50 Degree Conference on Engineering Mechanical Design and Stress Analysis (AIAS 2021)

IOP Conference Series: Materials Science and Engineering
Volume 1214

Online
1 - 3 September 2021

Editors:

Franco Furgiuele
Nicola Bonora
Filippo Cianetti
Paolo Livieri

ISBN: 978-1-7138-4673-4
ISSN: 1757-8981

Printed from e-media with permission by:

Curran Associates, Inc.
57 Morehouse Lane
Red Hook, NY 12571

Some format issues inherent in the e-media version may also appear in this print version.

Additional copies of this publication are available from:

Curran Associates, Inc.
57 Morehouse Lane
Red Hook, NY 12571 USA
Phone: 845-758-0400
Fax: 845-758-2633
Email: curran@proceedings.com
Web: www.proceedings.com

TABLE OF CONTENTS

Preface

N Bonora[1], F Cianetti[2], P Livieri[3]

[1]University of Cassino and Southern Lazio, Cassino I-03043, Italy
[2]University of Perugia, Perugia I-06123, Italy
3University of Ferrara, Ferrara, I-44121, Italy

bonora@unicas.it

The Italian Association for Stress Analysis (AIAS) was founded in 1971 by researchers from academia, research centers and industry. AIAS was intended as a community where to discuss, to share and to develop scientific knowledge related to all technical aspects of stress analysis. In the years, from an initial focus on experimental techniques, AIAS contributed considerably to the development of modern numerical methods and computational techniques for the mechanical engineering design. In 2015, AIAS turned in the Italian Scientific Society of Mechanical Engineering Design.

Today, AIAS is an institutional partner that supports the instances from academia in subject area of the mechanical engineering design. Every year, AIAS organizes a technical conference offering the possibility to present research updates, share new ideas and foster collaborations. The AIAS conference has become a fundamental event for all those interested in current developments in mechanical engineering design and stress analysis, where to meet researchers, testing equipment and software developers.

The 50th AIAS Conference edition was initially re-scheduled to be held in Genoa, Italy after it was switched as virtual conference in 2020 due to COVID-19 emergency. Unfortunately, due to the continuing pandemic emergency, it was decided to go virtual again to keep continuity with the tradition, confident that, despite the difficulties of the moment, united we stand.

As for the previous year, the response of researchers and students has been outstanding: over 180 oral contribution have been presented in synchronous, during the three days of the conference, with three simultaneous parallel sessions. In addition to the thematic sessions on AIAS traditional subjects, special sessions on *additive manufacturing*, *energetic methods for structural analysis* and *smart materials and MEMS* have been successfully organized with the contribution of the AIAS technical committees.

Among all contributions presented at the conference, 54 have been selected to be published after peer review, in this volume. This was made possible thanks to the active participation of all AIAS members, to the work of the AIAS Scientific Committee and Conference Papers Review panel (Profs Giovanni Meneghetti, AIAS Scientific Coordinator, Luciano Afferrante, Francesco Bucchi, Filippo Cianetti, Enrico Armentani, Marco Sasso). Their outstanding contribution is gratefully acknowledged.

Disclaimer
AIAS2021 Conference was changed to a virtual format for its second time due to continuing COVID-19 restrictions that did not allow to have it in presence.

AIAS Board of directors and conference organizers has decided to hold the meeting virtual instead of being postponed so as not to miss the chance to offer the opportunity to share researchers' own scientific results with the community of reference. The cancellation of the event would have represented a significant loss for researchers in the field of mechanical design.

The Virtual Conference dates and the conference organization scheme where the same of the initially scheduled conference. The conference was held on Sept. 1-3, 2021. The organizers operated remotely from two locations: the University of Cassino and Southern Lazio and the University of Ferrara. The Conference was organized for synchronous presentation: each author made its presentation live, as for the question and answer time. Presentations were previously uploaded and made available to all attendees. The presentation time slot was 15 min. Approximately 240 participants attended the conference during the three days event. The virtual conference was organized using the Google Suite tools available to AIAS. All sessions have been continuously monitored by the organizers to provide technical support in case of connection issues. The conference ran smoothly, and participants feedback was very positive.

Editors:

Franco Furgiuele, Nicola Bonora, Filippo Cianetti, Paolo Livieri

Scientific Committee:

Giovanni Meneghetti, Università di Padova,

Luciano Afferrante, Politecnico di Bari
Enrico Armentani, Università di Napoli "Federico II"
FIlippo Cianetti, Università di Perugia

Francesco Bucchi, Università di Pisa
Marco Sasso, Università Politecnica delle Marche

Peer Review Statement

All papers published in this volume have been reviewed through processes administered by the Editors. Reviews were conducted by expert referees to the professional and scientific standards expected of a proceedings journal published by IOP Publishing.

- **Type of peer review:** Single Anonymous
- **Conference submission management system:** AIAS proprietary abstract management system
- **Number of submissions received:** 184
- **Number of submissions sent for review:** 108
- **Number of submissions accepted:** 54
- **Acceptance Rate (Submissions Accepted / Submissions Received × 100):** 0
- **Average number of reviews per paper:** 2
- **Total number of reviewers involved:** 16
- **Contact person for queries:**
  Name: Nicola Bonora
  Email: bonora@unicas.it
  Affiliation: University of Cassino and Southern Lazio

Effect of process parameters on the mechanical properties of a Titanium alloy fabricated by Electron Beam Melting (EBM)

G Mirone, R Barbagallo and S Di Bella

University of Catania, DICAR - Department of Civil Engineering and Architecture, Via Santa Sofia 64 – 95125 – Catania, Italy

gmirone@dii.unict.it

Abstract. Electron Beam Melting is a layer-by-layer additive manufacturing technique which proved to be very efficient and versatile in the fabrication of medical devices and automotive or aerospace components. Due to the highly innovative character of the technology, its potential is not yet fully exploited and a comprehensive understanding of how the different processing parameters may affect the mechanical performance of the resulting materials still deserves to be achieved. Tensile tests on a Titanium alloy (Ti-6Al-4V-ELI) are ran at static rates for investigating the effects of different fabrication parameters on the material response, namely the speed function, the line offset, the focus offset, the number of contours and the horizontal-vertical growing direction of the specimens. A combination of such parameters is identified as a "process setup" and different setups are adopted for producing corresponding series of specimens, machined after fabrication according to ASTM F2924 and ASTM E8 regulations. The experimental procedure relies on enhanced true stress-true strain data based on the optical measurement of the neck cross section of the specimen, as an alternative to the traditional elongation-based approach to the true stress-true strain evaluation. This approach allows to better highlight the effects of the different process setups on the material performance, by focusing on the most stressed/strained material points within the overall specimen volume. The true curves allow to relate the material performance in terms of both strength and ductility to each process setup, allowing to outline the influence of single/combined fabrication parameters on the two above performance indicators.

1. Introduction

Electron Beam Melting is a layer-by-layer additive manufacturing technique that proved to be very efficient and versatile in the fabrication of, among others, medical devices and automotive/aerospace components. According to the ASTM 52900, EBM is a powder bed fusion process. This means that "thermal energy selectively fuses regions of a powder bed" [1][2]. EBM builds parts in a high vacuum chamber providing an ideal contamination-free atmosphere for the printing of reactive materials, such as titanium alloys, that have a high affinity to nitrogen and oxygen [3]. Another advantage is that the deposition occurs at elevated temperatures (build temperature is higher than 700 °C), reducing residual stresses in the final part [4]. The EBM process consists of continuous repetition of four fundamental steps: 1) Spreading of a powder layer through the rake system. 2) Preheating of the powder layer to grow up the temperature close the melting temperature of the powder. 3) Melting the powder through the electron beam. 4) Lowering the production table by a certain quantity [5].

In very recent years, numerous alloys were processed by EBM. Among them, titanium alloys, with an emphasis on Ti-6Al-4V ELI, are commonly used in the aerospace and bioengineering industries because of their good mechanical properties [6][7], low density, and good corrosion resistance [8]. Furthermore, non-toxicity makes this class of materials a reference for a great variety of biomedical applications [9][10]. On the other hand, one of the major drawbacks is that the as-built components are characterized by a high surface roughness [11]. It is known that the processing of Ti6Al4V ELI is of great interest for several industrial applications but, despite some efforts in the literature [12][13][14], the link between processing conditions and properties of the parts is not fully understood. A lot of studies show results of parts fabricated by ARCAM EBM equipment used with automatic process parameter control. In these conditions, the printer, through a series of algorithms, continuously adjusts the beam parameters to optimize the deposition. The only way to obtain full control of the EBM process is to use the manual operation mode where it is allowed to change the value of each process parameter like the speed function, the focus offset, the line offset, and the number of contours (last two shown in Figure 1).

The speed function adjusts the scanning speed using a proprietary (ARCAM) automatic algorithm. It controls and maintains the thickness of the melting pool constant during the melting process keeping the ratio between current and speed constant. The line offset represents the distance between two successive beam lines, the standard default value is 0.1 mm [15].

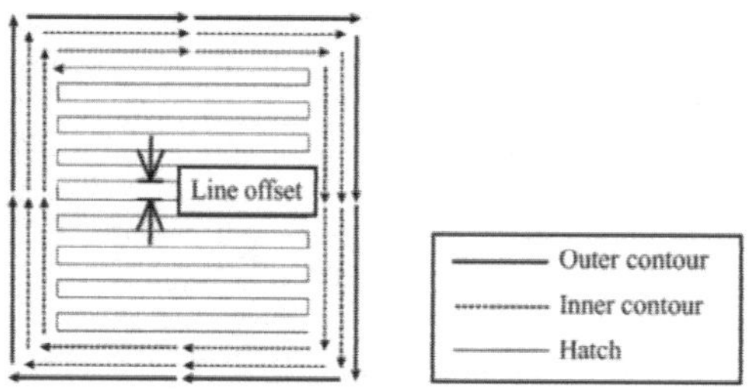

Figure 1. Line offset and number of contours.

The focus offset controls the size of the spot generated by the beam of electrons incident the powder bed. The smaller the spot, the deeper the melting pool is between the layers of powder.

The main objective of this paper is to assess the influence of speed function, focus offset, line offset e number of contours on the mechanical properties of the Ti6Al4V ELI made by EBM. The influence of orientation of the parts in respect to the production plane will be evaluated through a specific experimental campaign. Experimental tensile test will be performed at static rates on a set of specimens made by EBM production under a set of specified combination of production parameters. Despite some efforts have been performed by the scientific community to obtain local material information from global tensile elongations [16], the used experimental procedure will be based on true stress-true strain data obtained by optical measurement of the neck cross section of the specimen [17][18][19][20], as an alternative to the traditional elongation-based approach. The results will be to understand the effects of the different process setups on the material performance.

2. Materials and Methods

The samples were fabricated with an ARCAM Q10 EBM machine (property of MT Ortho srl) using pre-alloyed Ti6Al4V ELI (Titanium grade 23) powder provided by Arcam. The size of the powder particles was within the range of 45 to 150 μm.

As widely recognized, "as-built" specimens do not necessarily need after-print CNC machining but, very often, they have geometric defects and high surface roughness not suitable for the repeatability of the mechanical tests. Therefore, for a large-scale comparative study, they are probably inadequate. On the other hand, CNC machined specimens present advantages in terms of repeatability of production processes, results and performance of the material, but they can be characterized by possible modification of its response due to the mechanical processing. The choice fell on machined specimens that were obtained from a block of solid material fused through EBM technology.

Regarding the geometry to be used, we referred to the ASTM E8 standard related to powder metallurgy. The cylindrical samples were machined to obtain the ready-to-test samples shown in Figure 2, with two ISO M8 threads at the specimens' ends for appropriate coupling with the gripping system. Moreover, a slight tapering is enforced by machining the diameter of the cross section varying between 7 mm at the shoulders fillets and 6.9 mm at the mid length section.

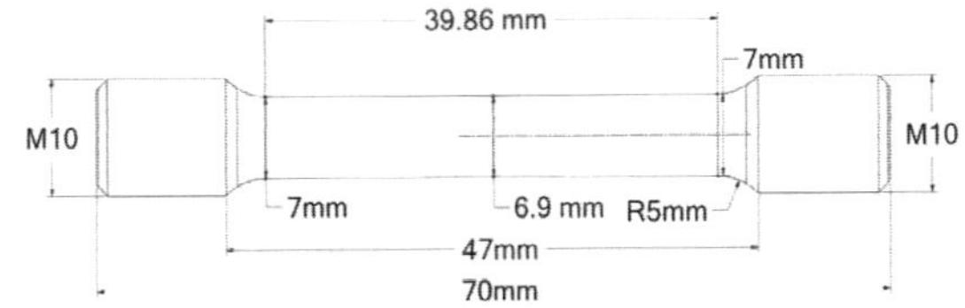

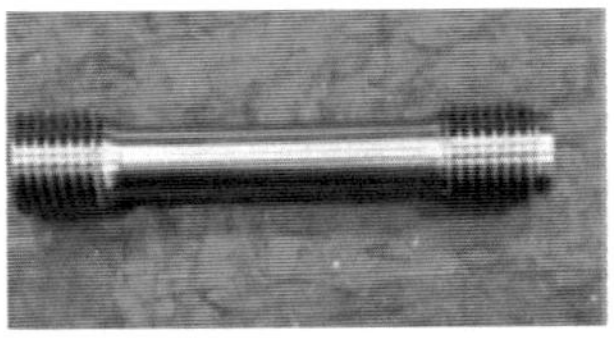

Figure 2. Machined sample for mechanical test **Figure 3.** Experimental set-up

Quasi-static tensile tests were carried out on a Zwick/Roell Z100 at a strain rate of 0.003 s^{-1}. A high-resolution video camera (1280x960) was used to record the evolution of the profile of the sample during the test, especially in the post-necking phase, while a clip gage has been used in order to evaluate more accurately the elastic properties of the material. The complete experimental setup is shown in Figure 3.

The first experimental campaign concerned the study of the influence on the material characteristics of the orientation of the specimens with respect to the production plane. Two cylindrical samples were built perpendicularly to the production plane and two other samples were built horizontally (Figure 4), using the default process parameters shown in Table 1. The thickness of the powder layer was set to 50 μm and the building plate, following the default procedure, was preheated to a temperature of about 730 °C before the printing process. The entire job was performed under a controlled vacuum of 10^{-3} mbar in the chamber. In particular, the purpose of this first group of tests was to evaluate which one of the two orientations allowed to obtain the best repeatability in terms of the mechanical performance of the material.

Table 1. Default Process parameter used for EBM

Speed Function	Focus offset (mA)	Line Offset (mm)	Number of contours
46	32	0.2	3

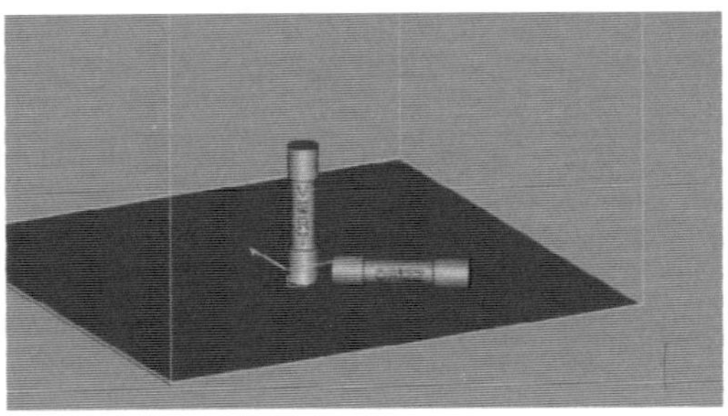

Figure 4. First Experimental campaign: Orientation

The second and third experimental campaigns were focused on the study of the influence on the material behaviour of the process parameters "speed function", "focus offset", "line offset", and "number of contours". For each configuration of process parameters, two identical specimens have been made to evaluate the repeatability of the results. The complete experimental campaign is shown in Table 2.

Table 2. Experimental campaign

Campaign	Specimen Code	Horizontal	Vertical	Speed Function	Focus Offset	Line Offset	N. of contours
1	TESTORIZZ	x		46	32	0.2	3
1	TESTVERT		x	46	32	0.2	3
2	SF36	x		36	32	0.2	3
2	SF56	x		56	32	0.2	3
2	FO22	x		46	22	0.2	3
2	FO42	x		46	42	0.2	3
2	LO05	x		46	32	0.5	3
2	LO075	x		46	32	0.75	3
2	NC5	x		46	32	0.2	5
2	NC7	x		46	32	0.2	7
3	SF30	x		30	32	0.2	3
3	SF60	x		60	32	0.2	3
3	LO01	x		46	32	0.1	3
3	LO08	x		46	32	0.8	3
3	SF56_LO075	x		56	32	0.75	3
3	SF56_LO05	x		56	32	0.5	3
3	SF36_LO075	x		36	32	0.75	3
3	SF36_LO05	x		36	32	0.5	3

For each test, the engineering and true stress-strain curves were obtained, together with the Young modulus, ultimate tensile strength, and elongation values.

To obtain the engineering curves, the well-known eqs. (1) and (2) have been used. Two nominal gage lengths are used: 10 mm for the mechanical extensometer and 30 mm for the optical extensometer; the mechanical measurement of the elongation was limited to the early plastic range, while the optical one was extended up to failure; optical elongation measurements were carried out be image scaling and pixel counting.

$$\sigma_{eng} = \frac{F}{A_0} \tag{1}$$

$$\varepsilon_{eng} = \frac{\Delta L}{L} = \frac{L - L_0}{L_0} = \frac{L}{L_0} - 1 \tag{2}$$

The true curves were obtained analysing optically the evolution of the diameter of the specimen, with the well-known eqs. (3) and (4).

$$\sigma_{true} = \frac{F}{A} \tag{3}$$

$$\varepsilon_{true} = 2\ln\frac{d_0}{d} \tag{4}$$

3. Results

3.1. Horizontal and vertical specimens

The objective of the first experimental campaign was to understand how the different positioning of the specimens in respect to the production plane affects the mechanical characteristics of the material and their repeatability.

The engineering stress-strain curves obtained from the horizontal and vertical specimens are shown in Figure 5. All the curves show a similar maximum tensile strength. Moreover, all the first elastic phases are well-matched. On the other hand, the horizontal specimens have a greater elongation-based engineering strain at fracture than the vertical ones. Furthermore, there is much more dispersion in the post-necking phase between the vertical curves than between the horizontal ones.

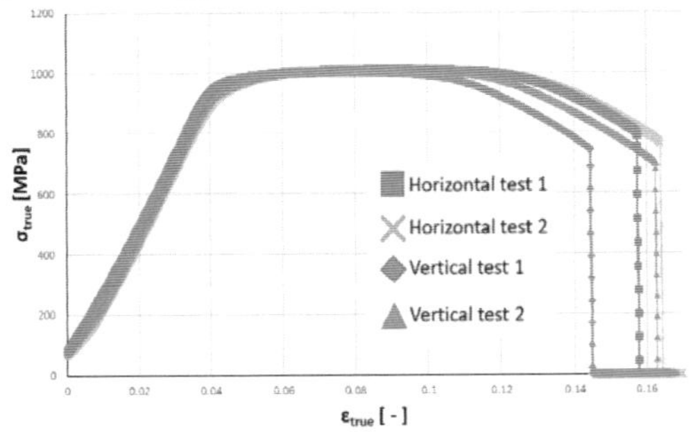

Figure 5. Horizontal Vs Vertical Engineering curves

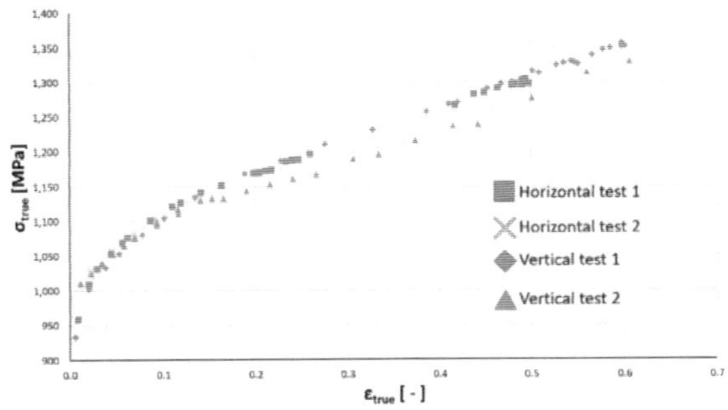

Figure 6. Horizontal Vs Vertical True curves

The diameter-based true stress-strain curves obtained from the horizontal and vertical specimens are shown in Figure 6. In this case, all curves are very similar to each other, with a maximum dispersion of 5%. The true deformations at fracture of the vertical specimens are greater than those of the horizontal specimens, which is exactly the opposite of what was found with the engineering curves. This fact confirms once again that the vertically printed specimens have greater variability of the quality of the printed material between the processing layers and, in turn, between the different single resistant sections. On the other hand, in the horizontal specimens, the differences between layers act uniformly on the whole length of the specimen and, therefore, do not determine single sections with different strengths affecting the overall specimen response. Another confirmation of this fact is that one of the vertical specimens did not show a centred necking (Figure 7), despite being all the specimens are slightly tapered with the minimum cross-section. Necking and failure stresses and strains of the horizontal and vertical specimens are shown in Table 3. Concluding the first experimental campaign, in order to have greater repeatability of the test results, only specimens printed with the horizontal orientation have been considered in the successive phases of the study.

Table 3. Horizontal Vs Vertical specimens tests results

		Necking		Failure	
Campaign	Specimen Code	σ_{true} [MPa]	ε_{true} [ ]	σ_{true} [MPa]	ε_{true} [ ]
1	Horizontal test 1	1077	0.062	1296	0.498
1	Horizontal test 2	1082	0.068	1282	0.544
1	Vertical test 1	1080	0.077	1351	0.601
1	Vertical test 2	1065	0.057	1302	0.648

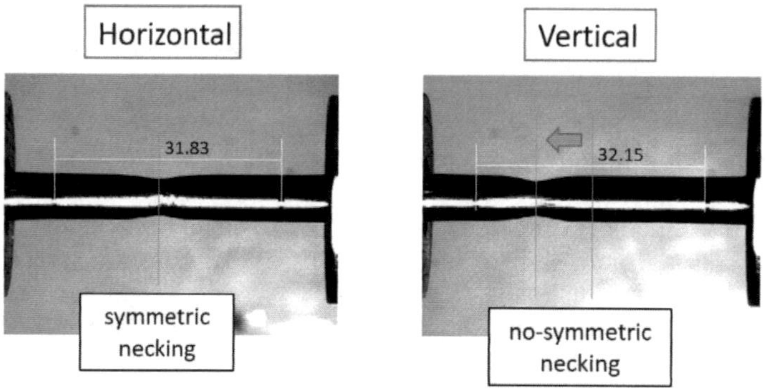

Figure 7. Specimens with horizontal Vs vertical orientation - Necking area

3.2. Changing the process parameters

Figure 8 and Table 4 show the obtained results with specimens printed changing only the speed function value and keeping all the other parameters as standard. In Figure 8 and in all the following ones, the yellow line represents the true stress-strain curve obtained with standard parameters. Increasing the speed function delivers improvements in both material strength and ductility, until a peak optimum value of the speed function equal to 56. Indeed, for speed function values higher or lower than 56, the obtained true curves are lower and shorter.

Table 4. Tests results of the specimens with different Speed Functions

Campaign	Specimen Code	UTS [MPa]	ε_{true} necking []	ε_{true} failure []
2	SF 30	947.83	0.069	0.413
2	SF 36	1011.72	0.068	0.492
2	SF 56	1060.87	0.066	0.569
2	SF 60	1002.92	0.071	0.496

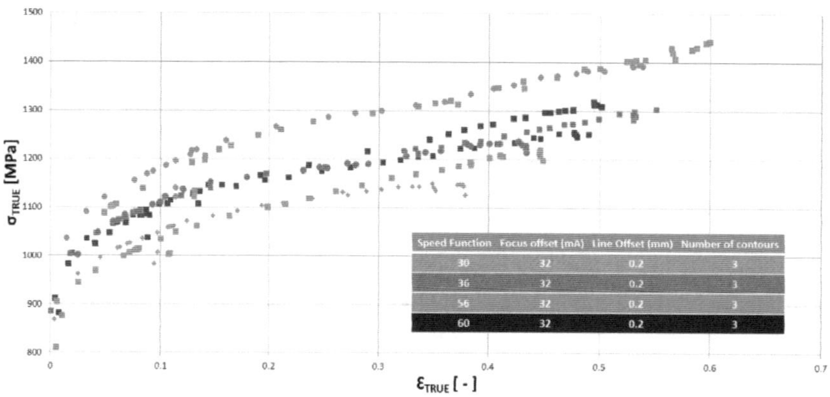

Figure 8. True curves of the specimens with different Speed Functions

Figure 9 and Table 5 show the obtained results with specimens printed only changing the line offset value and keeping all the other parameters as standard. Increasing the line offset delivers improvements of both material strength and ductility until a peak optimum value of the line offset equal to 0.75.

Table 5. Tests results of the specimens with different Line Offsets

Campaign	Specimen Code	UTS [MPa]	ε_{true} necking []	ε_{true} failure []
2	LO 0.1	970.80	0.070	0.572
2	LO 0.5	1007.41	0.071	0.587
2	LO 0.75	1055.60	0.060	0.619
2	LO 0.8	997.38	0.069	0.510

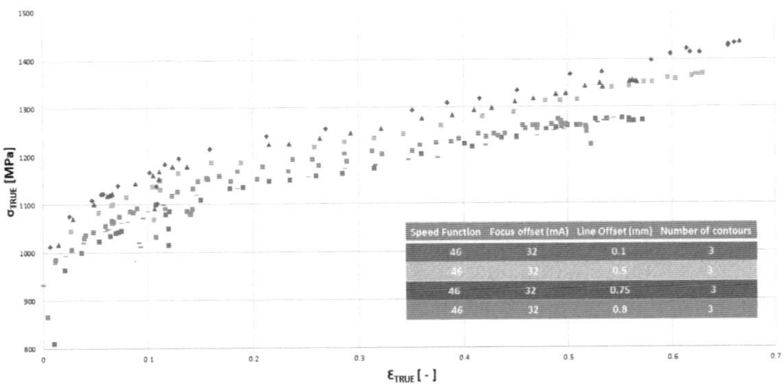

Figure 9. True curves of the specimens with different Line Offsets

Figure 10 and Table 6 show the obtained results with specimens printed changing only the focus offset value and keeping all the other parameters as standard. Increasing the focus offset of nearly 100% does not affect both the strength and the ductility of the resulting material.

Table 6. Tests results of the specimens with different Focus Offset

Campaign	Specimen Code	UTS [MPa]	ε_{true} necking []	ε_{true} failure []
2	FO 22	1040.23	0.071	0.476
2	FO 42	1032.40	0.065	0.514

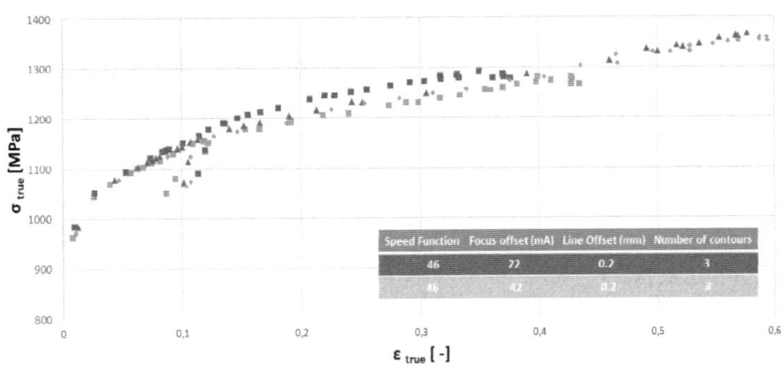

Figure 10. True curves of the specimens with different Focus offset

Figure 11 and Table 7 show the obtained results with specimens printed changing only the number of contours value and keeping all the other parameters as standard. Increasing the number of contours of nearly 50% does not affect both the strength and the ductility of the resulting material. This was highly expectable because specimens are machined.

IOP Conf. Series: Materials Science and Engineering 1214 (2022) 012001 doi:10.1088/1757-899X/1214/1/012001

Table 7. Tests results of the specimens with different Number of contours

Campaign	Specimen Code	UTS [MPa]	ε_{true} necking []	ε_{true} failure []
2	NOC 5	1046.21	0.044	0.588
2	NOC 7	1058.27	0.055	0.581

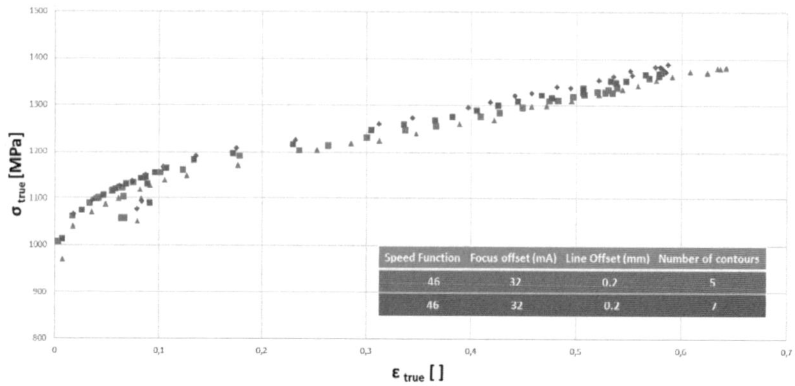

Figure 11. True curves of the specimens with different Nr. of contours

Figure 12 shows the obtained true curves with specimens printed changing both the speed function and line offset values and keeping all the other parameters as standard. With the same value of SF, the variation of line offset does not affect both the strength and the ductility of the resulting material. For different values of SF (with LO fixed) the graph shows an improvement of the material features at higher SF values.

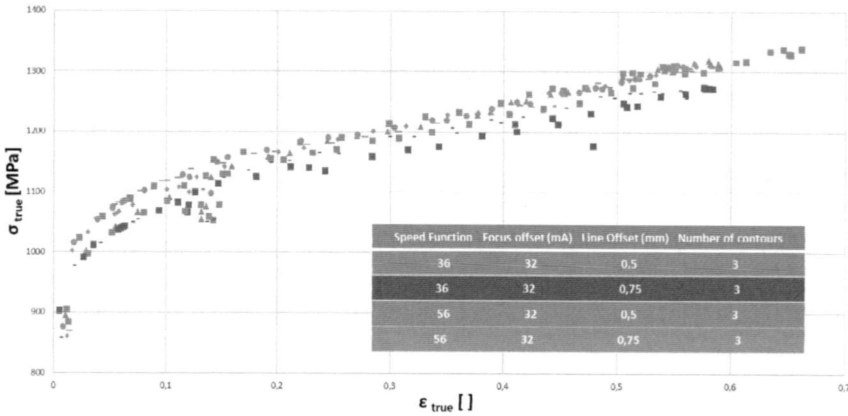

Figure 12. True curves of the specimens with different combined values of Speed Function and Line Offset

4. Conclusion

The results of the experimental campaigns can be summarized in the following points:

- The specimens printed perpendicularly and parallelly to the building plane give slightly different true stress-strain curves; moreover, perpendicular specimens present a higher dispersion in the test results, reasonably caused by the differences between the mechanical properties of different cross sections which, for the vertical specimen orientation, belong to different layers that are printed in different phases of the fabrication process under possibly slightly different printing conditions. Therefore, horizontally printed specimens are more suitable for a consistent analysis of the influence of the process parameters on the mechanical characteristics of the material.

- The speed function largely affects the mechanical properties of the material: starting from the default setting, greater values deliver better material strength and ductility until an optimum peak value of 56.
- The line offset affects the mechanical properties of the material: starting from the default one, greater values deliver better material strength and ductility until an optimum peak value of 0.75 mm.
- To change the focus offset and number of contours values in respect to the default ones shows no influence on the mechanical behaviour of the material.
- The curves obtained with the specimens printed changing simultaneously the line offset and speed function values shown that the speed function is the most decisive parameter among the studied ones regarding the mechanical performance of the resulting printed material.

References

[1] Frazier W J 2014, Metal additive manufacturing: a review. *J. Mater. Eng. Perform.* **23**(6), 1917–1928.
[2] Herzog D, Seyda V, Wycisk E and Emmelmann C 2016, Additive manufacturing of metals. *Acta Mater.* **117,** 371–392.
[3] Liu S and Shin Y C 2019, Additive manufacturing of Ti6Al4V alloy: A review. *Mater. Des.* **164**, 107552.
[4] Silvestri A T, Foglia S, Borrelli R, Franchitti S, Pirozzi C and Astarita A 2020, Electron beam melting of Ti6Al4V: Role of the process parameters under the same energy density. *J. Manuf. Process.,* **60**, 162-179.
[5] Körner C 2016, Additive manufacturing of metallic components by selective electron beam melting—a review. *Int. Mater. Rev.* **61**(5), 361-377.
[6] Mirone G, Barbagallo R, Corallo D and Di Bella S 2016, Static and dynamic response of titanium alloy produced by electron beam melting. *Procedia Struct. Integr.* **2**, 2355-2366.
[7] Mirone G, Barbagallo R, Giudice F and Di Bella S 2020, Analysis and modelling of tensile and torsional behaviour at different strain rates of Ti6Al4V alloy additive manufactured by electron beam melting (EBM). *Mater. Sci. Eng. A*, **793**, 139916.
[8] Chastand V, Quaegebeur P, Maia W and Charkaluk E 2018, Comparative study of fatigue properties of Ti-6Al-4V specimens built by electron beam melting (EBM) and selective laser melting (SLM), *Mater Charact*, **143**, 76-81.
[9] Elias C N, Lima J H C, Valiev R and Meyers M A 2008, Biomedical applications of titanium and its alloys. *JOM*, **60**(3), 46-49.
[10] Hao Y L, Li S J and Yang R 2016 Biomedical titanium alloys and their additive manufacturing. *Rare Met.*, **35**(9), 661-671.
[11] Galati M, Minetola P and Rizza G 2019, Surface roughness characterisation and analysis of the Electron Beam Melting (EBM) process, *Materials*, **12**(13), 2211.
[12] Prisco U, Astarita A, El Hassanin A and Franchitti S 2019, Influence of processing parameters on microstructure and roughness of electron beam melted Ti-6Al-4V titanium alloy, *Mater. Manuf. Process.,* **34**(15), 1753-1760.
[13] Wang X, Gong X and Chou K 2015, Scanning speed effect on mechanical properties of Ti-6Al-4V alloy processed by electron beam additive manufacturing, *Procedia Manuf.*, **1**, 287-295.
[14] Ge W, Guo C and Lin F 2014, Effect of process parameters on microstructure of TiAl alloy produced by electron beam selective melting, *Procedia Eng.,* **81**, 1192-1197.
[15] Shao M, Vijayan S, Nandwana P and Jinschek J R 2020, The effect of beam scan strategies on microstructural variations in Ti-6Al-4V fabricated by electron beam powder bed fusion.

Mater. Des., **196**, 109165.
[16] Mirone G, Verleysen P and Barbagallo R 2019, Tensile testing of metals: Relationship between macroscopic engineering data and hardening variables at the semi-local scale, *Int. J. Mech. Sci.*, **150**, 154-167.
[17] Zhang L, Gour G, Petrinic N and Pellegrino A 2020, Rate dependent behaviour and dynamic strain localisation of three novel impact resilient titanium alloys: Experiments and modelling, *Mater. Sci. Eng. A*, **771**, 138552.
[18] Mirone G, Barbagallo R and Giudice F 2019 Locking of the strain rate effect in Hopkinson bar testing of a mild steel. *Int. J. Impact Eng.* **130**, 97-112.
[19] Mirone G, Corallo D and Barbagallo R 2016 Interaction of strain rate and necking on the stress-strain response of uniaxial tension tests by Hopkinson bar. *Procedia Struct. Integr.*, **2**, 974-985.
[20] Mirone G and Barbagallo R 2021 How sensitivity of metals to strain, strain rate and temperature affects necking onset and hardening in dynamic tests. *Int. J. Mech. Sci.*, **195**, 106249.

Hydrogen diffusivity and tensile properties degradation in SLMed Inconel 718

F Aiello, M Beghini, L Bertini, G Macoretta, B D Monelli and R Valentini

Dipartimento di Ingegneria Civile e Industriale, Università di Pisa, Largo Lucio Lazzarino 2, 56122 Pisa, Italy

E-mail: giuseppe.macoretta@phd.unipi.it

Abstract. Besides elevated temperature applications, Inconel 718 is widely chosen for components subjected to heavy loads operating in aggressive environments, which can promote the production of hydrogen on the metal surface. Selective Laser Melting (SLM) is an emerging technology for the production of structural components in Inconel 718 or Inconel 625 alloys, allowing to create functionally optimized shapes and overcome the traditional manufacturing limitations. However, the SLM process introduces a peculiar microstructure and severe residual stresses that can affect hydrogen migration and accumulation. While the mechanical properties were deeply investigated in recent years, the resistance of SLMed Inconel 718 to the Hydrogen Embrittlement (HE) is still an open issue. The present work deals with a preliminary assessment of the hydrogen embrittlement susceptibility of the SLMed In718 in the as-built condition. Slow strain rate tests were carried out on unnotched specimens that were pre-charged by electrochemical cathodic charging. Fractographic analyses, carried out with a Scanning Electron Microscope (SEM), were used to identify the effects produced by the hydrogen intake on the material microstructure. The material resulted to be susceptible to hydrogen embrittlement, featuring a significant ductility reduction in presence of extremely elevated hydrogen concentrations. The hydrogen content was measured by employing the Hot Extraction Method (HEM), for each mechanical test. Due to a low diffusivity value, hydrogen penetrates only within a thin external layer of the specimen. Therefore, to identify the concentration profile, the effective hydrogen diffusion coefficient was identified and the hydrogen diffusion was simulated using Fick's equation. The loss in mechanical resistance was then correlated to the hydrogen concentration near the fracture onset.

1. Introduction

Nickel-based superalloys feature excellent mechanical properties, combining high strength and elevated ductility, both at room and high temperature, coupled with outstanding resistance to corrosion phenomena. These properties match the requirement of a wide range of industrial applications, from aero engines, power generation plants, to chemical plants and Oil&Gas applications.

Inconel 718 alloy is a widespread equiaxed nickel-based alloy characterized by high strength and excellent corrosion resistance, within a working temperature range between -250 °C and 800 °C. However, it is known to be susceptible to hydrogen embrittlement effects when exposed to elevated hydrogen concentrations [1, 2, 3, 4, 5]. Hydrogen can penetrate into the material during manufacturing processes or during service exposure to hydrogen-rich environments the surrounding environment during service exposure. In the case of aerospace applications, hydrogen can arise from fuel in rocket engines and its embrittlement effects were deeply studied [5].

Even if the Inconel 718 was developed for use in aircraft gas turbines, its combination of room-temperature strength and aqueous corrosion resistance made it a candidate for oilfield fasteners, valves, drill tools, and completion equipment, where hydrogen can arise from the conveyed fluids, [6]. Hydrogen embrittlement of the conventional Inconel 718 alloy was deeply investigated through SSR (Slow Strain Rate) tensile tests [2, 7, 8], along with the measurement of the hydrogen diffusivity coefficient [4, 9, 10]. Selective Laser Melting (SLM) is an emerging technology to produce structural components featuring complex geometries and it is particularly suitable for high-strength materials. Inconel 718 is the most widespread nickel-based superalloy produced by SLM, thanks to its wide range of applications and weldability properties. Even if the mechanical and fatigue properties of SLMed Inconel 718 are deeply investigated, the hydrogen diffusion properties and the resistance to hydrogen embrittlement phenomena are still open questions.

In the present paper, it is presented a preliminary investigation of the effects produced by cathodic hydrogen intake in SLMed Inconel 718 in the as-built condition.

The hydrogen diffusivity properties are conventionally evaluated by means of the Devanathan-Stachurski test method. However, in the case of a material featuring a very low diffusivity value, as the wrought Inconel 718, this test method requires a very thin metallic foil, namely featuring a thickness of about 50÷100 μm, to have experiment duration within two or four weeks at temperatures of 50 ℃, [9]. Producing a so thin foil by SLM is unfeasible, as its dimensions are comparable with the power size or laser beam diameter. On the other hand, machining a bar produced by SLM would alter the surface roughness, introducing a bias between the measured value and the behavior of a component produced by SLM and exposed to a hydrogen-rich environment. As a matter of fact, exploiting the potential of AM means producing complex geometries, where successive machining could be unfeasible or cost-ineffective. The hydrogen diffusion coefficient was thus measured indirectly by measuring the hydrogen intake of specimens exposed to the same surface hydrogen concentration for different durations.

Slow Strain Rate (SSR) tensile tests were carried out along with hydrogen concentration measurements and fractographic analyses were carried out to investigate the effects produced by the hydrogen intake. The results are analyzed within the framework of the hydrogen diffusion models developed in the literature. Further investigations will be performed based on the results of this preliminary investigation, studying the effects of the aging treatments and surface roughness.

2. Materials and methods

Specimens were produced by using the Renishaw RenAM 500E Selective Laser Melting (SLM) machine present in the *Metal Additive Manufacturing* laboratory of the University of Pisa.

The chemical composition of the Inconel 718 alloy powder, provided by Renishaw plc, is reported in Table 1.

The process chamber was filled with gas argon, reaching an oxygen concentration lower than 10 ppmw during the printing phase, guaranteeing to avoid oxidation phenomena. It was adopted a standard process parameters set, featuring a layer thickness of 60 μm. All the specimens were built in the vertical direction, namely with the longitudinal direction aligned with the direction of motion of the built plate, in order not to have traces of the support structures in the area of interest of the specimen and have uniform residual stresses on the specimen surfaces. The specimens were measured by a coordinate-measuring machine to check the occurrence of distortions due to the printing phase or the built plate removal. It resulted that the specimens presented no significant deviation, with values that are comprised in the typical range of accuracy

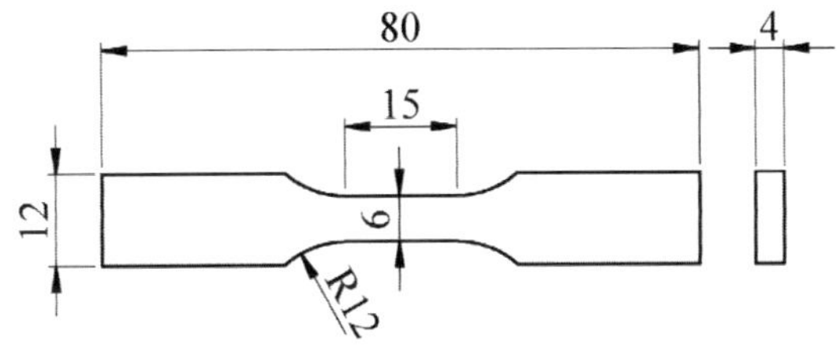

Figure 1. Specimen geometry and nominal dimensions. Dimensions in mm.

of the SLM process, namely 0.05 mm.

Table 1. Chemical composition of Inconel 718. Percentage mass values.

Ni	Cr	Fe	Nb + Ta	Mo	Ti
50÷55	17÷21	balanced	4.75÷5.5	2.80÷3.30	0.65÷1.15
Co	Al	Mn	Si	Cu	C
≤1.00	0.20÷0.80	≤0.35	≤0.35	≤0.30	0.02÷0.05

The specimen geometry, designed in agreement with the ASTM E8 standard, is characterized by a gauge length of 15 mm and by a rectangular cross-section, whose nominal dimensions, in the gauge section, are 4 mm × 6 mm, Figure 1. The specimens were tested in the as-built condition, preserving both the surface roughness and the microstructure produced by the SLM process. The effective cross-section dimensions of each specimen were measured before the mechanical tests.

Slow Strain Rate (SSR) tensile tests were carried out in air at room temperature on a servo-hydraulic machine having a load capacity of 50 kN (MTS Systems Corporation, Eden Prairie, MN USA). The machine was equipped with an extensometer having a gauge length of 10 mm (MTS 634.21-F25) and with a data acquisition system and control unit (RT3, Trio Sistemi e Misure S.r.l., Dalmine, Italy). SSR tests were performed by imposing a cross-head displacement of 1 μm/s and thus obtaining an almost constant strain rate of $1.5 \cdot 10^{-4}$ s^{-1}, comprised within the range provided by the G129-00 ASTM standard.

Microstructural investigations were carried out both in longitudinal and transverse directions by analyzing sections extracted from the gauge length of two specimens featuring the same geometry and SLM process history of the tested ones. Test samples were prepared following the ASTM E3-11 standard and etched with Kalling's II etchant (n.94 ASTM E407). The material porosity was characterized in terms of average material density and pore size and shape distribution by using the software ImageJ.

Scanning Electron Microscopy (SEM) analyses were carried out at the "Centro per l'Integrazione della Strumentazione Scientifica - Università di Pisa" (CISUP) using a FEI Quanta 450 FEG-SEM equipped with an EDX spectrometer Bruker QUANTAX XFlash Detector 6|10. Fractographic analyses were carried out by stereo microscope as well.

2.1. Hydrogen charging and measurement

The specimens were pre-charged before the SSR test by electrochemical cathodic charging, where a constant current density is imposed between the specimen and a platinum electrode. Before charging, all the specimens were cleaned by ultrasonic cleaning with acetone and rinsed with distilled water.

For the first three specimens, it was employed a 0.5 M aqueous solution of sulfuric acid (H_2S0_4), with the addition of 1 g/L of thiourea (CH_4N_2S) to increase hydrogen adsorption on the surfaces of specimens, [7, 11]. Hydrogen charging was performed at room temperature with a constant current density of 60 mA/cm^2 for durations between 45 and 66 hours. The room temperature, measured by a temperature data logger (Lascar EL USB-2) was 19±3 ℃.

To increase the hydrogen diffusivity coefficient and thus the hydrogen introduced into the material, a second batch of four specimens was charged at 70 ℃ by using a 3% aqueous solution of NaCl, containing 4 g/L of ammonium thiocyanate (NH_4SCN) to increase the hydrogen adsorption, [1, 12]. A constant current density of 20 mA/cm^2 was imposed. The hydrogen charging time ranged between 60 and 200 hours.

After each SSR test, the hydrogen content C_{HEM} was measured by Hot Extraction Method (HEM) employing the LECO® hydrogen determinator DH 603 (LECO, St. Joseph, MO, USA). The instrument measures the hydrogen flux that desorbs from the specimen heated at 265 ℃ and thus provides the average hydrogen content present in the analyzed coupon.

2.2. Diffusion coefficient determination

In the case of an ideal homogeneous material, where the distribution of hydrogen possible sites in the lattice is uniform, and a component without external loads and residual stresses, the hydrogen diffusion in metallic materials is known to be governed by Fick's law of diffusion:

$$\vec{J} = -\mathbf{D}\ \vec{\nabla} C + \vec{v} \cdot C, \tag{1}$$

$$\frac{\partial C}{\partial t} = -\vec{\nabla} \cdot \vec{J} + G, \tag{2}$$

where $\vec{J}$ is the diffusion flux density, $\mathbf{D}$ the diffusivity tensor, C the atomic concentration, $\vec{v}$ the transport velocity vector, and G the diffusing substance generation rate per unit volume. The presence of microstructural defects (such as dislocations, vacancies, grain boundaries, and precipitates) creates trapping sites, which provide an energetically favored environment for occupancy by the hydrogen, [13]. In the case of low occupancy in the reversible traps, Fick's first law can be used to model the diffusion of hydrogen in the alloy [13]. In this case, for an isotropic material, the diffusion equation takes the following form:

$$\frac{\partial C}{\partial t} = D_{eff} \frac{\partial^2 C}{\partial x^2}, \tag{3}$$

where D_{eff} is the effective diffusion coefficient. It varies with the temperature with an Arrhenius function:

$$D_{eff} = D_{eff,0} \cdot \exp\left(-\frac{E}{R\,T}\right), \tag{4}$$

where, in the present case, $D_{eff,0}$ is the pre-exponential parameter, and E is the activation energy equal to 48630 J/mol, [4].

Since conventional Inconel 718 alloy is known to present a low hydrogen diffusion coefficient, a direct measurement of D_{eff} is difficult due to unfeasible test durations. For this reason, D_{eff} was identified by measuring the H content (C_{HEM}) at different charging durations. The value of C_{HEM} as a function of time depends on the surface concentration and the effective diffusion coefficient D_{eff}. Therefore, it is possible to obtain the D_{eff} value through an optimization procedure.

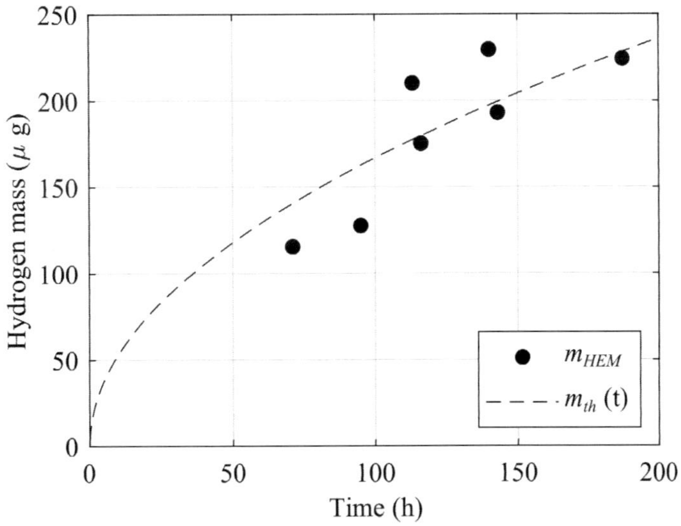

Figure 2. Effective diffusion coefficient identification. Measured hydrogen mass m_{HEM}, and theoretical trend m_{th} obtained using Equation 5 with the fitted parameters.

The theoretical value of the hydrogen mass m_{th}, at time t, as a function of the surface concentration C_s and D_{eff}, is obtained by solving Equation 3 and integrating the concentration profile in the volume:

$$m_{th}(C_s, D_{eff}, t) = \int_0^{l_p} C(C_s, D_{eff}, t)\ A\ dx, \tag{5}$$

where l_p and A are the specimen thickness and surface area, respectively. The D_{eff} value can be obtained by minimizing the following cost function:

$$F(C_s, D_{eff}) = \sum_{i=1}^{7} [m_{th}(C_s, D_{eff}, t_i) - m_{HEM}(t_i)]^2 . \tag{6}$$

Figure 2 shows the measured hydrogen mass m_{HEM} obtained from the electrochemically charged specimens as a function of charging times. The theoretical trend of the hydrogen mass m_{th}, obtained from Equation 5 using the fitted parameters, is reported with a dashed line.

Seven rectangular specimens having dimensions of 40 mm × 40 mm × 1.7 mm were electrochemically charged in the same conditions. It was adopted an electrolyte solution of NaCl 3.0% + NH_4SCN 0.4% (w/w), held at a constant temperature of 50 °C. The specimens were charged for different durations, with charging times (t_i) comprised between 60 and 200 hours. At the end of each charging time, the hydrogen mass was measured by HEM, by using the LECO® hydrogen determinator DH 603 (LECO, St. Joseph, MO, USA).

By solving Equation 6 for a fixed value of the effective diffusion coefficient D_{eff}, its value resulted to be equal to $5{\cdot}10^{-15}$ m^2/s at the test temperature of 50 °C. This value is very similar to the value obtained by Jebaraj [9] on the conventional wrought Inconel 718 in the solutionized condition, which can be considered as the most similar to the not heat-treated SLMed material.

2.3. Hydrogen profile within the specimens

The hydrogen concentration profile, reached in the specimen after the electrochemical charging, can be obtained by solving Fick's law, assuming a uniform initial concentration, in this case,

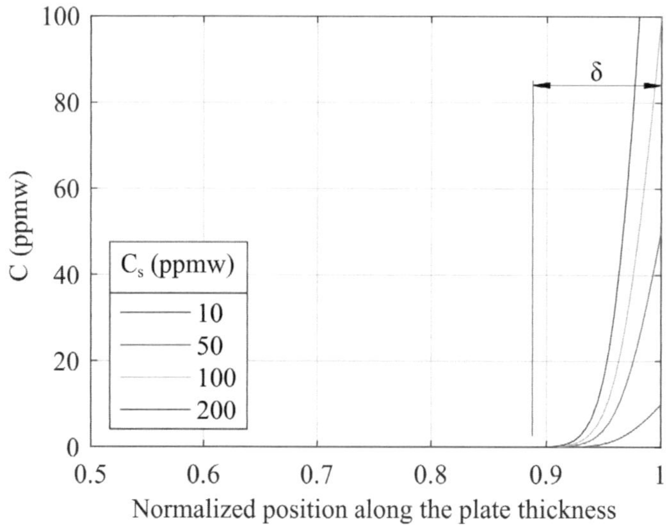

Figure 3. Sensitivity to surface concentration: hydrogen concentration profile within the specimen (charging time: 90 hours).

null, and imposing a boundary condition of constant and equal hydrogen concentration on the specimen walls.

In the case of a one-dimensional diffusion in a plane sheet, the solution is presented by Crank, [14]. Under these hypotheses, the hydrogen concentration profile depends only on the diffusivity concentration value, temperature-dependent, and on the hydrogen concentration on the specimen walls, which depends on the charging current and solution. Due to the low diffusion coefficient D_{eff}, the hydrogen concentration within the specimen does not reach a saturation value even after cathodic charging durations of hundreds of hours. The hydrogen penetration depth, calculated by using the approximate formulation derived for the case of a semi-infinite medium and presented by Crank [14], Equation 7, and employing the measured value of the effective diffusion coefficient ($5{\cdot}10^{-15}$ m^2/s at the test temperature of 50 °C), results to be about 0.26 mm after 150 hours of cathodic charging,

$$\delta = 4 \cdot \sqrt{D\,t} \tag{7}$$

The sensitivity of the concentration profile to the hydrogen concentration on the specimen surface at equal charging duration is reported in Figure 3. An increase in the hydrogen surface concentration produces only an increase in the average hydrogen content in the outer part of the specimen, while the penetration depth is influenced only by the charging duration and temperature. As shown in the schematic representation presented in Figure 4, after the cathodic charging, hydrogen is present only in the outer shell (black-filled area) of the specimen, while the middle part remains unaffected. Due to the non-uniform hydrogen concentration at the end of the cathodic charging, it is not possible to have a direct measure of the hydrogen concentration. As the HEM provides the average hydrogen concentration (C_{HEM}) present in the specimen, its measurement is strongly lower than the real value present near the specimen surfaces. However, based on the measured average hydrogen concentration content and the measured value of the hydrogen diffusivity D_{eff}, it is possible to calculate the hydrogen concentration on the specimen surface, C_s, and thus the average value, $C_{\mathrm{avg,\ s}}$, within the penetration depth through

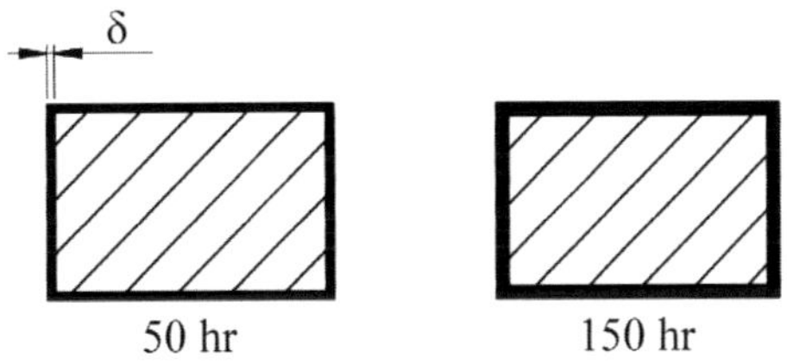

Figure 4. Hydrogen penetration within the specimen at increasing charging duration.

the integration of the hydrogen concentration profile, using the following equations:

$$C_{\text{HEM}} = \frac{1}{l_p} \int_0^{l_p} C(C_s, x) dx, \tag{8}$$

$$C_{\text{avg,s}} = \frac{1}{\delta} \int_{l_p - \delta}^{l_p} C(C_s, x) dx, \tag{9}$$

where l_p, and δ are the specimen thickness and width, and the hydrogen penetration depth, respectively. By applying this approach, an average hydrogen concentration of 10 ppmw, measured by HEM, corresponds to a surface concentration of 200 ppmw and an average concentration, within the hydrogen penetration depth, of 55 ppmw. It is worth noticing that it was neglected the contribution of the hydrogen desorption that happens between the end of the charging phase and the HEM measurement after the SSR test, thus underestimating the hydrogen content at the end of the cathodic charging.

3. Results and discussion

3.1. SSR test results

In Figure 5 the engineering stress-strain curves are reported as a function of the averaged hydrogen concentration measured by HEM. For the uncharged specimens it is reported only the average curve resulting from three tests carried out in identical conditions. In agreement with the results obtained on the conventional Inconel 718 alloy [1, 7, 11], it can be seen that the macroscopic effect of the hydrogen intake is an almost monotonic reduction of the material ductility, represented by the percentage elongation at fracture. For the lower values of the hydrogen content, the tensile strength of the material results to be unaffected by the hydrogen content as the crack happens after the occurrence of a necking phenomenon. As the hydrogen content increases, also the tensile strength decreases monotonically with the percentage elongation at fracture, due to a more consistent reduction of ductility. Within the investigated range, the hydrogen intake resulted not have an impact on the material elastic and strain-hardening phases, where the uncharged and charged curves are almost indistinguishable.

The embrittlement effect can be summarized by introducing a Hydrogen Embrittlement Index (HEI):

$$\text{HEI} = \frac{\varepsilon_{f,\text{uncharged}} - \varepsilon_{f,\text{charged}}}{\varepsilon_{f,\text{uncharged}}}, \tag{10}$$

where ε_f is the engineering fracture strain of the hydrogen-uncharged and charged specimens, respectively. As reported in Table 2, the HEI index reaches a value of 0.5 only for an average hydrogen concentration of 28 ppmw, corresponding to an estimated average value in the hydrogen affected part of the specimen of about 100 ppmw.

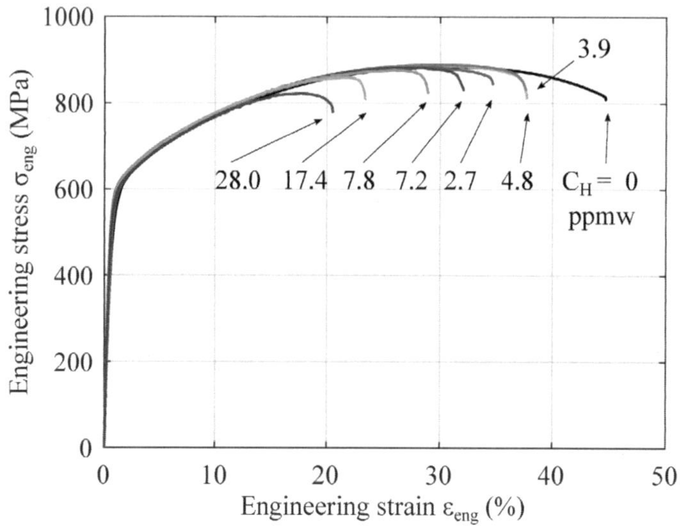

Figure 5. SSR tensile tests. For each curve it is reported the average hydrogen content measured by HEM, in ppmw.

Table 2. Hydrogen penetration depth, engineering fracture strain, and hydrogen embrittlement index values as a function of the hydrogen content.

C_{HEM} (ppmw)	0	2.7	3.9	4.8	7.2	7.8	17.4	28.0
δ (mm)	-	0.05	0.06	0.06	0.18	0.22	0.29	0.30
$C_{\text{avg,s}}$ (ppmw)	-	70	85	100	50	45	70	115
ε_f (%)	44.6	34.7	38.0	38.1	32.0	29.0	23.5	20.5
HEI	-	0.22	0.15	0.15	0.28	0.35	0.48	0.54

3.2. Fractographic investigations

The fractographic analysis, carried out by SEM, showed that the fracture surface can be divided into three regions: a central region, an intermediate region oriented at about 45° degrees to the loading axis, and a thin surface layer, Figure 6 (a). In the central region it is possible to see the traces of the SLM laser scanning path, suggesting that the final fracture took place by decohesion on the melt pool boundaries. Looking into detail to the outer part of the fracture surface, Figure 6 (b), it can be seen a clear difference between the part nearest to the specimen surface, a strip featuring a width of about 50 μm, and the more internal one. As the width of the outer strip is almost equal to the calculated hydrogen penetration depth in the investigated specimen, it suggests that the different appearance of the fracture surface is due to the effects of the hydrogen intake.

Moreover, it can be seen that there are no blisters or other cavities that can act as sites for the formation of gaseous hydrogen, suggesting that these phenomena do not affect a fully dense additively manufactured Inconel 718.

Fractographic analyses carried out by stereo microscope confirmed that the major difference between the hydrogen-free and charged material is represented by the external part of the fracture surface, within a thickness almost equal to the hydrogen penetration depth, Figure 7.

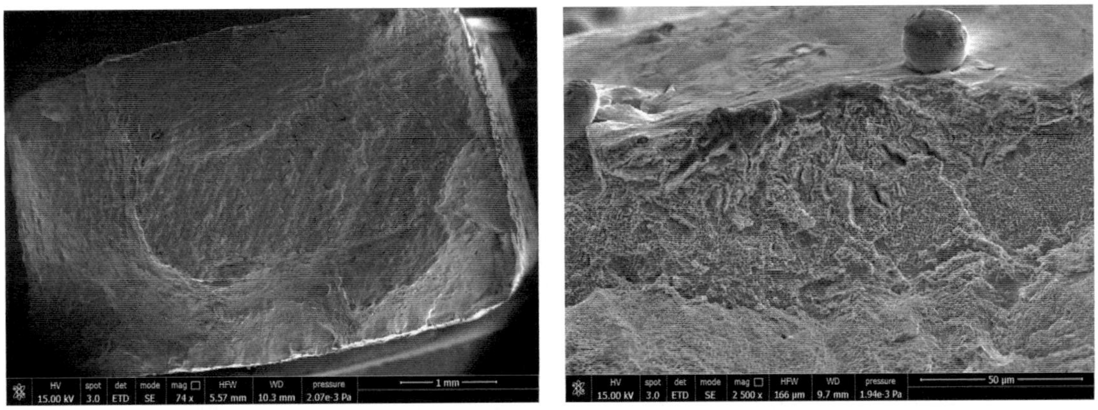

(a) (b)

Figure 6. SEM image of the whole fracture surface (a), and of detail near specimen wall (b).

(a) (b)

Figure 7. Fracture surfaces: specimens, not hydrogen charged (a), and with an average hydrogen concentration of 7.2 ppmw (b).

In the case of hydrogen charged specimens, the edges of the fracture surface appear to be sharper than the base material. In both cases in the central region, traces of the AM scan path are visible, slightly more marked for the hydrogen charged specimens.

While in the uncharged condition a visible necking occurs, charged specimens reached the fracture condition before the occurrence of necking, in particular at higher hydrogen concentrations, as shown in Figure 8.

As it is possible to observe in Figure 9, traces of layer delamination appeared along the gauge length of the specimen, confirming that the crack onset occurred on the specimen surface, and thus in the area affected by the hydrogen intake. The same traces can be observed also in Figure 7, in the case of the hydrogen-charged specimen. This effect can be considered peculiar of a SLMed material, which is built via a superposition of stacking layers.

3.3. Microstructural investigations

Microstructural investigations showed that the specimen microstructure is strongly influenced by the SLM process, with visible traces of the SLM scan paths, highlighted by segregations on the melt pool boundaries, and columnar grains, Figure 10, as it is typical for the as-built condition.

(a) (b)

Figure 8. Side view of the specimen fracture region: not hydrogen charged (a), and featuring an averaged hydrogen concentration of 17 ppmw (b).

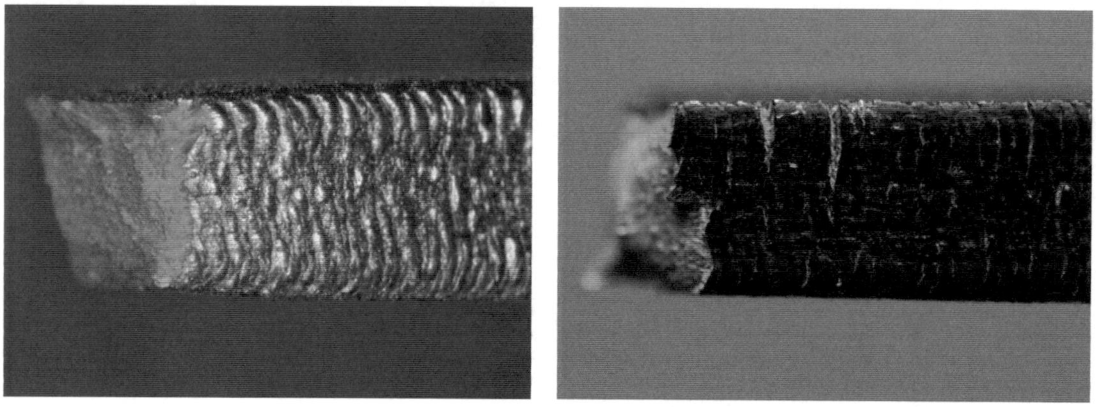

(a) (b)

Figure 9. Detail of the surface of an uncharged specimen (a), and of a specimen with an average hydrogen concentration of 7.8 ppmw (b), featuring visible traces of layer delamination.

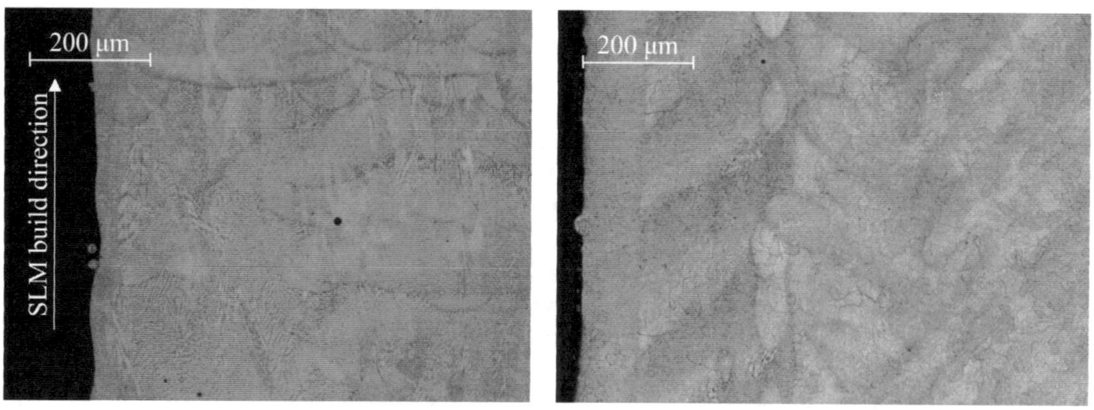

(a) (b)

Figure 10. Specimen microstructure in the longitudinal (a) and transverse direction (b), etchant Kalling's II.

The adopted SLM process parameters resulted to produce an almost full-dense material, having a density of 99.96%, if examined from sections extracted in longitudinal directions, and 99.94% in the transversal direction. The pores are round and small, the maximum length of the major axis is less than 60 μm, suggesting that they arise from gas bubbles entrapped during the SLM process.

4. Conclusions

A preliminary investigation of the hydrogen embrittlement effect on the Inconel 718 alloy produced by SLM was carried out. The effective hydrogen diffusivity within the material was indirectly evaluated by measuring the hydrogen content at different cathodic charging durations. It was obtained a value of $5 \cdot 10^{-15}$ m^2/s at a temperature of 50 °C, in line with the conventional wrought material in the solutionized condition. The tensile properties resulted to be degraded by the hydrogen intake, featuring a ductility reduction of about 50% in presence of average hydrogen content of 28 ppmw. As the hydrogen penetration depth is limited to the outer part of the specimen, the actual value of hydrogen concentration in the crack onset region is expected to be significantly greater. The susceptibility to hydrogen embrittlement of the SLMed material in the as-built conditions resulted to be in line with the conventional wrought Inconel 718.

Acknowledgments

The authors wish to thank Dr. Randa Anis Ishak Nakhla for her contribution to the SEM investigations and Mr. Roberto Minelli for the mechanical tests.

References

[1] Jothi S, Merzlikin S, Croft T, Andersson J and Brown S 2016 *Journal of Alloys and Compounds* **664** 664–681
[2] Li X, Zhang J, Akiyama E, Fu Q and Li Q 2019 *Journal of materials science & technology* **35** 499–502
[3] Hicks P and Altstetter C 1992 *metallurgical transactions A* **23** 237–249
[4] Xu J, Sun X, Liu Q and Chen W 1994 *Metallurgical and Materials Transactions A* **25** 539–544
[5] Jothi S, Sebald T, Davies H, Reese E D and Brown S 2016 *Materials & Design* **90** 101–114
[6] Debarbadillo J J and Mannan S K 2012 *Jom* **64** 265–270
[7] Fournier L, Delafosse D and Magnin T 1999 *Materials Science and Engineering: A* **269** 111–119
[8] Teus S *Physics of Metals and Advanced Technologies*
[9] Jebaraj J J, Morrison D J and Suni I I 2014 *Corrosion science* **80** 517–522
[10] Ehrlin N, Bjerkén C and Fisk M 2016 *AIMS Materials Science* **3** 1350–1364
[11] Li X, Zhang J, Fu Q, Akiyama E, Song X, Wang Y, Li Q and Zou N 2018 *international journal of hydrogen energy* **43** 20118–20132
[12] Zhang Z, Obasi G, Morana R and Preuss M 2016 *Acta Materialia* **113** 272–283
[13] Oriani R A 1970 *Acta metallurgica* **18** 147–157
[14] Crank J 1975 *New York* **19752** 1–21

Fracture mechanics of additive manufactured crack-like notches by digital image correlation

Giangiacomo Minak[1], Nenad Drvar[2]

[1]Alma Mater Studiorum – Università di Bologna, via Fontanelle 40, 47131 Forlì, Italy
[2]Topomatika d.o.o. Industrijska ulica 3, Novaki HR-10431, Sveta Nedelja, Croatia

giangiacomo.minak@unibo.it

Abstract. The rapid development of a variety of Additive Manufacturing (AM) techniques is witnessed by the large interest of the scientific community. Among the numerous new features offered by these techniques, very lately the possibility of including cracks of any shape, direction, and position inside AM samples to validate FEM models was explored. Therefore, the necessity arose of full-field measurement techniques that would allow the evaluation of the fracture mechanics parameters in the cases of both linear elastic and partially plastic materials, as a function of the fracture modes. Invented in the '80 of the XX century, the Digital Image Correlation (DIC) in recent years has taken place in nearly every laboratory. It is used to measure displacements, and by numerical differentiation or coupling with Finite Element Method, to calculate strains at different spatial scales. In particular, two procedures that were successfully employed on a specific kind of specimens, based on DIC at a small scale near the crack-tip, in this paper are shown in detail.

1. Introduction

Additive manufacturing in the last few years demonstrated unique capabilities and attracted much research interest. Industrial applications, which include automotive, aerospace, biomechanics, take advantage of the possibility of tailoring the characteristics of the products utilizing shapes not attainable using the classical subtractive manufacturing techniques.

The increasing use of AM components led to the study of the fracture mechanics of the relevant base materials, as a function of the process parameters and the built direction. An extensive analysis on the topic is reported in a recent review by Khosravani et al [1].

A new application for these techniques, proposed by Brugo et al. [2], consists in the embedding of a crack of any shape and position in AM components. In this way, the validation possibilities for numerical models can be greatly widened at least in terms of global stiffness and resistance. However, full-field measurement methods are needed to evaluate the local stress and strain at the crack-tip and to follow its evolution.

The use of the Digital Image Correlation technique to study fracture mechanics problems was introduced in the mid '80 [3,4], few years after the invention of the method in 1982 by Peters, Sutton et al. [5-6] and in the past few years emerged over other methods like the Object Grating [7].

In this paper, two procedures based on the (DIC) to evaluate the fracture behavior of AM components with embedded cracks are described.

Nowadays, the DIC is a relatively cheap, easy, and widely available optical technique used to measure the full displacement and strain field. It requires just one or two cameras, for 2D and 3D measurements respectively, of sufficiently high resolution and an algorithm to correlate the images. The correlation is performed by exploiting a speckle pattern (obtained by spray paint or ink powder) on the surface of the sample. This random pattern is a key point for the correlation algorithm, which matches subsets of pixels in consecutive images. By locating a point in the undeformed and the deformed image the displacement can be calculated.

While the displacement field is computed directly by DIC, the strain field is calculated by numerical differentiation of the displacement, with an inevitable introduction of noise. This is the reason why fitting methods usually are applied to the displacement field rather than to the strain one.

These displacement fields can be used to quantitatively characterize the fracture behavior of the material, and in particular, they can be exploited to estimate the stress intensity factors (SIFs).

Generally speaking, for linear elastic materials two main approaches to the fitting of theoretical models on the full displacement fields are shown in the literature. The first one is based on guessing a general form of an analytical function and fitting this to the displacement experimental data [8-10]. The second approach, which is applied in this paper, is based on the Williams' model [11-13], and the fitting on the experimental data is done by considering the first n term of the Williams expansion of the displacement.

The stress intensity factors (SIFs) are suitable parameters only in the case of a material that can be considered linear elastic. On the other hand, if the material exhibits an elastic-plastic behavior, the J-integral, originally proposed by Rice [14], can be successfully used to characterize the fracture mechanics behavior.

The J-integral represents the strain energy release rate or energy per unit fracture surface area of a material, and it is usually calculated from the integration of the experimental load-displacement curve.

However, this technique is suitable only for laboratory tests, in which the geometry factors for the specific specimen are known, and the load-displacement curve can be measured.

On the contrary, on a complex structure under various loads, it is generally impossible to compute the geometrical factors and the load-displacement boundary conditions for the specific position in which the crack is located.

To fit the purpose of analyzing cracks in a generic position of a real component, an alternative method consists of computing the J-integral from the displacement fields around the crack tip, measured by DIC. The DIC-based technique computes numerically the J-integral along a rectangular contour of the DIC grid [15-19].

To perform the J-integral calculation, displacement, strain, and stress fields must be known. The displacement and strain field are obtained from the DIC, while the stress-strain constitutive relation of the material must have been previously determined (e.g. by means of the Ramberg-Osgood model, as it is done in [20]).

In this paper the application of the two DIC based methods is shown in the case of two different materials, one Martensitic steel which can be considered linear elastic [21] and a Polyamide [22] which exhibits marked plasticity.

The experiments were performed on IASCB specimens [23-25] manufactured by Selective Laser Sintering (SLS). During the AM process, crack-like notches with different angles were embedded in the specimen and different values of support spans were considered in the test. Three repetitions were done for each specimen type and test condition.

The comparison with the results obtained employing standard methods for different fracture mode I-II mixity is shown.

2. Materials and Methods

2.1. Materials

The metallic specimens were manufactured by the EOSINT M 280 SLS machine (EOS GmbH), using the EOS Maraging Steel MS1 steel powder, with a chemical composition corresponding to the European 1.2709 classification. Other manufacturing details and material properties can be found in [21].

The polymeric specimens were produced by an EOS Formiga P100 SLS machine (EOS GmbH), using fine polyamide PA 2200 powder purchased from Prototal Plastic Design and Service (Götene, Sweden). The laser-sintered specimens were manufactured from mixed powder, recycled and new in equal proportions [22].

During the fabrication via the SLS technology, the specimens were oriented with the thickness in the built direction for both kinds of specimens.

2.2. IASCB Specimens

The IASCB specimen is a semicircular disk with an embedded radial edge-crack, tilted with respect to the load direction. The fracture test consists of the application of a vertical force, with the specimen located on two bottom supports, as shown in Figure 1. By varying the support spans and the crack tilt, it is possible to obtain a large variety of mixed I/II modes. In the cases under study, the thickness was 6 mm, the disk radius 60 mm and the crack length 24 mm. One of the spans was kept fixed at 42 mm and the other was varied, together with the crack angle, to change the mode mixity defined as $M = 2/\pi\, \mathrm{atan}(K_{II}/K_I)$ where K_I and K_{II} are the Sifs in mode I and II respectively.

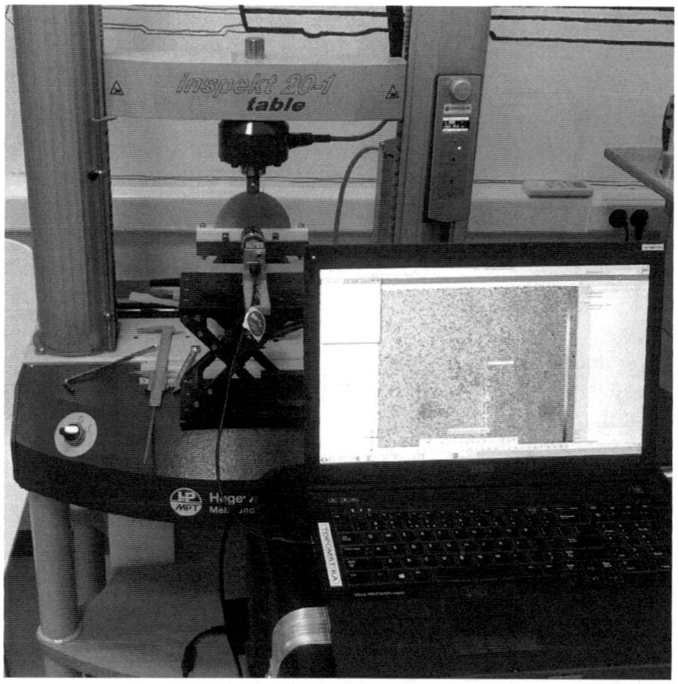

Figure 1: Experimental Setup

2.3. Digital Image Correlation

The 2D DIC system used to evaluate the full displacement field around the crack tip (in a window of 12 mm × 9 mm) consisted of a 10 MP Basler ace acA3800-14uc camera (Basler, Ahrensburg, Germany) equipped with Basler lens C125-2522–5 M-P f25 mm (Basler, Ahrensburg, Germany) and a custom led lamp.

A 2D measurement setup instead of 3D was chosen for a combination of measurement system reasons, specimen shape and loading conditions and postprocessing methodology. Since it is possible to assume that there are no large displacements that would cause the specimen to move significantly in the field of view, any errors that are the outcome of lens imperfections can be neglected considering that displacements are pretty much localized. When measuring small field of view (FOV) 2D setup gives more flexibility with FOV sizes, since the number of pixels in the camera is constant but just by changing the lens settings and distance from the object it is possible to adjust the 2D FOV practically to any desired size, compared to 3D DIC setups which tend to have the FOV predefined by the size of the available calibration objects. Care must be taken with 2D setups since the camera chip needs to be perfectly parallel to the specimen surface, and the specimen should have negligible out of plane displacement and rotations during the test since there is no possibility to compensate for out of plane rigid body motion, which is possible with 3D setups. Moreover, in 3D setups, the surface must be smooth enough not to cause problems to DIC when viewed from different sides (this should be considered for big magnifications and if 3D printed specimens with large layers or methods with visible layers are used).

The images were acquired by exploiting the GOM Snap 2D free software and processed by GOM Correlate (GOM GmbH, Braunschweig, Germany). GOM Snap can use USB cameras compatible with GenICam protocol, but as in the case of the present paper the loading application must be slow enough not to cause additional errors if using cameras with rolling shutters.

A black speckle pattern was spray-painted on the specimens (which naturally exhibits a white surface), as required by the digital correlation algorithm.

The DIC parameters were set according to the study of Palanca et al. [26], as follows: 19 pixels facet size, 16 pixels point distance, 8 facets spatial filter (median) and 3 time-step temporary filters (median). This parameter combination allows obtaining an optimal compromise between the spatial resolution and the precision of the displacement field measurement around the crack tip.

2.4. Evaluation of stress intensity factors by the DIC full displacement field

To model the displacement near the crack tip, the Williams' asymptotic formulation [27] was exploited. The u and v are the displacement along the crack axis (x-axis) and along an axis perpendicular to the former and contained in the crack plane (y-axis). Figure 2 shows the reference system used, with the origin located into the crack tip.

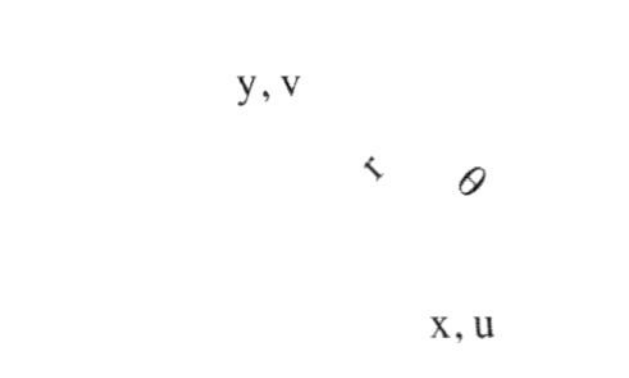

Figure 2: Reference system

In mode I, u and v are expressed as functions of the polar coordinates (r, θ), according to the series shown in Equations 1-2.

$$u_I = \sum_{n=1}^{\infty} \frac{r^{\frac{n}{2}}}{2\mu} a_n \left\{ \left[\kappa + \frac{n}{2} + (-1)^n \right] cos\frac{n\theta}{2} - \frac{n}{2} cos\frac{(n-4)\theta}{2} \right\} \tag{1}$$

$$v_I = \sum_{n=1}^{\infty} \frac{r^{\frac{n}{2}}}{2\mu} a_n \left\{ \left[\kappa - \frac{n}{2} - (-1)^n \right] sin\frac{n\theta}{2} + \frac{n}{2} sin\frac{(n-4)\theta}{2} \right\} \tag{2}$$

In mode II, u and v are computed as functions of the polar coordinates (r, θ), according to the series shown in Equations 3-4.

$$u_{II} = -\sum_{n=1}^{\infty} \frac{r^{\frac{n}{2}}}{2\mu} b_n \left\{ \left[\kappa + \frac{n}{2} - (-1)^n \right] sin\frac{n\theta}{2} - \frac{n}{2} cos\frac{(n-4)\theta}{2} \right\} \tag{3}$$

$$v_{II} = \sum_{n=1}^{\infty} \frac{r^{\frac{n}{2}}}{2\mu} b_n \left\{ \left[\kappa - \frac{n}{2} + (-1)^n \right] cos\frac{n\theta}{2} + \frac{n}{2} cos\frac{(n-4)\theta}{2} \right\} \tag{4}$$

where $\mu = E/(2+2\nu)$ is the shear modulus and $\varkappa = (3-\nu)/(1+\nu)$ for plane stress and $\varkappa = 3-4\nu$ for plane strain condition, respectively being E the elastic modulus and ν the Poisson ratio; a_n and b_n are the model parameters. The specimens are considered subjected to plane stress condition [28].

For mixed modes, the displacement fields can be obtained by the superimposition of the displacement of pure mode I and pure mode II, as shown in Equations 5-6

$$u = u_I + u_{II} \tag{5}$$

$$v = v_I + v_{II} \tag{6}$$

The values of K_I and K_{II} are calculated from the value of the Williams model parameters as shown in Equations 7-8.

$$K_I = a_1\sqrt{2\pi} \tag{7}$$

$$K_{II} = -b_1\sqrt{2\pi} \tag{8}$$

Four additional parameters (three translations and the rotation around the z-axis, perpendicular to the crack plane) were accounted to compensate rigid motions while performing the fitting, as done also in [28] and [29]. This approach is suitable when the material exhibits small-scale plastic deformation, on the other hand, it introduces non-linearity in the fitting parameters because the polar coordinates (r, θ) are modified as shown in Equations 9-10.

$$r = \sqrt{(x - x_0)^2 + (y - y_0)^2} \tag{9}$$

$$\theta = arctg(\frac{y - y_0}{x - x_0}) \tag{10}$$

2.5. J-integral computation by means of DIC

As said, the J-integral represents the strain energy release rate or energy per unit fracture surface area of a material. It is mathematically defined as a line-integral, as shown in Equation 11:

$$J = \int_{\Gamma} \left(W dy - t_i \frac{\partial u_i}{\partial x} ds \right) \tag{11}$$

where Γ is the integration path, W is the strain energy density, $t = \sigma\hat{n}$ is the surface traction vector ($\hat{n}$ is normal to the curve Γ and σ is the Cauchy stress tensor). The most important property of the J-Integral is that it does not depend on the path used for the integration. In our case, since the DIC data are defined on a rectangular discrete grid the definition of the J integral has to be suitably adapted. If the grid is sufficiently dense this is an acceptable approximation [16, 30, 31]. The integration is carried out by using the trapezoidal rule. The integrand function f in the discrete domain can be rewritten as shown in Equation 12, being Δl the path increment:

$$f \Delta l = W \Delta y - t_i \frac{\Delta u_i}{\Delta x} \Delta s \tag{12}$$

Expanding the term of the traction vector in the hypothesis of plane stress, and summing on the index i leads to Equation 13:

$$f \Delta l = W \Delta y - \left[\frac{\Delta u_x}{\Delta x} \left(n_x \sigma_x + n_y \tau_{xy} \right) + \frac{\Delta u_y}{\Delta x} \left(n_y \sigma_y + n_x \tau_{xy} \right) \right] \Delta s \tag{13}$$

The rectangular path is divided in five parts, as shown in figure 3, so that each part can be further simplified as in Equations 14-17.

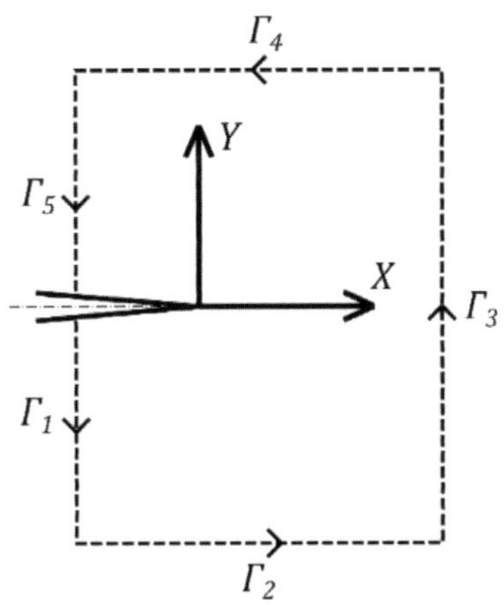

Figure 3: Rectangular path around the crack tip.

Path Γ1 and Path Γ5: $\hat{n} = (-1, 0)$, $\Delta s = -|\Delta gy|$, $\Delta y = -|\Delta gy|$ where $|\Delta gy|$ is the distance (constant and ≥ 0) between 2 points of the DIC grid along the y direction.

$$f \Delta l = \left(-W - \frac{\Delta u_x}{\Delta x} \sigma_x - \frac{\Delta u_y}{\Delta x} \tau_{xy} \right) |\Delta gy| \tag{14}$$

Path Γ2: $\hat{n} = (0, -1)$, $\Delta s = |\Delta gx|$, $\Delta y = 0$ where $|\Delta gx|$ is the distance (constant and ≥ 0) between two points of the DIC grid along the x direction.

$$f \Delta l = \left(\frac{\Delta u_x}{\Delta x} \tau_{xy} + \frac{\Delta u_y}{\Delta x} \sigma_y \right) |\Delta gx| \tag{15}$$

Path Γ3: $\hat{n} = (+1, 0)$, $\Delta s = |\Delta gy|$, $\Delta y = |\Delta gy|$

$$f\Delta l = \left(+W - \frac{\Delta u_x}{\Delta x}\sigma_x - \frac{\Delta u_y}{\Delta x}\tau_{xy}\right)|\Delta gy| \tag{16}$$

Path Γ4: $\hat{n} = (0, +1),\ \Delta s = -|\Delta gx|,\ \Delta y = 0$

$$f\Delta l = \left(\frac{\Delta u_x}{\Delta x}\tau_{xy} + \frac{\Delta u_y}{\Delta x}\sigma_y\right)|\Delta gx| \tag{17}$$

The stress components needed to compute f in each path are obtained from the strain components using the elastic-plastic constitutive relations, in our case the well-known Ramberg-Osgood relation, shown in Equation 18, is used.

$$\frac{\varepsilon_e}{\varepsilon_0} = \frac{\sigma_e}{\sigma_0} + \alpha\left(\frac{\sigma_e}{\sigma_0}\right)^n \tag{18}$$

where σ_e and ε_e are the equivalent stress and strain, σ_0 and ε_0 $(=\sigma_0/E)$ are the yield stress and strain, n, α and E the hardening exponent, the material constant, and the elastic modulus, respectively. The equivalent Von Mises strain shown in Equation 19. The strain along the z-direction can be computed according to Hooke's law in the case of plain stress from the x and y components as: $\varepsilon_z = \frac{-\nu}{(1-\nu)}(\varepsilon_x + \varepsilon_y)$.
The tangential stresses are null on the plane, and therefore ε_{xz} and ε_{yz} are null.

$$\varepsilon_e = \frac{2}{3}\sqrt{{\varepsilon_x}^2 + {\varepsilon_y}^2 + {\varepsilon_z}^2 - \varepsilon_x\varepsilon_y - \varepsilon_x\varepsilon_z - \varepsilon_y\varepsilon_z + 3{\varepsilon_{xy}}^2} \tag{19}$$

The Ramberg-Osgood parameters α, n and σ_0 are be obtained by an experimental characterization of the material.
Following the approach used in [17], the constitutive relation can be written as shown in Equation 20.

$$\varepsilon_{ij} = \frac{1+\nu}{E}s_{ij} + \frac{1-2\nu}{3E}\sigma_{kk}\delta_{ij} + \frac{3}{2}\alpha\varepsilon_0\left(\frac{\sigma_e}{\sigma_0}\right)^{n-1}\frac{s_{ij}}{\sigma_0} \tag{20}$$

where ν, δ_{ij} and s_{ij} are the Poisson's ratio, Kronecker delta and the stress deviator, respectively. The stress deviator is defined as shown in Equation 21:

$$s_{ij} = \sigma_{ij} - \frac{\sigma_{kk}}{3}\delta_{ij} \tag{21}$$

The strain energy density W can be computed, following Anderson [31], as shown in Equation 22.

$$W = \frac{1}{E}\left[\frac{1}{2}{\sigma_e}^2 + \frac{1}{2}(1-2\nu)\left(\sigma_x\sigma_y - {\tau_{xy}}^2\right) + \frac{\alpha n}{n+1}{\sigma_e}^2\left(\frac{\sigma_e}{\sigma_0}\right)^{n-1}\right] \tag{22}$$

3. Results

In this section some of the results of [21] and [22] are reported for sake of comparison of the two different DIC-based methodologies, the complete set of experimental data can be retrieved from the aforementioned papers.

3.1 Stress intensity factors by the DIC displacement field interpolation

In figure 5, the v components of the displacement as measured by DIC are reported for the Mode Mixity from pure Mode I to roughly 50%, while in figure 6 both the u and v displacements are shown in the case of nearly pure Mode II.

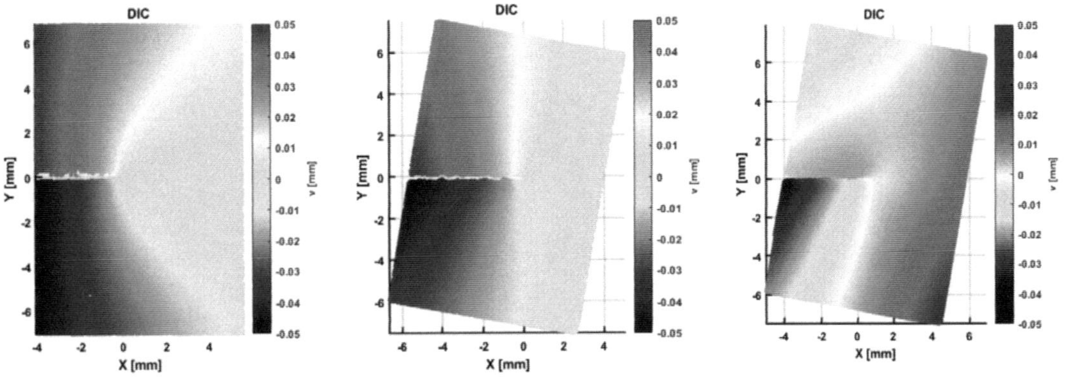

Figure 4: Measurement of the v displacement by DIC for Mode Mixity values (1, 0.77, 0.48)

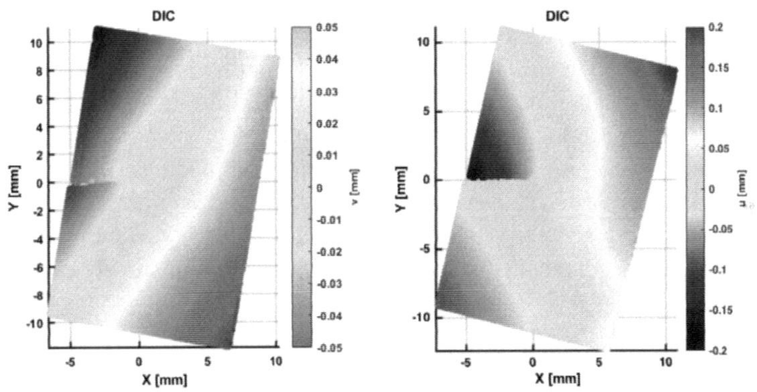

Figure 5: Measurement of the v and u displacement by DIC for Mode Mixity 0.12

The values of Stress Intensity Factors calculated according to the Equations 7 and 8 can be compared to those obtained from the Load-Displacement curve and the geometry factors defined for the IASCB specimens (PCR henceforth) as shown in Equations 23-24 (where P is the critical load, a is the crack length, R and t are the specimen radius and thickness and Y_I and Y_{II} are geometry factors calculated by FEM for the specific geometry [33].

$$K_I = \frac{P}{2Rt}\sqrt{\pi a Y_I} \tag{23}$$

$$K_{II} = \frac{P}{2Rt}\sqrt{\pi a Y_{II}} \tag{24}$$

In figure 6, the values of the SIFs obtained by the DIC procedure, and by the PCR method are compared for different values of Mode Mixity. A good agreement can be seen for Mode II while a large discrepancy is found for Mode I, in particular when it is pure.

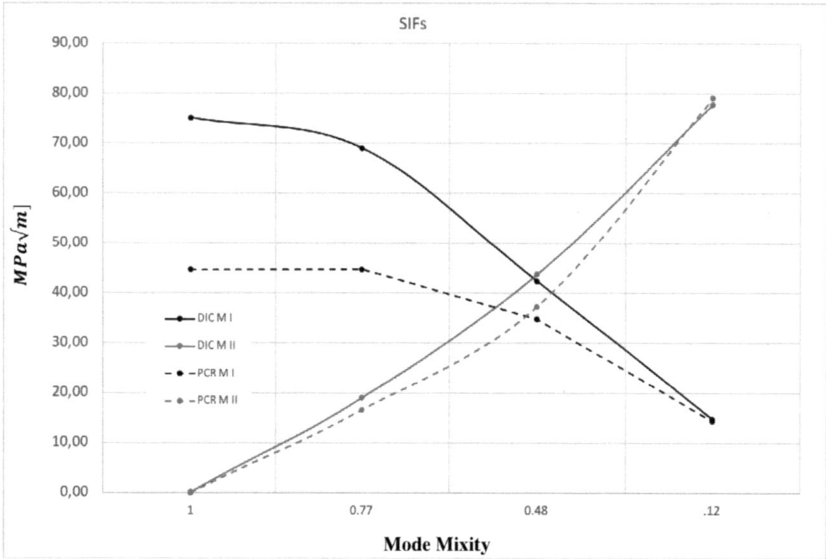

Figure 6: SIFs as a function of the mode mixity: comparison between DIC and PCR

This discrepancy cannot be provoked by the small plastic zone at the crack-tip, which by elementary fracture mechanics, could be responsible for not more than 5% of the difference. On the other side, it is well known, see Anderson, page 76 [31] that the K_I depends on the three-dimensional stress state at the crack tip. More specifically, it is higher for thin specimens, which are in-plane stress condition, while it decreases monotonically as specimen thickness increase until a plane-strain state is reached, and the outermost layers of materials become less important in the overall behavior.

In our case, the overestimation of the K_I obtained by DIC based technique is because the methodology evaluates the SIF from the displacement field measured on the outer surface, which is in-plane stress condition. This phenomenon becomes less important reducing the Mode Mixity until Mode II is reached.

3.2 . J-integral computed utilizing DIC

The J-integral was computed on different rectangular domains, similar to the one shown in figure 3, having removed the data within 0.3 mm (this value was determined empirically) from the crack axis because particularly noisy, due to the massive plasticization and to the crack opening. The value of the J-Integral was obtained by averaging on these paths a few millimeters long since, due to the data discretization, the integral was not strictly path-independent.

Figure 7 shows, as examples, the x and y displacement and strain maps obtained by the DIC two extreme configurations tested, pure Mode I and nearly pure Mode II. It can be observed that the strain in y and in the x-direction (in particular) increases going from Mode I to Mode II, which lead to a higher strain energy work to make the crack propagate.

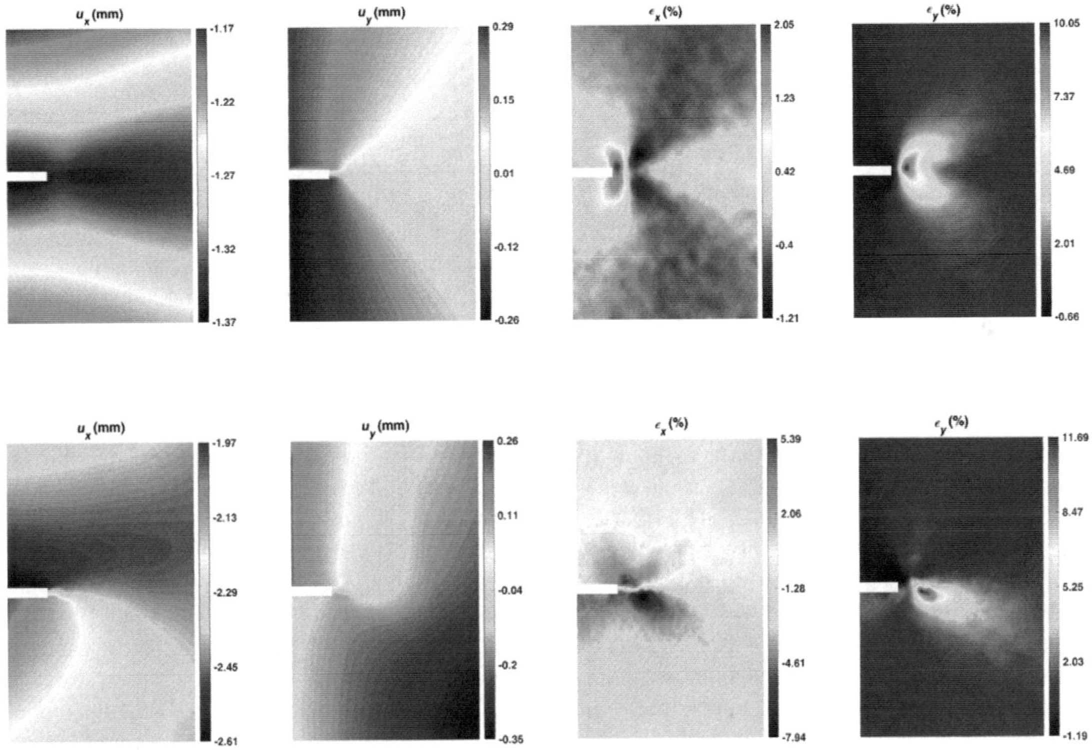

Figure 7: Measurement displacement and strain DIC for Mode Mixity values 1 and 0.12

As shown in [22] the J-integral can be derived by the Load-Displacement curve (LDC Method) once the indentation due to the support pressure has been removed and geometry factors dependent on the specimen shape and loading conditions are applied. In figure 8, the J-integral values computed through both methods are shown.

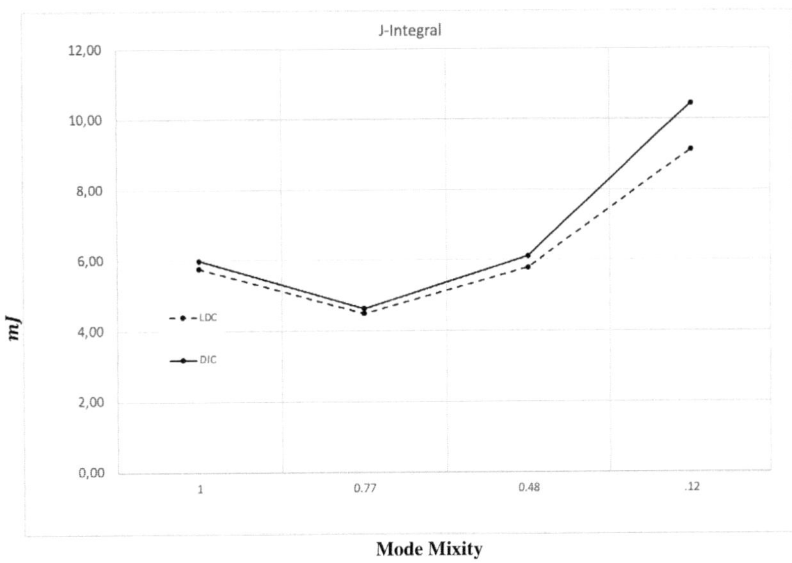

Figure 8: J-Integral versus the Mode Mixity as calculated by the DIC and LDC methodologies

Like in the previous case the DIC based technique tends to slightly overestimate the fracture mechanics parameter, but for all the Mode Mixity values. The J-integral is significantly higher in Mode II as it is well-known in literature [22].

4. Discussion

The two methods based on DIC which were shown, have some unique capabilities, but also some limitations to be considered. As recalled already in the title, by AM it is possible to produce crack-like notches, but not real cracks. The minimum radius of curvature depends on the technology and in the cases reported it was around 90 micrometers. The possibility to make this crack grow by fatigue pre-cracking is linked to the component geometry and loading. Moreover, it would be tricky to make the crack propagate exactly in the desired direction. References [21, 22] show the difference with respect to an actual crack in Mode I, which is more marked in the case of materials having higher-yielding strength.

On the other side, the possibility to embed crack-like notches of any shape, in any position and with any orientation in a model of a component can provide new design tools or validation benchmarks for numerical models. In the case of thick components, the evaluation of the fracture parameters by DIC on the surface can lead to an overestimation of the SIFs for high values of Mode Mixity, near to pure Mode I. To overcome this problem, a full 3D numerical analysis should support the DIC measurements (in general this task is quite straightforward since a detailed 3D model is necessary for the AM phase).

A very promising approach is the one based on the evaluation of the J-integral, which provides results very similar to the ones obtained by LDC in the case of specimens, where both are applicable.

The general limitation of these methods is the need to capture by two cameras in the most general 3D case the image of a surface in the immediate vicinity of the crack-dominated zone.

The methodologies described can be used also in the case of non-artificial cracks, with the same limitations above listed.

5. Conclusion

Two methodologies based on DIC have been shown in detail and compared in the case of IASCB specimens containing additive manufacture crack-like notches.

The main findings are:

- by DIC measurement of the displacement it is possible to evaluate the SIFs using the interpolation of the William's model.
- by DIC measurement of the displacement and by the derivation of the strain and stresses (using a material constitutive equation) it is possible to evaluate the average J-integral around the crack tip.
- The values obtained by the first method are overestimated in the case of pure Mode 1 with respect to the results obtained from the load-displacement curve.
- The values obtained by the second method are in good agreement with the results obtained from the load-displacement curve.

6. References

[1] Reza Khosravani M., Berto F-, Ayatollahi M.R., Reinicke T., 2020, Fracture behavior of additively manufactured components: A review, *Theor. Appl. Fract. Mech.* 109, 102763

[2] Brugo, T.; Palazzetti, R.; Ciric-Kostic, S.; Yan, X. T.; Minak, G.; Zucchelli, A., 2016, Fracture Mechanics of Laser Sintered Cracked Polyamide for a New Method to Induce Cracks by Additive Manufacturing. *Polym. Test.* **50**, 301–308.

[3] Peters, W. H., Ranson, W. F., Kalthoff, J. F., & Winkler, S. R., 1985,. A study of dynamic near-crack-tip fracture parameters by digital image analys. *J. Phys. Colloq*, **46** (C5), C5–631–C5–638.

[4] McNeill, S.R., Peters, W.H., Sutton, M.A., 1987, Estimation of stress intensity factor by digital image correlation. *Eng Fract Mech*, **28** (1), 101-112.

[5] Peters, W. H. and Ranson, W. F., 1982, Digital imaging techniques in experimental stress

analysis. *Opt Eng,* **21** (3), 427-431.
[6] Sutton, M.A., Wolters, W.J., Peters, W.H., Ranson, W.F., McNeill, S.R., 1983, Determination of displacements using an improved digital correlation method. *Image Vision Comput.* **1**, 133–139.
[7] Gubeljak, N., Semenski, D., Drvar, N., Predan, J., Kozak, D., Oblak, M., 2006, Object grating method application in strain determination on ctod tests // *Strain,* **42** (2) 81-87.
[8] Muskhelishvili, N.I., 1977, *Some basic problems of the mathematical theory of elasticity*, 4th edition, Noordhoff International, Leyden, The Netherlands
[9] Nurse, A. D., Patterson, E. A.,1993, Determination of Predominantly Mode II Stress Intensity Factors from Isochromatic Data. *Fat Frac Eng Mat Struct* ***16***(12), 1339–1354.
[10] Lopez-Crespo, P., Shterenlikht, A., Patterson, E. A., Yates, J. R., Withers, P. J., 2008, The Stress Intensity of Mixed Mode Cracks Determined by Digital Image Correlation. *J Strain Anal Eng* **43** (8), 769–780.
[11] Williams, J.G., Ewing, P.D., 1972, Fracture under complex stress – the angled crack problem. *Int J Fracture*, **8**:441–6.
[12] Papadopoulos, G. A., Poniridis, P. I., 1989, Crack Initiation under Biaxial Loading with Higher-Order Approximation. *Eng. Fract. Mech.,* **32** (3), 351–360.
[13] Silva, A. L. L., de Jesus, A. M. P., Xavier, J., Correia, J. A. F. O., 2017, Fernandes, A. A. Combined Analytical-Numerical Methodologies for the Evaluation of Mixed-Mode (I + II) Fatigue Crack Growth Rates in Structural Steels. *Eng. Fract. Mech*, **185**, 124–138.
[14] Rice, J.R., 1968, A path independent integral and the approximate analysis of strain concentration by notches and cracks. *J. Appl. Mech.* , **35**, 379–386.
[15] Vavrik, D., Jandejsek, I., 2014, Experimental Evaluation of Contour J Integral and Energy Dissipated in the Fracture Process Zone. *Eng. Fract. Mech.*, **129**, 14–25
[16] Kozłowiec, B. , 2017, Numerical Methods for Estimating J Integral in Models with Regular Rectangular Meshes. *Iop. Conf. Ser. Mater. Sci. Eng.* **175**, 012062,
[17] Breitbarth, E., Strohmann, T., Besel, M., Reh, S., 2019, Determination of Stress Intensity Factors and J Integral Based on Digital Image Correlation. *Frat. ed Integrita Strutt.*, **13**, 12–25.
[18] Catalanotti, G.,2010, Measurement of Resistance Curves in the Longitudinal Failure of Composites Using Digital Image Correlation. *Compos. Sci. Technol.*, **70**, 1986–1993.
[19] Shih, C.F., Moran, B., Nakamura, T., 1986, Energy Release Rate along a Three-Dimensional Crack Front in a Thermally Stressed Body. *Int. J. Fract.* **30**, 79–102.
[20] Gonzáles, G.L.G., González, J.A.O., Castro, J.T.P., Freire, J.L.F., 2017, A J-Integral Approach Using Digital Image Correlation for Evaluating Stress Intensity Factors in Fatigue Cracks with Closure Effects. *Theor. Appl. Fract. Mech.* **90,** 14–21.
[21] Campione, I., Brugo, T.M., Minak, G., Tomić, J.J., Bogojević, N., Kostić, S.Ć., 2020, Investigation by Digital Image Correlation of Mixed Mode I and II Fracture Behavior of Metallic IASCB Specimens with Additive Manufactured Crack-Like Notch. *Metals* **10**, 400.
[22] Brugo, T.M., Campione, I., Minak, G. ,2021, Investigation by Digital Image Correlation of Mixed-Mode I and II Fracture Behavior of Polymeric IASCB Specimens with Additive Manufactured Crack-Like Notch. *Materials* , ***14***, 1084.
[23] Saghafi, H., Monemian, S., 2010, A New Fracture Toughness Test Covering Mixed-Mode Conditions and Positive and Negative T-Stresses. *Int J Fract* , **165** (1), 135–138.
[24] Ayatollahi, M. R., Aliha, M. R. M.;,Saghafi, H. 2011, An Improved Semi-Circular Bend Specimen for Investigating Mixed Mode Brittle Fracture. *Eng. Fract. Mech.*, **78** (1), 110–123.
[25] Saghafi, H., Zucchelli, A., Minak, G., 2013, Evaluating Fracture Behavior of Brittle Polymeric Materials Using an IASCB Specimen. *Polymer Testing,* **32** (1), 133–140.
[26] Palanca, M., Brugo, T. M., Cristofolini, L., 2015, Use of Digital Image Correlation to Investigate the Biomechanics of the Vertebra. *J. Mech. Med. Biol.* **15** (02), 1540004.
[27] Williams, M. L., 1957, On the Stress Distribution at the Base of a Stationary Crack. *J Appl Mech*

, **24** (6), 109-114.
[28] Yates, J. R., Zanganeh, M., Asquith, D., Tai, Y. H., 2010, Quantifying Crack Tip Displacement Fields: T-Stress and CTOA. *Eng. Fract. Mech.*, **77** 2063–2076
[29] Yoneyama, S., Ogawa, T., Kobayashi, Y., 2007, Evaluating Mixed-Mode Stress Intensity Factors from Full-Field Displacement Fields Obtained by Optical Methods. *Eng. Fract. Mech.*, **74** (9), 1399–1412.
[30] Yamane, H.,Arikawa, S., Yoneyama, S.; Watanabe, Y., Asai, T., Shiokawa, K., Yamashita, M., 2014, J-Integral Evaluation for an Interface crack using digital image correlation, *J. of JSEM*, **14**, 122-127.
[31] Anderson, T.L. ,2017, *Fracture Mechanics: Fundamentals and Applications*; CRC Press: Boca Raton, FL, USA
[32] Darban, H. Haghpanahi, M. Assadi, A. , 2011, Determination of crack tip parameters for ascb specimen under mixed mode loading using finite element method. *Comput. Mater. Sci.*, **50,** 1667–1674.

Acknowledgments
The research presented in this paper is part of the A_Madam Project, funded by the European Union's Horizon 2020 research and innovation program under the Marie Skłodowska-Curie action, grant agreement No. 734455.

Experimental analysis on the resistance of fixing nails for spine cages

G La Rosa, R Mineo, R Barbagallo and C Tosto

Dept. of Civil Engineering and Architecture, University of Catania, Via S. Sofia 64, 95125 Catania – Italy

guido.larosa@unict.it

Abstract. Arthrodesis is one of the most common surgical treatment for the management of pathologies concerning the intervertebral discs. Special cages replace the intervertebral disc and realize the fusion between the vertebral body, often together with a fixing plate, screws, and rods. An alternative method is a cage that contains integrated screws, designed with suitable holes, it allows an easier insertion of fixing system during surgery. Mt Ortho company offers a cage equipped with two diverging locking nails in Ti6Al4V-ELI titanium alloy. These nails are placed directly inside the cage inclined by 30° with respect to the horizontal. This paper, performed in collaboration with Mt Ortho, reports the experimental tests aimed at evaluating the performance of locking nails under pull-out stresses. To evaluate the nails pull-out in a more physiological way, together to the standard condition, tests were performed on a composite layer made of wood and foam. Furthermore, the notch effects generated by the locking nails are qualitatively evaluated by photoelastic analysis, in view of a geometry able to reduce the stress concentration factor on the bone tissue, without reducing the pull-out resistance.

1. Introduction

One among the most important pathologies of the vertebral column concerns the intervertebral discs. The surgical treatment that can reduce these pathologies is arthrodesis, realized by the interbody (or intervertebral) fusion. Since its introduction in 1911, lumbar fusion has proven to have excellent clinical outcomes in treating several pathological spine conditions, such as spinal stenosis, spondylolisthesis, degenerative disc disease, spinal deformity, and trauma [1]. Nowadays intervertebral fusion is obtained distracting the intervertebral space enough to implant cages stabilizing the adjacent vertebral bodies until complete fusion occurs. The stretching of the annulus and ligaments provides stability and restore the disc height and spinal lordosis. The time for fusion varies, but is 6–12 months at minimum, therefore immediate stabilization could be required [2, 3, 4].

Special cages were designed to replace the intervertebral disc but, to allow the primary stability (biomechanical stability until the bone remodeling can assure a permanent locking) the cages are often mounted together with a fixing plate between the bodies of the two adjacent vertebrae, to prevent forces causing the ejection of the cage [5, 6].

Pedicle screws and rods are sometimes required for maintenance spinal stability until fusion occurs. However, some interbody devices used during anterior and lateral approaches may be combined with plate fixation or contain integrated screws that ensures primary stabilization and avoid the need for posterior patient repositioning [5, 6].

EBM (Electron Beam Melting) is a 3D printing technique in which a high energy source (concentrated beam of electrons) hits a bed of titanium powder causing its fusion. Mt Ortho designs and produces customized prostheses and orthopedic components in Titanium alloy by EBM additive manufacturing, to replace or substitute parts of the skeleton or as a support for surgical operations.

Aim of this study was to compare the pull-out forces of new cage integrated fixation system with traditional screw. The new fixation system was evaluated in two different manufacturing solutions.

2. Materials and methods

The traditional fixation element is a long screw with the appropriately shaped thread (Figure 1 (a)), used for mounting orthopaedic plates, in this study, two different fixation elements were compared to the traditional one: a nail with sawtooth ridges to prevent ejection (Figure 1 (b)), used in other fields of orthopedics and a similar nail made with EBM technology (Figure 1 (c)).

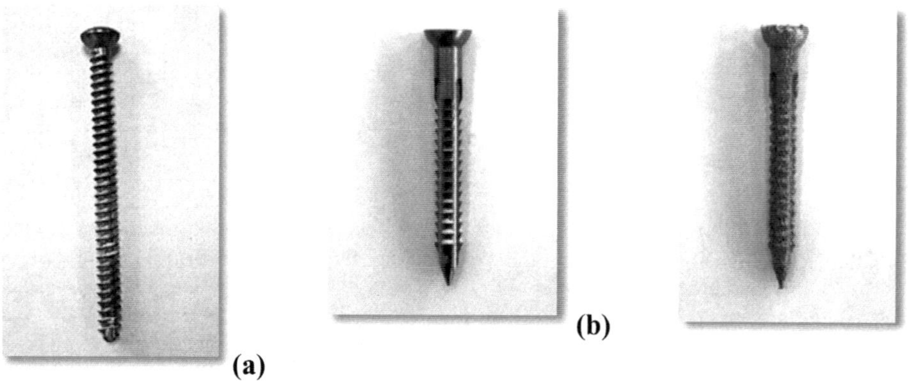

(a) (b) (c)

Figure 1. Specimens tested – (a) traditional fixation system, (b) traditional nail, (c) nail made with EBM technology

2.1. *Test procedure*

Standard *in vitro* pull-out tests require trials to be carried out on the fixation elements inserted in different foams simulating cancellous bone with different degrees of bone density [7, 8, 9]. To approach the physiological situation, in which the fixation elements cross the cortical part of the vertebral body, tests were also carried out on a mixed system consisting of a thin layer of fresh citrus wood (simulating the cortical part) overlapped on the foam (simulant the spongy part) as shown in Figure 2. Overall, several pull-out tests have been performed on the screw/nail inserted in a material simulating the bone:

- 3 foams with different density (grade 15, 20 and 40), simulating the cancellous bone, following the standard procedures.
- 3 combinations of foam (grade 15, 20 and 40) + thin wood layer, simulating more realistically the cortical (wood) + cancellous (foam) bone

2.2. *Introductory procedures*

Preliminary tensile tests were carried out on the foams and the wood, in order to define their mechanical characteristics. To avoid the errors of the measuring chain, the specimen displacements were defined by DIC measurements [10]. The tests were performed in static loading under displacement control on dog-bone specimens treated for the DIC measurements by an Instron 8501 testing machine.

Figure 3 shows the stress-strain behavior of the grade 40 foam, as an example, and the wood. The results highlighted the following characteristics for the foam: E = 0.5 GPa, σ_r = 14 MPa, similar to those of the cancellous bone in a healthy patient and E = 9.6 GPa, σ_r = 65 MPa for the fresh citrus wood, similar to those of the cortical bone. Moreover, the wood presented a noticeable laminar behavior.

IOP Conf. Series: Materials Science and Engineering 1214 (2022) 012004 doi:10.1088/1757-899X/1214/1/012004

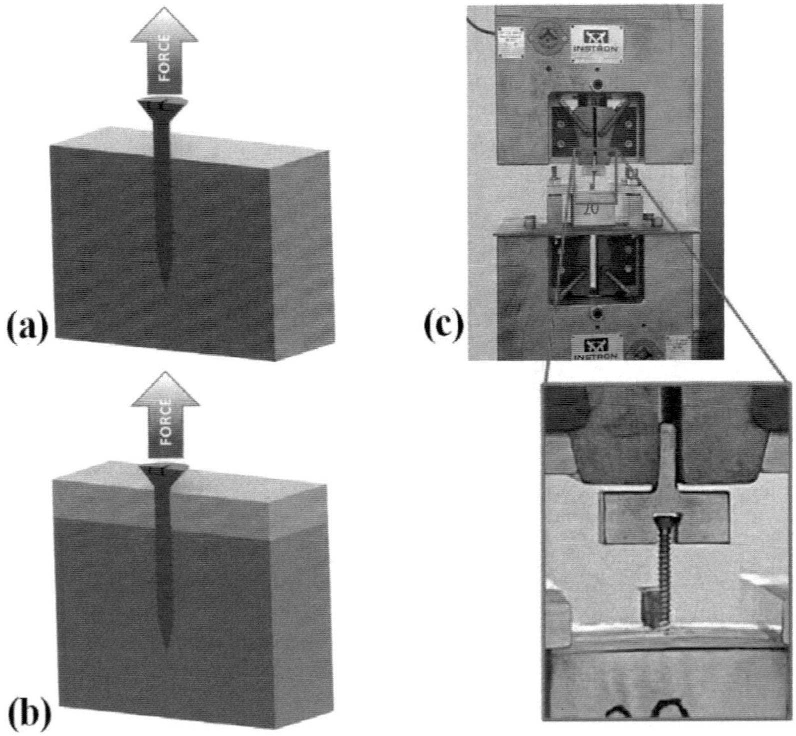

Figure 2. Pull-out tests on: (a) foam; (b) wood+foam; (c) experimental setup and clamping system.

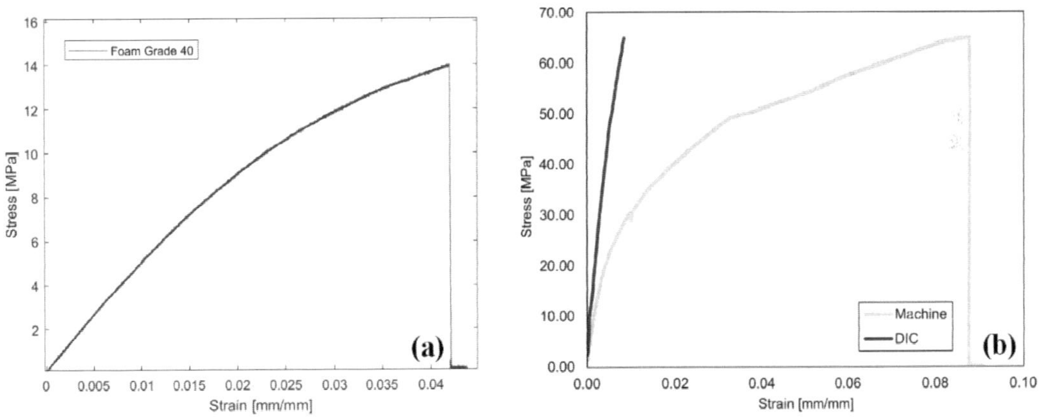

Figure 3. Mechanical characteristics of the grade 40 (a) foam and of the citrus wood (b).

2.3. Experimental setup

The tests were performed by means of an Instron electrical static testing machine with a 10 kN load cell. Figure 2 shows the experimental setup highlighting the upper and lower clamps in the testing machine.

The screws or nails were inserted respectively by screwing or by compression it on a preliminary hole with a diameter of 2 mm [11].

Screw and nails were also inserted in polycarbonate blocks with a thickness of 10 mm in order to evaluate the stress concentration factor around the tips due to the insertion using a photoelastic bench. Nails were inserted by compression in a previously made cylindrical hole with a diameter of 2 mm, screw was inserted by screwing on a previously made cylindrical hole with a diameter of 2.5 mm.

3. Results and discussion

As previously specified, the first tests were performed on the screw/nails inserted in the foam only. Five specimens were tested in order to mediate the results for each foam grade. Figure 4, as an example, shows the curves for the grade 40 foam for the two nails.

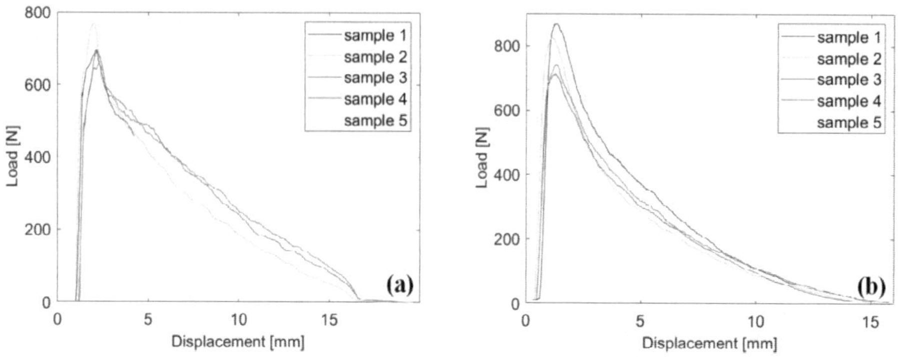

Figure 4. Curves force-displacement on grade 40 for the commercial nail (a) and EBM nail (b).

Figure 5 shows the pull-out forces for the three fixation components in the configuration with the thin wood layer with the different foam density. Tests performed show that the pull-out force is much greater for the screw respect to the nails but the weaker values for the nails, however, are large enough to resist to physiological loads.

Moreover, the insertion of a thin layer of citrus wood had more than doubles the pull-out forces in the grade 15 and 20, but it did not change the pull-out strength in grade 40. This is understandable due to the great density of the grade 40 foam, corresponding to a cancellous bone of high mechanical characteristics.

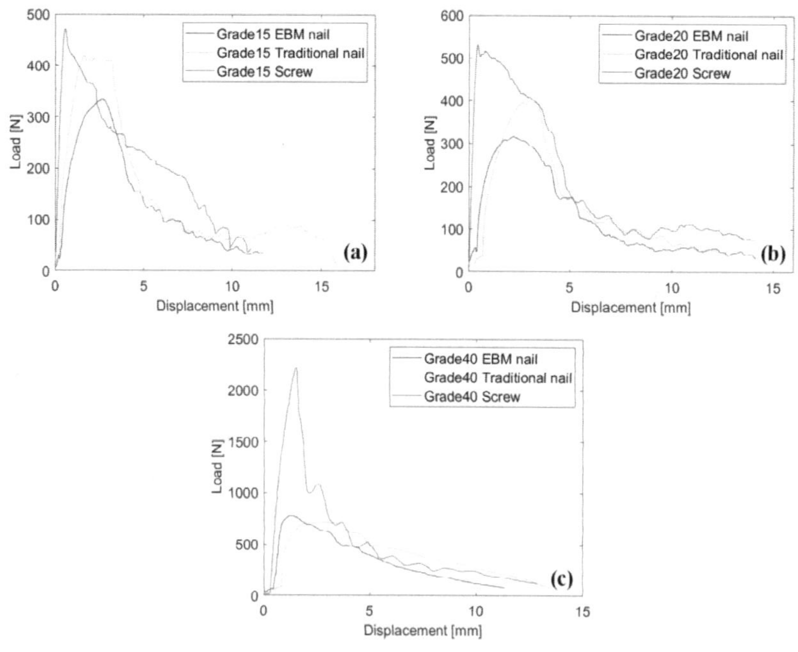

Figure 5. Curves force-displacement for the commercial nail and EBM nail on the composite wood+foam for the three different foam grades in the configuration with the wood layer: (a) grade 15; (b) grade 20; (c) grade 40.

The photoelastic images clearly show the high stress concentration factor of the screw and the traditional nail respect to the EBM nail, due to the rounded apex of the protrusions, then, the stress concentration factor is drastically reduced using the EBM nail.

4. Conclusions

Surgical requests were forwarded to verify the amount of nails instead of screws to fix the intervertebral cages, being the nails insertion easier and faster than the screws.

Tests were carried out to assess *in vitro* the pull-out forces for the two typologies of fixing systems., either on the foams (following the standard) or on a composite made of a thin layer of wood over the foam (more similar to the physiological condition).

Results verified that, even if the screw system has a greater resistance, the nail ones resist to a pull-out force large enough for the physiological use.

Among the nails, the EBM configuration show almost the same resistance than the traditional ones but a lower stress concentration effect.

References

[1] De Kunder SL, Rijkers K, Caelers I, De Bie RA, Koehler PJ and Van Santbrink H 2018 Lumbar Interbody Fusion: A Historical Overview and a Future Perspective *Spine (Phila Pa 1976)* **43**(16) 1161-1168

[2] Burkus JK, Heim SE, Gornet MF and Zdeblick TA 2003 Is infuse bone graft superior to autograft bone? An integrated analysis of clinical trials using the lt-cage lumbar tapered fusion device *J Spinal Disord Tech* **16**(2) 113-22

[3] Rodgers WB, Gerber EJ and Patterson JR 2010 Fusion after minimally disruptive anterior lumbar interbody fusion: Analysis of extreme lateral interbody fusion by computed tomography *SAS J.* **4**(2) 63-6

[4] La Rosa G, Gioè G and Fargione G 2019 Numerical simulation and experimental tests on a lumbar disc prosthesis *Mat. Today* **7** 586-592

[5] Szadkowski M, D'Astorg H, Bouhali H, Aleksic I, Ramos-Pascual S and Fière V 2020 Outcomes of stand-alone anterior lumbar interbody fusion of L5-S1 using a novel implant with anterior plate fixation *Spine J.* **20**(10) 1618-1628

[6] Baliga S, Treon K and Craig NJ 2015 Low Back Pain: Current Surgical Approaches *Asian Spine J.* **9**(4) 645-57

[7] ASTM Designation: F 543 – 02 Standard Specification and Test Methods for Metallic Medical Bone Screws

[8] ASTM Designation: F 1839 – 01 Standard Specification for Rigid Polyurethane Foam for Use as a Standard Material for Testing Orthopaedic Devices and Instruments

[9] La Rosa G, Clienti C, Mineo R and Audenino A 2016 Experimental analysis of pedicle screws *Procedia Struct. Integr.* **2** 1244-1251

[10] Bing P et al 2009 Two-dimensional digital image correlation for in-plane displacement and strain measurement: a review *Meas. Sci. Technol.* **20** (6)

[11] Daftari T.K., Horton W.C. and Hutton, W.C. 1994 Correlations Between Screw Hole Preparation, Torque of Insertion, and Pullout Strength for Spinal Screws *Journal of Spinal Disorders* **7** (2) 139-145

Numerical study on the influence of boundary conditions on the blast response of composite plates

L Lomazzi* and A Vescovini

Politecnico di Milano, Department of Mechanical Engineering, Via La Masa n.1, Milan, Italy

*Corresponding author: luca.lomazzi@polimi.it

Abstract. Blast loading represents a critical dynamic condition for both metal and composite structures. The blast response of metal materials has been extensively dealt with by researchers through experimental, analytical and numerical analyses, while composite materials appear to have been investigated in less detail. In this work, the blast response of a composite plate is numerically investigated employing the pure Lagrangian and the fully coupled Eulerian-Lagrangian (CEL) approaches using the LS-DYNA® software package. The latter methodology is set up to describe the effects determined by close-range explosions, such as the strongly non-uniform pressure distribution on the plate exposed area. Moreover, this work focuses on the characterisation of the influence of the boundary conditions on the plate response. The numerical methodology established in this work is set up and validated according to similar studies present in the literature.

1. Introduction

Composite structures are widely used in the structural engineering field and have several applications ranging from civil infrastructures, such as bridges and buildings, to mechanical systems, such as aircraft, ships and vehicles in general. Composite materials are also widespread in the military field, where they are commonly used both as part of high-performance protections, such as in protective panels for armoured vehicles, and for lightweight structures. Moreover, future trends in transportations are likely to move toward a larger use of this type of materials even for primary structures. Given the wide variety of applications, it is worth investigating the response of composite plate-like structures subjected to dynamic loading conditions, such as blast loading.

In the past fifty years many experimental tests have been performed to study the dynamic response of blast loaded plates with the purpose of formulating analytical and empirical predictive methods [1–5] or validating numerical models and methodological approaches [6–8]. The large number of studies published in the literature highlights that it is hard to accurately predict the dynamic response of plates subjected to blast loading, since it is determined by several multi-physics properties, such as the explosive material involved and its chemical composition and behaviour, the charge mass, the distance of the plate from the detonation point, the dimensions of the structure and the material it is made of.

Traditionally, most of the published works have only investigated the blast response of flat metallic monolithic plates. However, more recently, thanks to the increasing interest in advanced structural layouts and composite materials, experimental campaigns involving complex structures such as multi-layered plates [9] and structures made of composite materials [10–13] have been conducted. Similarly, also analytical methods to predict the blast response of composite plate-like structures have been recently propounded in the literature [14–17]. However, so far, the available analytical theories can only deal with simple boundary conditions and geometries and do not consider complex physical phenomena, such as the extremely non-linear effects rising in case of failure of composite plates.

Hence, finite element (FE) techniques are usually employed to deal with complex structures subjected to dynamic loading from nearby explosions [17]. This type of numerical simulations can also take into account the non-linear post-failure behaviour of composite materials, that reveals to be paramount to satisfactorily reproduce the blast loaded plate response [12,18,19]. However, to date, a limited number of numerical models employed to simulate composite plates under blast loading have been validated with experimental observations [17]. An exception is represented by the works in [12,13], that present a validated FE model of carbon fibre-polymer laminates and of woven glass/epoxy composite plates subjected to blast loading, respectively. Moreover, almost all the works present in the literature characterise the blast wave-structure interaction using the pure Lagrangian approach, that is known not to be accurate enough in close-range scenarios [20] and cannot predict eventual fluid-structure interaction (FSI) effects [21].

In this work, two methodological approaches to simulate the blast wave propagation and interaction with the target structure, i.e, the pure Lagrangian approach and the fully coupled Eulerian-Lagrangian (CEL) approach, are presented and the potentialities of each method are investigated by means of a case study taken from the experimental campaign presented in the work in [12], that involves carbon fibre-polymer laminates under blast loading. Moreover, the influence of the technique employed to model the plate boundary conditions on the structural response is investigated. This work is organised as follows. Section 2 presents and discusses the methodological approaches employed in this work to simulate the dynamic response of blast loaded composite plates. Section 3 presents a case study to investigate the potentialities of the two approaches and to characterise the influence of the modelling technique adopted to simulate the plate boundary conditions on the structural response. Finally, Section 4 discusses the main results of the case study and Section 5 draws out the conclusions from this work.

2. Methodological approach

Two approaches to the numerical simulation of blast loaded structures are investigated in this work, i.e., the pure Lagrangian approach and the fully coupled Eulerian-Lagrangian technique.

In this Section, the two approaches mentioned above are briefly described, along with the methodology set up to model the target structure. No mention to the specific solver used in this work is made in this Section, since all the information about the case study and the modelling procedure involved are reported in detail in Section 3.

2.1. *Pure Lagrangian approach*

A pure Lagrangian analysis consists of modelling the structure with Lagrangian elements and predicting the blast pressure time history exploiting empirical equations. The blast pressure is exerted on the structure as an analytical pressure load resorting to the specific equations implemented in the software package employed for the analysis. In particular, in this work the Kingery-Bulmash equations are exploited to predict the blast loading properties due to the explosion of some high explosive (HE) material [22]. Such equations, which were obtained via model fitting to experimental results, only require as input value the scaled distance *Z*, which is defined according to Equation 1 [23,24]:

$$Z=\frac{R}{W_{TNT}^{1/3}} \, (1) \qquad (1)$$

where *R* is the standoff distance, i.e., the distance from the centre of the HE spherical charge to a specific point of interest, while W_{TNT} is the TNT equivalent weight of the explosive material considered. The Kingery-Bulmash relationships employed in this work are valid at scaled distance values 0.147 $m/kg^{1/3}$<Z<40 $m/kg^{1/3}$, below this limit the predicted blast parameters may be inaccurate [25].

Instead, the analytical pressure applied to the target structure is typically computed according to Equation 2 [26]:

$$P = P_R \cdot \cos^2\theta + P_I \cdot (1 + \cos^2\theta - 2 \cdot \cos\theta) \tag{2}$$

where P_R and P_I are the reflected and the incident pressure values estimated by the Kingery-Bulmash equations, respectively, while θ is the angle of incidence. Note that Equation 2 is not valid for negative values of cosθ, i.e., for points on the surface not facing the explosion. In such a case, the simplified relationship $P=P_I$ is considered. The value of P is computed at each iteration of the solution for each loaded element, since (i) the variable θ varies during the structural deformation process and (ii) the values of P_R and P_I depend both on the spatial position of the target element with respect to the detonation point and on the time passed from the detonation of the HE material. No more information is reported in this work on the analytical prediction of blast loading. The interested reader is referred to the work in [27] to go deeper into the topic and for a critical discussion of Equation 2.

The blast loading prediction in the pure Lagrangian approach is not a computationally critical operation, since only empirical and analytical equations are involved, but no fluid structure interaction effects can be evaluated using this method, i.e., the effect of the deformation rate of the structure on the pressure loading is neglected.

2.2. *Fully coupled Eulerian-Lagrangian approach*

A CEL analysis consists of modelling both the target structure, exploiting the Lagrangian elements formulation, and the HE charge and the surrounding material, resorting to the Eulerian elements formulation. The thermodynamic state evolution of the HE material after the detonation event is determined employing the Jones-Wilkins-Lee (JWL) equation of state shown in Equation 3 [28]:

$$P(V,e) = A\left[1 - \frac{\omega \cdot V_0}{V \cdot R_1}\right] \cdot e^{-\frac{V \cdot R_1}{V_0}} + B\left[1 - \frac{\omega \cdot V_0}{V \cdot R_2}\right] \cdot e^{-\frac{V \cdot R_2}{V_0}} + \omega\rho e \tag{3}$$

where V is the inverse of density ρ, *e* the internal energy, A and B parameters with the units of pressure, ω the Grüneisen coefficient and R_1 and R_2 dimensionless parameters. The shock wave produced by the detonation of the explosive charge propagates in the surrounding material and reaches the target structure exerting on it a pressure loading. Hence, the surrounding material, i.e., the air in this work, needs to be modelled as well. In particular, the ideal gas equation of state shown in Equation 4 is typically assigned to the air domain [29]:

$$P = \rho \cdot (C_p - C_v) \cdot T \tag{4}$$

where C_p and C_v are the specific heat at constant pressure and at constant volume, respectively, while T is the temperature.

The thermodynamic state evolution of the HE material, the shock wave development and propagation in air and the prediction of the pressure load exerted on the structure are computationally expensive operations and usually represent the bottleneck of the CEL approach. Despite this issue, complex phenomena such as fluid structure interaction effects can be evaluated employing the fully coupled Eulerian-Lagrangian approach.

In this work, a hybrid modelling technique is presented that makes the CEL approach less computationally expensive. In particular, the explosive charge is not explicitly modelled, but only the air domain surrounding the target structure is considered. The shock wave is introduced in the simulation by loading the air domain with an analytical pressure time history determined resorting to the Kingery-Bulmash equations referred to in Section 2.1. This technique allows analytically propagating the blast wave to a region close to the target structure, saving the time required to numerically simulate the HE material detonation and the formation and propagation of the shock wave far from the region of interest. The proposed CEL methodology is shown in Figure *1*. Note that this hybrid technique may inaccurately characterise the blast loading if the scaled distance value Z evaluated considering as standoff distance R the distance R_{EUL} (Figure *1*) from the detonation point to the closest air domain point, i.e., $Z_{EUL}=R_{EUL}/W_{TNT}^{1/3}$, does not meet the requirement $0.147\ m/kg^{1/3}<Z_{EUL}<40\ m/kg^{1/3}$.

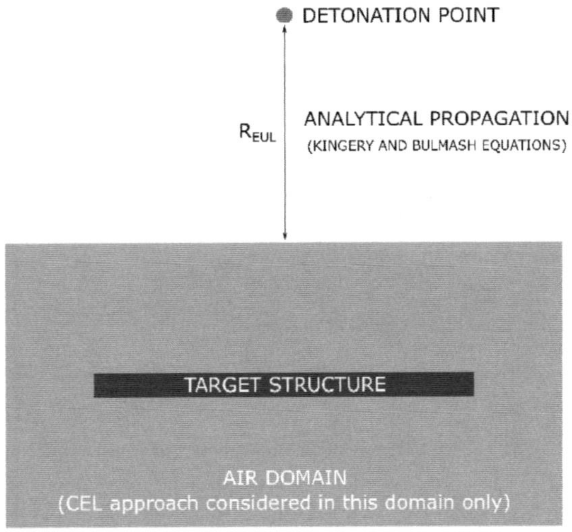

Figure 1. Proposed solution to speed up the CEL approach.

2.3. *Target structure modelling approach*

The target structure is modelled exploiting the Lagrangian elements formulation. Solid elements are preferred to shell elements in this work. This choice is made since the authors verified that the fully coupled Eulerian-Lagrangian approach works better with structures modelled as solid entities, even though shell elements would speed up the computations. The simulations proving this statement are not reported here for the sake of brevity. The composite material is modelled with a macro-homogeneous discretisation, i.e., each ply is modelled with a layer of solid elements. The intra-laminar and inter-laminar properties of the composite material are both retained in the analyses. The former is assigned to the solid elements representing each ply, while the latter is implemented as contact interaction between adjacent layers. More information about the specific properties of the structural elements is reported in the next Section with reference to the case study.

3. Case Study

The methodological approaches described in Section 2 are herein employed to simulate one of the experimental scenarios presented in the work in [12]. In this Section, the case study setup is briefly introduced and the numerical simulations are described in terms of main modelling parameters and results. Moreover, within the subsection dedicated to the CEL analysis, the influence of the modelling technique employed to apply the boundary conditions of the blast loaded structure is investigated. All the numerical simulations are performed employing the LS-DYNA® software package.

3.1. *Description of the scenario under assessment*

The scenario considered in this work involves the detonation of a 100g spherical Type 4 plastic explosive charge at 0.4m standoff distance from a flat quadrangular plate. The blast loaded plate is a Carbon-Polyester laminate composed of seven plies with thickness 0.6mm each and has dimensions

275x275mm^2. Each ply is made of plain-woven fabric. The plate is fixed to a support structure by means of a steel window frame leaving an exposed area of 250x250mm^2. Soft EPDM 414 foam is placed between the steel frame and the plate to create a simply supported-like boundary condition for the blast loaded structure. The scenario considered in this case study is shown in Figure *2*.

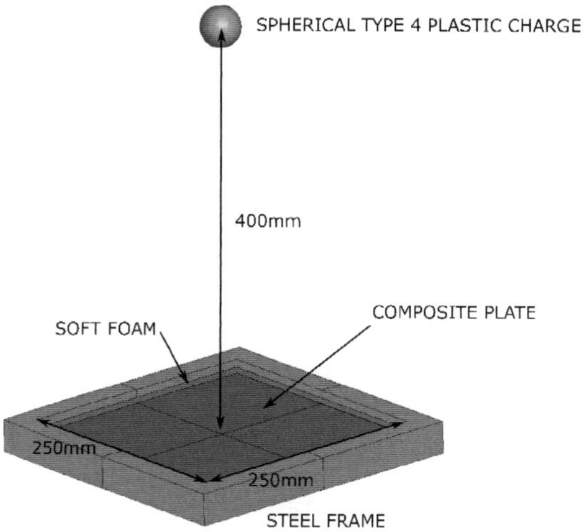

Figure 2. Scenario considered in the case study.

Note that the scenario described above is taken from the work in [12], where the experimental campaign is presented along with the numerical simulation performed employing the pure Lagrangian approach in the Abaqus® software package.

3.2. Modelling of the structural elements

In this subsection the modelling technique and parameters employed to model the steel window frame, the foam and the composite plate are presented.

The steel window frame is modelled using solid hexahedral elements with characteristic dimension 2.45mm and fully integrated formulation (elform=-1). Steel is modelled resorting to the elastic isotropic constitutive law implemented in LS-DYNA® with MAT_001 (*MAT_ELASTIC) using the parameters shown in Table 1, where RO identifies the material density, E the elastic modulus and PR the Poisson ratio.

Table 1. Parameters of the material of the steel frame [12].

Steel - MAT_001	
RO	7800 kg·m^{-3}
E	203 GPa
PR	0.3

The foam placed between the steel clamp and the composite plate is the soft EPDM 414 foam. However, in the work in [12] the material parameters employed in the constitutive law are not reported. Hence, since to the authors' best knowledge no information is present in the literature about the specific foam employed in the experimental setup, a different foam taken from the work in [30] is modelled in this case study. This choice is not considered to be critical for the numerical simulations since the foam is only employed to simulate simply supported-like boundary conditions and the stress state of this component is not investigated. Solid hexahedral elements with characteristic dimension 2.5mm are employed to model the foam material. Since this component is expected to undergo large deformations, a reduced integration element formulation (elform=1) is chosen to avoid possible negative volume errors preventing the normal termination of the numerical simulations. Moreover, to the same purpose, erosion is added to the material constitutive law to delete excessively distorted elements using the material card MAT_000 (*MAT_ADD_EROSION), setting the maximum effective strain at failure EFFEPS value to 5. The material constitutive law is implemented resorting to the LS-DYNA keyword MAT_057 (*MAT_LOW_DENSITY_FOAM), that is a law dedicated to highly compressible low density foams. The main material parameters are shown in Table *2*, where E identifies the Young's modulus used in tension, LCID the ID of the curve of the nominal stress versus strain, HU the hysteretic unloading factor, BETA the decay constant to model creep in unloading, DAMP the viscous coefficient to model damping effects, SHAPE the shape factor for unloading and KCON the stiffness coefficient for contact interface stiffness. The fields left to the default value are not shown in Table *2*, the interested reader is referred to the LS-DYNA® keyword user's manual (Vol. II) to go deeper into the topic [29].

Table 2. Parameters of the material of the foam [30,31].

Foam - MAT_057	
RO	63 kg·m^{-3}
E	8.4 MPa
LCID	Curve taken from [30]
HU	0.25
BETA	5.0
DAMP	0.5
SHAPE	5.0
KCON	1150 MPa

Each ply of the composite plate is modelled with solid hexahedral elements with characteristic dimension 2mm, as suggested in the work in [12], using one element in the thickness. A fully integrated element formulation (elform=-1) is employed to avoid hourglass effects. The intra-laminar properties are defined describing the composite material constitutive law employing the LS-DYNA® keyword MAT_054 (*MAT_ENHANCED_COMPOSITE_DAMAGE), that represents a built-in material model based on the Hashin failure criteria [32]. The main non-default material properties considered in this work are reported in Table 3, where EA, EB and EC are the Young's moduli in the three principal directions, PRBA is the in-plane Poisson ratio and GAB, GBC and GCA are the shear moduli in the three principal planes. Moreover, the parameter 2WAY=1 represents a flag to turn on the 2-way fibre action, DFAILT and DFAILC are the maximum strain values for fibre tension and compression, respectively, that are intentionally set to large values to avoid element deletion and to replicate the same conditions considered in the work in [12]. Finally, XC, XT, YC and YT are the strength values for the in-plane compressive (C) and tensile (T) modes and SC is the shear strength, that is intentionally set to a large value to neglect its contribution to the material failure [12].

Table 3. Material properties of the Carbon-Polyester laminate plies [12].

Carbon-Polyester laminate ply - MAT_054	
RO	1600 kg·m^{-3}
EA	55 GPa
EB	55 GPa
EC	7 GPa
PRBA	0.25
GAB	4.5 GPa
GBC	1.8 GPa

GCA	1.8 GPa
2WAY	1
DFAILT	1
DFAILC	-1
XC	240 MPa
XT	680 MPa
YC	240 MPa
YT	680 MPa
SC	1000 MPa

The inter-laminar properties are modelled employing the Cohesive Zone Model (CZM) approach using the keyword *CONTACT_AUTOMATIC_SURFACE_TO_SURFACE_TIEBREAK. This contact algorithm keeps the corresponding nodes between adjacent layers connected until failure occurs; after that, the interaction is turned into a simple surface-to-surface contact between the plies. Failure is accounted for by means of a quadratic criterion that considers both the normal and the shear interlaminar stresses. The maximum normal (NFLS) and shear (SFLS) stress values considered in this work are reported in Table 4.

Table 4. Properties of the contact interaction between adjacent plies [12].

NFLS	60 MPa
SFLS	60 MPa

The finite element model of the structural components is shown in Figure *3*, where the reference system considered in the analysis is also presented. The plate surface facing the blast loading lies in the xy plane characterised by z=0m and the plate central point on the same surface is at coordinates x=y=0m. Note that only a quarter of the model is considered to speed up the computations, accordingly introducing symmetry constraints on the x and y negative faces.

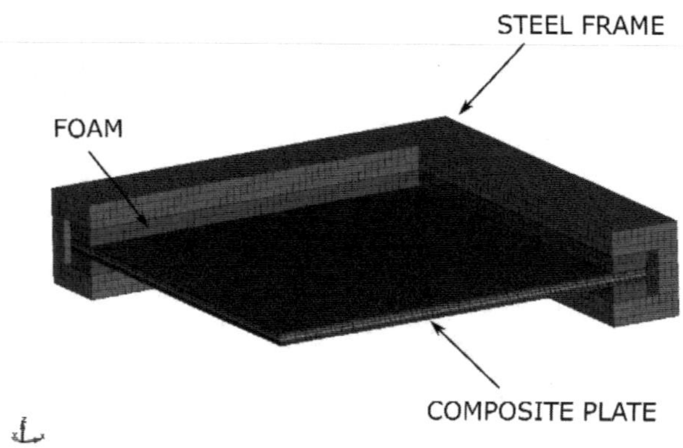

Figure 3. Finite element model of the structural components.

Contact algorithms are employed to consider in the numerical simulations the eventual interaction between the structural components. The foam is glued to the steel frame to reproduce the foam lined frame configuration referred to in the work in [12]. This is achieved introducing in the numerical analysis the keyword *CONTACT_TIED_SURFACE_TO_SURFACE with pinball segment based contact (SOFT=2) option. The contact between the foam and the plate is described using the algorithm *CONTACT_ERODING_SURFACE_TO_SURFACE with pinball segment based contact (SOFT=2) option and static and dynamic coefficients of friction FS=1 and FD=1, respectively [33,34]. This contact algorithm allows tracking the contact between the involved parts even if they undergo erosion, as it might happen to the foam elements as discussed above.

Boundary conditions are introduced to reproduce the effect of the experimental support structure. Two different models are set up to evaluate the effect of the modelling of the boundary conditions on the blast loaded plate response. The first model is hereinafter referred to as *CBC* (Complete Boundary Conditions) model, it consists of the whole setup shown in Figure *3* to which an encastre boundary condition is applied on the steel frame face opposite to the explosive charge. Instead, the second model, which is hereinafter referred to as *SBC* (Simplified Boundary Conditions) model, only includes the plate exposed area and the encastre boundary condition is applied to the edges of the plate. The two models are compared in Figure *4*.

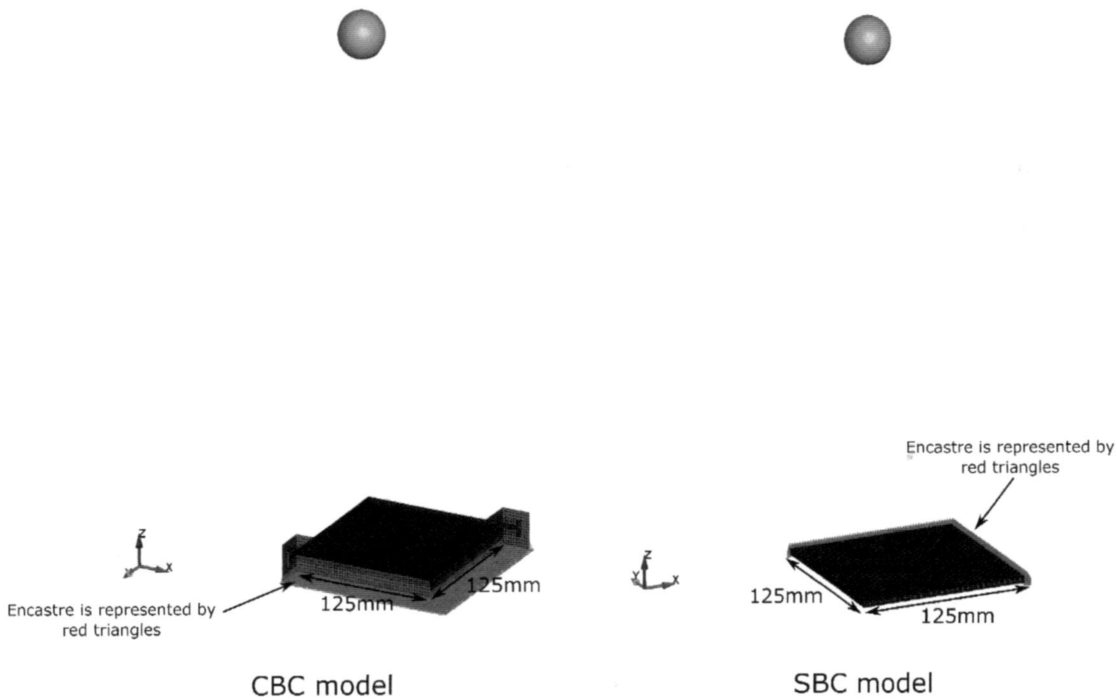

Figure 4. Boundary conditions modelling strategies.

3.3. Pure Lagrangian analysis

In the pure Lagrangian analysis the blast loading is applied on the plate exposed area facing the explosive charge according to the approach described in Section 2.1. The blast pressure loading is determined employing the keyword *LOAD_BLAST_ENHANCED with the parameters shown in Table 5, where M is the TNT equivalent weight of the explosive material (referred to as W_{TNT} in Section 2.1), XBO, YBO and ZBO the coordinates of the explosive charge in the reference system shown in Figure *4*, TBO the time of detonation, BLAST=2 the flag to specify that the explosion is a spherical free-air burst and NEGPHS=0 the flag to retain the blast pressure negative phase in the analysis.

Table 5. Parameters to compute the blast loading in the pure Lagrangian approach.

*LOAD_BLAST_ENHANCED	
M	168 g
XBO	0 m
YBO	0 m
ZBO	0.4 m
TBO	-0.13 ms
BLAST	2
NEGPHS	0

Looking at the data reported in Table *5*, the explosive charge is placed at a standoff distance of 0.4m from the plate surface facing the blast load and the equivalent TNT weight of the 100g spherical Type 4 plastic explosive charge is M=168g. This TNT equivalent weight value is determined employing the equations proposed in [35] that satisfy the equivalence in terms of blast pressure prediction. In particular, to stay on the conservative side, the upper bound value of the TNT equivalent weight is adopted in this work. According to the standoff distance and the TNT weight considered in the analysis, the scaled distance value associated to the explosion is $Z=0.72\ m/kg^{1/3}$, which satisfies the requirement for the validity of the Kingery-Bulmash equations reported in Section 2.1. Note that, to reduce the computational time, the arrival time (AT) of the blast wave is estimated using the Kingery-Bulmash equations and the analysis is started right before the wave impacts the structure. This is achieved by specifying a negative value in the field TBO so that $|TBO|<AT$.

The blast pressure and impulse exerted on the centre of the plate are shown in Figure *5*.

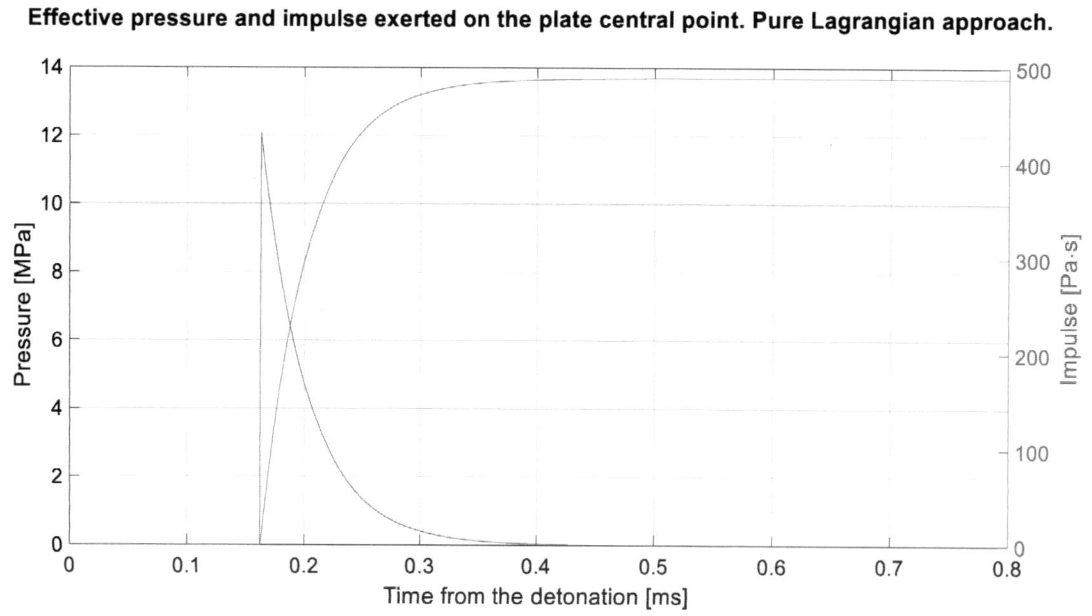

Figure 5. Effective pressure and impulse exerted on the plate central point. Pure Lagrangian approach.

The absolute value of the displacement of the plate centre is shown in Figure *6* for the only *CBC* model, since the *SBC* model is not assessed using the pure Lagrangian approach. Note that the absolute value of the displacement is plotted since, usually, positive values are shown in the literature, while according to the reference system shown in Figure *4* the plate gets deformed in the negative z direction in this case study.

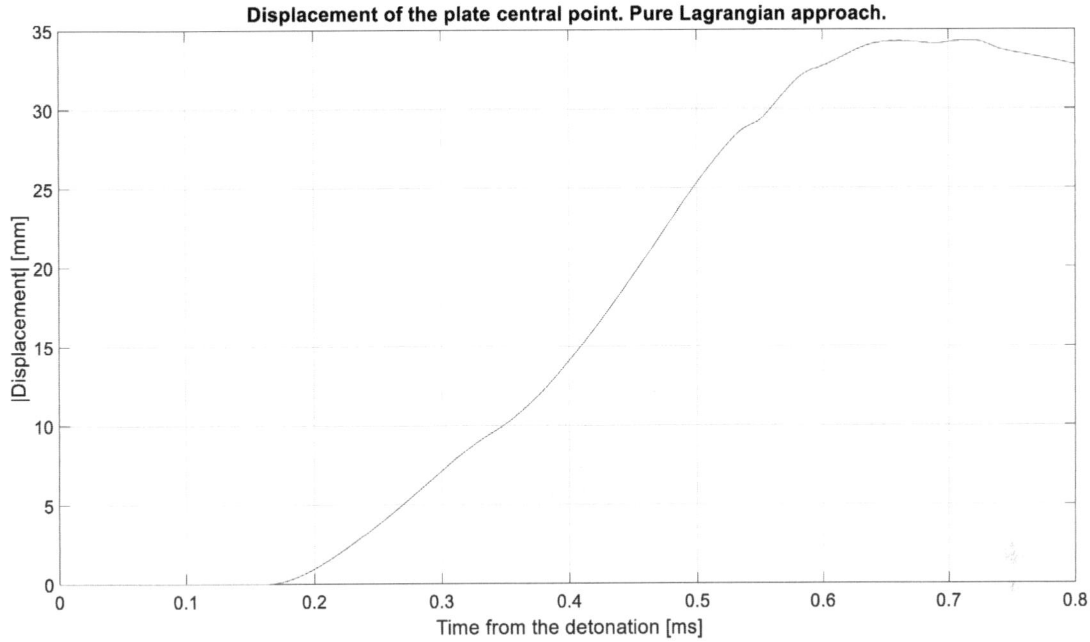

Figure 6. Absolute value of the displacement of the plate central point. Pure Lagrangian approach.

A summary of the maximum registered pressure (p_{max}^{Lag}), impulse (i_{max}^{Lag}) and displacement ($d_{max}^{Lag,\ CBC}$) at the central point of the plate is reported in Table *6*. The apex *Lag* identifies the results from the pure Lagrangian analysis, the apex *CBC* identifies the *CBC* model.

Table 6. Summary of the pure Lagrangian analysis results.

p_{max}^{Lag}	12.1 MPa
i_{max}^{Lag}	488.9 Pa·s
$d_{max}^{Lag,CBC}$	34.3 mm

3.4. CEL analysis

According to the CEL approach introduced in Section 2.2, the blast wave propagation and its interaction with the target structure are dealt with by explicitly modelling the air domain exploiting solid hexahedral elements with characteristic dimension at convergence 1mm. The formulation selected in this work is the solid section elform=5, which identifies 1 point Arbitrary Lagrangian-Eulerian (ALE) elements. The keyword *ALE_REFERENCE_SYSTEM_GROUP is employed to model the behaviour of the ALE elements. In particular, the air domain is assigned a mesh smoothing option dedicated to scenarios involving shock waves (PRTYPE=8) and the pure Eulerian behaviour is forced setting the initial mesh remapping factor (EFAC) to 1. Moreover, the card *CONTROL_ALE is included in the analysis to set the following global control parameters for the Eulerian calculations. The advection method employed in the analysis is the *donor cell + HIS* method (METH=3), which is a first order accurate method conserving total energy over each advection step. The ALE mesh smoothing is turned off (AFAC=-1) and flow out boundary conditions are specified at the Eulerian boundaries to avoid undesired reflections of the shock wave as it impacts the mesh boundary (EBC=0). Finally, the reference pressure value applied to the free surfaces of the ALE mesh boundary (PREF field) is set to 101,325 Pa. The CEL model is shown in Figure *7*.

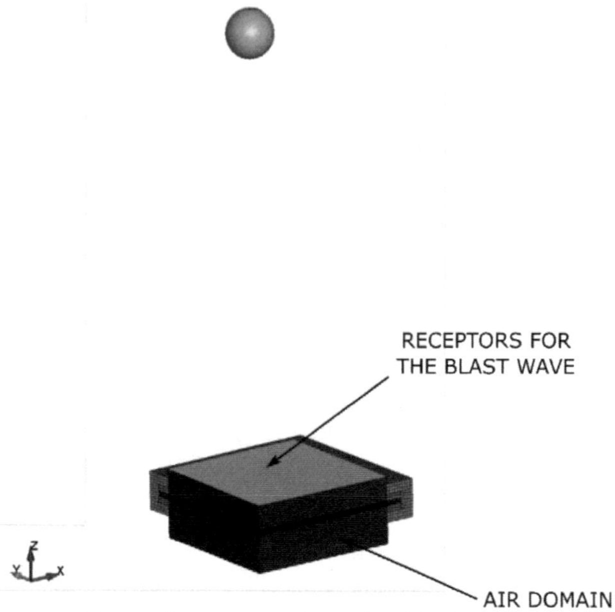

Figure 7. Finite element model involved in the CEL analysis.

The air behaviour is modelled using the material model MAT_009 (card *MAT_NULL) and the ideal gas equation of state (card *EOS_IDEAL_GAS). The non-default parameters included in the analysis are reported in Table 7, where MU identifies the dynamic viscosity, CV0 and CP0 the specific heat constants for the definition of C_v and C_p, respectively, V0 the initial relative volume and T0 the initial temperature value.

Table 7. Parameters to model the air behaviour [36,37].

Air - MAT_009	
RHO	1.225 kg·m^{-3}
MU	1.8·10^{-5} Pa·s
Air - *EOS_IDEAL_GAS	
CV0	717 J·Kg^{-1}·K^{-1}
CP0	1007 J·Kg^{-1}·K^{-1}
V0	1
T0	288.15 K

According to the procedure described in Section 2.2, the shock wave is initiated in the air domain by loading its upper surface with the analytical prediction from the Kingery-Bulmash equations. The elements acting as receptors for the blast wave are included into a segment set (see Figure 7) that is specified in the card *LOAD_BLAST_SEGMENT_SET. In this card, the field SFNRB is set to 1 to attenuate the shock waves reflected back to the ambient elements. The parameters included in the *LOAD_BLAST_ENHANCED card to define the blast load are the same as the parameters shown in Table 5. According to the standoff distance (R_{EUL}=380mm) and the TNT weight considered in the analysis, the scaled distance value associated to the explosion is Z=0.69 m/kg$^{1/3}$, which satisfies the requirement for the validity of the Kingery-Bulmash equations reported in Section 2.1.

Symmetry boundary conditions are applied to the x and y negative surfaces of the Eulerian domain, as also described in Section 3.2 for the structural components. The interaction between the shock wave propagating in the air domain and the composite plate is set up employing the card *CONSTRAINED_LAGRANGE_IN_SOLID. In this card, the air domain is set as master part, while the set of plies is specified in the slave field. Moreover, the fluid-structure coupling method CTYPE=4 is considered in the analysis, which is a penalty coupling for solid elements without erosion.

The blast pressure and impulse exerted on the centre of the plate are shown in Figure *8*.

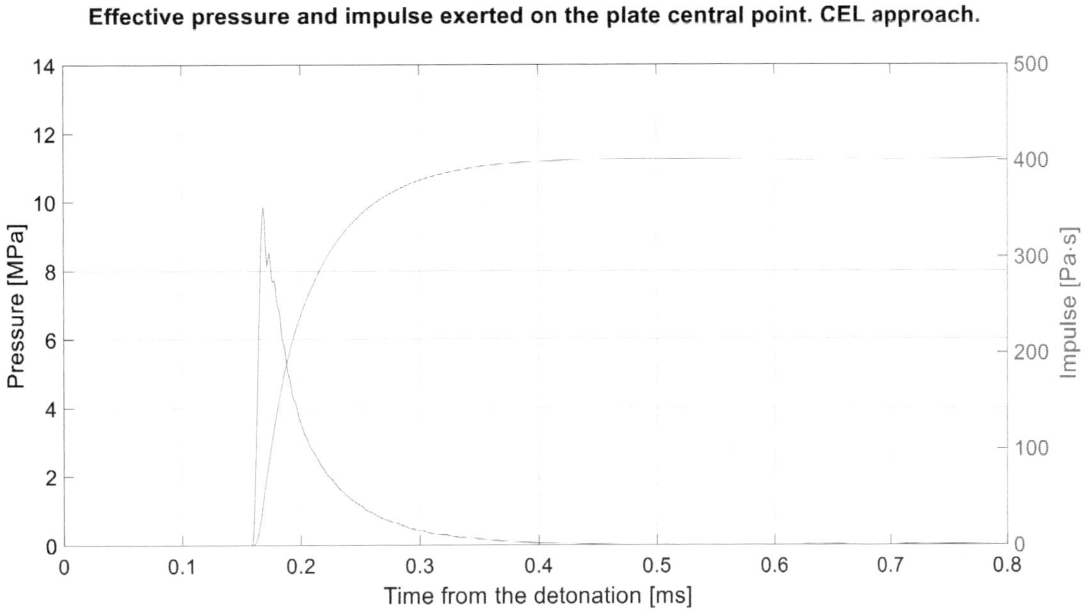

Figure 8. Effective pressure and impulse exerted on the plate central point. CEL approach.

The absolute value of the displacement of the plate centre is shown in Figure *9* for the *CBC* model (solid curve) and for the *SBC* model (dashed curve). The curve obtained with the *SBC* model is stopped as soon as the plate passes through the initial position.

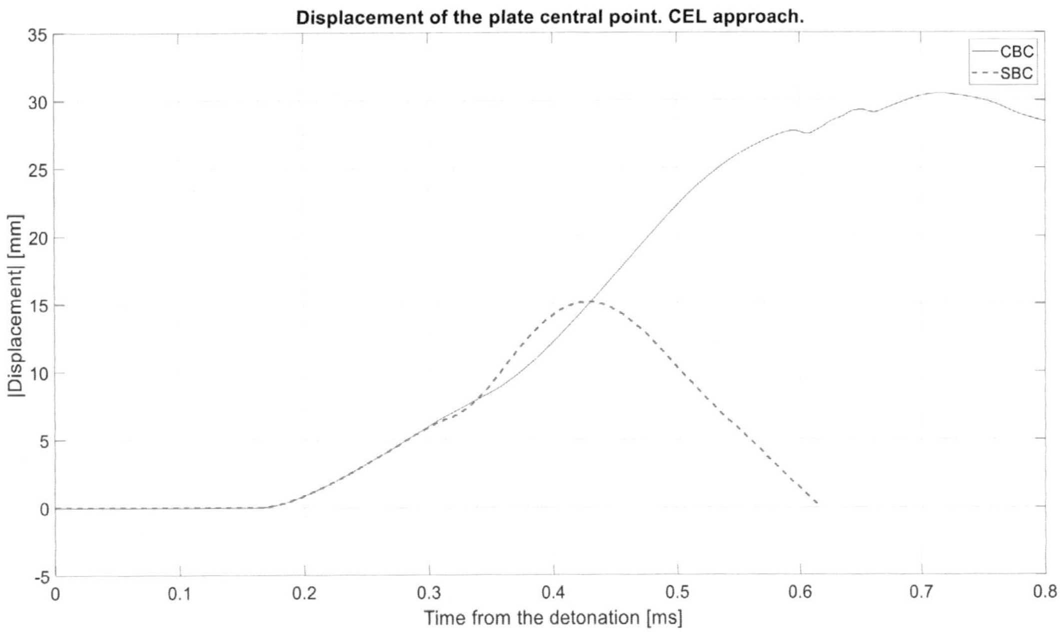

Figure 9. Absolute value of the displacement of the plate central point. CEL approach.

A summary of the maximum registered pressure (p_{max}^{CEL}), impulse (i_{max}^{CEL}) and displacement (d_{max}^{CEL}) at the central point of the plate is reported in Table *8*. The apex *CEL* identifies the results from the CEL analysis, the apex *SBC* identifies the *SBC* model.

Table 8. Summary of the CEL analyses results.

p_{max}^{CEL}	10.0 MPa
i_{max}^{CEL}	405.3 Pa·s
$d_{max}^{CEL,CBC}$	30.5 mm
$d_{max}^{CEL,SBC}$	15.1 mm

4. Discussion

In this Section, the results of the numerical simulations presented in Section 3 are discussed and compared.

The deflection time histories of the plate centre predicted by the numerical analyses presented in this work are shown in Figure *10*, along with the experimentally measured maximum displacement.

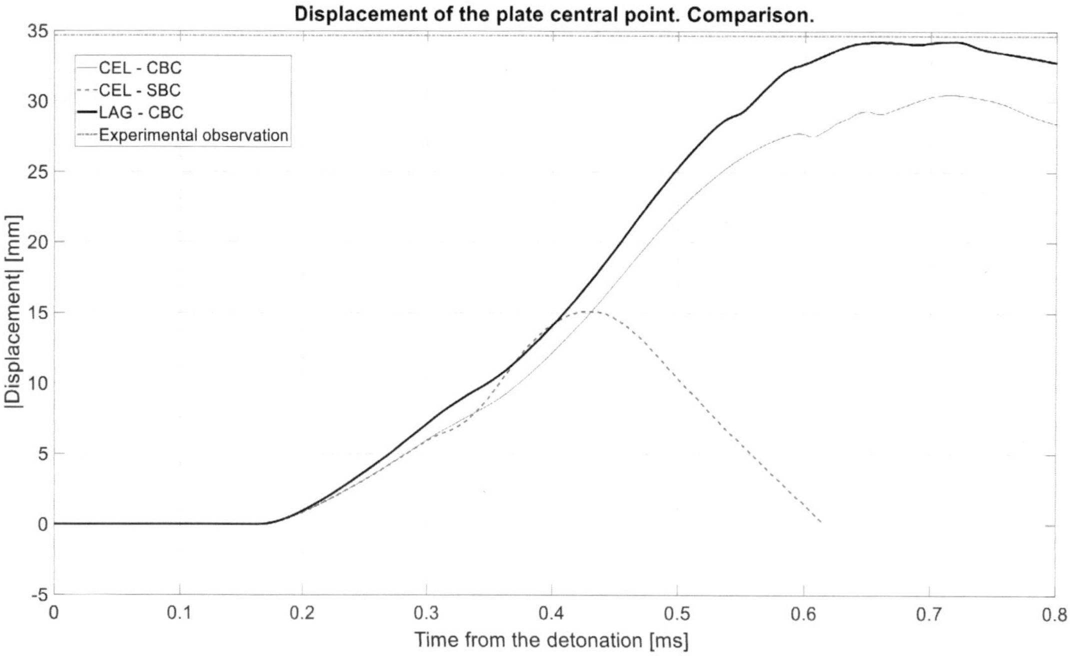

Figure 10. Displacement of the plate central point.

It turns out that the model with simplified boundary conditions (SBC) underestimates the deflection time history of the plate centre, while the CBC models satisfactorily reproduce the experimental observation. To further prove this, the maximum displacement values are compared in Table *9* considering the numerical models ($d_{max}^{Lag,CBC}$, $d_{max}^{CEL,CBC}$, $d_{max}^{CEL,SBC}$) and the experimental observation (d_{max}^{Exp}). The error affecting the *ith* numerical prediction is computed according to Equation 5, where the apex *i* identifies the numerical analysis of interest, i.e., *i* can assume the values *Lag,CBC*, *CEL,CBC* and *CEL,SBC*.

$$\text{Error}^i = \frac{d_{max}^i - d_{max}^{Exp}}{d_{max}^{Exp}} \cdot 100\% \tag{5}$$

Table 9. Comparison of the maximum deflection of the plate centre.

		Error
d_{max}^{Exp}	34.7 mm [12]	~
$d_{max}^{Lag,CBC}$	34.3 mm	-1.2%
$d_{max}^{CEL,CBC}$	30.5 mm	-12.1%
$d_{max}^{CEL,SBC}$	15.1 mm	-56.5%

The results shown in Table *9* confirm that the CBC models satisfactorily reproduce the experimentally observed maximum deflection of the plate centre, while the SBC setup strongly underestimates this value. This result suggests that accurately modelling the boundary conditions of the blast loaded structure is paramount to satisfactorily reproduce the real scenario.

Moreover, the Lagrangian analysis seems to provide more accurate results than the fully coupled Eulerian-Lagrangian approach. This result needs a deeper investigation to be explained, since the work in [12] labels the scenario assessed in the case study as near-field explosion (or close-range explosion), which usually requires the CEL approach for an accurate characterisation of the underlying physical phenomena involved. The near-field or far-field nature of an explosive scenario is determined by the value of the radial expansion of the shock front at the plate location ($\bar{r}_{plate}$), that is computed according to Equation 6:

$$\bar{r}_{plate} = \frac{\text{standoff distance}}{\text{charge radius}} \tag{6}$$

Typically, close-range effects are relevant for $1<\bar{r}_{plate}\leq 10$, while far-field conditions can be safely assumed for explosions characterised by $\bar{r}_{plate}>10$ [38]. The former case requires employing fully coupled Eulerian-Lagrangian analysis to satisfactorily reproduce the experimental observations, while the latter can be safely simulated employing the pure Lagrangian approach [20]. In the case study presented in this work, considering the explosive charge as a sphere of TNT with density 1610 $kg \cdot m^{-3}$ [39], the radial expansion of the shock front at the plate location value is $\bar{r}_{plate}=13.7$, that identifies a far-field scenario. Hence, the pure Lagrangian approach is suitable to simulate the explosive event and no improvement is expected by performing the CEL analysis. The smaller maximum deflection of the plate centre predicted using the fully coupled approach is determined by the lower peak effective pressure and impulse imparted to the blast loaded structure than the values determined by the pure Lagrangian approach, as shown in Figure *11* and Figure *12*, respectively.

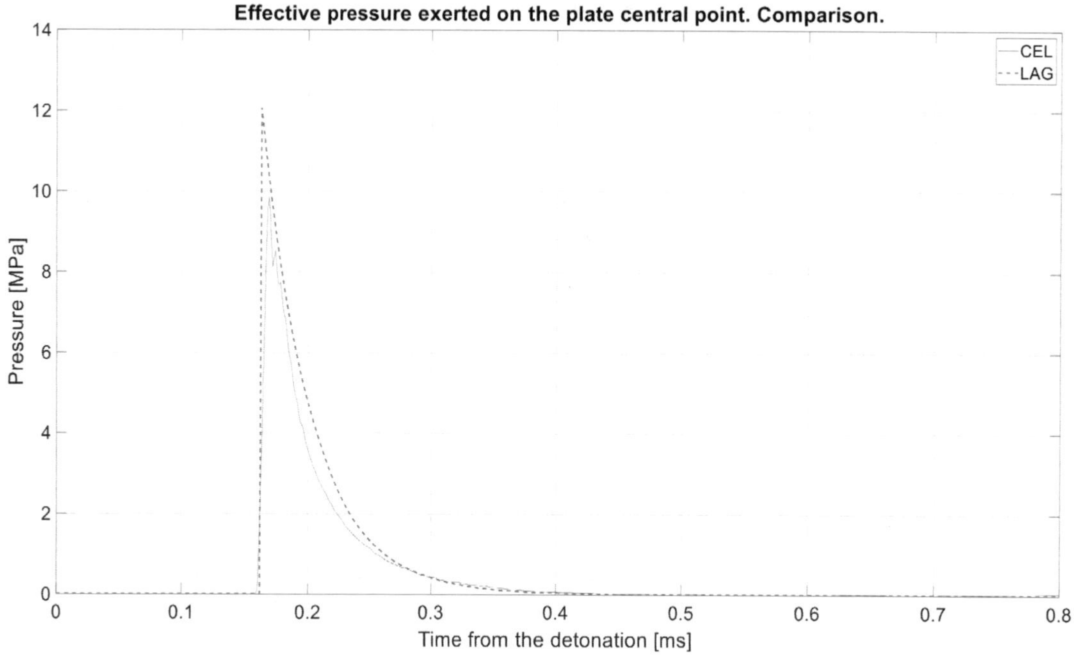

Figure 11. Effective pressure exerted on the plate central point in the analyses presented in Section 3.

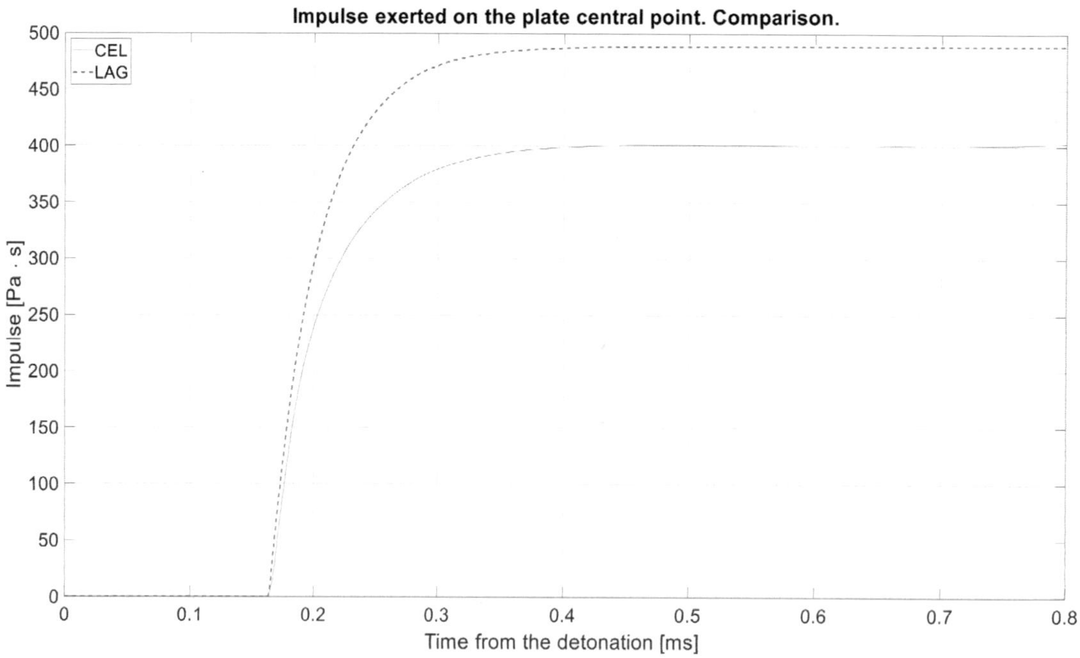

Figure 12. Impulse exerted on the plate central point in the analyses presented in Section 3.

It is worth highlighting that the peak effective pressure and the impulse imparted to the plate were not experimentally measured in the work in [12], hence no further considerations can be given about the corresponding numerical predictions presented herein.

5. Conclusions

In this work, two approaches to simulate blast loaded composite plates have been presented and compared. In particular, the pure Lagrangian approach and a hybrid CEL approach have been described in Section 2 and employed in the case study presented in Section 3. Moreover, the influence of the technique employed to model the plate boundary conditions has been investigated.

As discussed in Section 4, it turns out that explicitly modelling the boundary conditions of the blast loaded plate is paramount to satisfactorily reproduce the experimental observations. Moreover, considering the model characterised by the accurate reproduction of the boundary conditions, the pure Lagrangian approach appears to give slightly more satisfactory results than the predictions provided by the CEL analysis. This may seem unexpected at first sight, but it has been shown that the scenario assessed in the case study presented in Section 3 is a far-field explosion, thus no improvement is expected by employing the CEL approach. The smaller maximum deflection of the plate central point predicted by the CEL analysis is caused by the lower peak effective pressure and impulse imparted to the plate than the values determined by the pure Lagrangian approach. No further considerations have been reported about this dissimilarity, since experimental observations of the pressure loading are not available. More accurate numerical models might be set up to get a deeper insight into the explosive event presented in the case study, but this requires to have (i) experimental data of the pressure time history imparted to the plate, in order to set a more accurate TNT equivalent weight than the one assumed in this work, and (ii) the stress-strain curve of the foam material placed between the composite plate and the steel frame.

References

[1] Yao S, Zhang D and Lu F 2015 Dimensionless numbers for dynamic response analysis of clamped square plates subjected to blast loading *Arch. Appl. Mech. 2015 856* **85** 735–44

[2] Chung Kim Yuen S, Nurick G N, Verster W, Jacob N, Vara A R, Balden V H, Bwalya D, Govender R A and Pittermann M 2008 Deformation of mild steel plates subjected to large-scale explosions *Int. J. Impact Eng.* **35** 684–703

[3] Nurick G N and Martin J B 1989 Deformation of thin plates subjected to impulsive loading—a review Part II: Experimental studies *Int. J. Impact Eng.* **8** 171–86

[4] Nurick G N and Martin J B 1989 Deformation of thin plates subjected to impulsive loading—A review: Part I: Theoretical considerations *Int. J. Impact Eng.* **8** 159–70

[5] Chung Kim Yuen S, Nurick G N, Langdon G S and Iyer Y 2017 Deformation of thin plates subjected to impulsive load: Part III – an update 25 years on *Int. J. Impact Eng.* **107** 1339–51

[6] Langdon G S, Yuen S C K and Nurick G N 2005 Experimental and numerical studies on the response of quadrangular stiffened plates. Part II: localised blast loading *Int. J. Impact Eng.* **31** 85–111

[7] Yuen S C K and Nurick G N 2005 Experimental and numerical studies on the response of quadrangular stiffened plates. Part I: subjected to uniform blast load *Int. J. Impact Eng.* **31** 55–83

[8] Børvik T, Hanssen A G, Langseth M and Olovsson L 2009 Response of structures to planar blast loads – A finite element engineering approach *Comput. Struct.* **87** 507–20

[9] Ziya-Shamami M, Babaei H, Mostofi T M and Khodarahmi H 2020 Structural response of monolithic and multi-layered circular metallic plates under repeated uniformly distributed impulsive loading: An experimental study *Thin-Walled Struct.* **157** 107024

[10] Langdon G S, Cantwell W J, Guan Z W and Nurick G N 2014 The response of polymeric composite structures to air-blast loading: a state-of-the-art *http://dx.doi.org/10.1179/1743280413Y.0000000028* **59** 159–77

[11] Fallah A S, Micallef K, Langdon G S, Lee W C, Curtis P T and Louca L A 2014 Dynamic response of Dyneema® HB26 plates to localised blast loading *Int. J. Impact Eng.* **73** 91–100

[12] Gargano A, Das R and Mouritz A P 2019 Finite element modelling of the explosive blast response of carbon fibre-polymer laminates *Compos. Part B Eng.* **177** 107412

[13] Gunaryo K, Heriana H, Sitompul M R, Kuswoyo A and Hadi B K 2020 Experimentation and numerical modeling on the response of woven glass/epoxy composite plate under blast impact loading *Int. J. Mech. Mater. Eng. 2020 151* **15** 1–9

[14] Turkmen H S 2017 The Dynamic Behavior of Composite Panels Subjected to Air Blast Loading: Experiment and Theory *Explos. Blast Response Compos.* 57–84

[15] Schiffer A and Tagarielli V L 2015 The response of circular composite plates to underwater blast: Experiments and modelling *J. Fluids Struct.* **52** 130–44

[16] Baştürk S, Uyanik H and Kazanci Z 2014 An analytical model for predicting the deflection of laminated basalt composite plates under dynamic loads *Compos. Struct.* **116** 273–85

[17] Mouritz A P 2019 Advances in understanding the response of fibre-based polymer composites to shock waves and explosive blasts *Compos. Part A Appl. Sci. Manuf.* **125** 105502

[18] LeBlanc J and Shukla A 2010 Dynamic response and damage evolution in composite materials subjected to underwater explosive loading: An experimental and computational study *Compos. Struct.* **92** 2421–30

[19] Batra R C and Hassan N M 2008 Blast resistance of unidirectional fiber reinforced composites *Compos. Part B Eng.* **39** 513–36

[20] Lomazzi L, Giglio M and Manes A 2021 Analytical and empirical methods for the characterisation of the permanent transverse displacement of quadrangular metal plates subjected to blast load: comparison of existing methods and development of a novel methodological approach. *Int. J. Impact Eng.* 103890

[21] Aune V, Valsamos G, Casadei F, Langseth M and Børvik T 2021 Fluid-structure interaction effects during the dynamic response of clamped thin steel plates exposed to blast loading *Int. J. Mech. Sci.* **195** 106263

[22] Kingery C N and Bulmash G 1984 *Air Blast Parameters from TNT Spherical Air Burst and Hemispherical Surface Burst* (Aberdeen Proving Ground, Maryland: Ballistic Research Laboratories)

[23] Cranz K J, Eberhard O V. and Becker K E 1926 *Lehrbuch der Ballistik. Ergänzungen zum. Band II* ed Springer (Berlin)

[24] Hopkinson B 1915 *British ordnance board minutes, Report 13565* (London)

[25] Shin J, Whittaker A S and Cormie D 2015 Incident and Normally Reflected Overpressure and Impulse for Detonations of Spherical High Explosives in Free Air *J. Struct. Eng.* **141** 04015057

[26] Randers-Pehrson G, Bannister K A and Qxuaij L C 1997 *Airblast Loading Model for DYNA2D and DYNA3D*

[27] Lomazzi L, Giglio M and Manes A 2021 Analysis of the blast wave – structure interface phenomenon in case of explosive events *IOP Conf. Ser. Mater. Sci. Eng.* **1038** 012083

[28] Lee E L, Hornig H C and Kury J W 1968 *Adiabatic Expansion Of High Explosive Detonation Products* (Livermore, CA (United States))

[29] LIVERMORE SOFTWARE TECHNOLOGY CORPORATION (LSTC) 2018 *LS-DYNA® KEYWORD USER'S MANUAL - VOLUME II Material Models*

[30] Zhang T G, Satapathy S S, Dagro A M and McKee P J 2013 Numerical study of head/helmet interaction due to blast loading *ASME International Mechanical Engineering Congress and Exposition, Proceedings (IMECE)* vol 3 A (American Society of Mechanical Engineers (ASME))

[31] Fitek J and Meyer E 2013 *Design of a Helmet Liner for Improved Low Velocity Impact Protection*

[32] Hashin Z 1980 Failure Criteria for Unidirectional Fiber Composites *J. Appl. Mech.* **47** 329–34

[33] Tiwari A, Dorogin L, Tahir M, Stö K W, Heinrich G, Espallargas N and Persson B N J 2017 Rubber contact mechanics: adhesion, friction and leakage of seals *Soft Matter* **13** 9103

[34] Karger-Kocsis J, Mousa A, Major Z and Békési N 2008 Dry friction and sliding wear of EPDM rubbers against steel as a function of carbon black content *Wear* **264** 359–67

[35] Bogosian D, Yokota M and Rigby S 2016 TNT equivalence of C-4 and PE4: a review of

traditional sources and recent data *undefined*

[36] Touloukian Y S, Liley P E and Saxena S C 1970 *Thermophysical Properties of Matter, Vol. 3: Thermal Conductivity* (NY: IFI/Plenun)

[37] Touloukian Y S, Liley P E and Saxena S C 1970 *Thermophysical Properties of Matter, Vol. 11: Viscosity* (NY: IFI/Plenun)

[38] Shin J, Whittaker A S, Cormie D and Wilkinson W 2014 Numerical modeling of close-in detonations of high explosives *Eng. Struct.* **81** 88–97

[39] Koli S, Chellapandi P, Bhaskara Rao L and Sawant A 2020 Study on JWL equation of state for the numerical simulation of near-field and far-field effects in underwater explosion scenario *Eng. Sci. Technol. an Int. J.* **23** 758–68

Sound of a Composite Failure: An Acoustic Emission Investigation

C Barile[a], C Casavola[a], G Pappalettera[a] and P K Vimalathithan[a]

[a] Dipartimento di Meccanica, Matematica e Management, Politecnico di Bari, Via Orabona 4, 70126 – Bari, Italy

claudia.barile@poliba.it

Abstract. The failure progression characteristics of adhesively bonded Carbon Fiber Reinforced Polymer (CFRP) composites are investigated using Acoustic Emission (AE) technique. Different failure progression modes such as matrix cracking, fiber breakage, delamination and through-thickness crack growth releases AE waveforms in different frequency domains. The characteristic features of these different AE waveforms are studied in Mel Scale, which is a perpetual frequency scale of average human hearing frequency. The recurring noise in the recorded waveforms has been identified more efficiently when the waveforms are analysed in Mel Scale. The recorded AE signals from the adhesively bonded CFRP under static tensile loading are stretched to match the Mel filter banks. The sampling rate of the recorded signal is adjusted from 1 MHz to 20 kHz. Following that, the Mel spectrogram and its cepstral coefficients are used for identifying the different failure modes from which the AE signals are generated. A comprehensive comparison of the AE analysis in Mel scale with conventional waveform processing techniques such as Fast Fourier Transform (FFT), Continuous Wavelet Transform (CWT), Wavelet Packet Transform (WPT) and Hilbert-Huang Transform (HHT) has been made. The advantages and further applications of Mel Scale over traditional waveform processing techniques in defining the failure modes in the composites are also discussed.

1. Introduction

Any irreversible deformation in a material or a structure due to the external loads generates elastic waves [1]. The characteristic features of these elastic waves in the time-frequency domain carries the signature of their source. For instance, in a Fiber Reinforced Plastic (FRP) composite material, the sources of these elastic waves are matrix cracking, delamination, fiber breakage and so on. Each of the source generates elastic waves at a certain frequency, propagating for a certain wavelength and duration [2]. The propagation of these elastic waves within a solid, in this case, a composite material is known as Acoustic Emission (AE). Most of the energy of these elastic waves (or acoustic waves) lies within 1 kHz and 1 MHz frequency range while, most of the significant energy of the elastic waves can be said to be in the ultrasonic frequency range. Nonetheless, these acoustic waves are essentially sound waves.

Sounds waves are recognized by human auditory system not in a linear scale, but rather in a logarithmic scale. Mel scale, the word which is taken from 'melody' is the perpetual scale in which the human can recognize sounds [3]. If the frequency characteristics of the acoustic waves generated from the failure of a composite material, or general in a structure can be converted to into their Mel scales, perhaps, one may then have the sound of a failure. However, a question may arise on why the Mel scales must be used for analysing the AE waves from a composite failure. The objective of this research work is to characterize the time-frequency features of the AE signals generated from a CFRP composite failure using different waveform processing techniques and compare them with the Mel spectrogram. The advantages of using Mel scale and its applications are also discussed in this research work.

2. Experimental Procedure

The acoustic waves taken for this study are generated from a CFRP specimen in Single Lap Shear (SLS) configuration; subjected to a static tensile load. The dimensional configurations of the specimen are presented in Figure 1.

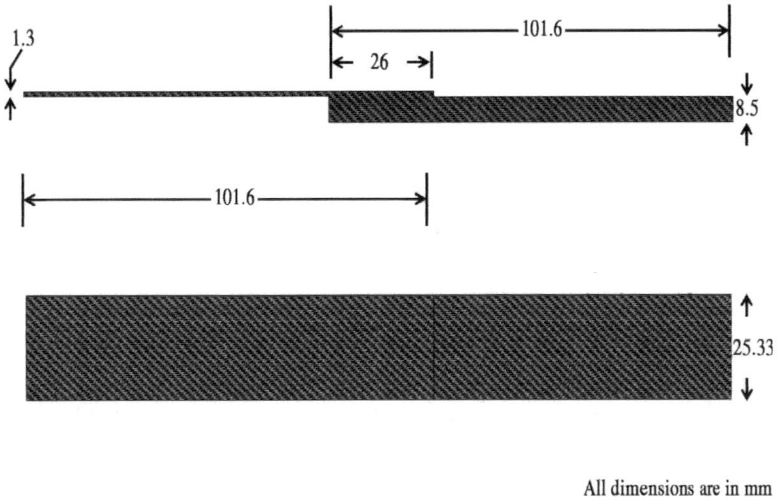

Figure 1. Schematic of SLS Specimens.

Two piezoelectric sensors with an operating frequency of 150 kHz to 400 kHz are used for recording the acoustic signals generated during the failure progression of the specimen [4]. The acoustic waveforms are sampled at a frequency on 1 MHz, which means the acoustic waves with frequency of 500 kHz are recorded according to the Nyquist criterium. The length of the waveforms recorded is of 3072 samples. Since this research work focuses on perceiving the failure progression in the CFRP as a sound, more importance is given to the subsequent sections, that are dedicated to the defined purpose.

3. Acoustic Emission Signal Processing

3.1. Selection of Acoustic Emission Signals

During the entire loading history of the SLS specimen, a total of 510 AE signals are recorded by both the piezoelectric sensors. The peak frequency of the AE signals as a function of the duration of the test is plotted in Figure 2.

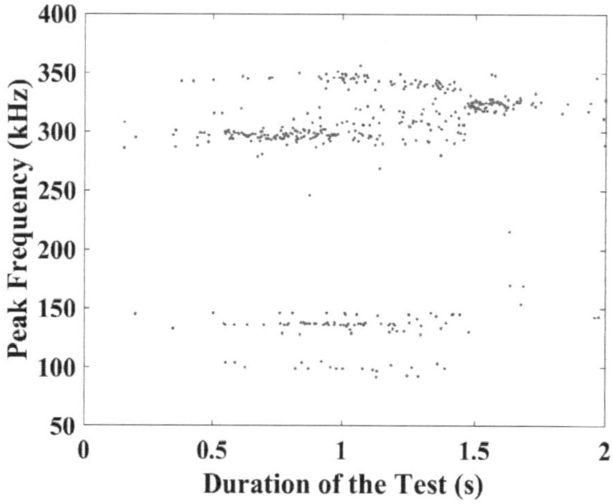

Figure 2. Peak Frequency distribution of the AE signals.

Analysing all the waveforms in the time-frequency domain is quite difficult owing to the large storage space and time required. Therefore, the characteristic acoustic waves representing different damage progression has been taken for this study. It has been extensively discussed in the literature that different failure modes generate acoustic signals with different peak frequency [2, 5]. It is illustrated in Table 1. Based on that, two AE waveforms representing each damage mode is taken from different loading stages and analysed further.

Table 1. Frequency characteristics of acoustic waves generated from different failure modes.

Mode of Failure Progression	Frequency of the Acoustic Waves
Matrix Cracking	150 kHz – 200 kHz
Debonding/Delamination	200 kHz – 300 kHz
Fiber Breakage Interlaminar Crack Growth	Above 300 kHz

3.2. Waveform Processing Techniques

The selected waveforms from the entire dataset are processed initially using different waveform processing techniques. The details and the parameters involving the waveform processing techniques used in this study are presented in this section.

First, Fast Fourier Transform (FFT) is used for processing the waveform and also for obtaining the frequency characteristics of the waveform. Next, the time-frequency characteristics of the waveform are studied using Continuous Wavelet Transform (CWT). One of the analytical wavelets named, Morse wavelet is used for CWT. The Fourier transform of this wavelet vanishes for negative frequencies, which makes it suitable for time-frequency studies. Morse wavelet with the symmetry parameter 3 and the time-bandwidth product of 60 is used with the number of octaves being 3 and number of voices per octave 32.

For Wavelet Packet Transform (WPT), the same Morlet wavelet is used, while the decomposition level is set at 3. The waveforms are decomposed to level 3 giving 8 frequency components.

Finally, for HHT, the signals are decomposed into 5 Intrinsic Mode Functions (IMF) before removing the noise, reconstructing them and analysing the time-frequency characteristics in the Hilbert spectrum.

3.3. Mel Spectrogram

Since the human auditory system can hear sounds only up to 20 kHz, the frequency rate of the acoustic waveforms is first rescaled from 1 MHz to 20 kHz. Then the waveforms are buffered into frames with a window length of 64. The frames are overlapped by 16 number of samples, followed by converting the frames into their time-frequency representation. Then each of these frames carrying the time-frequency representation is passed through 32 Mel filter banks. The Mel frequency (f_M) representation equivalent to the original frequency (f) can be given by Eqn. (1).

$$f_m = 2595\log\left(1+\frac{f}{700}\right) \tag{1}$$

4. Results and Discussions

For discussing the different waveform processing techniques, excluding the Mel scale frequency representation, the acoustic waveform generated from a matrix cracking event is taken. The waveform to be processed is presented in Figure 3.

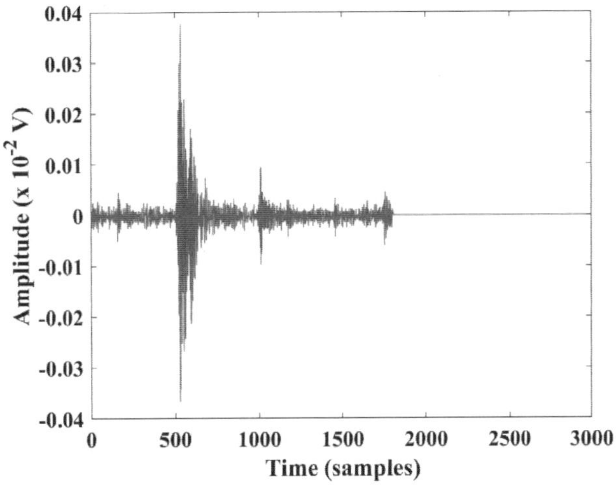

Figure 3. Acoustic Emission waveform generated from a matrix cracking event.

The frequency representation in FFT and the time-frequency representation in CWT, WPT and HHT are presented in Figures 4(a), 4(b), 4(c) and 4(d), respectively.

In Figure 4(a), the waveform seems to be a peak frequency around 150 kHz and also have some significant amount of spectral density around 275 kHz and 300 kHz. Nonetheless, it does not show any information about the time domain. On the other hand, CWT, WPT and HHT shows the information about the localization of different frequency band in time domain.

Now, for the same waveform, the Mel spectrogram is calculated by passing the frames of time-frequency values through Mel filters and is presented in Figure 5. In Mel spectrogram, the features of the waveform in the time-frequency domain are clearer and it shows the perpetual representation of how human ears would perceive the matrix cracking. This also gives a visual for how many frames the sound resonates and where its maximum power is centered.

Following this, a total of 8 waveforms are selected, each representing the different failure modes (indicated in Table 1) and are analysed. The results are discussed in this section.

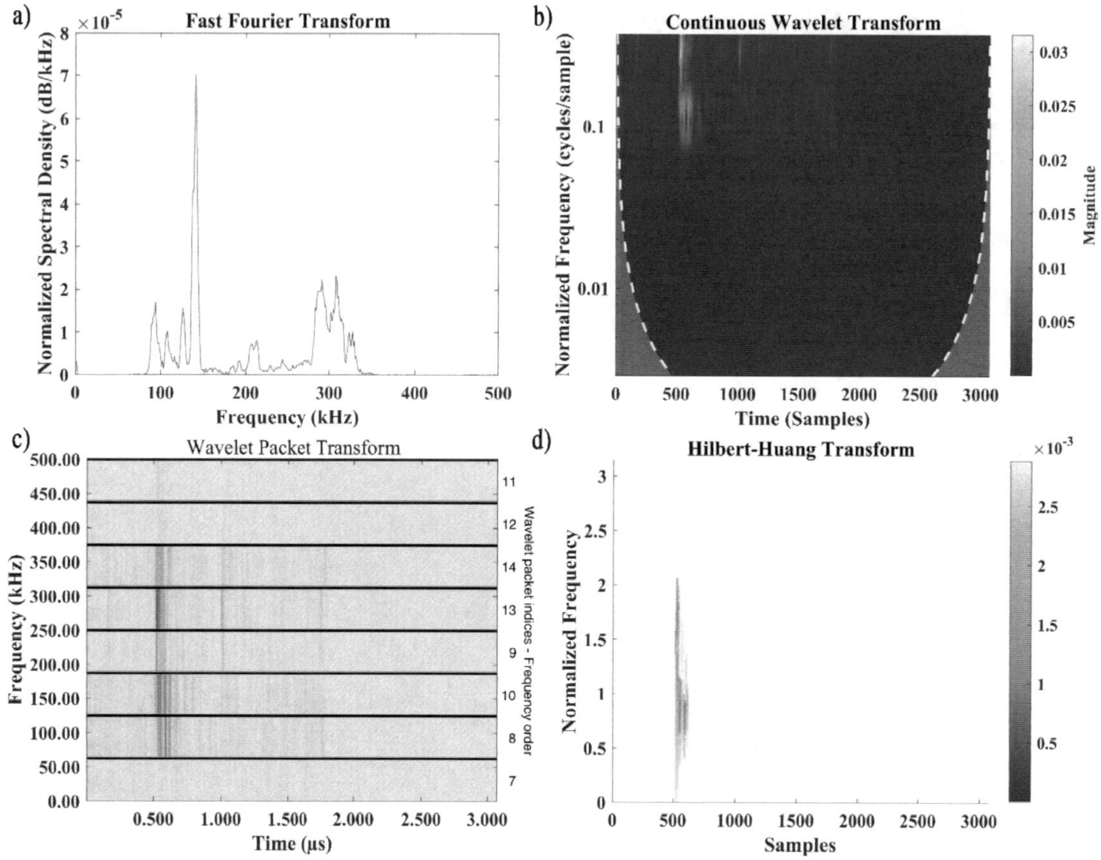

Figure 4. a) FFT, b) CWT, c) WPT and d) HHT of the Acoustic Emission waveform generated from a matrix cracking event.

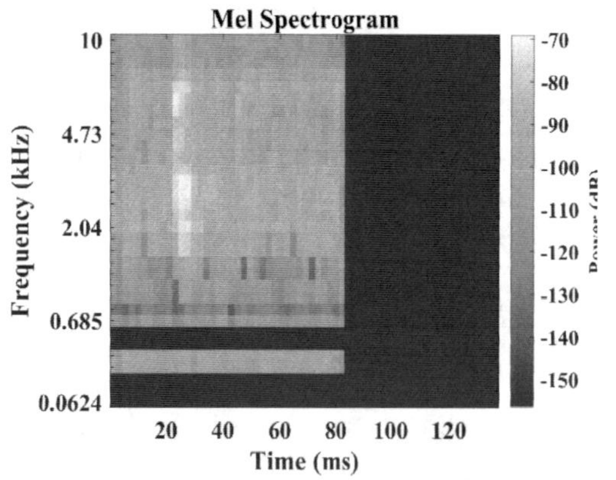

Figure 5. Mel Spectrogram of the acoustic wave generated from Matrix Cracking event.

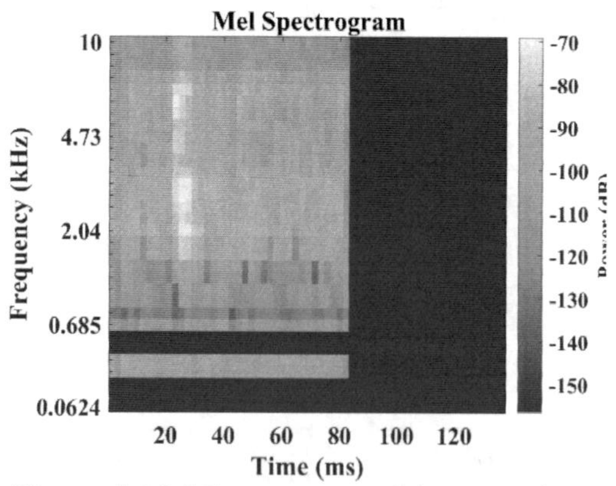

Figure 6. Mel Spectrogram of the acoustic wave n.2 generated from a Matrix Cracking.

Since Figure 5 already shows the Mel spectrogram of one of the matrix cracking events, the second one is presented in Figure 6. Similarly, the results of acoustic waves generated from delamination events are presented in Figure 7 and Figure 8.

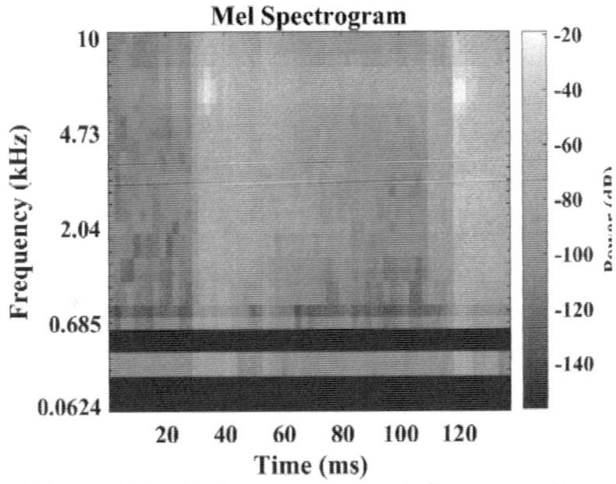

Figure 7. Mel Spectrogram of the acoustic wave n.1 generated from Delamination event.

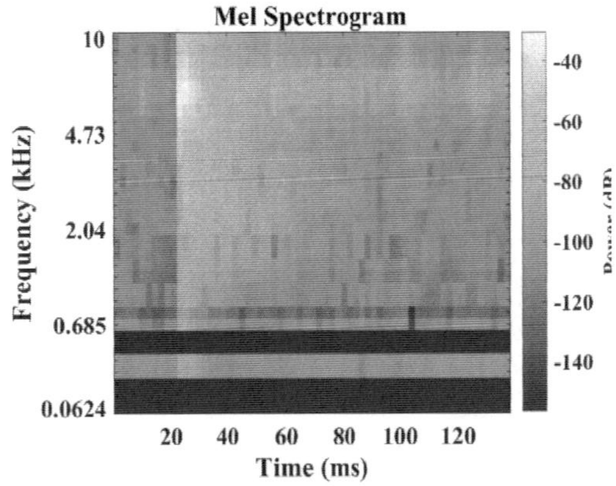

Figure 8. Mel Spectrogram of the acoustic wave n.2 generated from Delamination event.

Unlike the matrix cracking event, which resonates for a longer duration and its frequency in Mel scale is maximum between 1.93 kHz and 4.39 kHz, acoustic waves from delamination event have the maximum amplitude centered around 5 kHz in Mel scale. Particularly, the event presented in Figure 7 repeats its pattern after a certain period of time and is very much comparable to the acoustic wave presented in Figure 8, showing significant similarity between the two events.

The Mel spectrograms of the acoustic waves generated from interlaminar crack growth events are presented in Figure 9 and Figure 10. The Mel scale where these events centered are higher than the delamination event, while they show a repetitive pattern, with the acoustic signal in Figure 10 shows more repetition. Nonetheless, the resonance of the repetition seems to get smaller as the number of repetitions progresses.

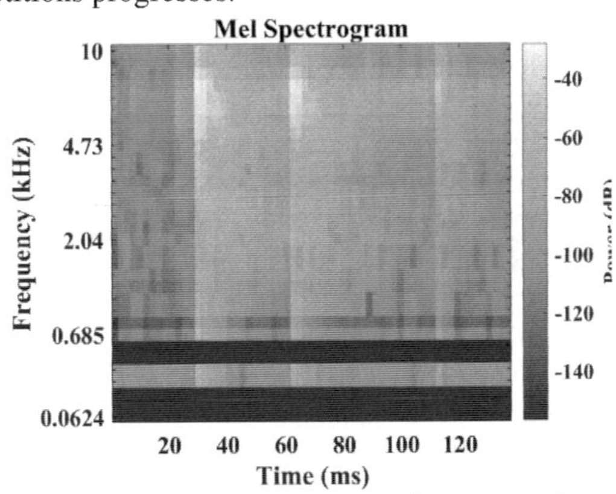

Figure 9. Mel Spectrogram of the acoustic wave n.1 generated from Interlaminar Crack Growth event

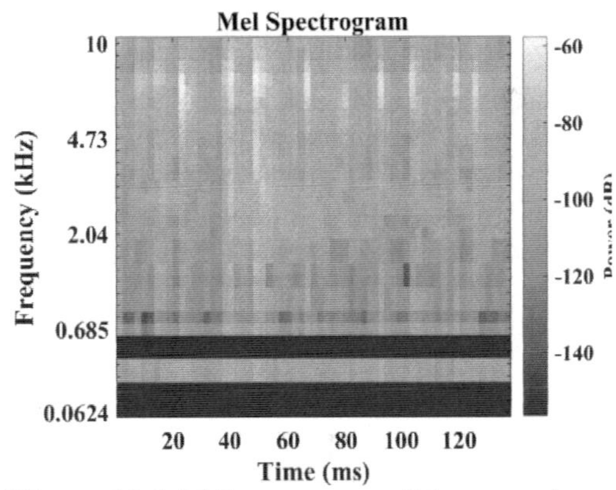

Figure 10. Mel Spectrogram of the acoustic wave n.2 generated from Interlaminar Crack Growth event

Finally, the Mel spectrograms of the acoustic events from fiber breakage events are presented in Figure 11 and Figure 12. Although the frequency in Mel scale with maximum power is almost the same as the events from interlaminar crack growth events, the signals from Fiber breakage are much shorter and they barely resonate. This is commensurable to the time-frequency analysis of fiber breakage and interlaminar crack growth events found in the literature. The peak frequencies of the acoustic signals from these two events mostly look similar, however, the fiber breakage event produces more transient signals between the two.

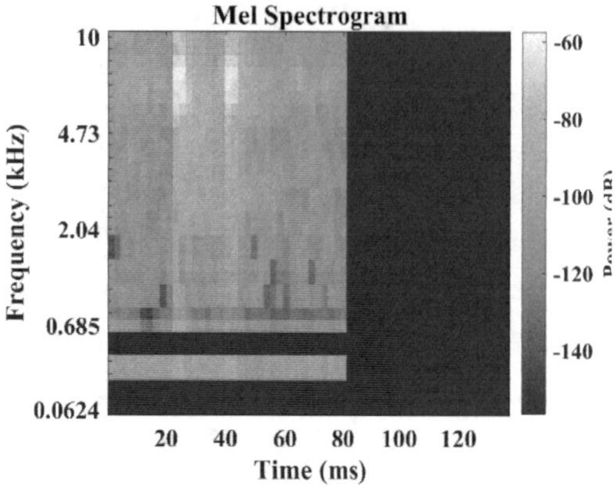

Figure 11. Mel Spectrogram of the acoustic wave n.1 generated from Fiber Breakage event

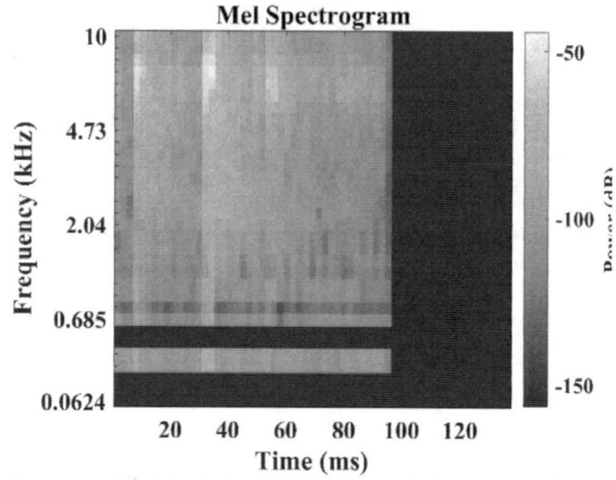

Figure 12. Mel Spectrogram of the acoustic wave n.2 generated from Fiber Breakage event

Figures 5-12 are essentially the apparent sounds of failure in the CFRP specimen tested for this study. What analysing the acoustic signals from different failure modes as 'sounds' accomplice? First, the repetitiveness of the frequency bands is more apparent in Mel scale. Even though CWT and WPT shows the time-frequency representations of the signal, it could not clearly magnify the repetitiveness in the time domain. Secondly, when these waveform features are processed for some other applications such as machine learning, CWT, WPT or HHT does not provide more features for classifications. On the other hand, Mel spectrogram has visually more features, which can be extracted and analysed for machine learning purposes. In fact, for most of the machine learning operations such as speech/words recognition, the Mel spectrogram is preferred over CWT or any other waveform processing techniques.

5. Conclusions

The acoustic emission signals generated from different failure modes in a SLS configuration of a CFRP specimen tested under static load is analysed. First, the waveforms are processed using conventional waveform processing tools such as FFT, CWT, WPT and HHT. The limitations in these conventional tools are explored. Then the acoustic waveforms from different failure modes are converted to their Mel scale equivalent in the time-frequency domain. The apparent sounds of failure in the Mel scale are analysed for find the characteristic differences between the signals generated from different failure modes. The results show that analysing the acoustic signal as 'sound' show clearer differences in defining the signals from different sources.

References

[1] Sause MG 2016 *In situ monitoring of fiber-reinforced composites: theory, basic concepts, methods, and applications* Springer (vol 242) p 131.

[2] De Groot PJ, Wijnen PA and Janssen RB 1995 Real-time frequency determination of acoustic emission for different fracture mechanisms in carbon/epoxy composites *Compos Sci Tech* **55(4)** 405-412.

[3] Wang HL, Song DZ, Li ZL, He XQ, Lan SR and Guo HF 2020 Acoustic emission characteristics of coal failure using automatic speech recognition methodology analysis. *Int J Rock Mech Min* **136** 104472.

[4] Barile C, Casavola C, Pappalettera G and Vimalathithan PK 2021 Multiparameter approach for damage propagation analysis in fiber-reinforced polymer composites *Appl Sci* **11(1)** 393.

[5] Barile C, Casavola C, Pappalettera G and Vimalathithan PK 2020 Application of different acoustic emission descriptors in damage assessment of fiber reinforced plastics: A comprehensive review *Eng Fract Mech* **235** 107083.

IOP Conf. Series: Materials Science and Engineering 1214 (2022) 012007 doi:10.1088/1757-899X/1214/1/012007

Damage Detection in Composite Materials by Flexural Dynamic Excitation and Accelerometer-Based Acquisition

G Loi[1], N Uras[2], M C Porcu[2] and F Aymerich[1]

[1] Department of Mechanical, Chemical and Materials Engineering, University of Cagliari, Piazza d'Armi, 09123 Cagliari, Italy
[2] Department of Civil-Environmental Engineering and Architecture, University of Cagliari, Piazza d'Armi, 09123 Cagliari, Italy

E-mail address: gabriela.loi@unica.it

Abstract. Composite materials provide many advantages over more conventional materials. However, their susceptibility to impact damage can question their use in critical load-bearing structures, and efficient methods are needed for early damage detection. To this purpose, the nonlinear vibro-acoustic modulation (VAM) technique applies a low-frequency pump excitation and a high-frequency probe excitation to exploit the onset of harmonic components around the probe frequency of the damaged structural response. The VAM technique has been widely studied on structures instrumented with piezoceramic transducers used for both actuation and sensing, but few attempts have been made to use equipment typical of modal testing, such as shakers and accelerometers. In this study, the VAM technique is applied to a composite laminate beam by employing an electro-dynamic shaker to generate low-frequency flexural excitation, a low-profile piezoceramic transducer to introduce the probe wave, and a micro-accelerometer to sense the structural response. Three resonance low frequencies and two acoustic frequencies are considered in different testing scenarios, at increasing levels of excitation amplitude. The results show a general good performance of the technique with the adopted experimental setup, the choice of the probe frequency and the higher level of the pump excitation having a significant impact on its sensitivity.

Keywords: Nonlinear Vibro-Acoustic Modulation, Structural Health Monitoring, Damage detection, composite materials

1. Introduction

Assessing the integrity of load-bearing structures is one of the major concerns in engineering applications and many non-destructive testing (NDT) methods have been developed to this purpose [1]. Some of them involve inspections by a range of observation techniques, such as x-ray radiography, thermography or shearography [2], while some others exploit the properties of transmission and reflection of ultrasonic waves into the materials [3]. These kinds of techniques are however expensive, time-consuming, and often unsuitable for testing large structural systems and/or advanced heterogeneous materials like sandwich panels or composite laminates. Structural Health Monitoring (SHM) methods based on embedding sensors within the structure may be a viable alternative for damage detection [4, 5]. Vibration-based SHM techniques assume that the modal properties of a damaged structure differ from those of the undamaged one. However, noticeable changes of modal properties are generally associated with rather high levels of damage, while the detection of comparatively small damage severities typically requires the adoption of more advanced approaches, such as those based on the analysis of the curvatures of the mode-shapes or of the frequency response function [6]. An elastic behavior of both the undamaged and the damaged system is assumed by such methods. In contrast, some SHM techniques focus on the nonlinear phenomena induced by damage. This makes them generally more sensitive to early-stage damage [7-9] whose detection is of primary interest when composite materials are concerned.

Indeed, the presence of defects or cracks perturbates the way an impinging wave propagates in a material, causing the onset of nonlinear features such as hysteretic behavior [10], wave resonance amplitude distortion [11], higher and sub-harmonics [12, 13] frequency mixing effects [14] and non-linearity in the elastic response of the system [15-18]. Among the techniques that exploit the non-linear features induced by damage is the Vibro-Acoustic Modulation (VAM) method, which has been proven to be effective in detecting fatigue cracks in steel beams [19, 20], aluminum specimens [21-23] and pipelines [24] as well as in wind turbine blades [25]. This technique was also applied to detect barely visible impact damage in composite materials [8, 26, 27] and composite sandwich panels [28]. The VAM technique applies simultaneously a low-frequency wave (pump excitation) and a high-frequency wave (probe or carrier excitation) to the test system. The interaction of these two waves leads to modulation effects. The appearance of modulation sidebands, i.e., spectral components around the main high-frequency harmonic, is an index of nonlinearity in the system response and reveals the presence of damage.

When carrying out VAM tests the excitation signals can be provided by means of different actuation methods and the system response captured through different kinds of sensors. Low-profile lead zirconate titanate (PZT) transducers are by far the most common choice for providing the high-frequency excitation [7-9, 26-28] and sometimes also the low-frequency modal excitation [29]. The latter is however more commonly applied by means of piezoceramic stack actuators [27, 28] or, more rarely, through shakers [7]. Shakers as low-frequency actuators in VAM tests were for example adopted in [7, 9, 23] to detect damage modes in aluminium and composite specimens. It can be also mentioned that an instrumented hammer was used in [30] to detect damage in steel specimens. As far as the sensors used to acquire the response in composite materials during VAM tests are concerned, PZT transducers [28] or scanning laser doppler vibrometers [8, 26, 27] are undoubtedly the most used.

An experimental setup including a shaker and an accelerometer, which are instruments typically used in modal vibration-based methods, is adopted in this study to apply the VAM technique for the detection of low-velocity impact damage in a composite laminate beam. In the experimental campaign, an electro-dynamic shaker and a piezoceramic transducer were used to generate, respectively, the low frequency flexural excitation and the carrier wave, while a micro-accelerometer was employed to acquire the structural response of the sample. Three resonance low frequencies and two acoustic frequencies were considered for different damage scenarios at increasing levels of excitation amplitude. It is worth noting that modal shakers may provide the power required to excite low-frequency vibration modes of systems, while accelerometers may efficiently measure the out-of-plane vibration response of the structure with the additional advantage of easy and fast removal and installation at different locations. The efficacy of the test setup and the effect of the excitation frequencies on the damage detection capability of the VAM method has been investigated in the study.

2. Vibro-Acoustic Modulation

Nonlinear Vibro-Acoustic Modulation (VAM), also referred to as Nonlinear Acoustics [5] or Nonlinear Elastic Wave Spectroscopy [11, 14], was proposed in [31] to detect the presence of closed cracks. In a VAM test the specimen is simultaneously excited by two waves: a low frequency modal/vibration (pump) wave and an acoustic high-frequency (probe) wave. The low-frequency pump wave aims to generate stress within the structure to perturb possible damage, while the high-frequency wave is expected to sense the variation of stiffness produced by the former, acting as a probing wave that can be modulated by the presence of damage.

As illustrated in figure 1, the spectrum of the signal response of a linear pristine material exhibits only two frequency components corresponding to the probe and pump excitation signals. In contrast, the presence of cracks, damage or discontinuities in the material induces wave perturbations that manifest themselves as additional frequency components (sidebands) around the probing carrier frequency. These sidebands typically occur in pairs (left and right sidebands) at the frequencies:

$$f_{SB_n} = f_{probe} \pm n\, f_{pump} \tag{1}$$

where f_{probe} and f_{pump} are the frequency of the pump and probe waves respectively and n is a positive integer number.

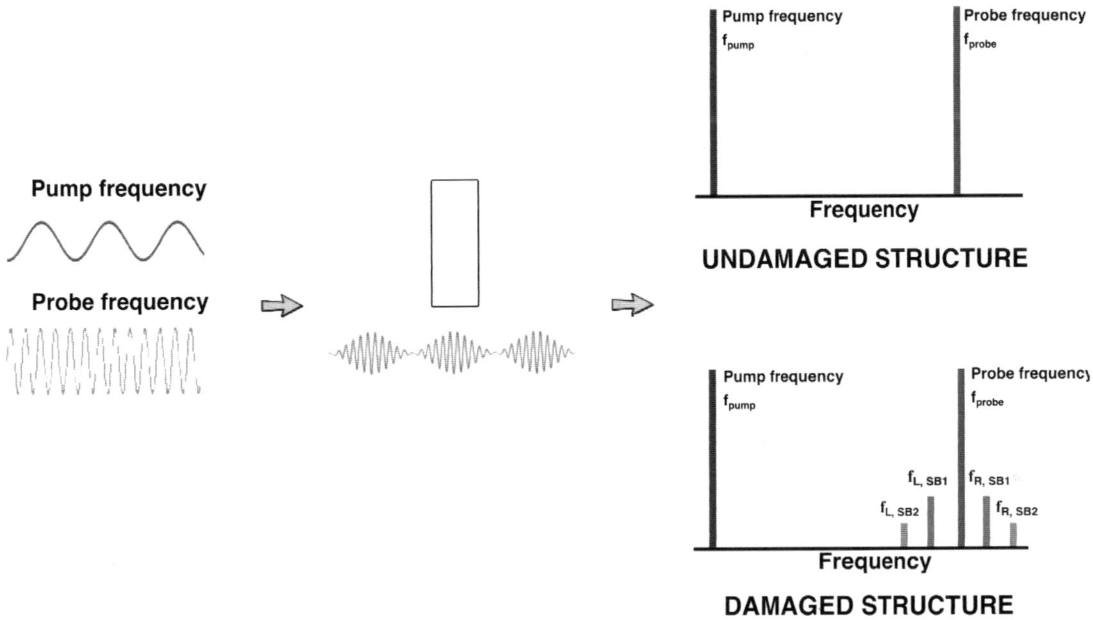

Figure 1. Schematic representation of the vibro-acoustic modulation (VAM) technique.

The onset of these modulated frequency components was found to depend on both the presence and the severity of damage and various damage indicators, ranging from the amplitude of the first pair of sidebands to parameters that relate the sidebands amplitude to the amplitude of the probe wave, have been proposed in the literature to monitor the degradation state of a structure. Commonly adopted damage parameters are based, for instance, on the ratio or on the difference between the sum of the amplitudes of the *i-th* pair of sidebands and the amplitude of the response at the probe frequency [3, 26, 27]. Further damage indicators involve the comparison between the amplitude of the sidebands and particular functions of the signal amplitudes at the pump and probe frequencies [30].

3. Experimental tests

Experimental tests have been conducted on a composite beam made of *Seal Texipreg ®HS160/REM* carbon/epoxy prepreg plies laminated with a $[0/90]_{3s}$ layup and consolidated in autoclave at a pressure of 8 bar and a maximum temperature of 160 °C. The examined sample was a 530 mm x 59 mm beam with a thickness of 2 mm. Prior to testing, the composite beam was ultrasonically C-scanned to exclude the presence of manufacturing defects.

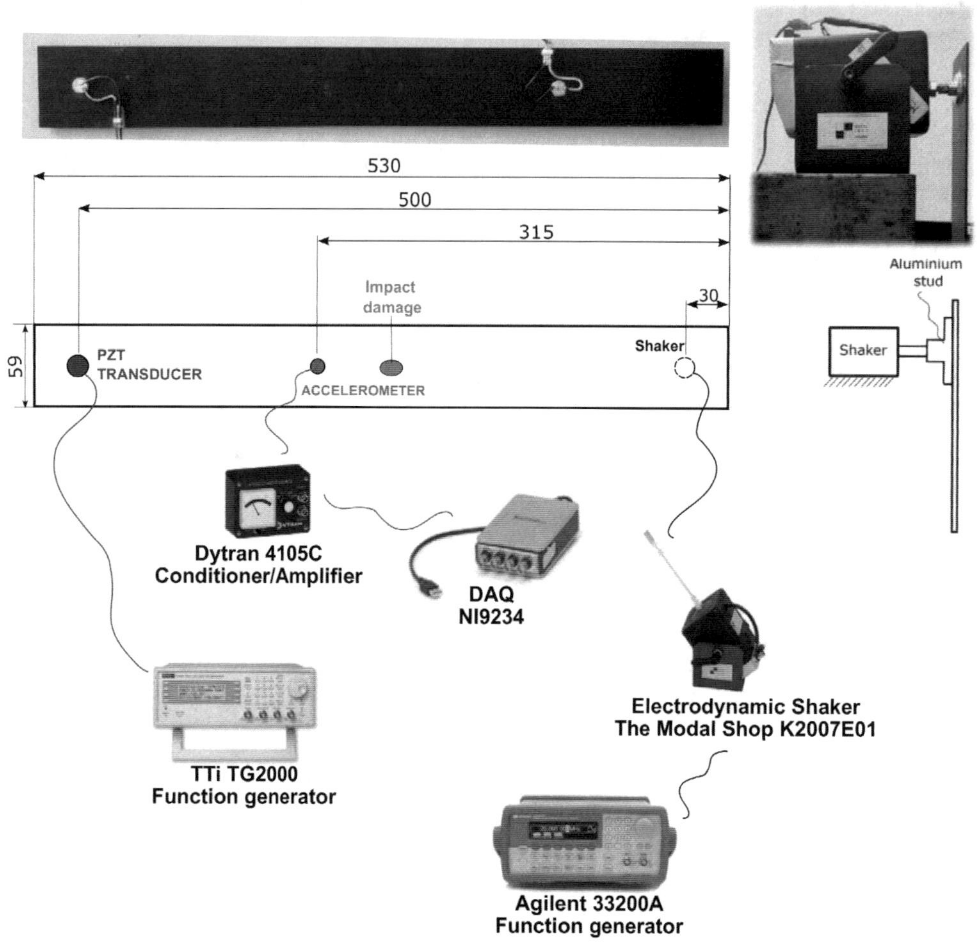

Figure 2. Schematic set-up of the vibro-acoustic modulation tests.

The beam was instrumented with a *PI Ceramic PIC 151* low-profile PZT transducer (10 mm diameter and 1 mm thickness) and a *DYTRAN 3224A1* micro-accelerometer, positioned as shown in figure 2. The PZT transducer was bonded to the surface of the beam by means of a two-component epoxy adhesive and wired through welded connectors, while the accelerometer was attached to the beam surface through a thin wax layer.

A drop-weight testing machine, equipped with an instrumented impactor provided with a hemispherical indenter of 10 mm diameter, was used to introduce damage in the sample. Two different levels of damage, referred to as D1 and D2 in the following sections, were obtained by impacting the composite beam with energies of 1.8 J and 2.4 J, respectively. Penetrant-enhanced X-radiographies were taken to characterize the extent and the nature of the two damage levels. As shown in Figure 3, the damage resulting from the impact loads is mainly a combination of matrix cracks and delaminations at various interfaces of the laminated beam. Some minor fibre breakage is however visible at the indentation area after the 2.4 J impact. The projected delamination areas corresponding to the two damage levels are respectively 31.5 mm^2 and 92 mm^2.

Experimental modal analyses were preliminarily carried out to identify the natural frequencies of the system for subsequent selection of pump and probe frequencies. To this purpose, the laminated composite beam was subjected to a sweep sine excitation, starting from 1 Hz and crossing 1000 Hz in 15 s, generated by a *Modal Shop K2007E01* electrodynamic shaker driven by an *Agilent 33200A* function generator (figure 2). The sample was coupled to the shaker by a threaded aluminium stud that

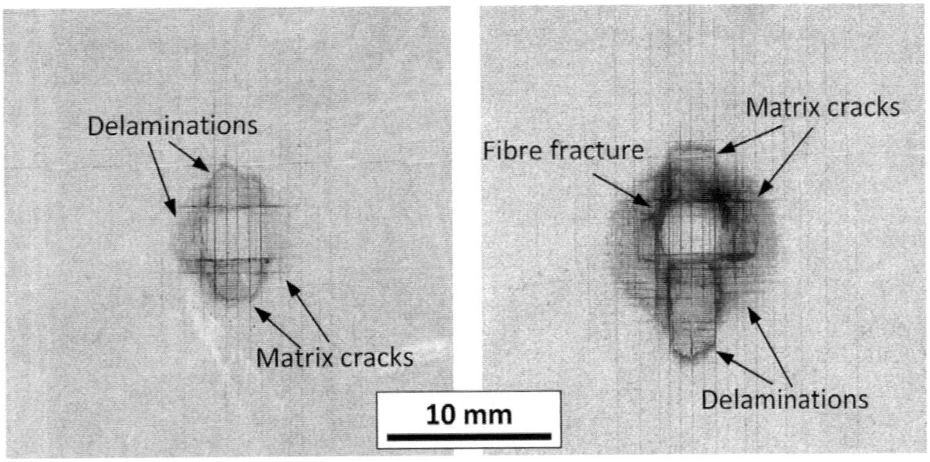

Damage level D1
(Impact energy = 1.8 J)

Damage level D2
(Impact energy = 2.4 J)

Figure 3. X-ray images of the damaged areas at the two examined damage severities.

was bonded to the beam by epoxy adhesive on one end and connected to the threaded hole of the shaker on the other end. The system response was monitored through a miniature accelerometer *DYTRAN 3224A1* mounted on the axis of the beam. The accelerometer output was fed to a *DYTRAN 4105C* conditioner (set to a 100x amplification) and then acquired by a *National Instrument NI9234* PC-controlled data acquisition unit. Ten sets of data were used to calculate the average Fast Fourier Transform (FFT) amplitude of the signal (figure 4) to increase the signal-to-noise ratio. Three frequencies, namely 155 Hz, 282 Hz and 494 Hz (evidenced in figure 4), were selected for the pump/modal excitation in the VAM experiments. A 3D Finite Element (FE) model of the composite sample was developed in *Abaqus* to evaluate the mode shapes associated to these three frequencies. The aluminium stud was modelled with 10-nodes solid tetrahedral elements, while the composite beam was discretized with 8-nodes brick elements. In order to correctly reproduce the stacking sequence, one solid element was used for each ply across the thickness. The mode shapes associated to the selected frequencies were found to be mainly characterized by out-of-plane bending, as shown in figure 5.

A similar procedure was applied to choose the frequencies of the probe wave. The electro-dynamic shaker was driven by the *Agilent 33200A* function generator using a sweep sine signal starting from 1 kHz and reaching 7.5 kHz in 15 s, while the system response was measured by the miniature

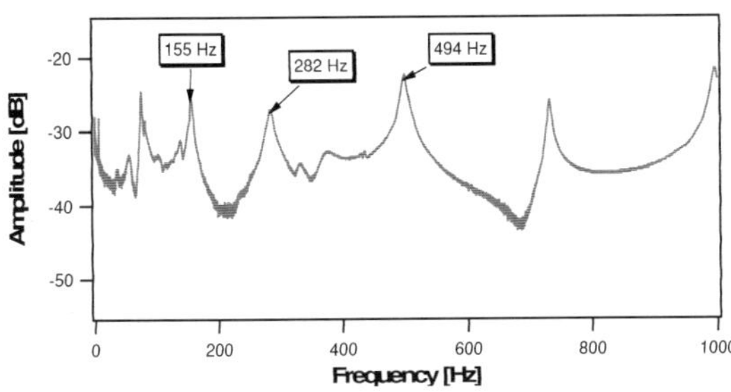

(a)

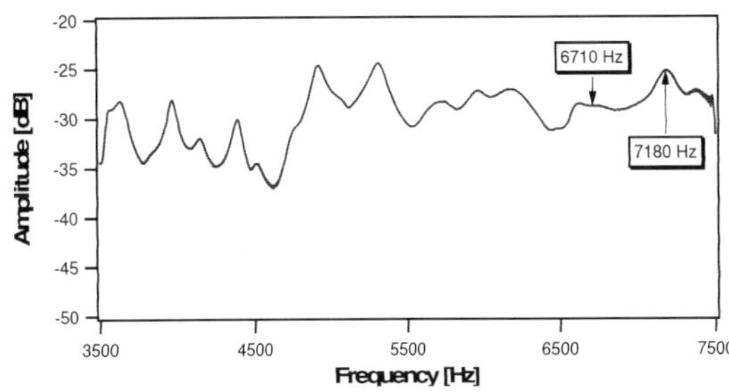

(b)

Figure 4. Power spectra of the accelerometer signal in the low-frequency (a) and high-frequency (b) ranges.

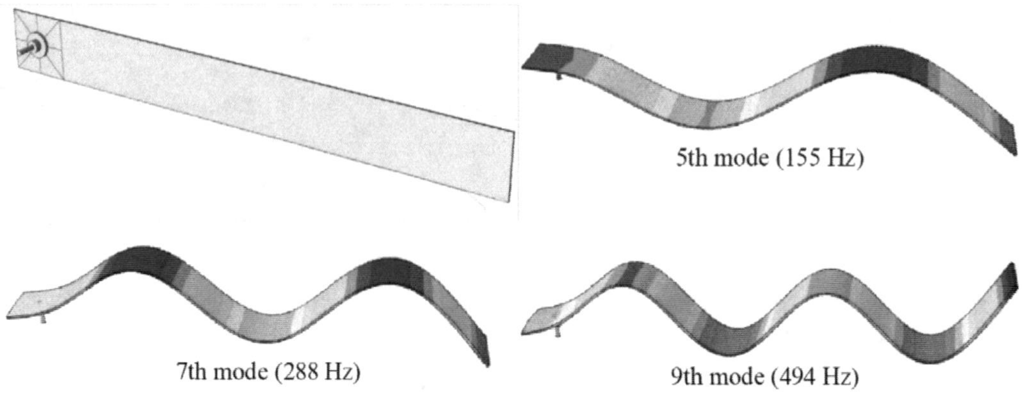

Figure 5. FEM model and numerically predicted mode shapes associated to the three selected pump frequencies.

accelerometer. The power spectrum of the acquired time signal is shown in fig. 4b. In this case, one resonance frequency (7180 Hz) and one generic frequency (6710 Hz) were selected for probe excitation.

Nonlinear vibro-acoustic modulation tests were then performed. A probe (high frequency) sinusoidal signal was applied to the PZT actuator and, simultaneously, the beam was excited with a pump (low frequency) sinusoidal wave by means of the electrodynamic shaker (Figure 2). The amplitude of the probe signal was set to 20 V_{pp} for all the tests, while different amplitudes of the low-frequency wave were considered to explore the role of the pump excitation level on the quality of the VAM indications. The experimental tests were carried out for six combinations of pump and probe frequencies, at various amplitudes of the pump excitation, for a total of eighteen different test scenarios. A summary of them is provided in Table 1.

Table 1. Test scenarios for different frequency combinations and pump excitation amplitudes

Probe frequency	Pump frequency	Pump amplitudes		
6710 Hz	152 Hz	0.5 V_{pp}	1.0 V_{pp}	2.0 V_{pp}
	282 Hz	2.0 V_{pp}	4.0 V_{pp}	6.0 V_{pp}
	494 Hz	1.0 V_{pp}	2.0 V_{pp}	3.0 V_{pp}
7180 Hz	152 HZ	0.2 V_{pp}	0.5 V_{pp}	1.0 V_{pp}
	282 Hz	2.0 V_{pp}	4.0 V_{pp}	6.0 V_{pp}
	494 Hz	1.0 V_{pp}	2.0 V_{pp}	3.0 V_{pp}

The probe excitation was kept fixed at 20 V_{pp}

The system response, sensed by the accelerometer, was acquired by the *National Instrument NI9234* digitizer at a 51.2 kSa/s sampling rate. Thirty data sets, each consisting of 256 kpoints, were saved for any of the different scenarios and then post-processed in *Matlab* to calculate the averaged power spectra of the acquired signals. The response spectra were finally zoomed on a window around the carrier wave frequency to retrieve the presence of modulation sidebands.

4. Results and discussion

Figures 6 and 7 show, as an example, power spectra around the probe frequency of the undamaged and damaged system responses obtained for two different combinations of vibro-acoustic excitations. The spectra reported in figure 6 correspond to testing conditions characterized by a pump excitation of 4 V_{pp} at a frequency of 282 Hz together with a simultaneous probe excitation at a frequency of 6710 Hz. Figure 7 shows the spectral response acquired with the beam subjected to the same pump vibration (282 Hz at 4 V_{pp}) but to a different probe excitation frequency (7180 Hz). It is worth recalling that the same amplitude of 20 V_{pp} was used in all testing scenarios to drive the PZT sensor for probe excitation.

When 6710 Hz is used as probe excitation frequency (figure 6), modulation sidebands appear in the spectral response of the beam after the introduction of the first damage (D1) and the amplitude of these sidebands increases after the beam is subjected to the second impact (damage level D2). In this case the average value of the amplitudes of the left and right sidebands increases of about 50% when advancing from damage level D1 to damage level D2. If we however examine the spectra acquired for a probe excitation frequency of 7180 Hz (Figure 7), we may see that while a sideband first appears on the left of the probe frequency for damage level D1, it then disappears when the severity of damage is increased to level D2. This suggests that, for the same values of excitation amplitudes, the effectiveness of this technique may depend on the selected excitation frequencies.

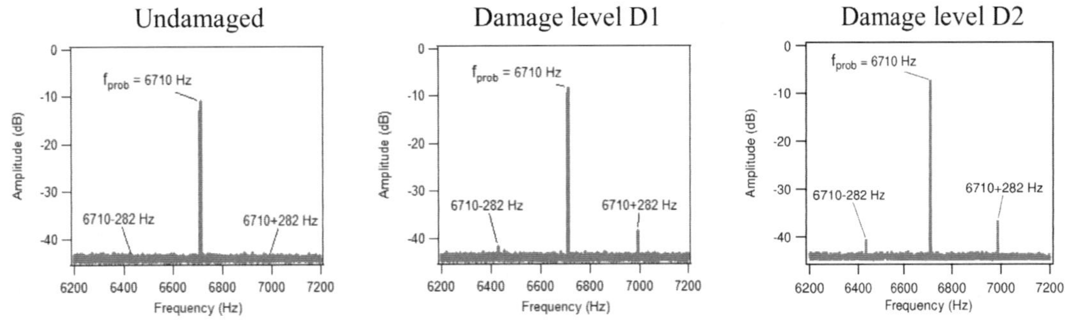

Figure 6. Power spectra of the response of undamaged and damaged sample: probe frequency 6710 Hz and pump frequency 282 Hz (4 V_{pp} amplitude).

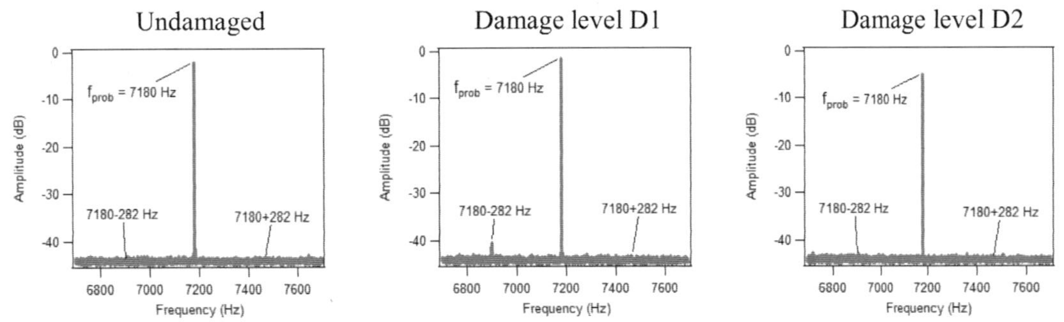

Figure 7. Power spectra for undamaged and damaged sample: probe frequency 7180 Hz and pump frequency 282 Hz (4 V_{pp} amplitude).

The graphs of figures 8 and 9 report a summary of the results obtained for all testing scenarios described in Table 1. The graphs compare the average values of the amplitudes measured at the left and right sideband frequencies for different pump excitation levels when examining the undamaged and the damaged beam. The average amplitude of the sidebands was assumed as an indicator of damage to explore the damage detection capabilities of the VAM approach for different pump and probe excitations.

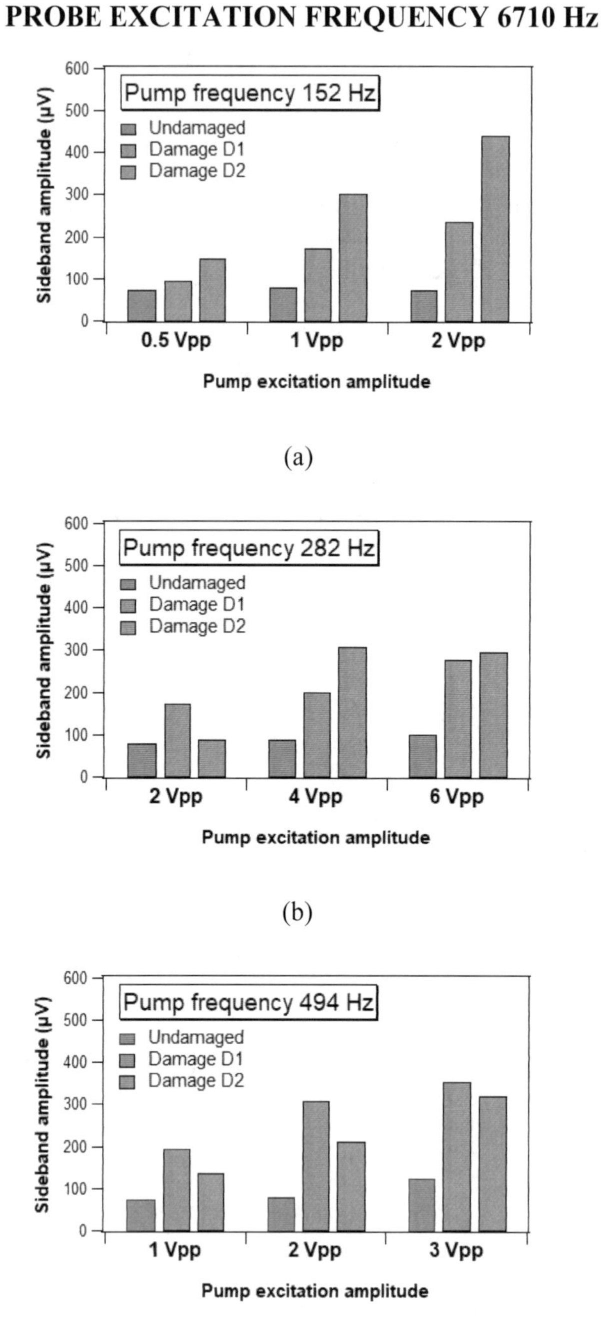

Figure 8. Average sideband amplitudes recorded for the different damage conditions at increasing pump excitation levels. The data plotted in the graphs were obtained using a probe frequency of 6710 Hz and pump frequencies respectively of 152 Hz (a), 282 Hz (b) and 494 Hz (c).

PROBE EXCITATION FREQUENCY 7180 Hz

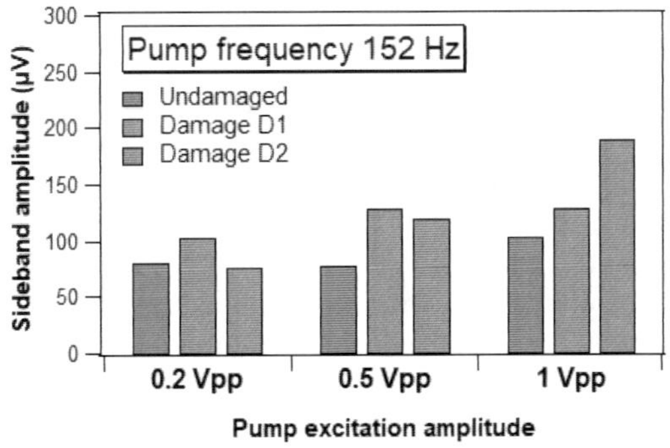

(a)

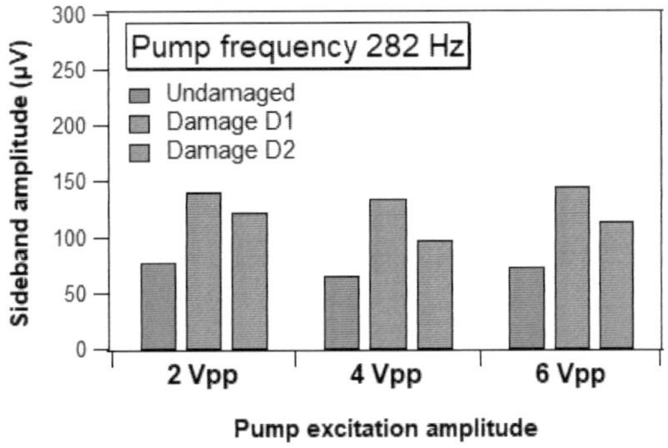

(b)

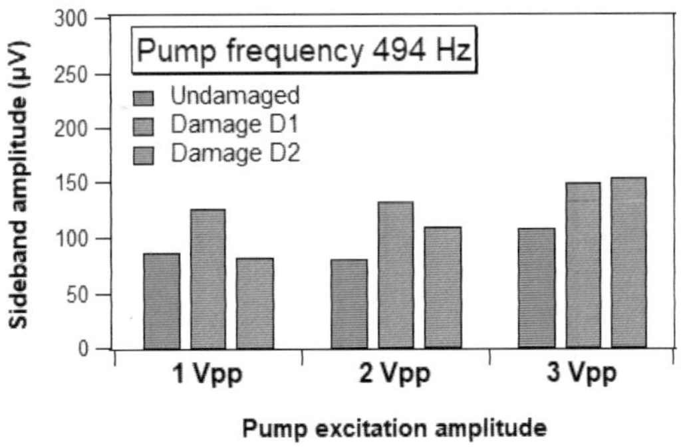

(c)

Figure 9. Average sideband amplitudes recorded for the different damage conditions at increasing pump excitation levels. The data plotted in the graphs were obtained using a probe frequency of 7180 Hz and pump frequencies respectively of 152 Hz (a), 282 Hz (b) and 494 Hz (c).

The data plotted in the graphs show that the VAM technique is generally capable of identifying the presence of the damage introduced in the composite beam by the low-velocity impacts, although, as mentioned before, its efficacy appears to be dependent on the choice of the pump and probe frequencies. The data also show that the choice of a resonance frequency for the probe excitation (7180 Hz) does not necessarily translate into improved detection performances. These indications confirm the findings of previous studies [9, 27], which indeed showed a clear effect of the excitation frequencies on the damage detection capabilities of nonlinear vibroacoustic approaches.

A clear influence of the pump amplitude on the performance of the VAM technique can be also inferred from the present results. If we focus our attention on the data acquired at the maximum pump excitation levels, we observe that the sideband amplitudes measured on damaged conditions are invariably higher than those measured on the undamaged beam. In some cases, however, (see figures 8c and 9b), the level of the sidebands recorded for the larger damage severity (D2) is lower than that recorded for the smaller damage (D1). A possible explanation of these results may be found in the small size of the damaged areas and in the similarity between the features of the two impact-induced damage scenarios, which are both dominated by matrix damage modes, and as such are not expected to lead to large differences in the associated nonlinear wave perturbation mechanisms.

A global comparison of the results obtained across the full range of the examined pump excitation levels shows that the sensitivity to damage of the method decays when the amplitude of the pump vibration is reduced. We notice that in several cases (see for example the graphs of figures 8b, 9a, 9c) the average sideband amplitude recorded for the undamaged beam is comparable or even higher than that measured for the D1 or D2 damaged beam. The loss of sensitivity at low pump excitation amplitudes can be related to the difficulty in the measurement of low-level signals, which can be close to or below the background noise floor, or to the need of attaining a minimum excitation energy for the activation of the mechanisms of nonlinear interaction between the pump and probe waves.

5. Conclusions

The detection of early-stage damage in composite materials plays a key role for their safe use in structural mechanics. The non-linear Vibro-Acoustic Modulation (VAM) is a non-destructive technique which can be effective for this purpose, even though some aspects still need to be better investigated to improve the effectiveness of such technique in practical applications. Among them are the role of frequency and amplitude of the pump and probe excitation waves as well as the impact of the testing setup. As a contribution to address this matter, the paper presents some experimental results obtained by testing a laminated composite beam, where two different levels of barely visible damage were introduced by low-velocity impacts. The beam was tested by means of an equipment typically adopted in experimental modal testing. It was simultaneously subjected to a harmonic probing excitation, provided through a piezoceramic disk, and a harmonic pump excitation provided by an electrodynamic shaker, able to supply the power required to excite low-frequency vibration modes in both pristine and damaged conditions. A micro-miniature accelerometer was used to sense the system response.

Six combination of pump/probe frequencies and three levels of pump amplitude were considered in the experimental campaign for a total of eighteen different test scenarios, each one involving the acquisition of data for the undamaged case and for the two damaged cases (D1 and D2). The following insights can be drawn from the results of the study.

i. The damage (typically delamination) associated to low-velocity impacts was found to cause the onset of a pair of sidebands at the left and right side of the probe frequency in the power spectra of the system response. The amplitude of these sidebands was found to generally increase with the severity of damage and the amplitude of the pump wave. However, some exceptions to this behavior called for an exploration of the role of the probe and pump frequencies as well as of the pump amplitude on the sensitivity of this technique. The average amplitude of the first two sidebands was adopted as modulation intensity index to investigate on this matter.

ii. The selection of the pump modal frequency seemed to affect slightly the results, although the best damage detection performances were generally associated with the lowest pump frequency (152 Hz) among the three considered in this study.
iii. The sensitivity of the technique was found to be mostly better at the highest pump excitation level for the different pump/probe frequency combinations. In four out of six cases, at the highest pump excitation amplitude the technique was able to correctly rank the severity of damage. The loss of sensitivity at low pump excitation amplitudes can be related to the difficulty in the measurement of low-level signals and/or to the need of reaching a minimum excitation level for the activation of nonlinear wave interaction mechanisms.
iv. In some of the considered cases the average amplitude of the sidebands recorded for the larger damage severity (D2) was lower than that recorded for the smaller damage (D1). This can be due to the small extent of the damaged area and to the comparable features of the examined damage conditions.
v. The choice of the probe frequency seemed to have a significant impact on the sensitivity of the VAM technique, as one of the two selected probe high frequencies (6710 Hz) led to generally better detection performances. Setting the probe excitation to a resonance frequency did not improve, however, the damage detection efficacy.
vi. The experimental setup adopted in this study, which included a shaker to apply the pump harmonic excitation and a micro-accelerometer to sense the response, was found to be suitable for the application of the VAM technique to detect damage in composite materials.

It can be concluded that the technique still presents some critical issues related to the choice of excitation frequencies and amplitudes that need further investigations. This suggests caution in interpreting VAM data for identification of damage of this nature and extent in this class of materials.

References

[1] Moore P O 2012 Nondestructive Testing Handbook vol 10, in Overview 3rd ed. (Columbus: American Society for Nondestructive Testing) p 594
[2] Rojas-Vargas F, Pasqual-Francisco J B and Hernandez-Cortés T 2020 Applications of shearography for non-destructive testing and strain measurement *International Journal of Combinatorial Optimization Problems & Informatics* **11**(3)
[3] Wronkowicz A, Dragan K and Lis K 2018 Assessment of uncertainty in damage evaluation by ultrasonic testing of composite structures *Comp. Struct.* **203** 71-84.
[4] Stepinski T, Uhl T and Staszewski W J 2013 *Advanced Strctural Damage Detection: from Theory to Engineering Applications* (Chichester: Wiley) p 352
[5] Kundu T 2019 N*onlinear Ultrasonic and Vibro-Acoustical Techniques for Nondestructive Evaluation* (Cham: Springer) p 773
[6] Porcu M C, Patteri D M, Melis S and Aymerich F 2019 Effectiveness of the FRF curvature technique for structural health monitoring *Constr. Build. Mater.* **226** 173-87
[7] Aymerich F and Staszewski W J 2010 Impact damage detection in composite laminates using nonlinear acoustics, *Compos. Part A-Appl. S.* **41** 1084-92
[8] Klepka A, Pieczonka L, Staszewski W J and Aymerich F 2014 Impact damage detection in laminated composite by nonlinear vibro-acoustic modulation *Compos. Part B-Eng.* **65** 99-108
[9] Klepka A, Staszewski W J, Jenal R B, Szwedo M, Iwaniec J and Uhl T, 2012 Nonlinear acoustics for fatigue crack detection – Experimental investigations of vibro-acoustic wave modulations *Struct. Health Monit.* **11**(2) 197–211
[10] Bentahar M, El Aqra H, El Guerjouma R, Griffa M and Scalerandi M 2006 Hysteretic elasticity in damaged concrete: quantitative analysis of slow and fast dynamics, *Phys. Rev. B* **73** 014116
[11] Van Den Abeele K E A, Johnson P A and Sutin A 2000 Nonlinear Elastic Wave Spectroscopy (NEWS) techniques to discern material damage, part I: nonlinear wave modulation spectroscopy (NWMS) *Res. Nondestruct. Eval.* **12** 17-30

[12] Lissenden C J, Liu Y, Choi G W and Yao X 2014 Effect of localized microstructure evolution on higher harmonic generation of guided waves *J. Nondestruc. Eval.* **33** 178-86.
[13] Shah A A and Ribakov Y 2009 Non-linear ultrasonic evaluation of damaged concrete based on higher order harmonic generation *Mater. Des.* **30** 4095-102
[14] Van Den Abeele K E A, Carmeliet J, Cate J A T and Johnson P A 2000 Nonlinear Elastic Wave Spectroscopy (NEWS) techniques to discern material damage. Part II: single mode nonlinear resonance acoustic spectroscopy. *Res. Nondestruct. Eval.* **12**(1) 31-42.
[15] Scalerandi M, Gliozzi A S, Bruno C L E, Masera D and Bocca P, 2008 A scaling method to enhance detection of a nonlinear elastic response *Appl. Phys. Lett.* **92** 101912.
[16] Porcu M C, Pieczonka L, Frau A, Staszewski, W J and Aymerich F 2017 Assessing the scaling subtraction method for impact damage detection in composite plates *J. Nondestruc. Eval.* **36** 33
[17] Loi G, Porcu M C and Aymerich F 2021 Impact damage detection in composite beams by analysis of non-linearity under pulse excitation *J. Compos. Sci.* **5** 39
[18] Loi G, Porcu M C, Pieczonka L, Staszewski W J and Aymerich F 2019 Scaling subtraction method for damage detection in composite beams *Procedia Struct. Integr.* **24** 118-26.
[19] Duffour P, Morbidini M and Cawley P 2006 A study of the vibro-acoustic modulation technique for the detection of cracks in metals *J. Acoust. Soc. Am.* **119**(3) 1463
[20] Yoder C and Adams D E 2010 Vibro-acoustic modulation utilizing a swept probing signal for robust crack detection *Struct. Health Monit.* **9**(3) 257–267
[21] Sohn H, Lim H J, DeSimio M P, Brown K and Derriso M 2014 Nonlinear ultrasonic wave modulation for online fatigue crack detection, *J. Sound Vib.* **333**(5) 1473–84
[22] Straka L, Yagpdzinskyy Y, Landa M and Hanninen H 2008 Detection of structural damage of aluminum alloy 6082 using elastic wave modulation spectroscopy *NDT E Int.* **41**(7) 554–63
[23] Dziedziech K, Pieczonka L, Adamczyk M, Klepka A and Staszewski W J 2018 Efficient swept sine chirp excitation in the non-linear vibro-acoustic wave modulation technique used for damage detection *Struct. Helath Monit.* **7**(3) 565–76
[24] Jiao J, Zheng L and Song G 2011 Vibro-acoustic modulation technique for micro-crack detection in pipeline *Proc. SPIE - Int. Soc. Opt. Eng.* **8321**(83213X)
[25] Kim S et al. 2014 Crack detection technique for operating wind turbine blades using vibro-acoustic modulation *Struct. Health Monit.* **3**(6) 660–70
[26] Pieczonka L, Zietek L, Klepka A, Staszewski W J, Aymerich F and Uhl T 2017 Damage imaging in composites using nonlinear vibro-acoustic wave modulations, *Struct. Control Health Monit.* **25**(2)
[27] Frau A, Pieczonka L, Porcu M C, Staszewski W J and Aymerich F 2015 Analysis of elastic nonlinearity for impact damage detection in composite laminates *J. Phys. Conf. Ser.* **628** (1)
[28] Pieczonka L, Ukowski P, Klepka A, Staszewski W J, Uhl T and Aymerich F 2014 Impact damage detection in light composite sandwich panels using piezo-based nonlinear vibro-acoustic modulations *Smart Mater. Struct.* **23**(10)
[29] Donskoy D M, Sutin A M and Ekimov A 2001 Nonlinear acoustic interaction on contact interfaces and its use for nondestructive testing *NDT Eval. Int.* **34**(4) 231–38
[30] Meo M and Zumpano G 2005 Nonlinear Elastic Wave Spectroscopy identification of impact damage on a sandwich plate *Comp. Struct.* **71**(3–4) 469–74
[31] Zaitsev V Y, Sutin A M, Belyaeva I Y and Nazarov V E 1995 Nonlinear interaction of acoustical waves due to cracks and its possible usage for cracks detection *J. Vib. Control* **1**(3), 335–44

Full-field monitoring methods for damage analysis on aeronautical CFRP specimens under fatigue loads

R Nobile[1], F W Panella[1], A Pirinu[1*] and A Saponaro[1]

[1]University of del Salento, Department of Engineering for Innovation, Viale per Monteroni, Lecce (LE), 73100, Italy

*alessandra.pirinu@unisalento.it

Abstract. The present paper is focused on full-field experimental monitoring procedures to be employed during HCF fatigue testing on two series of CFRP open hole samples. Two different experimental methodologies based on thermographic techniques and displacement measurements with Digital Image Correlation (DIC) analysis are employed for damage settlement and evolution to be detected up to failure, together with correspondent compliance analysis. Combined monitoring approaches, based on thermo-elastic and dissipative phenomena, together with stiffness properties variations, are claimed to offer precise damage state localization during tests in real time; in addition, DIC analysis is performed during low-frequency fatigue cycle is studied for better failure prediction and damage location. The thermal parameters and experimental compliance correlation seem to indicate similar signal variation during damage progress and after proper data elaboration; contemporaneous raw thermal measurements in critical zones of specimens under fatigue life offer a kind of delamination recognition at specific layer interface and location, as well as propagation before final failure. Sample under tensile load on the other hand reveal delaminations indirectly on the surface. In addition, non-destructive thermographic and ultrasound tests are performed at regular intervals during fatigue life.

Keywords: CFRP, damage prediction, delamination, DIC, Fatigue, TSA.

1. Introduction

Composite structures showed a growing employment in aerospace and automotive field due to their elevated mechanical properties, such as strength-to-weight and stiffness-to-weight ratios in case of Carbon fiber-reinforced polymers (CFRP), as well as improved challenging geometries to be achieved with advanced manufacturing and elevated fatigue resistance in comparison to classical metallic alloys [1]. However, the non-homogeneous behaviour and different interface conditions with other parts seem to exhibit critical mechanical response behaviour, showing a continuous degradation of mechanical properties before the final structural failure with alternate stressed during service life [2].

The detection and prevention of damage evolution, employing a no-contact and less expensive monitoring technique represents an ongoing challenge for the Structural Health Monitoring (SHM) of composite aircraft parts under fatigue with relatively high frequency levels [3, 4]. Non-Destructive Inspections (NDIs) and preventive replacement are required by the scheduled maintenance program to ensure stable and reliable operating condition during service life, involving consequent considerable costs. Among the different SHM techniques, optical and non-contact ND methods such as Thermoelastic

Stress Analysis [5] (TSA) and Digital Image Correlation [6] (DIC) seem to provide the full-field stress states and damage induced effects of monitored structures [1]. Several authors demonstrate the damage evolution analysis, employing the Infrared Thermography (IRT), which provides a real-time continuous monitoring procedure on inspected elements under cyclic loads, unlike the local assessing of strain gauges. Thermoelastic stress analysis (TSA) is a non-contact sub-surface technique based on the thermoelastic effect which correlate the stress distribution to temperature variations detected in inspected component subjected to cyclic loads employing a sensitive infrared camera [1, 5]. Therefore, recent researches highlight the damage detection capability and monitoring potentials of micro-bolometric infrared camera showing the more cost effective and commercial availability solutions for TSA approach without an expensive thermographic equipment [7, 8].

In this work, two different Carbon Fiber/epoxy (CFRP) composite specimens in the form of Flat Open Hole laminated elements are tested under tension-tension cyclic loading, realized in pre-preg autoclave curing or through the Liquid Resin Infusion technology, to study fatigue behaviour in presence of hole.

Experimental tests are in-real time monitored with thermal acquisitions, exporting reveal effects of localized tensile loads and notched zones as source of delamination initiation, whilst delamination growth is experimentally investigated before failure and correlated to fatigue residual life expectancies. However, the present proposed work efforts consist of better damage localization (well observed with visual and thermographic inspection) and quantification of propagation severity, compared with the well-established compliance analysis. Continuous damage accumulation is monitored with infrared thermal camera and DIC measurements at specified intervals during fatigue life, coupled with mechanical stiffness degradation during whole fatigue life for CFRP elements after High Cycle fatigue (HCF) loading.

2. Materials and Methods

2.1. CFRP elements and experimental procedures

Both sample series are CFRP composites in Figure 1(a) in Flat Open Hole configuration, and all static and cyclic tests are performed in load control mode on the servo - hydraulic INSTRON 8850 axial-torsion test machine, equipped with a load cell of ± 250 KN and governed by a computer equipped with INSTRON WAVE MATRIX software.

(a)

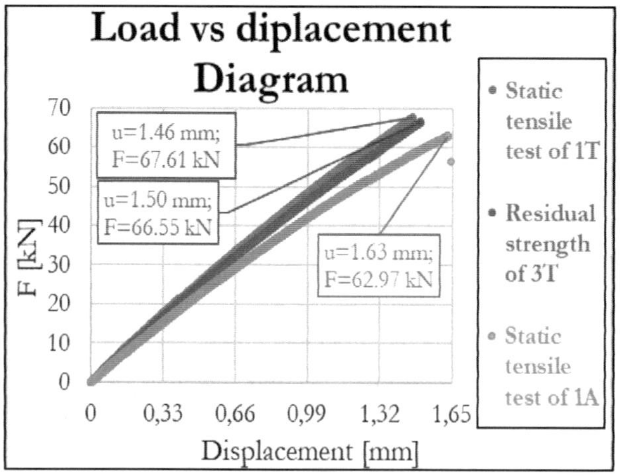

(b)

Figure 1. (a) Example OHT specimens of 'T-S' (left) and 'A' (right) series; (b) load-displacement diagram for OHT series.

The first batch (denotated 'T-S' type) consist of flat samples [Dim. 177.8 × 28.8 × 6.55 mm] with a central hole of 4.74 mm with countersink 2 × 45°, realized in autoclave of composite, aided to reproduce tensile fatigue behaviour in presence of hole for aeronautic bolted junctions. These specimens consist of 20 woven fabrics of thickness 0.327 mm with a stacking sequence of [±45, 0/90, 0/90, 0/90, ±45]2s symmetrically arranged.

The second case of study includes two CFRP elements [Dim. 130 × 39 × 6.78 mm] with a flat central hole of 6 mm, denotated as 'A-type' samples, laminated with unidirectional layers in a specific layup [+ 45, - 45, 0, + 90]$_{2s}$ symmetrically arranged in angle-ply configuration of 16 plies, manufactured through Liquid Resin Infusion process.

For both OHT batches, two preliminary static tests are done for assessing the reference ultimate tensile strengths (UTS), to establish subsequent cyclic loads level. Figure 1(b) shows example diagrams of static test for each series, where an average maximum static load (F_{static}) of 67.61 kN and 62.97 kN have been measured for 'T-S' samples and of 'A' specimens, respectively. Successively, three 'T-S'-type specimens and two 'A'-type samples are tested HCF selecting specific load levels, test parameters and resulting data scheduled in Table 1 in terms of endurance.

The static and cyclic tests follow the standard practices ASTM D5766 M for open-hole tensile strength and ASTM D7615 M for open-hole fatigue response of polymer matrix composite laminates, respectively. The fatigue tests are conducted on CFRP samples at a frequency load of 9 Hz employing R = 0.1 as load ratio under sinusoidal waveform. In Table 1, fatigue parameter F_{max_norm} represents the normalized maximum load percentage of each fatigue test and damaged zones and failure modes are displayed in each CFRP series with similar aspect, as shown in Figure 2(a).

Table 1. Experimental fatigue tests for 'A' and 'T-S' OHT series.

Sample batch	ID fatigue sample	Stress ratio (R)	Load frequency [Hz]	F_{max_norm} [%]	Number of cycles (N_f)	*Damage zone and type*
'A' Series	A2			87.33	204050	Crack at 1st and delamination at 8th ply
	A3			79.39	126225	Crack at 1st and delamination at 8th ply
'T-S' Series	3T	0.1	9	85.00	Stop at 10^6	Undamaged
	4T			91.66	737481	Crack and delamination around countersunk layers
	5T			93.88	318797	Crack and delamination around countersunk layers

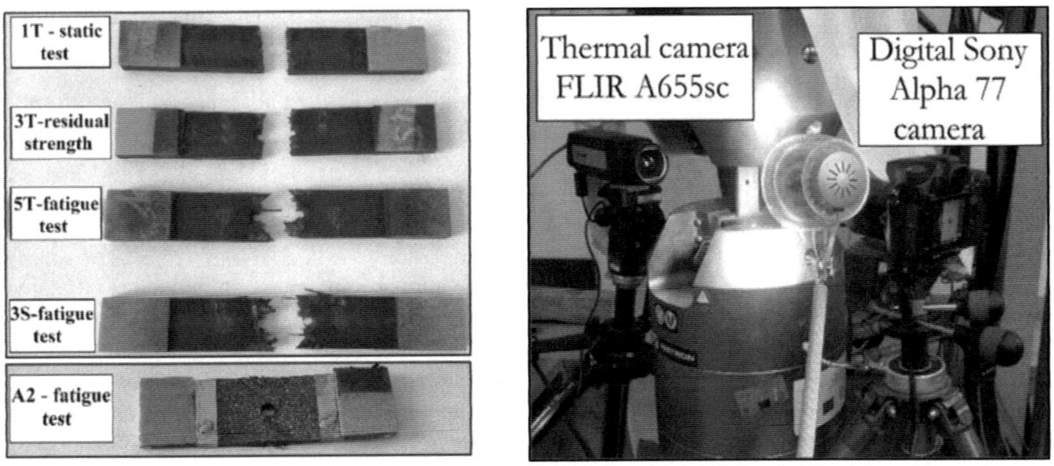

(a) (b)

Figure 2. (a) Example of damaged samples of 'T-S' and (b) 'A' series; (b) experimental and monitoring setup of OHT samples for fatigue damage evaluation with IR and optical cameras.

In the 'T-S' series, delamination initiations, matrix cracking and finally fibre breakage are observed through fatigue cycle, in the layers at the onset of the countersunk region, while the shear cracks at 45°

to fibre directions on 1st and 16th layers with delamination observed in 'A' series during cyclic loading, with central ply delamination observed in the final stages.

A residual strength test is carried out on the undamaged 3T after fatigue loads and similar load-displacement trends and damage modes are detected for static 1T and 3T samples, as shown in Figures 1 (b) and 2 (a). In fatigue test, the specimen stiffness variation is commonly employed to express the damage state of polymer composites [9] and a cumulative damage D_K parameter is defined as [10]:

$$D_K = \left[1 - \frac{K}{K_0}\right] * 100 \quad (1)$$

where K and K_0 represent the actual and initial stiffnesses, respectively. The experimental results include the compliance analysis where the stiffness behaviour is evaluated during all fatigue life until final failure.

2.2. Damage monitoring and inspection procedures

A monitoring strategy consists of thermographic and digital image correlation acquisitions to evaluate damage initiation and progression during cyclic loading. Specimens are positioned inside the test machine clamps and monitored at regular intervals with a micro-bolometric infrared thermal camera and digital high-resolution camera, as shown in the setup of Figure 2(b).

The FLIR A655sc thermal camera is employed, equipped with an uncooled vanadium oxide micro-bolometer detector that provides 14-bit data up to 50 frames per second. For each CFRP elements, thermal signals are recorded at regular intervals of 5000 cycles as a function of the expected fatigue life with a frame of 640 × 240 pixels, using a frame rate of 100 Hz and observation time of 10s, in controlled environment temperature between 23÷26 °C. The thermal camera is MATLAB compatible and controlled by a computer equipped with the Research-IR software, with which the thermal sequences are acquired and converted in '.mat' or '.csv' file for the post processing analysis.

According to thermoelastic theory, the temperature change between the peak and valley of the cyclic load on an orthotropic material could be correlated to the stress change in an adiabatic environment following the Equation [10]:

$$\frac{\Delta T}{T} = -\frac{1}{\rho * C_p} \sum_{k=1}^{3} \alpha_k * \Delta\sigma_k \quad (2)$$

Where ΔT [K] represents the temperature oscillation and T [K] is the absolute temperature at the mean load level on inspected specimen under fatigue loading, ρ [kg/m^3] the material density, C_p [J/(kg*K)] is the specific heat capacity at constant pressure, α_k [K^{-1}] represents each directional coefficient of thermal expansion (CTE), and $\Delta\sigma_k$ represents the amplitude of the principal stresses [1, 10]. The ratio $\alpha_k/(\rho * C_p)$ is denotated as thermoelastic constant.

As demonstrated by different authors [1, 10], thermal dissipation could be related to damage progression and TSA based approach could be successfully employed as an indicator of damage accumulation during fatigue loading. During low frequency fatigue test, the average temperature of sample seems to increase with cycles, as illustrated in Figure 3(b). Thermal variation in terms of amplitude oscillation is related to reversible thermoelastic heating and cooling during cycling loads [10]. Therefore, when the material becomes deformed or damaged, the material volume always followed by heat releases and a part of energy necessary to damage propagation transformed in an irreversible heat source [11]. Since, both the damage initiation and damage propagation up to final failure result extremely difficult to accurately detect with TSA approach, still widely used for rapid fatigue limit estimation tests [13]. In fact, several authors proposed different thermoelastic approaches successfully applied for rapid estimation of the fatigue limit of homogeneous or orthotropic materials in stepwise cyclic tests with increasing stress amplitude levels [12 - 16].

In previous works [20, 22], the authors employed an alternative TSA analytical model proposed by Galietti et al. [13, 23] to extract thermal signal information from thermographic sequences acquired during each fatigue test. In particular, the thermal signal is processed in the time domain based on the following equation implemented in MATLAB [30]:

$$T(i,j,t) = T_0 + bt + T_{the} * \sin(\omega * t + \pi + \varphi) + A * cos(2 * \omega * t + \psi) \tag{3}$$

Where the sum $T_0 + b * t$ (denotated as T_{sum}) describes the mean increase in the average dissipative thermal signal during fatigue test; ω is the angular frequency with which the imposed load signal varies; T_{the} and φ are respectively the amplitude and the phase of first harmonic component related to thermoelastic signal; A and ψ represent respectively the second component amplitude and the phase of thermal signal obtained at a double frequency [16, 23]. Algorithm allow to extract the various signal information from each selected ROI to obtain as output six (T_0, b, T_{the}, A, φ and ψ) signal parameter reported in equation (3), evaluated during fatigue tests.

The aim of this study consists of a different damage fatigue correlation evolution, introducing TSA contrast parameters based on reversible and irreversible thermoelastic sources. In fact, different methods are proposed in case of ND control thermographic inspections to provide enhancing contrast for defect detection and to reduce various noises and other error sources [17, 18].

Thermal processing techniques lie in thermal contrast evaluation, comparing relative temperature behavior during observation time in selected Regions of Interest (ROIs) in different defective and defect-free zones [18]. In this work, the research focus lies on the possible introduction of a TSA contrast-based method for the cumulative damage identification and quantification in an established zone. The first term of Equation (2) could be combined to evaluate a first normalized Thermoelastic Contrast parameter for specific damage evaluation from reversible heating source during the fatigue life (N_{tot}) as:

$$\begin{aligned} D_{rev}(N) &= \left[\left(\frac{\Delta T}{\bar{T}} - \left(\frac{\Delta T}{\bar{T}}\right)_0\right) / \left(\frac{\Delta T}{\bar{T}}\right)_0\right]_d - \left[\left(\frac{\Delta T}{\bar{T}} - \left(\frac{\Delta T}{\bar{T}}\right)_0\right) / \left(\frac{\Delta T}{\bar{T}}\right)_0\right]_i = \\ &\left[\frac{\Delta T}{T} / \left(\frac{\Delta T}{T}\right)_0\right]_d - \left[\frac{\Delta T}{T} / \left(\frac{\Delta T}{T}\right)_0\right]_i \end{aligned} \tag{4}$$

Where $\left(\frac{\Delta T}{T}\right)_d$ and $\left(\frac{\Delta T}{T}\right)_i$ represent the TSA ratio of Eq. (1) for damaged and undamaged regions (ROIs) at a given time of fatigue life (N/N_{tot}), normalized with respect to the same relative TSA ratio at initial fatigue life, respectively. The aim of this parameter is the analysis of a damage evolution associated to temperature oscillation and normalized to actual temperature and referred to undamaged reference zone.

In addition, assuming initially undamaged conditions, the cumulative damage zone elated to dissipative heat release could also provide interesting information on material degradation. The authors propose an alternative damage analysis coefficient to the previous one, to be evaluated in function of fatigue life (N/N_{tot}) with a second parameter:

$$D_{diss}(N) = \left[\left.\frac{T - T_0}{T_0}\right|_d - \left.\frac{T - T_0}{T_0}\right|_i\right] = \left[\left.\frac{T}{T_0}\right|_d - \left.\frac{T}{T_0}\right|_i\right] * 100 \tag{5}$$

Where the TSA parameter D_{diss} represents damage progression associated with a normalized contrast evaluated with temperature ratio of the mean load level in selected damaged and undamaged zones respectively. Therefore, preliminary thermal analysis consists of a suitable selection of proper damaged and undamaged zones in the raw temperature images. Then, in the second processing phase, each sequence of 3D thermal ROIs is post-processed in MATLAB environment following Equations (4) and/ or (5) as function of fatigue life and compared with stiffness behaviour.

For each OHT specimen, the authors optimize experimental monitoring setups to achieve a suitable ROI of the most critical zones where interesting thermal data, displacement fields and in-situ deformations could arise, as illustrated in Figures 2(b) and 3(a) with selected regions of interest (ROIs) on reduced section. The reference intact zones are selected calculated in several ROIs without recorded significative variations on the proposed damage factors. Since, the damage presented multiple defective spots due to widespread delamination, therefore the authors considered only the defective ROIs close to the hole.

Both sample surfaces are preventively prepared to create the proper pattern with a high random contrast for DIC analysis employing on one side of acrylic paint and with high-temperature black paint for thermal analysis on the opposite one. In Figure 1 (a), an example of surface preparation for DIC analysis is reported for 'A' type samples.

During static and fatigue test, a Digital Sony Alpha 77 camera with macro lens is utilized to acquire the high-resolution images of 4000 × 6000 pixels at regular intervals of a low-speed test. A test rate of 0.5 mm/min and a load frequency of 0.05 Hz is employed during static and fatigue tests respectively to maximize the high resolution of acquired images. The Ncorr open source 2D codes [24, 25] implemented in MATLAB is employed for the full-field post-processing DIC analysis of acquired images for different experimental tests, previously used in other work [20].

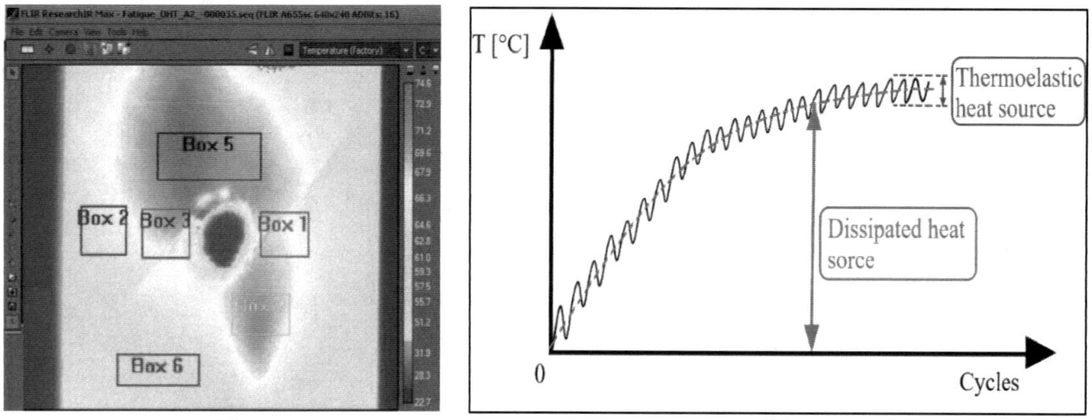

(a) (b)

Figure 3. (a) Thermal map at 69.6% of fatigue life for A3 coupon with ROIs; (b) schematic diagram of thermal behaviour during fatigue test.

In addition, non-destructive infrared thermographic (IRT) and ultrasonic (UT) tests are performed during suitable fatigue test interruption [19]. As example, the ND results for 5T sample are reported at 184,000 cycles (57.7 % of fatigue life) to allow UT and IRT inspections and check material state. Non-destructive evaluation employing pulsed thermography is done with the FLIR A655sc thermal imaging camera and a halogen lamp 1000 W. The IR technique provides a full-field inspection without sample removal from machine actuators. IR thermal acquisitions have a duration of about 150 seconds with a frame rate of 6.25 Hz and heating times of 20-30s is selected.

The UT inspections are performed with Omni-scan MX control unit employing a Phased Array OLYMPUS 2.25L64 probe with 64 transducers. A Plexiglas base was applied to the probe to optimize the wave beam between probe and sample surface with a coupling gel to ensure good transmission of the UT signal [20].

3. Results and discussion

3.1. TSA results

Preliminary compliance analysis is done for each fatigue test. Comparative analysis of stiffness behaviour as function of fatigue life is presented in Figure 4(a) for 'T-S' type and 'A' type elements. Stiffness data are normalized to specific maximum initial value (K_0) and show a clear different trend between the two-sample series, motivated mostly by different manufacturing process effects and load history.

In Figure 4(a), experimental data of 5T and 4T samples highlight a slow decreasing trend during initial and intermediate test phase, achieving a final stiffness degradation around 4 ÷ 8 % at 99% of fatigue life. Small *K* deviations are followed by a sudden decay before final failure. Specimen 3T results undamaged after the selected load (10^6 cycles) and damage absence obviously shows an almost flat stabilized trend with a slightly reduction of 2% of stiffness. The A2 sample, representative of 'A' series, presents significative stiffness degradation with a final reduction of 43 % at 99% of fatigue life. In this series, the 40% of total life seems to produce a compliance decrease around 7% with evident damages around the critical zone. After 60% of fatigue life, internal damage initiations seem to accelerate until final failure providing the collapse when internally the induced macro defects start to propagate through the whole specimen section. Thus, in the fatigue damage assessment composites, the classical approach based on compliance analysis confirms a robust reference for data evaluation.

As anticipated, fatigue damage initiation and failure prediction could be determined in terms of signal perturbations of raw thermal sequences or TSA processed data acquired during fatigue test [20].

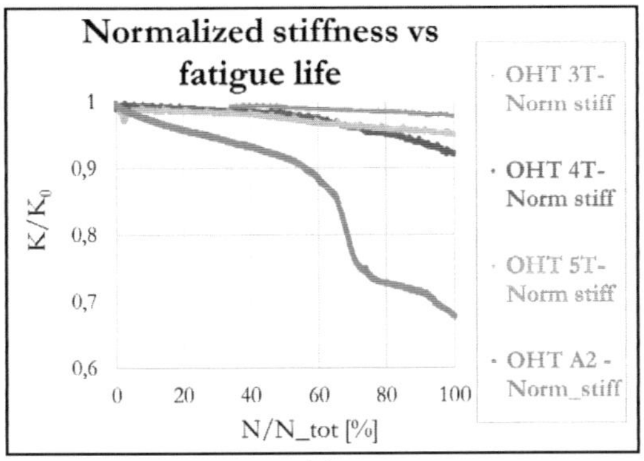

(a)

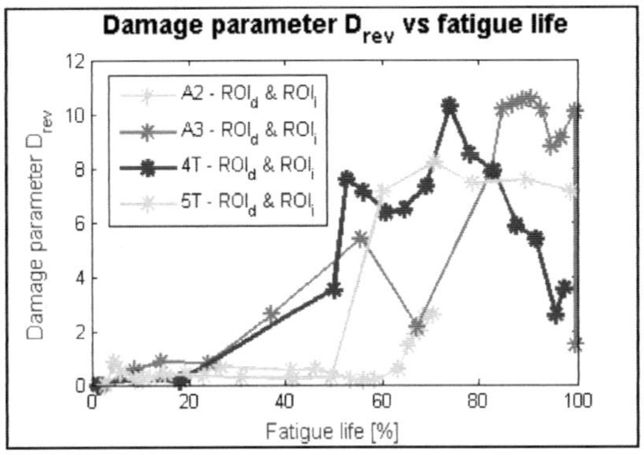

(b)

Figure 4. (a) Normalized stiffness and (b) reversible damage D_{rev} vs fatigue life for OHT series.

In fact, thermographic approach could be successfully employed for damage monitoring, providing as output results from equations (4) and (5) that should result relate to thermoelastic effects to cyclic loads or to intrinsic energy dissipation. In Figures 4(b), 5(a) and 5(b), the processed thermal profiles of selected ROIs are reported for damage evaluation with stiffness degradation (D_K), reversible (D_{rev}) or dissipative heating (D_{diss}) sources during fatigue life respectively. The comparative analysis of damage parameters defined as Equation (4) and (5) provides different diagrams of TSA contrast data at different fatigue cycles for each sample series. Specific analysis ROIs are chosen for each specimen by maximizing the damage parameter were analysed. The intact area was calculated in several zones to evaluate the influence of ROI selection on damage factors values. In Figure 4(b), the D_{rev} data show high level of noise (scattering) and a certain trend variability due to ROI choice has been noted, but parameter values seem to increase in each specimen. It appears that D_{rev} behaviour follows the damage regime even if it suffers a lot of noise and critical damage propagation is clearly distinguished, highlighting similar behaviour between A3, 5T and 4T samples. In general, the D_{rev} parameter seems characterized by an initial phase, that starts form 20% of fatigue life without an evident growth of parameter trend, and by a second phase, where an evident curve growth over 1 seems to occur in all the specimens after 20% and at 50-55% life in 'A' and 'T-S' specimens' series, respectively.

In Figures 5(a) and 5(b), both diagrams show comparative analysis between damage diagrams evaluated with stiffness degradation (D_K) and behaviour with the dissipative thermal parameter (D_{diss}) profiles. D_{diss} data show a damage evolution with an approximately exponential trend with foible data variability. Even the variation of stiffness does not differ much from D_{diss} curves, which look very similar and common steep severe damage increase in case of A2 and A3 specimens manufactured via Infusion technique, whilst three autoclave produced samples show a damage behaviour more gradually increasing up to failure. In the first stage, thermal contrast profiles elaborated through D_{diss} parameter provide slight data variation due to internal micro-cracking between fibers and matrix and to damage nucleation that begin to propagate in initial phases of fatigue test. In the next phase, the temperature reaches an almost stable increasement due to slow damage propagation and initial delamination, highlighting the different damage evolution between CFRP sample typology. Finally, this increasing trend is followed by a rapid contrast rise, leading to the final rupture. D_K and D_{diss} diagrams appear interlocking and intersecting each other, resulting in very similar correspondence and there is a better agreement between the proposed damage parameter D_{diss} and the stiffness measurement for all the specimens, even for the A2 sample for which we have not retrieved the final data.

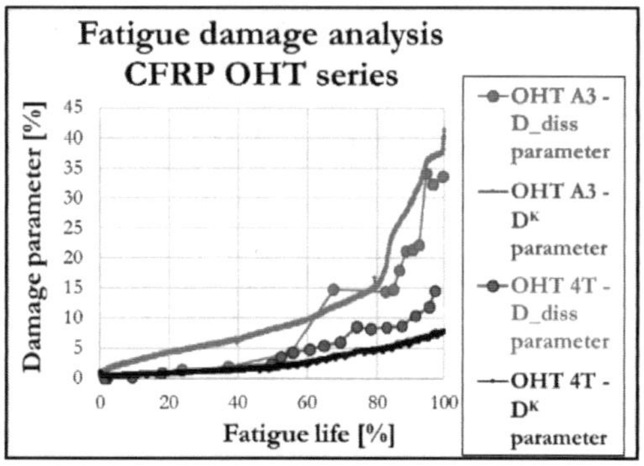

(a)

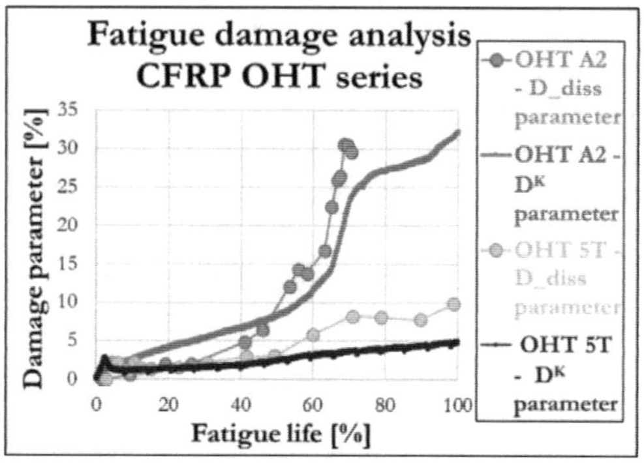

(b)

Figure 5. (a) – (b) Comparative analysis between D_K and dissipative damage D_{diss} parameter vs fatigue life for OHT series.

Assuming initially undamaged conditions, the proposed damage parameter evaluated by D_{diss} data seems to provide an interesting comparable behaviour, as observed in cumulative damage D_K profiles related to stiffness degradation as function of fatigue life, validating initial assumptions on raw damage initiation proposal, based on TSA approach, and implemented according to a contrast-based method. This comparative analysis validates the new damage parameter definition previously described for realistic fatigue damage estimation using TSA contrast-based approach contained in the D_{rev} coefficient and the alternative thermal dissipation D_{diss} coefficient, proposed for damage propagation.

3.2. DIC analysis

In fatigue of composite materials, fibre-reinforced polymer composites are matrix cracking, fibre–matrix debonding, ply delamination and fibre breakage represent the most common failure mechanisms. The strain field on surface should provide important aspects of fatigue damage propagation prior to final failure in terms of displacement variations. Thus, the authors employ digital image processing to analyse strains and displacements for selected ROIs onto specimens, performed with Ncorr software implemented in MATLAB environment.

Six Virtual Extensometers (VE) of 16 ÷ 22 mm in length are selected for strain and displacement analysis in static and fatigue test cases. As shown in the example Figure 6(a), the employed VE consist of four extensometers parallel to applied load and the other two normal to load.

In Figure 6(b), the diagram shows the σ_y / ε_y trends of the VE-1 and VE-4 (located as shown in Figure 6(a)) for the residual strength test of 3T sample post fatigue cycles, where higher strain values are observed in the critical section at the countersunk zone. Specific ROIs are selected for different test cases for a proper evaluation in critical areas where maximum stress, displacements and strain concentration should be observed. In the first phase, the DIC analysis is carried out for the static test case (as seen in the 2.2 paragraph) in an area of interest relatively large area of around the hole zone. In the second phase, the DIC data acquired at the various regular intervals are evaluated during entire fatigue to estimate displacement and strain variation.

Figures 7 to 8 report an example comparative analysis the angular deformation γ_{xy} maps for 5T and A3 respectively at two three different instants (1% to 79% ÷ 88%), highlighting in both test a high strain state around the hole's zone and similar cross deformation state since test beginning. As damage seems to evolve during fatigue life, the strain values increase in the critical zones for a specific cyclic load, as seen for the TSA results. As expected, in Figures 7, the γ_{xy} data of 5T shows an evident asymmetrical deformation state due to countersunk bolted joining caused by the load transfer through the bolted plates that may lead to higher strain levels.

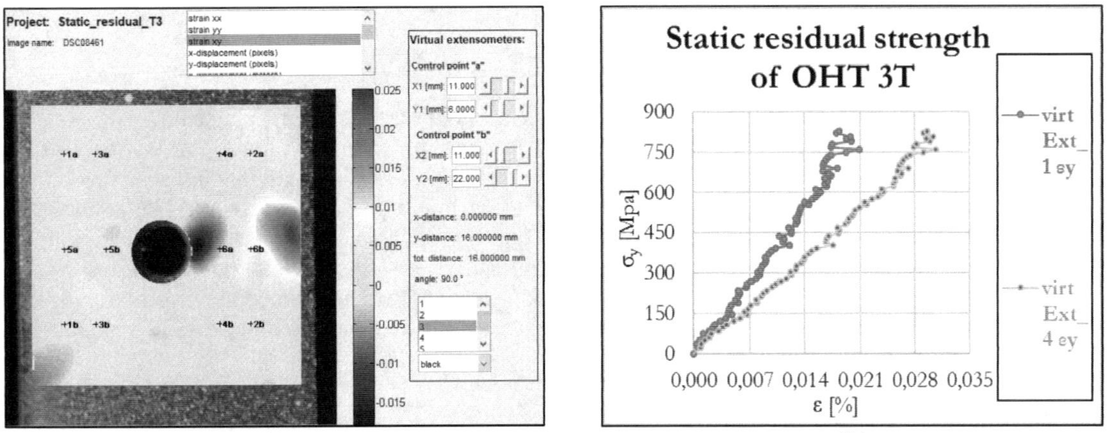

(a) (b)

Figure 6. (a) Angular deformation γ_{xy} map and (b) σ_y / ε_y diagram virtual extensometer VE-1 and VE-4 for residual strength test of 3T sample.

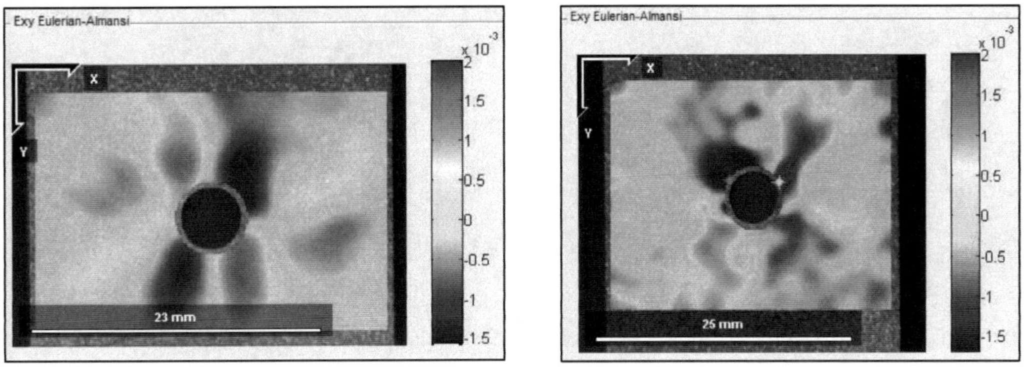

(a) (b)

Figure 7. (a) Angular deformation γ_{xy} maps at 1% vs (b) 88% fatigue life for 5T sample.

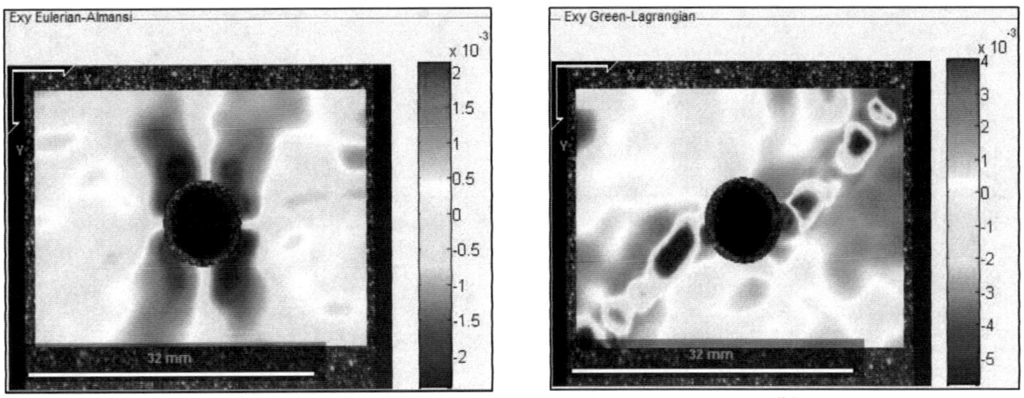

(a) (b)

Figure 8. (a) Angular deformation γ_{xy} maps at 1% vs (b) 79% fatigue life for A3 sample.

In Figures 8, higher deformation of critical section could be clearly observed in 'A' sample during cyclic test, along with shear coupling deformation along the fibre orientation of unidirectional first ply specimen. Therefore, the processed DIC images seem to establish complementary information also useful for a qualitative evaluation of damage initiation and location on inspected part during fatigue phases.

3.3. ND evaluation results

Figure 9(a) shows thermal distribution of 5T sample at 1.5 cooling seconds after a heating time of 20s carried out during short test interruption at 57.7 % of fatigue life, where the hottest zones seem to be localized in the middle section of specimen when final damage degradation arises, as described from D_{diss}/D_K parameters. As expected, the static ND evaluation highlights two widespread delaminations around the critical zone of countersink hole, as reported in Figure 9(a).

In previous works, the authors introduced a contrast-based (Local Boundary Contrast, LBC) algorithm implemented in a MATLAB GUI that avoids defining the free-defect zones when computing SNR, providing different contrast maps that highlights defective zones boundaries and location, as shown in Figure 9(b) [21].

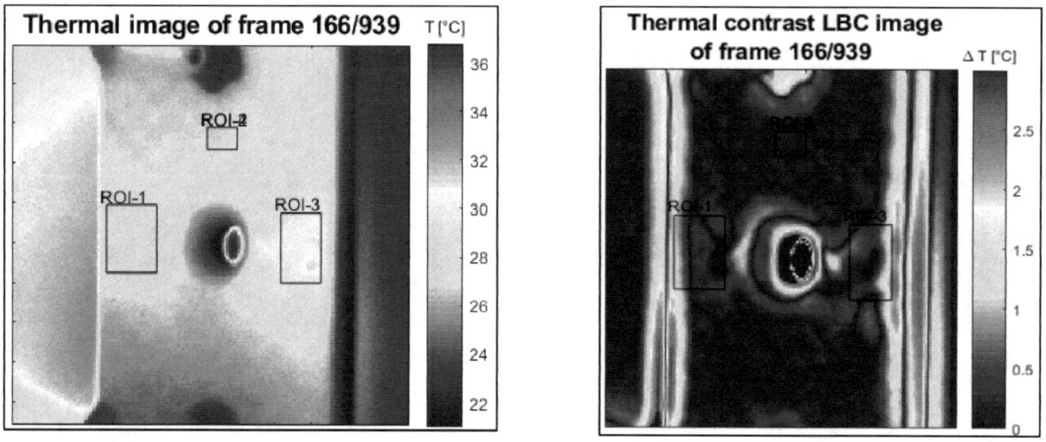

(a) (b)

Figure 9. (a) Thermal vs (b) LBC maps at 1.5 cooling seconds after a heating time of 20s of 5T sample.

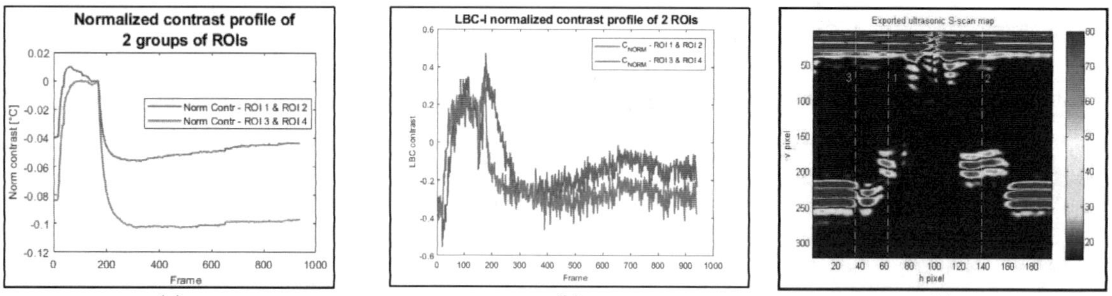

(a) (b) (c)

Figure 10. (a) Thermal vs (b) LBC contrast profiles and (c) S-Scan (0°) for post-fatigue 5T sample.

The same MATLAB GUI provide a manual selection of Region of Interest (ROIs) employed for a comparative analysis of normalized contrast evaluation of raw thermal data and LBC, as shown in Figures 10(a) and 10(b) respectively.

As shown in Figure 10(c), the post-fatigue damage on the reduced section is also qualitatively evaluated in the ultrasonic S-scan map, where the presence of both delaminations can be observed affecting the first plies from the side of the detached hole in accordance with TSA degradation of real-time IRT signal detected during the fatigue test. Damage extension seems to be greater on right side compared to the left one, in perfect accordance with IRT data.

4. Conclusions

The composite damage initiation is not easy to be detected during fatigue tests before final failure. Therefore, in this proposed study the authors provide to perform a synergic and suitable combination of full-field monitoring methods for a possible damage analysis based on open hole CFRP specimens under cyclic loading.

The comparative analysis between the stiffness degradation and the post-processed thermal results obtained using TSA approach during fatigue tests is discussed for two batches of OHT elements, showing an interesting correlation between damage evolution evaluated with experimental results and TSA post-processed data. The comparative proposes a fatigue damage evaluation using TSA contrast-based approaches.

In addition, thermographic inspections highlight the presence of internal delamination after 57.7 % of fatigue life in perfect accordance with the well-established UT data.

5. References

[1] Vassilopoulos AP and Keller T 2011 Experimental Characterization of Fiber-Reinforced Composite Materials. In: Vassilopoulos AP, Keller T., editors. Fatigue of Fiber-reinforced Composites, Springer, 25 - 67.

[2] Subramanian S, Reifsnider K L and Stinchcomb W W 1995. A cumulative damage model to predict the fatigue life of composite laminates including the effect of a fibre-matrix interphase. *Int J Fatigue* 17 (5), 343 – 51. 10.1016/0142-1123(95)99735-S.

[3] Güemes A, Fernandez-Lopez A, Pozo A R and Sierra-Pérez J 2020 Structural Health Monitoring for Advanced Composite Structures: A Review, *J. Compos. Sci.*, **4**, 13, 10.3390/jcs4010013

[4] de Castro B A, Baptista F G and Ciampa F 2019 Comparative analysis of signal processing techniques for impedance-based SHM applications in noisy environments, *Mechanical Systems and Signal Processing* **126**, 326 – 340, 10.1016/j.ymssp.2019.02.034

[5] Middleton C A, McCrory J P, Greene R J, Holford K and Patterson E A 2019 Detecting and Monitoring Cracks in Aerospace Materials Using Post-Processing of TSA and AE Data, *Metals* **9** (7) 748, 10.3390/met9070748

[6] Rodrigues F de Sá, Marques R, Tabrizi I E, Kefal A, Qasim Ali H, Yildiz M and Suleman A 2021 A new methodology for thermoelastic model identification in composite materials using digital image correlation. *Opt and Las in Eng* **146** 106689

[7] Thatcher J, Crump D, Devivier C, Bailey P and Dulieu-Barton J 2020 Low cost infrared thermography for automated crack monitoring in fatigue testing, *Opt. Lasers Eng* **126** :105914

[8] Pitarresi G, Cappello R and Catalanotti G 2020 Quantitative thermoelastic stress analysis by means of low cost setups. *Opt. Lasers Eng* **134**: 10615

[9] Subramanian S, Reifsnider K L and Stinchcomb W W 1995. A cumulative damage model to predict the fatigue life of composite laminates including the effect of a fibre-matrix interphase. *Int J Fatigue* **17** (5), 343 – 51. 10.1016/0142-1123(95)99735-S.

[10] Toubal L, Karama M and Lorrain B 2006 Damage evolution and infrared thermography in woven composite laminates under fatigue loading. *Int J Fatigue* **28**, 1867 – 1872. 10.1016/j.ijfatigue.2006.01.013

[11] Chrysochoos A 2002 Infrared thermography, a potential tool for analysing the material behaviour. *Mécanique & Industries* **3** (1), 3-14.

[12] La Rosa G and Risitano A 2000 Thermographic methodology for rapid determination of the fatigue limit of materials and mechanical components, *Int J Fatigue* **22** (1), 65-73

[13] Galietti U, Demelio G P, Palumbo D and De Finis R 2016 A new rapid thermographic method to assess the fatigue limit in GFRP composites, *Compos B*. **103**: 60 - 67.

[14] Colombo C, Bhujangrao T, Libonati F and Vergani L 2019 Effect of delamination on the fatigue life of GFRP: A thermographic and numerical study. *Composite Structures* **218**, 152-161. 10.1016/j.compstruct.2019.03.023

[15] Ricotta M, Meneghetti G, Atzori B, Risitano G and Risitano A 2019. Comparison of Experimental Thermal Methods for the Fatigue Limit Evaluation of a Stainless Steel. *Metals* **9** (6): 677, DOI: 10.3390/met9060677

[16] Meneghetti G, Ricotta M and Pitarresi G 2019 Infrared thermography-based evaluation of the elastic-plastic J-integral to correlate fatigue crack growth data of a stainless steel. *Int J Fatigue* **125**, 149–160. 10.1016/j.ijfatigue.2019.03.034

[17] Fleuret J, Ebrahimi S, Ibarra-Castanedo C and Maldague X P V 2021 Independent Component Analysis Applied on Pulsed Thermographic Data for Carbon Fiber Reinforced Plastic Inspection: A Comparative Study May 2021 Applied Sciences. *Applied Sciences* **11**(10): 4377, 10.3390/app11104377.

[18] Vavilov V P and Burleigh D D 2015 Review of pulsed thermal NDT: Physical principles, theory and data processing. *NDT Int.* **73**, 28–52, 10.1016/j.ndteint.2015.03.003.

[19] Marzo F 2021 Analisi e monitoraggio del danneggiamento a fatica di provini CFRP aeronautici. *Master Thesis in Mechanical Engineering*, University of Salento, Lecce (73100).

[20] Dattoma V, Panella F, Pirinu A and Castriota A 2019 Fatigue damage on CFRP plates under bending by thermographic and UT analysis, aided with FEM-DIC prediction. *Pro Stru Inte* **24**, 978 – 987. 10.1016/j.prostr.2020.02.085.

[21] Dattoma V, Nobile R, Panella F, Pirinu A and Saponaro A 2018 Advanced NDT procedures and thermal data proceessing on CFRP aeronautical components. In Proceedings of the IRF 2018 of 6th International Conference Integrity-Reliability-Failure, Lisbon, Portugal, 22–26 July 2018.

[22] F. W. Panella and A. Pirinu 2021 Fatigue behaviour and damage evaluation on aeronautical CFRP elements under tension-tension and bending loads: two cases of study, *Int J Fatigue* **152** (13): 106403, 10.1016/j.ijfatigue.2021.106403.

[23] Galietti U, Demelio G P, Palumbo D, De Finis R 2017 Study of damage evolution in composite materials base on the Thermoelastic Phase Analysis (TPA) method, *Compos B* **117**, 49 - 60.

[24] Blaber J, Adair B and Antoniou A 2015 Ncorr: Open-Source 2D Digital Image Correlation Matlab Softwar,e *Exp. Mech.* **55** (6): 1105–1122, 10.1007/s11340-015-0009-1

[25] Nežerka V Antoš J, Sajdlová T and Tesarek P 2016 Use of Open Source DIC Tools for Analysis of Multiple Cracking in Fiber-Reinforced Concrete, *Appl Mech and Mater* **827**: 336 – 339, 10.4028/www.scientific.net/AMM.827.336

Acknowledgments
Authors would like to thank the Prof. U. Galietti for sharing the thermographic equipment and the student Francesco Marzo for his work in the experimental activities.

Comparison of fatigue damage criteria for a CFRP aeronautical joint subjected to random fatigue load

A Castriota[1], **V Dattoma**[1] **and R Nobile**[1]

[1] Department of Engineering for Innovation, University of Salento, 73100, Lecce, Italy

Corresponding author: alessandro.castriota@unisalento.it

Abstract. The fatigue life prediction of components in CFRP subjected to random fatigue load was carried out considering different fatigue damage models. The selected component is a filled hole tension specimen, which is representative of the typical aeronautical riveted joint. The stress state of the riveted section was determined through a FEM model, created with CODE ASTER. Subsequently, the fatigue load history was introduced, calculating the stress field for the different load levels. Applying the constant life diagrams, fatigue cycles were counted and the damage was evaluated. Finally, the damage that occurs in the different ply of the specimen was calculated using different fatigue damage criteria present in literature. The creation of a parametric FEM model offers the possibility of optimizing the study and constitutes a useful tool in order to choose the most suitable damage criteria for the particular geometry and loading mode of the component. Moreover, different severity of the load history and different load sequence were introduced to show their effect in the fatigue failure.

Keywords: CFRP, damage, fatigue, FEM

1. Introduction

Within the aeronautical and aerospace industry, the use of composite materials is assuming increasingly important positions: the trend is to increase the percentage of composite materials used in the construction of an aircraft, exploiting the high mechanical characteristics that these materials offer to front with a lower specific weight than metallic materials; in fact, a composite material structure weighs less than a light alloy structure of equal strength [1, 2, 3]. The progressive replacement of the material used to make the structures and sub-structures of an aircraft has introduced, in addition to many advantages of a structural nature, various problems related to different aspects such as design and production. In particular, the use of a composite material introduces novelties regarding the failure modes and the fatigue life that differ considerably from what happens in metallic materials, mainly for two reasons: the different behaviour, given that the composite material is anisotropic, at the most orthotropic, while a metal is isotropic, and the inhomogeneity of the composite material with respect to the metal material. These two aspects introduce a whole series of structural problems different from metallic materials [4].

A very important aspect is that of the junctions: the different structures that make up an aircraft are joined together through junctions and are also made up of smaller parts joined together in different ways

IOP Conf. Series: Materials Science and Engineering 1214 (2022) 012009 doi:10.1088/1757-899X/1214/1/012009

depending on the type of loads, of the form and function they must perform [1, 2]. Composite materials basically have two ways of joining: bonded joints and riveting and / or bolting joints, in which rivets or bolts are used, usually in titanium alloy, to mechanically join two or more parts together. The difference between the two types of joint is that the first type of joint cannot be dismantled except by damaging the parts connected between them, while the second type allows the joint to be disassembled without damaging the connected parts. These two types of connection between the parts introduce complications in the behaviour of a structure. In particular, the presence of holes, necessary for making a connection with rivets or bolts, introduces a concentration of stresses on the edge of the hole which tend to reduce the static and fatigue resistance of the component [1, 4, and 5]. This effect of reducing the resistance, due both to a reduction of the resistant section and to the cutting effect at the edges of the hole, interacts with the presence of the rivet inside the hole, which tends to modify the rigidity of the component due to the obstacle action that it carries out against the tendency of the hole to reduce its dimension in a transverse direction with respect to the load. Evaluations of this type are very important in the fatigue resistance, a fundamental aspect in the service life of the joint. In particular, the damage due to a fatigue stress in components of this type is substantially modified by the presence of a foreign body such as a rivet. Consequently, it is important to evaluate how the damage is calculated, which can have different values depending on the criterion used for its count.

In the bibliography there are some works that have dealt with similar issues. Payan et al. [6] developed a cumulative nonlinear damage model in order to evaluate laminated carbon/epoxy composites under static and fatigue loading. Instead, Subramanian et al. [7] have defined a new parameter in the evaluation of the fatigue damage that occurs at the interface of composite specimens subject to tensile fatigue, calling it "interface efficiency", which evaluates the degradation of the interface between the sheets of a specimen subject to fatigue. Van Paepegem et al. [8] have introduced the skip-the-cycle approach to calculate the damage in fiberglass specimens to overcome the problem of redistribution of stresses within the specimen as the damage progresses and limit the computational cost of numerical simulations on fatigue. Also, Papanikos et al. [9] developed a numerical model with three-dimensional elements on ANSYS that evaluates the fatigue life of fiber-reinforced plastic components, using the Hashin damage criterion and the Ye delamination criterion. On the joints, Zhao et al. [10] have developed a numerical model to evaluate the degradation of material properties as the imposed fatigue cycles advance.

It is evident that it is necessary to be able to numerically evaluate the damage that occurs in components subjected to fatigue loads. The present work comes from this consideration: the numerical results given by the application of different criteria for the calculation of the damage due to the application of random fatigue loads are compared, in order to evaluate which of them is more appropriate in estimating the damage and how this is affected by the load history sequence. The applied load history is derived from a load spectrum multiplied by a load constant. In this way it is also possible to define the absolute level of the load history needed to produce a unitary damage in the specimen. This document reports a preliminary work for the evaluation of the damage in a CFRP specimen that simulates a riveted joint.

2. Geometry and material

The specimen consisted of 16 plies in CFRP with a stacking sequence $[45/0/45/0/0/45/0/45]_S$. In the centre there is a hole with a conical part, in which a titanium alloy Ti6Al4V rivet is housed, in order to evaluate the interaction of composite and rivet of an aeronautical joint. The rivet has some clearance inside the hole. In addition, the specimens are equipped with tabs at the ends for gripping the test machine (Figure 1 and Figure 2).

(a) (b)

Figure 1. Top view (a) and bottom view (b) of the specimen with the rivet

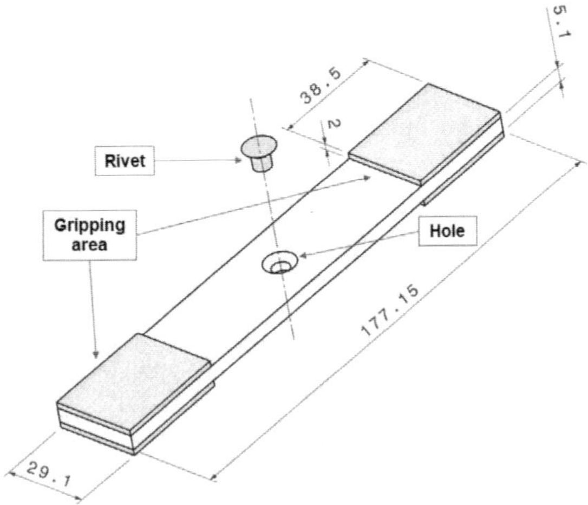

Figure 2. Axonometric view of the riveted specimen

The mechanical characteristics of CFRP material are shown in Table 1, indicating the ones that have been hypothesized assuming they are similar to the mechanical characteristics of other similar materials.

Table 1. Mechanical characteristics of specimen material

E_{11}	60535.97	MPa	available
E_{22}	58398.59	MPa	available
E_{33}	8756	MPa	unavailable
G_{12}	4378.17	MPa	available
G_{13}	2700	MPa	unavailable
G_{23}	2700	MPa	unavailable
ν_{12}	0,34		unavailable
ν_{13}	0,545		unavailable
ν_{23}	0,545		unavailable
UTS_1	743.53	MPa	available
UCS_1	739.66	MPa	available
E_{c_11}	58881.22	MPa	available
UTS_2	625.35	MPa	available
UCS_2	694.99	MPa	available
E_{c_22}	55847.53	MPa	available

3. Numerical model

The numerical model of the central area of the specimen has been built, as damage occurs in this area due to the stress concentration induced by hole. The useful length reported in the numerical model is 60 mm in total, with the hole positioned at the centre of this length. The numerical model was developed with the FEM Code Aster, which uses Salome - Meca as a pre-processing environment and Paravis as a post – processing environment.

Modeling is based on the use of 3D elements to reproduce each lamina and the rivet. The elements for composite material have a quadratic form function, while linear elements have been used for the rivet. Two different meshes were performed: mesh mapped for the laminas that make up the composite material and tetrahedral free mesh for the rivet. This choice was made because the results related to the rivet in terms of stress and strain are not relevant, while a mapped mesh for the edges allows to obtain more robust results and to better manage the post processing phase. The model thus produced has a total number of nodes equal to 11177 and a total number of three-dimensional elements equal to 2818, of which 574 tetrahedral elements relating to the rivet and 2244 hexahedral elements relating to the CFRP specimen (Figure 3). The numbering of the plies start from above, with the ply 1 corresponding to the rivet head.

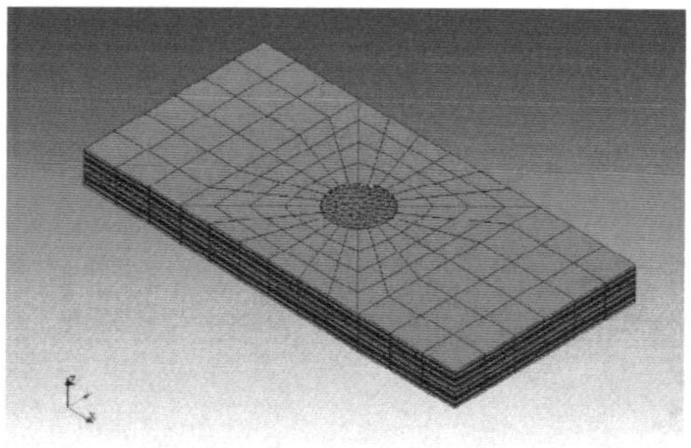

Figure 3. Numerical model

3.1. Constraints and load history

The specimen was fully constrained to one end blocking all DOFs. On the load application side, transverse displacements with respect to the direction of application of the load have been blocked and the relative motions between the nodes of the load application face. Furthermore, the motion of the rivet in the z direction (parallel to the axis of the rivet) and the transverse motion have been blocked, in order to have congruence in the simulation of the contact between the internal surface of the hole and the rivet (Figure 4).

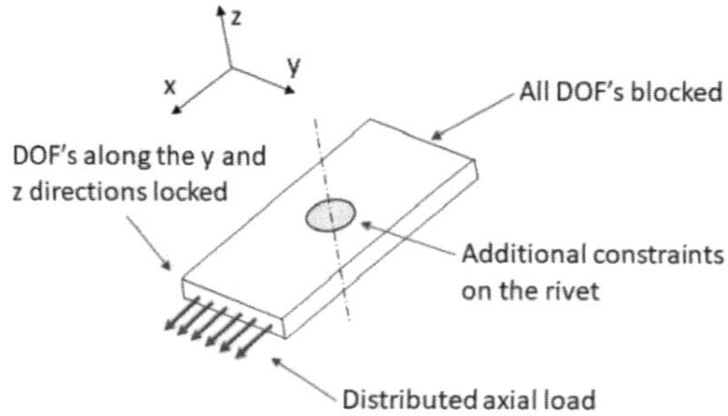

Figure 4. Constraints and load for the numerical model

The load was applied on the opposite end to the constrained one, using a surface force. The load consists of a load history that simulates the actual stress state that the component would receive during its service. There are several load histories used in the literature to evaluate the random fatigue life of

CFRP components. Schön et al. [11] used a load history with different R load ratios, modifying the load history based on the damage that was achieved in representative bolted joint specimens. In the aeronautical field, especially military, the FALSTAFF spectrum [12] is widely used, which is a particular pseudo-random spectrum equivalent to 200 aircraft flights. In order to be general, it is a scalable load spectrum based on the application for which it is used.

Observing the load spectrum reported in the work conducted by Schön et al. [11], a simple load spectrum was considered for the case in question, in order to highlight the influence that the sequence of application of the load levels has in the generation of fatigue damage. In particular, a load spectrum was designed consisting of five blocks formed by two load levels each, with a load ratio $R = 0.1$ and prevailing traction for all the blocks, as shown in Figure 5a. Taking up the distribution of loads in the spectrum reported in the same work, we have given a higher frequency of occurrence at low loads and a lower frequency of occurrence at higher loads (Figure 5b).

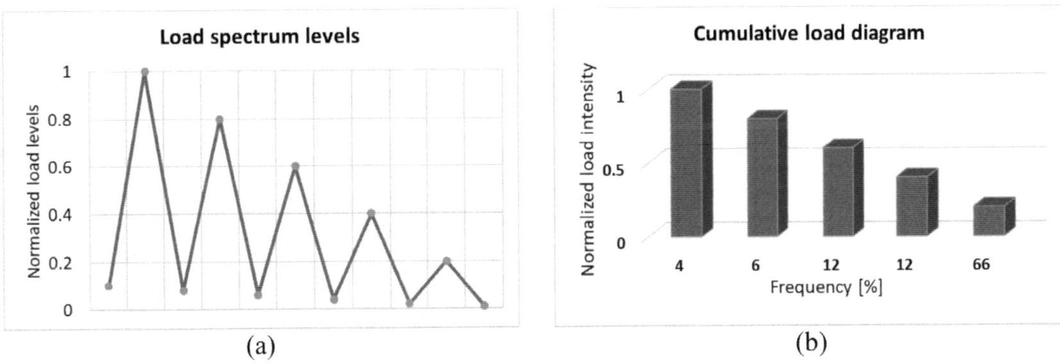

Figure 5. Example of load spectrum levels (a) and cumulative load diagram (b)

Being a spectrum, in order to have the load history to be applied and all the load levels, it is necessary to multiply the levels of the spectrum by a scale factor. In this way it is possible to evaluate for which combination of loads there is the unitary damage that marks the failure of the component.

The calculation of the fatigue damage was considered over a total number of cycles equal to 10^6 cycles. Knowing this data, three load histories have been constructed, with the characteristic of having a normal distribution of the intervention frequency of each single block constituting the load history itself. By identifying the blocks with the numbers 1, 2, 3, 4, 5, the following load histories were generated.

Table 2. Load spectrum 1

Block ID	1	2	3	4	5
Number of repetitions	660000	120000	120000	60000	40000

Table 3. Load spectrum 2

Block ID	5	4	3	2	1
Number of repetitions	40000	60000	120000	120000	660000

Table 4. Load spectrum 3 [a]

Block ID	1	2	3	4	5	4	2	3	1	5	2	4	1
Number of repetitions	220	60	60	20	10	20	40	60	300	30	20	20	140

[a] This load spectrum has a total of 1000 cycles. It will be repeated 1000 times to reach 10^6 expected cycles

3.2. Contact

The presence of the rivet provides for the simulation of the contact between the external surface of the rivet and the internal surface of the hole. This is because the rivet is not always in adherent contact with the internal surface of the sample hole, there is no type of adhesion between the surfaces. The presence of the rivet intervenes in the overall stiffness of the specimen since, depending on the type of stress to which the specimen is subjected, some areas of the rivet surface come into contact or not with the internal surface of the hole.

For example, in the case of traction, the rivet is in contact with the internal surface of the hole in a transverse direction with respect to the direction of application of the load while it has no contact in the longitudinal direction, as can be seen in Figure 6.

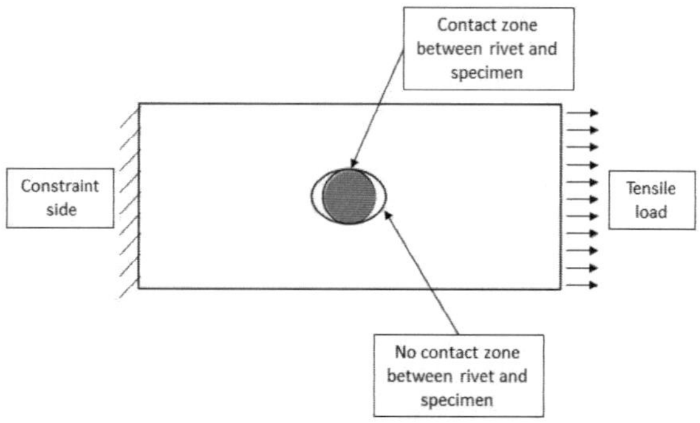

Figure 6. Contact zone between specimen and rivet for tensile load

The modelling of the contact introduced some geometric non-linearities which required the resolution of the problem in the non-linear field.

4. Damage criteria

The stress state resulting from the FE model was used to evaluate the fatigue damage induced by the three different load sequences. For this reason it is useful to calculate the damage that occurs after the application of a random load history for a total of 10^6 cycles. Among the different criteria present in the literature, three criteria have been chosen for the evaluation of damage. The results of the chosen criteria will be compared, in order to evaluate which of them best approximates the calculation of the damage to the material. The choice of criteria was made on the basis of the data available on the material of the specimen.

The first criterion chosen is the classic Palmgren - Miner [13, 14], which expresses the damage as:

$$D = \sum_i D(\sigma_{ai}) \tag{1}$$

where D indicates the cumulative damage given by the sum of the single damages that are generated for the cycles performed at a stress amplitude equal to σ_{ai}. The single damage due to these n_i cycles compared to the total number of cycles N_i executable for that load level can be evaluated according to the following equation:

$$D(\sigma_{ai}) = \frac{n_i}{N_i} \tag{2}$$

The second criterion chosen is that of Bond - Farrow [15], which is a generalization of the damage criterion of Owen - Howe [16]. The calculation model is expressed by the following equation:

$$D = \sum_i \left[A \left(\frac{n_i}{N_i} \right) + B \left(\frac{n_i}{N_i} \right)^C \right] \tag{3}$$

Parameters A, B and C depend on the mechanical properties of the material, i.e. UTS, UTC and Young's modulus under compression E_c (Table 1). The terms for calculating the parameters change depending on whether you are in prevailing traction or prevailing compression conditions.

Both these criteria do not consider the effect of the sequence of application of the load levels.

The third criterion is that of Hashin - Rotem [17]. In this case, the authors approached the problem from a different point of view: in fact their method is based on the introduction of the concept of residual life time which, unlike the damage parameter D, is a measurable quantity. The damage is evaluated later. This approach can be summarized in the equation that expresses the damage D_i achieved at the end of the i-th loading step in a multi-step sequence:

$$D_i = D_{i-1}{}^{\frac{1-s_i}{1-s_{1-1}}} + \frac{n_i}{N_i} \tag{4}$$

where $s_i = \sigma_i / \sigma_s$, with σ_i is the i-th stress and σ_s the static failure stress.

The correct application of these criteria require the knowledge of the curves with constant life for each load ratio R, which have been assumed on the basis of the non-dimensional constant life curves reported by Harris et al. [18]. Since these curves are dimensionless with respect to the maximum tensile and compressive failure stresses, they have been reconstructed numerically with a Python routine for the material in question. By plotting different lines for different R, the necessary parameters were calculated to be able to evaluate the number of limit fatigue cycles N for each R needed to calculate the damage. Figure 7 shows the reconstruction of the constant life curves for the specimen material.

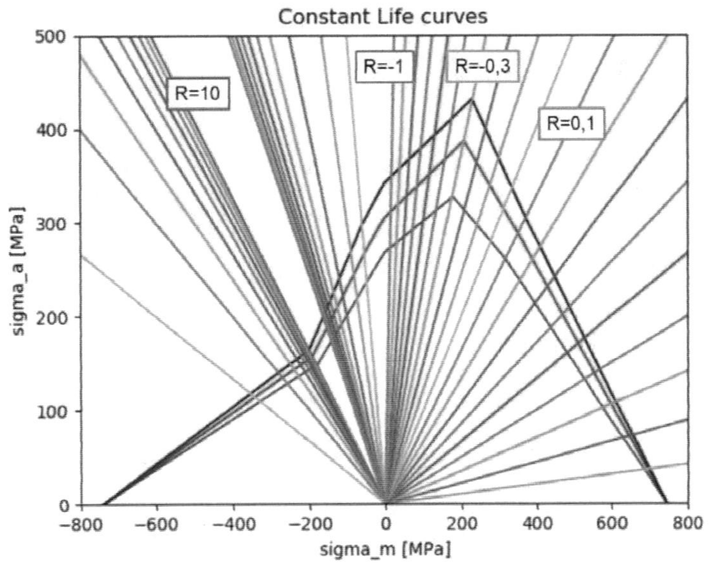

Figure 7. Constant life curves for principal direction 1 of the specimen material rebuilt with Python

5. Discussion of results

Numerical simulations of the static type and for the calculation of the damage were carried out. The first simulation was carried out with a static tensile load to determine which areas of the specimen were most stressed both in tension and in compression. Starting from the results obtained statically for a maximum load level, the stress values for the nodes belonging to each ply in the most stressed areas were extracted. These tensions, extracted for each load level constituting the load history, were used for the calculation of the damage developed through Python routines for each of the considered criteria.

5.1. Static numerical result

A numerical simulation with static load was performed to evaluate the stress state of the specimen and to evaluate the behavior of the contact in the interface between specimen and rivet. The load applied for the execution of the static simulation is approximately equal to 70% of the static resistance in tension.

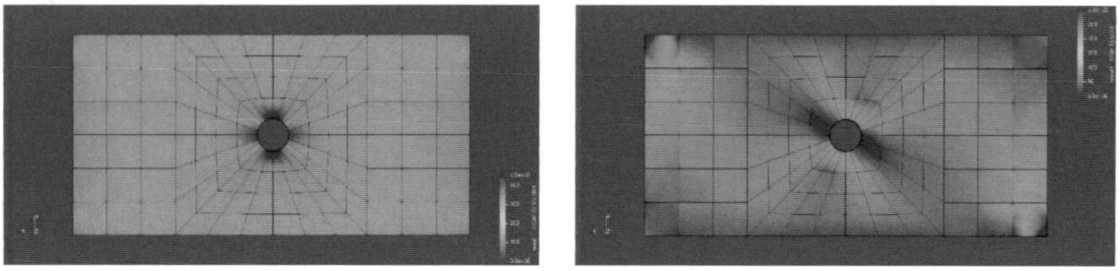

(a) (b)

Figure 8. Stress field for first direction for 0° ply (a) and 45° plies (b)

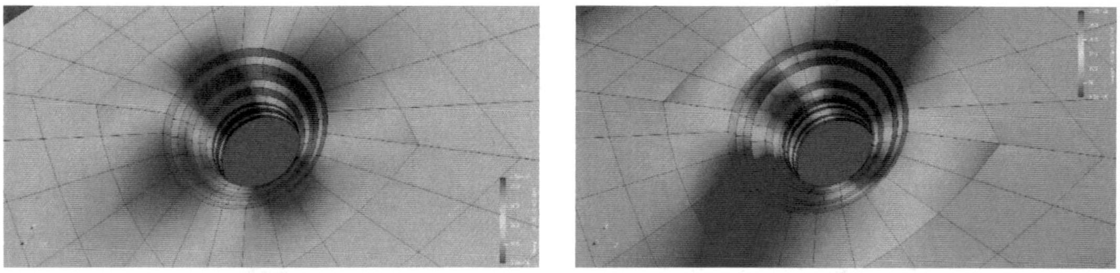

(a) (b)

Figure 9. Detail view of the stress field for first direction around the hole for 0° plies (a) and 45° plies (b)

By observing the stress field of the specimen, it can be seen how this changes as a function of the orientation of the ply considered. In particular, in the case of 0° plies it was found that the areas with the highest stress values are in correspondence of the hole edges (Figure 8a and 9a), in transverse direction (along the y direction), with a symmetrical distribution of the stress field in the x direction, while in the case of the 45° plies the area with the greatest stresses are always at the edges of the hole but in a position rotated by about 45°(Figure 8b and 9b), with an anti-symmetrical distribution respect to what is observed for the 0° plies. This observation holds for both stresses in the x direction and for stresses in the y direction. These are the directions that have been considered for the subsequent fatigue damage calculation. These observations are independent of the ply considered, net of the calculated tension value, as they depend only by the fibres orientation. The following Figure 10 shows a detail view of the stress field of the specimen near the hole, which highlights the different behaviour between the plies based on the orientation.

Figure 10. Detail view of the specimen stress field in first direction near the hole

The static calculation shows that the highest stress is obtained in correspondence of ply 7, near the hole. This is attributable both to the stress concentration effect that occurs in abrupt changes in shape and to the fact that the hole has a countersunk part which determines, even in the cross section, a concentration of stresses due to the change in shape.

5.2. Comparison of the damage assessment

Starting from the calculated stress state, the damage for some plies have been calculated and reported in Table 5, considering as the area of interest that near the hole, which already statically present high levels of tension. Figure 11 shows the nodes around the hole for which the damage assessment was conducted according to the chosen criteria.

Figure 11. Nodes near the hole where the damage was calculated

The calculation of the damage according to the three chosen criteria was performed with routines written in Python. The choice of plies to be considered was made by choosing those that presented a significant level of damage in tension and compression. Given the considerable amount of data produced, only the damages calculated for the gifts belonging to ply 7 and ply 8 are reported. The nodes considered are those most stressed both in tension and compression. Their position is shown in Figure 12.

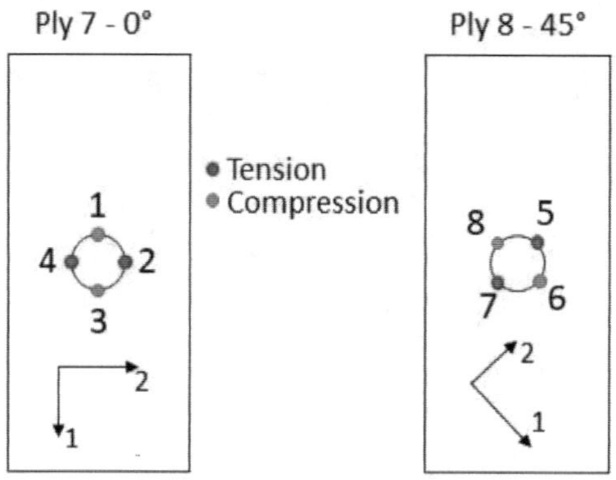

Figure 12. Location of the nodes whose damage values are reported

Table 5. Damage values calculated according to the chosen criteria – Load History 1

Ply	Node	Direction	Palmgren -Miner	Bond - Farrow	Hashin –Rotem
7	1	1	3.0662927 e-52	4.6016950048 e-52	7.65221368 e-57
		2	1.6364842 e-18	3.210328717 e-16	4.0824303644 e-23
	2	1	0.26207761	0.505624788	0.259993539
		2	1.214698356 e-37	1.956178897 e-33	3.030936211 e-42
	3	1	3.0662927 e-52	4.60169501 e-52	7.652213678 e-57
		2	1.6224767 e-18	3.18558058 e-16	4.04748196 e-23
	4	1	0.2585824614	0.498864494117	0.2599934518
		2	1.49219233 e-37	2.35378396 e-33	3.7237949436 e-42
8	5	1	8.165656835 e-34	1.328126355 e-33	2.0372336714 e-38
		2	4.6172964965 e-35	4.098411942 e-31	1.152780466 e-39
	6	1	6.92191788 e-11	1.261282077 e-10	1.72217459 e-15
		2	6.30057584 e-19	1.3590426965 e-16	1.57218388 e-23
	7	1	7.92911904 e-34	1.289576451 e-33	1.9781601 e-38
		2	4.0815362406 e-35	3.6681135 e-31	1.0189561922 e-39
	8	1	6.888041184 e-11	1.2587402238 e-10	1.718178515 e-15
		2	6.027049168 e-19	1.305836585 e-16	1.503891228 e-23

Table 6. Damage values calculated according to the chosen criteria – Load History 2

Ply	Node	Direction	Palmgren -Miner	Bond - Farrow	Hashin –Rotem
7	1	1	3.0662927 e-52	4.6016950048 e-52	2.356030225286 e-07
		2	1.6364842 e-18	3.210328717 e-16	0.0002728305
	2	1	0.26207761	0.505624788	0.763544067
		2	1.214698356 e-37	1.956178897 e-33	2.95013302 e-07
	3	1	3.0662927 e-52	4.60169501 e-52	2.3560302252 e-07
		2	1.6224767 e-18	3.18558058 e-16	0.000271713
	4	1	0.2585824614	0.498864494117	0.7613872544
		2	1.49219233 e-37	2.35378396 e-33	3.4028543124 e-07
8	5	1	8.165656835 e-34	1.328126355 e-33	5.733791302 e-07
		2	4.6172964965 e-35	4.098411942 e-31	1.478562482 e -06
	6	1	6.92191788 e-11	1.261282077 e-10	0.00945034
		2	6.30057584 e-19	1.3590426965 e-16	0.00027582025
	7	1	7.92911904 e-34	1.289576451 e-33	5.620221511 e-07
		2	4.0815362406 e-35	3.6681135 e-31	1.367493257 e -06
	8	1	6.888041184 e-11	1.2587402238 e-10	0.0094435922
		2	6.027049168 e-19	1.305836585 e-16	0.000270093

Table 7. Damage values calculated according to the chosen criteria – Load History 3

Ply	Node	Direction	Palmgren -Miner	Bond - Farrow	Hashin –Rotem
7	1	1	3.0662927 e-52	4.6016950048 e-52	9.6469702666 e-78
		2	1.6364842 e-18	3.210328717 e-16	1.33141237 e-43
	2	1	0.26207761	0.505624788	0.7623882
		2	1.214698356 e-37	1.956178897 e-33	6.047121243 e-63
	3	1	3.0662927 e-52	4.60169501 e-52	9.6469703 e-78
		2	1.6224767 e-18	3.18558058 e-16	1.33141237 e-43
	4	1	0.2585824614	0.498864494117	0.7621986
		2	1.49219233 e-37	2.35378396 e-33	6.047121243 e-63
8	5	1	8.165656835 e-34	1.328126355 e-33	5.32483807 e-59
		2	4.6172964965 e-35	4.098411942 e-31	3.971592214 e-61
	6	1	6.92191788 e-11	1.261282077 e-10	3.034053386 e-36
		2	6.30057584 e-19	1.3590426965 e-16	2.613035679 e-44
	7	1	7.92911904 e-34	1.289576451 e-33	5.324838073 e-59
		2	4.0815362406 e-35	3.6681135 e-31	3.971592215 e-61
	8	1	6.888041184 e-11	1.2587402238 e-10	3.034053386 e-36
		2	6.027049168 e-19	1.305836585 e-16	2.613035679 e-44

By observing the data reported in the tables for ply 7 and ply 8, it can be observed that the damage is relevant only for some nodes. It is evident that the damage begins near the edge of the hole and then spreads to the rest of the section.

Finally, a comparison was made for different load history of the damage criteria considering the damage that occurs in node 4 belonging to layer 7, which is the node that presents the highest damage at the change in shape of the cross section and of the hole.

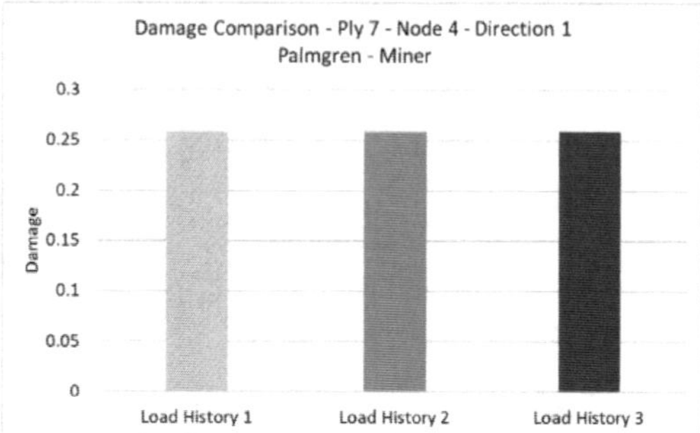

(a)

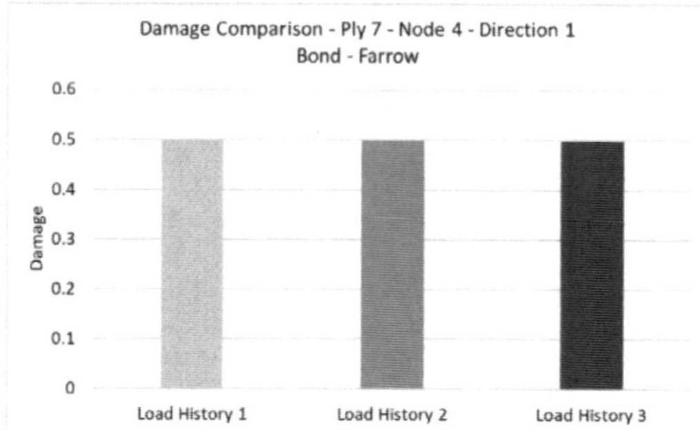

(b)

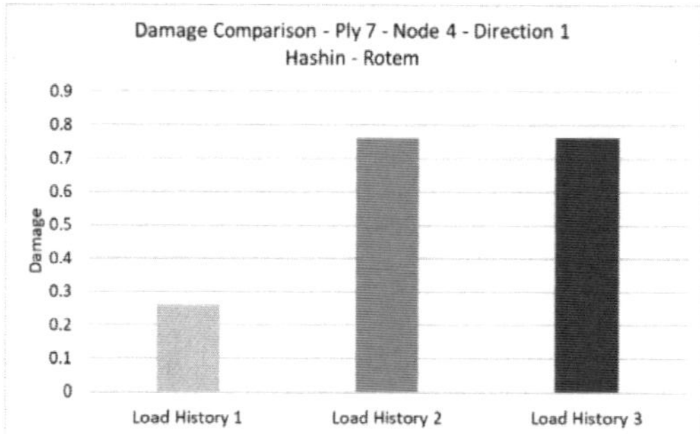

(c)

Figure 13. Comparison of damage levels calculated according to the criteria chosen for the three load histories

From the histograms shown in Figure 13a and 13b it can be seen that the damages calculated with the Palmgren - Miner and Bond - Farrow criteria always show the same values as the load application sequence varies. Instead, the Hashin - Rotem criterion (Figure 13c) highlights a damage value that is substantially affected by the load application sequence.

6. Conclusion

In the field of the study of the fatigue behavior of composites, it is of fundamental importance to be able to predict with a certain accuracy the damage that occurs when a component is stressed with a random load history. The calculation of the damage can be carried out according to criteria that evaluate the progress of the same as the load history advances, basing it on the sequence of stresses that the load history produces.

To evaluate the influence that the load application sequence has on the calculated damage, a numerical model of a CFRP specimen with a titanium alloy rivet, representative of an aeronautical joint, was created. Three load histories have been assigned with the same load levels but with a different load application sequence. From the static analysis of the specimen, it was found that the most stressed area is near the edge of the hole, due to the double change in shape of the section due to the presence of the hole itself and to the profile of the hole, which houses a rivet with head flared. The load level at which there was a maximum stress slightly lower than the ultimate tensile stress due to fatigue is equal to 26kN. The damage was then calculated according to three criteria found in the literature and chosen on the basis of the information available on the material of the specimen.

In conclusion, it was found that among the three criteria considered, that of Hashin - Rotem is the one that best evaluates the damage in the case of different load application sequence, allowing a better assessment of the damage. Furthermore, it must be remembered how this work was carried out without taking into account the damages that occur in the composites, such as delamination, and the redistribution of stresses in the case of damage in one point, which greatly affect the effective fatigue resistance.

References

[1] Mallick P K 1993 *Fiber-reinforced composites: materials, manufacturing and design* (New York: Dekker)
[2] Megson T H G 1999 *Aircraft structures for engineering students* (Amsterdam: Elsevier)
[3] Summerscales J, Short D 1978 *Carbon fibre and glass fibre hybrid reinforced plastics* Composites
[4] Jones R M 1999 *Mechanics Of Composite Materials* 2nd ed. (Philadelphia PA: Taylor & Francis)
[5] Jollivet T, Peyrac C, Lefebvre F 2013 Damage of composite materials 5th Fatigue Design Conference Fatigue Design 2013 Elsevier *Procedia Engineering* **66** 746 – 758
[6] Payan J, Hochard C 2002 Damage modelling of laminated carbon/epoxy composites under static and fatigue loadings Elsevier *International Journal of Fatigue* **24** 299 - 306
[7] Subramanian S, Reifsnider K L, Stinchcomb W W 1995 A cumulative damage model to predict the fatigue life of composite laminates including the effect of a fibre-matrix interphase Elsevier *International Journal of Fatigue* **17-5**, 343 - 351
[8] Van Paepegem W, Degrieck J, De Baets P 2001 Finite element approach for modelling fatigue damage in fibre-reinforced composite materials Elsevier *Composite: Part B* **32**, 575 - 588
[9] Papanikos P, Tserpes K I, Pantelakis S 2003 Modelling of fatigue damage progression and life of CFRP laminates Blackwell Publishing Ltd. Fatigue *Fract Engng Mater Struct* **26** 37 – 47
[10] Zhao L, Shan M, Hong H, Qi D, Zhang J, Hud N 2017 A residual strain model for progressive fatigue damage analysis of composite structures Elsevier *Composite Structures* **169** 69 – 78
[11] Schön J, Nyman T 2002 Spectrum fatigue of composite bolted joints Elsevier *International Journal of Fatigue* **24** 273–79
[12] Aicher W 1976 Description of a Fighter Aircraft Loading Standard For Fatigue Evaluation: FALSTAFF
[13] Palmgren A 1924 Die lebensdauer von kugellagern (The service life of ball bearings) *Zeitschrift des Vereinesdeutscher Ingenierure (VDI Journal)* **68**(14) 339 – 41
[14] Miner M A 1945 Cumulative damage in fatigue *Journal of Applied Mechanics* **67** 159 – 64
[15] Bond I P and Farrow I R 2000 Fatigue life prediction under complex loading for XAS/914 CFRP incorporating a mechanical fastener *Int. J. Fatigue* **22** 633–44
[16] Owen M J, Howe R J 1972 The accumulation of damage in a glass-reinforced plastic under tensile and fatigue loading *J. Phys. Appl. Phys.* **5** 1637
[17] Hashin Z and Rotem A 1978 A cumulative damage theory of fatigue failure *Mater. Sci. Eng.* **34** 147–60
[18] Harris B, Gathercole N, Lee J A, Reiter H, Adam T 1997 Life-prediction for constant-stress fatigue in carbon fibre-composites *Phil. Trans. R. Soc. Lond. A* Mathematical, Physical and Engineering Sciences. **355** 1259–1294.

Mechanical and fatigue behaviour of Sandwich elements with ABS core cells and GFRP plies, as function of cellular geometry and process parameters

F W Panella[1*], A Pirinu[1], A Saponaro[1] and M Santoro[2]

[1]University of Salento, Department of Engineering for Innovation, Viale per Monteroni, Lecce (LE), 73100, Italy
[2]Jindal Films Europe Brindisi Srl, Via Enrico Fermi, 25bis, Brindisi (BR), 72100, Italy

[*]francesco.panella@unisalento.it

Abstract. The present work aim is to fully apply the potential of FDM (Fused Deposition Modelling) technology with economic printing machine, in order to create performant ABS cores, to be subsequently used as shaped moulds for sandwich plate assembly with GFRP skins, representing simple prototypes to validate production concept and its structural behaviour. The further task of this study is to empirically optimize the geometry of the core as function of required thickness and technological building parameters, in order to achieve light and mechanically performing structures with extremely reduced costs. After compression experimental tests on core cell geometries, fatigue tests under bending are also executed on reinforced GFRP plates in order to validate best compromise between ABS core design and GFRP laminated skin properties and stiffness requirement, highlighting influence of ABS wire performances as deposited in the form of beam-based lattice geometry. In addition, the authors employ the digital image correlation technique to perform the strain and displacement field of lattice structures and sandwich interfaces.

Keywords: DIC analysis; fatigue bending test; FDM 3D printing; lattice ABS structure; sandwich panel.

1. Introduction

Engineering advances often must combine the light-weight requirement of mechanical structures with a suitable preservation in terms of stiffness and resistance, highlighting the common solution of a proper employment a different material with superior properties (i.e., composite materials) [1]. However, the problem seems to be not often resolved due to advanced material cost and mainly due to the possibility to combine the structural performance and lower density.

A promising strategy consist of adopting the new approach of topology optimization that provides the design tool to gain new unit cells for lattice material [2]. Recently, the research development gained an important focus on lattice structures obtained with the Additive Manufacturing (AM) process due to their distinctive specific properties such as high specific stiffness and strength, proper thermal and acoustic properties resulting a suitable heat exchanger and energy absorber [3]. Lattice structures consist of a spatial array arranged periodically to form specific unit cells with edges and faces, often related to cellular solids [4]. Cellular solids consist of stochastic (such as open-cell and closed-cell foams) and non-stochastic structures (2D and 3D lattice structures) with different mechanics and physical properties depending directly by structure and shape of unit cell itself [3, 4]. Traditional manufacturing process such as melt foaming or powder metallurgy using blowing agents could provide only the stochastic structures, whilst the additive manufacturing allows the introduction of the ordered non-stochastic components [5]. As described by different authors [2, 4, 6], the structural optimization seems to highlight the interdependencies between the application of AM and lightweight strategies such as topology optimization or latticing for structural optimisation.

The Sandwich-type structures seem to combine the well-known lightweight and high mechanical properties for the employment of stochastic or non-stochastic structures as material constituent of core [7]. Therefore, sandwich structures with different core material and geometries require simple but optimised manufacturing phases for reliable construction and proper structural design, especially in initial the prototyping stage. At same time, the realization of sandwich structures reinforced with GFRP skin may result to be rather convenient from the point of view of manufacturing times and costs. In fact, the main idea is to achieve complex geometry of the component by laminating external shell fabrics directly on the previously shaped core geometries, without using die and other equipment, with only exception for use of simple vacuum bag method.

In the present work, the authors propose a different GFRP (Glass fibre reinforced polymer) / ABS (Acrylonitrile butadiene styrene) lattice sandwich structures for aeronautical and biomedical field, manufacturing with the consolidated technologies of 3D printing and the vacuum bag techniques, as alternative to very low-weight core material as expanded polyurethane.

The employed core is an ABS lattice structures manufactured with FDM (fused deposition modelling) technology, employed as a 'base' for lamination of GFRP skins the external surfaces in ABS. Preliminary compressive tests are carried out on five selected lattice structures to evaluate different mechanical specific properties for a suitable selection of lattice configuration for sandwich panels. In the second phase of study, four-points bending static tests are executed firstly in static conditions to compare the effects of core lattice geometries, aided by digital image correlation (DIC) analysis to select the proper sandwich lattice structures for a fatigue campaign [8]. In addition, a simple numerical model is employed to simulate bending behaviour of the selected GFRP/lattice core geometry, verifying structural performance before static test.

In fact, numerous studies are present in literature, regarding mechanical behaviour and damage modes under static loads of lattice-based structures, but the recent interest is focused also on fatigue assessment and stiffness behaviour under cyclic loading [8, 9]. Therefore, fatigue assessment under bending loads is also performed after static tests to validate the selected compromise between core design and GFRP skin reinforcement, highlighting influence of ABS performances as deposited in the flat sandwich specimens. Experimental results and damage modes are described in the following work.

2. Materials and Methods

2.1. Lattice elements

For this research work, two sample series are employed for mechanical evaluation of 3D cell topologies analysis. Effect of extrusion width in terms of stress-strain curves are investigated, i. e. extrusion width of 0.4 mm or 0.2 mm. As shown in Figure 1a, smaller values of extrusion width seem to give better morphological accuracy, but longer printing time.

Then, the final configuration of lattice panels is optimized based on effect of extrusion width, as shown in Figure 1b. Core manufacturing parameters and cell geometries are studied and optimized to achieve a cellular structure with sufficient geometric accuracy as function minimum cell size. The influence of ABS 3D building direction and printing flow parameter on strength was also studied, as shown in example diagram of Figure 2a.

Therefore, each samples set is manufactured in the same 3D printer CREALITY ender 3 pro with the same process parameters (i. e. printing speed, nozzle temperature and heated bed temperature). Process parameters are selected following best practices recommended on printer manual. The first batch of samples consists of five lattice structures manufactured with a filament SUNLU type in ABS with a diameter of 1.75 mm. The selected lattice unit are the BCC (basic centred cubic), FCC (face centred cubic), SFCC, SFBCC and OCTET [9].

Each sample batch consists of 3 cubic specimens with 27 cell units (3 × 3 × 3) subjected to the compression test; these samples are manufactured with the upper and bottom flat surfaces to improve load conditions and the contact with test machine.

Cubic lattice samples present thickness and dimensions of the cell unit reduced in such a way as to create specimens ensuring enough units in thickness, such as to create functional components with a complex geometry. An overall summary of specific geometrical and physical properties is reported in Table 1.

(a) (b)

Figure 1. (a) Macroscopic effects of extrusion width on morphological accuracy and (b) final configuration of lattice panel.

Table 1. General overview of employed lattice structures for compressive test.

ID LATTICE STRUCTURE	BCC	FCC 50_45	SFCC	SFBCC	OCTET
Beam thickness [mm]	1.2	1.2	1.2	1.2	0.4
Cell size [mm]	7	7	7	7	10
ρ [g/cm³]	0.29	0.45	0.35	0.42	0.53
Lattice sample view				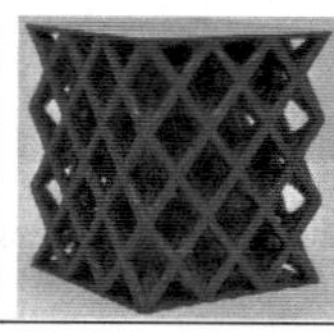	

The second case of study includes hybrid sandwich elements realized by ABS lattice panels laminated with GFRP skins. Operational steps for the manufacturing of GFRP / ABS lattice sandwich panels consist of surface preparation of lattice panels, the lamination process of GFRP skins, vacuum bag polymerization and cutting of sandwich panels; the final sandwich samples is reported in Figure 2b.

The manual lay-up process in steel or carbon molds represents the mainly common manufacturing method of sandwich structures with complex geometry with additional plastic deformation and machining processes to achieve the final shape of designed element. After lamination process, the resulting structures of external skins and core is assembled during cure cycle in autoclave for uniform transfer of resin and additives used for bonding. In this sample batch, the ABS lattice panels are the sandwich core constituent employing their wider external surfaces as 'base' for lamination process of GFRP skins. A single ply of twill e-glass 0/90 fabric (160 g/mm^2) is employed for each GFRP skin.

Table 2. Geometric properties of GFRP/ABS lattice sandwich elements for 4p static bending tests.

ID GFRP/ LATTICE ELEMENT	Lattice structure	Glass fabric of skin	N. of samples	Thickness [mm]	Width [mm]	Length [mm]
SFCC 1.2 × 7 P4n	SFCC	twill e-glass 0/90	6	22	23	129
SFBCC 1.2 × 7 P4n	SFBCC	twill e-glass 0/90	5	22.5	28	129
FCC 50_45 1.2 × 7 P4n	FCC	twill e-glass 0/90	5	20	30	152

Data results of compression test provide specific mechanical properties of employed lattice structures, resulting more performing for the FCC 50_45, SFCC and SFBCC geometry, utilized as lattice core for sandwich panels, avoiding the employment of octet structure as core due to higher density and complex geometry. Geometrical properties of sandwich elements are summarized in Table 2.

Static bending tests are performed to evaluate specific mechanical behaviour of employed sandwich structures, showing that the SFCC lattice panels provide the best pattern between mechanical properties and simplified manufacturing geometry. In addition, three fatigue tests are performed on lattice-GFRP sandwich structures (344.5 × 26 × 22.5 mm^3).

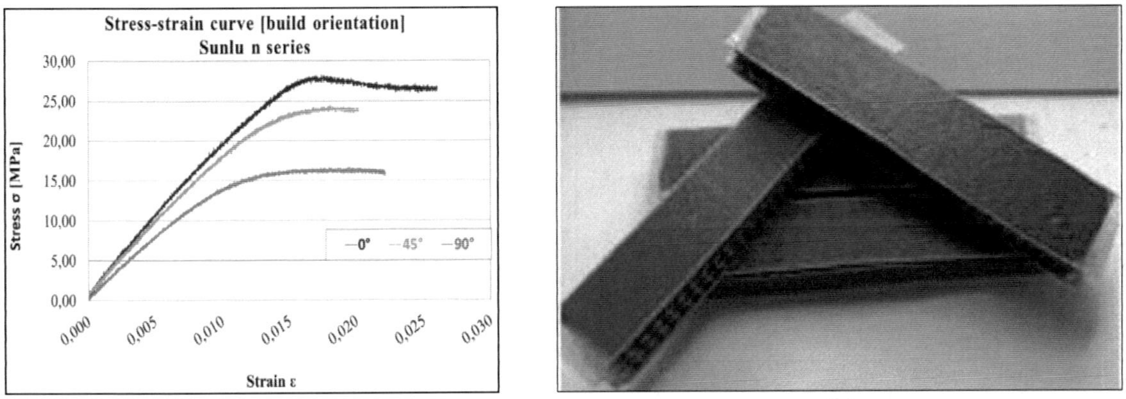

(a) (b)

Figure 2. (a) Influence evaluation of ABS 3D building direction and (b) final sandwich elements.

2.2. Experimental and monitoring procedures

The compressive and bending static tests are carried out are performed in displacement control mode using an MTS 810 servo-hydraulic axial testing machine. The test system is equipped with a load capacity of ±100 kN and governed by a computer equipped with Station Manager 3.A 900 software in the Experimental Mechanic Laboratory at University of Salento (Lecce, Italy). All experimental parameters (time, strain, load and displacement) are recorded during whole test.

(a) (b)

Figure 3. (a) Experimental setups for compressive tests and (b) for four-points bending tests.

Strain data of specimens are detected by an MTS S / N 1067668 extensometer and the authors select the load speeds of 0.2 mm/min and 1 mm/min are employed during compressive and bending static tests respectively to maximize the high resolution of acquired images for a proper evaluation of damage and for a suitable DIC analysis. As shown in Figures 3a and 3b, two specific experimental setups are performed employing specific steel supports for compressive and both static and fatigue bending tests respectively. As shown in Figure 3a, two steel plates are gripped in the machine clamps for the correct transmission of axial compressive loads on lattice cubic specimens during whole test.

The four-points bending test is performed with appropriate steel supports applied to simulate a four-point bending, following the ASTM C393-00 "Standard test method for Flexural Properties of Sandwich Constructions" to ensure correct constraint and load conditions determining determination final flexural properties. As shown in Figure 3b, sandwich elements are positioned on two fixed steel rollers with elastomeric blue slices and loaded with two compression rollers fixed at a determined distance from specimen centre.

In addition, during bending static test, a Digital Sony Alpha 77 camera with macro lens is employed to acquire the high-resolution images of 4000 × 6000 pixels during whole low-speed test. The Ncorr and Ncorr_post open source 2D codes [9, 10] implemented in MATLAB are employed for a full-field post-processing DIC analysis of acquired images of experimental tests, previously used in other works [11, 12]. The lattice structural configuration seems to highlight a high random white-black contrast on the inspected samples. Therefore, the authors decide to avoid the surface preparation with the speckle pattern for DIC analysis on the Region of Interest (ROI). In fact, as shown in Figure 3b, the non-stochastic structure could be directly used as random speckle pattern for strain and displacement mapping of lattice elements, as observed also for stochastic foams [13, 14].

2.3. Numerical model

In this work, a numerical model is performed on the bending test of GFRP/ABS sandwich element with more performing lattice FCC 50_45 to optimize sandwich panel design, skin thickness and critical zones. FEM model is built using ANSYS software, simulating only a quarter of sandwich specimen. Table 3 report the mechanical and physical properties of GFRP / ABS sandwich constituents with FCC 50_45 1.2 × 7 core geometry, implemented in the FEM model. In the analysis, core isotropy is adopted, related to a bilinear behaviour with a tangent modulus ET equal to about 44% of elastic modulus is adopted, whilst for GFRP fabric 0/90°, experimental data were obtained from static test.

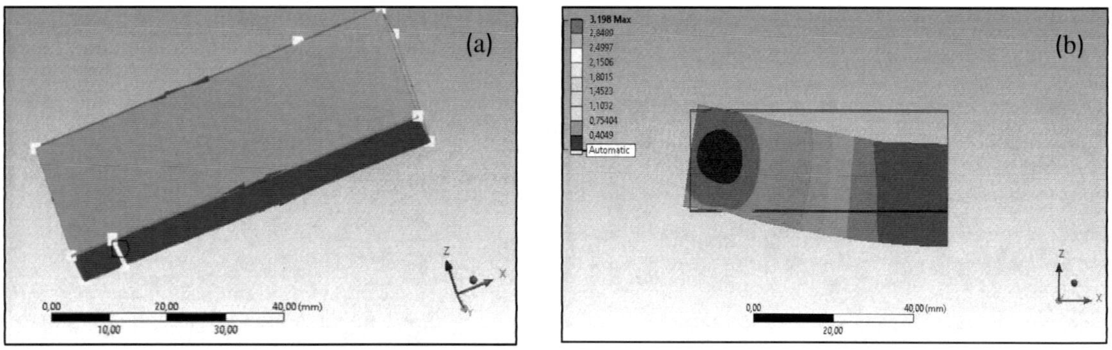

Figure 4. (a) FEM boundary conditions for bending test of GFRP/lattice core sandwiches elements and (b) deformed configuration of sandwich specimen under bending loads.

Table 3. Element constituent properties for FEM model of GFRP/ABS FCC 50_45 sandwich panel.

	Material properties						
Element constituent	**Young's modulus [Mpa]**	**Shear modulus [Mpa]**	**Tangent Modulus [MPa]**	**Yield strength [MPa]**	**Poisson's ratio**	**Density [g/m^3]**	**Thickness [mm]**
ABS_FCC 50_45 P4n	201.59	74.55	161.27	7.58	0.35	550000	21.7
GFRP	10000	5500	/	/	0.22	/	0.5

Constraint conditions are applied for the supports, with load conditions on the contact lines of internal supports, imposing progressive displacement along the Z axis, as reported in Figure 4a. A non-linear analysis is defined since core elements result heavily distorted. The equilibrium equations in non-linear field are solved using the modified Newton Rapson method, capable of improving the solution convergence. Figure 4b shows the deformed condition of the sandwich panel subjected to a four-point bending load condition.

3. Results and discussion

The mechanical properties of cell solids are influenced by base material, relative density, topology, and shape of unit cell. In particular, the relative density represents the ratio between the density of cellular unit and the base material density, calculated as

$$\rho_r = \frac{\rho_{cu}}{\rho_{bm}} \tag{1}$$

Where ρ_{cu} is the density of specific cellular unit (empirically evaluated) and ρ_{bm} represents density of base material (available in datasheet) of stochastic elements. In the present study, the same 3D-Printer Filament ABS is employed for manufacturing process of lattice materials. The authors evaluate specific mechanical properties normalized respect to ρ_r in both test cases.

3.1. Compression test results

In this case, an evident stiffness variation and damage evolution is observed for cubic lattice tested elements under compressive loading for each experimental test. Comparative analysis of mechanical behaviour in terms of specific stress-strain is presented in Figure 5a, where the example of static compression behaviour could be observed for different lattice configurations.

After elastic phase, all σ / ε curves seem to exhibit an increasing trend due to continuous densification process of cellular units that obviously arises after the first collapse of specific lattice unit.

In Figure 5a, the FCC, SFCC and SFBCC structures seems to provide an almost similar behavior in the initial linear phase, whilst different behaviors could be observed for OCTET and BCC structures. In fact, the octet configurations result the most rigid structures due to the consistent increase in density. The BCC structure also exhibits an approximating constant stress state with a slightly increase of stresses during the static test. After the first collapse of specific lattice unit, the SFCC core shows a severe decrease in the stress state with a successive slightly augmentation before final failure, whilst both SFCC and FCC structures provide an almost constant behavior.

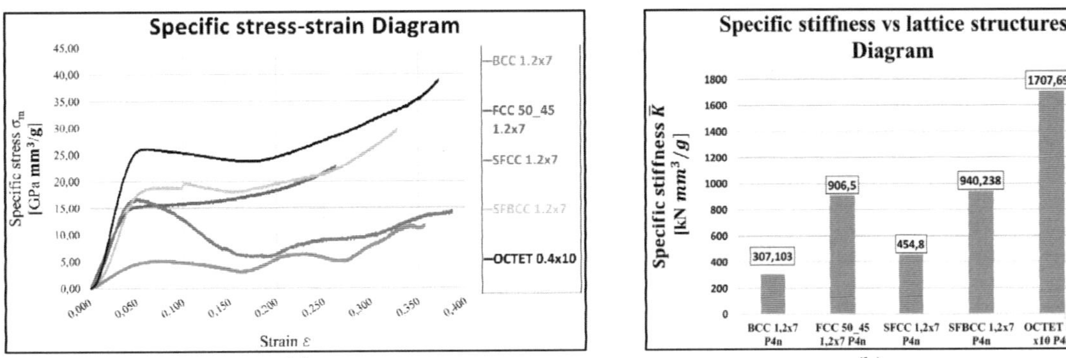

(a) (b)

Figure 5. (a) Comparison of specific Stress-strain and (b) specific stiffness behaviours for different cellular structures under compressive loads.

Example comparative analysis of compressive properties is also reported in the histogram of Figure 5b. The stiffness results of the five structures are normalized to relative density of specific lattice unit and show evident similar values for FCC_50_45 and SFBCC elements, whilst higher and lower data are observed for OCTED and BCC cellular geometries, respectively. By visual monitoring, damage evolution could be observed during compression test, as reported in example Figures 6a – 6d. The acquired images show damage initiation and progression of a cubic element with BCC_1.2×7 lattice configuration where the structure collapse seems to begin at the nodes of central units and subsequently the structure is affected by a continuous and gradual densification during compression test.

(a) (b) (c) (d)

Figure 6. (a) – (d) Example of damage evolution during compression test of BCC element.

3.2. Results and discussion of static bending tests

A monitoring strategy for these tests consists of compliance and digital image correlation analysis, employed in this work to evaluate damage progression during bending loading of sandwich samples, shown in Figure 2b.

The specimens are positioned on steel supports of test machine clamps, following the specific ASTM regulation, and monitored in real time with digital high-resolution Nikon camera for damage evaluation, as shown in Figure 7a.

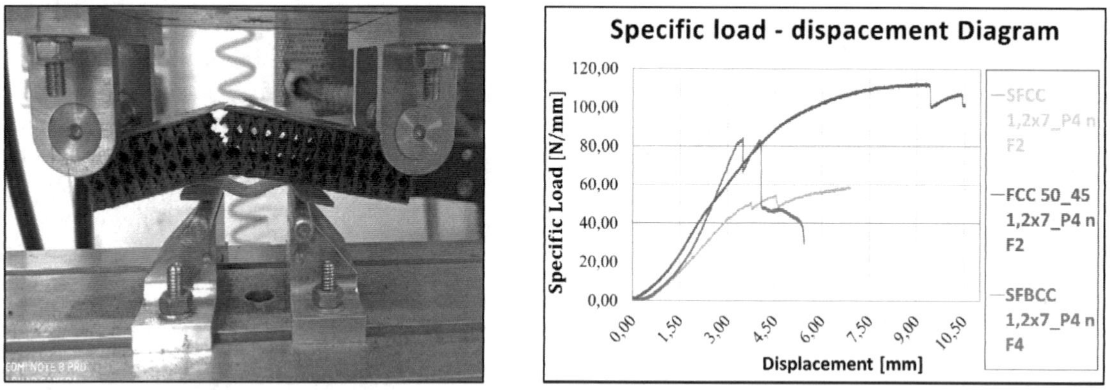

(a) (b)

Figure 7. (a) Example of failure mode for SFCC F5 n after static load; (b) example of specific load-displacement curves of SFCC, FCC 50_45 and SFBCC sandwich series.

For each test case, specific load-displacement diagrams are analysed showing better results in terms of specific mechanical parameters for FCC 50_45 lattice sandwich elements, as reported in the example curves of Figure 7b. The SFCC F2n specimen is affected by a first detachment of the laminate subject to compression, at the first peak, followed by a break in the tensile laminate, therefore in agreement with theoretical previsions.

In SFBCC case, not considering the intensity of the load per unit width, all specimens exhibit a similar behaviour characterized by a first detachment of laminate on compressed side arisen at the first peak of specific curve (as shown in Figure 7b), followed by a total separation of laminate on the opposite tensile side and by a final core shear failure.

In the FCC 50_45 elements, an almost flat stabilized is observed in the final test phase, where a slightly reduction of 2% between the residual and initial stiffness. The flexural properties are calculated according to ASTM C393-00 "Standard Test Method for Flexural properties of Sandwich Constructions" and consists of Core Shear Stress τ [MPa], Facing Bending Stress σ [MPa] and Sandwich Beam Deflection Δ [mm]. For each lattice core configuration, Table 4 reports the average flexural properties, where $\overline{Pw}$ represents the maximum axial force normalized to specimen width.

Samples damages seem to be affected highlight two modes of failure, the detachment of laminate subject to compression load with shear break of core and detachment of laminate subjected to compression load followed by a break in laminate subject to traction with final break of core. The last failure mode results perfectly aligned with the theoretical foundations, demonstrating the proper design of sandwich panels, as reported in Figures 8a - 8c.

In Figure 9, a comparison of the expected (real) stress values and those obtained from numerical simulations is reported. Given the geometric approximations used, the limits that characterize a non-linear analysis, the low number of experimental tests and the approximations introduced relating to the behaviour of the material, the numerical models created cannot be considered acceptable due to the high discrepancy between the values obtained experimentally, as shown in Figure 9.

Table 4. Average flexural properties of each GFRP/ABS lattice sandwich elements.

ID GFRP/ LATTICE SANDWICH SAMPLE	Lattice structure	$\overline{Pw}$ [N / mm]	$\bar{\tau}$ [MPa]	$\bar{\sigma}$ [MPa]	$\bar{\Delta}$ [mm]
SFCC 1.2 × 7 P4n	SFCC	54.45	1.31	115.40	2.84
SFBCC 1.2 × 7 P4n	SFBCC	92.41	2.09	104.13	2.74
FCC 50_45 1.2 × 7 P4n	FCC	110.43	2.76	137.30	4.94

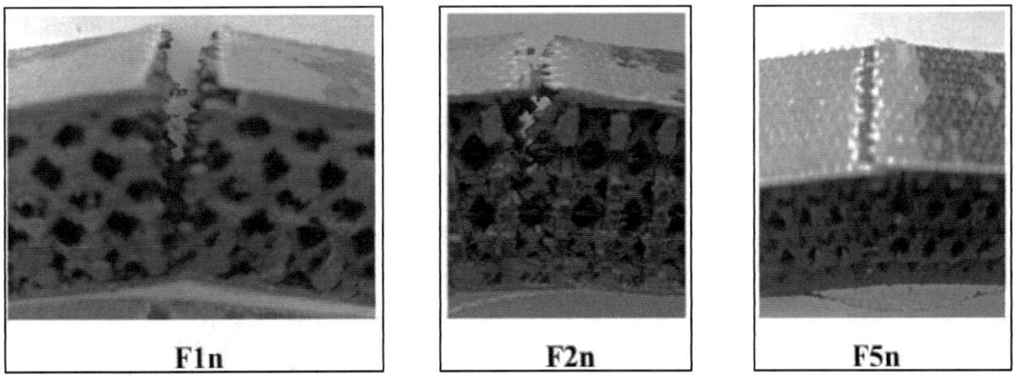

F1n F2n F5n

(a) (b) (c)

Figure 8. (a) – (b) Failure modes of GFRP / ABS panels with an SFCC 1.2x7 geometry core.

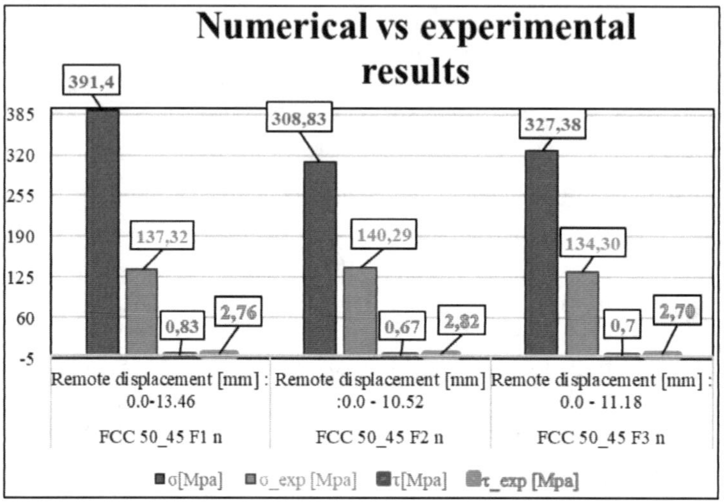

Figure 9. Numerical-experimental comparison for GFRP/ FCC_50-45 sandwich.

In the cellular solids, the strain field should provide the most important aspect of bending damage prediction and propagation prior to final failure; thus, the authors employ the digital image processing to analyse strains and displacements for selected ROIs of specimens, performed with Ncorr software implemented in MATLAB environment. Specific ROIs are selected for different test cases for a proper evaluation in critical areas where maximum stress, displacements and strain concentration should be observed. The DIC data are analysed at the various regular intervals evaluating during entire test to estimate displacement and strain variation. The Figures 10 to 11 report an example comparative analysis the angular deformation γ_{xy} and displacement V along Y axis maps respectively for four – points bending test of SFCC F5n lattice sandwich sample at two different instants (50% and 95%). In Figures 10a and 10b, high angular deformations are observed near the bending supports on the compressed lattice unit.

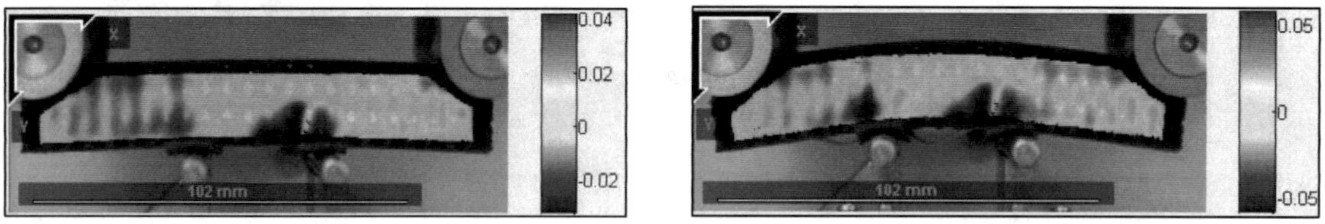

(a) (b)

Figure 10. (a) Angular deformation γ_{xy} maps at 50% and (b) 95% bending test for SFCC F5n lattice sandwich sample.

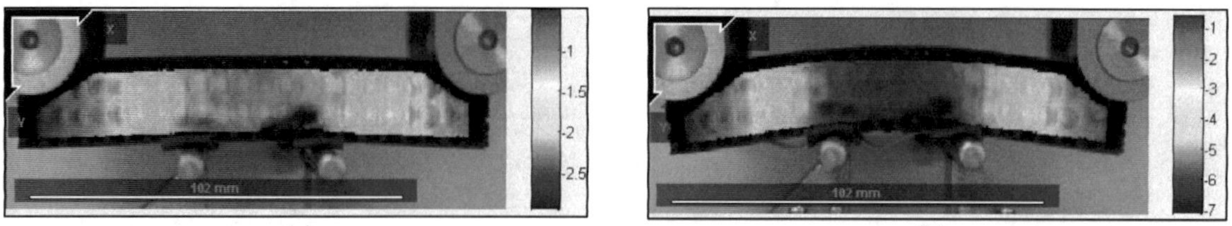

(a) (b)

Figure 11. (a) Axial displacement V maps at 50% and (b) 95% bending test for SFCC F5n lattice sandwich sample.

3.3. Preliminary fatigue results

Table 5 reports the experimental overview of fatigue results on SFCC lattice sandwich panels, where the maximum fatigue load F_{MAX} to specific sandwich width [N/mm] and the amplitude load to specific sandwich width AMP [N/mm] are normalized according to the average maximum axial force over sandwich width $\overline{Pw}$ [N/mm].

Table 5. GFRP/ABS lattice sandwich elements for 4p fatigue bending tests.

ID GFRP/ LATTICE ELEMENT	Weight [g]	Density [g/cm³]	Stress ratio R	$F_{MAX}/\overline{Pw}$ [%]	AMP / $\overline{Pw}$ [%]	Cycles to failure
SFCC 1.2 × 7 P4n F7	30	0.353	0.2	37.19	14.87	41430
SFCC 1.2 × 7 P4n F8	30	0.353	0.1	33.05	14.87	19162
SFCC 1.2 × 7 P4n F9	31	0.365	0.1	28.92	13.01	53113

Figure 12a shows normalized stiffness results against fatigue life of three tests. By the general trend of the curves, an almost flat behaviour of stiffness is observed from 0% to approximately 50% ÷ 60% of fatigue life. Subsequently, a rapid decrease appears before the final failure of lattice element. As shown in Figure 12b, fatigue damage seems highlight two failure modes, the detachment of laminate subject to compression load with shear break of core and break of core at bending support followed by laminate detachment.

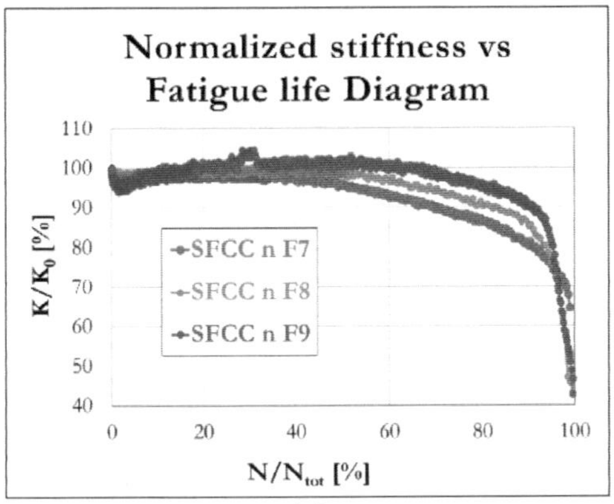

(a)

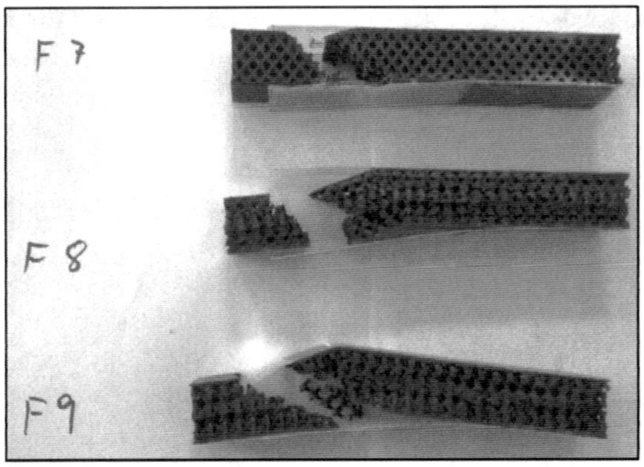

(b)

Figure 12. (a) Normalized stiffness vs fatigue life; (b) tested sandwich elements after fatigue loads.

4. Conclusions

In the present study, a comparative analysis is carried out to evaluate mechanical behavior of different sandwich structures with lattice unit core and GFRP skins.

Preliminary compressive tests are conducted on cubic lattice elements manufactured in ABS with five different reticulated geometries for a suitable selection of final cellular solid configuration.

The FCC 50_45, SFCC and SFBCC lattice unit are employed for the manufacturing of sandwich panel cores due to their proper mechanical properties combined to lower density. The octet structure provides better results but present the high density and difficult complex geometry.

From bending tests results, GFRP/ABS sandwich elements seem to provide good mechanical behavior as stress absorber due to large deformations, despite loss integrity. These interesting results seem to be perfectly aligned to initial assumption of a possible employment of ABS lattice core sandwich structures due to higher specific mechanical properties combined to lower costs.

Reference

[1] Rajak D K, Pagar D, Kumar R and Pruncu C 2019 Recent progress of reinforcement materials: a comprehensive overview of composite materials. *J of Mater Rese and Tech* **8** (6), 10.1016/j.jmrt.2019.09.068

[2] Gao W, Zhang Y, Ramanujan D, Ramani K, Chen Y, Williams C B, Wange C C L, Shina Y C, Zhang S and Zavattierif P 2015 The status, challenges, and future of additive manufacturing in engineering, *Computer-Aided Design* **69**, pp 65 – 89

[3] Tao W and Leu M C 2016 Design of lattice structure for additive manufacturing, *International Symposium on Flexible Automation (ISFA)*, 16544329, 325-332, 10.1109/ISFA.2016.7790182

[4] Seharing A, Azman A H and Abdullah S 2020 A review on integration of lightweight gradient lattice structures in additive manufacturing parts, *Advan in Mechan Engin* **12** (6): 168781402091695, 10.1177/1687814020916951

[5] Lehmhus D, Vesenjak M, De Schampheleire S and Fiedler T 2017 From Stochastic Foam to Designed Structure: Balancing Cost and Performance of Cellular Metals, *Mater* **10** (8): 922, 10.3390/ma10080922

[6] Nagesha B K, Dhinakaran V, Shree V M, Kumar K P M, Chalawadi D and Sathishc T 2020 Review on characterization and impacts of the lattice structure in additive manufacturing *Mater Today: Proc* **21**, Part 1, 916-919, 10.1016/j.matpr.2019.08.158

[7] Austermann J, Redmann A J, Dahmen V, Quintanilla A L, Mecham S J and Osswald T A 2019 Fiber-Reinforced Composite Sandwich Structures by Co-Curing with Additive Manufactured Epoxy Lattices, *J of Comp Sci* 3(2):53, 10.3390/jcs3020053

[8] Sutton M A, Orteu J J and Schreier H 2009 Image Correlation for Shape, Motion and Deformation Measurements: Basic Concepts, Theory and Applications. 1st ed. Springer Publishing Company, Incorporated, 10.1007/978-0-387-78747-3

[9] Ling C, Cernicchi A, Gilchrist M D and Philip Cardiff 2019 Mechanical behaviour of additively- manufactured polymeric octet-truss lattice structures under quasi-static and dynamic compressive loading. *Materials & Design* **162** (15) 106 - 118, 10.1016/j.matdes.2018.11.035

[10] Blaber J, Adair B and Antoniou A 2015 Ncorr: Open-Source 2D Digital Image Correlation Matlab Software. *Exp. Mech.* **55** (6): pp 1105–1122, 10.1007/s11340-015-0009-1

[11] Nežerka V Antoš J, Sajdlová T and Tesarek P 2016 Use of Open Source DIC Tools for Analysis of Multiple Cracking in Fiber-Reinforced Concrete. *Appl Mech and Mater* **827**: 336 – 339, 10.4028/www.scientific.net/AMM.827.336

[12] Dattoma V, Panella F, Pirinu A, Castriota A. 2019 Fatigue damage on CFRP plates under bending by thermographic and UT analysis, aided with FEM-DIC prediction. *Pro Stru Inte* **24**: pp 978– 987.

[13] Jung A, Al Majthoub K, Jochum Ch, Kirsch S M, Welsch F, Seelecke S and Diebels S 2019 Correlative digital image correlation and infrared thermography measurements for the investigation of the mesoscopic deformation behaviour of foams. *J of the Mech and Phy of Sol* **130**, pp 165-180

[14] Huo X, Sun G, Zhang H, Xiaojiang Lv and Qing Li 2019 Experimental study on low-velocity impact responses and residual properties of composite sandwiches with metallic foam core. *Composite Structures* **223**, 110835

IOP Conf. Series: Materials Science and Engineering 1214 (2022) 012011 doi:10.1088/1757-899X/1214/1/012011

Fatigue damage evolution using NDT techniques for Health Monitoring of CFRP laminates

R Nobile[1], F W Panella[1], R Antonazzo[1], A Pirinu[1], A Saponaro[1, 2]

[1]Department of Engineering for Innovation, University of Salento, Lecce 73100, Italy

[2]Corresponding author's e-mail: andrea.saponaro@unisalento.it

Abstract In this work, an experimental method based on ultrasonic measurements with piezoelectric transducers (PZTs) was proposed for monitoring the evolution of fatigue damage on a batch of carbon fabric Open Hole specimens. The amplitudes of the recorded ultrasonic sinusoidal waves in packets were evaluated with pitch-catch modality during the different phases of the fatigue life and correlated to the applied loads.
The examination of the ultrasonic data showed high sensitivity of the UT signal respect to stiffness decay and to fatigue damage associated with delamination near the hole. Occasional ultrasound measurements with the consolidated pulse-echo Phased Array technique were carried out by the authors at regular time intervals during the tests to evaluate the state of damage in relation to the degradation of the signal detected with the PZT sensors.
The results have shown the potential capability of the applied experimental technique for real-time detection of delamination on composite elements subjected to time-varying loads.

Keywords: CFRP, fatigue, damage, Non-Destructive Evaluation.

1. Introduction

Damage due to fatigue is one of the main problems of aeronautical structures. This phenomenon is more pronounced in areas of stress concentration such as the connection of components by means of bolts or rivets [1, 2]. For the study of these structures, specimens with different geometric configurations of the hole are used. The possibility of real-time monitoring of composite aeronautical structures subject to fatigue, to assess the state of health in service, is a very important objective to be achieved in a logic of Structural Health Monitoring SHM [3, 4]. Therefore, it is essential to develop structural health monitoring methods to detect fatigue damage within composite structures and continuously monitor its evolution [5-7].

In the literature there are several methods to investigate the presence of damage inside the materials and components: for example, the variation of the ultrasound propagation speed [8, 9] was used to determine the fatigue damage; alternatively, the variation of the electrical resistance of a material subjected to cyclic fatigue loads [10-12] has been proposed; these methods could be useful for a variety of SHM purposes.

A further technique is linked to the use of Lamb waves, which are ultrasound waves that propagate on the surface of the component and are generated by specific kind of excitations [13, 14]. This method is based on the emission of ultrasonic wave packets in the component to be analysed and on the reading, at a position of the receiver, of the value of the received ultrasound signal. Lamb waves were used as NDT in the studies conducted by Kessler et al. [15], who used this technique to investigate the damage within composite materials. Most of the current active structural monitoring systems based on the phenomenon of guided wave propagation describe the changes in the characteristics of the signal in time domain with respect to the basic signal [16].

The present work proposes one experimental method, based on ultrasonic measurements with piezoelectric transducers (PZTs) for monitoring fatigue damage evolution on a batch of carbon fabric Open Hole specimens. The ultrasonic signals acquired in wave packets with pitch-catch mode, during the different phases of the fatigue life, were exported and processed in Matlab with an appropriate analysis algorithm, based on the Fast Fourier Transform (FFT) and correlated to the applied loads. The examination of the ultrasonic data showed high sensitivity of the UT signal respect to stiffness decay and to fatigue damage associated with delamination near the hole.

The results showed the effective capacity of the applied technique for real-time detection of delamination on composite elements subjected to time-varying loads.

2. Materials and methods

2.1. Ultrasonic measurements on CFRP with countersunk hole

Three series of carbon fabric Open Hole Tension (OHT) specimens named S, T and V with different countersunk fastener hole geometry (Figure 1a) were considered. The specimens consist of carbon/epoxy prepregs with different woven patterns built with twenty-layer laminates for a total nominal thickness of 6.55 mm and stacking sequences symmetrical with respect to the mid-plane [+45°/-45° | 0°/90° | 0°/90° | 0°/90° | +45°/-45°]$_{2S}$. The nominal mechanical properties of each lamina are summarized in Table 1.

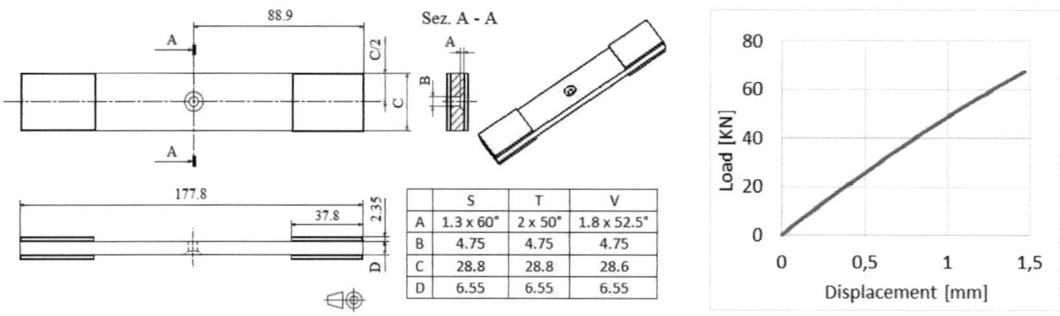

(a) (b)

Figure 1. (a) Geometries of the specimens used for the fatigue tests (dimensions in mm); (b) Load versus displacement curve of the specimen 1T.

Table 1. Mechanical properties of pre-impregnated fabric lamina.

E_1	[MPa]	71000
E_2	[MPa]	67900
E_3	[MPa]	7100
υ_{12}	-	0.1
$\upsilon_{13} = \upsilon_{23}$	-	0.34
$G_{12} = G_{13} = G_{23}$	[MPa]	4600
Ply thickness	[mm]	0.327

A preliminary static test was carried out on a specimen of the T series according to the ASTM D5766 M standard (Figure 1b), which allowed to evaluate an ultimate load equal to 67.6 kN and used as reference value for fatigue tests.

The set-up adopted during the fatigue tests includes a signal generator (HAMEG Instruments HMF 2550), a commercial Control Unit OmniScan® MX, two piezoelectric ceramic discs 7 x 0.2 mm and a servo-hydraulic testing machine - INSTRON Model 8850 with a load cell of 250 kN (Figure 2). This measurement system allowed to record and to analyse the ultrasonic signals in transmission and reception (Pitch-Catch method) without removing the specimen from the testing machine.

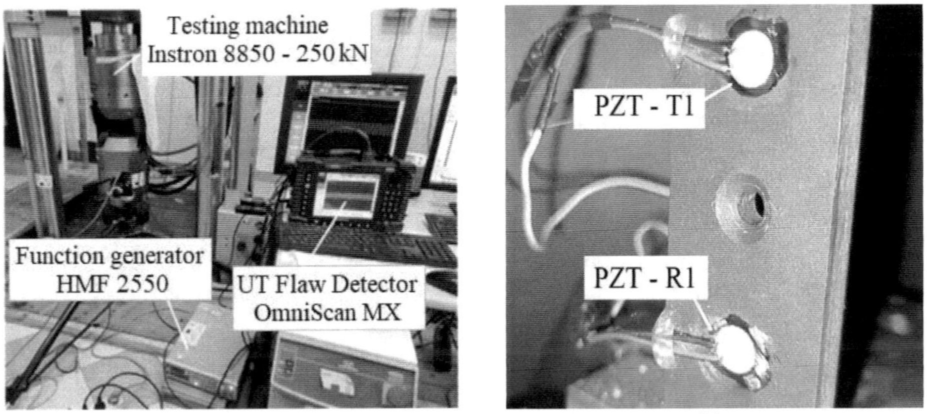

(a) (b)

Figure 2. (a) Experimental set-up of fatigue test and UT measurements devices; (b) piezo disks transducers bonded on the specimen surface.

The application of continuous sinusoidal Lamb wave during fatigue test was performed using the two piezoelectric sensors (PZTs) bonded to the surface of the specimen from the countersink side, as previous tests indicated this area is subjected to delamination around the countersunk seat of the hole.

The sensors used in this work have a resonant frequency centered at 300 kHz ±10 kHz (radial mode vibration). The position of the two sensors is shown in Figure 3 where the transmitter sensor T1 and the receiver R1 have been positioned on the longitudinal axis of the specimen at a distance of 40 mm between them to cover the entire area where the countersunk hole is present. A block diagram of the connections is shown in Figure 3a.

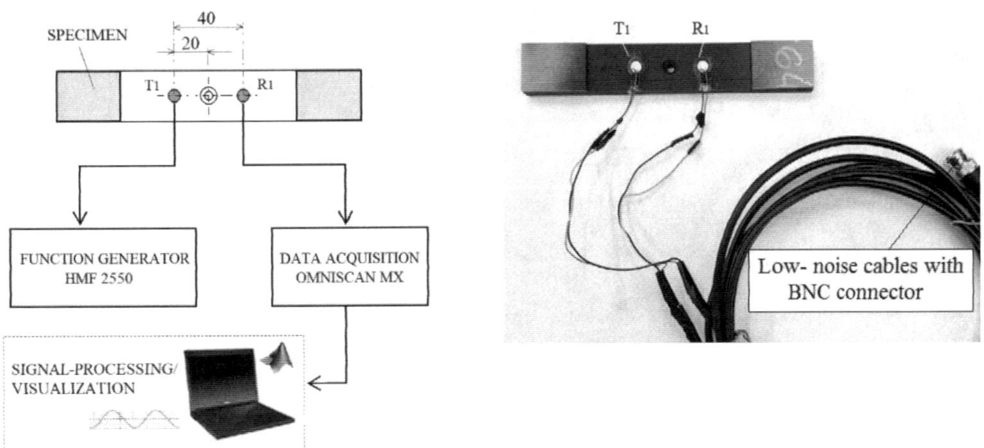

(a) (b)

Figure 3. (a) Block diagram of ultrasonic measurements systems (dimensions in mm); (b) location of PZT sensors on the specimen.

The input signal supplied to the T1 transmitter at 300 kHz (in accordance with the resonant frequency of the sensor) has a voltage of $20V_{pp}$ (peak-to-peak voltage). The signal received in real time by the sensor R1, amplified by 40 dB and averaged 16 times with the OmniScan® MX unit, was recorded at predetermined time intervals, according to the expected fatigue life and exported in Matlab for data processing with a specific routine based on the Fast Fourier Transform (FFT) to obtain the amplitude of the fundamental frequency.

The batch of the specimens for the fatigue tests are subjected to both tension-tension and tension-compression loading in force-controlled fatigue tests with a sinusoidal waveform, frequency of 9 Hz and stress ratio of $R = \sigma_{min}/\sigma_{max} = 0.1$ and $R = -0.5$. Seven different load levels were considered. Table 2 shows the experimental parameter used and the resulting number of cycles to failure (N_f). If no failure occurred before 10^6 cycles, test was interrupted.

Table 2. Scheduled experimental parameters for each fatigue tests.

ID Specimen	Frequency of load (Hz)	Stress ratio (R)	Max load level (% UTS)	Mean load (kN)	Amplitude load (kN)	Number of cycles to failure (N_f)
OHT 4T	9	0.1	90.18	33.55	27.45	732673
OHT 5T	9	0.1	93.86	34.92	28.57	318797
OHT 2S	9	0.1	87.22	32.45	26.55	10^6 (run-out)
OHT 3S	9	0.1	91.66	34.1	27.9	737481
OHT 2V	9	-0.5	44.35	7.5	22.5	10^6 (run-out)
OHT 3V	9	-0.5	66.52	11.25	33.75	45012
OHT 4V	9	-0.5	59.14	10	30	138167

Further ultrasonic ND analysis was implemented on specimens 2S, 4T and 5T with the purpose to analyse composite delamination evolution during fatigue life. The equipment used for conventional ultrasonic technique consist of a Control Unit OmniScan® MX and one flat Phased Array probe with 64 elements and a frequency of 2.25 MHz, which is optimized for investigating composite laminates with a total thickness of 3-8 mm [17-20] (Figure 4a-b). These ultrasonic measurements required to interrupt the fatigue test at regular intervals, to remove the specimen from the testing system and to record UT data of the S-scan maps (0°) at different probe positions (Figure 4b). Conventional ultrasonic technique presents the limitation of difficult development of real-time measurements on specimens due to induced cyclic deformation state of specimen during fatigue.

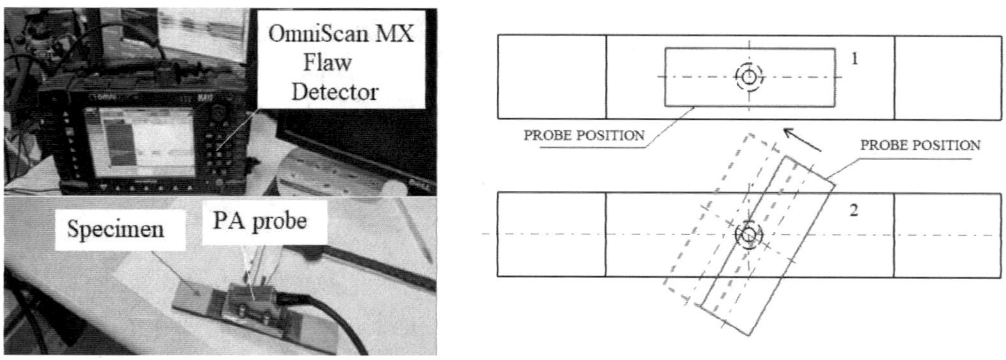

(a) (b)

Figure 4. (a) Experimental set-up during fatigue test for conventional Non-Destructive damage control.; (b) Probe location on the specimen's surface.

Particular attention was paid to the presence of secondary lobes in addition to the main ones displayed on the A-Scan during the specimen ultrasound scans.
These lobes are generated by the diffusion of the acoustic energy from the probe with angles different from the main path (Figure 5), they can be reflected by the surfaces of the element being inspected and be the cause of spurious indications in A-Scan or S-Scan analysis.

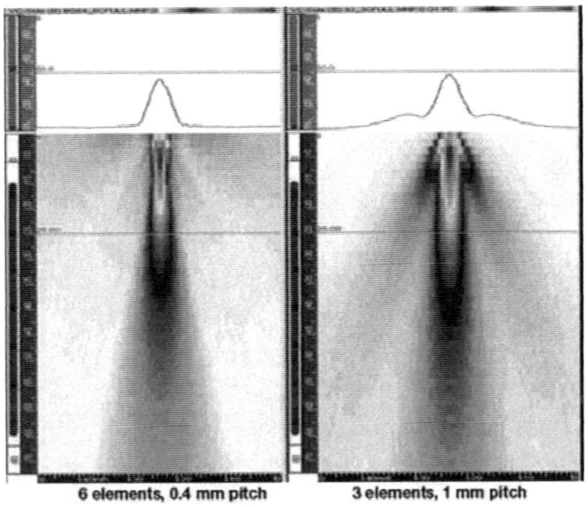

Figure 5. Example of main lobes and secondary lobes.

To identify the cause of the presence of the secondary lobes it is necessary to change the number of active elements of the probe, the pitch of the elements, the depth/distance of focus point and the position of the first and last active element (and various combinations of these parameters).

From the analysis of the A-Scans and S-Scans acquired the investigation was carried out keeping active all the elements of the Phased Array probe (16 x 64), a depth of focus equal to 2 mm and a pitch of 0.5 mm scan to obtain the best possible definition for the type of probe used.
Measurements were conducted by varying the beam opening angle by acting directly from the "focal law" section of the Phased Array instrumentation adopting, after appropriate checks inspection, to use a linear beam with at 0° also in virtue of the fact that possible defects were orthogonal to the surface being inspected.

3. Results and discussion

3.1. UT monitoring results – Analysis of damage
This section describes the ultrasound results performed on the different types of specimens subjected to fatigue. The amplitude of the received signal ΔA_{pp} and the fundamental frequency were selected to monitor the fatigue damage initiation and evolution. Amplitude changes in recorded signals are widely used to detect damage [8].

Before fatigue tests, the effects of the static load in tension-tension and tension-compression mode on the received signal were experimentally verified. The results show that the UT signal approximately linearly depending on the static load in the range of 0-12 kN and 2-16 kN in tension and compression test respectively (Figure 6a-b). The amplitude received signals at 300 kHz showed this linear dependence on the applied load (and then on the stress) both in traction and in compression load. This result is also confirmed in the literature [21].

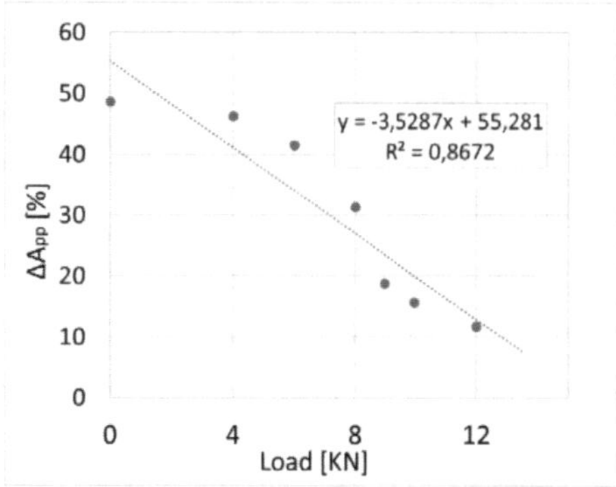

(a)

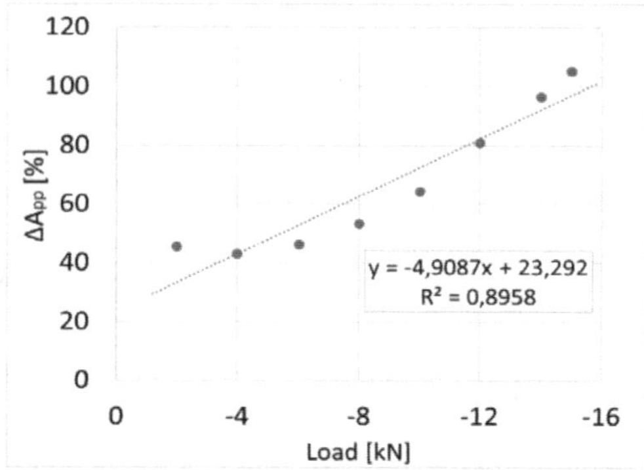

(b)

Figure 6. Effect of tension static load (a) and compression static load (b) on ultrasonic signal amplitude for specimens 4T and 2V.

Figure 7a shows an example of signals received in time domain at various fatigue steps for specimen 2S. The graph shows interesting variations in the amplitude of the UT signal which is reduced respect to baseline signal in intact specimen, due to continuous and early damage evolution. From Figure 7a, a time shift of the signal was observed at 960000 cycles. The raw data acquired in the time domain were subsequently converted into the frequency domain by means of an appropriate analysis routine based on the Fast Fourier Transform (FFT). Figure 7b shows the fundamental amplitude of the signal recorded at 300 kHz, highlighting an evident frequency attenuation of the reference signal of the undamaged specimen at 0 cycles with the progressive increase of fatigue cycles.

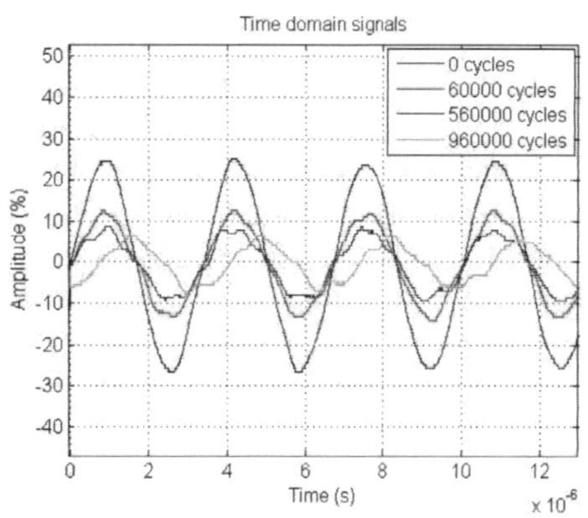

(a)

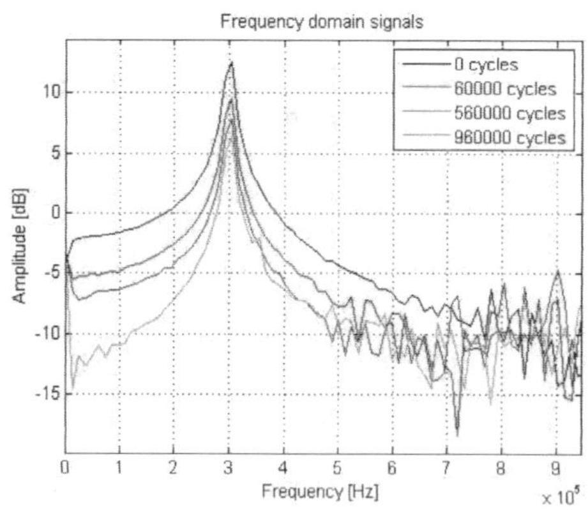

(b)

Figure 7. Example of received continuous sine guided waves at 300 kHz in time domain (a); (b) Zoom on fundamental frequency spectrum (FFT) in dB at various stage of fatigue for the specimen 2S.

By processing fatigue data, recorded with a frequency of 9 Hz, the stiffness of the tested specimens was determined by varying the number of load cycles. Figure 8 show the variation of the normalized stiffness with respect to its maximum value as the number of cycles varies, normalized with respect to the total of cycles (N / N_{tot}) both for the tests with R = - 0.5 and for those with R = 0.1.

From the general trend of the curves, a decrease in stiffness of 5-8% compared to the initial value can be observed. The specimens stressed with R = -0.5 showed a stiffness reduction of 12% for the 3V specimen and 20% in the case of the 4V. The 2V specimen shows a stabilized trend and only a slight reduction in the final phase before the interruption of the test at 10^6 cycles.

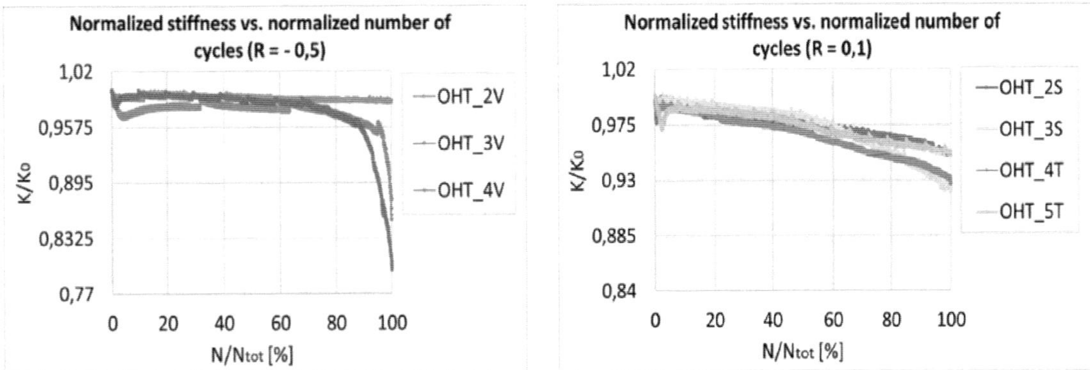

(a) (b)

Figure 8. Normalized stiffness against number of cycles, normalized respect to total cycles for each specimen tested with stress ratio R= - 0.5 (a) and R = 0.1 (b).

The peak-to-peak amplitude of the received signal was related to the stiffness of the material as the number of load cycles varies. To make a comparison between the data collected for the tested specimens, the UT signal of each of them was normalized respect to the reference signal relative at 0 cycles. From the general trend of the curves of Figures 9 and 10, a marked behaviour on the material is observed for the specimens stressed in tension-compression with R = -0.5. In particular, the specimens tested with the load levels indicated in Table 2 are damaged earlier and their fatigue life is reduced compared to those tested in tension-tension (R = 0.1).

Figure 9a shows a considerable sensitivity of the ultrasound signal to detect delaminations in the initial phases of fatigue affecting the first plies near the surface, on the side where the countersunk hole is present, while the stiffness slowly decreases with the progressive increase of the number of load cycles.

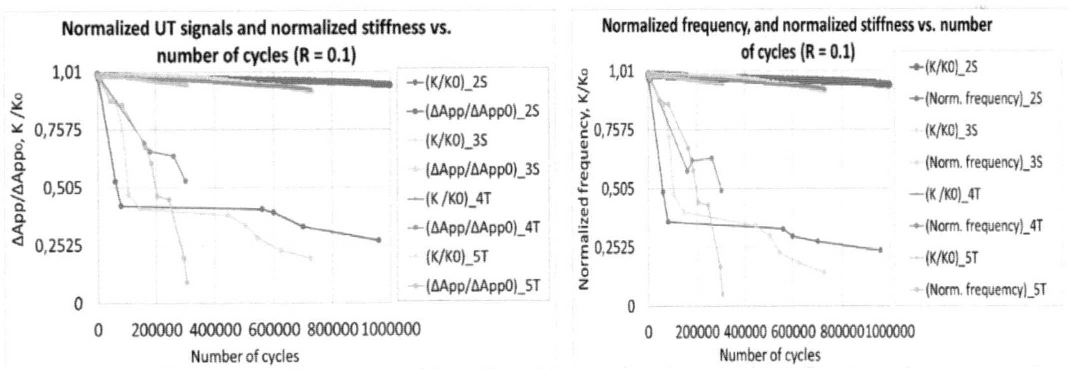

(a) (b)

Figure 9. Normalized peak-peak amplitude and stiffness vs. number of cycles (a); (b) normalized fundamental frequency and stiffness vs. number of cycles for all specimens during fatigue life with stress ratio R = 0.1.

The highest reduction of the signal with respect to the reference signal at 0 cycles was obtained for specimen 5T, which was subjected to load cycles with a maximum load of 63 kN. Due to the high level of the applied load the change of the signal started from the first cycles and were originated by the delamination phenomena on the first layers. Specifically, from 0% (0 cycles) to 26% (84000 cycles) of the fatigue life, the amplitude of the signal is reduced from 110.59% to 96%. Starting from approximately 50% of the fatigue life, the signal is reduced in amplitude (ΔA_{pp} = 75.29%) due to the progressive damage accumulation and delamination propagation. Subsequently, the amplitude is further reduced (ΔA_{pp} = 52%) at approximately 63% of the fatigue life. In the final stages, at 94% of the fatigue life, before the final failure of the specimen, the signal decreases rapidly (ΔA_{pp} = 10.2%); it is interesting to note that these relevant changes of the ultrasonic signal does not correspond to mechanical stiffness changes, which seems to vary with a quite uniform trend.

The specimen 3S, which fatigue cycles are characterized by a maximum force of 62 kN, showed a decrease in the amplitude of the UT signal from 0% (0 cycles) to approximately 20% (144500 cycles) of the fatigue life from 85.49% to 35.29% respectively, due to accumulation of damage in the laminate located around the hole on the countersunk side. At 60% of the fatigue life, the amplitude of the signal is reduced (ΔA_{pp} = 39.93%). Subsequently, starting at approximately 74% of the fatigue life (544500 cycles), the signal rapidly reduces from 24.31% to 19.6% before final failure. The specimen 4T, which fatigue cycles are characterized by a maximum load level of 61 kN, presents an initial amplitude reduction of the recorded signal from 101.17% to 0% of the fatigue life (0 cycles) to 70.6% at 22% of the life (160000 cycles). Starting from about 35% of the fatigue life (260000 cycles) up to 41% of the life, the amplitude is rapidly reduced from 65% to 54.1%. This quickly reduction of the signal is consistent with the experimental reduction of the stiffness of the sample due to the fatigue progressive damage.

The specimen 2S, unlike the other specimens tested with R = 0.1, did not show a final failure due to the low level of load applied (F_{max} = 59 kN). However, delaminations were observed in the first plies from the side of the countersunk hole which obviously did not affect the entire thickness of the laminate. The amplitude of the recorded signal is reduced from 51.7% at 0 cycles to 22% at 80000 cycles. Subsequently it presents a stable trend up to 560000 cycles (ΔA_{pp} = 21.2%) (Figure 9a). In the final phase, before the interruption of the test occurred at 10^6 cycles, the UT signal is rapidly reduced to 17.2% at 700000 cycles, in accordance with a more rapid decrease of the stiffness.

A similar behaviour was observed also considering the trend of the fundamental frequency, normalized with respect to its initial value, as the number of cycles varies (Figure 9b). The comparison shown in the graph allows to observe, also in this case, that the attenuation of the fundamental frequency, such as the amplitude of the signal, is higher than the decrease in the stiffness of the material; ultrasonic measurements with Lamb waves are therefore a very sensitive tool for damage monitoring. The greater attenuation of the UT signal and the fundamental frequency with respect to the reduction of stiffness is due to delaminations that occur during the first phases of fatigue, which affect approximately a thickness of 3-4 mm of the total thickness of the specimen, and this determines a lesser influence on the decrease in total stiffness of the sample.

Figure 10a shows the results of the specimens subjected to tension-compression load (R = - 0.5).

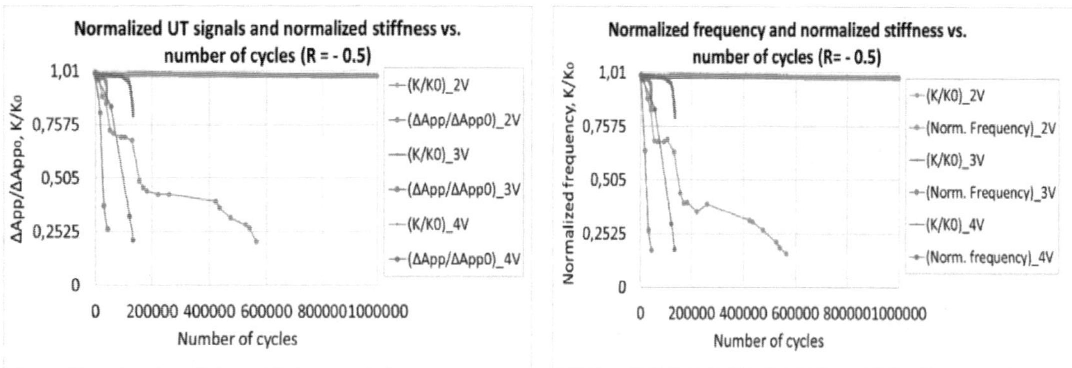

(a) (b)

Figure 10. Normalized peak-to-peak amplitude and stiffness vs. number of cycles (a); (b) normalized fundamental frequency and stiffness vs. number of cycles for all specimens during fatigue life with stress ratio R = - 0.5.

By the graph, it is possible to observe that the amplitude of the sine wave packet decreases rapidly for specimen 3V, subjected to a maximum load of 45 kN compared to the other two tested specimens 2V and 4V, due to delaminations that occur starting from the first plies near the countersunk hole and which subsequently affect an increasingly larger area until the failure. Specifically, from 0% (0 cycles) to 38% of the fatigue life (170000 cycles), the amplitude of the signal is reduced from 50.19% to 40.78%. Subsequently, at about half of the useful life of the sample (66% of the fatigue life), the detected amplitude of the signal is equal to 18.82%, reduced by 47.18% compared to the previous measurement. At 95% of the fatigue life (43000 cycles), the amplitude is further reduced before the final failure, in accordance with the rapid decrease in the stiffness of the sample.

The specimen 4V, stressed with a lower load (F_{max} = 40 kN), has a lower decay of the signal amplitude associated with the damage which is less than specimen 3V. The amplitude of the signal is reduced from 0% (0 cycles) to 40.53% (56000 cycles) from 84.7% to 71.37% respectively. Subsequently, at 87% of the useful life (120,000 cycles), the amplitude is reduced to 27.44%. In the final phase of failure (97% of the fatigue life), the signal is significantly reduced in amplitude (ΔA_{pp} = 18.03%) due to accumulation of damage with the number of load cycles. This result is coherent with the rapid decay of the stiffness of the material.

Specimen 2V did not show a final rupture. However, the sample was damaged and the delaminations observed affected the first plies near the countersunk hole. The amplitude of the signal decreases from 0 cycles (ΔA_{pp} = 49.4%) to 156000 cycles (ΔA_{pp} = 24.31%). Subsequently a reduction is observed to 221000 cycles (ΔA_{pp} = 21.7%) and finally, at 556000 cycles, before the interruption of the test occurred at 10^6 cycles, the amplitude decreases further (ΔA_{pp} = 10.19%) caused by large delaminated area. The reduction in stiffness was found to have little effect on specimen damage. The same behaviour was observed by plotting the fundamental frequency versus fatigue cycles (Figure 10b).

To achieve a quantitative evaluation of the progressive fatigue damage, in which ΔA_{pp} is the peak-peak amplitude of the signal recorded in the different phases of the test and ΔA_{pp0} is the amplitude of the signal referred to the intact specimen, a Damage Indicator (DI) was defined by the following equation:

$$DI = \left| \frac{\Delta A_{pp} - \Delta A_{pp_0}}{\Delta A_{pp_0}} \right| \quad (1)$$

Figure 11a-b show the trends of the damage indicator for R = 0.1 and R = - 0.5 of the tested specimens as a function of the number of cycles. From the graph in Figure 11a, the specimen 5T, stressed with a maximum load of 63.5 kN, has a damage index that rapidly increases to about 51% of the fatigue life, compared to the increase in damage observed at 22% of the useful life in the specimen 4T, stressed with a maximum load of 61 kN.

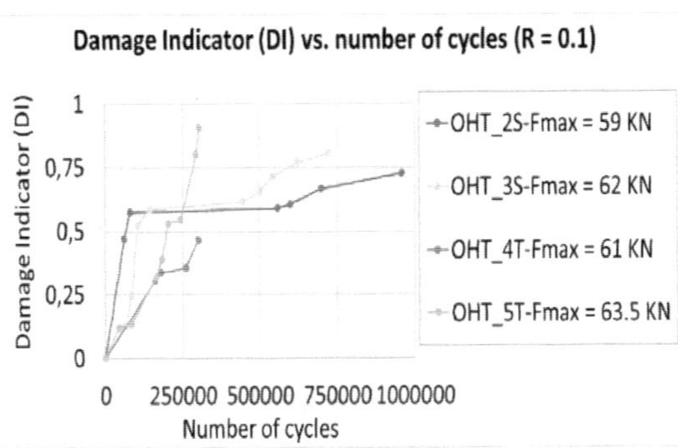

(a)

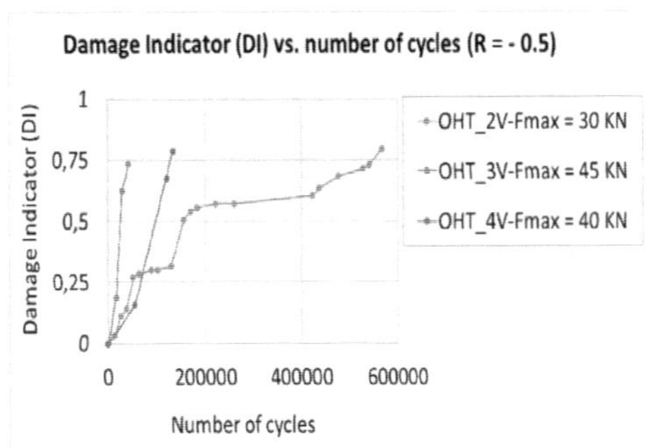

(b)

Figure 11. Trend of Damage Indicator (DI) for specimens with stress ratio R = 0.1 (a) and R = - 0.5 (b) against number of cycles.

For the specimens 2S and 3S, the damage indices increase most rapidly in the initial stages of the test up to approximately 20% of the fatigue life for the 3S specimen and approximately 10% for the 2S specimen. Subsequently, they show an almost constant trend with the number of cycles and then further increase to about 60% of the fatigue life with a greater increase observed in 3S specimen compared to 2S.

From general trend of the curves relating to the samples stressed in tension-compression (Figure 11b), a higher increase in the damage index is observed as the applied load increases, as the number of load cycles varies.

Figure 12a show an example of damaged section of specimen 2S with clearly visible delaminations after fatigue test observed with the stereo microscope, while Figure 12b shows an example of delamination during the test for the specimen 2V.

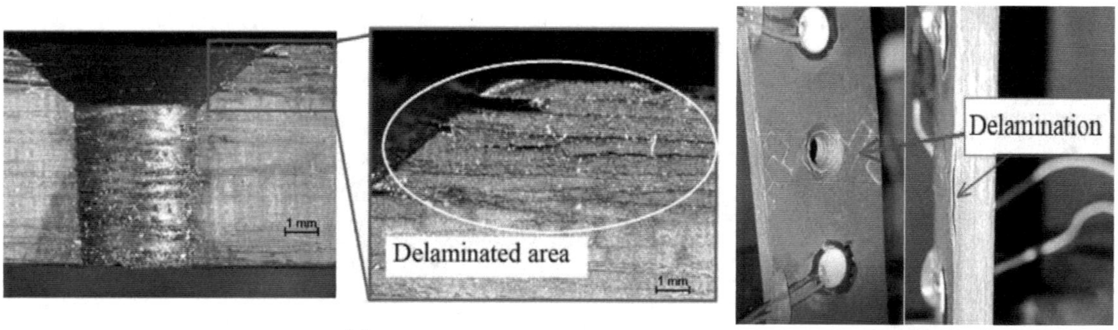

(a) (b)

Figure 12. (a) Example of damaged section analysis observed by stereo microscope after 10^6 fatigue cycles for specimen 2S; (b) Example of delamination during fatigue test for specimen 2V.

In order to confirm the reliability of the technique, authors monitored at regular intervals the specimens 2S, 4T and 5T with conventional UT control acquisitions. Figure 13 shows an example of delaminations for the specimen 5T, according to probe position 1 on the specimen surface (Figure 4b), involving the first plies from the side of the countersunk hole in accordance with what was detected in real-time with the PZT sensors by the degradation of the signal during the fatigue test.

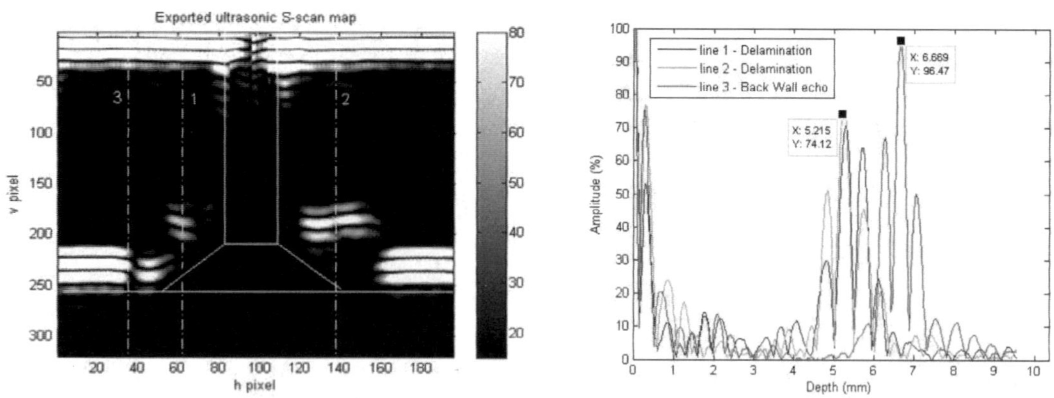

(a) (b)

Figure 13. (a) Example of delamination in S-scan map (0°); (b) amplitude profiles along directions 1, 2 and 3 for the specimen 5T.

From figure 13a, it is evident that the extent of the damage showed in the grey scale image is higher on the right side of the S-scan than on the left. The amplitude of the ultrasonic signal relating to line 1 is less high than that line 2, due to the greater extension of the delaminated area (see figure 13b).

Before fatigue test, UT scans was performed on the intact specimen 5T. Figure 14 shows the S-scan map (0°) where in the upper part the signal due to surface of the specimen from the no-countersunk side is observed, instead in the lower part, the signal of the back wall echo of the surface where is evident the countersunk hole. Between the signals of the two surfaces, in the thickness of the specimen, no pre-existing defect is observed.

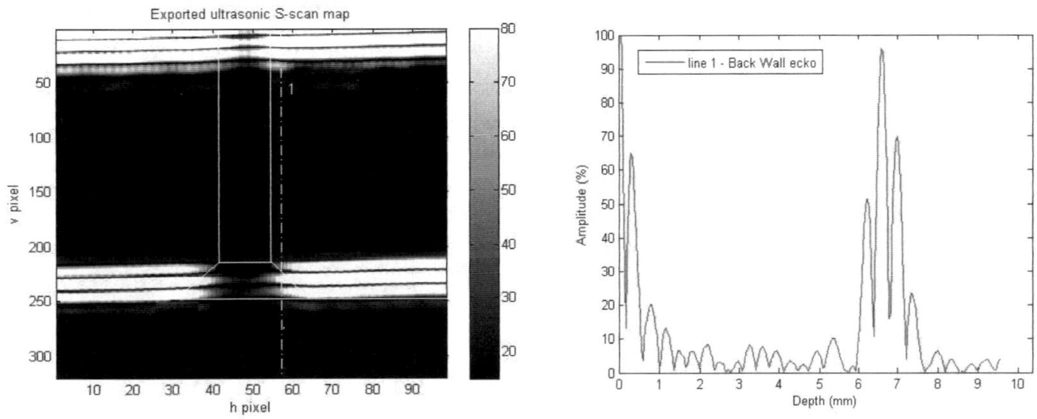

(a) (b)

Figure 14. (a) Example of S-scan map (0°) on un-damaged specimen before fatigue test; (b) amplitude profile along direction 1 for specimen 5T.

4. Conclusions

In this work, an experimental procedure for the progressive monitoring of fatigue damage of CFRP elements, based on the ultrasonic propagation of Lamb waves, was applied to monitor in real-time the onset and the progressive evolution of the damage up to failure. The acquired ultrasound signal was compared with the reference signal at the beginning of the fatigue test. From the measurements carried out on the batch of tested specimens, the attenuation of the wave packet amplitude of the received signal and of the fundamental frequency are much more sensitive than the reduction of stiffness. This result was explained by the fact that the Lamb wave received, in the pitch-catch mode configuration, can be sensitive to detect the first delaminations occurring on the side of the countersunk hole and affecting the first plies. This determines a signal degradation already from the initial stages of the fatigue life, while the stiffness is reduced very slowly with the number of fatigue cycles until the swiftly failure, due to the considerable thickness of the sample. The comparison with conventional ND monitoring methods were employed for damage analysis during the tests. The results obtained are satisfactory, though the first phases of damage affecting the first plies are not easily detectable because of noise and other factors.

In conclusion, with reference to the parameters studied in this work, the results demonstrated the effective potential of the technique adopted for real-time detection of the delamination onset at the interface and its evolution to predict fatigue damage.

Acknowledgements

The authors would like to thank the student Francesco Marzo for his contribution in the UT research carried out during the academic internship at the Experimental Mechanical laboratory of University of Salento – Department of Engineering for Innovation.

References

[1] Skorupa M, Machniewicz T, Skorupa A, et al. 2017 Fatigue life predictions for riveted lap joints *Int J Fatigue* **94** 41–57
[2] Grilo T J, Paulo R M F, Silva C R M, Davim J P 2013 Experimental delamination analyses of CFRPs using different drill geometries *Compos. Part B Eng.* **45** 1344–1350
[3] Palit Sagar S, Das S, Parida N, Bhattacharya D K 2006 Non-linear ultrasonic technique to assess fatigue damage in structural steel *Scripta Materalia* **55 (2)** 199-202
[4] Ciampa F, Meo M Impact detection in anisotropic materials using a time reversal approach 2012 *Structural Health Monitoring* **11(1)** 43–49
[5] Reis P, Ferreira J, Richardson M 2011 Fatigue damage characterization by NDT in polypropylene/glass fibre composites *Appl. Compos. Mater.* **18** 409–419
[6] Shin C S, Chiang C C 2006 Fatigue damage monitoring in polymetric composites using multiple fiber bragg gratings *Int. J. Fatigue* **28** 1315–1321
[7] Ling Y, Mahadevan S 2012 Integration of structural health monitoring and fatigue damage prognosis *Mech. Syst. Signal Process.* **28** 89–104
[8] Omari M A, Sevostianov I 2013 Estimation of changes in the mechanical properties of stain-less steel subjected to fatigue loading via electrical resistance monitoring *International Journal of Engineering Science* **65** 40-45
[9] Dattoma V, Nobile R, Panella F W, Saponaro A 2019 Real-time monitoring of damage evolution by nonlinear ultrasonic technique *Procedia Structural Integrity* **24** 583-592
[10] Nobile R, Saponaro A 2020 In-situ measurements of fatigue damage evolution by electrical resistance method *Procedia Structural Integrity* **28** 1321-1328
[11] Park J-M, Lee S-I, DeVries K L 2006 Nondestructive sensing evaluation of surface modified single-carbon fiber reinforced epoxy composites by electrical resistivity measurement *Compos Part B Eng.* **37** 612-626
[12] Nobile R, Saponaro A 2021 Real-time monitoring of fatigue damage by electrical resistance change method *International Journal of Fatigue* **151** 106404
[13] Giurgiutiu V 2014 Structural health monitoring with piezoelectric wafer active sensors *Elsevier Academic Press* New York 2nd edition
[14 Diamanti K, Soutis C 2010 Structural health monitoring techniques for aircraft composite structures *Progress in Aerospace Sciences* **46** 342–352
[15] Kessler S, Spearing S, Soutis C 2002 Damage detection in composite materials using lamb wave methods *Smart materials and structures* **11(2)** 269
[16] Michaels J E 2008 Detection, localization and characterization of damage in plates with an in-situ array of spatially distributed ultrasonic sensors *Smart Mater. Struct.* **17** 035035.
[17] Antonazzo R 2015 Ottimizzazione controlli ND ultrasonori per elementi in CFRP *University of Salento Master thesis*
[18] Dattoma V, Nobile R, Palano F, Panella F W, Pirinu A, Saponaro A 2021 Ultrasonic and thermographic fatigue monitoring on a full-scale CFRP aeronautical component after repairing *IOP Conf. Ser.: Mater. Sci. Eng.* **1038** 012027
[19] Marzo F 2021 Analisi e monitoraggio del danneggiamento a fatica di provini CFRP aeronautici *University of Salento* Master thesis
[20] Palumbo D, De Finis R, Saponaro A, Nobile R, Panella F, Galietti U 2020 Study of Damage Behavior of T-Joint Components by Means of Different Non-destructive Techniques Proc. SPIE Thermosense: Thermal Infrared Applications XLII **11409**
[21] Yan X, Qiu L, Yuan S 2017 Experimental study on the effect of propagation characteristics of guided waves under uniaxial static stress 9th *International Symposium on NDT in Aerospace* 1-7

IOP Conf. Series: Materials Science and Engineering 1214 (2022) 012012 doi:10.1088/1757-899X/1214/1/012012

Performance comparison between electromechanical and electro-hydrostatic regenerative shock absorbers

R Galluzzi[1], S Circosta[2], N Amati[2] and A Tonoli[2]

[1] School of Engineering and Sciences, Tecnologico de Monterrey, Mexico City 14380, Mexico
[2] Department of Mechanical and Aerospace Engineering, Center for Automotive Research and Sustainable Mobility@PoliTO, Politecnico di Torino, 10129 Turin, Italy

E-mail: `renato.galluzzi@tec.mx`

Abstract. Controllable damping solutions are increasingly finding their way into passenger cars and other vehicles. They improve ride quality by enhancing comfort and handling performance. Semi-active solutions based on controlled valves or variable-rheology fluids are by far the most common technologies in the automotive industry. However, regenerative alternatives that employ electric machines can be seen as a desirable evolution of the suspension market. In this context, the present paper studies two regenerative solutions: electro-hydrostatic and rotary electromechanical dampers. Transmission interfaces are particularly relevant in these technologies. Adequate choices are needed to guarantee proper packaging, performance, structural integrity, and high efficiency. This work compares both technologies in terms of design and packaging. Furthermore, experimental results are analysed to understand the performance of damping solutions. It is demonstrated that the rotary electromechanical device outperforms its electro-hydrostatic counterpart in important features, such as conversion efficiency, envelope, mass and dynamic behaviour.

1. Introduction

Current trends in the automotive field foresee a strong push towards the electrification of power-train and chassis devices. This change has been driven by the search of more sustainable energies, and the inclusion of smart features. The so-called *x-by-wire* technologies are finding their implementation through the addition of numerous mechatronic solutions.

In ground vehicles, smart actuators have been proposed to control different parameters of the vehicle suspension. Since the seminal work proposed by Karnopp [1], damping has been regulated predominantly through semi-active means using solenoid valves or fluid rheology, thus allowing comfort or handling enhancement. However, recent efforts have proposed the use of electromagnetic damping solutions. They exploit an electric machine—linear or rotary—to control suspension forces. From an energy standpoint, actuators of this type are intrinsically bidirectional. This means that electric machines can absorb or exert power. Hence, they provide active or damping forces to the suspension. Furthermore, when damping, they offer regenerative capability, i.e. vibrational energy can be partially transformed into electricity. Active suspensions are particularly attractive for the autonomous vehicle paradigm, as they provide improved vibration suppression for the chassis, so passengers can focus on secondary activities comfortably while travelling.

The aforementioned benefits have led to numerous research efforts in the field of regenerative damping solutions [2]. The use of electric machines brings the benefit of linearity, simple, well-known dynamics and improved controllability with respect to other technologies. However, a key challenge of regenerative solution consists in identifying and implementing a proper transmission system to interface the electric machine with the vehicle suspension.

Although linear motors [3] fit well with the natural stroke of conventional shock absorbers, their limited force-to-mass ratio makes them unsuitable for many vehicles. Thereby, linear-to-rotary conversion is needed. In this regard, numerous attempts have been made with different mechanisms: ball screw [4, 5], rack-pinion [6] and rotary with lever [7, 8, 9]. In an attempt to inherit the use of fluid from passive solutions, electro-hydrostatic actuators with hydraulic pistons and fixed-displacement pumps have been also proposed for this task [10, 11].

Despite the numerous efforts, there is no clear indication as to which trend to follow: electro-hydrostatic or electromechanical. Comparison attempts have been carried out to identify the most suitable technology. However, many of these technologies are developed *ad hoc* for specific applications, thus hindering direct, one-to-one technology comparisons.

To address this shortcoming, the present paper provides two detailed design alternatives for specific regenerative suspension requirements. Specifically, electro-hydrostatic (EHA) and rotary electromechanical (REM) versions of a regenerative shock absorber are designed, built and tested. Design aspects for each technology have been already discussed in previous works [7, 10]. Nevertheless, the aim of this paper is to present them in a comparative manner in terms of design choices, prototype features and their experimental response.

This work is organised as follows. Section 2 describes the damping requirements. Subsequently, section 3 outlines the design for the offered damping solutions to reach the final layout of both actuators. Then, section 4 presents experimental results. In section 5, solutions are compared. Finally, section 6 concludes the work.

2. Requirements

For the present comparative analysis, a B-class vehicle is taken as a reference. In the suspension, a maximum-damping piecewise characteristic is desirable (see figure 1). Thus, a first damping coefficient $c_1 = 10\,\mathrm{kNs/m}$ is needed within damping speeds $|v_d| \leq 0.1\,\mathrm{m/s}$. Above $0.1\,\mathrm{m/s}$, significantly lower coefficients are expected: $c_2 = 0.5\,\mathrm{kNs/m}$. Besides damping, maximum damping force ($F_{d,max} = 2\,\mathrm{kN}$) and speed ($v_{d,max} = 2\,\mathrm{m/s}$) are relevant design specifications.

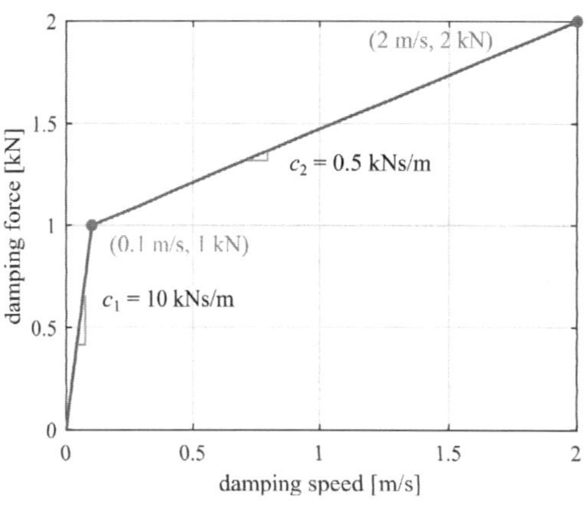

Figure 1. Target maximum damping specification.

Although the damping characteristic already highlights important design parameters, simulations are needed to understand the expected duty cycle of the shock absorber. This is particularly relevant to size the electric machine in regenerative solutions. To this end, vertical dynamics can be analysed through a quarter-car model (figure 2).

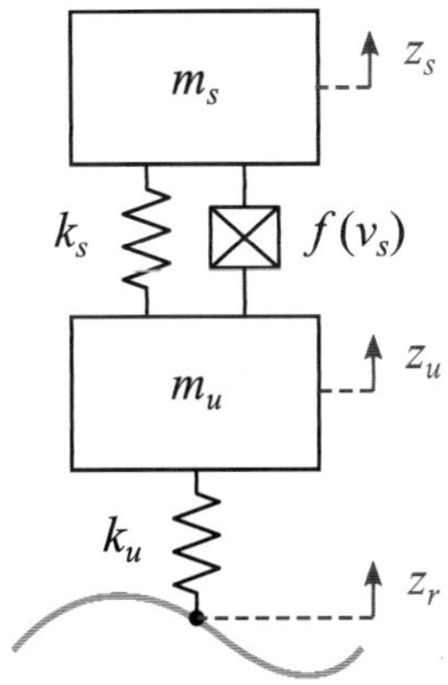

Figure 2. Quarter-car model to evaluate damper duty cycle.

The chassis of the vehicle (sprung mass m_s) is connected to the wheel hub (unsprung mass m_u) by means of a suspension spring (stiffness k_s) and shock absorber (piecewise damping $f(v_d)$). In turn the wheel hub receives the road unevenness vibrations through the tyre (stiffness k_u).

To reproduce realistic damping load profiles, the model in figure 2 must be simulated with the damping characteristic curve presented in figure 1 and a realistic road profile for $z_r(t)$. These conditions lead to a root-mean-square (RMS) damping force $F_{d,rms} \cong 504\,\mathrm{N}$ when the parameters in table 1 are used.

Table 1. Quarter-car model parameters.

Description	Symbol	Value	Unit
Sprung mass	m_s	417	kg
Unsprung mass	m_u	40	kg
Suspension stiffness	k_s	23	$\mathrm{kN/m}$
Tyre stiffness	k_u	226	$\mathrm{kN/m}$
Vehicle cruise speed	v_{car}	70	$\mathrm{km/h}$
Road roughness index (B class, ISO 8608)	G_r	$6.4 \cdot 10^{-7}$	$\mathrm{m \cdot cycle}$

3. Design

3.1. Overview

The aim of the design is to size the transmission and the electric machine of the actuator. The transformation from the rotary domain to the linear one is obtained by means of the transmission ratio τ_t $[\mathrm{m/rad}]$. In an EHA solution, this relation is ideally attained by closing a hydrostatic circuit with a pump unit (fixed displacement D_p) and a hydraulic cylinder (piston cross section A_p), thus

$$\tau_t = \frac{D_p}{A_p} \tag{1}$$

Likewise, for a REM solution, this transformation is achieved through a linkage with transmission ratio τ_l $[\mathrm{m/rad}]$. Usually, an additional gearbox multiplier (τ_g) is required to achieve the target performance:

$$\tau_t \cong \tau_l \tau_g \tag{2}$$

Disregarding the type of actuator, the definition of the transmission ratio allows designing the electric machine. From a practical standpoint, a feasible transmission system able to guarantee τ_t is needed. This process is usually iterative, as a high degree of integration among subsystems is desirable.

3.2. Electric machine

The design of the electric machine must maximise the machine output torque (T_m), while satisfying working and geometric constraints. To simplify this step, a candidate machine can be designed either by analytic means or through the finite-element method. To match a desired envelope for the actuator, the stator outside radius r_{so} must be constrained. A design procedure in two dimensions yields the torque per unit length r_{Tl} $[\mathrm{Nm/m}]$. Subsequently, the machine active length l_a can be adjusted to match the desired output torque.

The design of electric machinery is a research area *per se* and thus, is not covered thoroughly in this work. Nevertheless, some considerations can be followed for compact actuators. In particular, a three-phase permanent-magnet brushless machine is proposed to maximise torque density. Concentrated winding solutions are preferred due to their lower Joule losses, shorter end turns and simpler winding layouts. For the target machine size, a configuration with 12 slots and 10 magnets is selected. The machine geometry is optimised to yield maximum torque while avoiding saturations on the magnetic circuit. The obtained machine design is shown in figure 3 and its parameters are listed in table 2.

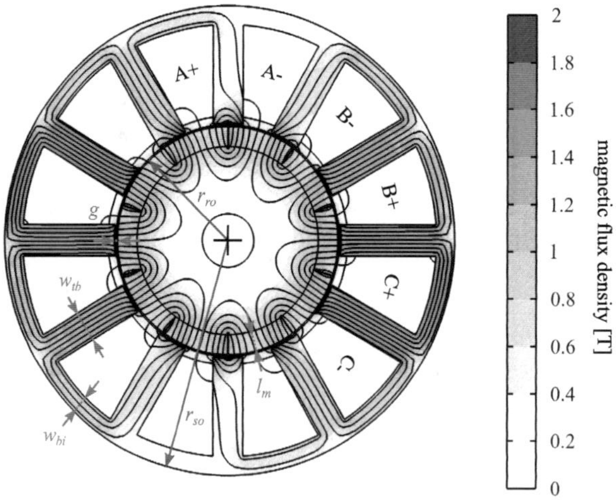

Figure 3. Electric machine cross section. The colour map indicates the flux density norm distribution, whereas the contour lines denote the out-of-plane magnetic vector potential. Letters in the slots indicate the winding layout for the three phases (A,B,C). Dimensions are reported in table 2.

Once that the machine has been designed in two dimensions, an effective approach should size the transmission ratio and the output torque of the machine, given that, in static conditions

$$F_d = \frac{T_m}{\tau_t} \tag{3}$$

$$\omega_m = \frac{v_d}{\tau_t} \tag{4}$$

Table 2. Electric machine parameters.

Description	Symbol	Value	Unit
Phasc rcsistance	R	20	$\mathrm{m\Omega}$
Inductance (d and q axes)	L_{dq}	80	$\mu\mathrm{H}$
Phase back-EMF constant	K_e	23	$\mathrm{mVs/rad}$
Nominal torque	T_n	0.7	Nm
Maximum torque	T_{max}	2.3	Nm
DC link voltage	V_{dc}	48	V
Base speed	ω_n	10	krpm
Maximum generator efficiency	η_g	0.88	$-$
Stator outside radius	r_{so}	35	mm
Rotor outside radius	r_{ro}	14	mm
Stator back iron width	w_{bi}	2.8	mm
Stator tooth base width	w_{tb}	4.7	mm
Air gap	g	0.5	mm
Permanent magnet length	l_m	3	mm
Active length	l_a	22	mm

Thus, assuming constant electric machine torque, the damping force can be maximised by reducing τ_t as much as possible. However, this poses a bi-fold effect, as it would lead to very high angular speed (ω_m) given a fixed damping speed. Moreover, in linear-to-rotary transmissions, the inertia of rotating parts J_m and viscous damping c_m are transported to the linear domain as

$$c_{eq} = \frac{c_m}{\tau_t^2} \tag{5}$$

$$m_{eq} = \frac{J_m}{\tau_t^2} \tag{6}$$

These equivalent parameters exhibit a quadratic inverse relation with the transmission ratio.
A proper selection of τ_t and l_a depends on specific criteria, as listed below.

3.2.1. Continuous operation. For machines cooled by natural air convection, continuous operation is attained with an RMS wire current density of $6\,\mathrm{A/mm^2}$ [12]. In this case, the motor should be able to guarantee the RMS damping force obtained from the simulations in section 2 ($F_{d,rms} \cong 504\,\mathrm{N}$):

$$F_{d,rms} = \frac{T_{m,cont}}{\tau_t} = r_{Tl,cont}\frac{l_a}{\tau_t} \tag{7}$$

3.2.2. Impulsive operation. In transient conditions, air-cooled machines can attain $20\,\mathrm{A/mm^2}$ of current density for few seconds. The machine should comply with maximum force requirements in this condition ($F_{d,max} = 2\,\mathrm{kN}$):

$$F_{d,max} = \frac{T_{m,max}}{\tau_t} = r_{Tl,max}\frac{l_a}{\tau_t} \tag{8}$$

3.2.3. Damping coefficient. For electromagnetic devices, the maximum viscous damping is attained in short circuit and it is inversely proportional to the phase resistance:

$$c_{eq,max} = \frac{3K_e^2}{2\tau_t^2 R} \tag{9}$$

where K_e is the motor back-electromotive force (EMF) constant and the phase resistance is given by

$$R \cong \frac{2N_c N_t \rho_{Cu}}{A_w} l_a \tag{10}$$

being N_c the number of series coils per phase, N_t the number of turns per coil, A_w the wire cross section and ρ_{Cu} the resistivity of copper.

The combination of the most conservative force constraint (equation 7 or 8) with the maximum damping requirement (expressions 9 and 10) lead to an optimal choice of motor active length l_a and total transmission ratio τ_t. For the present study, a transmission ratio $\tau_t = 1.3\,\mathrm{mm/rad}$ and an active length $l_a = 22\,\mathrm{mm}$ are determined.

3.3. Motion transmission

Up to this point, both solutions share the same electric machine and total transmission ratio. However, there are significant differences regarding how τ_t is achieved.

3.3.1. Electro-hydrostatic transmission. This transmission is achieved by coupling a hydraulic cylinder with a pump. To reuse the passive shock absorber tube and piston, a cross section $A_p = 4.2\,\mathrm{cm}^2$ is preserved. According to equation 1, a fixed volumetric displacement $D_p = 3.6\,\mathrm{cm}^3/\mathrm{rev}$ is required to match the target transmission ratio.

To favour low friction and extended lifespan, gerotor pump is selected for the task. A seven-chamber device is designed with trochoidal gear profiles and maximum eccentricity to minimise its envelope. With a pressure range up to $60\,\mathrm{bar}$, gerotor clearances are tuned to fulfil a suitable trade-off between leakages and mechanical losses [13]:

- Total axial clearance: $20\,\mu\mathrm{m}$
- Outer gear outside diameter clearance: $80\,\mu\mathrm{m}$
- Inner gear inside diameter clearance: $20\,\mu\mathrm{m}$
- Tooth tip clearance: $40\,\mu\mathrm{m}$

The EHA solution depicted in figure 4 packs a motor (4,7) and a pump (1,2) within the same pressure-tight oil medium. The inner gear (1) of the pump is directly coupled to the electric motor shaft (10). The outer gear (2) is supported hydrostatically by a journal bearing machined on the casing (9). The rotor of the unit is supported by two ball bearings (3) in a face-to-face arrangement. The unit is coupled to the chambers of the damper by means of a hydraulic manifold (11) welded onto the shock absorber tube. The rotor position is measured by means of a Hall sensor array (7) that measures the magnetic field of permanent magnets on one end of the rotor shaft. The back of the motor cover (8) presents four sealing cable glands (5) that are used to access power and signal wirings. A drain cap (6) is used to ease oil filling.

3.3.2. Lever and multiplier transmission. In this case, the total transmission ratio τ_t is split between a linkage (τ_l) and a gearbox multiplier (τ_g). Since the lever system is mainly conceived for new suspension architectures, it is defined in an initial step. In the present study, we refer to a four-bar kinematic assembly in which the end rod is fixed onto the passive shock absorber tube. The latter is preserved for structural

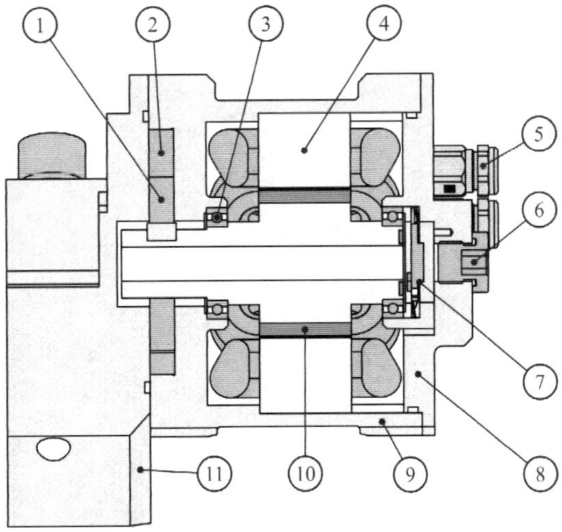

Figure 4. Electro-hydrostatic damper actuation unit. (1) gerotor inner gear; (2) gerotor outer gear; (3) motor ball bearing [×2]; (4) stator assembly; (5) cable gland [×4]; (6) drain cap; (7) position sensor; (8) motor cover; (9) casing; (10) rotor assembly; (11) manifold.

purposes only. In the future, the shock absorber tube will be removed and the rod end will be connected to other suspension parts. In this layout, a linkage with $\tau_l = 115\,\mathrm{mm/rad}$ is obtained.

To fulfil $\tau_t = 1.3\,\mathrm{mm/rad}$, a two-stage planetary multiplier is designed. Each stage presents a ratio of $1/9.375$, thus leading to a total ratio of $\tau_g = 1/87.89$. Gear profiles are carefully defined to match load and fatigue considerations from the ISO 6336 method B standards.

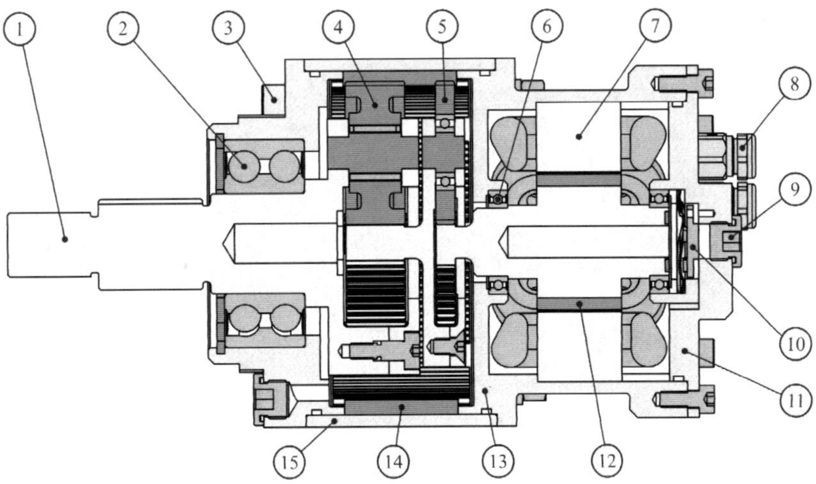

Figure 5. Rotary electromechanical damper actuation unit. (1) input shaft; (2) input bearing; (3) gearbox cover; (4) first stage assembly; (5) second stage assembly; (6) motor ball bearing [×2]; (7) stator assembly; (8) cable gland [×4]; (9) drain cap [×2]; (10) position sensor; (11) motor cover; (12) rotor assembly; (13) motor casing; (14) outer ring gear; (15) gearbox casing.

The final assembly embodies the same electric machine (7,12) as the EHA counterpart. In contrast, the rotor tip presents a sun gear profile that interfaces directly to the second stage (5) of the multiplier (2). An input splined shaft (1) is connected to the linkage. Axial and radial forces are supported by a double-row angular contact ball bearing arrangement (2). The same input shaft serves as a carrier of the first stage (4). An outer ring gear (14) is shared by both stages in a gearbox casing (15). Gearbox components are axially constrained by a gearbox cover (3) and a motor casing (13). As the EHA unit, the motor side presents a motor cover (11), a feedback sensor (10), ball bearings for the rotor shaft (6) and cable glands (8). Drain caps (9) help in this case to introduce gearbox lubricants.

Final versions of both prototypes are shown in figures 6 and 7.

Figure 6. Picture of the electro-hydrostatic damper prototype installed on a shock absorber.

Figure 7. Picture of the rotary electromechanical damper prototype with its linkage.

4. Experimental characterisation

The goal of the experimental characterisation is to understand the power contributions involved in multiple operating points and define the efficiency behaviour of the regenerative damper prototypes. This allows for a proper comparison of both technologies.

4.1. Test procedure

During testing, prototype inputs must be driven by means of a physical flow variable, while they impose a damping torque. To this end, their electric machine is current-controlled with field-oriented techniques using custom power and logic hardware. Specifically, the motors are characterised in the regenerative (damping) quadrant with a quadrature current ramp profile $I_q(t)$ increasing from zero to $25\,\mathrm{A}$ at a rate of $1\,\mathrm{A/s}$. This profile allows spanning multiple operating points in regeneration.

To drive the EHA unit, an additional motor (Kollmorgen™ DBL5) and a pump (Parker™ MGG2) are directly coupled to the prototype pump by means of a manifold and two circuit lines, as seen in figure 8. The driving motor imposes steady angular speed ω_d, which in turn becomes a constant flow rate on the hydraulic ports of the characterised device through the driving pump (fixed displacement of $11.43\,\mathrm{cm^3/rev}$). During testing, the hydraulic power is calculated by means of the pressure drop Δp_h measured by two GEFRAN™ TK sensors, and flow rate Q_h sensed by a KEM™ HM11 meter. The

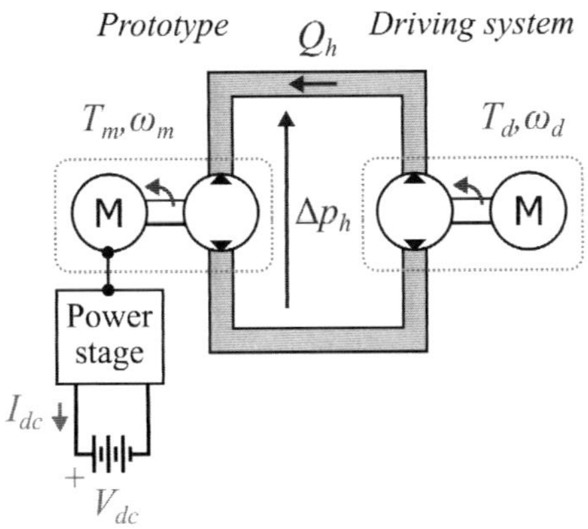

Figure 8. Scheme of the electro-hydrostatic damper prototype test bed.

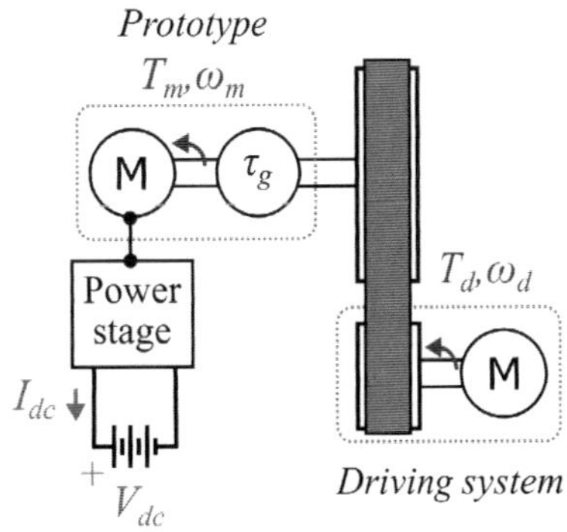

Figure 9. Scheme of the rotary electromechanical damper prototype test bed.

mechanical power is obtained from the prototype speed feedback (ω_m) and the torque estimated from the current ($T_m \cong 3K_eI_q/2$). Finally, the regenerated power is obtained from the DC link battery voltage (V_{dc}) and current (I_{dc}) signals.

Similarly, the REM unit is tested by means of the belt-pulley system schematised in figure 9. The transmission is driven by an electric machine (Kollmorgen™ DBL5), which again imposes constant angular speed ω_d through a pulley ratio of 2. The input mechanical power is obtained from the driving motor torque (T_d) and speed (ω_d), while the regenerated power is measured from DC link variables.

In both test setups, signals were synchronously acquired using an LMS™ SCADAS system at a sample rate of 12.8 kHz.

4.2. Results

Power contributions and conversion efficiency results are reported in figures 10 and 11 for EHA and REM prototypes, respectively. In both cases, results are reported for different constant speeds of the driving unit.

Both prototypes were able to span a significant portion of the force-speed plane. The EHA unit could range from 300 to 1.4 kN; these forces were calculated as $F_d = \Delta p_h A_p$. The REM solution yielded lower forces between 110 and 876 N. This difference can be attributed to constructive features of the two test beds, which fundamentally operate under diverse physical principles. More importantly, the EHA solution showed larger friction losses, thus limiting the minimum attainable force, even in the absence of electromagnetic damping. To confirm this point, figure 12 depicts the characterised points in the force-speed plane, where the target maximum damping curve is reported for comparison purposes.

A further difference is seen in the entity of power contributions. The REM device presents a mechanical input power and an output electrical power at the DC link. Conversely, the EHA device shows an additional hydraulic power term. While mechanical power usually increases the effort output due to friction, hydraulic power involves flow losses. In other words, an EHA solution presents leakages that eventually reduce the motion transmission. By comparing the hydraulic and mechanical power terms of the EHA device, one can observe a net decrease in the latter at high forces. In fact, high forces imply larger pressure drops, which in turn amplify leakages in the pump gaps. These phenomena are also present in the gaps between the piston and the damper tube, but these components did not take part of the experimental campaign. In turn, power terms from the REM solution show steadier behaviour due to an ideally rigid motion ratio without flow (speed) reduction. Leakage losses are also advisable in figure 12, where the effective damping speed of the EHA device decreases as force ramps up.

The presence of multiple conversion stages hinders the EHA device total conversion efficiency,

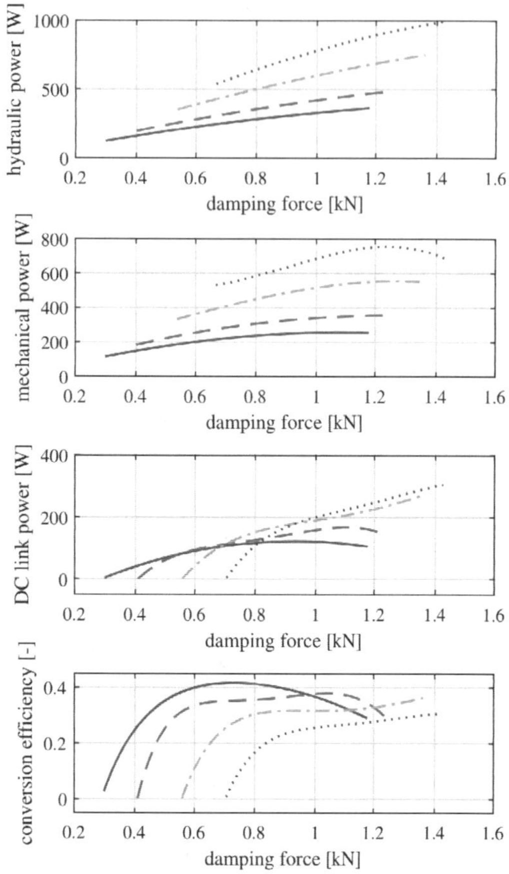

Figure 10. Electro-hydrostatic damper prototype test results. Power contributions and conversion efficiency at constant speed of the driving motor and pump: 1 (solid), 1.2 (dashed), 1.6 (dash-dotted) and 2 krpm (dotted).

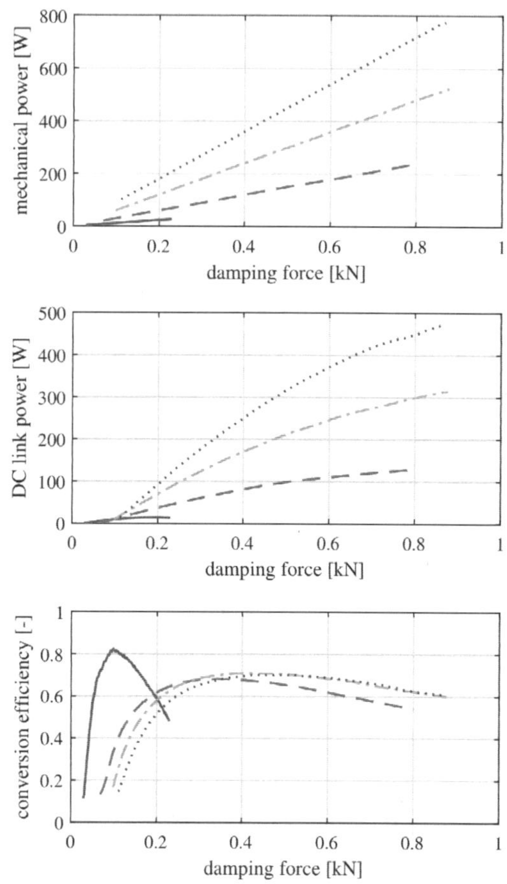

Figure 11. Rotary electromechanical damper prototype test results. Power contributions and conversion efficiency at constant speed of the prototype input shaft: 10 (solid), 25 (dashed), 50 (dash-dotted) and 75 rpm (dotted).

calculated as

$$\eta_{tot} = \frac{V_{dc} I_{dc}}{\Delta p_h Q_h} \tag{11}$$

which in the tests involving the lowest driving speed, led to a maximum of 0.417. At higher-valued inputs, the efficiency drops below 0.36 and presents a decreasing value as pressure escalates due to leakage effects.

By converse, the REM prototype shows more favourable results in this regard. With

$$\eta_{tot} = \frac{V_{dc} I_{dc}}{T_d \omega_d} \tag{12}$$

a maximum is found at 0.823, where the driving speed is the lowest setting (20 rpm on the driving motor, 10 rpm on the prototype input shaft). Higher speeds induce larger damping forces, but in this case, the efficiency finds maximum values close to 0.7, with very slight decays as force increases.

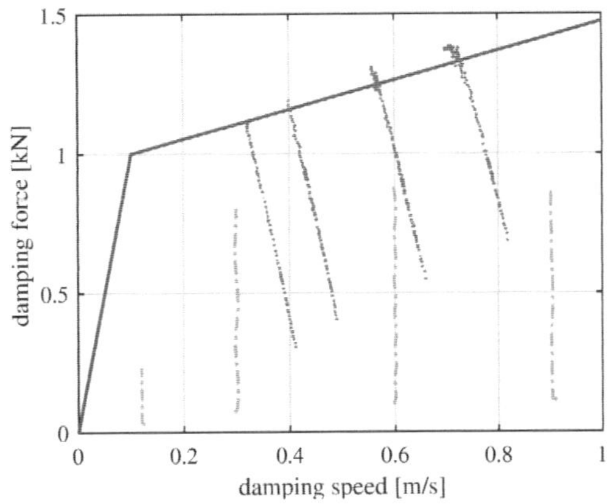

Figure 12. Characterised points in the force-speed plane for the electro-hydrostatic (dotted) and the rotary electromechanical (dash-dotted) prototypes. The target maximum damping characteristic (solid) is shown for comparison purposes.

5. Discussion

In the following, key features for both prototypes are compared. Table 3 summarises the comparison in qualitative terms. However, for some of the listed features, quantification is possible.

Table 3. Qualitative comparison among regenerative damper technologies.

Feature	Electro-hydrostatic	Rotary electromechanical
Mechanical robustness and fatigue	●	◐
Noise	◐	○
Oil-free	○	●
Conversion efficiency	◐	●
Packaging	◐	●
Actuator mass	◐	●
Equivalent inertia	◐	●
Total score units	●●●◐	●●●●●◐

Regarding mechanical robustness and fatigue, the EHA solution presents an advantage due to the intrinsic lubrication of the rotating parts, thus reducing tribology issues. Furthermore, gerotor pumps are well-known for inducing very low relative angular speeds among the parts in contact. The REM prototype, on the other hand, is lubricated with transmission oil, but the amount of parts in contact is much larger than in the EHA counterpart. In terms of noise, the presence of fluid dampens sound pressure emissions for the EHA solution. This is still applicable but less evident for the REM alternative.

The fluid presence becomes a downside when ageing, leakages and fluid contamination are accounted for. For this reason, many hydraulic devices in the automotive industry are being dismissed in favour of oil-free alternatives.

The conversion efficiency was demonstrated to be much lower and presented a narrow region for the EHA damper. This aspect is dominated by larger friction due to tight clearances between the gerotor gears and the pump casing. If clearances are increased to reduce friction, leakages would dominate significantly the behaviour.

When considering only the actuation units, the EHA device presents a maximum diameter of 88 mm and an axial length of 117 mm. The REM device is larger at 93 mm of diameter and 194 mm of axial length considering its input shaft. The mass of the REM prototype with linkage bars is 3.5 kg, while the EHA device and the shock absorber weigh 5.5 kg. Nevertheless, the use of a rotary electromechanical system could imply a total rethinking of the vehicle suspension with collateral benefits on the geometric envelope and distribution of masses. In addition, the REM prototype mass is added to the sprung component of the quarter-car model. The EHA, by converse, sits on the unsprung mass, thus having a more significant impact on vertical dynamics.

In terms of dynamic behaviour, the EHA system presents an equivalent inertia of 23 kg (equation 6). The REM device presents a lower inertial contribution of 15.9 kg. Despite sharing the same electric machine and overall transmission ratio, the moment of inertia of both gerotor gears is larger than that of rotating carriers and planets inside the gearbox. This inertial term is relevant as frequency escalates, as it tends to lock the suspension and eventually requires an active attenuation using the control system.

6. Conclusions

The present paper dealt with the comparison of electro-hydrostatic and rotary electromechanical regenerative damping technologies. A unified design for the electric machine and transmission ratio was motivated. Then, two different approaches led to the definition of specific transmission systems for both technologies. Details regarding packaging and assembly were discussed. The resulting prototypes were built and tested using dedicated test beds. Despite having different architectures, a similar testing methodology led to comparable results in terms of power contributions and conversion efficiencies. Finally, a discussion on qualitative and quantitative features of both technologies was carried out.

From the obtained results, it is clear that the rotary electromechanical solution outperforms the electro-hydrostatic one in terms of key metrics. However, the implementation of a rotary electromechanical damper requires important efforts to integrate the actuator into a conventional suspension, while electro-hydrostatic actuators fit in traditional layouts.

References

[1] Karnopp D, Crosby M J and Harwood R A 1974 *Journal of Engineering for Industry* **96** 619–626
[2] Abdelkareem M A A, Xu L, Ali M K A, Elagouz A, Mi J, Guo S, Liu Y and Zuo L 2018 *Applied Energy* **229** 672–699
[3] Gysen B, Paulides J, Janssen J and Lomonova E 2010 *IEEE Transactions on Vehicular Technology* **59** 1156–1163
[4] Tonoli A, Amati N, Detoni J G, Galluzzi R and Gasparin E 2013 *Vehicle System Dynamics* **51** 1186–1199
[5] Kawamoto Y, Suda Y, Inoue H and Kondo T 2008 *Vehicle System Dynamics* **46** 1053–1063
[6] Li Z, Zuo L, Kuang J and Luhrs G 2013 *Smart Materials and Structures* **22** 025008
[7] Galluzzi R, Circosta S, Amati N and Tonoli A 2021 *Mechatronics* **77** 102580
[8] Todmal P E and Melzi S 2020 *SAE International Journal of Passenger Cars - Mechanical Systems* **13** 06–13–01–0002
[9] Yu M, Dini D and Cleaver G D 2018 *IEEE/ASME Transactions on Mechatronics* **23** 2066–2077
[10] Galluzzi R, Xu Y, Amati N and Tonoli A 2018 *Applied Energy* **210** 16–27
[11] Guo S, Xu L, Liu Y, Guo X and Zuo L 2017 *IEEE/ASME Transactions on Mechatronics* **22** 2684–2694
[12] Dorrell D G, Hsieh M F, Popescu M, Evans L, Staton D A and Grout V 2011 *IEEE Transactions on Industrial Electronics* **58** 3741–3757
[13] Tessari F, Galluzzi R and Amati N 2020 *Journal of Mechanical Design* **142** 063501

IOP Conf. Series: Materials Science and Engineering 1214 (2022) 012013 doi:10.1088/1757-899X/1214/1/012013

A comparison on the Energy Release between traditional and Additive Manufactured AISI 316L steel during static tensile test

D D'Andrea, E Guglielmino, G Risitano and D Santonocito*

University of Messina, Department of Engineering, Contrada di Dio, 98166 Messina Italy

dsantonocito@unime.it

Abstract. Additive Manufacturing (AM) allows the creation of mechanical components and biomechanical devices with low weight and reduced material consumption compared to traditional manufacturing techniques. Despite design opportunities increase, on the other hand, also the process parameters to consider when dealing with AM are relevant on the mechanical performance. It is of fundamental importance to assure the structural integrity and reliability of such components. Energy release could give relevant information on the mechanical performance of the material; hence the Static Thermographic Method (STM) has been proposed as a rapid test procedure to assess the initiation of damage within the material monitoring the temperature evolution during static tensile tests. The aim of the present work is to compare the energy release of a stainless steel during static tensile tests. Specimens made of AISI 316L were produced with the same geometry, both with traditional and AM techniques. A comparison on the microstructure has been also performed in order to assess the possible cause of failure under static loads.

1. Introduction

Additive manufacturing (AM) is a very promising rapid prototyping technique which is moving to the production of final mechanical components. Industrial applications in aerospace [1] or biomedical field [2,3] are still increasing. Primary developed for polymers, 3D printing also allow the realization of components made in steel [4], with the possibility of exploring new design thanks to topology optimization [5]. One of the most common 3d-printed metal is the stainless steel AISI 316L, a low carbon steel with excellent mechanical properties. However, the mechanical properties of 3D-printed materials are severally affected by process parameters [6], as well as their microstructure. In high energy deposition process, the laser beam power, scanning speed and beam size play an important role in the finished part [7,8].

As the mechanical characteristics change with the printing parameter, it is necessary to perform several tests to retrieve both static and fatigue properties. This would lead to performing a huge number of tests with a high material consumption. On the hand, for traditional engineering materials rapid test procedure based on the energy release of the material have been developed in the last thirty years. The first researcher who apply the infrared thermography to monitor the superficial temperature of a specimen subjected to fatigue load was Risitano [9]. The development of the Thermographic Method (TM or Risitano TM) [10] leads to a rapid assessment of the fatigue limit of the material and of the SN curve [11]. Other researcher adopted also thermography and other energy-based techniques to assess the fatigue properties of a wide class of materials. In 2013 Risitano and Risitano proposed a rapid test procedure to assess the beginning of damage within the material monitoring the superficial temperature during a static tensile test [12], i.e. the Static Thermographic Method (STM). A limit stress could be identified when a deviation from the linear thermoelastic trend of the temperature signal is noticed. The Static Thermographic Method has been applied to several kind of materials and compared both with traditional fatigue test and TM, showing good agreement [13–16]. Santonocito, for the first time applied the STM to 3D-printed PA12 [17] and monitored the temperature trend of AISI 316L specimens obtained by SLM [18].

In this work, a comparison between traditional AISI 316L specimens and AM specimens of the same material, but with two different printing configurations, has been carried out monitoring the energy release during static and fatigue tests.

2. Theoretical background

2.1. Thermographic Method

As observed by La Rosa and Risitano [10], during a fatigue test, performed at a stress level above the fatigue limit σ_0 of the material and at a given stress ratio R and test frequency f, the temperature evolution exhibits three phases (Figure 1a). In the first phase (Phase I), there is an increment until the temperature stabilize at a value equal to ΔT_{st} (Phase II). As the material approaches to fail, temperature experiences a very rapid increment (Phase III), compared to the previous one. Fatigue limit can be identified in a rapid way as the first stress level at which the stabilization temperature is noticeably higher compared to the previous value. For each constant amplitude (CA) fatigue test, it is possible to evaluate the energy parameter Φ as the subtended area of the temperature versus number of cycle curve (ΔT-N). Generally, the higher the applied stress, the higher the stabilization temperature, but the energy parameter could be assumed as material property, at a given stress ratio and test frequency. It is also possible to perform a stepwise fatigue test (Figure 1b), increasing the applied stress level and registering the relative stabilization temperature. As the specimen fail, it is possible to evaluate the energy parameter Φ and assess the number of cycles to failure for each stress level, as the specimen would be stressed at that stress level with CA tests, simply dividing the energy parameter for the different stabilization temperatures and neglecting Phase I and III, usually smaller compared to Phase II. In this way, knowing the N-σ values, it is possible to obtain the complete SN curve of the material with a very limited number of tests.

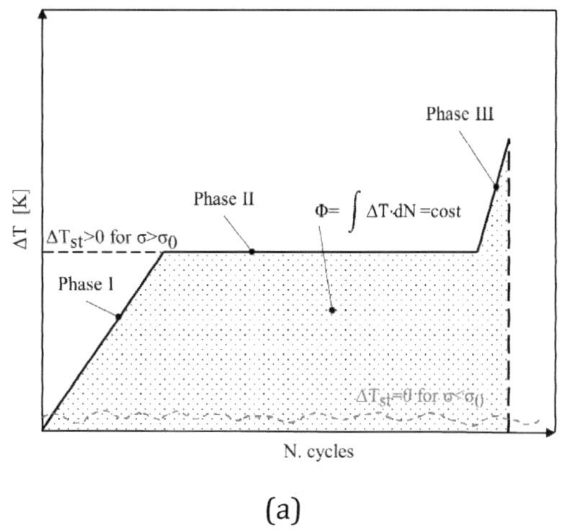

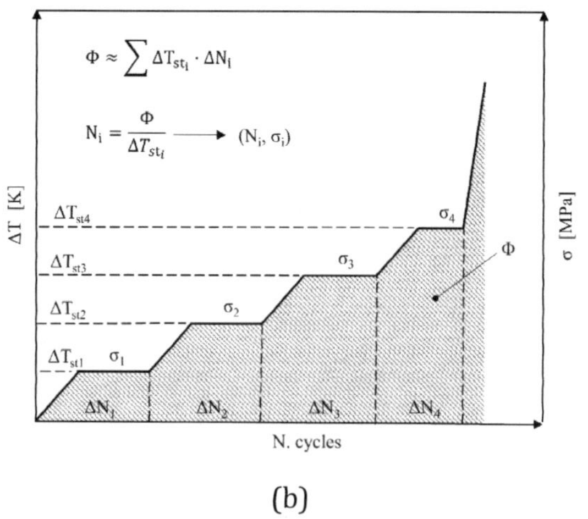

Figure 1. a) Temperature trend during a constant amplitude fatigue test; b) Temperature trend during a stepwise fatigue test.

2.2. *Static Thermographic Method*

In 2013, Risitano and Risitano [12] proposed a very rapid procedure to assess the first damage within the material monitoring its temperature evolution during a uniaxial tensile test. During a static tensile test of common engineering materials, the temperature evolution, detected by means of an infrared camera, is characterized by three phases (Figure 2): an initial approximately linear decrease due to the thermoelastic effect (obeying to Lord Kelvin's law, Phase I), then the temperature deviates from linearity until a minimum temperature value (Phase II), therefore it experiences a very high further increment until material failure (Phase III).

Under uniaxial stress state and in adiabatic test conditions, Equation 1 can be simplified as:

$$\Delta T_s = -K_m T \sigma_1 \tag{1}$$

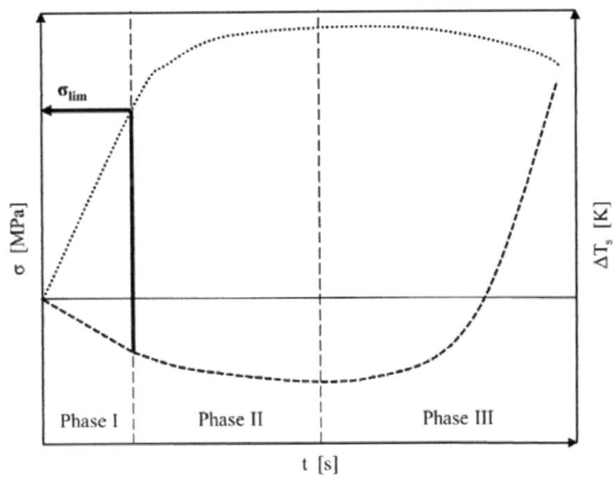

Figure 2. Temperature trend during a static tensile test.

The use of high precision IR sensors allows to define experimental temperature vs. time diagram during static tensile test in order to define the stress at which the linearity is lost. In 2010, Clienti et al. [19] for the first time correlated the damage stress σ_{lim} related to the first deviation from linearity of ΔT temperature increment during static test (end of phase I) to the fatigue limit of plastic materials. If it is possible during a static test to estimate the stress at which the temperature trend deviates from linearity, that stress could be related to a critical macro stress σ_{lim} which is able to produce in the material irreversible micro-plasticity. This critical stress is the same stress that, if cyclically applied to the material, will increase the microplastic area up to produce microcracks, hence fatigue failure.

3. Materials and Method

Three set of three specimens each, made of low carbon stainless steel AISI 316L, were tested. The geometry of the specimens was of the type "Continuous radius between ends", according to the ASTM E466 standard (Figure 3a). The first set of specimens was realized by traditional manufacturing, processing a bar of material by turning. The second and third set of specimens were produced by Selective Laser Melting (SLM) printing process, adopting AISI 316L powder. The specimens were built along Z direction, in nitrogen atmosphere, adopting a laser power of 230W and a laser scanning speed of 1400 mm/s for the second set; while 190W and 1200 mm/s for the third set and tested "as built".

Static tensile tests were performed under stress control, adopting a stress rate equal to 3 MPa/s, with a servo-hydraulic loading machine MTS 810 with a maximum load capacity of 250 kN (Figure 3b). The stress rate must be adopted in order to assure adiabatic conditions during the tensile tests, i.e. the specimen must not have the time to exchange heat with the surrounding environment.

During the tests, the specimen surface temperature was monitored with an infrared camera FLIR A40, recording the temperature evolution with a frequency of 5 Hz. The maximum value of a rectangular area placed on the specimen was recorded and post processed with a rlowess filter, with a data span of 5%, to exclude outliers and enhance the linear trend.

Stepwise fatigue tests were performed on four specimens with the same loading machine, adopting two stress ratio, R= -1 and R= 0.1, and a testing frequency of 20 Hz. During the fatigue test the temperature trend was monitored with the IR camera.

Figure 3. a) AISI 316L specimens obtained by SLM; b) Experimental setup.

4. Results and discussion

4.1. Static tensile test on Traditional steel

Tensile tests were performed on three specimens made of AISI 316L and manufactured by turning. The tests were performed under stress control, adopting a stress rate of 3 MPa/s, able to assure a low heat exchange between the specimen and the surrounding environment. The rupture instant of the specimen has been chosen to sync the temperature and load data. The applied stress has been reported versus the filtered temperature signal in Figure 4.

It is possible to distinguish a first linear thermoelastic trend, then the temperature trend shows a deviation from the linearity till a minimum point. Suddenly, it experiences a high increment, due to prevalent plastic deformation, till specimen failure. To assess the limit stress, it is possible to perform two linear regressions and make their intersection. The first regression has been performed with the ΔT_1 points, corresponding to the linear thermoelastic trend (Phase I); while the second regression has been performed with the ΔT_2 temperature points of the plateau region (Phase II). Temperature points near the knee have been neglected.

The first two specimens (Figure 4a-b) show a clear temperature behaviour, with a value of limit stress σ_{lim} equal to 233.7 and 265.7 MPa. Also the third specimen shows a deviation from the linearity, but for a stress level lower compared to the previous specimens, i.e. 207.3 MPa (Figure 4c).

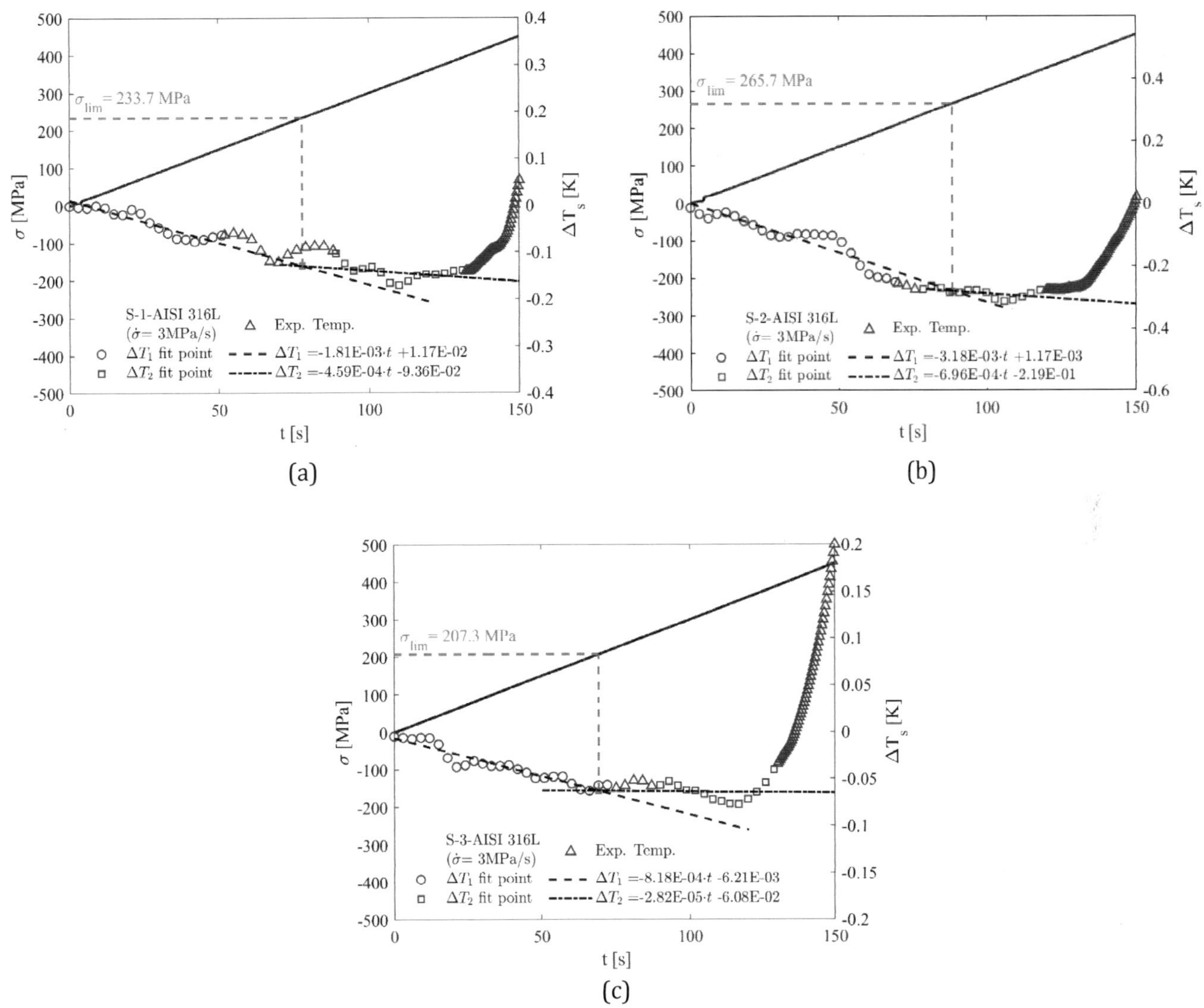

Figure 4. Temperature trend vs. applied stress of Traditional AISI 316L specimens.

4.2. Static tensile test on AM steel

Static tensile tests were performed on specimens, made of AISI 316L powder, produced by SLM printing process. Two laser powers were adopted: 230W and 190W. The tests were performed under stress control, with the same stress rate of the traditional material.

In Figure 5 are reported the temperature trend of the three specimens built along Z direction, with a laser power of 230W. As done for the traditional material, it is possible to perform two linear regressions with the temperature points of the linear and plateau regions. Compared to the traditional material, the deviation from the linearity is slightly evident. For the first two specimens, the limit stress is equal to 202.6 and 199.3 MPa, while for the third specimen it is equal to 171.1 MPa.

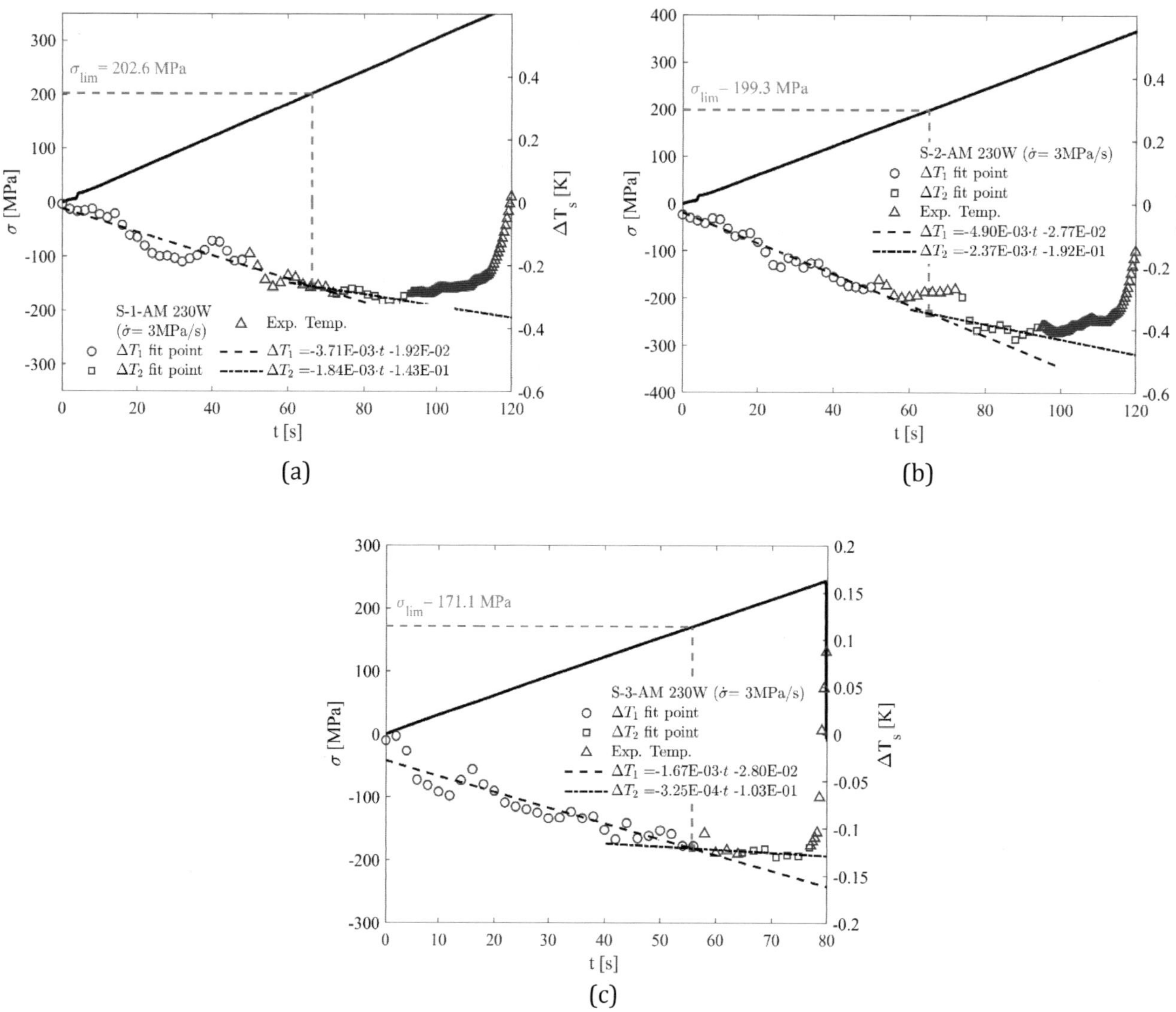

Figure 5. Temperature trend vs. applied stress of AM AISI 316L specimens (230W, 1400 mm/s).

As regard the specimens built with a laser power of 190W, the temperature signal it is not very clear. For the first test (Figure 6a) the deviation from the thermoelastic trend has been observed for a stress level equal to 174.7 MPa. For the second specimen (Figure 6b), a very noisy temperature signal has been obtained, hence it was not possible to assess a value for the limit stress.

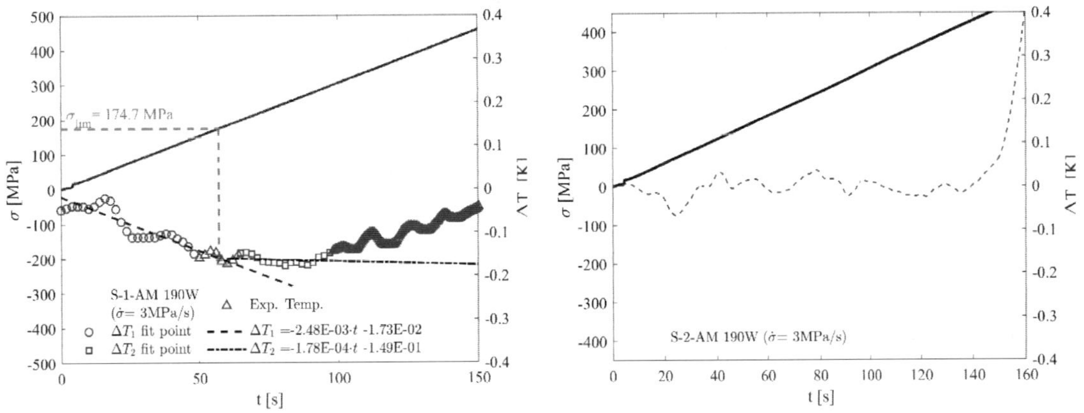

(a) (b)

Figure 6. Temperature trend vs. applied stress of AM AISI 316L specimens (190W, 1200 mm/s).

4.3. Comparison

In Table 1 are reported, for each tested specimen, the value of the ultimate strength σ_U and the limit stress σ_{lim} as assessed by STM, with their average estimation. While the traditional specimens show a little standard deviation, the AM specimens exhibits a high variability in terms of ultimate strength.

Table 1. Comparison of the mechanical properties of AISI 316L specimens.

Specimen type	No. Specimen	σ_U [MPa]	σ_{lim} [MPa]	$\sigma_{U\ ave}$[MPa]	$\sigma_{lim\ ave}$[MPa]
AISI 316L Traditional	1	707	234	714 ± 11	236 ± 30
	2	709	266		
	3	727	207		
AISI 316L AM (230W)	1	370	203	329 ± 74	191 ± 17
	2	373	199		
	3	244	171		
AISI 316L AM (190W)	1	614	175	580 ± 48	175
	2	546	-		

4.4. Fatigue behaviour

In order to compare the value of the limit stress with the value of fatigue limit of the traditional material, four fatigue tests were performed with a test frequency of 20 Hz and two stress ratio R, 0.1 and -1.

In Figure 7a the temperature trend vs. the applied stress level, for a specimen tested at R= -1 with a stress increase of $\Delta\sigma$= 10 MPa and a number of cycles per block ΔN= 20.000, is reported. For each stress level the corresponding value of the stabilization temperature has been estimated. As it is possible to observe, the temperature experiences a very high further increment when the stress level raise from 250 to 260 MPa. For the last stress level of 270 MPa, temperature does not stabilize, increasing up to specimen failure. It is reasonable to consider the fatigue limit for a stress ratio of R= -1 in the range 250÷260 MPa, where the higher temperature increment is noticed. It is worth noting that the limit stress found with the STM is within the previous stress range assessed with fatigue test.

The stepwise fatigue test with stress ratio R= 0.1 was performed with $\Delta\sigma$= 20 MPa and ΔN= 10.000, ranging from 200 MPa to 560 MPa (Figure 7b) and recording the stabilization temperature.

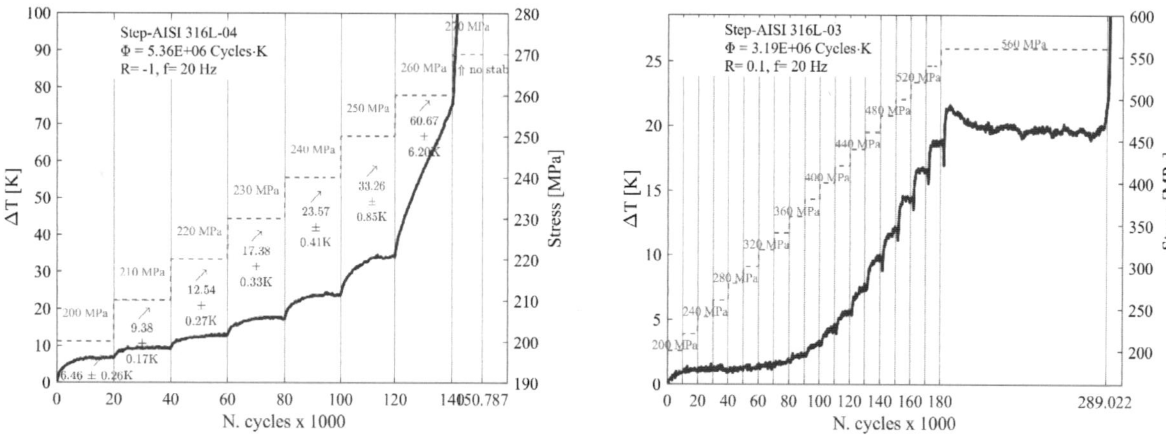

(a) (b)

Figure 7. Stepwise fatigue test of Traditional AISI 316L specimens: a) R= -1; b) R=0.1.

Reporting the stabilization temperature vs. the relative stress level (Figure 8), it is possible to observe a bilinear trend. For stress levels below the fatigue limit of the material, with a stress ratio R= 0.1, a slight temperature increment is noticeable. On the other hand, temperature increments are clearly higher for stress levels above the fatigue limit. By performing the linear regression of the lowest set of points and of the highest, then making their intersection, it is possible to assess the fatigue limit at R= 0.1 with the TM, which is equal to $\sigma_{0\ R0.1}$= 388 MPa. Considering the same applied maximum stress, it is evident how the fatigue limit for stress ratio of 0.1 is higher compared to the one for R= -1. In facts, this stress condition, in the tension-compression domain, is the more damaging for the materials, leading also to higher temperature increment.

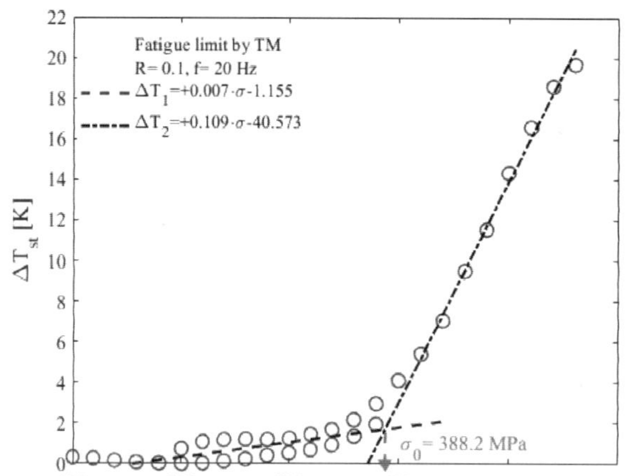

Figure 8. Fatigue limit of traditional AISI 316L at R= 0.1 by Thermographic Method.

Fatigue tests on AM AISI 316L specimens, printed by SLM with a laser power of 230W and laser scanning speed of 1400 mm/s, were performed by Santonocito et al. [18] on the same specimen geometry of the present work (Figure 9). In that work, the authors shown how the fatigue life, for a stress ratio R= 0.1, of such AM material is shorten compared to other literature data [20]. From the TM, the fatigue limit at R= 0.1 for traditional AISI 316L is equal to 388 MPa, a value considerably higher compared to the value of circa 80 MPa at 10^6 cycles, as reported by Solberg et al. (green triangles of Figure 9). In addition, under stepwise fatigue tests (red and black circles in Figure 9), specimens fail for stress levels below the value of the limit stress (191 ± 17 MPa and 175 MPa) in an unpredictable way and fatigue life seems depending also on the test frequency.

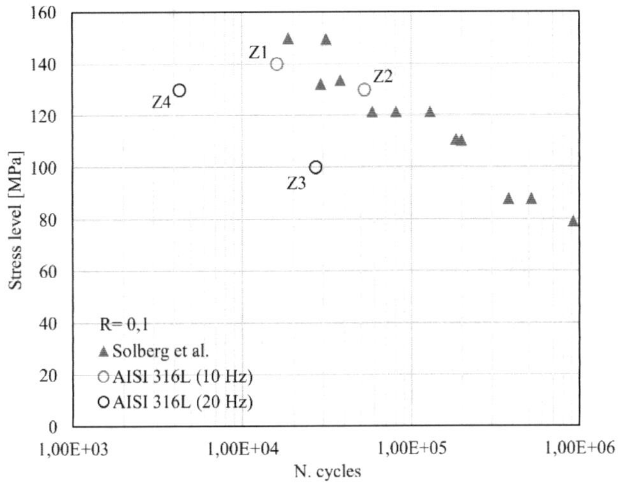

Figure 9. S-N curve for AM AISI 316L, from [18].

4.5. Microstructure

To confirm the mechanical behaviour of AISI316L steel obtained through AM and by traditional method, microscopies were carried out on the fracture surface suitably treated with a lapping machine.

As is clearly shown in Figure 10a and Figure 10b, the surfaces have a different structure which justifies the different behaviour of the samples when subjected to static and fatigue tests. In fact, the specimen obtained by traditional method has a uniform surface and does not show any macroscopic defects.

On the other hand, the sample made by AM technique has a surface characterized by numerous defects, such as inclusions and incomplete fusions. In particular, as shown in Figure 10c, the microstructure defects take on considerable dimensions even in the order of hundreds of microns, becoming the cause of possible crack initiation.

As anticipated in the materials and methods section, the sample production methodology is shown in Figure 10d, in which characteristic streaks are highlighted indicating the laser melting direction.

Andreau et al. [21] performed a fatigue test campaign on SLM specimen, made of 316L stainless steel, fabricated using three different hatching strategies, varying the scanning speeds in the internal portions of the parts, in order to assess the impact of the inner porosity rate on the fatigue behaviour. Samples with lack of fusion near the surface showed reduced fatigue properties with increasing pore size distribution. A high amount of internal porosity was found to not impact the fatigue properties below 10% of the areal layer porosity, while they exhibit premature cracking for values above. Fatigue life is mostly affected by small porosities at the specimen surface.

Shrestha et al. [22] found that lack of fusion in subsequent layers is responsible for fatigue crack initiation. The presence of large internal defect due to lack of fusion has a preponderant effect on fatigue cracks compared to surface roughness. Internal defects can severally affect the ultimate strength when the pore diameter is about 2400 μm and elongation when they are about 1800 μm [23].

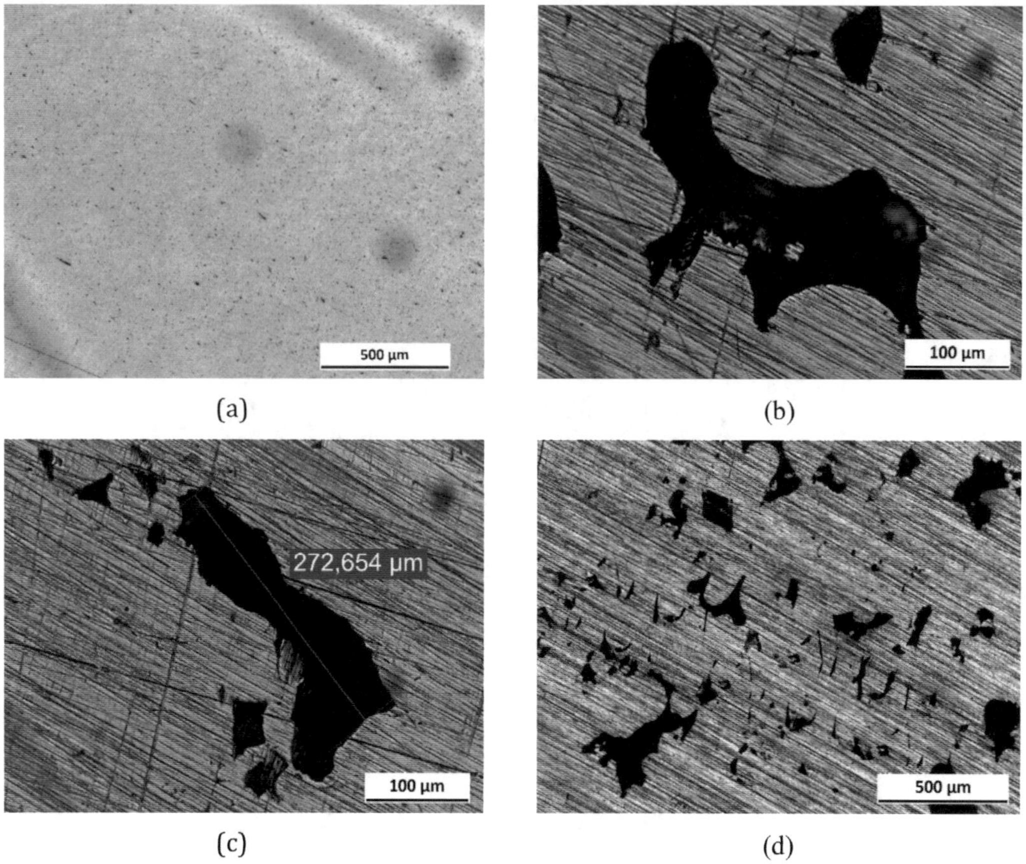

Figure 10. a) Micrography of Traditional AISI 316L; b-c) Defect of AM specimens; d) Micrography of AM AISI 316L.

5. Conclusion

In this work, the energy release of traditional and AM stainless steel AISI 316L has been monitored during static tensile and stepwise fatigue test. The mechanical properties of the different set of specimens have been assessed and compared in terms of ultimate strength and limit stress. AM specimens have shown a higher dispersion compared to the traditional material. In addition, by applying the Static Thermographic Method, the limit stress of traditional material is considerably higher than the AM one. To verify the limit stress, conceivable as the first damage initiation within the material, stepwise fatigue tests have been performed on traditional specimens with two stress ratios. The limit stress is in good agreement with the range of fatigue limit assessed by Thermographic Method for a stress ratio of R= -1. For a stress ratio of R= 0.1, traditional AISI 316L exhibit better fatigue life performances compared to the AM counterpart.

Energy methods, such as Static Thermographic Method and Thermographic Method, can be a useful aid to assess in a rapid way the fatigue properties of AM materials, where the process parameters strictly affect the mechanical properties of the produced mechanical components.

References

[1] Pradeep P I, Kumar V A, Sriranganath A, Singh S K, Sahu A, Kumar T S, Narayanan P R, Arumugam M and Mohan M 2020 Characterization and Qualification of LPBF Additively Manufactured AISI-316L Stainless Steel Brackets for Aerospace Application *Trans. Indian Natl. Acad. Eng.* **5** 603–16

[2] Tilton M, Lewis G S, Bok Wee H, Armstrong A, Hast M W and Manogharan G 2020 Additive manufacturing of fracture fixation implants: Design, material characterization, biomechanical modeling and experimentation *Addit. Manuf.* **33** 101137
[3] Chen R K, Jin Y an, Wensman J and Shih A 2016 Additive manufacturing of custom orthoses and prostheses-A review *Addit. Manuf.* **12** 77–89
[4] Sing S L, Tey C F, Tan J H K, Huang S and Yeong W Y 2019 *3D printing of metals in rapid prototyping of biomaterials: Techniques in additive manufacturing* (Elsevier Ltd.)
[5] Cucinotta F, Guglielmino E, Longo G, Risitano G, Santonocito D and Sfravara F 2019 *Topology optimization additive manufacturing-oriented for a biomedical application* vol 1 (Springer International Publishing)
[6] Bartolomeu F, Buciumeanu M, Pinto E, Alves N, Carvalho O, Silva F S and Miranda G 2017 316L stainless steel mechanical and tribological behavior—A comparison between selective laser melting, hot pressing and conventional casting *Addit. Manuf.* **16** 81–9
[7] Vastola G, Zhang G, Pei Q X and Zhang Y W 2015 Modeling and control of remelting in high-energy beam additive manufacturing *Addit. Manuf.* **7** 57–63
[8] Saboori A, Toushekhah M, Aversa A, Lai M, Lombardi M, Biamino S and Fino P 2020 Critical Features in the Microstructural Analysis of AISI 316L Produced By Metal Additive Manufacturing *Metallogr. Microstruct. Anal.* **9** 92–6
[9] Curti G, La Rosa G, Orlando M and Risitano A 1986 Analisi tramite infrarosso termico della temperatura limite in prove di fatica *Proc. XIV Convegno Naz. AIAS* 211–20
[10] La Rosa G and Risitano A 2000 Thermographic methodology for rapid determination of the fatigue limit of materials and mechanical components *Int. J. Fatigue* **22** 65–73
[11] Fargione G, Geraci A, La Rosa G and Risitano A 2002 Rapid determination of the fatigue curve by the thermographic method *Int. J. Fatigue* **24** 11–9
[12] Risitano A and Risitano G 2013 Determining fatigue limits with thermal analysis of static traction tests *Fatigue Fract. Eng. Mater. Struct.* **36** 631–9
[13] Corigliano P, Cucinotta F, Guglielmino E, Risitano G and Santonocito D 2020 Fatigue assessment of a marine structural steel and comparison with Thermographic Method and Static Thermographic Method *Fatigue Fract. Eng. Mater. Struct.* **43** 734–43
[14] Risitano G, Guglielmino E and Santonocito D 2018 Evaluation of mechanical properties of polyethylene for pipes by energy approach during tensile and fatigue tests *Procedia Structural Integrity* vol 13 (Elsevier B.V.) pp 1663–9
[15] Cucinotta F, D'Aveni A, Guglielmino E, Risitano A, Risitano G and Santonocito D 2021 Thermal emission analysis to predict damage in specimens of high strength concrete *Frat. ed Integrita Strutt.* **15** 258–70
[16] Crupi V, Guglielmino E, Risitano G and Tavilla F 2015 Experimental analyses of SFRP material under static and fatigue loading by means of thermographic and DIC techniques *Compos. Part B Eng.* **77** 268–77
[17] Santonocito D 2020 Evaluation of fatigue properties of 3D-printed Polyamide-12 by means of energy approach during tensile tests *Procedia Struct. Integr.* **25** 355–63
[18] Santonocito D, Gatto A and Risitano G 2021 Energy release as a parameter for fatigue design of additive manufactured metals *Mater. Des. Process. Commun.* 1–7
[19] Clienti C, Fargione G, La Rosa G, Risitano A and Risitano G 2010 A first approach to the analysis of fatigue parameters by thermal variations in static tests on plastics *Eng. Fract. Mech.* **77** 2158–67
[20] Solberg K, Guan S, Razavi S M J, Welo T, Chan K C and Berto F 2019 Fatigue of additively manufactured 316L stainless steel: The influence of porosity and surface roughness *Fatigue Fract. Eng. Mater. Struct.* **42** 2043–52
[21] Andreau O, Pessard E, Koutiri I, Penot J D, Dupuy C, Saintier N and Peyre P 2019 A competition between the contour and hatching zones on the high cycle fatigue behaviour of a 316L stainless steel: Analyzed using X-ray computed tomography *Mater. Sci. Eng. A* **757** 146–

59
[22] Shrestha R, Simsiriwong J and Shamsaei N 2019 Fatigue behavior of additive manufactured 316L stainless steel parts: Effects of layer orientation and surface roughness *Addit. Manuf.* **28** 23–38
[23] Wilson-Heid A E, Novak T C and Beese A M 2019 Characterization of the Effects of Internal Pores on Tensile Properties of Additively Manufactured Austenitic Stainless Steel 316L *Exp. Mech.* **59** 793–804

Combined experimental methods to assess the fatigue limit

G Fargione, F Giudice, R Barbagallo, D Corallo and G La Rosa

Dept. of Civil Engineering and Architecture, University of Catania, Via S. Sofia 64, 95125 Catania – Italy

guido.larosa@unict.it

Abstract. Purpose of this paper is to perform an energetic analysis on metal specimens under fatigue testing using the simultaneous application of different non-destructive techniques: thermography (TH), acoustic emission (AE) and digital image correlation (DIC). They assure the better evaluation, being contactless methods, usable in real time and, even if TH and DIC are related to the specimen surface, representative of the inside state of the materials. In particular, the cascade of hits and the acoustic energy were recorded to assess the parameters linked to the fracture of the materials. At the same time, the surface temperature, detected by a thermocamera, and the displacements induced on the specimen, using the DIC on the free zone of the specimens, were recorded. The DIC analysis allows the evaluation of the correct displacement by a pattern realized on the opposite face respect to TH. AE data and hysteresis curves (derived by DIC) were compared with the measures acquired by TH and reveal a reliable correlation between them in terms of fatigue limit.

1. Introduction

Energetic approaches are widely used to preview structure failures, often preceded by different energy emissions in the form of heat, sounds or vibrations. They could be detected during the fatigue life, following the crack nucleation and propagation. In the present study the coupling of different energy methodologies was considered: Thermographic analysis (TH), hysteresis area by Digital Image Correlation (DIC) at high velocity and Acoustic Emission (AE).

Many researchers have performed different methodologies to detect the fatigue parameters using thermography in rapid tests and, consequently many procedures were defined to assess the fatigue and fracture limit, applying single load or a sequence of incremental loading steps [1-19].These procedures allow assessing the fatigue life [5], the cumulative damage [14], and finally, the fatigue limit under the application of simple static loading also [20-23]. The latter method is based on the limit of the linear thermoelasticity as the starting of the fatigue process [24, 25].

Digital Image Correlation (DIC) is widely used to assess the correct displacements in specimens under static or dynamic loading and can be usefully exploit to define the area of hysteresis, linked to the state of damage of the specimen [23, 26-36].

Acoustic Emission (AE) was proposed as a control methodology, becoming one of the more effective Non-Destructive Techniques in the industrial field. It can be also used to define the fatigue parameters, detecting the hits and their acoustical energetic amount to highlight the fracture propagation [37-40].

Following some studies assessing fatigue parameters by coupling different methodologies [23, 34-40], the aim of the present paper is the comparison about the results found by coupling TH, DIC and AE to define the fatigue parameters and the crack growing in static and dynamic testing. In particular, the paper describes the evaluation of the fatigue limit using the three methodologies and the comparison among them.

2. Materials and methods

Several tensile and fatigue tests have been performed on dog bone shaped specimens made of S275JR steel with a cross section and a gauge length of 20x6.5 mm, respectively.

2.1. Experimental setup

In order to perform at the same time TH, AE and DIC acquisitions, the specimens were accurately prepared before the tests. In particular, black paint, for the TH, and a random black and white pattern, for the DIC, were applied on the opposite sides of the specimen, respectively. Moreover, the two EA sensors were positioned and fixed symmetrically at the opposite ends of the gage length, with a small layer of grease interposed by the specimen and the sensors themselves, useful to transmit the acoustic emissions. The two piezoelectric sensors, connected to the corresponding preamplifiers, allow localizing the emission source along the axis of the specimen.

For the detection of the AE signals, it was used a AMSY4-MC6 Vallen System, with a central unit equipped with four ASIPP acquisition cards and four channels and an Analog Digital Converter (ADC) sampling frequencies up to 10 MHz and with a resolution of 16 bits by eliminating the lower frequencies. The AE sensor were placed at a distance of 75 mm, symmetrically respect the middle point of the specimen. A Hsu-Nielsen test was carried out for the calibration of the acoustic signal. As a result, a suitable threshold value of about 43 dB (for the static tests) and 50 dB (for the fatigue tests) with a 34 dB gain of the acoustic signal have been set. Rise time, duration, maximum amplitude, energy and number of hits were acquired.

A Flir ThermaCAM X6540 SC thermocamera was used, operating in the infrared field having a spatial resolution of 320x240 pixels, thermal resolution of 20 mK. The thermal images were acquired and processed using the Flir ThermaCAM Researcher Professional 2.10 software.

Both the static tensile tests and the dynamic fatigue tests have been analysed by means of the DIC technique. In the static tests, the images were acquired by a standard Image Source DMK23G445 monochromatic camera. In the dynamic tests, in order to evaluate the hysteresis cycle at different instants of the tests, the images were acquired by a high-speed Phantom V711 rapid camera. The DIC analysis has been performed using the Gom Correlate software.

The tests have been performed by means of an Instron 8501 hydraulic testing machine with a 100 kN load cell both in cyclic loading (under loading control) and in static loading (under displacement control). Fig. 1 shows the experimental setup with the thermal camera on the right, the high-speed camera on the left, and the specimen with the acoustic sensors clamped in the testing machine in the middle.

2.2. Testing procedure

Static tests were run with a test speed of 1 mm/min. During the tests, the acoustic emission parameters and the thermal gradient were recorded. From the acoustic signal it is possible to derive and plot the cascade of hits trend and the cumulative acoustic energy. The cascade of hits ranks in a single event all the "hits" that exceed the threshold value set in a time interval.

As already known, basing on the thermoelastic phenomenon, the fatigue limit can also be estimated from the slope change on the thermoelastic curve [20]. The deviation by the linear thermoelastic behaviour, in fact, highlights the beginning of the thermoplastic phenomenon, strictly linked to the damage process and, then, to the fatigue limit.

Fatigue tests were run with R=-1, frequency of 10 Hz and loading trains of 1000 cycles each, starting from ±10 kN to ±30 kN with steps of ±2 kN to ±5 kN each. The tensile-compressive stress, as a consequence, varied from ±80 MPa to ±240 MPa.

IOP Conf. Series: Materials Science and Engineering 1214 (2022) 012014 doi:10.1088/1757-899X/1214/1/012014

Fig. 1. Experimental setup

3. Results and discussion

An example of the AE and TH responses as a function of stress are reported in figure 2 for the fatigue loading. The number of hits and the AE energy were processed evaluating their behaviour at the different stress applied (figure 2a), as well as the thermal increments respect to the initial temperature (figure 2b). It is possible to put in evidence as both the methodologies converge to the same value of stress (about 80 MPa).

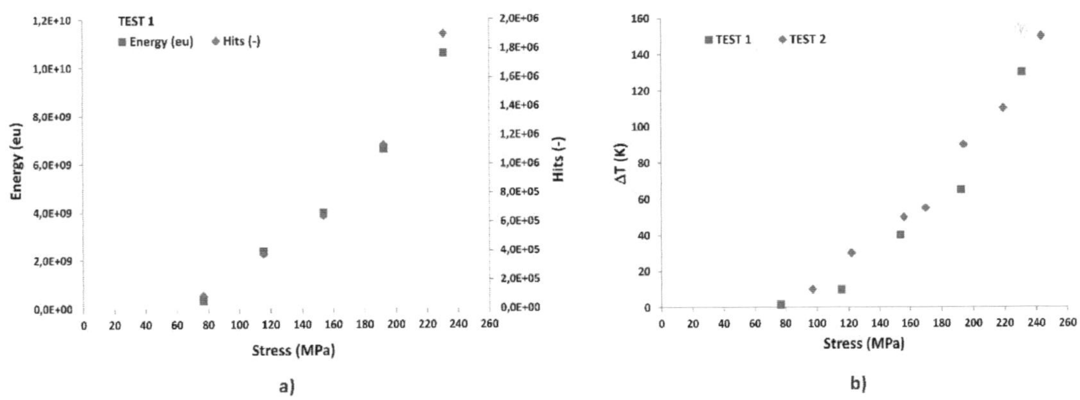

Fig. 2. AE energy and hits (a), and TH responses (b), both as functions of the stress applied.

At the same time, using the displacements obtained by the DIC processing, it was possible to determine the hysteresis curve, thanks to the acquisition by the rapid camera. It was necessary to allow a load frequency high enough to assure an easy acquisition of the thermal variations (proportional to the frequency) and to control local plastic phenomena. An example of the hysteresis curve for a load step is shown in figure 3a. The amount of the mechanical energy (expressed by the hysteresis areas) is shown in figure 3b as a function of the applied stress. Once again, the curve, representing the mechanical plastic energy, converges to the same value of the AE and TH responses, demonstrating that also the measure of the hysteresis areas can be used to assess the fatigue limit. In particular, the values obtained by the different techniques show a close agreement between AE (80.5±1.1 MPa) and DIC (77.8±2.9 MPa). The fatigue limit assessed by TH provides affine values (90.0±1.2 MPa), even if they tend to be higher.

These results, as a whole, confirm the potential of multi-technique approaches, but also suggest the need to deepen the investigation on the comparison between the three techniques used also for other materials and to compare it with that obtained using other traditional or innovative methodologies.

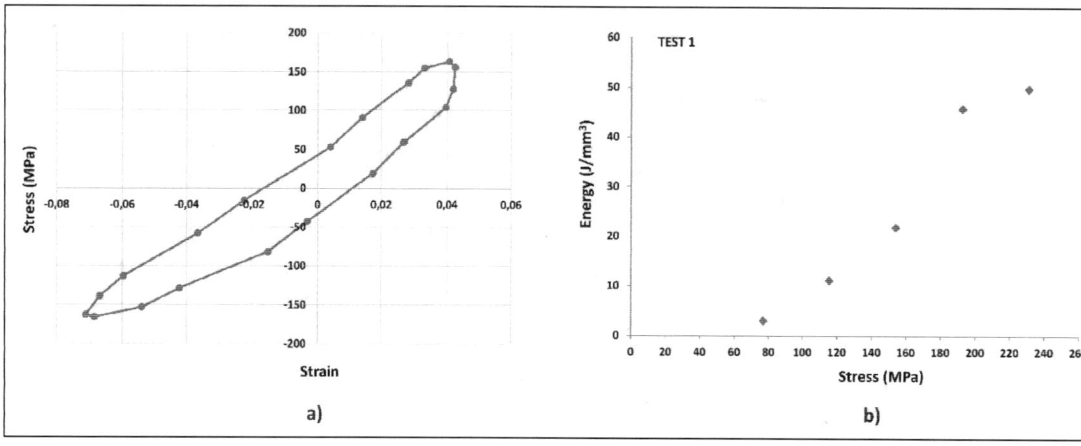

Fig. 3. Example of hysteresis curve detected by DIC (a), and hysteresis area as a function of the stress applied (b).

4. Conclusions

Series of tests were carried out on specimens in S275JR steel, coupling the methodologies of acoustic emission, digital image correlation and thermography, in order to verify the possibility to detect the fatigue limit by the comparison of the three forms of energy: thermal, mechanical and acoustic. The tests were performed under static tensile loading and under cyclic loading at R=-1.

The results obtained show that the acoustic emission is able to define the fatigue limit, either in terms of cascade hits or in terms of released energy, as well as the thermographic analysis, already tested by many authors. By the DIC displacements the hysteresis areas were calculated and, once again, the curve of the plastic mechanical energy converges to the same fatigue limit. Then, the acoustic emission parameters as well as the hysteresis areas can be used to define the fatigue limit using a methodology similar to that applied by thermography. The approach needs to be better study in deep to verify if the energetic amount detected by acoustic emission could be better linked to other fatigue parameters and, in this case, translate the results acquired by thermography to the other methodologies.

The results obtained show as all the methods converge to the same fatigue limit, demonstrating that all the forms of detected plastic energy (thermal, acoustic and mechanical) can be used (separately or combined) to determine the fatigue behaviour.

Following these results, the authors intend to prosecute the analysis on a larger series of specimens to better investigate the correlation between the energy detected by the different methodologies, in order to formulate a reliable procedure based on a combined data analysis able to predict the fatigue parameters.

References

[1] Bodner SR, Davidson DL and Lankford J 1983 *Eng. Fract. Mech.* **17** 189
[2] Luong MP 1988 *Mech. Mater.* **28** 155
[3] Kaleta J, Blotny R and Harig H 1990 *J. Test. Eval.* **19** 326
[4] La Rosa G and Risitano A 2000 *Int. J. Fatigue* **22** 65
[5] Fargione G, Geraci A, La Rosa G and Risitano A 2002 *Int. J. Fatigue* **24** 11
[6] Klingbeil NW 2003 *Int. J. Fatigue* **25** 117

[7] Boulanger T, Chrysochoos A, Mabru C and Galtier A 2004 *Int. J. Fatigue* **26** 221
[8] Meneghetti G 2007 *Int. J. Fatigue* **29** 81
[9] Plekhov OA, Saintier N, Palin-luc T, Uranov SV and Naimark O 2007 *Mat. Sci. Eng. A* **462** 367
[10] Charkaluk E and Constantinescu A 2009 *Mech. Mater.* **41** 483
[11] Maquin F and Pierron F 2009 *Mech. Mater.* **41** 928
[12] Naderi M, Amiri M and Khonsari MM 2010 *P. R. Soc. A* **466** 423
[13] Crupi V, Chiofalo G and Guglielmino E 2011 *P. I. Mech. Eng. C-J. Mech.* **225** 833
[14] Risitano A and Risitano G 2013 *Int. J. Fatigue* **48** 214
[15] Risitano A, Fargione G, Giudice F and Patanè G 2015 *Procedia Eng.* **109** 346
[16] Fargione G, Giudice F and Risitano A 2017 *Theor. Appl. Fract. Mech.* **88** 97
[17] Meneghetti G and Ricotta M 2018 *Int. J. Fatigue* **115** 107
[18] De Finis R, Palumbo D and Galietti U 2019 *Fatigue Fract. Eng. Mater. Struct.* **42** 267
[19] Colombo C and Vergani L 2019 *Procedia Struct. Integr.* **24** 658
[20] Clienti C, Fargione G, La Rosa G, Risitano A and Risitano G 2010 *Eng. Fract. Mech.* **77** 2158
[21] La Rosa G and Risitano A 2014 *Key Eng. Mater.* **577/578** 69
[22] Risitano A, Fargione G and Guglielmino E 2014 *Frat. Integrita Strutt.* **30** 201
[23] Giudice F, La Rosa G, Fargione G and Barbagallo R 2019 *Procedia Struct. Integr. 24* 706
[24] Caglioti G 1982 The thermoelastic effect: Statistical mehanic and thermodynamics of he elastic deformation *Mechanical and Thermal Behaviour of Metallic Materials* eds G Caglioti and A Ferro Milone (Amsterdam; North Holland) chapter 1 pp 1–34
[25] Melvin AD, Lucia AC, Solomos G, Volta G and Emmony DC 1990 *J. Mater. Sci. Lett.* **9** 752
[26] Wattrisse B, Chrysochoos A, Muracciole JM and Némoz-Gaillard M 2001 *Eur. J. Mech. A-Solid* **20** 189
[27] Kanchanomai C, Yamamoto S, Miyashita Y, Mutoh Y and McEvily AJ 2002 *Int. J. Fatigue* **24** 57
[28] Sánchez-Arévalo FM and Pulos G 2008 *Mater. Charact.* **59** 1572
[29] Hunady R, Hagara M and Schrötter M 2012 *Procedia Eng.* **48** 242
[30] Roy SC, Goyal S, Sandhya R and Ray SK 2013 *Procedia Eng.* **55** 165
[31] Li L, Muracciole JM, Waltz L, Sabatier L, Barou F and Wattrisse B 2016 *Opt. Laser Eng.* **8** 1
[32] La Rosa G, Clienti C and Marino Cugno Garrano A 2016 *Procedia Struct. Integr.* **2** 2140
[33] La Rosa G, Clienti C, Marino Cugno Garrano A and Lo Savio F 2017 *Eng. Fract. Mech.* **183** 53
[34] La Rosa G, Clienti C, Marino Cugno Garrano A and Lo Savio F 2018 *Procedia Struct. Integr.* **13** 1583
[35] La Rosa G, Lo Savio F, Giudice F, Clienti C and Marino Cugno Garrano A 2020 *Fatigue Fract. Eng. Mater. Struct.* **43** 2597
[36] Barbagallo R, Fargione G, Giudice F and La Rosa G 2021 *IOP Conf. Ser.: Mater. Sci. Eng.* **1038** 012050
[37] Kordatos EZ, Aggelis DG and Matikas TE 2012 *Compos. Part B-Eng.* **43** 2676
[38] Kordatos EZ, Dassios KG, Aggelis DG and Matikas TE 2013 *Mech. Res. Commun.* **54** 14
[39] La Rosa G, Clienti C and Lo Savio F 2014 *Procedia Eng.* **74** 261
[40] Giudice F, La Rosa G, Lo Savio F and Clienti C 2019 *Procedia Struct. Integr.* **18** 886

IOP Conf. Series: Materials Science and Engineering 1214 (2022) 012015 doi:10.1088/1757-899X/1214/1/012015

A first quantitative approach for detecting volumetric defects in additive manufactured metal samples by using active thermographic technique

Ester D'Accardi, Davide Palumbo, Vito Errico, Andrea Fusco and Umberto Galietti

Politecnico di Bari - Dipartimento di Meccanica, Matematica e Management, Via Orabona 4, 70125 Bari, Italy

E-mail: ester.daccardi@poliba.it

Abstract. This work is focused on the quantitative analysis of defects produced in metal samples of AISI 316L with Laser Powder Bed Fusion (L-PBF) additive manufacturing (AM) process by means of active thermographic techniques. A simple and common set-up with 2 flash lamps and a cooled sensor has been used to analyse the differences between planar and volumetric defects. In particular, the presence of imprinted spheres containing inside non-sintered metallic powder has been compared with those of similar cylinders in size and depth, induced in similar samples. Furthermore, the differences in terms of thermal contrasts have been studied between the cylinder manufactured as an internal defect with powder inside, against the same geometry, but with a small channel for the discharge of the non-melted material. The analysis of the thermal signal and the application of post processing algorithms allowed us to identify the thermal features suitable for describing the behaviour of different kind of defects, typical of the process.

1. Introduction

Pulsed thermography is a fast, contactless, full field non-destructive technique used for detecting defects in composite and metals materials [1-8]. Commonly, it involves the use of flash heating lamps to inspect the surface and monitoring the thermal behaviour with an infrared camera, both during heating and cooling phases [1]. The most important applications include the detection and quantification of imposed defects, as flat bottom hole and inserts in composite materials to simulate delaminations and voids [1-8]. However, only few applications regard real defects [9-13], such as delaminations, impact damage, detachment, joint defects or even porosity, a typical and widespread defect especially in components produced using additive technology [9-10]. Recently, some works [9-10] demonstrated the capability of pulsed thermography to detect volumetric irregularly shaped defects, in comparison with micro computed tomography (μCT) investigation used as reference results.

As known, μCT is the more accurate NDT technique for characterizing volumetric defects, but it is time consuming and thus not suitable for industrial applications, as the additive manufacturing ones, in which many components need to be controlled [14-15]. However, it is known that diagnosing and quantifying a defect are two profoundly different things that are both necessary for obtaining the complete non-destructive control of components. Like all non-destructive testing techniques, thermography also needs calibration procedures, to define limits, advantages, and disadvantages of the techniques.

There are different works and references addressed to quantitative thermography, both for the depth [16-23] and the size characterization [24-26]. For the depth measurement, all the above-mentioned works are based on 1D models.

Lau et al. [16], following Carslaw and Jaeger's solution [8], developed a relation for the surface temperature history of finite uniform thickness plate subjected to a Dirac pulse. Later, Almond and Pickering [17] modified the 1D analytical model to consider the finite size of defects, and then to obtain the prediction of peak contrast amplitudes and times for defects with different aspect ratios.

Starting from these models, the relationship between the decay of temperature on the inspected surface and time is widely used to estimate defects depth and many methods are based on the time [3] or frequency domain [5-6] analyses.

There are different methods that start from the study of the contrast between a sound and a defect area, such as the peak-slope time method (PST) and the Pulsed Phase Thermography (PPT). There are also methods without a reference point as sound area include, for example, logarithm second derivative method (LSD) [3], by means of the well-known algorithm of the thermal signal reconstruction (TSR), and the least-squares fitting method (LSF) [20-22].

These methods assume that the thermal wave reflection coefficient γ is 1, which is not true for most real situations [22], as the case of a volumetric defect or a defect failed with powder and not air. These methods have been validated in previous works by means of numerical simulations and experimental data on composite and plastic materials [20-22]. Moreover, these methods are greatly influenced by the noise and the amplitude of the thermal contrast, making the application more difficult in the case of metals and defects due to lack of fusion, typical of additives, with powder trapped inside.

In this work, a series of AISI 316L stainless steel samples were obtained by using the Selective Laser Melting (SLM) process, according to a complete plan, providing for the insertion of imposed defects of different types and sizes directly in the CAD. Two samples were printed with cylindrical defects inside, planar with respect to the examined surface, considering in the first case the creation of suitable channels to eliminate the non-fused powder trapped inside, while in the other one leaving it. On the other hand, in a complete second plan, twelve samples have been printed with internal volumetric defects of spherical shape, with dimensions (distance from the surface and maximum dimension) similar to those of the cylinders. Also in this case, the very compact unfused powder, of the same material, was left inside.

The aim of the work is to investigate the different thermal behaviour of planar and volumetric defects and the effect of the presence of unfused powder inside.

For this reason, the samples were inspected by means of pulsed thermography technique, using a set-up in reflection configuration with two flash lamps for a nominal energy value of 6 kJ and a cooled sensor. The material properties characterization required to assess the diffusivity value according to the ASTM standard applying the well-known Parker method [27].

The analysis of the raw thermal data has been carried out starting from the study of 1D thermal models present in the literature [17], studying the simple thermal signal and the defect-sound contrast and then applying the non-linear least-squares fitting method (NLSF). The raw thermal data were then post-processed, applying the well-known algorithm of the Pulsed Phase Thermography (PPT) [5-6], looking for the differences between a planar defect and a volumetric defect in the frequency domain and in particular in the depth – blind frequency correlation.

The obtained results show the lowest values for the reflection coefficient in the case of volumetric defects, and lower values comparing cylindrical defects with and without metal powder inside. The errors in evaluating both the size and depth have been calculated considering the nominal defect dimensions. The biggest errors (>20%) have been obtained for the volumetric defects, as expected, and, in general, for the defects with the smallest diameter (3 mm) and therefore far from the hypothesis of a large reflector.

2. Theory

In its simplest form, the pulsed thermographic technique in reflection configuration sees the typical configuration of one or two flash lamps and a microlometer or cooled sensor to stimulate the investigated material with a short and high energy light pulse and observe the thermal decay by means of infrared thermography. A temperature response of an adiabatic semi-infinite body subjected to flash heating (theoretical Dirac pulse) is described with the well-known equation (1) [1]:

$$T(0,t)=\frac{Q}{\sqrt{\pi\rho ckt}}\exp\left(-\frac{d^2}{4\alpha t}\right) \tag{1}$$

where *d* represents the defect depth, *t* is time and *T* is the temperature value. *ρ*, *c*, *k*, *α* indicate the thermophysical material properties, respectively equal to density, heat capacity, thermal conductivity and the thermal diffusivity, equal in fact to the ratio *k/ρc*. *Q* takes into account the energy source for unit of surface transferred directly to the sample.

Following the models in [16], the presence of an inhomogeneity within the medium alters the transient temporal decay of temperature, depending on the difference of effusivities between the defect and the background material, and so directly according to the reflection coefficient γ. Specifically, the collected thermal decay is:

$$T(0,t)=\frac{Q}{\sqrt{\pi\rho ckt}}\left[1+2\sum_{n=1}^{\infty}\gamma^n\exp\left(-n^2\frac{d^2}{\alpha t}\right)\right] \tag{2}$$

Almond et al. [17], proposed a simple analytical model in which the effect of the defect aspect ratio on the signal contrast is considered, starting from the analysis of the thermal contrast between the defect and related sound area, over the centre of the detected defect:

$$T_c(0,t)=\frac{2Q}{\sqrt{\pi\rho ckt}}\left[\sum_{n=1}^{\infty}\gamma^n\exp\left(-n^2\frac{d^2}{\alpha t}\right)\right]\left[1-\exp\left(-\frac{(Pd)^2}{16m\alpha t}\right)\right] \tag{3}$$

where *P=D/d* is the defect aspect ratio, *m* is the ratio of in-plane to through thickness thermal and *D* is the diameter of a circular defect. Equation (3) can be used for estimating size (diameter) and depth of defects with a known aspect ratio and setting *γ=1*. However, equation (3) does not consider the shape or typology of defects.

Later, Sirikham et al. [21] proposed a modification that allows the equation (3) to be applicable to any segment of collected data. The proposed model is reported in equation (4):

$$T_C(0,t+t_s)=\frac{2Q}{\sqrt{\pi\rho ck(t+t_s)}}\left[\sum_{n=1}^{\infty}\gamma^n\exp\left(-\frac{(nd)^2}{\alpha(t+t_s)}\right)\right]\left[1-\exp\left(-\frac{D^2}{16m\alpha(t+t_s)}\right)\right] \tag{4}$$

where t_s is the starting time of sampling.

It is necessary to repeat that the analytical models in thermal NDT are typically 1D, and so they mainly refer to planar defects oriented in a parallel direction with respect to the sample surface [22]. In fact, the expressions above are approximate and allow for the correcting 1D analytical solutions by finite size of defect, typically flat bottom hole with different aspect ratio.

As known, the study of the simple thermal contrast doesn't allow to characterize and quantify properly a defect, especially when this value is too weak, or the defect have a small aspect ratio.

For this reason, there are several algorithms of post processing used to improve the quality of the raw thermal data, and certainly the oldest one is the Pulsed Phase thermography (PPT), that allow to study the thermal response of the inspected surface in the frequency domain, passing through the appropriate use of the Discrete Fourier Transform (DFT), which is not always very simple to apply correctly. In particular, the correct application of this algorithm required the choice of the right truncation window size and the frame rate, to discretize the data in the correct way and by dividing the energy into the right number of components. Thanks to this type of analysis, it is therefore possible to have a rough estimation of the defect depth, considering the phase maps and the phase contrast for the detected defect. The main mathematical steps are enclosed in the following equations and in the last one, equation (7), which links the depth of the defect to the blind frequency, or the frequency for which the phase contrast is equal to zero, when the thermophysical properties of the material are known.

$$\text{DFT } F_n=\Delta t\sum_{k=0}^{N-1}T(k\Delta t)e^{-\frac{2j\pi nk}{N}}=Re_n+Im_n \tag{5}$$

$$\text{phase} \quad \varphi_n = \tan^{-1}\left(\frac{Im_n}{Re_n}\right) \tag{6}$$

$$\text{blind frequency } d = C_1 \sqrt{\frac{\alpha}{\pi f_b}} + C_2 \tag{7}$$

In these equations, *i* is the imaginary number, *Re* and *Im* are, respectively, the real and imaginary parts of the transform, and the subscript *n* represents the frequency increment. Phase maps are then computed with the equation (6), by repeating this process for all pixels. With *N* thermograms in the available sequence, *N*/2 frequency values can be obtained (due to the symmetry of the Fourier transform). Considering the sequence related to the phase maps, a detected defect and the related sound zone, it is possible to define the phase contrast and then the equation (7), that correlate the blind frequency f_b to the defect depth d; here, C_1 and C_2 represent two calibration constants, and their value depends on the material properties and type of the defect.

3. Experimental setups and materials

3.1. Experimental setup and materials for the SLM process

A series of AISI 316L stainless steel samples (figure 1) were obtained by using a commercially available Concept Laser M1 Selective Laser Melting (SLM) machine. The latter is equipped with a Nd: YAG laser source characterized by a wavelength of 1064nm, a maximum laser power equal to 100W and a laser spot diameter equal to 200 μm. AISI 316L stainless steel powder, a commercial spherical powder with a particle size in the range of 15-53 μm, produced through gas atomization by Cogne Acciai Speciali S.p.a (Italy), was used as powder metal. Table 1 shows the nominal compositions of the powder as certified by the manufacturer.

Figure 1. SLM samples.

Table 1. Chemical composition of the AISI 316L stainless steel powder (weight %).

Alloying element	Cr	Ni	C	Mn	Si	S	Mo	N	P	O	Cu	Fe
Nominal	16.0-18.0	11.0-13.0	0.03	2.0	0.75	0.010	2.0-3.0	0.10	0.025	0.10	0.50	Balance

The samples were fabricated with defects of known sizes within them. As already said, two different shapes were chosen for the defects: spherical and cylindrical (figure 2). Some of the samples produced were manufactured with holes in the direction of the defects to allow the removal of the metal powder trapped in the defect during printing.

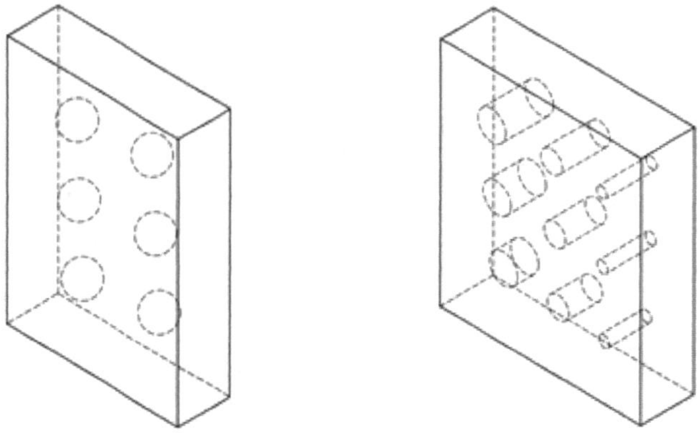

(a) (b)

Figure 2. Internal defects of the manufactured samples: (a) spherical volumetric defects; (b) cylindrical planar defects.

The samples were constructed with the same process parameters and scanning strategy optimized in previous studies [28-29]. A 5x5 mm^2 island scanning strategy was used (figure 3). The order in which these islands are scanned is random, a strategy patented by Concept Laser; the combination of these scanning factors allows to reduce the residual thermal stresses [30]. Samples were rotated 45° on the construction platform to avoid problems with powder coating from the blade. The process parameters considered as constant are presented in table 2.

Table 2. Process parameters.

Parameter	Unit	N	Value
Laser power	W	P	100
Scanning speed	mm s^{-1}	v	200
Laser spot diameter	μm	d	200
Layer thickness	μm	LT	30
Hatch distance	μm	h_d	140

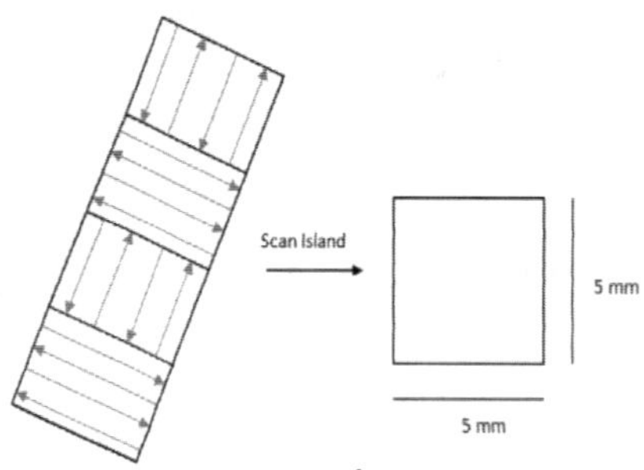

Figure 3. Random island scanning strategy: 5x5mm^2 islands with scan angle variation of 90° between adjacent islands.

The experimental plan is resumed in the following tables (3, 4 and 5). It is necessary to point out that not all defects have been used in this work. The full analysis is intended for future work. In figure 4 are instead reported the technical drawings of the samples 2 and 3, used to show the intended geometry.

Table 3. Experimental plan, sample 1 – cylinders no powder (planar defects), diameters (D), depths (d) and defect aspect ratio (D/d).

SAMPLE 1 (Cylinders no powder)			
DIAMETER (D) [mm]	8,00	ASPECT RATIO (D/d)	DRILLED
DEPTH (d) [mm]	0,50	16,00	
	2,00	4,00	
	4,00	2,00	
DIAMETER (D) [mm]	6,00		
DEPTH (d) [mm]	0,50	12,00	
	1,50	4,00	
	2,50	2,40	
DIAMETER (D) [mm]	3,00		
DEPTH (d) [mm]	0,20	15,00	
	0,40	7,50	
	1,00	3,00	

Table 4. Experimental plan, sample 2 – cylinders filled with powder (planar defects), diameters (D), depths (d) and defect aspect ratio (D/d).

SAMPLE 2 (Cylinders filled with powder)			
DIAMETER (D) [mm]	8,00	ASPECT RATIO (D/d)	Z LENGHT [mm]
DEPTH (d) [mm] SIDE A	0,50	16,00	13,50
	2,00	4,00	10,00
	4,00	2,00	6,00
DEPTH (d) [mm] SIDE B	1,00	8,00	13,50
	3,00	2,67	10,00
	5,00	1,60	6,00
DIAMETER (D) [mm]	6,00		
DEPTH (d) [mm] SIDE A	0,50	12,00	13,50
	1,50	4,00	11,50
	2,50	2,40	9,50
DEPTH (d) [mm] SIDE B	1,00	6,00	13,50
	2,00	3,00	11,50
	3,00	2,00	9,50
DIAMETER (D) [mm]	3,00		
DEPTH (d) [mm] SIDE A	0,20	15,00	14,20
	0,40	7,50	13,80
	1,00	3,00	12,00
DEPTH (d) [mm] SIDE B	0,60	5,00	14,20
	0,80	3,75	13,80
	2,00	1,50	12,00

Table 5. Experimental plan, sample 3, 4 and 5 – spheres filled with powder (volumetric defects), diameters (D), depths (d) and defect aspect ratio (D/d).

SAMPLE 3 x4			**SAMPLE 4 x4**			**SAMPLE 5 x4**		
DIAMETER (D) [mm]	8,00	(D/d)	DIAMETER (D) [mm]	6,00	(D/d)	DIAMETER (D) [mm]	3,00	(D/d)
DEPTH (d) [mm] SIDE A	0,50	16,00	DEPTH (d) [mm] SIDE A	0,50	12,00	DEPTH (d) [mm] SIDE A	0,20	15,00
	1,00	8,00		1,00	6,00		0,40	7,50
	2,00	4,00		1,50	4,00		0,60	5,00
	3,00	2,67		2,00	3,00		0,80	3,75
	4,00	2,00		2,50	2,40		1,00	3,00
	5,00	1,60		3,00	2,00		1,20	2,50
DEPTH (d) [mm] SIDE B	5,00	1,60	DEPTH (d) [mm] SIDE B	2,80	2,14	DEPTH (d) [mm] SIDE B	2,00	1,50
	4,50	1,78		2,30	2,61		1,80	1,67
	3,50	2,29		1,80	3,33		1,60	1,88
	2,50	3,20		1,30	4,62		1,40	2,14
	1,50	5,33		0,80	7,50		1,20	2,50
	0,50	16,00		0,30	20,00		1,00	3,00

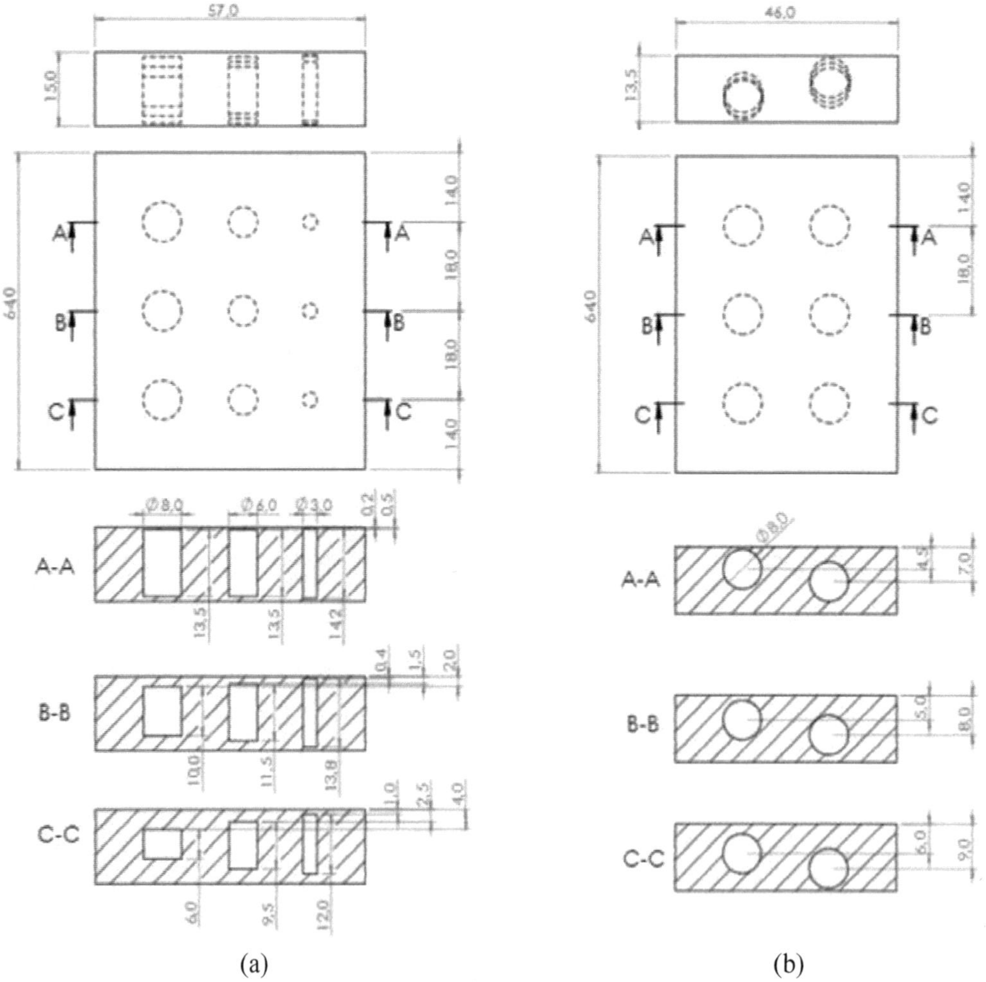

Figure 4. Technical drawings reported as an example for the sample 2 (cylinders, planar defects) and sample 3 (spheres, volumetric defects).

3.2. Experimental setup for Pulsed Thermography

The experimental setup (figure 5) in reflection configuration used to perform the pulsed thermographic tests included a cooled sensor MWIR FLIR A6751 *(FLIR Systems, Wilsonville,* OR, USA) with a lens of 50 mm and a Hensel EH Pro 6000 (Hensel, Fairfield, NJ, USA) flash tube (optical pulse energy 6 kJ, pulse duration 3-5 ms). The main testing parameters are resumed in the following table. The contemporary need for a high frame rate and a good spatial resolution required the use of a no full frame window, with a consequent total number of tests equal to approximately 250, considering the complete plan and therefore the replicated samples.

Table 6. Main experimental setup parameters.

Frame rate (Hz)	Acquisition time (s)	Calibration temperature range (°C)	Integration time (μs)	Windowing (pixels)	Spatial resolution (mm/pixel)
600	10	15.7-85 °C	595.43	320x156	0.14

(a) (b)

Figure 5. Experimental setup: thermal source flash lamps and IR cooled sensor FLIR A6751.

4. Methodology and first results

4.1. 1D model and NLSF optimization procedure

As suggested in previous publications [20-22], it is possible to fit experimental data with the analytical solution using equation (3), choosing properly the number of variables and the variables themselves.

In this work, one of the main aims is to understand if the modification of the coefficient γ can describe the behaviour of a volumetric defect, and if the powder inside the defect reduces its value.

Another question is related to the depth d and diameter D indicated as result or input parameters of the fitting, being the same variables in the case of a sphere, as a volumetric defect, based on the position of the pixels used and the cooling time analysed (truncation window size).

Besides, as already specified, the fitting required the previous assessing of the sound area and the defect one.

Here, considering the diagram and the result shown as an example in figure 6, a total area of 100 pixels is selected to evaluate the sound and 9 pixels for the defect area, then the thermal contrast is calculated, as represented in figure 6, passing from the thermal trends reported in figure 7 to the contrast, for the defects with the same nominal sizes diameter 8 mm and depth 0.5 mm, in the case of a cylinder without powder (blue line), a cylinder filled with powder (green line) and a sphere (red line). The figure 6 reports a thermal map taken, as an example, within the sequence at the time of maximum contrast.

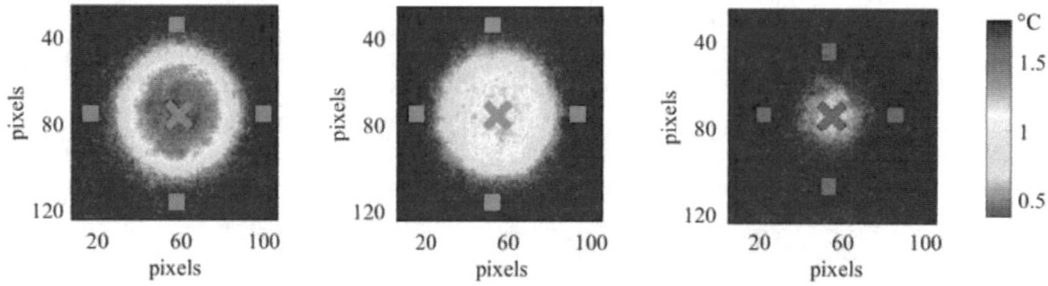

(a) (b) (c)

Figure 6. Thermal contrast maps and diagram for sound and defect selection: (a) cylinder no powder, (b) cylinder filled with powder, (c) sphere filled with powder – D=8mm, d=0.5 mm.

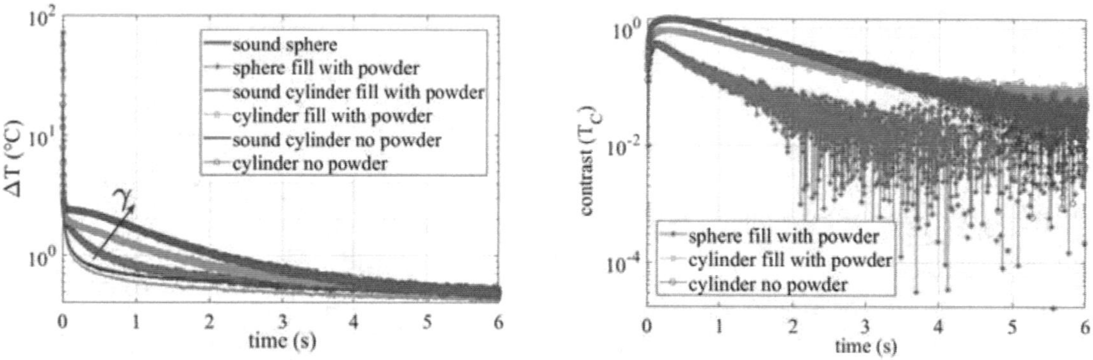

(a) (b)

Figure 7. (a) Thermal trends and (b) contrasts, considering cylinder no powder (blue line), cylinder filled with powder (green line), sphere filled with powder (red line) – D=8mm, d=0.5 mm.

In this study, the free parameters for the fitting will be γ, D, d and t_s, following, as model, the modification proposed by Sirikham et al. (equation 4), m was assumed equal to 1, considering the material thermally isotropic, instead the number of iterations was assumed equal to 10, since the higher summing terms have appeared negligible. The fitting involved a precise and limited truncation window size of the raw thermal data, considering the signal relative to the standard deviation of the sound area, and therefore as the last frame the one for which the thermal contrast, taken at the centre of the defect, exceeded the chosen threshold by 2 times. To reduce the noise, a Gaussian filter equal to 1 was applied to the raw thermal data.

The NLF algorithm used the Matlab *lsqnonlin* least-square function to find the parameters, following an optimization procedure, following the equation (8):

$$\min_{\gamma,D,d,t_s}\left|T_{c_{est}}(t)-T_{c_{exp}}(t)\right| \tag{8}$$

considering the contrast between defect and sound, estimated T_{c_est} and experimental T_{c_exp}. The values of γ, d, t_s, D varied as follows:

$$x_0=\left[0.6,\ 2.5\text{x}10^{-3},\ 0.005,\ 5\text{x}10^{-3}\right];$$
$$l_b=\left[0.1,\ 0.2\text{x}10^{-3},\ 0,\ 0.3\text{x}10^{-3}\right];\ u_b=\left[1,\ 5\text{x}10^{-3},\ 0.05,\ 8\text{x}10^{-3}\right] \tag{9}$$

considering x_0 the initial value, and l_b (lower band) with u_b (upper band) the extremes of the chosen interval. The used model is the one reported in the equation (10), where the parameters left as free in the fitting are reported in square brackets:

$$T_C[\boldsymbol{\gamma},\mathbf{D},\mathbf{d},\mathbf{t_s}]=\frac{2Q}{\rho c\sqrt{\pi\alpha(t+\mathbf{t_s})}}\left[\sum_{n=1}^{10}\gamma^n\exp\left(-\frac{(n\mathbf{d})^2}{\alpha(t+\mathbf{t_s})}\right)\right]\left[1-\exp\left(-\frac{\mathbf{D}^2}{16\alpha(t+\mathbf{t_s})}\right)\right] \tag{10}$$

The unknown parameters $Q/\rho c$ and α are estimated, separately. In particular, for $Q/\rho c$, the same model reported in equation (10) was used, calibrating with the biggest defect in terms of aspect ratio, with no powder inside, and therefore closer to the condition of the 1D model; in this case, the defect sizes are considered as known, and respectively equal to 8 mm for the diameter *D* and 0.5 mm for the depth *d*, fitting only the portion of the curve between t_s and the time of the maximum contrast, in order to have a correct estimate of the energy input *Q*.

For the rough estimation of the thermal diffusivity α, reference was made to the ASTM standard [27], applying the Parker method, with a typical setup in transmission, cooled thermal camera and flash lamp, using a sample printed as per standard, of the same material with dimensions of 15x15x3 mm^3, averaging the result on 5 different tests.

4.2. Application of the Pulsed Phase Thermography (PPT) algorithm

The raw thermal data are then analysed, by means of the PPT algorithm. A truncation window size of 4096 is analysed, obtaining 2048 phase maps. The typical trend of the phase contrast is reported in figure 8, considering the defects with a diameter equal to 8 mm and a depth of 0.5 mm, in the case of a cylinder without powder (blue line), a cylinder filled with powder (green line) and a sphere (red line). As it is possible to appreciate from the reported zoom, the value related to the blind frequency is different, and it corresponds to a discrete value near to the theoretical one equal to 0. In this work, the value closest to 0 was automatically chosen by comparing the last negative and the first positive. 8 different defects are used to have an estimation of the constants C_1 and C_2, with the same nominal sizes, to calibrate the result and so to have a rough estimation of the defect depth.

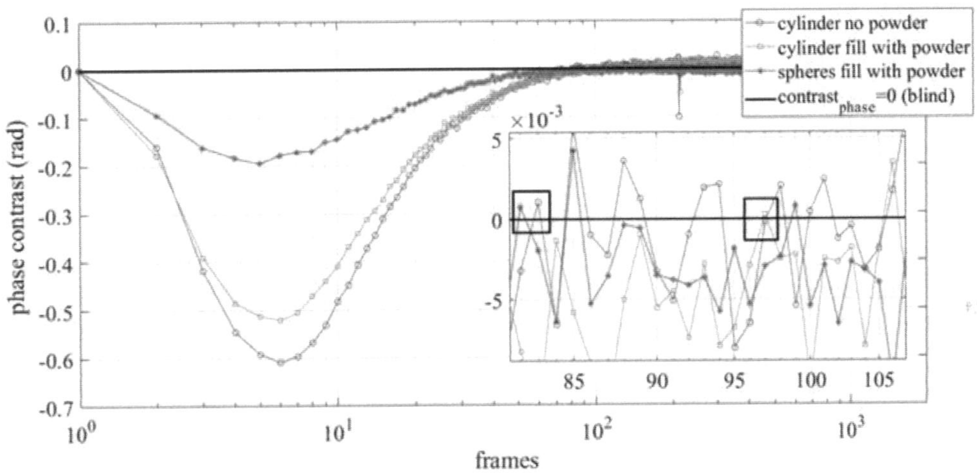

Figure 8. Phase contrast trends, considering cylinder no powder (blue line), cylinder filled with powder (green line), sphere filled with powder (red line) – D=8mm, d=0.5 mm.

The quantitative analysis consists in estimating the depth (d_{est}) and diameter (D_{est}) of the defects with both methods, calculating the percentage error respect to the nominal values (d_{nom}, D_{nom}) and so the equation (11):

$$\%Error_d=\frac{d_{est}-d_{nom}}{d_{nom}}x100;\ \%Error_D=\frac{D_{est}-D_{nom}}{D_{nom}}x100 \tag{11}$$

5. Quantitative results and discussion

5.1. 1D model: assessing the reflection coefficient value and defect sizes estimation

Following what is reported in the previous paragraph 4.1, some of the available thermographic data have been analysed by means of the NLSF method. As mentioned above, the use of the NLF technique requires a priori knowledge of material thermal diffusivity, that was evaluated using the Parker method considering a sample with standard dimensions and obviously of the same material and lot of powder. After five different measurements, considering an average area in the middle of the sample with homogeneous and uniform heating, and so the mean and standard deviation value of the same, the estimate obtained is in any case in the range of the literature and equal to $3.25 \times 10^{-6} \mp 0.02 \left[\frac{m^2}{s}\right]$.

Following the equation 11 and the NLSF method, it is possible to obtain an estimation of the energy input for this material, equal to $\frac{Q}{\rho c}$=0.0015 [K*m], with a contrast trend over the defect centre and the relative fitting shown in figure 9. For this case, the reflection coefficient value was determined by fitting equal to *γ=0.84*.

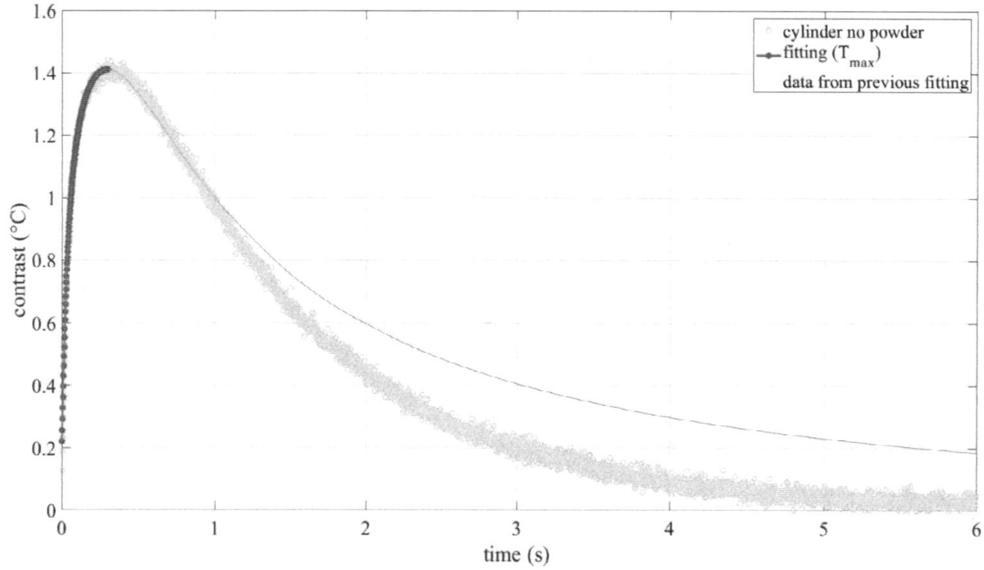

Figure 9. Contrast trend (cyan line), data used for the fitting with NLSF method (blue points) and result of the fitting considering the case of a cylinder with no powder – D=8mm, d=0.5 mm.

The 1D model application concerns the defects visible from the analysis of the thermal contrast, in any case with a value higher than 2 times the standard deviation of the relative sound signal. The result for the empty cylinders is reported in figure 10, for the cylinders filled with powder in figure 11, while for the spheres in figure 12. As it is easily to see, the chosen criterion allows a better fitting for lower contrasts.

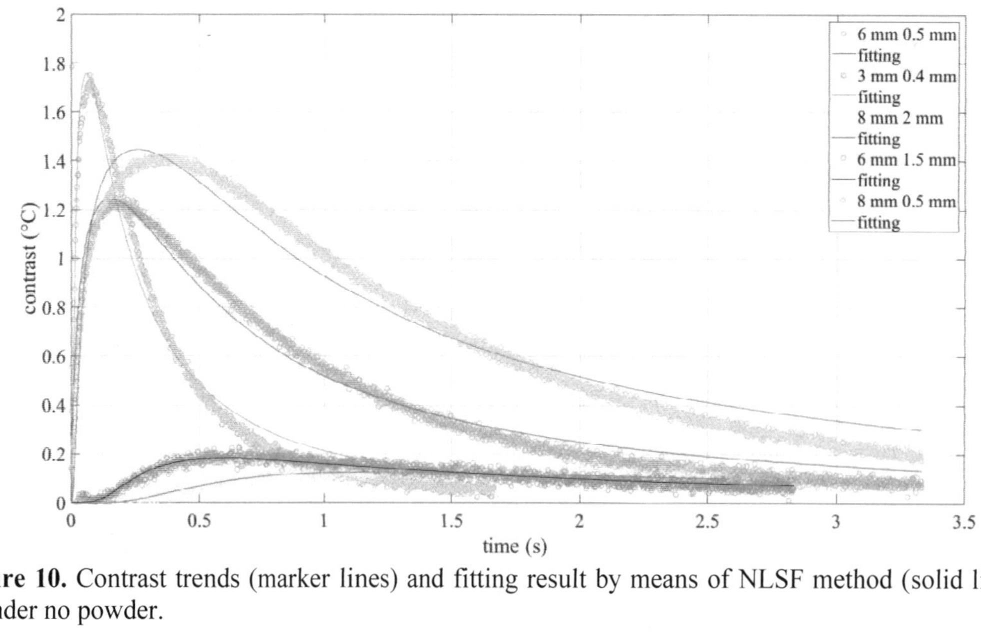

Figure 10. Contrast trends (marker lines) and fitting result by means of NLSF method (solid line) – cylinder no powder.

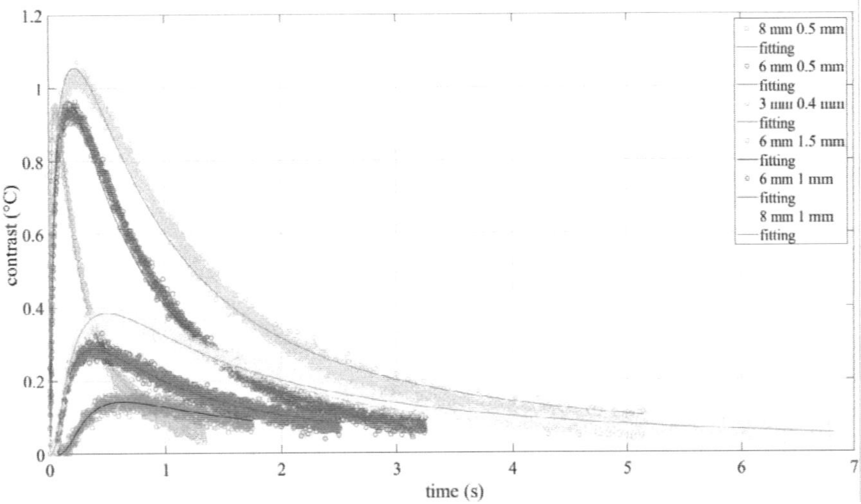

Figure 11. Contrast trends (marker lines) and fitting result by means of NLSF method (solid line) – cylinder filled with powder.

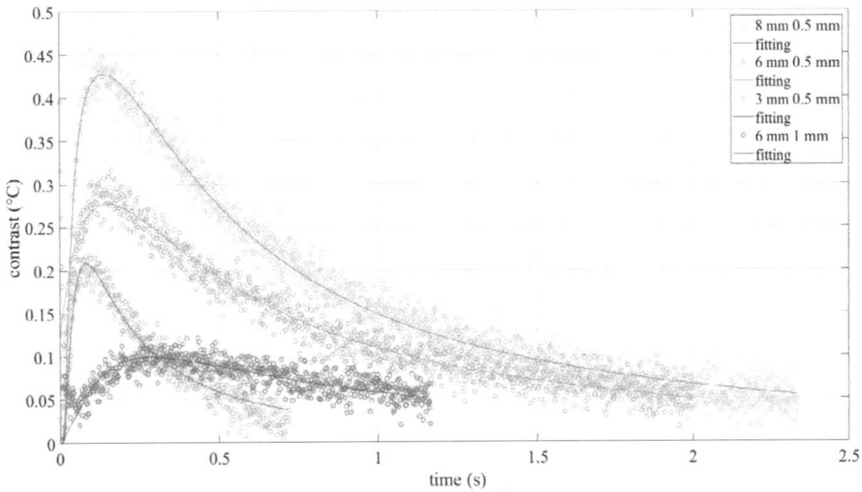

Figure 12. Contrast trends (marker lines) and fitting result by means of NLSF method (solid line) – spheres filled with powder.

The following table 7 instead summarizes the main results of the fitting, as comparative indexes to distinguish the types of defects and to have an estimation of the errors carried out in determining their size.

Table 7. Quantitative results in size estimation by means of NLSF method.

Nominal depth (d_n) [mm]	Nominal Diameter (D_n) [mm]	γ	Estimated depth (d) [mm]	t_s [s]	Estimated Diameter (D) [mm]	% depth error	% Diameter error	R^2
5,00E-04	8,00E-03	0,69	4,32E-04	0,00	7,66E-03	-13,55	-4,19	0,86
5,00E-04	6,00E-03	0,65	4,22E-04	0,02	6,24E-03	-15,63	3,98	0,84
1,50E-03	6,00E-03	0,65	1,38E-03	0,00	6,10E-03	-7,68	1,75	0,93
2,00E-03	8,00E-03	0,62	1,93E-03	0,05	8,00E-03	-3,37	0,00	0,96
4,00E-04	3,00E-03	0,63	2,56E-04	0,00	3,55E-03	-35,94	18,35	0,78

5,00E-04	8,00E-03	0,42	6,15E-04	0,05	5,80E-03	23,02	0,76
5,00E-04	6,00E-03	0,38	5,77E-04	0,05	6,37E-03	15,37	0,93
4,00E-04	3,00E-03	0,36	4,98E-04	0,02	3,20E-03	24,55	0,95
1,00E-03	6,00E-03	0,36	1,13E-03	0,05	7,09E-03	12,59	0,73

By means of the reflection coefficient γ, it is possible to discern the thermal behaviour between metal-air $\gamma = (0.7/0.6)$, metal-powder metal $\gamma = (0.6/0.5)$. This difference is even more evident in the case of the spheres (metal-powder metal) $\gamma = (0.4/0.3)$.

The higher errors (>20%) in the estimation of the sizes are obtained in the case of spheres or also planar defects with lower diameter (3 mm), because they are further away from the hypotheses of 1D model, and they are highlighted in yellow in table 7.

The column of the errors committed in estimating the diameter in case of volumetric defects is left blank because the size depends on the section of the sphere and therefore from the analysed truncation window size.

5.2. Quantitative results by means of the application of the Pulsed Phase Thermography (PPT) algorithm

The raw thermal data have been processed by means of PPT. The analysis returns different phase maps in the frequency domain with result quality and thermal resolution that depends on the testing and analysis parameters, and in particular on the frame rate and the truncation window size.

Each frequency result of this analysis theoretically refers to a precise defect depth, and the two quantities are inversely proportional. It follows that for each frequency, in the case of a planar defect, there is always the same thermal imprint, as shown for example in figure 13 (a, b, c) for the defect with sizes D=8 mm and d=0.5 mm considering 3 different frequencies. Instead, in the case of a volumetric defect, the typical result is the one reported in figure 13 (d, e, f), for the same frequencies and the sphere with the analog sizes of the previous cylinder.

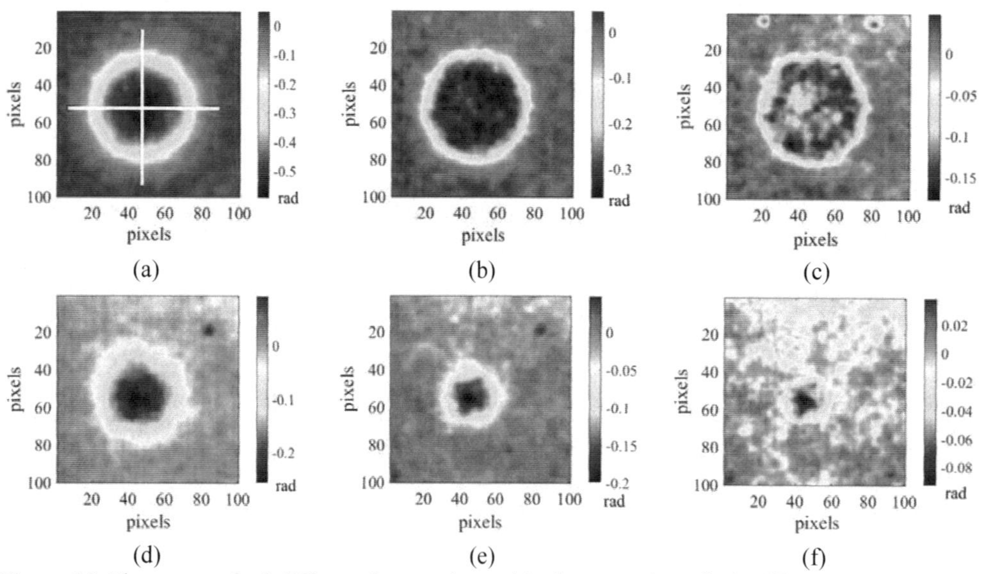

Figure 13. Phase maps for 3 different frequencies and in the case of a cylinder filled with powder and the analogous sphere D=8mm, d=0.5 mm; (a) f=0.59 Hz – cylinder, (b) f=1.46 Hz – cylinder, (c) f=3.22 Hz – cylinder, (d) f=0.59 Hz – sphere, (e) f=1.46 Hz – sphere, (f) f=3.22 Hz – sphere.

Considering a profile in both directions (figure 13 a, with white lines), it is possible to recognize the same typical trend, therefore characteristic of planar defects and volumetric defects. This effect is obviously less pronounced for small defects. Figure 14 reports the case for planar and volumetric defects filled with metal powder, considering 3 different diameters (8 mm, 6 mm and 3 mm) and 3 different depths (0.5 mm, 0.5 mm and 0.2 mm).

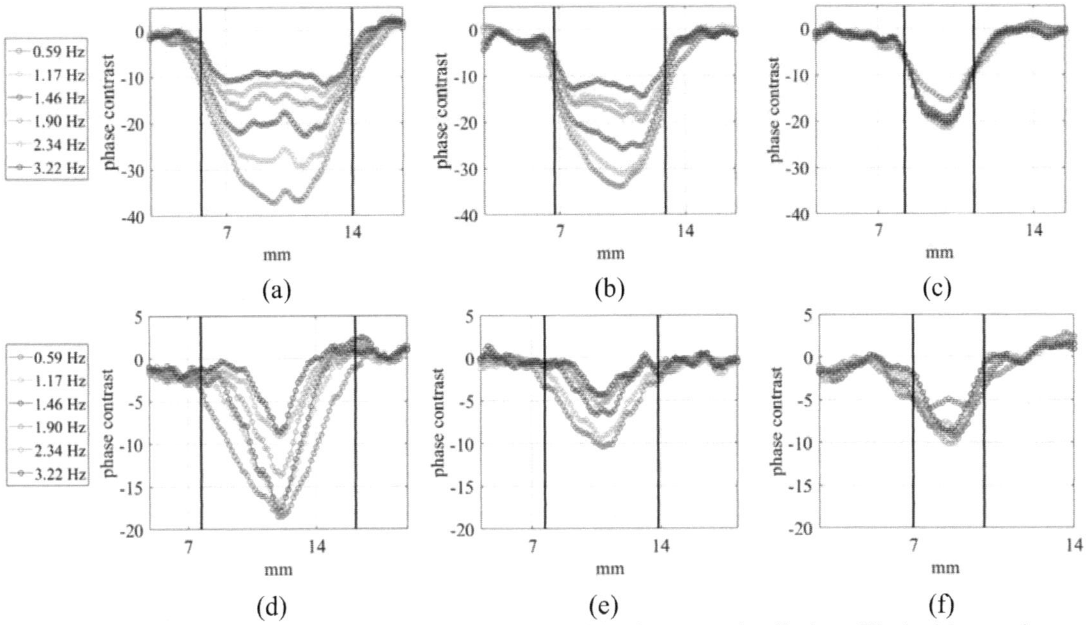

Figure 14. Phase trends for 6 different frequencies and in the case of cylinders filled with powder and the analogous spheres for 3 different dimensions; (a) cylinder D=8 mm, d=0.5 mm, (b) cylinder D=6 mm, d=0.5 mm, (c) cylinder D=3 mm, d=0.2 mm, ; (d) sphere D=8 mm, d=0.5 mm, (e) sphere D=6 mm, d=0.5 mm, (f) sphere D=3 mm, d=0.2 mm.

Finally, it is possible to note a characteristic thermal trend for the two types of defects by plotting the surface at a certain frequency. There is therefore a phase contrast with an almost constant value for the entire surface of the defect, showing an example in figure 15 a for the cylinder filled with powder D=8 mm – d=0.5 mm, and the analogous sphere figure 15 b with a phase peak in the centre of the defect, considering obviously the same scale.

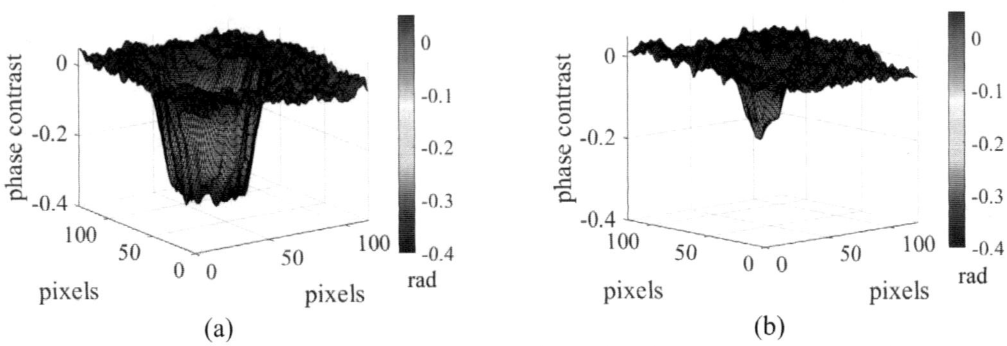

Figure 15. Phase surface maps in the case of a cylinder filled with powder (a) and the analogous sphere (b) D=8mm, d=0.5 mm for a f=1.46 Hz.

Considering the equation 6 and so the phase contrast trends for 8 different defects (see table 8 and rows 1 and 2 for the nominal sizes) and all the cases (cylinders without powder (blue line), cylinders filled with powder (green line) and spheres (red line)), the blind frequency value for each defect was assessed, following the criterion explained in the section 4.2. The calibration curve (equation 6) seems to be different changing the defect type, and in particular the calibration constant C_2. The constant C_1 is instead very similar and in fact the calibration curves are almost parallel.

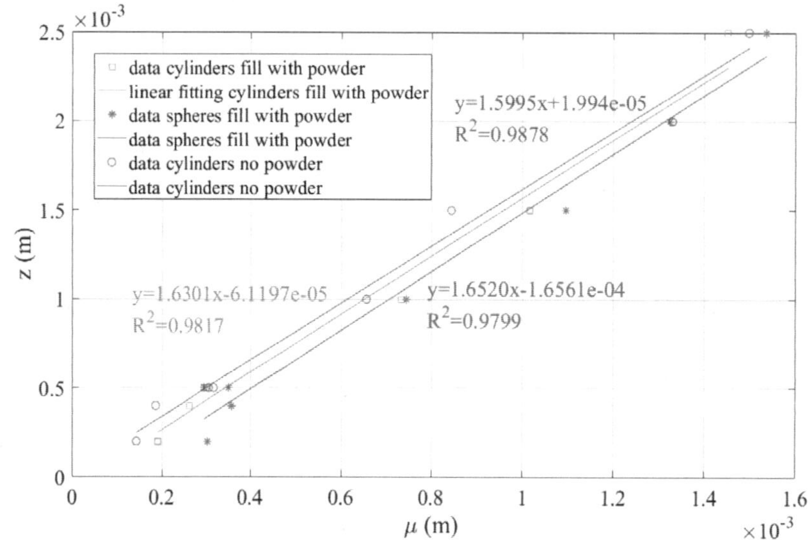

Figure 16. Calibration curves by means of PPT (equation 6); cylinders without powder (blue line), cylinders filled with powder (green line) and spheres (red line).

The errors carried out in estimating the defect depths by means of the previous equations and the calibration curves are reported in table 8, with their respective colours. Again, the highest errors (>20%) are recorded in the case of volumetric defects and for the smallest planar defects, especially when the wrong curve is used to determine the error on the depth (the type of defect is not always known a priori). The obtained curves are logically still valid for the defects not considered for the calibration, as reported for example in last column of the table 8, in bold.

Table 8. Quantitative results in size estimation by means of PPT.

Nominal Diameter (D_n) [mm]	3,00E-03	6,00E-03	6,00E-03	6,00E-03	8,00E-03	8,00E-03	3,00E-03	3,00E-03	**6,00E-03**
Nominal depth (d_n) [mm]	2,00E-04	2,50E-03	1,50E-03	5,00E-04	2,00E-03	5,00E-04	4,00E-04	1,00E-03	**1,00E-03**
Estimated depth (d) [mm]	2,48E-04	2,41E-03	1,37E-03	5,29E-04	2,14E-03	5,11E-04	3,17E-04	1,07E-03	
% depth error	24,23	-3,48	-8,77	5,72	7,24	2,27	-20,68	6,81	
Estimated depth (d) [mm]	3,24E-04 2,50E-04	2,34E-03 2,30E-03	1,64E-03 1,59E-03	5,00E-04 4,29E-04	2,14E-03 2,10E-03	4,95E-04 4,23E-04	4,39E-04 3,67E-04	1,19E-03 1,13E-03	**1,07E-03** **1,00E-03**
% depth error	62,15 24,77	-6,46 -7,91	9,60 6,30	0,08 -14,20	7,24 5,24	-1,06 -15,36	9,82 -8,33	18,86 13,04	**6,52** **0,46**
Estimated depth (d) [mm]	5,07E-04 4,35E-04 3,38E-04	2,47E-03 2,44E-03 2,37E-03	1,77E-03 1,72E-03 1,64E-03	5,83E-04 5,13E-04 4,16E-04	2,14E-03 2,10E-03 2,02E-03	4,94E-04 4,22E-04 3,24E-04	5,93E-04 5,23E-04 4,27E-04	1,21E-03 1,15E-03 1,06E-03	**7,77E-04** **5,97E-04** **5,23E-04**
% depth error	153,34 117,71 68,84	-1,06 -2,41 -5,24	18,08 14,94 9,58	16,54 2,58 -16,76	6,85 4,84 1,07	-1,25 -15,55 -35,14	48,24 30,83 6,69	20,68 14,89 6,08	**-2,28** **-3,22** **-12,28**

The analysis of the error will be the subject of future works and thanks to replicated samples, it will allow us to obtain the confidence bands and the other defects available in the experimental plan.

The results demonstrate the importance of adopting a suitable calibration procedure for obtaining an accurate estimation of the size and depth, above all for volumetric defects.

6. Conclusions and outlooks

The main conclusions of this work can be summarized by the following bulleted list:

- Assessment of quantitative indexes to discern the thermal behaviour between metal-air γ=(0.7/0.6), metal-powder metal γ=(0.6/0.5);
- Assessment of quantitative indexes to discern the thermal behaviour between planar $\gamma = (0.7/0.5)$ and volumetric defect γ=(0.4/0.3);
- Evaluation of the errors done in defects characterization by using 1D methods (1D model with the diameter correction and the blind frequency-PPT);
- First indication to define the limit of detection of volumetric defects by active thermography $\left(\frac{D}{d} \geq \frac{6}{2.5} = 2.4; \frac{D}{d} \geq \frac{3}{1} = 3\right)$ and the adopted set-up → 2 flash lamps $\frac{Q}{\rho c}$=0,0015 [K*m], MWIR cooled thermal sensor 600 Hz, acquisition time 7s.

The results presented in this first work are to be understood as preliminary results of a larger activity, which aims to the following future developments:

- Completing all the analyses provided for in the experimental plan and evaluating the probability of detection (POD) for volumetric defects by means of different algorithms;
- Carrying out a FEM model to study the 3D effects produced by the volumetric defects on the thermal signal comparing with the experimental tests (detection limit for volumetric defects);
- Using volumetric defects for calibrating the thermographic data and then quantifying porosity as typical defects of metal additive manufacturing (micrographic analysis as a reference).

Conflicts of Interest: the authors declare no conflict of interest.

References

[1] Maldague X P V 2001 *Theory and practice of infrared technology of non-destructive testing*. John Wiley & Sons, Inc. ISBN 0-471-18190-0.

[2] Oswald-Tranta B 2017 Time and frequency behaviour in TSR and PPT evaluation for flash thermography. *Quantitative InfraRed Thermography Journal*. **14(2)** 164-184.

[3] Shepard S M, Lhota J R, Rubadeux B A, Wang D and Ahmed T 2003 Reconstruction and Enhancement of Active Thermographic Image Sequences. *Optical Engineering*. **42(5)** 1337-1342.

[4] Vavilov V P and Burleigh D D 2015 Review of pulsed thermal NDT: physical principles, theory and data processing. *NDT & E International*. **73** 28–52.

[5] Ibarra-Castanedo C 2005 Quantitative subsurface defect evaluation by pulsed phase thermography: depth retrieval with the phase. Collection Mémories et thèses électroniques.

[6] Ibarra-Castanedo C, Maldague X P 2004 Defect depth retrieval from pulsed phase thermographic data on plexiglas and aluminum samples. In *Thermosense XXVI International Society for Optics and Photonics*. **5405** 348-357.

[7] Krapez J C, Lepoutre F and Balageas D 1994 Early detection of thermal contrast in pulsed stimulated thermography. *J. Phys. IV*, 4, C7-47–C7-50.

[8] Carslaw H S and Jaeger J C 1959 *Conduction of Heat in Solids* (Oxford University Press: Oxford, UK).

[9] D'Accardi E, Altenburg S, Maierhofer C, Palumbo D and Galietti U 2019 Detection of Typical Metal Additive Manufacturing Defects by the Application of Thermographic Techniques. *Multidisciplinary Digital Publishing Institute Proceedings*. **27** 1-24.

[10] D'Accardi E, Ulbricht A, Krankenhagen R, Palumbo D and Galietti U 2021 Capability of active thermography to detect and localize pores in Metal Additive Manufacturing materials. *IOP Conference Series: Materials Science and Engineering*. **1038 (1)** 012018.

[11] D'Accardi E, Palano F, Tamborrino R, Palumbo D, Tatì A, Terzi R and Galietti U 2019 Pulsed phase thermography approach for the characterization of delaminations in CFRP and comparison to phased array ultrasonic testing. *Journal of Nondestructive Evaluation*. **38(1)** 20.

[12] Dell'Avvocato G, Palumbo D, Pepe R and Galietti U 2021 Non-destructive evaluation of resistance projection welded joints (RPW) by flash thermography. *IOP Conference Series: Materials Science and Engineering.* **1038 (1)** 012003.
[13] Maierhofer C, Röllig M, Ehrig K, Meinel D and Céspedes-Gonzales G 2014 Validation of flash thermography using computed tomography for characterizing inhomogeneities and defects in CFRP structures. *Composites Part B: Engineering.* **64** 175-186.
[14] DebRoy T, Wei H L, Zuback J S, Mukherjee T, Elmer J W, Milewski J O, ... and Zhang W 2018 Additive manufacturing of metallic components–process, structure and properties. *Progress in Materials Science.* **92** 112-224.
[15] Mani M, Feng S, Lane B, Donmez A, Moylan S and Fesperman R 2017 Measurement science needs for real-time control of additive manufacturing powder bed fusion processes. CRC Press.
[16] Lau S K, Almond D P and Patel P M 1991 Transient thermal wave techniques for the evaluation of surface coatings. *Journal of Physics D: Applied Physics.* **24(3)** 428.
[17] Almond D P and Pickering S G 2012 An analytical study of the pulsed thermography defect detection limit. *Journal of applied physics.* **111(9)** 093510.
[18] Beemer M F and Shepard S M 2018 Aspect ratio considerations for flat bottom hole defects in active thermography. *Quantitative InfraRed Thermography Journal.* **15(1)** 1-16.
[19] Han X, Favro L D, Kuo P K and Thomas R L 1996 Early-time pulse-echo thermal wave imaging. *Review of Progress in Quantitative Nondestructive Evaluation.* Springer, Boston, MA.
[20] Sun J G 2006 Analysis of pulsed thermography methods for defect depth prediction. 329-338.
[21] Sirikham A, Zhao Y and Mehnen J 2017 Determination of thermal wave reflection coefficient to better estimate defect depth using pulsed thermography. *Infrared Physics & Technology.* **86** 1-10.
[22] Moskovchenko A I, Švantner M, Vavilov V P and Chulkov A O 2021 Characterizing depth of defects with low size/depth aspect ratio and low thermal reflection by using pulsed IR thermography. *Materials.* **14(8)** 1886.
[23] Müller J P, Dell'Avvocato G and Krankenhagen R 2020 Assessing overload-induced delaminations in glass fiber reinforced polymers by its geometry and thermal resistance. *NDT & E International.* 102309.
[24] Meola C, Carlomagno G M and Giorleo L 2004 Geometrical limitations to detection of defects in composites by means of infrared thermography. *Journal of Nondestructive Evaluation.* **23(4)** 125-132.
[25] D'Accardi E, Palumbo D, Tamborrino R and Galietti U 2018 Quantitative analysis of thermographic data through different algorithms. *Procedia Structural Integrity.* **8** 354-367.
[26] Ratsakou A, Skarlatos A, Reboud C and Lesselier D 2020. Shape reconstruction of delamination defects using thermographic infrared signals based on an enhanced Canny approach. *Infrared Physics & Technology.* **111** 103527.
[27] ASTM E1461-13 2013. *Standard test method for thermal diffusivity by the flash method.* (West Conshohocken, PA).
[28] Casalino G, Campanelli S L and Ludovico, A D 2013 Laser-arc hybrid welding of wrought to selective laser molten stainless steel. *Int. J. Adv. Manuf. Technol.* **68** 209-216.
[29] Casavola C, Campanelli S L and Pappalettere C 2009 Preliminary investigation on distribution of residual stress generated by the selective laser melting process. *J. Strain Anal. Eng. Des.* **44** 93-104.
[30] Lu Y, Wu S, Gan Y, Huang T, Yang C, Junjie L and Lin J 2015 Study on the microstructure, mechanical property and residual stress of SLM Inconel-718 alloy manufactured by differing island scanning strategy. *Optics & Laser Technology.* **75** 197-206.

IOP Conf. Series: Materials Science and Engineering 1214 (2022) 012016 doi:10.1088/1757-899X/1214/1/012016

A direct methodology for the calibration of ductile damage models from a simple multiaxial test

Gabriele Cortis[1], Filippo Nalli[1], Marcello Piacenti[2]

[1]Sapienza Università di Roma - Dipartimento di Ingegneria Meccanica e Aerospaziale, Via Eudossiana 18, 00184 Roma, Italia,
[2]Sapienza Università di Roma, Piazzale Aldo Moro 5, 00185, Roma, Italia

gabriele.cortis@uniroma1.it

Abstract. In the present work a straightforward calibration procedure of ductile damage models is proposed. The direct methodology involves the use of a simple multiaxial specimen, to be tested with a universal testing machine, capable to reproduce different stress states in the material. The specimen geometry was the one proposed by Driemeier et al. [1]. In addition, a numerical-analytical procedure was devised for the identification of material strains to fracture and corresponding stress states, directly from experimental tests. This allowed to overcome the use of Finite Element Analysis and inverse methods usually adopted to retrieve the local parameters representative of the material ductility.

1. Introduction

There are many cases where a correct description of the structural performance of the material and an accurate prediction of its ultimate strength is crucial. For instance, this is of major importance in all forming processes, where plastic formability limits and post-manufacturing residual stresses must be determined with great accuracy. Also, the knowledge of the material ultimate resistance is crucial in understanding how a component will withstand potential accidental overloads without reaching failure. For this purpose many damage models for ductile materials have been developed over the years. A ductile damage model is a mathematical model describing how the material gets damage with deformation and enables us to predict the onset of material failure and to know how much the material is damaged when subjected to a load. There are different classes of damage models based on different criteria. In particular, the empirical criteria-based models assume the damage accumulates with the equivalent plastic strain growth weighted on a function of the stress state [2, 3, 4]. These damage models need to be calibrated for the specific material and the calibration procedure is not trivial. It consists of two phases: the first is an experimental campaign in which several tests are carried out to induce different stress states into the material investigated. At this stage the global quantities, such as force and displacement at failure, are collected. The second phase involves numerical finite element simulation of the experimental tests to retrieve local quantities descriptive of the stress state and not directly measurable.

In particular, to perform an accurate calibration it is often necessary to induce several complex stress states in the material, especially with non-linear formulations [5, 6], which require a dedicated multi-axis machines and a custom specimens that are difficult to manufacture. Because of this complexity, these models are not yet used if not marginally by the industries, which often does not have the know-how or the right equipment to perform these types of tests or manage the subsequent post-processing stage.

In this context, the aim of this work regards the development of a straightforward calibration procedure for ductile damage models in order to simplify and promote the use of damage criteria also on an industrial scale. The new procedure is based on the use of a simple multiaxial specimen, to be tested with a common tensile machine, which is able to generate different stress states, such as pure shear or combined tensile-shear states. The specimen geometry was the one proposed by Driemeier et al [1]. Moreover, to overcome the demanding phase of Finite Element Analysis, a numerical-analytical model was developed for the identification of the local quantities that describe the ductility of the material, directly from the experimental tests.

2. Theoretical background

It is well known that a wide class of ductile damage models consider the accumulated damage depending on the increase of plastic strain, weighted on a function of the stress state, as reported in equation (1). The function of the stress state can be described using two scalar parameters: the triaxiality T and the Lode parameter X. In turn these two quantities are function of the invariant of the stress tensor as stated in equation (2) and (3):

$$D = \int_0^{\varepsilon_f} f(T, X) d\varepsilon_p \tag{1}$$

$$T = \frac{1}{3} \frac{I_1}{\sqrt{3J_2}} \tag{2}$$

$$X = \frac{27}{2} \frac{J_3}{(\sqrt{3J_2})^3} \tag{3}$$

where I_1 is the first stress invariant, J_2 and J_3 are the second and the third deviatoric stress invariant, respectively. The fracture onset occurs when D reaches a critical value conventionally set equal to 1. Under the assumption of proportional loading T and X are constant and posing D equal to 1, the equation (1) can be inverted obtaining the equation (4).

$$\varepsilon_f = f^{-1}(T, X) \tag{4}$$

This formula describes a surface called "Fracture Locus" in the space (ε_p, T, X) for which each pair of (T, X) is coupled to a unique value of plastic strain at incipient failure. Usually, T and X and are not perfectly constant, exhibiting limited variation, therefore it is commonly accepted to employ the average value of the two parameters as shown in equations (5) and (6):

$$T_{avg} = \frac{1}{\varepsilon_f} \int_0^{\varepsilon_f} T(\varepsilon) d\varepsilon_p \tag{5}$$

$$X_{avg} = \frac{1}{\varepsilon_f} \int_0^{\varepsilon_f} X(\varepsilon) d\varepsilon_p \tag{6}$$

Over the years, several forms of the function $f^{-1}(T, X)$ have been proposed in the literature. For instance, equation (7) shows the analytical expression of the damage model proposed by Coppola, Cortese and Folgarait (CCF) [3].

$$\varepsilon_f = \frac{1}{C_1} e^{-C_2 T} \left(\frac{cos\left[\beta\frac{\pi}{6} - \frac{1}{3}\arccos(\gamma)\right]}{cos\left[\beta\frac{\pi}{6} - \frac{1}{3}\arccos(\gamma X)\right]} \right)^{\frac{1}{n}} \tag{7}$$

Where C_1, C_2, β and γ are material parameters to be tuned through a calibration procedure and n is the hardening exponent of the Hollomon's power law.
Instead, the equation (8) represents the damage model devised by Bai and Wierzbicki [4].

$$\varepsilon_f = \left\{ \frac{A}{C_2} \left[\left(\frac{1+C_1^2}{3} \right)^{\frac{1}{2}} \cos\left(\frac{\pi}{6} - \theta \right) + C_1 \left(T + \frac{1}{3} \sin\left(\frac{\pi}{6} - \theta \right) \right) \right] \right\}^{-\frac{1}{n}} \quad (8)$$

Where C_1 and C_2 are the parameters to be identified when the Von Mises yielding condition is adopted and n is the hardening exponent. Testing different specimen's geometries permits the identification of the four points in the ε_p, T, X space. An example of fracture locus is illustrated in figure 1 for a titanium alloy (SLM Ti6Al4V) [7].

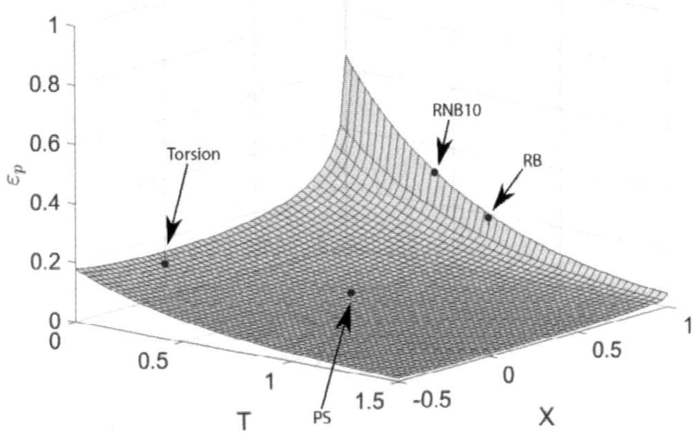

Figure 1. SLM Ti6Al4V: Fracture surfaces calibrated using the CCF model. The blue dots represent the experimental calibration points.

3. The "Driemeier" specimen

The starting point of this work was to define the design criteria that led to the choice of specimen. They can be summarised as follows: the easiness of fabrication, the capability of generating both a pure shear state and a multiaxial stress states like tensile-shear with uniaxial tensile machines, the capacity to focus the maximum plastic strain in the test area and the ability to maintain the parameters T and X as constant as possible to satisfy the quasi-proportional loading condition.

After the definition of the design criteria, a literature study was carried out, which for the sake of brevity is not reported, in which several specimens available in the literature designed for shear or tensile-shear tests were analysed. The geometry that most closely meets the design criteria is the one proposed by L. Driemeier et al. [1, 8] shown in figure 2A, a specimen used to perform shear tests with which is possible to obtain information of stress-triaxiality dependence on the material behaviour in the low triaxiality regime. The specimen, hereafter referred to as the "Driemeier", was considered the most suitable because it concentrates the maximum plastic strain value in its critical point in the gauge centre, it can be produced with conventional machine tools and it is possible to obtain both pure shear and combined tensile-shear, using a universal standard machine by simply varying the position of the two central holes. This is achieved by rotating the centres of the two holes by an alpha angle (α) on a circumference centred on the specimen, as illustrated in the figure 2B. In this way, a range of different couple of T_{avg} and X_{avg} can be achieved in a low triaxiality regime.

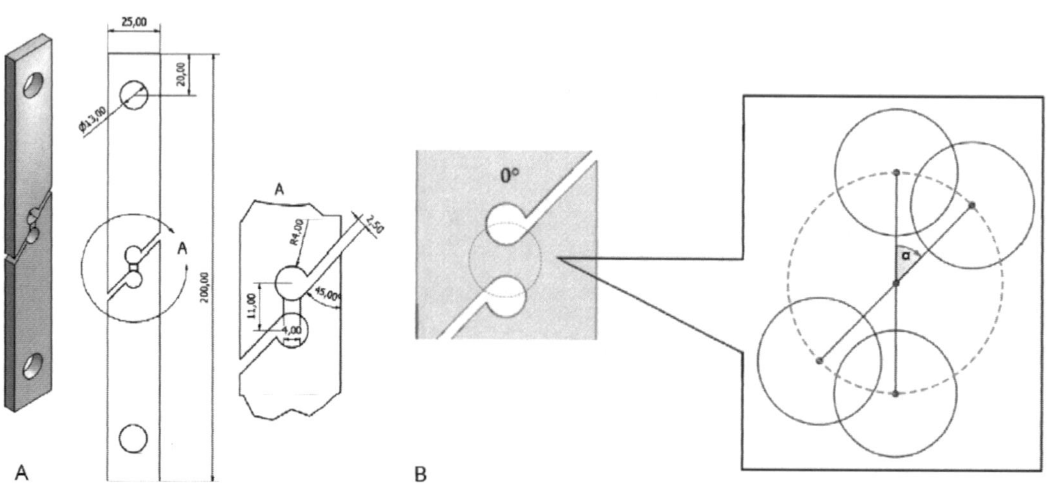

Figure 2. (A) Original geometry of the "Driemeier" specimen; (B) Representation of alpha angle (α).

The only limitation of the specimen is the constancy of the parameters T and X for high level of strain. In fact, if an excessive displacement is applied, the test area tends to rotate aligning itself with the machine's axis of pull, turning the test into a tensile test for high plastic strains. To reduce excessive rotation and hence comply with the quasi-proportional loading conditions, a sensitivity analysis was carried out to optimize the geometry.

The final specimen dimensions are shown in figure 3, and they were scaled to be machined from metal bars with a diameter of 20 mm. In the present work, three different configurations of the central holes were chosen, corresponding to three different alpha angles (α) of -10°, +10° and +30°. The -10° angle was chosen to further reduce the rotation of the test area in order to generate a stress state close to pure shear, i.e. T_{avg} and X_{avg} close to zero. Instead, the other two angles were chosen to generate combined tensile-shear stress states and to achieve different values of the Lode parameter at low triaxiality.

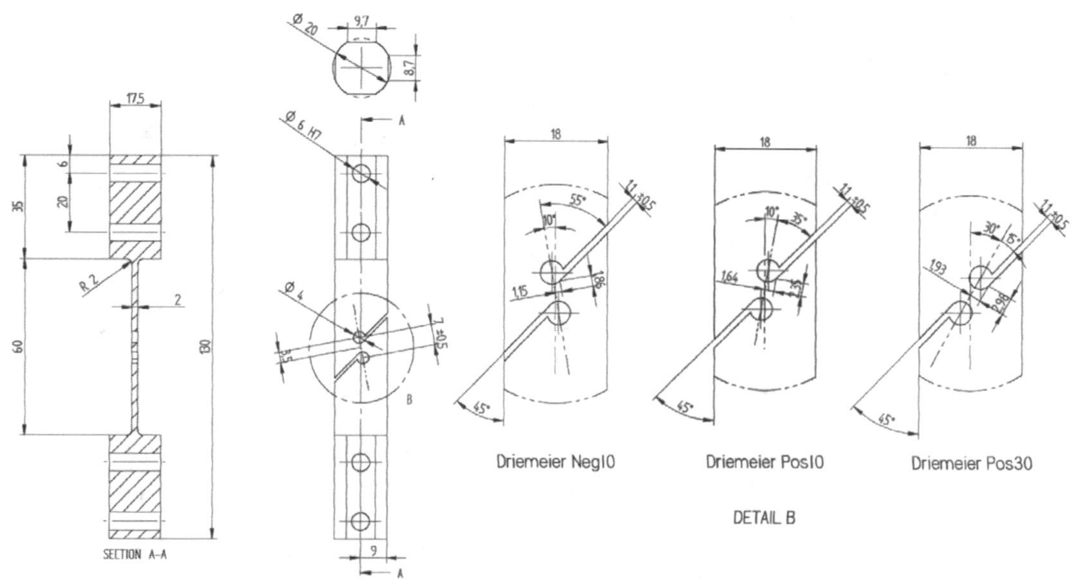

Figure 3. Technical drawing of the optimised geometry for the three configurations: -10°, +10°, +30°.

All the analyses carried out in this study were performed through finite element simulations using the commercial software ANSYS 2019. The models were meshed with brick elements and the function "*large displacement*" of the solver was activated. The displacement was applied to the upper part of the specimen, while the lower part was fixed. The stainless steel 17-4PH was chosen as benchmark material, since it was fully characterised in a study proposed by Nalli et al. [7]. Its constitutive behaviour was modelled using both the multilinear and bilinear stress-strain constitutive law, as shown in figure 4.

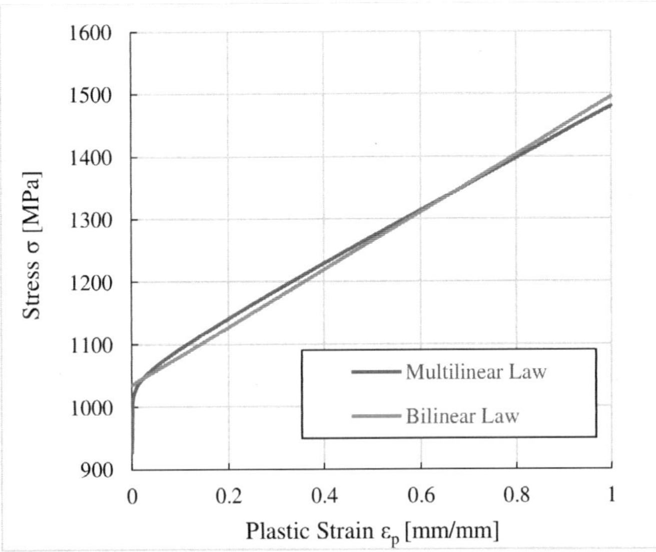

Figure 4. Stress-strain constitutive law for 17-4PH Wrought.

4. Results

The present section reports the results of the simulations used to verify the stress state at low triaxiality regime generated by the three different configurations introduced above -10°, +10°, +30°. Figure 5 shows the history of actual values of T and X, as a function of accumulated plastic strain. It is evident that suitably different pairs of T and X values were obtained, and this allows a proper characterisation of the fracture locus at low triaxiality regime. However, these values can only be considered approximately constant for plastic strain smaller than about 0.65, complying with the quasi-proportional loading hypothesis. Above this threshold, there is an increase of T and X due to the excessive rotation of the section.

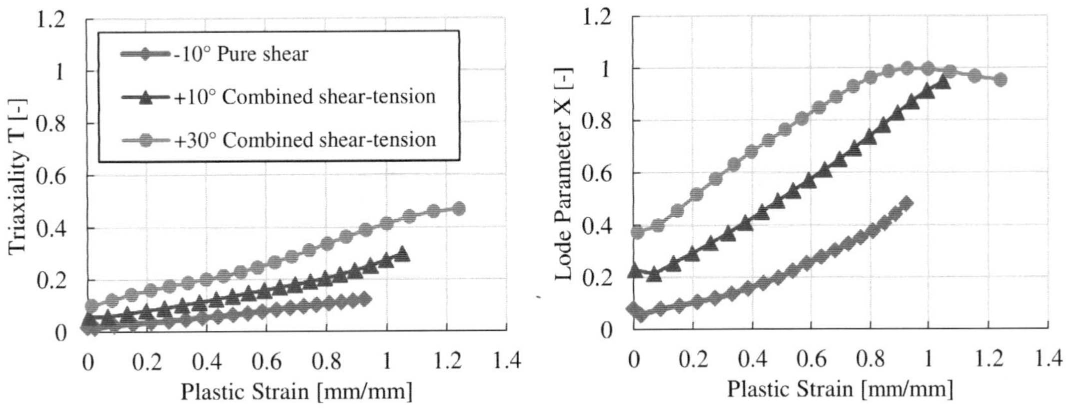

Figure 5. History of the instantaneous values of the Triaxiality and Lode parameters.

Considering a value of plastic strain of about 0.63, within the threshold, and using the formulae (5-6), it was possible to derive T_{avg} and X_{avg} for the three configurations as reported in table 1. In addition, the value of the displacement, applied to the specimen in the finite element simulation, required to obtain a deformation of 0.63 is reported.

Table 1. T_{avg} and X_{avg} values for different configurations.

Configuration	*d [mm]*	ε_p *[mm/mm]*	T_{avg} *[-]*	X_{avg} *[-]*
$\alpha = -10°$ Pure shear	2.4	0.63	0.05	0.15
$\alpha = +10°$ Combined shear-tension	1.87	0.63	0.11	0.38
$\alpha = +30°$ Combined shear-tension	1.5	0.63	0.18	0.60

It is clear that the configuration at -10°, characterised by a couple of T and X with values close to zero, generates a stress state nearly equal to that of pure shear ($T = 0$, $X = 0$) as anticipated above. The other two couples represent a good combination of values from which to derive information of stress state at low triaxiality for the calibration procedure, and thus improve the prediction accuracy of the damage models. Another important result, that fulfils one of the design requirements, is the presence of the maximum plastic strain value in its critical point in the gauge centre. A proof of this result is illustrated in figure 6A where only the contour plot of the -10° configuration is shown. Moreover, figure 6B and figure 6C display the contour plots of T_{avg} and X_{avg}. These two quantities, and therefore the corresponding graphs, are not present in the standard set of available outputs, but ANSYS allows the implementation of a custom script, i.e. textual commands in the "ANSYS Parametric Design Language" (APDL), with which to numerically derive and graphically represent an user-defined result. Therefore, the contour plots in figures 6B and 6C were obtained through the implementation of a custom APDL script devised to represent the behaviour of T_{avg} and X_{avg}.

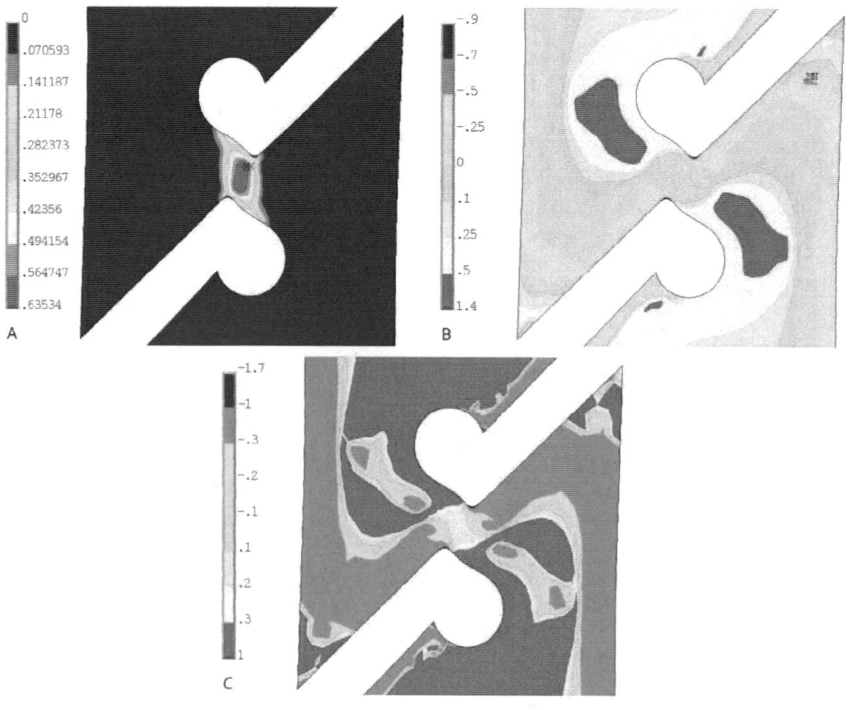

Figure 6. Driemeier -10°, contour plot of: (A) Plastic Strain, (B) Triaxiality, (C) Lode parameters.

IOP Conf. Series: Materials Science and Engineering 1214 (2022) 012016 doi:10.1088/1757-899X/1214/1/012016

In addition to the significant results obtained with the Driemeier test specimen, a mathematical model was created to derive the local parameters describing the stress state directly from the experimental tests without using the Finite Element Method (FEM). Namely, a set of formulae was developed for each type of experimental test to derive ε_p, T_{avg} and X_{avg} at the most critical point, requiring only the knowledge of the displacement at failure d_f from the experimental tests and the knowledge of the bilinear constitutive law of the material, i.e. yield point σ_y and tangent modulus E_t. These formulae were developed as a result of an extensive simulation campaign performed via FEM with the aim of analysing the behaviour of ε_p, T_{avg} and X_{avg} as a function of a general displacement d applied to the specimen and as function of the material represented by σ_y and E_t.

The procedure used to derive these analytical expressions is the following. Once the three Driemeier specimen configurations were selected, extensive ranges of σ_y, E_t and d were chosen within which to perform the study to consider a wide spectrum of ductile materials, namely σ_y 200-1250 [MPa], E_t 100-1200 [MPa] and d 0.15-4 [mm] respectively. Subsequently, many finite element simulations were carried out for each configuration and for each combination of σ_y and E_t for different values of displacement d. In particular the effect of σ_y and E_t was studied individually, i.e. σ_y was studied keeping E_t fixed for a certain value of d and vice versa. From now on, for the sake of brevity, the only focus will be on the Driemeier configuration -10°, being the procedure the same for the other geometries. For example, according to figure 7, the functions that best fit the effects of σ_y and E_t on plastic strain are the logarithmic function and the decreasing exponential function respectively.

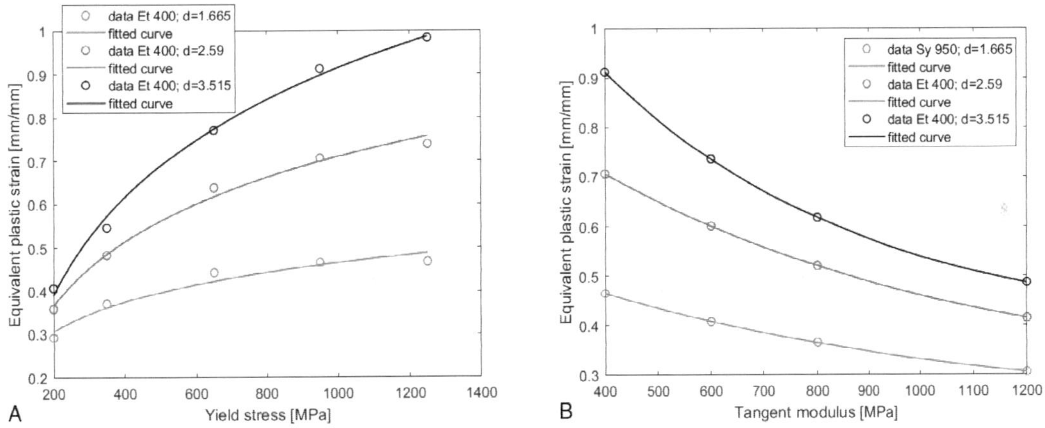

Figure 7. Strain-Yield stress relationship (A), Strain-Tangent modulus relationship (B), each for different values of displacement, keeping constant the Tangential modulus and the Yield stress in the first and second case respectively. The circle points represent the data acquired via FEM.

Thus, as a result, the equation (9) represents the formula that best approximates the plastic strain behaviour. In particular, the terms c_1, c_2, c_3 and c_4 are dependent on the imposed displacement according to the equations (10-13). The same analysis procedure was performed for T_{avg} and X_{avg}. The equations (9, 14, 15) of ε_p, T_{avg} and X_{avg} graphically represent a 3D surfaces as a function of d, σ_y and $\mathrm{E_t}$. For example, still regarding the configuration -10°, in figure (8) three surfaces that describe ε_p are reported at three different displacement values: 1.665, 2.59 and 3.515 mm.

$$\varepsilon_p = c_1 e^{(-c_2 E_t)} + c_3 \ln(\sigma_y) + c_4 \tag{9}$$

$$c_1 = 0.3738\, d - 0.1214 \tag{10}$$

$$c_2 = -1.37 * 10^{-4}\, d + 2.894 * 10^{-3} \tag{11}$$

$$c_3 = 0.1163\, d - 0.1002 \tag{12}$$

$$c_4 = -0.6778\, d + 0.7481 \tag{13}$$

$$T_{avg} = 0.04 \tag{14}$$

$$X_{avg} = -c_1 E_t + c_2 \sigma_y + c_3 \tag{15}$$

$$c_1 = 3.728 * 10^{-6}\, e^{0.927d} + 2.341 * 10^{-6} \tag{16}$$

$$c_2 = 1.849 * 10^{-5}\, d + 5.123 * 10^{-5} \tag{17}$$

$$c_3 = 0.0478\, d - 0.0316 \tag{18}$$

Notice that, for the configuration -10°, in equation (14), the value of T_{avg} was assumed equal to 0.04, because the variation with the material and displacement was observed to be minimal, thus it was considered constant. Considering all the simulation with different combinations of d, σ_y and E_t, the value of T_{avg} varied within the range 0.02-0.08, thus a value of 0.04 was considered a fair compromise. The equations (16-18) represent the parameters of the X_{avg} taking into account the dependence of displacement.

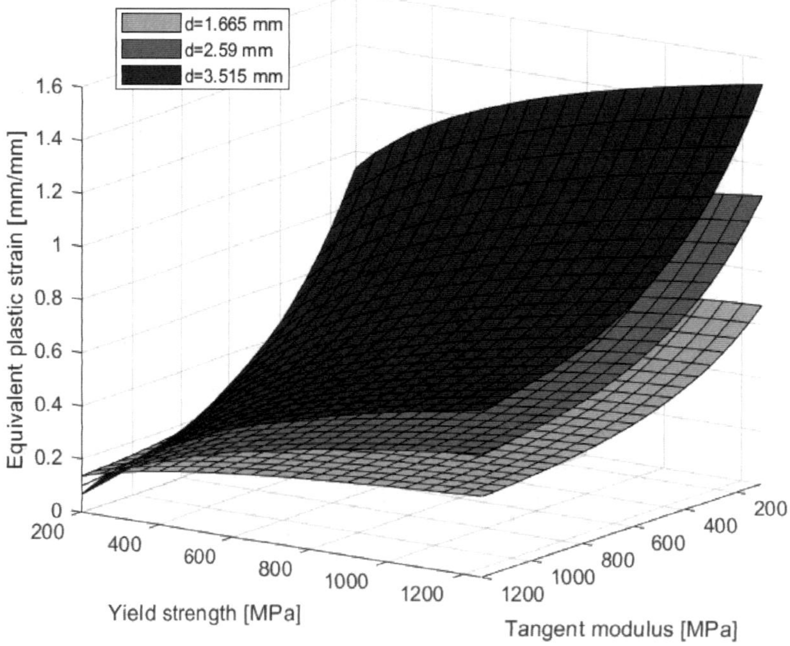

Figure 8. Three surfaces that describe the plastic strain in the space $(E_t, \sigma_y, \varepsilon_p)$ at three different displacement values.

Regarding the traditional method, the simulation of the experimental test is required and to find the triad (ε_p, T_{avg}, X_{avg}) it is necessary to stop the simulation at the same experimental condition in which the material failure occurs, i.e. at the same value of the displacement at failure. Then ε_p and the history of T, X are exported to calculate T_{avg} and X_{avg}. On the contrary, the use of the mathematical model allows to overcome the finite element analysis and the inverse method normally adopted to derive the local parameters representative of the material ductility.

A detailed example of how to use the mathematical model is provided below. Consider the Driemeier specimen at alpha -10° configuration and the corresponding results obtained via FEM for a displacement of 2.4 mm shown in table 1 and figure 6. Knowing only the applied displacement and the bilinear constitutive law of the material, it is possible to apply the formulae (9-18) and determine ε_p, T_{avg} and X_{avg}. For the 17-4PH material, values of 1035 and 460 MPa for yield strength and tangent modulus respectively were chosen to define the bilinear elastic-plastic constitutive law of the material, figure 4. table 2 compares the results obtained through both methods, i.e. via FEM and via Mathematical Model. The values obtained are very close to those obtained via FEM, demonstrating the goodness of the direct method.

Table 2. Comparison of values obtained with both FEM and mathematical model for the -10° configuration applying a displacement of 2.4 mm.

	FEM	**Math. Model**
ε_f [mm/mm]	0.63	0.60
T_{avg} [-]	0.05	0.04
X_{avg} [-]	0.15	0.16

For a more general overview of the results, table 3 provides a comparison of the triad values (ε_p, T_{avg}, X_{avg}) obtained from both the Finite Element Method and the mathematical model for several displacement d. The parameters evaluated through mathematical model are very close to those acquired via FEM, meaning that the validation of the model is positive.

Regarding the approximation committed using the mathematical model, in table 4 the average of the relative error in percentage (err_{rel}^{avg}) and the standard deviation (Std) for each test configuration are provided. The error committed is acceptable considering the benefits of the mathematical method: a simplification of the calibration operation and a significant reduction in computational time since the use of FEM is no more required.

Table 3. Comparison between the parameters obtained using the mathematical model and via FEM.

	Driemeier -10° ; d=1.665 mm			**Driemeier -10° ; d=2.59 mm**		
	ε_p [mm/mm]	T_{avg} [-]	X_{avg} [-]	ε_p [mm/mm]	T_{avg} [-]	X_{avg} [-]
FEM	0.438	0.036	0.109	0.679	0.049	0.155
Math. Model	0.415	0.040	0.123	0.651	0.040	0.175
	Driemeier +10° ; d=1.53 mm			**Driemeier +10° ; d=2.38 mm**		
	ε_p [mm/mm]	T_{avg} [-]	X_{avg} [-]	ε_p [mm/mm]	T_{avg} [-]	X_{avg} [-]
FEM	0.488	0.090	0.329	0.747	0.116	0.419
Math. Model	0.488	0.097	0.335	0.732	0.125	0.422
	Driemeier +30° ; d=1.35 mm			**Driemeier +30° ; d=2.1 mm**		
	ε_p [mm/mm]	T_{avg} [-]	X_{avg} [-]	ε_p [mm/mm]	T_{avg} [-]	X_{avg} [-]
FEM	0.515	0.167	0.556	0.804	0.207	0.669
Math. Model	0.547	0.173	0.569	0.822	0.213	0.680

Table 4. Mean relative error (%) and the standard deviation committed using the mathematical model.

	$err_{rel}^{avg}(\varepsilon_{pl})$	$err_{rel}^{avg}(T_{avg})$	$err_{rel}^{avg}(X_{avg})$	$Std(\varepsilon_p)$	$Std(T_{avg})$	$Std(X_{avg})$
- 10°	6.5%	30.8%	9.6%	0.026	0.015	0.015
+ 10°	2.2%	8.8%	1.6%	0.021	0.010	0.008
+ 30°	6.0%	3.0%	1.9%	0.063	0.008	0.013

5. Conclusion

The aim of this work was to develop a straightforward procedure to simplify the calibration of ductile damage models to make these failure criteria widespread on an industrial scale. The first step in the roadmap was to choose a suitable specimen from literature that met the following requirements: easiness of production, capability to confine the maximum value of plastic strain in the test area, capacity to generate pure shear or shear-tension states through a uniaxial testing machine and the capacity to maintain the quasi-proportional loading conditions. The chosen specimen that met all the requirements is the one proposed by L. Driemeier. The geometry of the specimen was optimised with respect to the original to keep the Triaxiality and Lode parameters as constant as possible. In addition, three different geometry configurations (-10°, +10°, +30°) generating respectively one pure shear stress state and two combined tensile-shear states were verified via FEM. The last step to further simplify the procedure was to develop a mathematical model that permits to derive locally in the critical point the plastic strain, T and X directly from the experimental test knowing the material and the displacement at failure. Since this numerical-analytical procedure only requires the use of a spreadsheet, it is no longer necessary to derive parameters using the finite element method, which means a saving in computational time and a straightforward procedure. The objective of the future work is to produce these specimens and to carry out experimental tests in order to calibrate several damage models for different materials using the new procedure presented in this work.

References

[1] Driemeier L, Brünig M, Micheli G and Alves M 2010 Experiments on stress-triaxiality dependence of material behavior of aluminum alloys *Mechanics of Materials* **42** 207-17
[2] Bai Y and Wierzbicki T 2008 A new model of metal plasticity and fracture with pressure and Lode dependence *International Journal of Plasticity* **24** 1071-96
[3] Coppola T, Cortese L and Folgarait P 2009 The effect of stress invariants on ductile fracture limit in steels *Engineering Fracture Mechanics* **76** 1288-302
[4] Bai Y and Wierzbicki T 2010 Application of extended Mohr–Coulomb criterion to ductile fracture *International Journal of Fracture* **161** 1
[5] Cortese L, Nalli F and Rossi M 2016 A nonlinear model for ductile damage accumulation under multiaxial non-proportional loading conditions *International Journal of Plasticity* **85** 77-92
[6] Papasidero J, Doquet V and Mohr D 2015, Ductile fracture of aluminum 2024-T351 under proportional and non-proportional multi-axial loading: Bao–Wierzbicki results revisited *International Journal of Solids and Structures* **69–70** 459-474
[7] Nalli F, Cortese L and Concli F 2021 Ductile damage assessment of Ti6Al4V, 17-4PH and AlSi10Mg for additive manufacturing *Engineering Fracture Mechanics* **241**, 107395
[8] Brünig M, Chyra O, Albrecht D, Driemeier L and Alves M 2008 A ductile damage criterion at various stress triaxialities *International Journal of Plasticity* **24** 1731-55

IOP Conf. Series: Materials Science and Engineering 1214 (2022) 012017 doi:10.1088/1757-899X/1214/1/012017

Development of statistical methods to highlight differences between endurance limit of re-rolled steel sheets in an assumed homogeneous population

M Beghini[1], B D Monelli[1], M Barsanti[1,*], G Macoretta[1] and A Piccinocchi[2]

[1] Department of Civil and Industrial Engineering, University of Pisa, Largo Lazzarino 2, 56122 Pisa, Italy

[2] Toscotec S.p.A., Viale Europa 317/F, 55012 Marlia (LU), Italy

E-mail: marco.beghini@unipi.it bernardo.disma.monelli@unipi.it giuseppe.macoretta@phd.unipi.it michele.barsanti@unipi.it alessandro.piccinocchi@toscotec.com

Abstract. In this paper, we present the results of the activities carried out to estimate the axial fatigue strength of ASTM E566 Gr. 70 sheet metal. We describe the experimental setup, followed by the collected data. To highlight a possible inhomogeneity in some of the material properties, we estimate the endurance limit for the investigated material using various statistical approaches on the set of test articles. The data analysis was carried out both by considering as a whole the set of test articles and by dividing the set into two groups. As a new analysis technique, we propose a method introducing a bimodal distribution (Gaussian mixture).

1. Introduction

The present paper shows the results of the activities carried out in the framework of a collaboration with Toscotec S.p.A. The paper is divided into four sections. After this short introduction, section 2 is dedicated to the presentation of the investigated material and theadopted testing procedure. In section 3 the experimental results are presented, and the estimations of the material fatigue strength by the Staircase Procedure and the Maximum Likelihood Principle are reported. A new analysis technique is finally proposed, in which the population distribution is a linear combination (mixture) of two Gaussian distributions. In section 4 some concluding remarks are presented and a suggestion for the choice of the endurance limit of the material is proposed.

2. Materials and methods

In this section we describe some of the mechanical properties of the steel under investigation and the geometry of the test articles made of this material. We finally describe how the test campaign was carried out.

*Corresponding Author

2.1. Material properties
In this work we study the fatigue strength limit of steel SA 516 - Gr. 70. The supplier has sent the material needed to manufacture the test articles in two sheets, called sheet A and sheet B, which are nominally the same material, therefore with the same fatigue strength limit. However, it could have occurred that the two sheets were obtained in two separate castings. Therefore it might be interesting to try to understand if differences can be highlighted by considering the test articles extracted from the two sheets as two different samples.

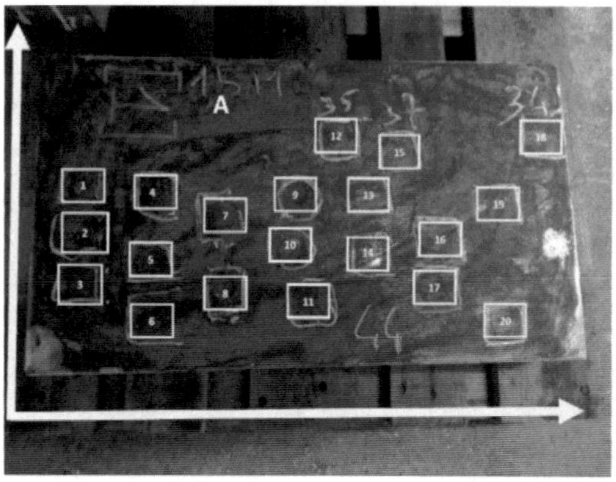

Figure 1. Sheet A, 310 ×600×90 (mm). The extraction positions of the test articles are highlighted in yellow.

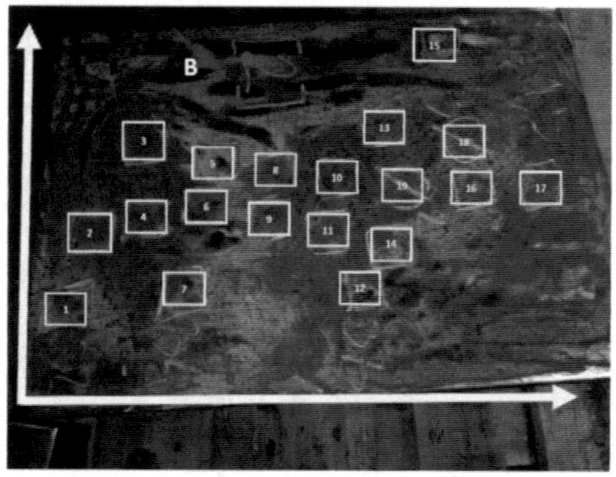

Figure 2. Sheet B, 410 ×620×90 (mm). The extraction positions of the test articles are highlighted in yellow.

Table 1. Main tensile properties of steel SA 516 - Gr. 70

Property	Value
Ultimate Tensile Strength	(520±10) MPa
Yield Strength	(320±10) MPa
Elongation after fracture	(23±3)%
Reduction of area	(67±3)%

The main tensile properties of the material under investigation are listed in table 1. The reported values have been obtained averaging 4 measurements (using test articles extracted from the two sheets). We estimated the dispersion by calculating the standard deviation of these 4 values. The results shown in table 1 do not suggest that a difference can be noted between the mechanical properties of the two sheets, in accordance with the supplier's declaration.

2.2. Description of test articles
The test articles have been extracted from the two sheets through thickness. The extraction positions are highlighted in figures 1 and 2. The geometric dimensions of the test articles are shown in figure 3

We obtained 20 test articles from sheet A (named A1,..., A20) and 19 test articles from sheet B (named B1,..., B19).

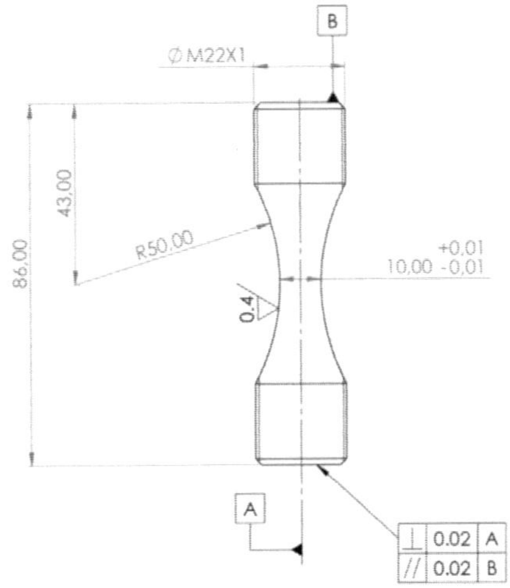

Figure 3. Test article design.

2.3. Experimental setup and test conditions

A campaign of high cycle number fatigue tests was carried out using a resonance testing machine. The test conditions are shown in table 2. The test campaign was conducted following the UNI 3964 standard [1] having the foresight to select test articles from sheet A in the first part of the test campaign and from sheet B in the second and final part. In each of the two parts, the selection was random. This method of conducting the test campaign allows highlighting any differences between the mechanical fatigue behaviour of the test articles extracted from sheet A compared to the ones extracted from sheet B. In fact, the number of test articles present in each of the two parts of the test allows statistically reliable values to be obtained considering separately the data obtained by analyzing the test articles extracted from each of the sheets as well.

Table 2. Test conditions.

Parameter	Value or condition
Loading configuration	Uniaxial tension
Load ratio ($\sigma_{\min}/\sigma_{\max}$)	0
Runout	5×10^6 cycles
Criterion for choosing the stress level $\Delta\sigma$	Staircase procedure
Testing machine	Rumul resonant machine 150 kN
Test frequency	175-176 Hz
Test control	Force
Test temperature	Room temperature
Test stop condition	Frequency drop greater than 1 Hz

The fatigue test facility, with the resonant testing machine, is shown in figure 4. Figure 5 shows the detail of the assembly of the test article in the testing machine.

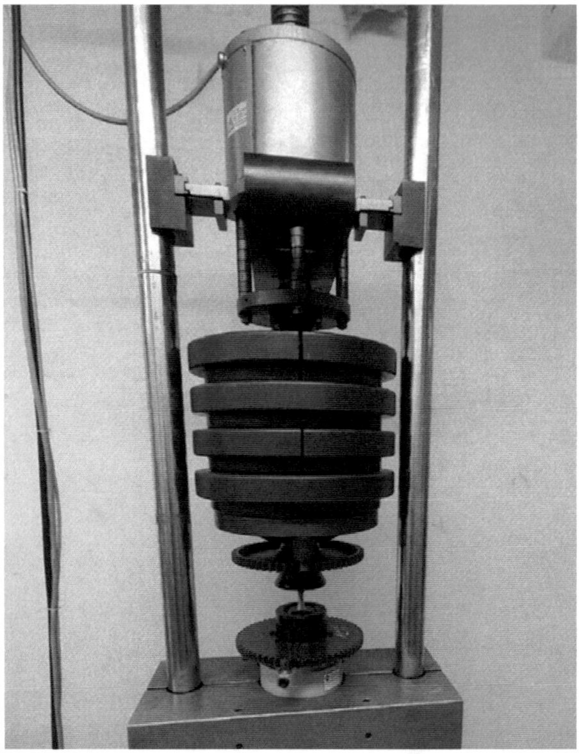

Figure 4. Fatigue test facility.

Figure 5. Detail of the test article assembly.

3. Fatigue strength estimation

It is well known that the fatigue strength is not a physical quantity measurable as such. On the contrary, we can obtain the mean value of this quantity (for a statistically homogeneous population) by carrying out a statistical analysis of a binary random variable that indicates the outcome of each of the tests in the campaign (runout/failure). In the present work, we present three methods of statistical analysis. Two of them are applied in two different ways: the test campaign is either considered as a whole or divided into two sub-campaigns.

3.1. Outline of the Staircase Procedure

We briefly recall the data processing method reported in the UNI 3964 standard [1] for determining the fatigue strength. We recall that this method was presented by Dixon and Mood [2] in a famous work in which they show, for the staircase experimental procedure, a smart approximation for the determination of the likelihood maximum. We will describe the "exact" likelihood maximization method in subsection 3.3. In table 3 we list the quantities used in the elaboration of the calculation. The method of analysis that we describe below is applicable in the hypothesis that the sample (set of test articles) is extracted from a Gaussian population distribution.

With the symbols listed in table 3, it is possible to compute the mean value $\Delta\sigma_{50\%}$ of the fatigue strength S as

$$\Delta\sigma_{50\%} = \Delta\sigma_0 + d\left(\frac{A}{N} \pm \frac{1}{2}\right) \qquad \begin{array}{l} + \text{ if the less frequent outcome is the runout} \\ - \text{ if the less frequent outcome is the failure} \end{array} \tag{1}$$

Table 3. Nomenclature used in staircase procedure.

Symbol	Nomenclature
d	Staircase stress step
$\Delta\sigma_0$	Lowest stress amplitude for the less frequent event in the test results
N	Number of the less frequent event
i	Stress level index, starting from 0
n_i	Number of samples at the stress level i
A	$\sum i \cdot n_i$
B	$\sum i^2 \cdot n_i$
$\Delta\sigma_{50\%}$	Mean value of the fatigue strength $\Delta\sigma$ (50% survival probability)
$\Delta\sigma_{10\%}$	Fatigue strength corresponding to 10% survival probability
$\Delta\sigma_{90\%}$	Fatigue strength corresponding to 90% survival probability
SD	Standard deviation of $\Delta\sigma$

and the standard deviation SD of this mean value as

$$SD = 1.62d\left(\frac{NB - A^2}{N^2} + 0.029\right) \qquad \text{valid if } \frac{NB - A^2}{N^2} > 0.30. \tag{2}$$

The extremes of the 90% confidence interval for $S_{50\%}$ can be obtained, again assuming the validity of Gaussian population, computing

$$\Delta\sigma_{10\%} = \Delta\sigma_{50\%} + 1.96 \cdot SD \qquad \Delta\sigma_{90\%} = \Delta\sigma_{50\%} - 1.96 \cdot SD. \tag{3}$$

3.2. Elaboration of the staircase outcomes

In figure 6 we show the outcomes of the staircase procedure, obtained starting at an initial stress level of 235 MPa and with a staircase stress step $d = 12$ MPa. It should be noted that, although the knowledge of the mechanical properties of the material under examination allowed to establish a reasonable stress value for the first test, three test articles were "wasted" in determining the correct initial stress level (two consecutive different outcomes are needed to start the staircase procedure).

		Actual tests sequence																											
ID		A14	A16	A17	A19	A20	A10	A8	A11	A9	A7	B4	B15	B3	B13	B16	B5	B10	B14	B8	B11	B2	B17	B6	B9	B12	B7	B19	B18
Δσ (MPa)	271																		×		×								
	259																	•		•		×							
	247																•						×		×				
	235	×		×												•								•		×		×	
	223		•		×										•												•		×
	211					×				×				•															
	199						×		•		×		•																
	187							•				•																	
N (Mcycle)		0.24	5.00	1.37	0.85	3.63	0.88	5.00	5.00	0.98	0.33	5.00	5.00	5.00	5.00	5.00	5.00	5.00	0.11	5.00	0.34	0.33	0.70	5.00	0.23	1.29	5.00	0.32	2.11

Figure 6. Staircase outcomes assuming the sample of test articles as a unique set.

The statistical analysis carried out using equations 1, 2 and 3 provides the results presented in table 4.

Table 4. Fatigue strength assuming the sample of test articles as a unique set.

$\Delta\sigma_{10\%}$ [MPa]	$\Delta\sigma_{50\%}$ [MPa]	$\Delta\sigma_{90\%}$ [MPa]	SD [MPa]	$SD/\Delta\sigma_{50\%}$ [%]
326	228	130	77	33.6

We note that SD is very large when compared with $\Delta\sigma_{50\%}$. Generally, in fact, the ratio $SD/\Delta\sigma_{50\%}$ falls in the interval $[0.080, 0.16]$. We also note that, when the results obtained with this procedure are inserted in the relationship $\Delta\sigma_{50\%} - 6 \cdot SD$ for the determination of the endurance limit, we obtain a negative value. This result would make this material useless in the mechanical design. Observing that all the first seven outcomes carried out using test articles from sheet B were runouts, we analyzed separately the results involving test articles extracted from different sheets. The outcomes are shown in figures 7 and 8. Note that the first six outcomes of the test articles extracted form sheet B cannot be considered in the statistical analysis. We aim to highlight a possible difference between the average values of the mean fatigue stength in the two groups.

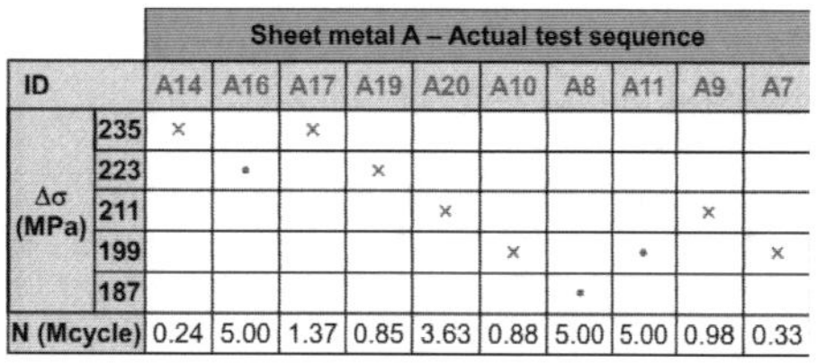

ID		A14	A16	A17	A19	A20	A10	A8	A11	A9	A7
		Sheet metal A – Actual test sequence									
Δσ (MPa)	235	×		×							
	223		•		×						
	211					×				×	
	199						×		•		×
	187							•			
N (Mcycle)		0.24	5.00	1.37	0.85	3.63	0.88	5.00	5.00	0.98	0.33

Figure 7. Staircase outcomes considering only test articles from sheet A.

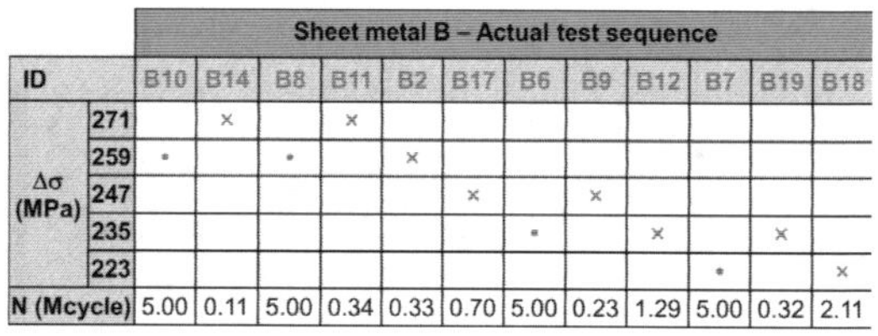

ID		B10	B14	B8	B11	B2	B17	B6	B9	B12	B7	B19	B18
		Sheet metal B – Actual test sequence											
Δσ (MPa)	271		×		×								
	259	•		•		×							
	247						×		×				
	235							•		×		×	
	223										•		×
N (Mcycle)		5.00	0.11	5.00	0.34	0.33	0.70	5.00	0.23	1.29	5.00	0.32	2.11

Figure 8. Staircase outcomes considering only test articles from sheet B.

The statistical analyses carried out following equations 1,2,3 for the computation of the fatigue strenngth limit provides the results presented in tables 5 and 6.

Table 5. Fatigue strength analyzing only test articles from sheet A.

$\Delta\sigma_{10\%}$ [MPa]	$\Delta\sigma_{50\%}$ [MPa]	$\Delta\sigma_{90\%}$ [MPa]	SD [MPa]	$SD/\Delta\sigma_{50\%}$ [%]
248	209	170	31	14.7

Table 6. Fatigue strength analyzing only test articles from sheet B.

$\Delta\sigma_{10\%}$ [MPa]	$\Delta\sigma_{50\%}$ [MPa]	$\Delta\sigma_{90\%}$ [MPa]	SD [MPa]	$SD/\Delta\sigma_{50\%}$ [%]
293	250	207	33	13.3

Comparing the results reported in tables 5 and 6 with the results of table 4, we observe that the average values for the two groups are significantly different from each other and that the standard deviation of each of the two groups is about half the standard deviation obtained when the results from the test articles extracted from the two sheets are analyzed together.

3.3. Fundamentals of the Maximum Likelihood Principle
To enforce the result shown above, we analyzed the same test campaign outcomes using likelihood maximization. Maximum Likelikood Estimators (MLE) must be obtained using a numerical procedure to obtain the values of the arguments at whiuch maximum is attained. In this section, we briefly outline the parameter estimation procedure based on maximization of likelihood (ML). The random variable $\Delta\sigma$, assumed to be Gaussian, cannot be sampled. Given a sample made of n test articles, for the $k-$th test ($k = 1, \ldots, n$), a binary response random variable e_k is sampled: $e_k = 0$ for runout and $e_k = 1$ for failure.

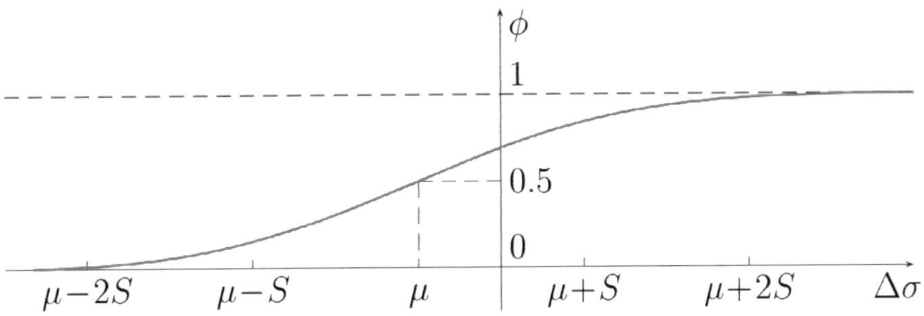

Figure 9. An example of Gaussian cumulative distribution function.

Let μ and S be the mean and standard deviation of the random variable $\Delta\sigma$. If we introduce the cumulative distribution function ϕ (see figure 9), i.e. the probability $\mathbb{P}$ (for given values of μ and S) that the random variable $\Delta\sigma$ is less than the value of $\Delta\sigma_k$ used in the $k-$th test, as

$$\mathbb{P}(\Delta\sigma < \Delta\sigma_k) = \phi(\Delta\sigma_k|\mu, S) = \frac{1}{\sqrt{2\pi}S}\int_{-\infty}^{\Delta\sigma_k} \exp\left(-\frac{(x-\mu)^2}{2S^2}\right) dx \tag{4}$$

we can define $L(\mu, S)$ as the logarithm of the likelihood function:

$$L(\mu, S) = \sum_{k=1}^{n} [e_k \log(\phi(\Delta\sigma_k|\mu, S)) + (1 - e_k)(1 - \log(\phi(\Delta\sigma_k|\mu, S)))]. \tag{5}$$

We can obtain the MLEs of μ and S taking the couple of values $\hat{\mu}$ and $\hat{S}$ at which L attains its maximum value.

3.4. Maximum Likelihood estimates and comparison with staircase results
In figure 10 we show the outcomes of the staircase procedure used to obtain the ML estimators. It should be noted that the three initial failure outcomes can be inserted in the calculations, increasing the value of n.

The statistical analysis carried out using maximization of 5 provides the results presented in table 7.

		Elaborated samples																													
ID		A5	A6	A13	A14	A16	A17	A19	A20	A10	A8	A11	A9	A7	B4	B15	B3	B13	B16	B5	B10	B14	B8	B11	B2	B17	B6	B9	B12	B7	B19	B18
Δσ (MPa)	271	×																				×		×								
	259		×																		•		•		×							
	247			×																•						×		×				
	235				×		×												•								•		×		×	
	223					•		×										•												•		×
	211								×				×				•															
	199									×		•		×		•																
	187										•				•																	
N (Mcycle)		0.23	0.63	0.14	0.24	5.00	1.37	0.85	3.63	0.88	5.00	5.00	0.98	0.33	5.00	5.00	5.00	5.00	5.00	5.00	5.00	0.11	5.00	0.34	0.33	0.70	5.00	0.23	1.29	5.00	0.32	2.11

Figure 10. Outcomes used in MLE calculations, considering the test articles all together.

Table 7. MLE estimators of fatigue strength assuming the sample of test articles as a unique set.

$\Delta\sigma_{10\%}$ [MPa]	$\Delta\sigma_{50\%}$ [MPa]	$\Delta\sigma_{90\%}$ [MPa]	SD [MPa]	$SD/\Delta\sigma_{50\%}$ [%]
295	218	141	60	27.5

We note that the ML estimate of SD is smaller than the Dixon-Mood estimate. A comparison of the estimates of $\Delta\sigma_{50\%}$ and SD obtained wwith different methods is reported in [3]. The bias of Dixon-Mood SD estimators has been shown in [4] and is not generally considered critical, since it tends to decrease the endurance limit. Notwithstanding this, we observe that, when these results are inserted in the relationship $\Delta\sigma_{50\%} - 6 \cdot SD$ for the determination of the endurance limit, we again obtain a negative value. Again we analyzed separately the results involving test articles extracted from different sheets. The outcomes are shown in figures 11 and 12, for which we can again use all the outcomes.

		Sheet metal A – Elaborated samples												
ID		A5	A6	A13	A14	A16	A17	A19	A20	A10	A8	A11	A9	A7
Δσ (MPa)	271	×												
	259		×											
	247			×										
	235				×		×							
	223					•		×						
	211								×				×	
	199									×		•		×
	187										•			
N (Mcycle)		0.23	0.63	0.14	0.24	5.00	1.37	0.85	3.63	0.88	5.00	5.00	0.98	0.33

Figure 11. Outcomes used in MLE calculations, considering only test articles from sheet A.

		Sheet metal B – Elaborated samples																	
ID		B4	B15	B3	B13	B16	B5	B10	B14	B8	B11	B2	B17	B6	B9	B12	B7	B19	B18
Δσ (MPa)	271								×		×								
	259							•		•		×							
	247						•						×		×				
	235					•								•		×		×	
	223				•												•		×
	211			•															
	199		•																
	187	•																	
N (Mcycle)		5.00	5.00	5.00	5.00	5.00	5.00	5.00	0.11	5.00	0.34	0.33	0.70	5.00	0.23	1.29	5.00	0.32	2.11

Figure 12. Outcomes used in MLE calculations, considering only test articles from sheet B.

The statistical analyses carried out maximizing likelihood defined in equation 5 provides the results presented in tables 8 and 9.

Table 8. ML estimators of fatigue strength obtained analysing only test articles from sheet A.

$\Delta\sigma_{10\%}$ [MPa]	$\Delta\sigma_{50\%}$ [MPa]	$\Delta\sigma_{90\%}$ [MPa]	SD [MPa]	$SD/\Delta\sigma_{50\%}$ [%]
228	196	166	25	12.6

Table 9. ML estimators of fatigue strength obtained analysing only test articles from sheet B.

$\Delta\sigma_{10\%}$ [MPa]	$\Delta\sigma_{50\%}$ [MPa]	$\Delta\sigma_{90\%}$ [MPa]	SD [MPa]	$SD/\Delta\sigma_{50\%}$ [%]
285	244	203	32	13.1

Comparing the results reported in tables 8 and 9 with the results of table 7, we observe that also for this statistical technique the average values for the two groups are significantly different from each other. This result confirm that the difference between the two mean values is not due to a limitation of the Dixon-Mood approximation, but is a consequence of a real difference between some of the mecanical properties of the two sheets, only nominally equal.

3.5. Maximum Likelihood estimation of a mixture of Gaussian distribitions

We present here an extension of the ML method using, as a distribution function, a mixture of two Gaussians. We define a linear combination ψ of two Gaussian cumulative distribution function with two differerent mean values μ_1 and μ_2 and (for simplicity) the same standard deviation S for both components as

$$\psi(\Delta\sigma_k|\mu_1,\mu_2,S,\alpha) = \alpha\phi(\Delta\sigma_k|\mu_1,S) + (1-\alpha)\phi(\Delta\sigma_k|\mu_2,S). \tag{6}$$

where $\alpha \in (0,1)$ is the relative weight of the two components. In this way, a new objective function (to be maximized) $L'(\mu_1,\mu_2)$ can be defined as

$$L'(\mu_1,\mu_2) = \sum_{k=1}^{n}[e_k \log(\psi(\Delta\sigma_k|\mu_1,\mu_2,S,\alpha)) + (1-e_k)(1-\log(\psi(\Delta\sigma_k|\mu_1,\mu_2,S,\alpha)))] \tag{7}$$

in which, for simplicity, the same standard deviation S has been kept fixed (30 MPa, a reasonable value obtained by previous analyses), and also the weights α and $1-\alpha$ have been kept fixed as well (0.42 and 0.58, values of the relative numbers of test articled from the two sheets). The numerical maximization of L' provides $\mu_1 = 191$ MPa and $\mu_2 = 269$ MPa confirming a separation of the two mean values more than twice a reasonable estimation of the standard deviation. These results do not vary much if S and α are varied within a reasonable range of values (20-40 MPa, 0.3-0.6). However, in the current state of research, it is not possible to show reasonable values of the estimate of S using a Gaussian misture using the ML method because the estimation of S is lowered until it reaches very small and meaningless values. The probability density function obtained with this method is shown in figure 13.

4. Conclusions

In this paper, we have shown the differences that can occur if we introduce the hypothesis that the sample is not distributed like a Gaussian. This hypothesis, which is usually assumed to be true in the analysis of fatigue tests, is very strong and conditions the interpretation of a series of fatigue tests. In staircase tests, 90% confidence intervals for the means show that $\Delta\sigma_{50\%}$ seems to be different for sheets A and B nominally made of the same steel. The standard deviation S is slightly overestimated in the Staircase Procedure with respect to the Maximum Likelihood estimation. A new procedure for the estimation of means of a mixture of two Gaussians confirms the difference between the values of $\Delta\sigma_{50\%}$ for the two sheets. The Staircase method tends to be conservative, and even ML estimates do not differ much. Perhaps the development of

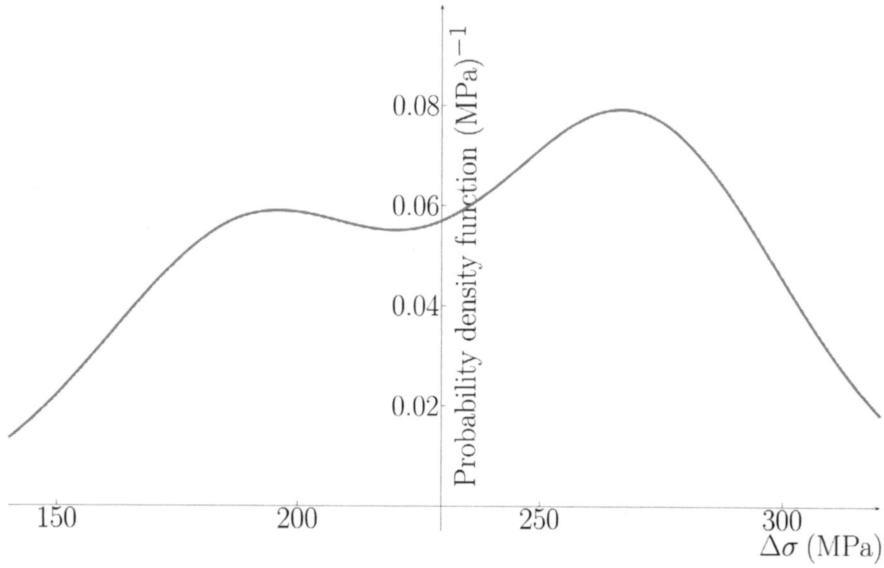

Figure 13. The probability distribution function of the population, modellized as a mixture of gaussian distributions.

new test/analysis methodologies could provide results that improve component design. For the estimation of the endurance limit for this material (to be used in component design), we propose for $\Delta\sigma_{50\%}$ the lowest of the two mean values (209 MPa) and for SD the corresponding value (31 MPa) as obtained from the analysis, carried out according to [1], of test articles extracted from sheet A only, which has the worst mechanical fatigue characteristics. For the endurance limit, we obtain therefore $\Delta\sigma_{50\%} - 6 \cdot SD = 23\,\text{MPa}$. Although small, it is still a value that can be used in mechanical design. In future developments of this work, we will investigate the characteristics of the Gaussian mixture method also using Monte Carlo simulations, to obtain, for instance, the confidence intervals of the estimates. It should be as well interesting a comparison of MLEs with other estimators or other statistical techniques (Fisher Information, Bayesian analysis).

References

[1] UNI 3964 Standard 1985 Prove meccaniche dei materiali metallici. Prove di fatica a temperatura ambiente. Principi generali. Tech. rep. UNI - Ente Nazionale Italiano di Unificazione 20123 Milano
[2] Dixon W J and Mood A M 1948 *Journal of the American Statistical Association* **43** 109 – 126
[3] Amorena M, Barsanti M, Gubinelli M, Manfredi E, Moretto I and Vitali M 2002 Sull'impostazione di programmi di prove per ingranaggi Proceedings of AIAS2002 - 31st Conference on Stress Analysis and Mechanical Engineering Design, 18-21 September 2002, Parma, Italy
[4] Pollak R, Palazotto A and Nicholas T 2006 *Mechanics of Materials* **38** 1170 – 1181

IOP Conf. Series: Materials Science and Engineering 1214 (2022) 012018 doi:10.1088/1757-899X/1214/1/012018

Analysis of fracture behaviour in active materials for lithium ion batteries

D Clerici[1], F Mocera[1] and F Pistorio[1]

[1] Department of Mechanical and Aerospace Engineering, Politecnico di Torino, Corso duca degli Abruzzi 24, 10129 Torino, Italy

E-mail: davide.clerici@polito.it

Abstract. Several strong points make lithium ion battery one of the most widespread energy storage system. Nevertheless, one of the biggest drawbacks is the progressive damage which affects active materials, and influences cycle life as well. The hosting process of lithium ions causes the rise of mechanical stress in active material, which ultimately leads to the propagation of micro-flaws already present in fresh material. Finally, the damage of active material and solid-electrolyte interphase growth caused by cracks propagation result in capacity drop. The distribution of Mode I stress intensity factor is calculated along the semi-elliptical crack front on the outer surface and in the core of a three-dimensional spherical active material particle. A 3D and 2D finite element method analysis is performed in ANSYS Mechanical APDL starting from the mechanical stress state in active material computed with the electrochemical-mechanical model presented in previous works. The model is built using collapsed singular elements along the crack front, the not-singular version of these elements is used to model the outlying region of the crack area. The dependence of stress intensity factor on geometry size is deepened to evaluate the most critical condition. Moreover, the influence of current rate on stress intensity factor is investigated, in order to identify a current threshold beyond stress intensity factor is greater than the toughness of active material, and cracks start to propagate.

1. Introduction

Lithium ion battery (LIB) is nowadays the most common energy storage system, it meets the needs of a widespread field of applications, from small electronics up to heavy duty vehicles [1,2], thanks to its good energy and power density, great capacity and safe. Nevertheless, its nominal performance decreases during cycle life [3] and eventually it needs to be replaced after a certain number of cycles.

The aforementioned degradation is mainly correlated to mechanical damage of the electrodes [4]: lithium storage in electrode materials causes the rise of mechanical stress in the electrode itself, and leads to crack propagation ultimately, as deepened in previous studies [5,6].

Crack growth in electrode micro-structure causes two main problems: electrode isolation and SEI growth. Electrode isolation occurs when a portion of active material is detached and it can not host lithium ions anymore. On the other hand, cracks propagation causes the creation of new fresh surfaces of active material, where solid electrolyte interphase (SEI) precipitation can occur. The isolation of some portion of active material causes capacity fade and impedance rise, on the other hand, SEI precipitation consumes lithium ions that are no longer available to be cycled, then capacity fade occurs.

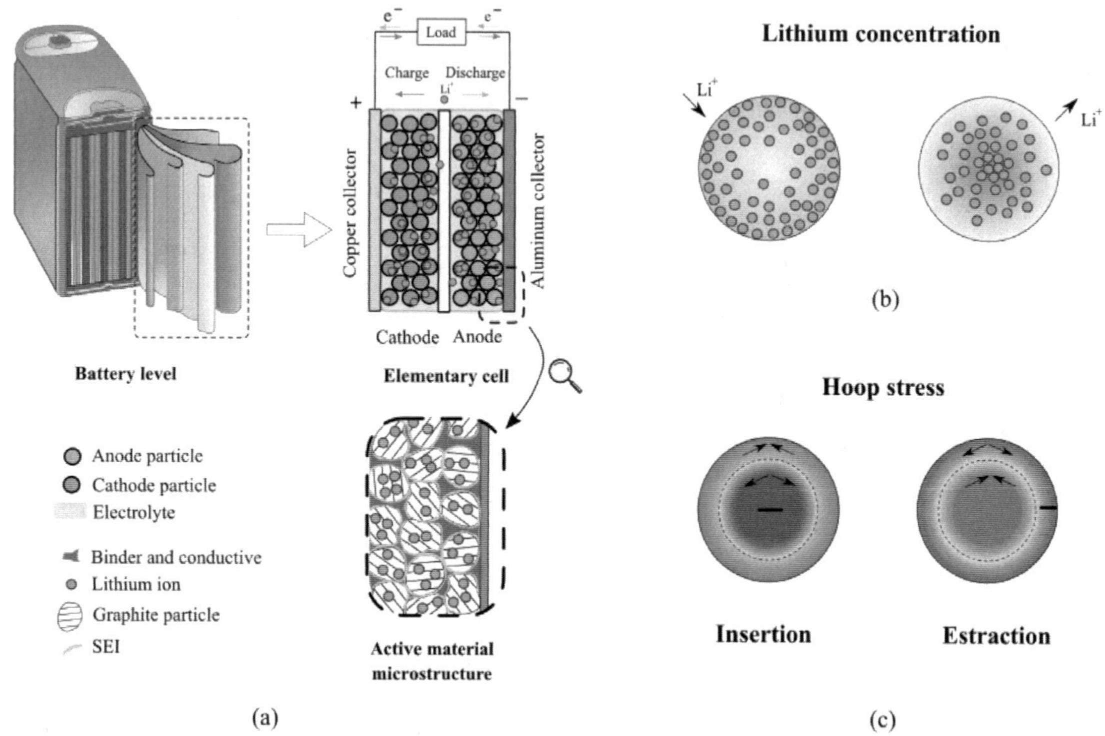

Figure 1: Battery model. (a) Internal structure of a typical LIB, from battery scale to micrometre scale. (b) Lithium concentration during intercalation and deintercalation: as lithium diffuses inside the particle, a concentration gradient is established, which leads to strain mismatch and stress. (c) Hoop stress distribution during intercalation and deintercalation. Hoop stress is compressive in the core and tensile on the surface during deintercalation, vice-versa during intercalation. Areas are marked with red where cracks can propagate and with green where cannot, as crack growth occurs just where hoop stress is tensile, according to mode I. As a consequence, internal cracks are likely to propagate during lithium insertion, and surface cracks during extraction.

A sketch of a typical LIB is reported in Figure 1a: battery is composed by several "elementary cells" that are made in turn by positive and negative current collectors, cathode, anode and separator. Each of the aforementioned components is a sheet few tens of micrometers thick. Cathode and anode are composed in turn by a mixture of several components, as shown in Figure 1a: Active material - a powder of quasi-spherical particles which occupies about 50% of the volume; Conductive materials and binder - about 20% in volume; finally 30% of volume is void, and is then occupied by liquid electrolyte. For this reason they are referred as porous electrodes.

Graphite is the most popular active material in negative electrode, whereas several metal oxides are used as positive electrode, such as: lithium iron phosphate (LFP), lithium cobalt oxide (LCO), lithium manganese oxide (LMO), Lithium nickel manganese cobalt oxide (NMC), Lithium nickel cobalt aluminium (NCA). These materials are known as intercalation materials, as lithium ions are intercalated into the interstices of the crystal structure. Lattice parameters vary according the lithium content, causing a volume increase during lithium intercalation, vice-versa during deintercalation. This behaviour may be seen in analogy with thermal strain: greater concentration (or temperature) gradient causes greater strain, following a linear trend, where the proportionality coefficient is represented by fraction molar volume (or coefficient of thermal expansion).

Lithium ions diffuse inside the active material particles, causing an inhomogeneity in lithium concentration, as highlighted in Figure 1b, and thus a strain mismatch arises in the particle, leading to the so called diffusion induced stress (DIS) [5].

Mechanical stress caused by concentration gradient is the primary driving force of crack propagation [7–9]. Opening mode I, caused by a tensile stress normal to crack surfaces, is considered. Then, hoop stress is the driving force of crack propagation, as clarified in Figure 1c. Hoop stress is tensile on the surface during lithium deintercalation, and in the core during intercalation, as found in previous studies [5, 6]. This trend makes the cracks in the core to propagate during lithium intercalation, and those on the surface during deintercalation, as highlighted in Figure 1c.

Another effect which causes crack growth is phases transition [10, 11]. Some materials, such as LFP, present a strong boundary between li-rich and li-poor rather than a smooth gradient. The existence of two phases with a marked difference in lithium content causes a strong mechanical stress at the interface, which causes crack growth as well.

Fracture in active materials is affected by several factor: such as size and shape of the particle, size of pre-existing cracks and current rate [12]. Size and shape of the particle influences mechanical stress [13], indeed it was observed that small particles are less subjected to fracture because of smaller stress [12]. Current rate has a negative influence on cracks, as greater current rate causes greater stress, as well as longer pre-existing cracks makes their growth easier.

Crack modeling in LIB electrodes is still under debate in literature [12], several approaches are followed, such as phase field [14], cohesive model [15], energetic approaches [16]. In this work, a finite elements (FE) analysis in ANSYS Mechanical APDL is carried out on a spherical active material particle with superficial and internal cracks. Stress intensity factor is computed for different boundary conditions - namely current rates - to identify the boundary beyond unstable crack growth occurs. Several geometric configurations are considered to clarify their influence on crack propagation.

2. Fundamentals

2.1. Electrochemical-mechanical model

Fundamentals of electrochemical-mechanical modeling of active materials for LIB was treated in details in previous works [5, 6, 13], however a brief explanation is reported here as well.

Mechanical stress arises in the micro-structure of active material as a consequence of lithium inhomogeneity, caused by its diffusion through active material particle. As a consequence, this analysis involves two different fields: mechanical and electrochemical.

Mechanical model is expressed by Equations (1)-(3) according to the axisymmetric approach, where radial and circumferential are the directions needed to describe the problem. Constitutive Equations (1) show that a chemical strain $\frac{\Omega C}{3}$ is added to elastic strain, in analogy with what is usually done with thermal strain αT. It is straightforward to understand that an inhomogeneous concentration field can produce mechanical stress, exactly as temperature field does. At this stage, concentration field is needed to be solved in order to calculate mechanical stress.

Diffusion Equations (4)-(6) solve the concentration fields considering chemical potential, lithium flux and mass conservation equations. Flux of lithium ions goes from areas with high potential (high concentration) to areas with low potential (low concentration). This unbalanced concentration, and thus potential, is due to boundary flux, which is directly proportional to the current rate of the battery. Negative flux makes the lithium ions enter inside the particle, and diffuse towards the core, where concentration is lower. On the contrary, positive flux retrieves lithium ions from the surface and makes the ones in the core to diffuse towards the outer areas, where concentration is lowering.

Once concentration field is solved with a proper boundary flux, which can be expressed as current density I as well (Equation (7)), mechanical field is solved in turn (Equation (8)-(10)) considering radial stress to vanish on the external surface (free expansion), and constraining the central point of the particle in order to prevent rigid body motion. More complex surface constraints and contact characteristics between particles were deepened in a previous study [13].

Table 1: DIS model equations [5].

Mechanical equations		
Constitutive	$\sigma_r = \frac{E\left[\left(\varepsilon_r-\frac{\Omega c}{3}\right)(1-\nu)+2\nu\left(\varepsilon_c-\frac{\Omega c}{3}\right)\right]}{(1+\nu)(1-2\nu)} \quad \sigma_c = \frac{E\left[\left(\varepsilon_c-\frac{\Omega c}{3}\right)+\nu\left(\varepsilon_r-\frac{\Omega c}{3}\right)\right]}{(1+\nu)(1-2\nu)}$	(1)
Congruence	$\varepsilon_r = \frac{du}{dr} \quad \varepsilon_c = \frac{u}{r}$	(2)
Equilibrium	$\frac{d\sigma_r}{dr} + \frac{2}{r}\left(\sigma_r - \sigma_c\right) = 0$	(3)
Diffusion equations		
Chemical potential	$\mu = \mu_0 + R_g T ln(c)$	(4)
Flux	$J = -Mc\nabla\mu = -D\nabla c$	(5)
Mass conservation	$\frac{\partial c}{\partial t} + \nabla \cdot (\nabla J) = 0$	(6)
Solutions		
Concentration	$c(r,t) = c_0 + \frac{IR}{F_a D}\left[3\tau + \frac{1}{2}\left(\frac{r}{R}\right)^2 - \frac{3}{10} - 2\frac{R}{r}\sum_{n=1}^{\infty}\left(\frac{sin(\lambda_n r/R)}{\lambda_n^2 sin(\lambda_n)}e^{-\lambda_n^2\tau}\right)\right]$	(7)
Displacement	$u(r) = \frac{\Omega}{3(1-\nu)}\left[(1+\nu)\frac{1}{r^2}\int_0^r c(r)r^2\,dr + 2(1-2\nu)\frac{r}{R^3}\int_0^R c(r)r^2\,dr\right]$	(8)
Radial stress	$\sigma_r(r) = \frac{2\Omega}{3}\frac{E}{1-\nu}\left[\frac{1}{R^3}\int_0^R c(r)r^2\,dr - \frac{1}{r^3}\int_0^r c(r)r^2\,dr\right]$	(9)
Hoop stress	$\sigma_c(r) = \frac{\Omega}{3}\frac{E}{1-\nu}\left[\frac{2}{R^3}\int_0^R c(r)r^2\,dr + \frac{1}{r^3}\int_0^r c(r)r^2\,dr - c(r)\right]$	(10)

2.2. Linear elastic fracture mechanics

Linear elastic fracture mechanics (LEFM) theory predicts the stress field singularity ahead the crack in terms of stress intensity factor and energy release rate G. This assumption holds whether the material is linear elastic or characterized by a small-scale yielding at the crack tip, i.e. the plastic zone extension at the crack tip is small relative to the crack length. The stress intensity factor K_I for a crack of characteristic size a and subjected to uniform tension σ in pure mode I loading is:

$$K_I = Y\sigma\sqrt{\pi a} \tag{11}$$

where Y is the dimensionless geometric factor. The stress field describing the singularity near the crack tip in mode I loading can be expressed as follows:

$$\sigma_{ij} = \frac{K_I}{r_c^{1/2}} f_{ij}(\theta) \tag{12}$$

where σ_{ij} are the elastic stress components, r_c and θ are polar coordinates with origin at the crack tip, and $f_{ij}(\theta)$ is a dimensionless shape function. For a single fracture mode, the energy release rate G is a function of the stress intensity factor K_I:

$$G = \begin{cases} \dfrac{1}{E} K_I^2 & \text{plane stress} \\ \\ \dfrac{1-\nu^2}{E} K_I^2 & \text{plane strain} \end{cases} \tag{13}$$

where E is the Young's modulus and ν is the Poisson's ratio. According to LEFM theory, the J-integral is a convenient way to calculate the energy release rate G. The 2D path-independent J-integral on an arbitrary counterclockwise path around the crack tip Γ, introduced by Rice [17], is defined as follows:

$$J = \int_\Gamma \left(W dx_2 - \mathbf{T} \frac{\partial \mathbf{u}}{\partial x_1} \right) ds \tag{14}$$

where W is the strain energy density, $\mathbf{T}$ is the traction vector on the path Γ of the outward unit normal vector $\mathbf{n}$, $\mathbf{u}$ is the displacement vector, x_1 and x_2 are the coordinate directions taken parallel and perpendicular to the crack direction, respectively. Using the divergence theorem, the J-integral can also be generalized to 3D by considering a tubular surface around the crack front. This domain integral method is used to evaluate contour integrals automatically in the FE analysis. The stress intensity factor K_I can be therefore computed from the J-integral by equating Equations (14) and (13).

According to LEFM, the onset of crack growth depends on the relative values of stress intensity factor and material-specific fracture toughness. The crack will propagate when the stress intensity factor exceeds the material fracture toughness. In the case of pure mode-I crack opening fracture occurs when:

$$K_I \geq K_{Ic} \tag{15}$$

where K_{Ic} is the fracture toughness relative to mode I loading.

3. Method

Fracture behaviour of active material particles induced by the insertion or extraction of lithium ions is studied in spherical particles of radius R with pre-existent internal or superficial cracks. The internal crack is disk-shaped with a diameter of $2a$, and is located in the center of the particle where the tensile stress is larger during lithiation, as shown in Figure 2a. Figure 2b shows the semi-elliptical crack, located on the particle surface where the strongest tensile stress occurs during delithiation, with semi-minor and semi-major axis a and c, respectively. The crack depth corresponds to the semi-minor axis a, instead the half width of the crack at the particle surface c' is slightly lower than the semi-axis c. However, the original semi-major axis c is considered in subsequent analyses, as c can be approximated by c' for small cracks, i.e. $R \gg a$. For sake of simplicity, the analyses are restricted to superficial circular cracks with $c = a$.

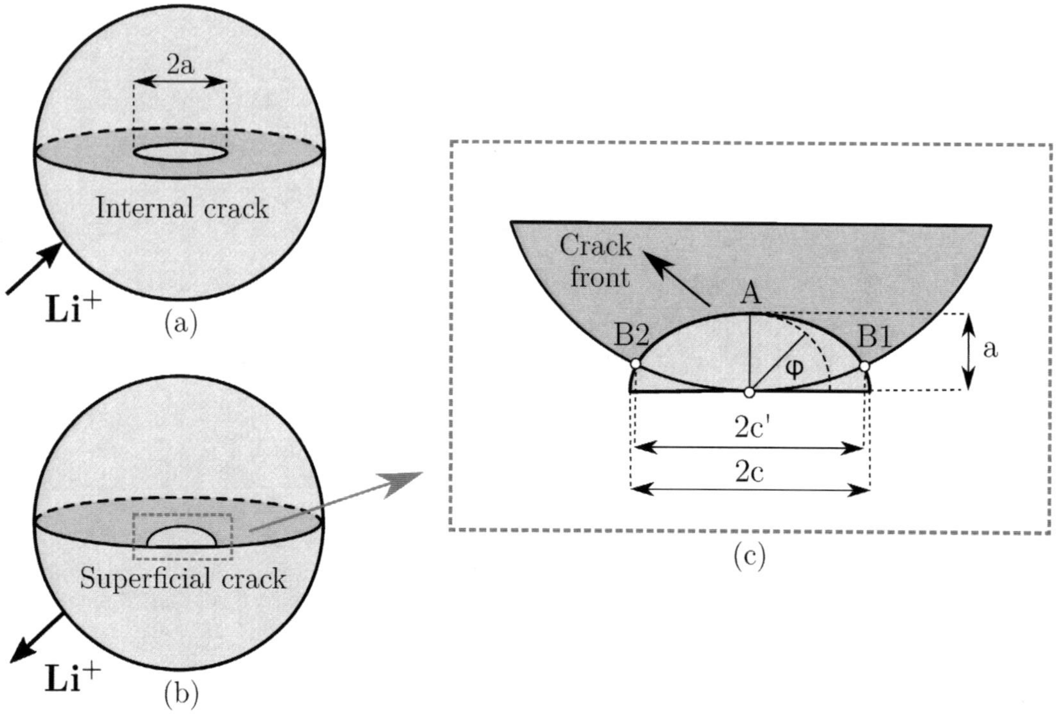

Figure 2: Sketch of a 3D spherical active material particle with pre-existent flaws: (a) disk-shaped crack located at the particle centre during lithiation, (b) semi-elliptical superficial crack during delithiation and (b) detail of the semi-elliptical superficial crack geometry in the sphere mid-plane.

The tensile hoop stress induced by the diffusion process during lithium ions insertion and extraction can drive internal cracks to grow towards the surface and surface cracks to penetrate further towards the core according to mode I, respectively. Since mode II and III toughness values, K_{IIc} and K_{IIIc}, are generally much larger than K_{Ic}, the corresponding crack face loadings can be neglected when analyzing the fracture behavior of active material particles. Based on the LEFM theory, unstable crack propagation occurs when the stress intensity factor K_I along the crack front exceeds the fracture toughness of the material K_{Ic} as shown in Equation 15. In general, K_I depends on the stress distribution over the crack area induced by the concentration gradient, which is a function of intercalation (or deintercalation) time and current density, crack geometry and the investigated position along the crack front in turn, whereas K_{Ic} is an intrinsic material parameter and can be evaluated experimentally.

3.1. FEM model

Fracture behaviour of spherical active material particles is analyzed performing a FEM-based analysis in ANSYS Mechanical APDL. The axisymmetry of the problem is exploited (Figure 3a) in order to reduce the complexity of the study in case of spherical particle with an initial central crack, whereas a 3D model is build when the initial crack is located on the particle surface (Figure 4a).

The spherical particle with central crack is modelled with 2D 8-node plane element PLANE183 in axisymmetric mode (Figure 3d). Singular elements, which are produced by collapsing one side and shifting the respective midside nodes to a quarter position (Figure 3e), are adopted in the FE model to improve the numerical results close to the crack tip [18], where the stress

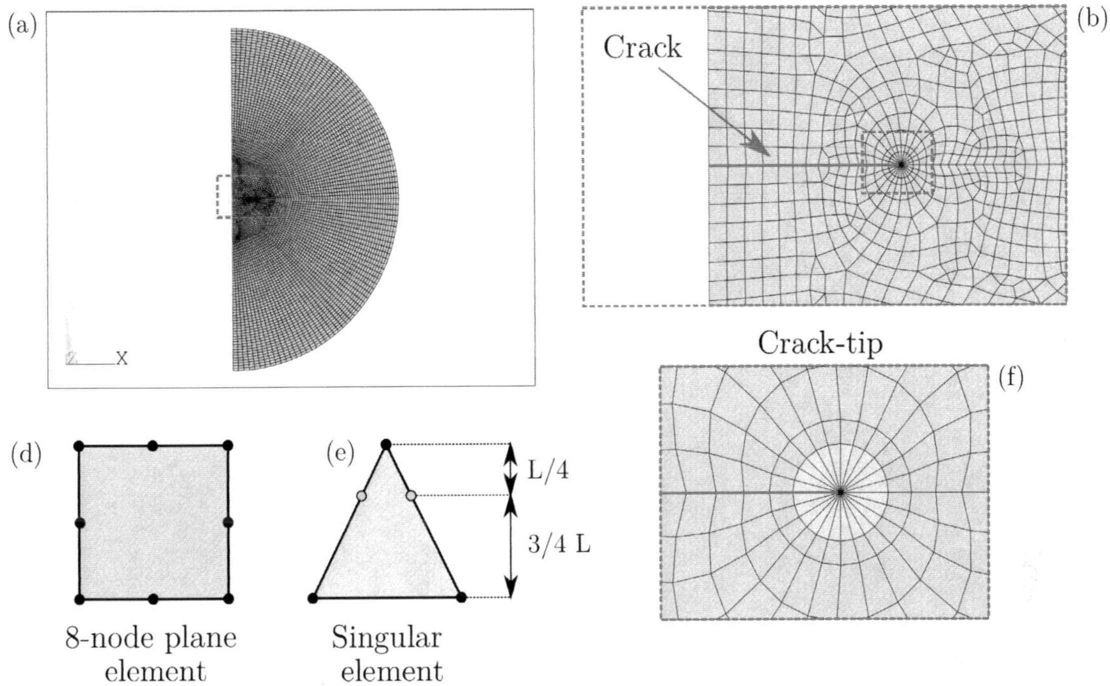

Figure 3: The 2D axisymmetric spherical particle with central crack. (a) Meshed FEM-based model with detail of (b) crack-tip region, (c) elements around crack-tip region: (d) 8-node plane element and (e) collapsed singular element with midside nodes (represented in green) at quarter-point positions.

field is singular. Since the crack tip causes a stress concentration, stress and strain gradients become larger as the crack tip is approached, therefore the elements are chosen to be small near the crack region, and their size is gradually increased when moving away from it (Figure 3b). The command of "KSCON" is used to automatically generate 12 singular elements around the specified crack-tip (Figure 3b) .

Only one half of the spherical particle in 3D model is considered, due to the symmetry of the problem. A refined mesh with collapsed quarter-point singular elements is employed in a tubular region around the crack front, to obtain a well-established singular stress field close to the crack front. Singular elements are constructed from the 15-node version of 3D isoparametric brick SOLID186 elements collapsed to wedges by shifting the mid-side nodes one-quarter away from the crack front (Figure 4e). On the top of the singular elements layer, seven additional layers, consisting of the 20-node version of isoparametric brick elements SOLID186 (Figure 4d) are added. A torus is created around the crack front to control the mesh generation. A spider web mesh made of 8 concentric rings is created on one of two meridian torus surfaces using MESH200 elements, with the innermost ring containing 12 singular elements in circumferential direction generated through the KSCON command (Figure 4g). The meshed area is then swept generating brick elements with one layer of singular elements around the crack front. A minimum of 56 elements along the crack front is used (Figure 4g). The rest of the model is meshed with 10-node tetrahedral version of SOLID186 elements (Figure 4f). Figures 4a and b show the finite element mesh of the model implementd in ANSYS. This particular mesh topology is able to accommodate different crack shapes and allows a smooth change from a refined mesh near the crack front to a larger mesh far from the crack area.

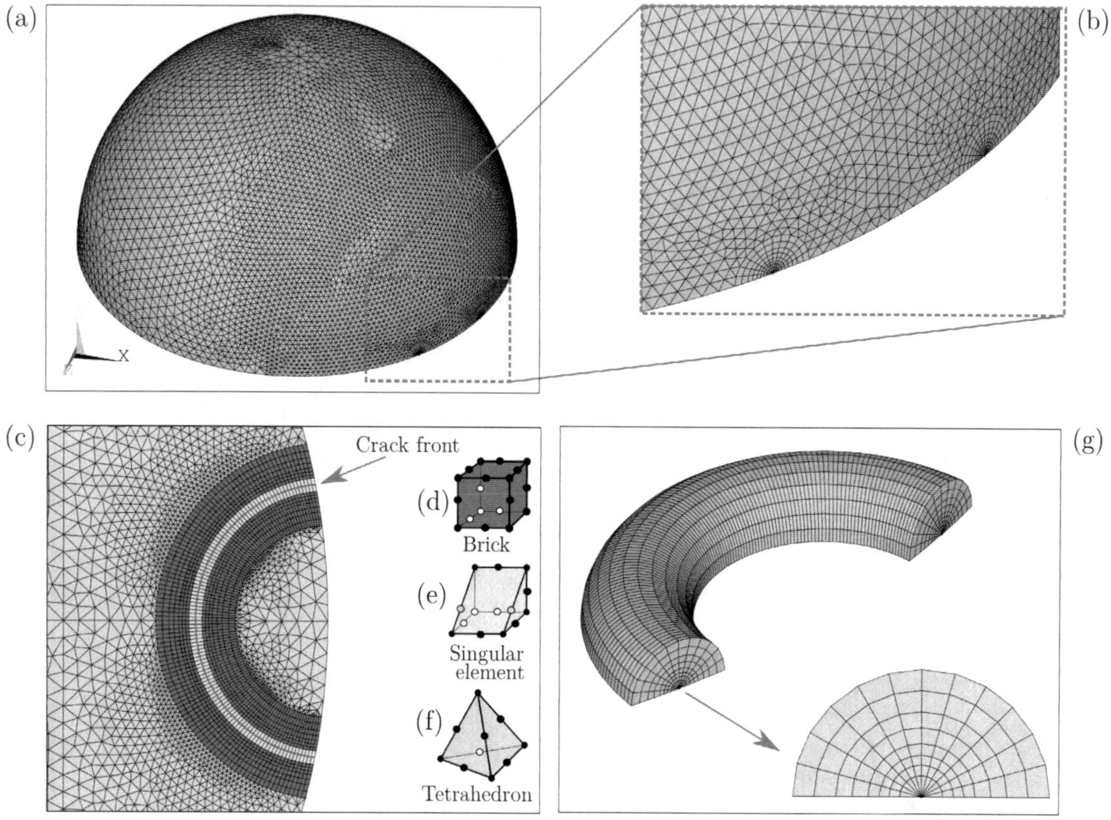

Figure 4: The 3D half spherical particle with semi-elliptical superficial crack. (a) Meshed FEM-based model with (b) detail of the crack region. (c) Bottom view of mesh around the crack region with focus on (d) 20-node brick element, (e) 15-node collapsed singular element with nodes at quarter-point positions (represented in green) and (f) 10-node tethraedral element. (g) Elements breakdown near the crack front with detail of the spider web pattern.

3.2. Simulation procedure

According to previous works [5, 6], the governing equations of lithium ions diffusion problem are similar to that of heat transfer. Exploiting the analogy between diffusive and thermal equations, the multi-physics diffusion-mechanical problem presented in section 2.1 can be reformulated with an equivalent thermal-mechanical problem by replacing the concentration distribution in the spherical particle with an equivalent temperature distribution. This allows an easier implementation in the commercial finite element software ANSYS^{TM}. In order to obtain the solution of the diffusion-mechanical problem in a faster and simpler way, the lithium concentration distribution inside the particle is computed analytically with proper mechanical and chemical boundary condition, neglecting the influence of the mechanical stress on the concentration itself [5]. The equivalent temperature gradient is shown in Equation (16) equating diffusion strain induced by the inhomogeneous chemical concentration with thermal strain.

$$\alpha(T - T_{ref}) = \frac{\Omega}{3}(C - C_{ref}) \Longrightarrow T = T_{ref} + (C - C_{ref}) \tag{16}$$

Where α is the thermal expansion coefficient, $\frac{\Omega}{3}$ is the coefficient of volume expansion due to concentration gradient, T_{ref} and C_{ref} are temperature and concentration referred to zero strain. The concentration distribution is interpolated from the centre to external surface of the particle using a polynomial function. Based on the concentration polynomial function, the temperature field computed using Equation (16) is mapped on FE model nodes including the surface crack. The thermal induced stress and strain states, which are similar to the ones induced by the lithium diffusion inside the spherical particle, are then computed and used for the fracture mechanics analysis. A static fracture analysis is performed to evaluate the stress intensity factor K_I and to determine the subsequent crack propagation process. By assuming plane strain conditions at the tip of the central crack and at all positions along the front of the superficial crack, the K_I is computed from the J-integral value, which corresponds to the energy release rate G according to the LEFM theory:

$$K_I = \sqrt{\frac{E \cdot J}{(1-\nu^2)}} \tag{17}$$

LMO is chosen as case study to analyse the fracture in cathode intercalation materials. The LEFM theory is valid as LMO can be considered a brittle material. Lithiation and delithiation of LMO particles produce tensile hoop stresses that cause the opening of central and surperficial pre-existent crack, respectively. Lithium insertion and extraction are simulated adopting galvanostatic operation consisting of a constant lithium flux at the edge of the spherical particles. Different particle radii (R = 5, 10, 15, 20 μm), initial normalized crack depths (a/R= 0.025, 0.05, 0.1) and current densities over the particle surface (I = 1, 2, 3, 4, 5 A/m^2) are analyzed. The LMO material properties are summarized in table 2.

Table 2: LMO material properties (cathode).

Parameter	Symbol	Value	Unit	Reference
Young Modulus	E	10	GPa	[19]
Possion ratio	ν	0.3	-	[19]
Diffusion coefficient	D	$7.08 \cdot 10^{-15}$	m^2/s	[19]
Partial molar volume	Ω	$3.497 \cdot 10^{-6}$	m^3/mol	[19]
Maximum Concentration	C_{max}	$2.29 \cdot 10^{4}$	mol/m^3	[19]
Fracture toughness	K_{Ic}	0.24	MPa $\cdot$ m$^{1/2}$	[20]

4. Simulation results and discussion

4.1. Concentration and stress analysis

The lithium ions concentration field inside the LMO spherical particle without initial defects is computed based on the analytical model introduced in section 2.1. Exploiting the thermal analogy, diffusion induced radial and hoop stresses are then evaluated with the 2D and 3D FE models in ANSYS and compared with the analytical solution. Figure 5 shows the comparison of the results of concentration and stress state inside a particle of radius R = 10 μm during galvanostatic insertion assuming a null initial concentration within the particle and a lithium flux equivalent to a surface current density of 1 A/m^2. As previously stated, the radial stress is tensile and vanishes on the particle surface, as the interaction with surroundings is neglected and traction-free condition is assumed on the outer surface of the particle. On the contrary, the hoop stress is compressive on the surface and the largest tensile hoop stress is found to occur in the core of the particle, where pre-existing defects could lead to fracture. Figure 6 shows the comparison among FE and analytical results of concentration and stress state inside a spherical particle of radius R = 10 μm during galvanostatic extraction assuming an initial concentration

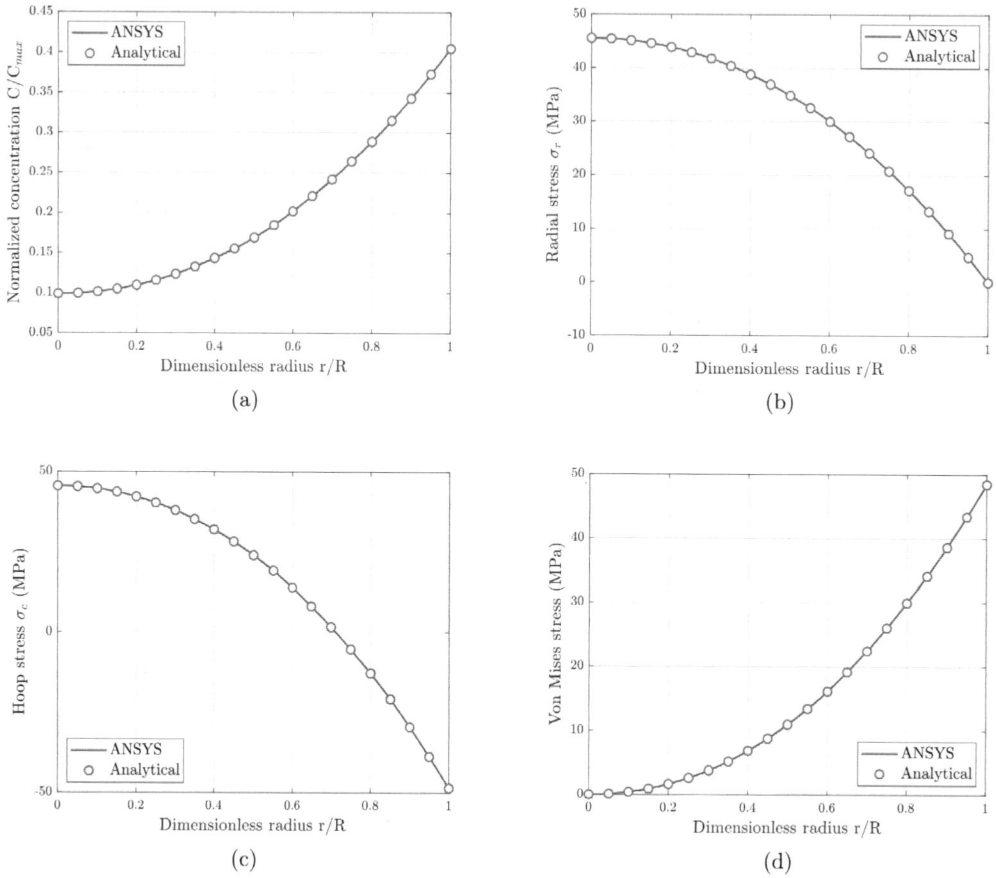

Figure 5: Comparison between analytical and FEM-based simulation results in galvanostatic insertion: (a) normalized lithium concentration C/C_{max}, (b) radial stress, (c) hoop stress and (d) Von Mises stress within a spherical particle of radius $R = 10$ μm at 1 A/m^2 current density and null initial concentration. The simulation time is equal to 2000 s.

within the particle equal to C_{max} and surface current density equal to -1 A/m^2. Radial stress is compressive and vanishes on the particle surface, on the contrary the hoop stress is compressive in the core and reaches the maximum tensile value on the particle surface, being the driving force for crack propagation.

The evolution of the hoop stress at different location along the particle radius during the lithium ions insertion and extraction processes is shown in Figure 7a and b, respectively. The hoop stress at the centre and on the surface of the particle increases monotonically at the beginning of lithiation and delithiation respectively, then it keeps constant after reaching the maximum value. Since the particle experiences the most severe loading condition at this time, it is used to compute the stress intensity factor K_I and predict if the fracture propagation occurs. Moreover, since the stress inside the particle keeps constant after an initial transient, the same stress intensity factor K_I can be computed in correspondence of almost the whole state of charge (SOC) range [5].

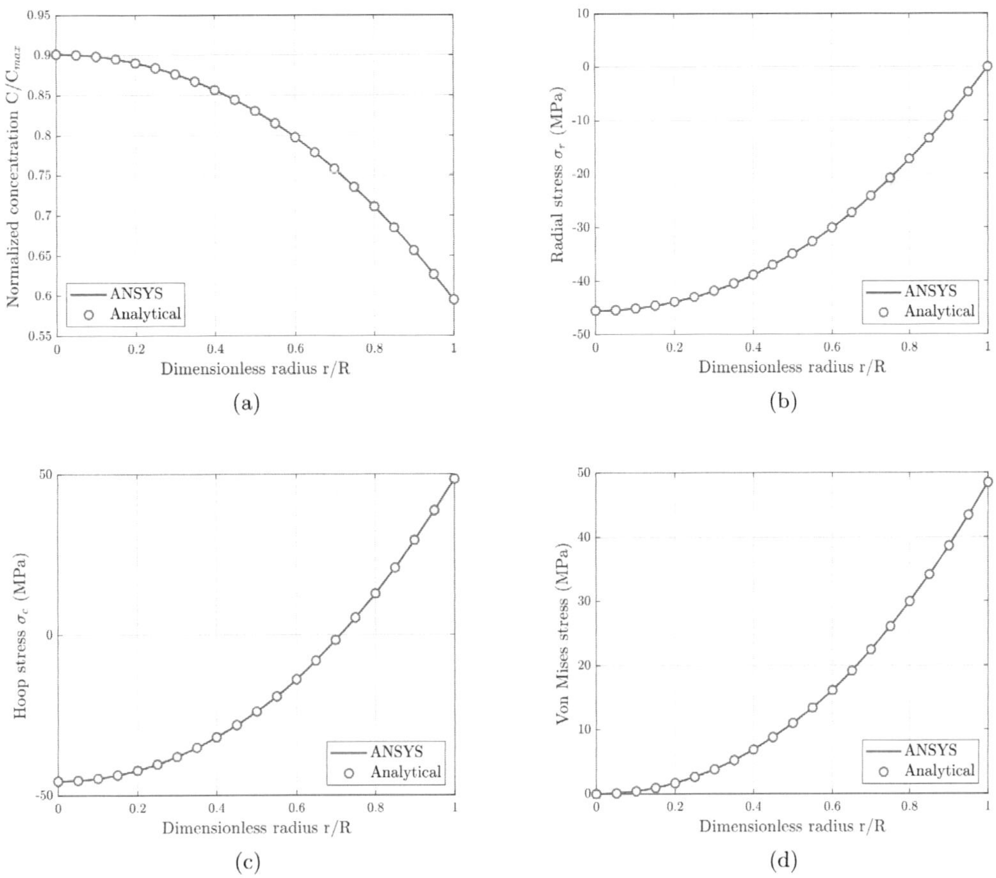

Figure 6: Comparison between analytical and FEM-based simualation results in galvanostatic extraction: (a) normalized lithium concentration C/C_{max}, (b) radial stress, (c) hoop stress and (d) Von Mises stress within a spherical particle of radius $R = 10\ \mu$m at -1 A/m^2 current density and initial concentration $C_0 = C_{max}$. The simulation time is equal to 2000 s.

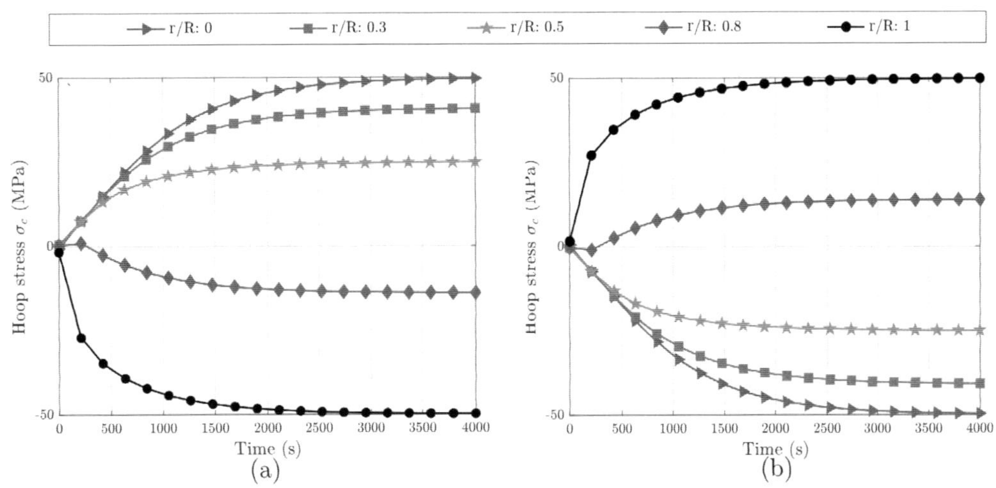

Figure 7: Hoop stress within a spherical particle of radius $R = 10\ \mu$m as a function of diffusion time. Each curve represents a different radial position in the particle: $r/R = 0$ is the particle center and $r/R = 1$ is the particle surface. (a) galvanostatic insertion at 1 A/m^2 current density and null initial concentration and (b) galvanostatic extraction at -1 A/m^2 current density and initial concentration $C = C_{max}$.

Figure 8a shows the normalized concentration field during galvanostatic insertion, analytically computed and mapped as equivalent temperature in the 2D axisymmetric FE model of the spherical particle with central crack. Figure 8c shows the diffusion induced hoop stress distribution inside the spherical particle. Near the outer layer the particle, i.e. far from the crack region, the hoop stress distribution is compressive and relatively unaffected by the crack. On the contrary, Figure 8d shows that the particle experiences a stress singularity at the particle center close to the crack tip, according to LEFM theory.

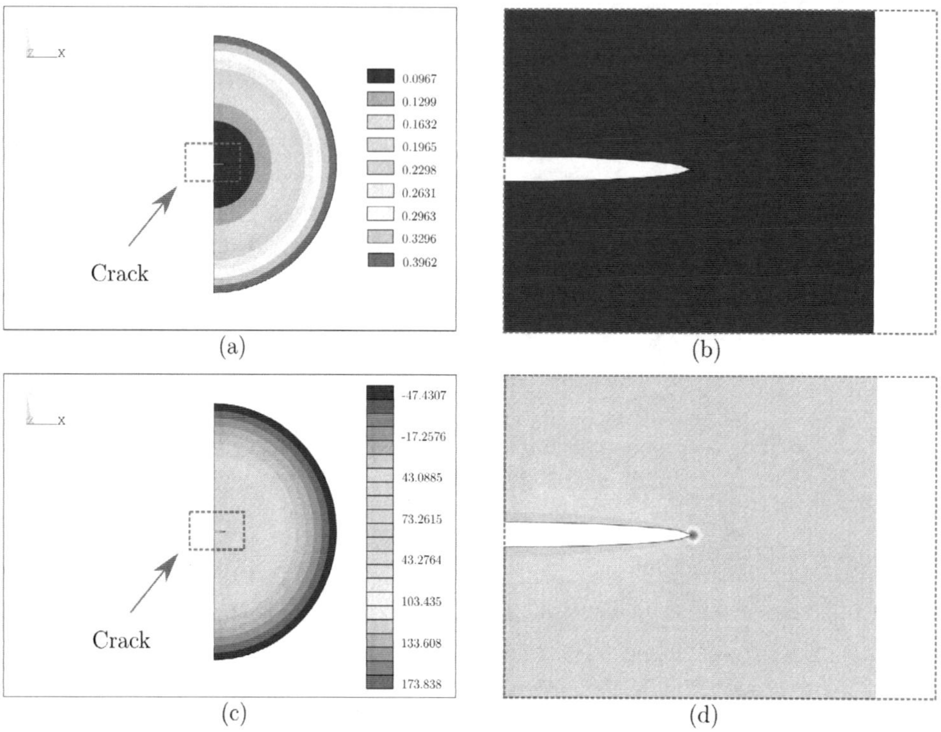

Figure 8: FE simulation results for the spherical particle of radius $R = 10\ \mu\text{m}$, $a/R = 0.1$ during galvanostatic insertion at 1 A/m^2 current density and null initial concentration, obtained with thermal/chemical analogy. The simulation time is equal to 2000 s. (a) Normalized concentration distribution C/C_{max} with (b) detail of concentration near the crack tip region, (c) hoop stress distribution σ_c (MPa) with (d) detail of stress near the crack tip region.

Similarly, crack propagation in spherical particle with superficial pre-existing crack during lithium extraction is analysed carrying out a 3D FEM-based simulation and exploiting the thermal analogy. Lithium ions diffusion is not influenced by the crack, and the concentration distribution, shown in Figure 9a, varies from lower values in the centre of the particle to higher values on the surface resulting in higher expansion of the outer layer of the particle. Figures 9b-d show the hoop stress distribution induced by inhomogeneous volume changes inside the spherical particle. The hoop stress field away from the crack is spherically symmetric, and it is not influenced by the crack, whereas a singular stress field occurs in correspondence of the crack front.

IOP Conf. Series: Materials Science and Engineering 1214 (2022) 012018 doi:10.1088/1757-899X/1214/1/012018

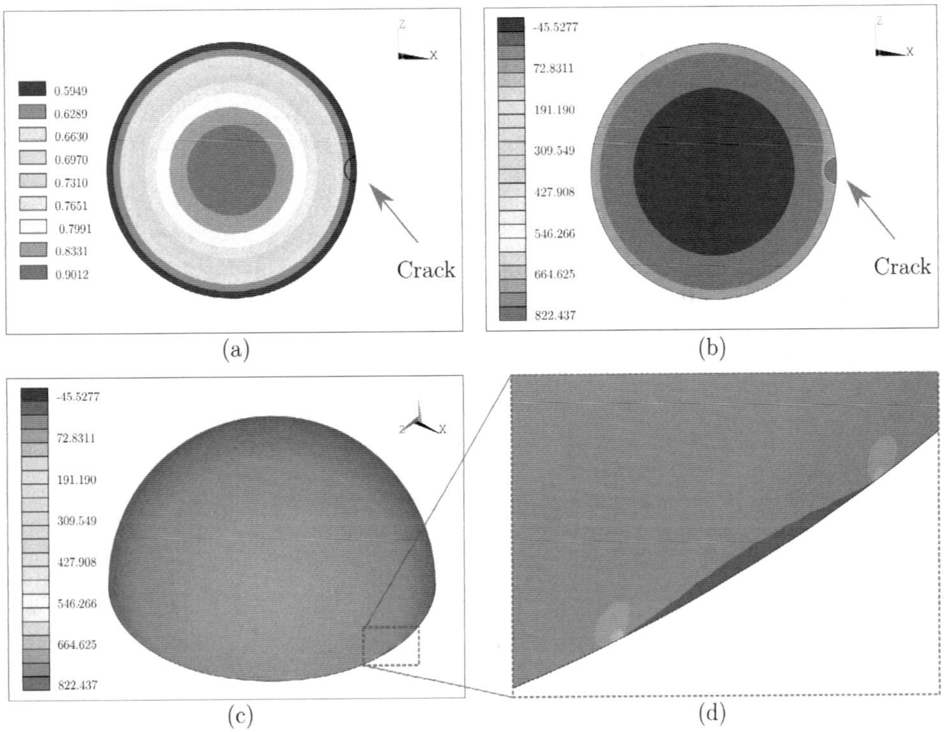

Figure 9: FE simulation results for the spherical particle of radius $R = 10$ μm, $a/R = 0.1$ during galvanostatic extraction at -1 A/m^2 current density and initial concentration equal to C_{max}, obtained with thermal/chemical analogy. The simulation time is equal to 2000 s. (a) Normalized concentration distribution C/C_{max}, hoop stress distribution σ_c (MPa) (b) inside the particle and (c) at the particle surface with (d) detail of the crack region.

4.2. Validation of the FEM model

The stress distribution in the spherical particle is used as input to compute the stress intensity factor K_I at the crack tip of central and surface cracks.

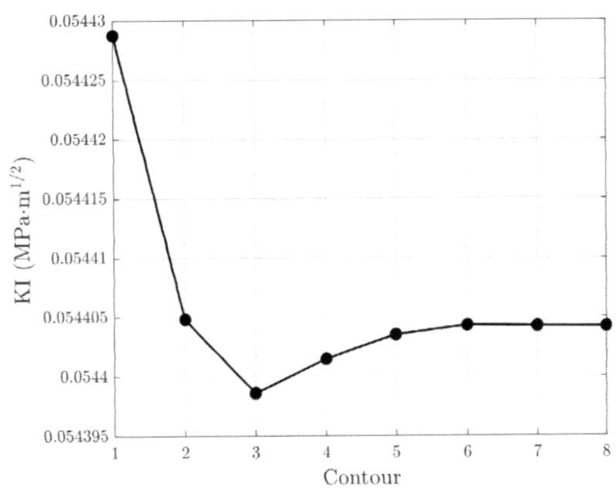

Figure 10: Stress intensity factor K_I evaluated over 8 contours in galvanostatic insertion (2000s). Particle radius $R = 10$ μm, central crack length $a/R = 0.1$ and current density $I = 1$ A/m^2.

Figure 10 shows the K_I values obtained computing the J-integral over eight different contours characterized by an increasing distance from the crack tip. The results show that K_I convergences after the fifth contour to a constant value, meaning that K_I is path-independent. The results obtained from the last contour are considered for the subsequent evaluations.

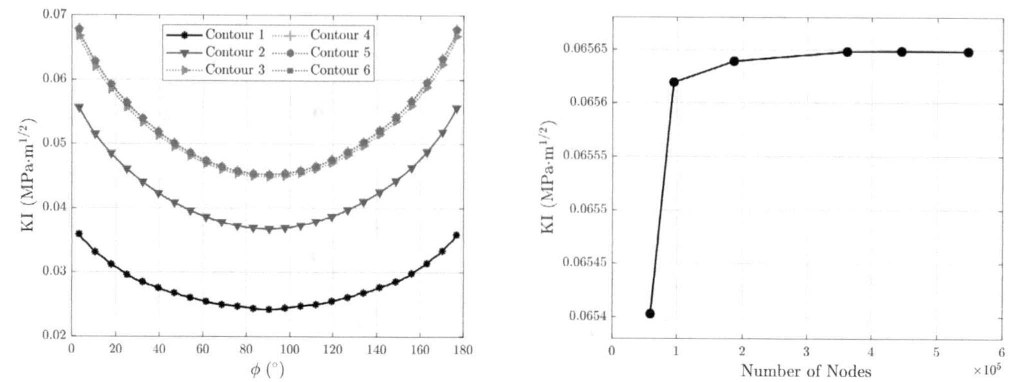

(a) (b)

Figure 11: Study of stress intensity factor K_I convergence for a spherical particle of radius $R =$ 10 μm and superficial crack length $a/R = 0.1$ during galvanostatic extraction ($I = -1$ A/m^2, $t = 2000$ s): (a) K_I distribution along the crack front for six different contours. (b) K_I at point B_1 (referring to Figure 2) as a function of the number of nodes.

Figure 11a shows the stress intensity factor K_I distribution along the crack front obtained from the FEM model in galvanostatic extraction. The stress intensity factors K_I is evaluated over six contours with an increasing distance from the crack tip. The results show that K_I keeps constant when computed from the third contour, which means that the path-independence of J-integral is verified sufficiently distant from the tip. Moreover, the K_I distribution along the crack front is not uniform, but the peak value occurs at points B_1 and B_2 on the spherical surface where the hoop stress reaches its maximum value. It is highlighted that the points characterized by maximum K_I are denoted as B_1 and B_2 even if they are not directly on the spherical surface, but slightly below it. This can be explained by the existence of the so-called vertex singularity close to a corner point, i.e. where a crack front intersects the free surface in a 3D body [21], which changes the distribution of the stress field. The singularity exponent of the stress field (Equation 12) close to the free surface differs from $-1/2$. Moreover, it can be shown that the stress singularity in an elastic material must vary as $r^{-1/2}$ so that J-integral does not vanish [22]. Since K_I distribution along the crack front is computed from the J-integral, which lose their meaning on the particle surface, it is at least reasonable to assume that KI evaluated at points B_1 and B_2 have the same values as the first neighbouring crack-tip node.
In order to evaluate the 3D model reliability, a convergence test of the stress intensity factor K_I in point B_1 is performed as a function of the number of nodes. The results shown in figure 11b clearly indicate that as the number of nodes increases, the K_I converges to a constant value.

4.3. Influence of geometric factors and current density

Different simulations are performed by varying the particle radius R, normalized crack depth a/R and current density I, in order to evaluate the effect of the diffusion induced hoop stress on the particle fracture behavior and the conditions that cause the crack to propagate as well. Both central and superficial crack are analyzed. Figure 12 shows the variation of K_I as a function of particle radius R at different initial crack depths a/R.

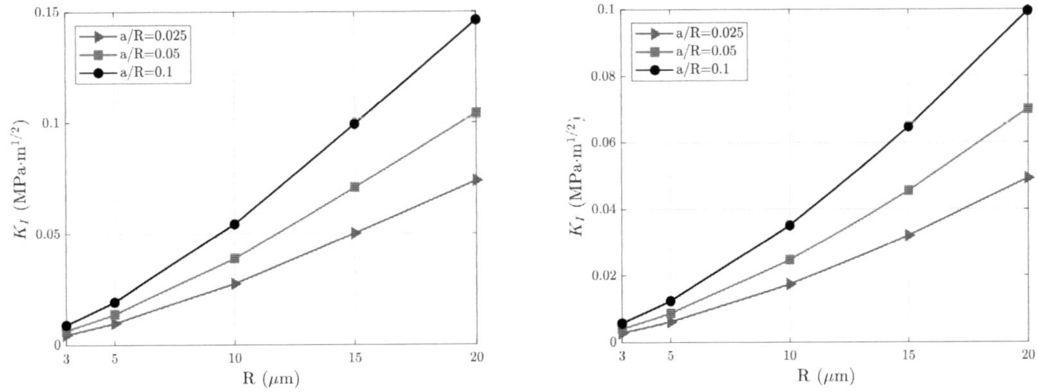

(a) (b)

Figure 12: Stress intensity factor as a function of particle radius R and for different normalized crack depths a/R. (a) Central crack and galvanostatic insertion with current density $I = 1$ A/m^2. (b) Superficial crack and galvanostatic extraction with current density $I = -1$ A/m^2.

The stress intensity factor K_I is computed for the most severe loading condition, namely when hoop stress reaches its maximum. K_I increases monotonically with the particle radius for both galvanostatic insertion and extraction, considering a constant a/R. Smaller particles are characterized by a lower concentration gradient and consequently lower diffusion induced stress. This results in a lower K_I and greater robustness than larger particles from the fracture mechanic point of view. Moreover, for the same particle radius, the stress intensity factor K_I increases with the crack depth a/R. Indeed, a larger a/R leads the hoop stress near the crack region to increase, facilitating the fracture.
Figure 13 shows stress intensity factor computed in a particle with central crack with varying current densities I and particle radii R in galvanostatic insertion.

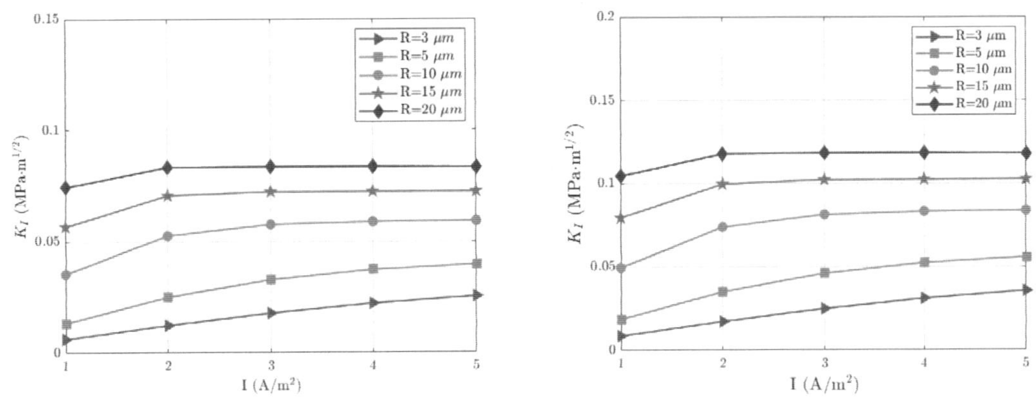

(a) (b)

Figure 13: K_I as a function of current density I in a spherical particle with central crack during galvanostatic insertion for different particle radii R. Normalized crack depth is (a) $a/R = 0.05$ and (b) $a/R = 0.1$. Simulations are carried out for a SOC level equal to 30 %.

The current density I applied on the external surface of the particle affects the stress intensity factor as larger current densities induce higher tensile hoop stresses. As a consequence, higher

current densities are more dangerous from the crack propagation point of view, and this trend is even amplified in bigger particle. Indeed, previous works have confirmed that greater particle size results in higher stresses [5,6], leading to higher K_I in turn. As previously highlighted, the stress intensity factor K_I always becomes greater with increasing the normalized crack depth a/R.
An increase of current density over the surface of smaller particles, namely with radius R equal to 3 and 5 μm, makes the K_I grow monotonically, whereas for particles with larger size the K_I keeps almost constant increasing the current density. This can be explained by the fact that in larger particles with higher lithium flux, i.e. current density, the concentration distribution at the particle center, where the crack is located, is fairly flat compared to the surface. Therefore, the hoop stress at the particle centre does not increases with greater current densities, as this region is barely not affected by the applied flow at the edge of the particl,e and the concentration on the surface quickly reaches the saturation value. Potentiostatic operation is needed to evaluate crack propagation during insertion properly.
Figure 14 shows the stress intensity factor K_I computed in a particle with superficial crack as a function of current density I, with different particle radii R and normalized crack depths a/R.

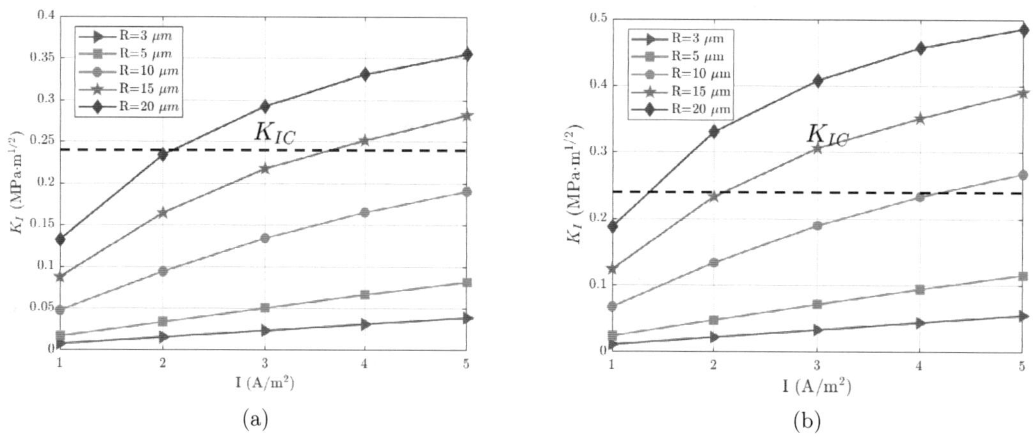

Figure 14: K_I as a function of current density I in a spherical particle with superficial crack during galvanostatic extraction for different particle radii R. Normalized crack depth is (a) $a/R = 0.05$ and (b) $a/R = 0.1$. Simulations are carried out for a SOC level equal to 30 %. The fracture toughness value K_{Ic} of LMO is reported with dashed lines.

The stress intensity factor increases with the increase the normalized crack depth a/R and the applied current density I. However, smaller particles are subjected to lower stresses, and thus lower K_I, which relieve crack propagation. Spherical particles with radius greater than 10 μm characterized by superficial crack may experience fracture, as K_I exceeds the fracture toughness K_{Ic} of LMO for certain current densities.

5. Conclusions

Fracture behaviour of active materials for lithium ion batteries is studied in this work. Mechanical stress which arises in active material due to lithium inhomogeneity are known to be the driving force of crack propagation and subsequent damage of electrode material. Starting from the electrochemical-mechanical model presented in previous works, a finite element model is built up to compute stress intensity factor in active material particles subjected to fracture. Mode I is considered, as is the one characterized by lower fracture toughness and it is the most likely mode to cause crack propagation. Then, tensile circumferential stress is the necessary condition to have crack propagation, which occurs when stress intensity factor overcome fracture toughness of the material. Superficial and internal cracks are considered, according to the stress which is able to make the crack propagate. Indeed, during lithium insertion tensile hoop stress is present in the core of the particle, then internal cracks are likely to propagate. On the other hand, tensile hoop stress is present on the external surface during lithium deintercalation, then superficial cracks are likely to propagate. Lithium manganese oxide is chosen as case study material, considering its brittle characteristic it is suitable to LEFM hypotesis of the finite element model. Several simulations with different geometries and boundary conditions are carried out in order to evaluate when unstable propagation occurs. For what concerns geometric aspect, the results show that small particles are less likely to fracture, in agreement with the stress trend: smaller particles are characterized by lower stress as well. Then, different boundary conditions are considered, namely different current densities, proportional to the current delivered by the whole cell. Results show the combinations of particle size and current density which are safe (no unstable crack propagation) and the one that could be subjected to fracture. Potentiostatic operation will be needed to evaluate properly the propagation of central crack in bigger particles.

References

[1] Mocera F, Vergori E and Somà A 2019 Battery performance analysis for working vehicle applications *IEEE Transactions on Industry Applications* **56** 644–653
[2] Somà A, Bruzzese F, Mocera F and Viglietti E 2016 Hybridization factor and performance of hybrid electric telehandler vehicle *IEEE Transactions on Industry Applications* **52** 5130–5138
[3] Mocera F, Somà A and Clerici D 2020 Study of aging mechanisms in lithium-ion batteries for working vehicle applications *2020 Fifteenth International Conference on Ecological Vehicles and Renewable Energies (EVER)* (IEEE) pp 1–8
[4] Deshpande R, Verbrugge M, Cheng Y T, Wang J and Liu P 2012 Battery cycle life prediction with coupled chemical degradation and fatigue mechanics *Journal of the Electrochemical Society* **159** A1730
[5] Clerici D, Mocera F and Somà A 2020 Analytical solution for coupled diffusion induced stress model for lithium-ion battery *Energies* **13** 1717
[6] Clerici D and Mocera F 2021 Micro-scale modeling of lithium-ion battery *IOP Conference Series: Materials Science and Engineering* vol 1038 (IOP Publishing) p 012007
[7] Raghavan R S, Kumar A M and Narayanrao R 2015 Intercalation induced surface cracking in electrode particles *ZAMM-Journal of Applied Mathematics and Mechanics/Zeitschrift für Angewandte Mathematik und Mechanik* **95** 845–858
[8] Woodford W H, Chiang Y M and Carter W C 2010 “electrochemical shock” of intercalation electrodes: a fracture mechanics analysis *Journal of the Electrochemical Society* **157** A1052
[9] Klinsmann M, Rosato D, Kamlah M and McMeeking R M 2016 Modeling crack growth during li extraction and insertion within the second half cycle *Journal of power sources* **331** 32–42
[10] Chen B, Zhou J and Cai R 2016 Analytical model for crack propagation in spherical nano electrodes of lithium-ion batteries *Electrochimica Acta* **210** 7–14
[11] Deshpande R, Cheng Y T, Verbrugge M W and Timmons A 2011 Diffusion induced stresses and strain energy in a phase-transforming spherical electrode particle *Journal of the Electrochemical Society* **158** A718
[12] Li P, Zhao Y, Shen Y and Bo S H 2020 Fracture behavior in battery materials *Journal of Physics: Energy* **2** 022002
[13] Clerici D, Mocera F and Somà A 2021 Shape influence of active material micro-structure on diffusion and contact stress in lithium-ion batteries *Energies* **14** 134
[14] Zuo P and Zhao Y P 2015 A phase field model coupling lithium diffusion and stress evolution with crack propagation and application in lithium ion batteries *Physical Chemistry Chemical Physics* **17** 287–297
[15] Bower A F and Guduru P 2012 A simple finite element model of diffusion, finite deformation, plasticity and fracture in lithium ion insertion electrode materials *Modelling and Simulation in Materials Science and Engineering* **20** 045004
[16] Stamps M A, Eischen J W and Huang H Y S 2016 Particle-and crack-size dependency of lithium-ion battery materials lifepo 4 *AIMS Materials Science* **3** 190–203
[17] Rice J R 1968 A path independent integral and the approximate analysis of strain concentration by notches and cracks *J. Appl. Mech.* **35** 379–386

[18] Barsoum R S 1976 On the use of isoparametric finite elements in linear fracture mechanics *International journal for numerical methods in engineering* **10** 25–37
[19] Zhang X, Shyy W and Sastry A M 2007 Numerical simulation of intercalation-induced stress in li-ion battery electrode particles *Journal of the Electrochemical Society* **154** A910
[20] Mughal M Z, Amanieu H Y, Moscatelli R and Sebastiani M 2017 A comparison of microscale techniques for determining fracture toughness of limn2o4 particles *Materials* **10**
[21] Bažant Z P and Estenssoro L F 1979 Surface singularity and crack propagation *International Journal of Solids and Structures* **15** 405–426
[22] Anderson T L 2017 *Fracture Mechanics: Fumdamental and Applications* ISBN 9781498728133

IOP Conf. Series: Materials Science and Engineering 1214 (2022) 012019 doi:10.1088/1757-899X/1214/1/012019

Structural Health Monitoring of civil structures through FEM high-fidelity modelling.

G Mirasoli[1], C Brutti[1], C Groth[1], L Mancini[1], S Porziani[1] and M E Biancolini[1*]

[1] Department of Enterprise Engineering "Mario Lucertini", University of Rome "Tor Vergata", Via del Politecnico 1, 00133 Roma, Italy

*biancolini@ing.uniroma2.it

Abstract. FEM high-fidelity modeling is a useful method for evaluating the dynamic behavior of civil structures and their state of damage through the processing of real-time vibration data in the frame of the general Structural Health Monitoring. The development of reliable digital twin involves the use of high-fidelity models, which must be validated with experimental data. In this study, a prototype of a bridge, composed by a frame structure of steel beams and a slab made in concrete and asphalt, was considered. Two of the main beams have a bolted connection that have been used to simulate a local damage of the structure. FEM models have been validated with experimental data, thus they are able to evaluate both static and dynamic behavior of the bridge and its state of health, referred to benchmark structure. The bridge was degraded considering the breakage of one or more plates of the bolted connection, or an accidental overload capable of creating plastic deformations. The proposed models identify damage of the structure through two different methods: shift of natural frequencies, and evaluation of frequency response of the structure (FRF) at the sensor nodes. The second method is more suitable since the variations of observed variables is more consistent. Therefore, with these models, it is possible to carry out monitoring operations for the predictive maintenance of the structure and defining the limits that the observed variables can undergo in order to still consider the structure suitable for the operating loads.

1. Introduction

It is well known that unexpected damage of structure, mechanical systems or plant components can origin relevant safety risk or loss of money. This can be avoided by means of a smart maintenance based on continuous and accurate monitoring during the operating life coupled with models able to identify the damaging process and to predict its time progress [1][2]. In the last decades, Structure Health Monitoring (SHM) has attracted the interest of researchers and many papers and books on this topic are available in the scientific literature [3][4].

Generally, a robust SHM system should have some requirements that can be summarized as follows: able to identify very early the damage and locate where it is occurring. In addition, it is necessary to estimate its severity, predict the remaining useful life basing on objective procedures without involving the engineering judgement of the user. This last feature is the prognostic part of SHM [5]. All these features must be automated in order to manage lot of data collected on field. In recent years, thanks to the possibility of automatically handling large amounts of data at an acceptable cost, the creation and use of a digital twin of the structure under observation has been widely used in many fields of application, especially considering how complex can be the physical phenomena involved [6][7][8].

IOP Conf. Series: Materials Science and Engineering 1214 (2022) 012019 doi:10.1088/1757-899X/1214/1/012019

Without any doubt the heart of the Structural Health Monitoring is the selection of the methods to identify the behavior of the structure and its variation that can indicate a possible damage [9][8]. There are two strategies for structures monitoring: static monitoring, in which some parameters are detected in a long time, and dynamic monitoring oriented towards the evaluation of dynamic parameters, like natural frequencies, damping coefficients and modal shape. Consequently to this selection, the sensors, the acquisition chain, the managing software can be stated. The modern experimental technology offers several methods to investigate a structure as visual automated inspection, acoustic emission, static deformation, vibration behavior [10]. All the experimental data can compared with those extracted by a numerical model (the digital twin) in the same operative condition with the aim to identify and locate the damaging process.

Relating to the dynamic behavior of a structure, it is important to note that large-scale civil or mechanical structures are normally stressed by various loads time varying that can cause vibration as earthquakes, wind, temperature, or human-made excitation. On the other end both the sensors and numerical simulation methods are widely used so that this kind of monitoring is the main system for SHM. Several types of approaches are now available to monitor how a structure is changing [9][10] as frequency [11], mode shapes and frequency response functions [12], flexibility and compliance, damping and so on [9].

In this paper the change of frequency and of frequency response is used to identify the damage in a structural part of a prototype of bridge. In general, bridges are infrastructure very relevant [13] as they connect different regions and originate social and economic benefits. The health monitoring of these is essential to maintain properly adapting to changing conditions of use, to climate-driven or environmental events [14][15].

In this study a prototype of bridge built in the frame of the research project SHAPE was used to develop a digital twin FEM based, in order to define some type of damaging process and to evaluate the change in the dynamic behavior as frequency response function. In particular, as the prototype has in two of the main beams a bolted connection, on these elements has been simulated a local damage both of the bolts and of the connecting plates. As first step of the investigation, the FEM model was tuned basing on the natural modes measured on the prototype using, in the part where the damage is located, a high degree of definition. This allowed not only to study accurately how the dynamic response change due to the damage but also to reduce the computational times and costs allowing in this way to simulate several damaging scenarios.

2. SHM of civil structures: bridge example.

The laboratory structure investigated is a bridge 1:4 scale using an orthotropic steel-concrete deck and two concrete piles Figure 1. The deck consists of five profiles IPE 300 for the main structure made of steel S275, while three of them were used as main beam, the other two were used as head transverse beam Figure 2. The in-plane stiffness was increased by using a series of coupled UPN 140 profiles as a transverse beam and "L" shaped profile as cross-bracing. This small steel grillage works together with a composite slab made of a steel corrugated sheet and covered with concrete. The steel deck rests on a rubber layer of 15 mm placed below each beam, each support rest on the two lateral abutments made of concrete.

Figure 1. View of the steel-concrete bridge.

As above noted, two of the main beams have a bolted connection where the simulated damage has been located. The bolted joint is place at one third of the length (Figure 2).

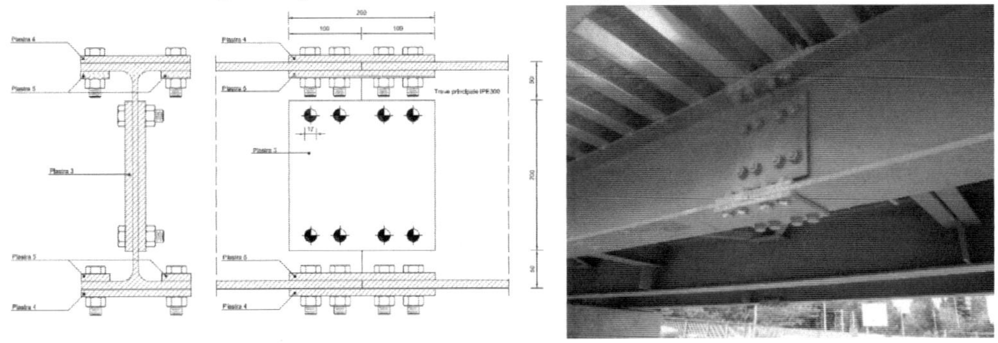

Figure 2. Bolted connection at one third of the length.

Experimental dynamic investigation on the bridge was performed extracting the mechanical parameters of concrete and the stiffness of supports [16]. Table 1 shows the results obtained for the dynamic investigation executed placing accelerometers on the deck and on the beams. As excitation source a hammer below the beam was used in different positions.

Table 1. Results of the dynamic investigation.

	Order Type	Frequency [Hz]
1	1° Flexural	17.50
2	1° Torsional	20.71
3	2° Flexural	56.78
4	2° Torsional	59.63
5	1° Plate	64.42

3. Simplified model identification and calibration.

Basing on the data above reported, the FE model for the code NASTRAN NX [17] equipped with pre-post processor FEMAP was implemented. The model is composed by shell elements (equal to 76 mm or 3 inches) reproducing the slab of the deck, truss elements modelling the cross bracing, and finally beam elements are used to model the main and transverse beam. During the modelling phase, the rubber layer below each beam has been modelled by replacing the rigid restraints with three elastic springs along the three main direction. Starting with the initial values of the density, the elastic modulus of each material and preliminary values of the springs' stiffness, these parameters have been updated to better fit the experimental frequencies.

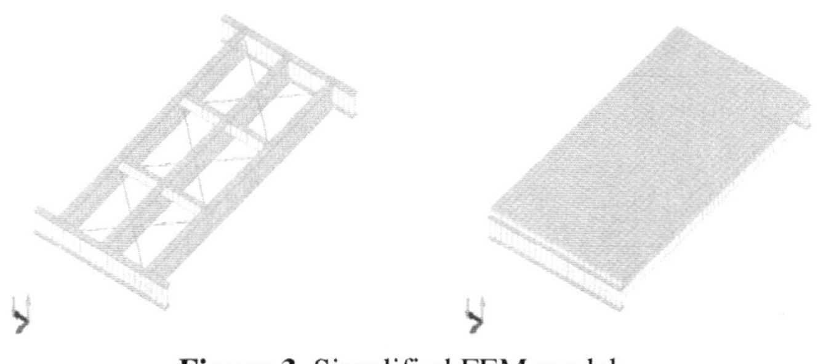

Figure 3. Simplified FEM model.

Table 2 shows the values of mechanical properties of the model, while Table 3 shows the properties of the restraint springs for the FEM.

Table 2. Mechanical properties assigned to the steel bridge.

	Thickness [m]	Elastic modulus [MPa]	Density [kg/m^3]
Steel S275	-	210000	7860
Concrete	0.073	30000	2450
Asphalt	0.03	6000	1800

Table 3. Properties of restraint springs.

Spring direction	Direction	Elastic modulus [kN/m]
X	Deck axis	$1*10^4$
Y	Vertical	$1*10^5$
Z	Transversal	$1*10^6$

After identification of material properties, a comparison between the natural frequencies of FE model and experimental ones has been carried out. Relative error between the two values has been computed and collected in Table 4.

Table 4. Comparison between experimental results and numerical results.

	Order Type	Experimental [Hz]	FEM [Hz]	Err %
1	1° Flexural	17.50	18.26	4.34
2	1° Torsional	20.71	19.76	4.59
3	2° Flexural	56.78	55.37	2.48
4	2° Torsional	59.63	58.35	2.15
5	1° Plate	64.42	64.41	0.02

Figure 4 shows the mode shapes for the bridge. They are in good agreement with those identified experimentally by using Frequency Domain Decomposition procedure.

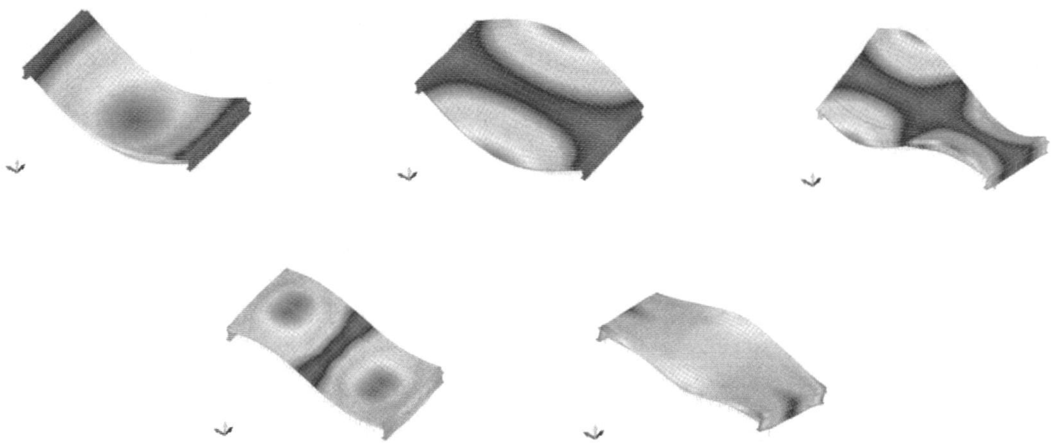

Figure 4. Mode shapes.

IOP Conf. Series: Materials Science and Engineering 1214 (2022) 012019 doi:10.1088/1757-899X/1214/1/012019

4. Static and dynamic behaviour of bolted joint.

The second step of the investigation was to study the mechanical behaviour of the steel beam with bolted connection. Figure 5 shows the beam section geometry. Both beam and plates of connection are modelled with solid elements, while bolts are modelled with beam elements. RBE2 rigid elements, available in NASTRAN NX, simulate the presence of nut and lock nut. Bolt preloads are evaluated starting from tightening torque of 281 Nm, according to the values suggested by EUROCODE. Contacts between plates and beam are modelled as frictional contacts. In this case, the continuity between the two parts is linked to the effect of bolts tightening. The friction coefficient is set to $\mu = 0.4$.

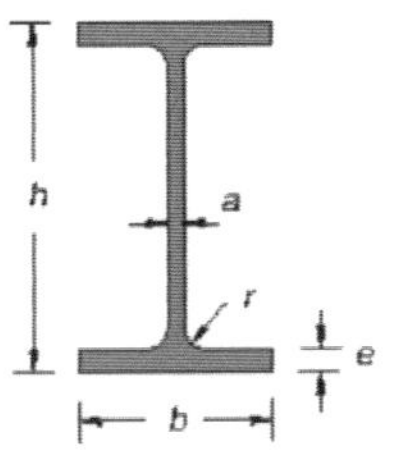

h = 300 mm; b = 150 mm; e = 10.7 mm; a = 7.1 mm ; A = 5188 mm^2

Ixx = $7.996 \cdot 10^7$ mm^4; Iyy = $6.027 \cdot 10^6$ mm^4; Jt = $1.557 \cdot 10^5$ mm^4

Wxx = $5.331 \cdot 10^5$ mm^3; Wyy = $8.036 \cdot 10^4$ mm^3;

Wt = $1.456 \cdot 10^4$ mm

Figure 5. Beam section geometry.

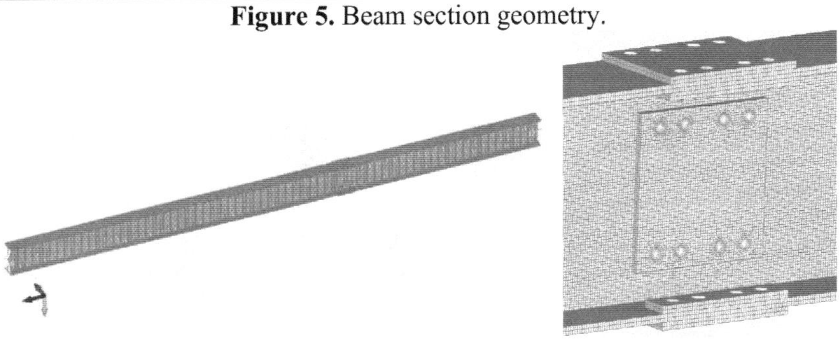

Figure 6. FE mesh.

4.1. Prestressed Modal analysis.

It is customary to perform modal analysis on mechanical systems without due regards to their stress state. This approach is of course well accepted in general but can prove inadequate when dealing with cases like beam with a bolted connection. It is believed that the stress stiffening can change the response frequencies of a system which impacts both modal and transient dynamic responses of the system. This is explained by the fact that the stress state would influence the values of the stiffness matrix [18].To takes into account of bolt preloads, preliminary nonlinear static analysis is performed for calculating differential stiffness to be included in the normal mode solution.

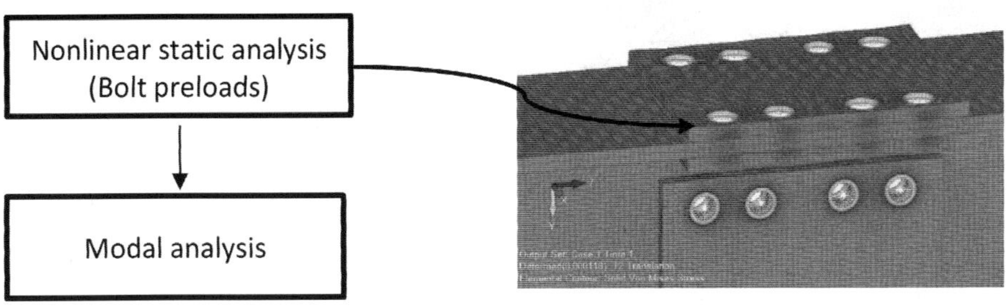

Figure 7. Prestressed modal analysis.

Prestressed modal analysis is performed using SOL401 solution in NASTRAN NX. SOL 401 - NLSTEP. SOL 401 is a multistep, structural solution which supports a combination of static (linear or nonlinear) subcases and modal (real eigenvalue) subcases. The primary solution operations are time increments, iterations with convergence tests for acceptable equilibrium error, and stiffness matrix updates. The iterative process is based on variations of Newton's method. The stiffness matrix updates are performed to improve the computational efficiency. SOL401 consider the internal force computation as follows [5]:

$$F = \int_V \bar{B}^T \sigma dV \tag{1}$$

the element matrix B is defined from the strain definition as

$$d\varepsilon = \bar{B} du \tag{2}$$

in which B could be divided into two parts (linear and nonlinear), i.e.

$$\bar{B} = B_L + B_N \tag{3}$$

upon differentiation of Eq.(1), we have

$$dF = \int_V \bar{B}^T d\sigma dV + \int_V d\bar{B}^T \sigma dV \tag{4}$$

σ represents stresses. Substituting Eq(3) and

$$d\sigma = D d\varepsilon = D\bar{B} du \tag{5}$$

Eq(4) becomes

$$dF = [K_L + K_R + K_\sigma] du \tag{6}$$

with

$$K_L = \int_V B_L^T D B_L dV \tag{7}$$

$$K_R = \int_V B_L^T D B_N + B_N^T D B_N + B_N^T D B_L dV \tag{8}$$

$$K_\sigma du = \int_V dB_N^T \sigma dV \tag{9}$$

in which K_L represents the usual linear stiffness matrix, K_R a stiffness due to large rotation, and K_σ a geometric stiffness dependent on the initial stress level.

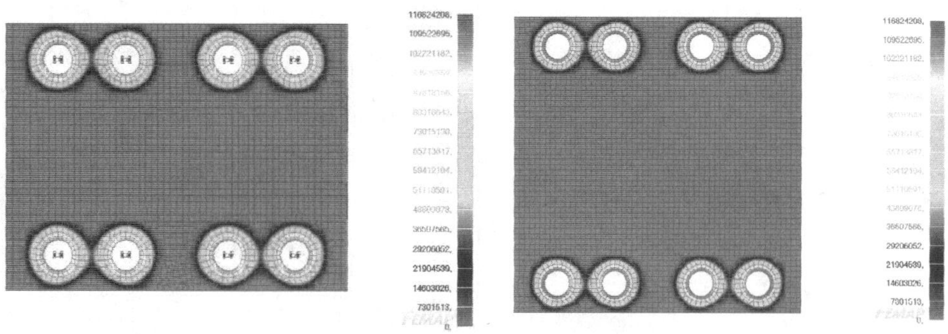

Figure 8. Contact pressure on plates.

IOP Conf. Series: Materials Science and Engineering 1214 (2022) 012019 doi:10.1088/1757-899X/1214/1/012019

Figure 8 shows contact pressure on plates caused by bolt preloads. Beam is simply supported at both ends. Table 5 shows the results of the modal analysis in different grip conditions of bolted joint:

- C0: design conditions.
- C1: joint with one free beam wing.
- C2: joint with free beam wings.
- C3: joint without core plates.
- C4: joint without core plates and one free beam wing.

Table 5. Modal analysis results.

	C0 [Hz]	C1[Hz]	C2 [Hz]	C3 [Hz]	C4 [Hz]
1	26.24	24.75	22.35	26.60	4.59
2	102.06	98.82	95.08	101.51	78.28
3	206.92*	199.70*	195.39	204.15*	155.87
4	219.25	219.02	216.89	207.49	197.27
5	343.42	-	348.26	353.33	314.04

Later, a comparison of FE results with theoretical solution of a beam with a concentrated mass representing the bolted joint (Figure 8) is reported.

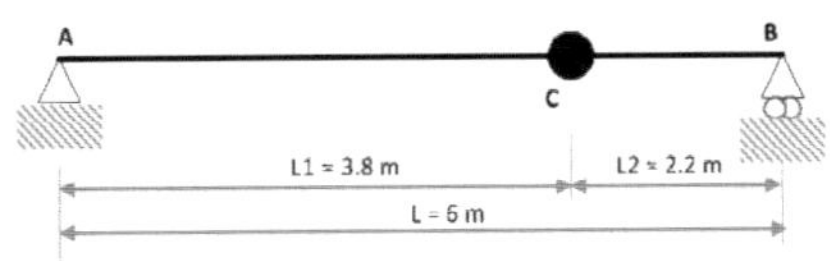

Figure 9. Simply supported beam.

Natural frequencies of a beam without joint can be obtained from Eq. (10):

$$f_i = \frac{K_i^2}{2\pi L^2}\sqrt{\frac{EI}{\rho A}} = 28.11 \text{ Hz} \tag{10}$$

where E = 210 GPa, is the Young modulus of material; $I_{xx} = 7.996e^{-5}$ m^4, represent moment of inertia of the section in the vertical deformation plane; A is the section area; density of material ρ = 7800 kg/m^3; L the distance between supports. K_i values depends on boundary condition and frequency order. Considering a simply supported beam and the first natural frequencies $K = \pi$. The bolted joint, including plates and bolts, weigh 17 kg. The application of superposition principle using Dunkerlay method allow the evaluation of the first natural frequency.

$$\frac{1}{f_{tot}} = \sqrt{\left(\frac{1}{f_t}\right)^2 + \left(\frac{1}{f_m}\right)^2} = 26.53 \text{ Hz} \tag{11}$$

Percentage error between numerical and theoretical is 1.09%. FE model is validated also considering axial deformation:

$$f = \frac{1}{4L}\sqrt{\frac{E}{\delta}} = 216.20 \text{ Hz} \tag{12}$$

Compared with FE result of 206.90 Hz, relative error is around 4.30%. The influence of different joint configurations on the dynamic behavior of the structure are shown in Table 5. The major contribution to the strength of the joint is given by the connections on the wings as they are more distant from the neutral axis (C1, C4).

4.2. Nonlinear static analysis.

Nonlinear static analysis is performed to study the mechanical performance of bolted joint. The applied loads are:

- Gravity load.
- Overload which causes yielding in the midsection of the beam.

For each step of the applied load the following quantities are calculated:

- Maximum deflection.
- Wingspan on the traction side of beam section.
- Bolt preload.

Figure 10 shows these quantities function of applied load.

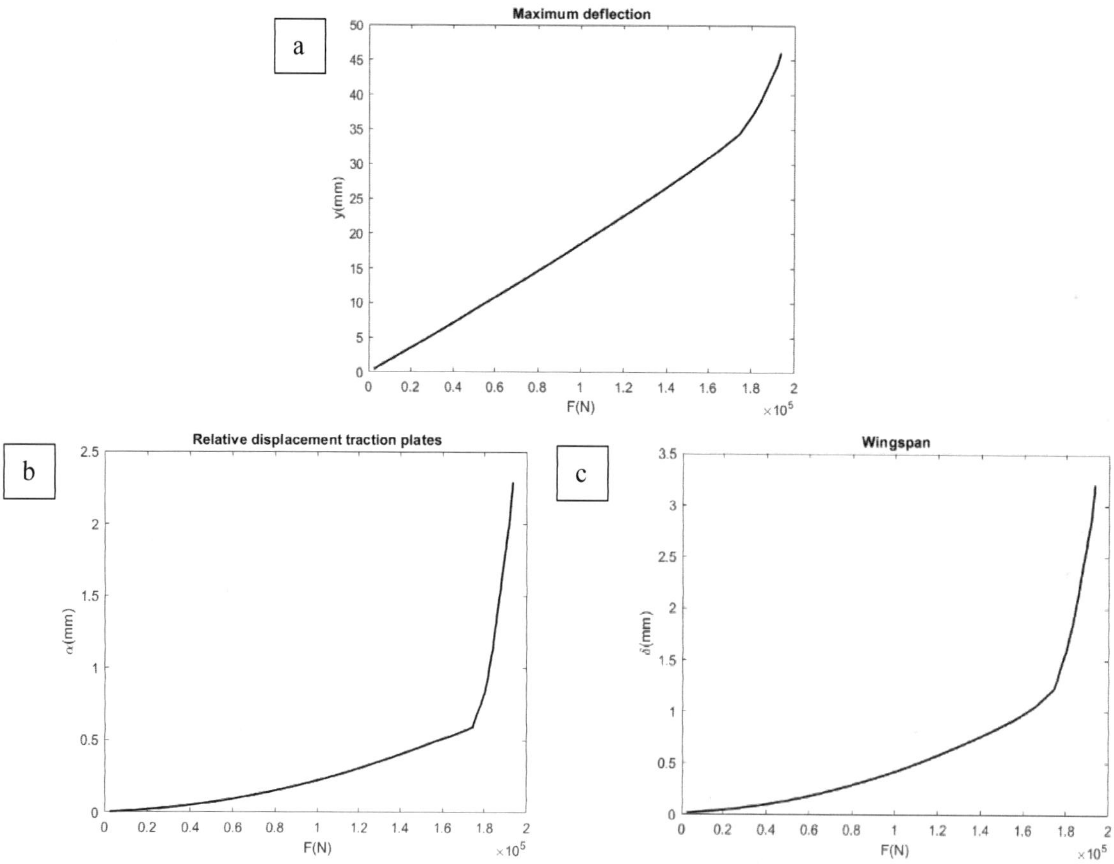

Figure 10. (a) Maximum deflection, (b) Relative displacement traction plates, (c) Wingspan.

No regular trends are linked to friction and local deformability. Friction between plates involves a variation in bolt preloads and a relaxation of the connection. This effect shows local deformability of the joint plates (Figure 11).

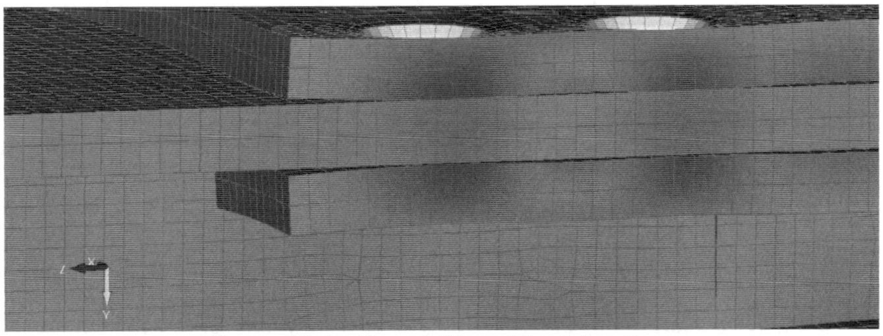

Figure 11. Deformability of the joint plates.

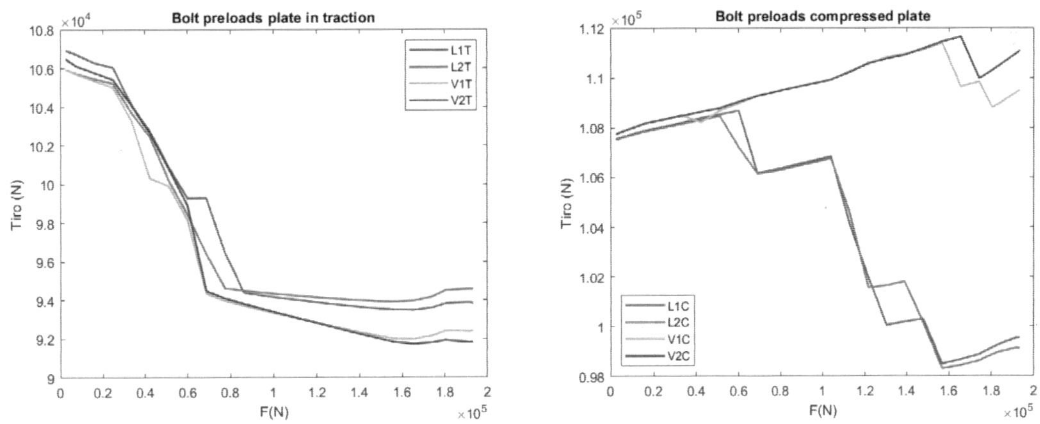

Figure 12. Bolt preloads wing plates.

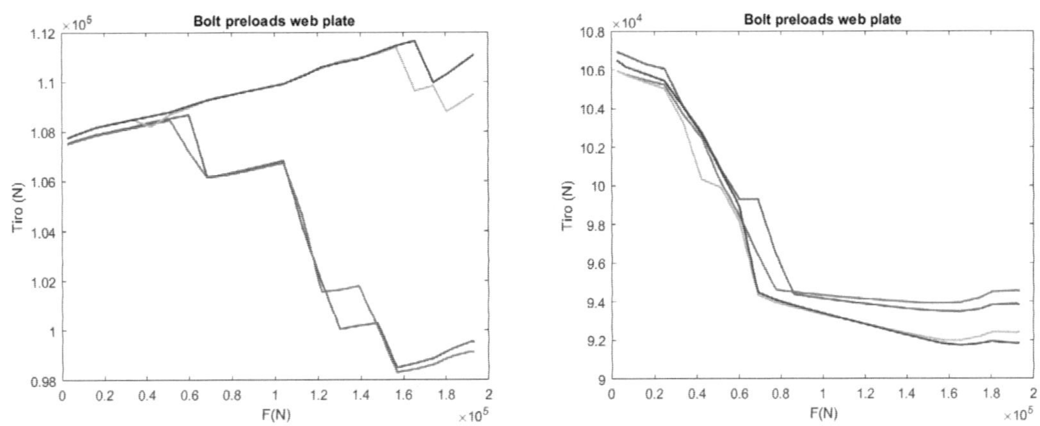

Figure 13. Bolt preloads core plates.

Damage of bolted connection is simulated removing lower plate of the joint (Figure 14).

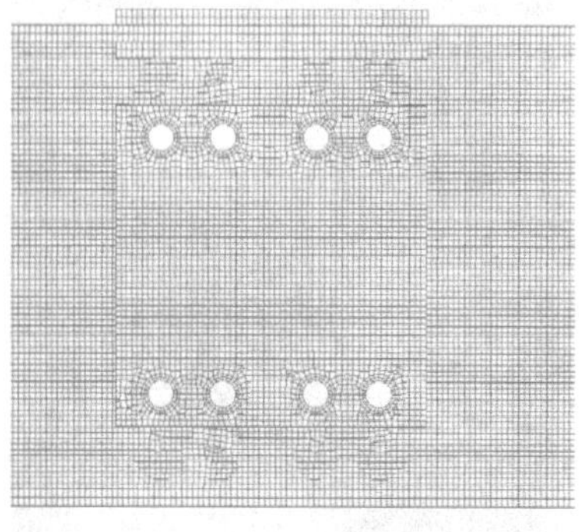

Figure 14. Bolted joint without lower plate.

Another nonlinear static analysis is performed in this configuration. The damaged joint has a lower stiffness, as expect. In this case, maximum deflection of the structure is at the joint (Figure 15).

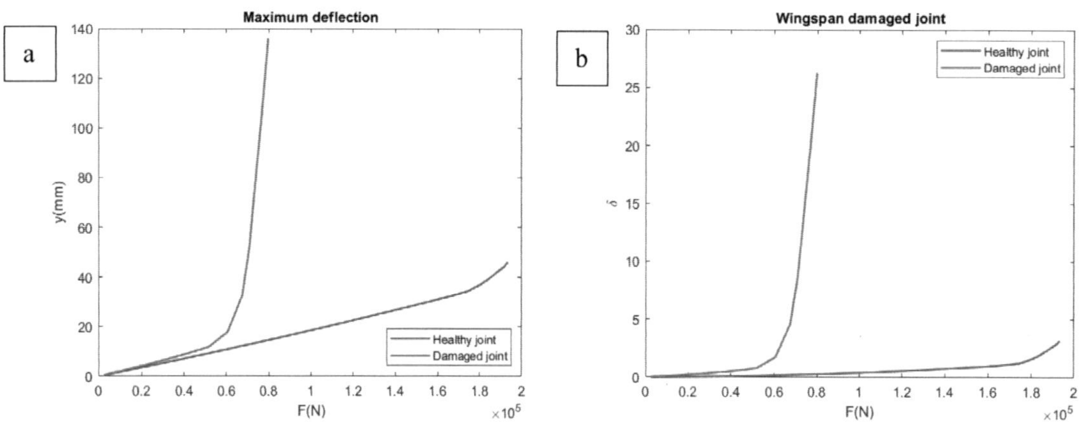

Figure 15. Damaged joint: (a) Maximum deflection, (b) Relative displacement traction plates, (c) Wingspan.

5. FEM model: Hybrid formulation.

Next step of this study concerns the development of a FEM hybrid model, where one of the main beams and its bolted joint are modelled with solid elements. In this way, the actual behaviour of bolted joint and its influence on the dynamic behaviour of the bridge is considered. Figure 16 shows the mesh used in this model. Glued contacts are set between solid beam and surrounding structure, in order to guarantee structural continuity.

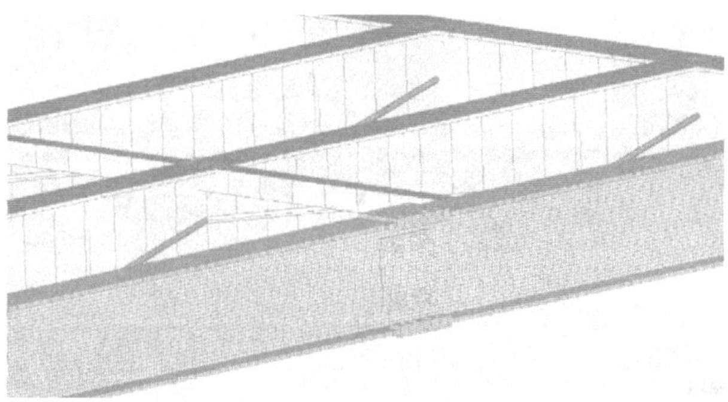

Figure 16. Hybrid model with solid beam modelled.

Dynamic analysis is performed in the same manner as previous analysis of steel beam with bolted joint shown in Chapter 4. Preliminary nonlinear static analysis takes into account of bolt preloads and its influence on the next normal mode solution. Table 6 shows the results of modal analysis.

Table 6. Comparison between experimental results and numerical results.

	Order Type	Experimental [Hz]	FEM [Hz]	FEM hybrid [Hz]
1	1° Flexural	17.50	18.26	16.89
2	1° Torsional	20.71	19.76	18.63
3	2° Flexural	56.78	55.37	53.14
4	2° Torsional	59.63	58.35	54.47
5	1° Plate	64.42	64.41	59.97

The results are in good agreement with experimental one. Relative errors are lower than 10%. Later, bridge is degraded considering the breakage of one or more plates of the bolted connection. Damage is identified through two different ways: shift of natural frequencies, and evaluation of frequency response of the structure (FRF) at the sensor nodes.

5.1. Shift of natural frequencies.

A modal analysis is performed, and the first twenty frequencies of the structure evaluated. Figure 17 shows the shift of each mode. Six different types of damage are considered:

1. Core plate.
2. Upper plate.
3. Lower plate
4. Upper plate + core plate.
5. Upper plate + lower plate.
6. Lower plate + core plate.

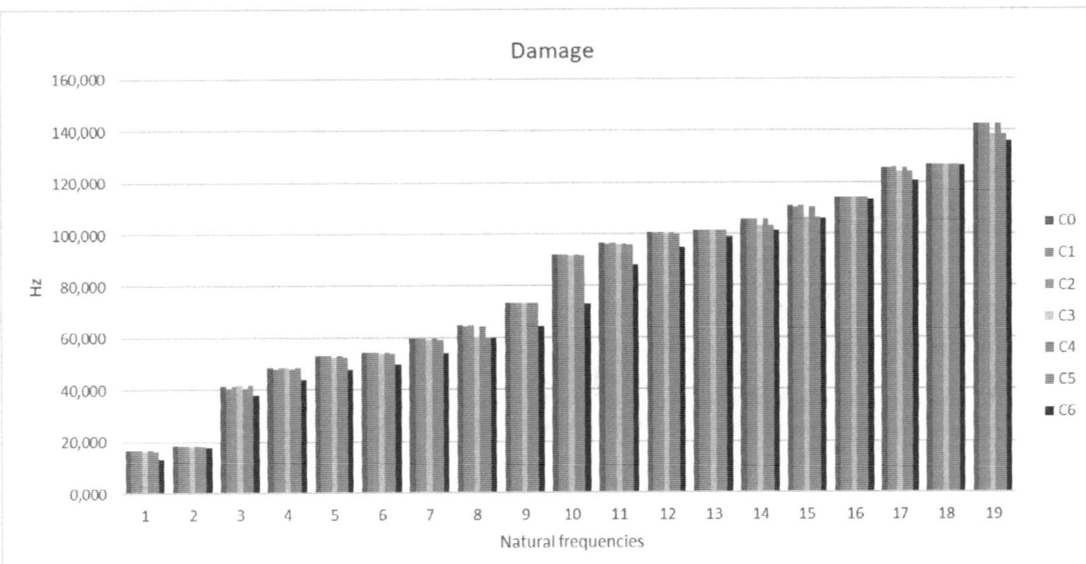

Figure 17. Shift of natural frequencies.

The greatest variation is observed in the absence of the bottom plate and core plates (Level 6). These frequencies are associated with modes shape in which the joint has high curvature.

Table 7. Structure natural frequencies for each damage level and relative variation respect to design condition.

DAMAGE LEVEL	1	2	3	4	5	6	7	8	9	10	11	12	13	14	15	16	17	18	19
0	16,886	18,631	41,586	48,390	53,141	54,468	59,971	64,972	73,648	92,112	96,534	100,566	101,453	105,597	110,910	114,072	125,572	126,852	142,575
1	16,848	18,599	40,487	48,174	52,995	54,374	59,886	64,637	73,596	91,818	96,190	100,463	101,419	105,566	110,240	114,047	125,343	126,772	142,523
2	16,897	18,644	41,581	48,391	53,156	54,480	59,985	64,970	73,647	92,118	96,517	100,588	101,449	105,568	110,907	114,080	125,606	126,825	142,576
3	16,426	18,322	41,746	48,427	52,475	54,206	59,142	59,992	73,501	91,565	95,956	100,091	101,398	103,263	106,190	113,915	124,081	126,714	138,123
4	16,860	18,612	40,485	48,174	53,012	54,384	59,898	64,636	73,595	91,823	96,167	100,486	101,413	105,538	110,235	114,055	125,380	126,741	142,523
5	16,443	18,331	41,740	48,429	52,499	54,213	59,139	60,007	73,501	91,577	95,942	100,113	101,392	103,249	106,169	113,924	124,125	126,682	138,121
6	13,331	17,919	38,053	44,072	47,775	49,699	53,997	59,830	64,402	73,270	88,086	94,631	99,114	101,392	105,770	113,382	120,545	126,436	135,819
1	0,225	0,173	2,645	0,447	0,275	0,173	0,142	0,516	0,071	0,319	0,357	0,103	0,034	0,029	0,604	0,022	0,182	0,063	0,037
2	-0,063	-0,069	0,013	-0,002	-0,028	-0,023	-0,023	0,004	0,001	-0,007	0,017	-0,022	0,005	0,028	0,003	-0,007	-0,027	0,021	0,000
3	2,726	1,659	-0,383	-0,076	1,253	0,480	1,382	7,665	0,200	0,594	0,599	0,473	0,054	2,211	4,256	0,138	1,187	0,109	3,123
4	0,155	0,103	2,650	0,447	0,242	0,154	0,123	0,518	0,072	0,313	0,381	0,080	0,040	0,056	0,608	0,015	0,153	0,087	0,036
5	2,629	1,615	-0,368	-0,079	1,208	0,469	1,387	7,643	0,200	0,580	0,613	0,450	0,061	2,224	4,275	0,130	1,153	0,134	3,125
6	21,055	3,824	8,496	8,924	10,098	8,755	9,961	7,914	12,555	20,455	8,751	5,902	2,306	3,983	4,634	0,605	4,003	0,328	4,739

5.2. FRF analysis.

Frequency response is the quantitative measure of the output spectrum of a system or device in response to a stimulus and is used to characterize the dynamics of the system. It is a measure of magnitude and phase of the output as a function of frequency, in comparison to the input. The frequency response diagrams were obtained using an input-output methodology named SISO (Single-Input-Single-Output). Figure 18 shows load node at the centre of slab and the sensor node located at the joint. Frequency range is 0 – 150 Hz. Structural damping is set to 5%, usually used in civil structures.

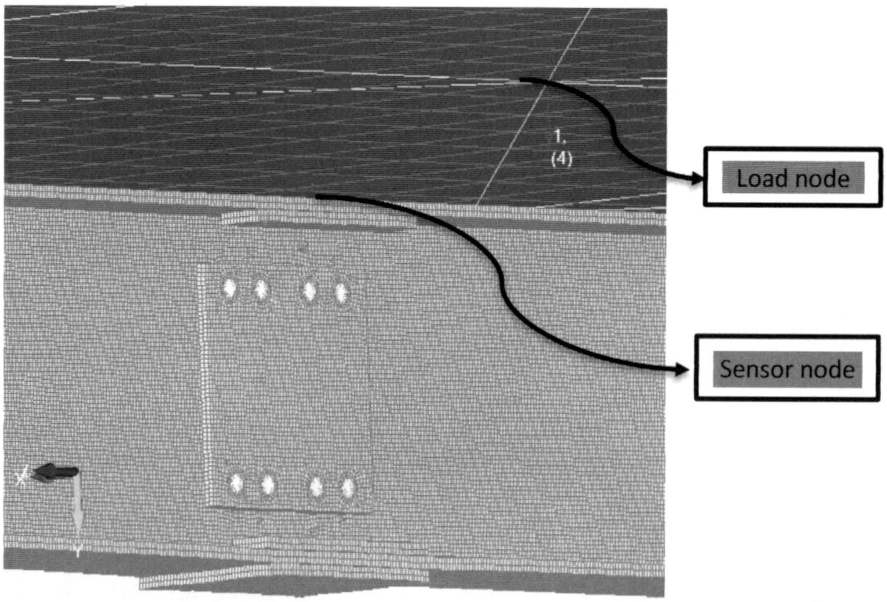

Figure 18. Load node and sensor node location.

Figure 19 display frequency response functions (inertness) for six different types of damage.

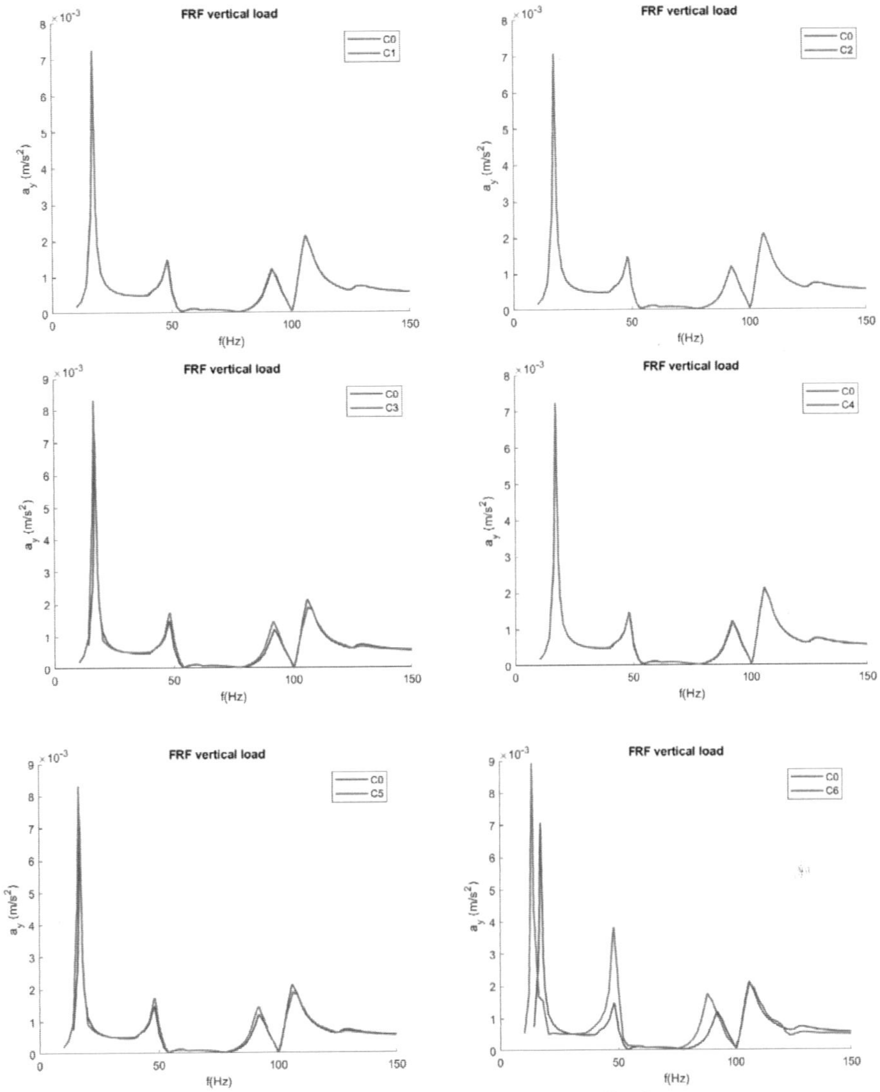

Figure 19. Frequency response amplitude.

Table 8 show the shift of first peak of measured acceleration for different levels of damage scenarios. In this case, amplitude variations are more consistent than frequency variations, compared with the same damage configuration. Considering the absence of core plates, amplitude variation is around 2.7%, while frequency variation is only 0.2%. For this reason, observing amplitude variations allow a better identification of structural damage.

Table 8. Shift of FRF amplitude (1° peak).

	Δa %	Δω %
1	2.72	0.22
2	0.35	0.06
3	17.57	2.73
4	2.31	0.16
5	17.57	2.63
6	26.42	21.07

FRF analysis is also performed considering a wind load in the horizontal direction.

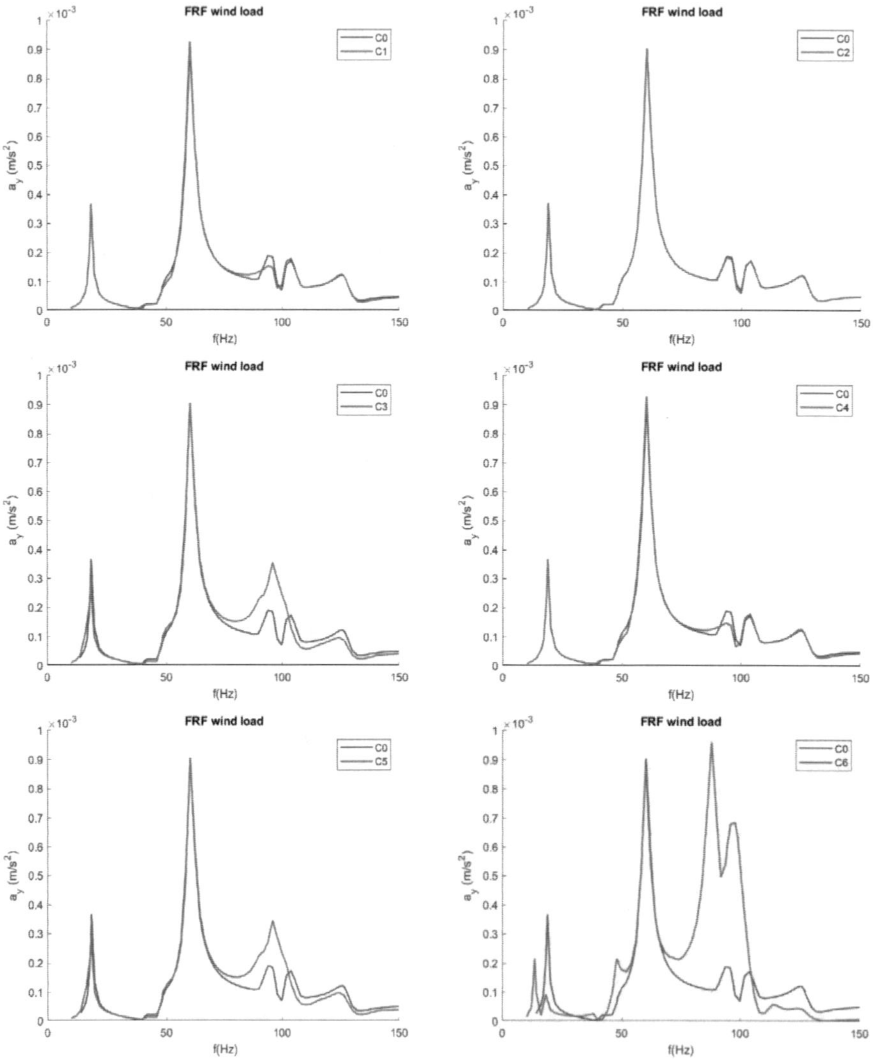

Figure 20. Frequency response amplitude horizontal load

As seen in Figure 20, frequency response functions for C1,C2 and C4 configurations have the peak amplitude at around 60 Hz, and its value is higher then benchmark configuration; while for C3, C5, C6 amplitude at 60 Hz is lower then undamaged structure but amplitude at around 100 Hz increases and for C6 configuration become the highest.

Conclusions

In this study, a digital twin model of a prototype bridge structure was built, with the aim of evaluating both static and dynamic behaviour. Two of the main beams of the bridge have a bolted connection used to simulate a local damage of the structure. The novel approach consists in modelling only one beam of the structure and its bolted connection with a high degree of definition. This allowed a reduction in computational times and costs and the possibility of simulating a damage to the joint considering its actual behaviour. Damage was identified through two different methods: shift of natural frequencies, and evaluation of frequency response of the structure (FRF) at the sensor nodes. Amplitude variations of FRF allow a better identification of structural damage since variations were more consistent. Therefore, with these models it is possible to carry out monitoring operations for the predictive maintenance of the structure and defining the limits that the observed variables can undergo in order to still consider the structure suitable for the operating loads.

Acknowledgement
The present work was developed as part of the research project "SMART MAINTENANCE OF INDUSTRIAL PLANTS AND CIVIL STRUCTURES BY 4.0 MONITORING TECHNOLOGIES AND PROG-NOSTIC APPROACHES - MAC4PRO", sponsored by BRIC-2018 of the National Institute for Insurance against Accidents at Work - INAIL (Italy).

References
[1] Balageas D, Frutzen CP and Güemes A eds 2010. Structural health monitoring **90**. Jhon Wiley & Sons
[2] Chen HP Ni YQ 2018 Introduction to Structural Health Monitoring. John Wiley & Sons.
[3] Sohn H, Farrar CR, Hemez FM, Shunk DD, Stinemates DW, Nadler BR and Czarnecki JJ 2003. A review of structural health monitoring literature: 1996–2001. Los Alamos National Laboratory, USA, 1.
[4] Sony S, Laventure S and Sadhu A 2019. A literature review of next-generation smart sensing technology in structural health monitoring. *Structural Control and Health Monitoring*, **26**(3), p.e 2321.
[5] Jones D, Snider C, Nassehi A, Yon J and Hicks B 2020. Characterising the Digital Twin: A systematic literature review. *CIRP Journal of Manufacturing Science and Technology*, ***29***, pp.36-52.
[6] Zonzini F, Aguzzi C, Gigli L, Sciullo L, Testoni N, De Marchi L, Di Felice M, Cinotti TS, Mennuti C and Marzani A 2020. Structural Health Monitoring and Prognostic of Industrial Plants and Civil Structures: A Sensor to Cloud Architecture. *IEEE Instrumentation & Measurement Magazine*, ***23***(9), pp.21-27.
[7] Rizzo P, Yu LL and Marzani A 2012. Structural Health Monitoring of Civil Structures: New Methodologies and Field Applications 2012.
[8] Errandonea I, Beltrán S and Arrizabalaga S 2020. Digital Twin for maintenance: A literature review. *Computers in Industry*, **123**, p.103316.
[9] Dos Santos JA, Maia NMM, Soares CM and Soares CM 2008. Structural damage identification: A survey. *Trends in computational Structures technology*, pp. 1-24.
[10] Fan W and Qiao P, 2011. Vibration-based damage identification methods: a review and comparative study. Structural health monitoring, **10**(1), pp. 83-111.
[11] Wang SQ and Li HJ 2012. Assessment of structural damage using natural frequency changes. *Acta Mechanica Sinica*, ***28***(1), pp.118-127.
[12] Bandara RP, Chan TH and Thambiratnam DP 2014. Structural damage detection methodusing frequency response functions. *Structural Health Monitoring*, ***13***(4), pp.418-429.
[13] Ye C, Butler L, Calka B, Iangurazov M, Lu Q, Gregory A, Girolami M and Middleton C 2019. A digital twin of bridges for structural health monitoring.
[14] Farrar CR and Jauregui DA 1998. Comparative study of damage identification algorithms applied to a bridge: I. Experiment. *Smart materials and structures*, ***7***(5), p.704.
[15] Farrar CR and Jauregui DA 1998. Comparative study of damage identification algorithms applied to a bridge: II. Numerical study. *Smart materials and structures*, ***7***(5), p.720.
[16] Shape: PREDICTING STRENGTH CHANGES IN BRIDGES FROM FREQUENCY DATA, SAFETY, HAZARD AND POLY-HARMONIC EVALUATION – Final report
[17] NX NASTRAN *Multi-Step Nonlinear User's Guide*
[18] Harrys C, Piersol P 1993 *Harrys shock and vibration handbook* (New York McGraw&Hill)

On the statistical nature of fatigue crack-growth through Monte Carlo simulations and experimental data

V. Giannella

Department of Industrial Engineering, University of Salerno, via Giovanni Paolo II, 132, 84084, Fisciano (SA), Italy

vgiannella@unisa.it

Abstract. Understanding factors that contribute to scatter in crack-growth rates data is fundamental to reduce risk of unexpected structural failures. This document discusses the statistical nature of material scattering observed in fatigue crack propagation tests through a comparison between Monte Carlo simulations and experimental test data. The case study consists of constant amplitude tests on central cracked 2024-T3 aluminium alloy plate specimens loaded in pure Mode-I with stress ratio R equals to 0.2. It was found that results of simulations matched the experimental fatigue lives in terms of both mean value and dispersion around it. Based on these outcomes, the crack-growth behaviour can be described in a statistical manner with the expectation to reduce the inaccuracy present in life predictions.

1. Introduction

Life prediction of structures undergoing variable loads necessitates deep knowledge about the fatigue phenomenon. As it is known today, the metal fatigue process is complex, still not fully understood, with many variables which can influence the life of a structure, such as: material, loading, geometric characteristics, pre-existing defects, etc. [1-4]. Quantifying and controlling these variables is essential to enhance the competitiveness in designing fatigue-safe products, e.g. for turbine engines, railway axles, etc. [1, 3, 5-6].

Engineering components generally fail due to irreversible micro-plastic deformation from which cracks are nucleated, generally from notches or at critical locations [6-8]. Considering the inhomogeneous microstructure and non-uniform mechanical properties of metal materials, crack tips and defects may encounter diversiform microstructures, in turn demonstrating variable strength against Fatigue Crack-Growth (FCG), as if "*the cracks would pass through a different material possessing different properties*" [9]. As a consequence, it can be stated that FCG has inevitably stochastic characteristics, at least from a macroscopic point of view.

According to numerous researches [2-3, 9-13], scatter of material data seems to represent the most relevant contribute to the variability of life predictions of engineering structures. This investigation developed statistical concepts to FCG behaviour by considering only the material scattering as source of variability for the definition of the lifespan distribution for a given case study, i.e. a plate specimen made of AA2024-T3, loaded in pure Mode-I, see Figure 1. More in detail, this document discusses the statistical nature of material scattering observed in fatigue crack propagation tests performed in [9, 13], through a comparison between Monte Carlo simulations and the literature test data. The document also attempted to describe crack-growth behaviour in a statistical manner with the expectation that this statistical description could be able to reduce the large amount of error currently present in numerous life prediction methods.

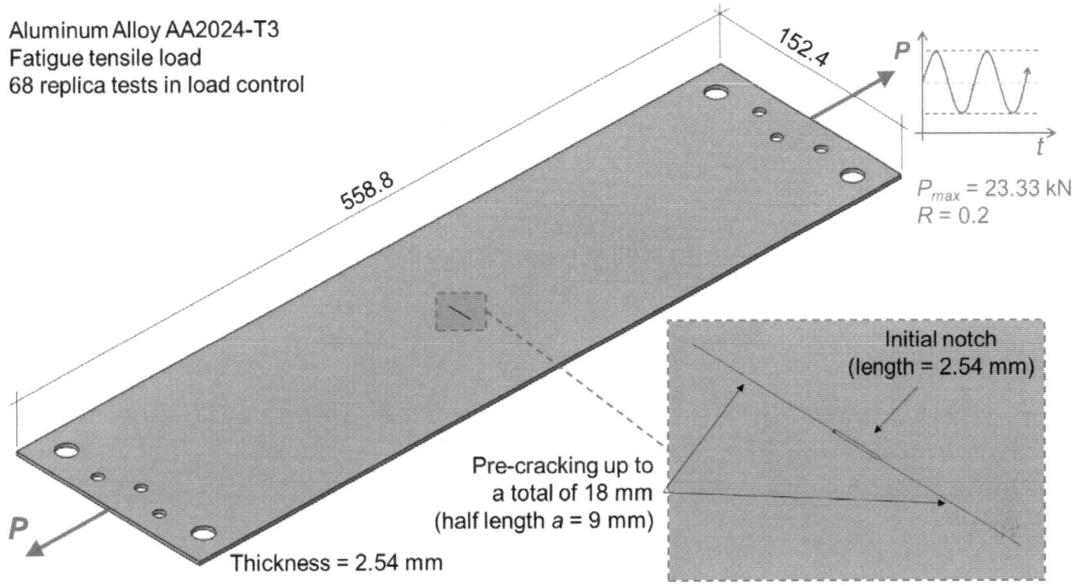

Figure 1. Case study description.

2. Materials and Methods

The study of fatigue crack propagation behaviour has been widely conducted for some time in an effort to understand metal fatigue more fully. To this aim, experimental tests on specimens are generally used in a scale approach from small uniaxial laboratory samples to larger and complex multiaxially loaded mechanical components. The information obtained from crack propagation studies on specimens is then used in estimating the fatigue life of structures and components. The current investigation has the main purpose of applying statistical concepts and theory to the study of fatigue crack propagation behaviour for laboratory tests on specimens. This experimental campaign was carried out in [9, 13] and was here briefly recalled (Figure 1). Although [9, 13] do not represent recent researches, these were considered as the perfect benchmark where to cross-validate the up-to-date stochastic framework. In particular, the experimental results of [9, 13] in terms of Crack-Growth Rates (CGRs), i.e. da/dN vs. ΔK (see Figure 2), were here used to perform life predictions for the same geometry, material and loading conditions in a probabilistic way through Monte Carlo simulations. Such a case study sounded perfect to this aim because consisting of 68 replica tests in which only the source of variability coming from material properties could be identified. Therefore, this represented the perfect benchmark through which derive quantifications based only on the variability of material data.

The experimental campaign consisted of 68 identical constant amplitude tests on central cracked AA2024-T3 aluminium alloy plate specimens loaded sinusoidally (load amplitude P) in pure Mode-I with stress ratio $R=0.2$. A rigorous pre-cracking phase was carried out for all the 68 replica tests so as to obtain a precise half crack length of 9 mm from which start the cycle counting. The aim of this experimental campaign was the same of the current investigation, namely to investigate the material scattering and how it can affect the fatigue crack propagation process.

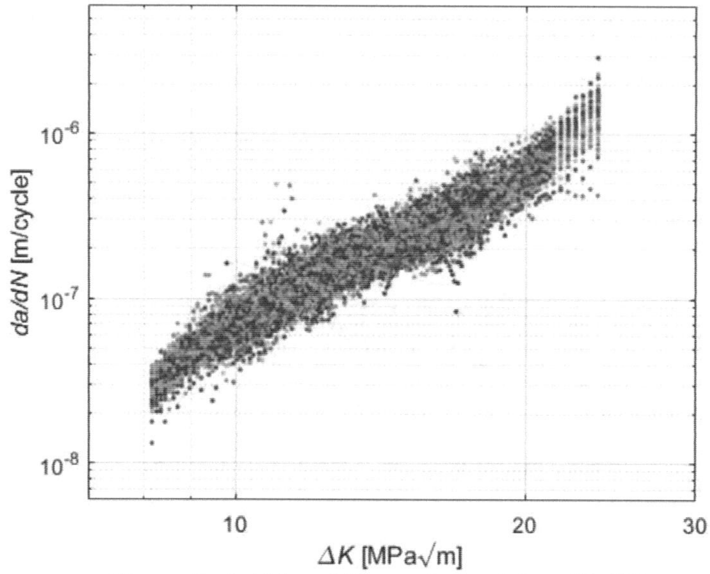

Figure 2. *da/dN* vs. ΔK experimental data. [9, 13]

Material data of Figure 2 are generally adopted in order to calibrate a FCG law useful to perform life estimates of the cracked components [14-16]. A Paris-like FCG law, based on Equations 1-3, was used in this work and implemented in a stochastic way. This law was selected among those available in literature since material data of Figure 2 presented a mostly perfectly linear tendency. More sophisticated FCG law were instead adopted in [3, 11] through a framework similar to that proposed here.

Thanks to the stochastic implementation of the FCG law, no material parameters needed to be directly calibrated through CGR test data. The two main parameters C and m of Equation 1 were derived through Monte Carlo simulations as detailed in the followings, see Figure 3. The plane stress/strain constraint factor α was kept as fixed to 1.5, as suggested in [3] with reference to relatively low-strength materials. σ/σ_0 ratio was calculated according to the applied load P ($\sigma = P/Wt$) and the flow stress σ_0 of the material (439 MPa). W is the width of the plate (i.e. 152.4 mm), t is the thickness of the plate (i.e. 2.54 mm), whereas a is the continuously increasing half crack length (varying from 9 to 49.8 mm, as for the experimental campaign). More details about how to calibrate these parameters can be found in [3].

Equations 1-3 were directly inserted in the MATLAB programming platform [17], in which a vast series of (C, m) sets were randomly sampled within the grid shown in Figure 3 (top left graph). A CGR curve was drawn for each set and compared with all the test data (in Figure 2). Each CGR curve was then judged as acceptable only if the related root mean square error was lower than a cut-off threshold (set to 2%). CGR curves that insufficiently correlated with test data were immediately discarded, whereas the "valid" ones were used to compute a fatigue life prediction according to Equation 4 [18], with this latter used to calculate the K values along the crack advancing.

These Monte Carlo simulations were performed until that a total of 1e5 CGR good curves (and corresponding life predictions) were derived (~30 min of run on a regular laptop).

$$\frac{da}{dN} = C\left[\left(\frac{1-f}{1-R}\right)\Delta K\right]^m \tag{1}$$

$$f = \frac{K_{op}}{K_{max}} = \begin{cases} max(R, A_0 + A_1 R + A_2 R^2 + A_3 R^3), & R \geq 0 \\ A_0 + A_1 R & , \quad -2 \leq R < 0 \end{cases} \tag{2}$$

$$\left. \begin{aligned} A_0 &= (0.825 - 0.34\alpha + 0.05\alpha^2)\left[\cos\left(\frac{\pi}{2} s_{max}/\sigma_0\right)\right]^{1/\alpha} \\ A_1 &= (0.415 - 0.071\alpha)\, s_{max}/\sigma_0 \\ A_2 &= 1 - A_0 - A_1 - A_3 \\ A_3 &= 2A_0 + A_1 - 1 \end{aligned} \right\} \tag{3}$$

$$K = \sigma\sqrt{\pi a} * \left[1 - 0.025\left(\frac{2a}{W}\right)^2 + 0.06\left(\frac{2a}{W}\right)^4\right]\sqrt{\sec\left(\frac{\pi a}{W}\right)} \quad , \tag{4}$$

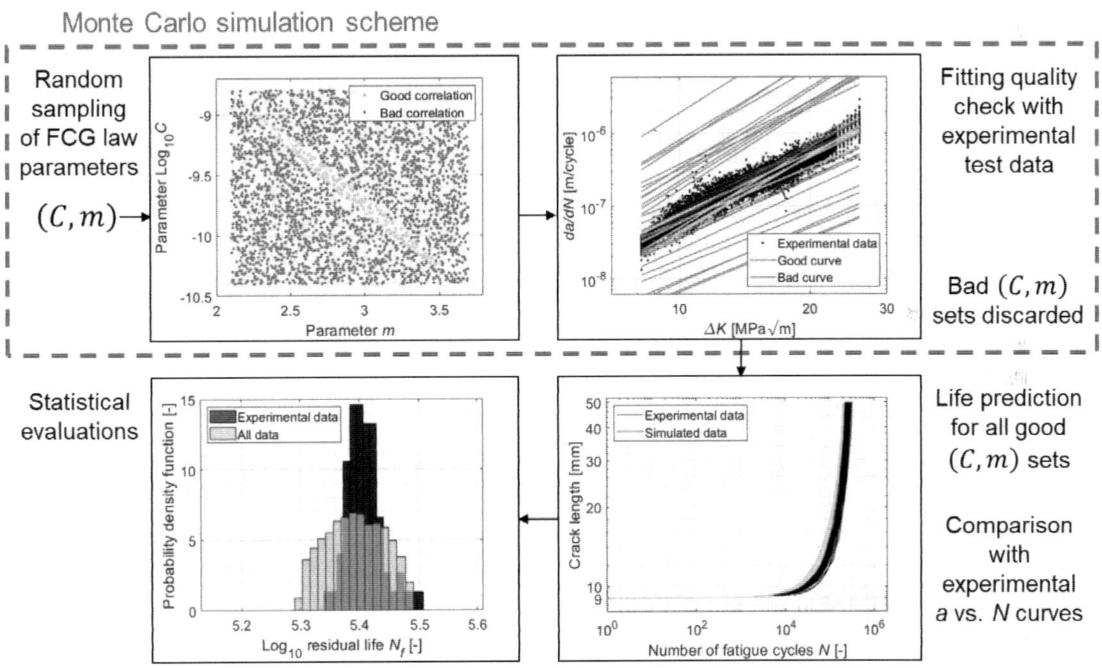

Figure 3. Monte Carlo simulation scheme.

3. Results and discussion

Ideally, it is desirable that the estimated life will exactly predict the actual life. Unfortunately, there are many variables which influence predictions and some of them are still not well understood. One of the most important of these variables is how well the FCG laws (Paris, NASGRO, …) calibrated onto experimental data actually represent the observed crack propagation behaviour.

Thanks to the Monte Carlo framework here adopted (Figure 3), the authentic distributions of the main FCG law parameters C and m can be derived. These are arranged in the correlation plot reported in Figure 4. A first key aspect to notice is that these parameters have rather large ranges of equally acceptable values. Namely, parameter m accepts a range of variability from 2.25 to 3.5, whereas $Log_{10}C$ can vary from -8.9 to -10.3 (thus C can vary from 1.3e-9 to 5e-11). However, their variation is not arbitrary since $Log_{10}C$ and m are mostly linearly correlated. This has been here assessed with reference to an aluminium alloy but was also observed in [3, 11] for a high-strength steel and already discussed from more theoretical standpoints, see e.g. [19-20].

Figure 5 reports all the CGR curves that were obtained from the Monte Carlo simulations. All these CGR curves returned a root mean square error lower than 2% and, therefore, they can be all equally considered as "good-fit" for the test data of Figure 2. The corresponding life predictions are reported in Figures 6-7 where it can be noticed a very good agreement between the simulation results and the outcomes measured experimentally in [9, 13].

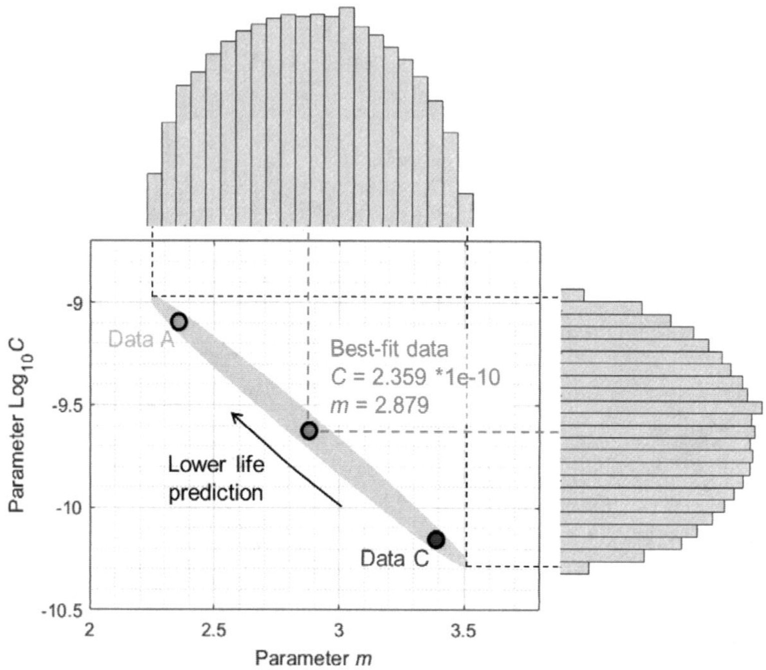

Figure 4. Correlation plot for FCG law parameters C and m.

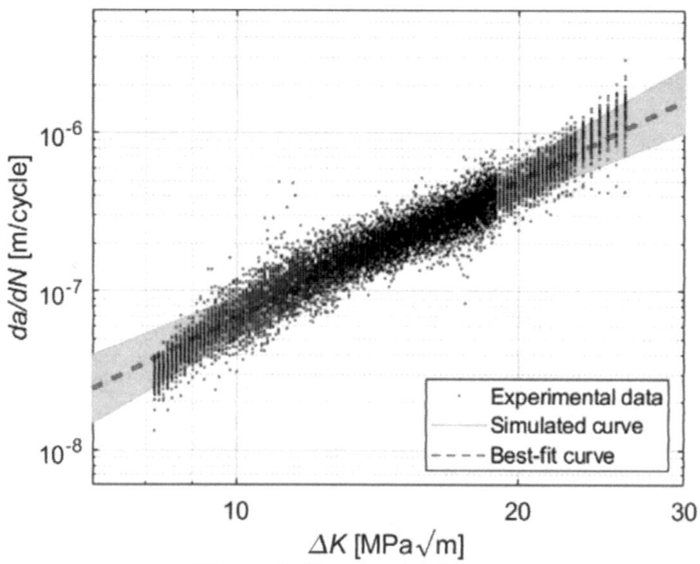

Figure 5. Simulated CGR curves.

In Figure 5, the CGR curve that returned the lowest root mean square error against the test data (i.e. the "best-fit" curve) is also highlighted, whereas the related predictions are also highlighted in Figures 6-7. "Best-fit" outcomes reported in Figures 4-7 represent what can be obtained by using a traditional deterministic approach. Although perfectly predicting the mean value of the lifespan distribution, a "best-fit" deterministic calculation cannot add any substantial information to the prediction itself, e.g. the likelihood to reach the predicted life, the likelihood of failure for a given number of cycles, etc. Furthermore, also the conservativeness or the non-conservativeness of this prediction cannot be guaranteed a priori [3, 11]. On the contrary, the stochastic approach here described and used to account for the material scattering properly captured both the mean value of experimental lives and its variance.

A further observation can be made with reference to Figure 8, in which three sets of equally acceptable FCG laws and corresponding life predictions were extracted (data sets shown in Figure 4). It is worth noting that high values of parameter *C*, associated with low values of parameter *m* (required to reduce the steepness of the curve so as to fit test data), produce a high *da/dN* in the initial part of propagation, in turn returning a reduced life prediction. Finally, it can be noticed that, although all these three lines well fit the test data, the variability of their life predictions can be non-negligible, returning a prediction for "Data C" more than 50% higher than that of "Data A" (310k vs. 200k cycles respectively). Due care has to be made when defining the FCG law parameters to perform fatigue crack-growth predictions, especially if considering that the results presented here refer to laboratory specimens, thus mostly taking into account of material scattering only, whereas numerous further contributors of uncertainty and variability have to be added when working on real structures.

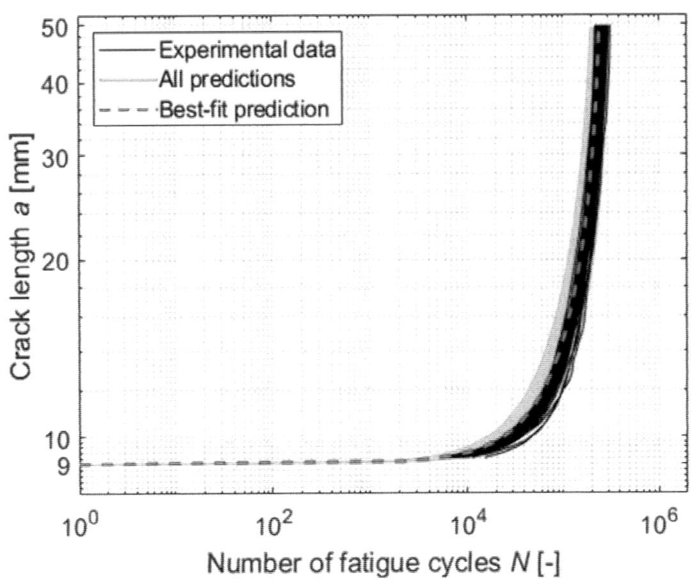

Figure 6. Comparison of experimental and simulated *a* vs. *N* curves.

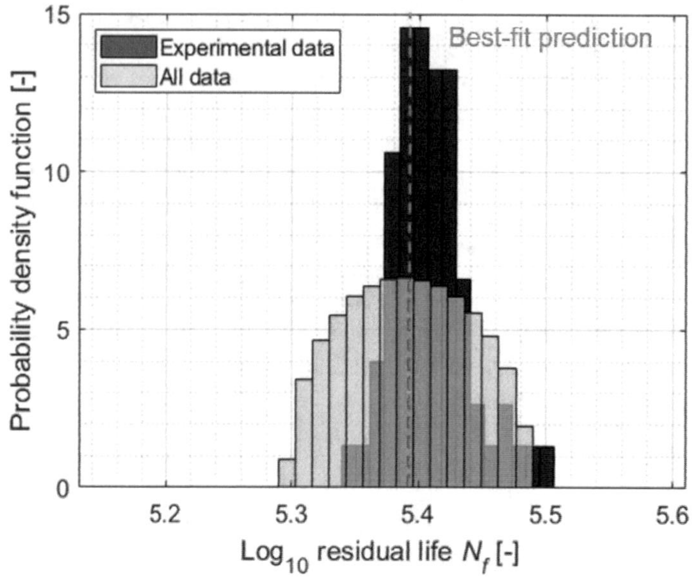

Figure 7. Histogram of the experimental and simulated life predictions.

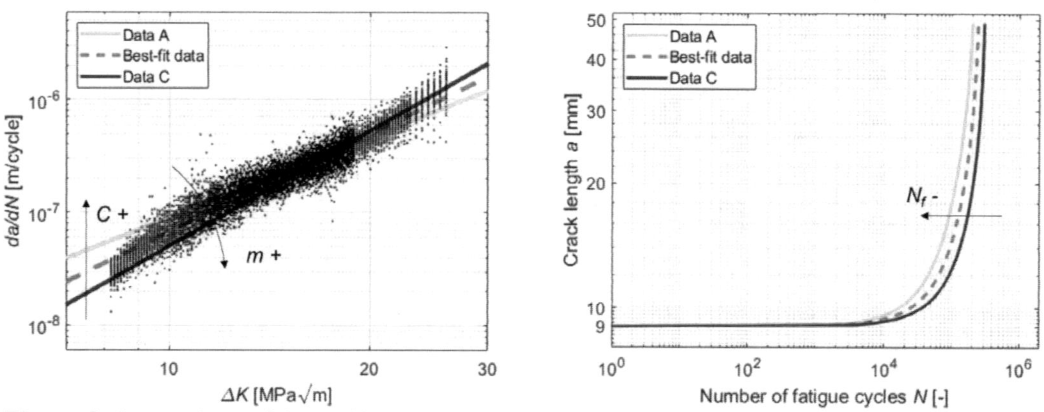

Figure 8. Comparison of three CGR data sets and corresponding predictions (data sets highlighted in Figure 4).

4. Conclusions

The main purpose of this work was to apply statistical concepts to the study of fatigue crack propagation behaviour. In doing this, this study involved the determination of the effects of the material properties variability on the life prediction for plate specimens made of AA2024-T3 loaded in Mode-I.

Monte Carlo simulations were run to derive the distributions of the two main FCG law parameters (C and m) and a vast series of CGR curves were drawn and used to predict the life of the specimens, together with an analytical equation used to calculate the K values at various crack lengths.

A very good match was achieved between experimental and simulated life predictions in terms of both mean value and dispersion around it. This demonstrated that the Monte Carlo simulations well captured the ranges of variability of parameters C and m of the FCG law. As a consequence, the whole stochastic framework demonstrated to correctly capture the scattering observed in experimental data. The study attempted to describe crack-growth behaviour in a statistical manner with the expectation that this stochastic framework can be used to reduce the uncertainty of life predictions by providing a solid range of variability to the predictions.

References

[1] Zhu S -P, Liu Q, Peng W and Zhang X -C 2018 Computational-experimental approaches for fatigue reliability assessment of turbine bladed disks *Int. J. Mech. Sci.* **142-143** 502-517.

[2] Zhu S -P, Foletti S and Beretta S 2017 Probabilistic framework for multiaxial LCF assessment under material variability *Int. J. Fatigue* **103** 371-385.

[3] Giannella V 2021 Stochastic approach to fatigue crack-growth simulation for a railway axle under input data variability *Int. J. Fatigue* **144** 106044.

[4] Romano S, Beretta S, Miccoli S and Gschweitl M 2020 Probabilistic Framework for Defect Tolerant Fatigue Assessment of Additively Manufactured Parts Applied to a Space Component, in Structural Integrity of Additive Manufactured Parts, ed. Shamsaei N, Daniewicz S, Hrabe N, Beretta S, Waller J, Seifi M, West Conshohocken, PA: ASTM International, 526-539.

[5] Citarella R, Giannella V, Vivo E and Mazzeo M 2016 FEM-DBEM approach for crack propagation in a low pressure aeroengine turbine vane segment *Theoretical and Applied Fracture Mechanics* **86** 143-152.

[6] Giannella V, Sepe R, Borrelli A, De Michele G and Armentani E 2021 Numerical investigation on the fracture failure of a railway axle *Engineering Failure Analysis* 129 105680.

[7] Giannella V, Citarella R, Fellinger J and Esposito R 2018 LCF assessment on heat shield components of nuclear fusion experiment "Wendelstein 7-X" by critical plane criteria *Procedia Structural Integrity* **8** 318-331.

[8] Sepe R, Greco A, De Luca A, Armentani E and Berto F 2019 Experimental and FEM numerical assessment of multiaxial fatigue failure criteria for a rolling Stock's seats structure *Engineering Failure Analysis* **102** 303-317.

[9] Virkler D A, Hillberry B M and Goel P K 1979 The Statistical Nature of Fatigue Crack Propagation *ASME J. Eng. Mater. Technol* **101**(2) 148-153.

[10] Beretta S, Foletti S, Rusconi E, Riva A and Socie D 2016 A log-normal format for failure probability under LCF: Concept, validation and definition of design curve *Int. J. Fatigue* **82** 2-11.

[11] Giannella V 2021 Fatigue crack-growth predictions for a railway axle under material data variability *IOP Conf. Ser: Mate.r Sci. Eng.* **1038** 012062.

[12] Beretta S and Carboni M 2006 Experiments and stochastic model for propagation lifetime of railway axles *Eng. Frac. Mech.* **73** 2627-2641.

[13] Virkler D A 1978 The Statistical Nature of Fatigue Crack Propagation, MS thesis, *Purdue University*.

[14] Giannella V, Fellinger J, Perrella M and Citarella R 2017 Fatigue life assessment in lateral support element of a magnet for nuclear fusion experiment "Wendelstein 7-X" *Engineering Fracture Mechanics* **178** 243-257.

[15] Beretta S and Carboni M 2004 Application of fatigue crack growth algorithms to railway axles and comparison of two steel grades *Proc. Instn. Mech. Engr. Part F: J. Rail and Rapid Transit* **218**(4) 317-326.

[16] Giannella V, Dhondt G, Kontermann C and Citarella R 2019 Combined static-cyclic multi-axial crack propagation in cruciform specimens *Int. J. Fatigue* **123** 296-307.

[17] MATLAB R2020a. Natick, Massachusetts: The MathWorks Inc. 2019.

[18] Tada H, Paris P C and Irwin G R, The Stress Analysis of Cracks Handbook. ASME Press, New York, 2000.

[19] Cortie M B and Garrett G G 1988 On the Correlation Between the C and m in the Paris Equation for Fatigue Crack Propagation *Eng. Frac. Mech.* **30**(1) 49-58.

[20] Bergner F and Zouhar G 2000 A new approach to the correlation between the coefficient and the exponent in the power law equation of fatigue crack growth *Int. J. Fatigue* **22**(3) 229-239.

Composite-to-metal multi-bolt joints: a simplified FE analysis method

V G Belardi[1], P Fanelli[2] and F Vivio[1]

[1]Department of Enterprise Engineering, University of Rome Tor Vergata, Via del Politecnico, 1, Rome, 00133, Italy.

[2]Department of Economics, Engineering, Society and Business Organization (DEIM), University of Tuscia, Largo dell'Università, Viterbo, 01100, Italy.

E-mail: valerio.belardi@uniroma2.it

Abstract. This paper presents a modified version of the Composite Bolted Joint Element (CBJE) that can be employed for the finite element analysis of double-lap joints made up of metal and composite material plates. The CBJE makes use if the theoretical definition of composite bolted region that is solved to analytically define the behavior of the joint. The solution is utilized to obtain the stiffness of the bolted region and the bolt-hole contact stiffness that are transferred to the CBJE. The definition based on the analytical solution allows to balance the accuracy and the computational effort of the solution. The results of CBJE for hybrid double-lap joints is compared with a 3D FE solid demonstrating a highly remarkable matching.

1. Introduction

Numerous engineering applications require the connection of composite and metal components. Bolted joints represent the most utilized connection method to deal with flanges made up of different materials. The reason of this important spread mainly lies in the high load bearing capabilities of bolted joints, along with the ease of disassembling the joint. Additionally, bolted joints, if properly designed, are characterized by progressive and non-catastrophic failure modes, which allow bolts-load redistribution during damage progression. These features are compatible with scheduled maintenance inspections and, therefore, bolted joints are considered a reliable technique. These aspects explain the relevance of hybrid composite-to-metal bolted joints structural assessment.

The realization of prototypes and experimental studies provide detailed results about the behavior of bolted joints [1, 2]; anyway, the experimental campaigns can be expensive and unlikely performed on a wide range of configurations.

A fast assessment of the bolted joints can be obtained with the analytical methods. Basically, the joint is studied by means of a mass-spring model obtaining the governing equations [3, 4]. The main drawback of these approaches is the level of simplifications that are introduced and that can be reason of high approximations.

Nowadays, this task is primarily performed exploiting the finite element method. Three-dimensional models allow a detailed and accurate analysis of the joint [5, 6]. Nevertheless, a complete and detailed three-dimensional modeling approach is arduous, and the analysis is time consuming for both the modeling and the computational aspects, especially considering that non-linear effects must be included. This particularly affects the preliminary design stage when different configurations must be explored and improved towards the optimization.

Conversely, the simplified FE approaches aim at the simulation of the bolted joint with a lower amount of computational resources [7, 8, 9]. To this end, a simplified modeling approach is outlined, it is funded on the solution of the bolted joint analytical model [10, 11, 12, 13]. The region surrounding the bolt is ideally split in angular sectors whose elastic properties are evaluated and assigned to a set of radially arranged beam-shaped stiffness matrix elements, that replace the angular sectors [14, 15, 16]. Also, a beam element is employed to model the bolt-shank, whereas two spring elements allow to consider: the friction effects, the bolt-hole clearances and the contact between the bolt and the plates. The proposed custom FE methodology allows the combination of an accurate definition of the joint stiffness properties along with a low computational burden. Furthermore, it is successfully compared with highly detailed 3D FE models.

2. FE modeling strategy for double-lap composite-to-metal joints

The FE modeling strategy is based on the concept of spot joint element defined by Vivio in Ref. [17]. The spot joint element makes use of a group of user-defined finite elements that are inserted in a global shell model to simulate a spot joint. The key idea of the methodology is the definition of a theoretical model including the joint and an annular portion of the surrounding plate, Figure 1. The model is clamped on the outer edge and connected to a rigid core representing the bolt. The action on the theoretical model of four typical load conditions is considered: (1) in-plane load, (2) transversal load P, (3) radial bending moment, (4) torsional moment. Then, the reference model is analytically solved, and the analytical solution is utilized to assess the stiffness of the region.

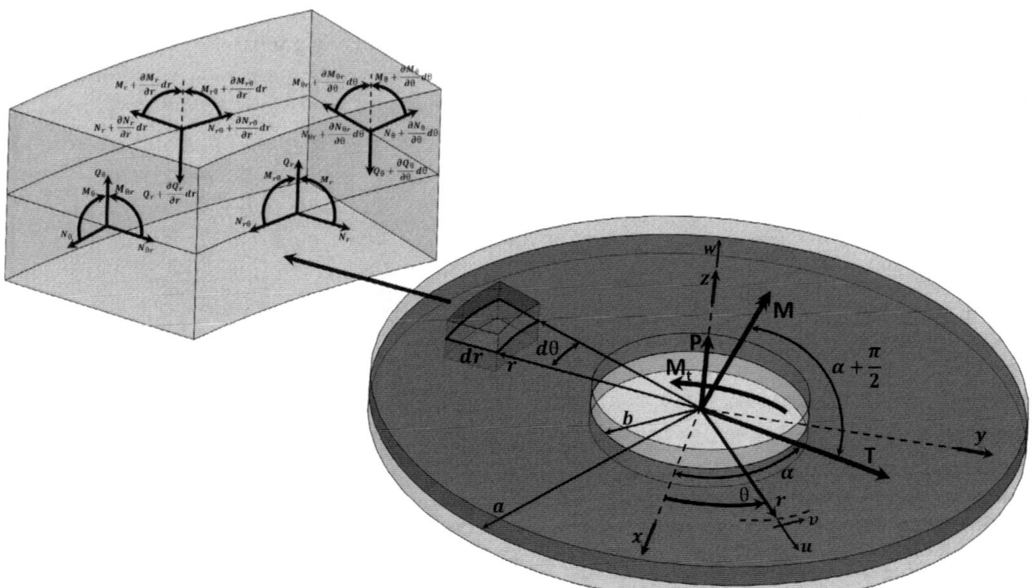

Figure 1. Theoretical reference model of composite bolted joint [16].

In particular, the theoretical model is split into N angular sectors, the stiffness properties of each sector are evaluated and transferred to custom beam elements, see Figure 2, one for each sector, determining a set of radial beam elements.

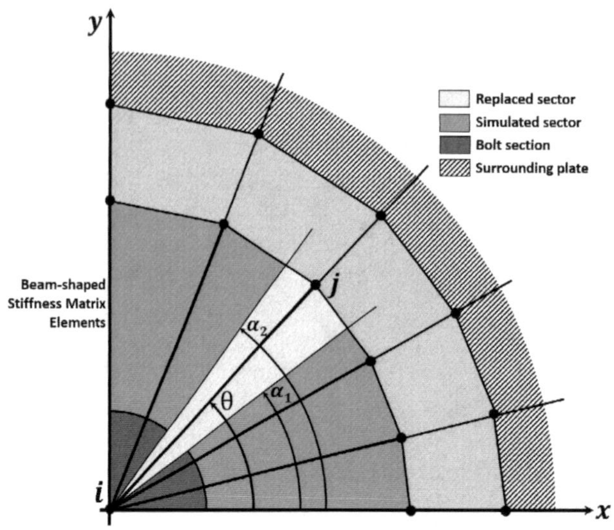

Figure 2. Structure of the Composite Bolted Joint Element [15].

Moreover, each beam-shaped stiffness matrix element is equivalent in stiffness to the corresponding angular sector. The stiffness properties are evaluated through the ratio between the stress resultants and the generalized displacements, these quantities are obtained from the analytical solution of the theoretical model. In the case of metal plate, the theoretical framework exploits the hypothesis of isotropic material and it allows the obtainment of a closed form analytical solution. This solution is described in Ref. [17] along with the evaluation of the stiffness properties of the custom elements to be employed for metal plates. On the other hand, the theoretical model with the plate made up of composite material features rectilinear orthotropic material properties, i.e. stiffness properties variable with the angular coordinate θ. In this case, a closed form solution is not achievable and the solution strategy is based on the application of the Ritz method to the principle of virtual displacements, as described in Refs. [10, 11, 12, 13]. Then, the upgraded version of the spot joint element for composite bolted joints, the Composite Bolted Joint Element (CBJE), was presented in Ref. [16] with the relative description of the stiffness matrix of the custom beam elements.

The configuration of the CBJE to be employed for the structural analysis of composite-to-metal double-lap bolted joints is shown in Figure 3. Specifically, the configuration of the double-lap joint is characterized by two metal outer plates and an inner composite plate. It follows that the stiffness matrix of the upper and lower groups of custom beam elements differs from the central one since the theoretical models are different.

Furthermore, there are other components building up the CBJE. The bolt shank is modeled with beam elements; these elements are connected to the upper and lower groups of custom beam elements through two rigid beam elements. This is necessary since the plates are meshed on their mid-surface. In addition, the connection between the beam elements of the bolt shank and the rigid beams is realized with spring elements. The spring elements k_r are needed to simulate the interaction between the bolt head and the shank. The non-linear spring elements k_c^m are necessary to model the bearing contact between the bolt shank and the holes of the metal plates and, in addition, the relative sliding occurring when the static friction force is reached. Then, a further non-linear spring k_c^c with analogous properties connects the middle node of the bolt shank to the group of custom beam elements of the composite plate.

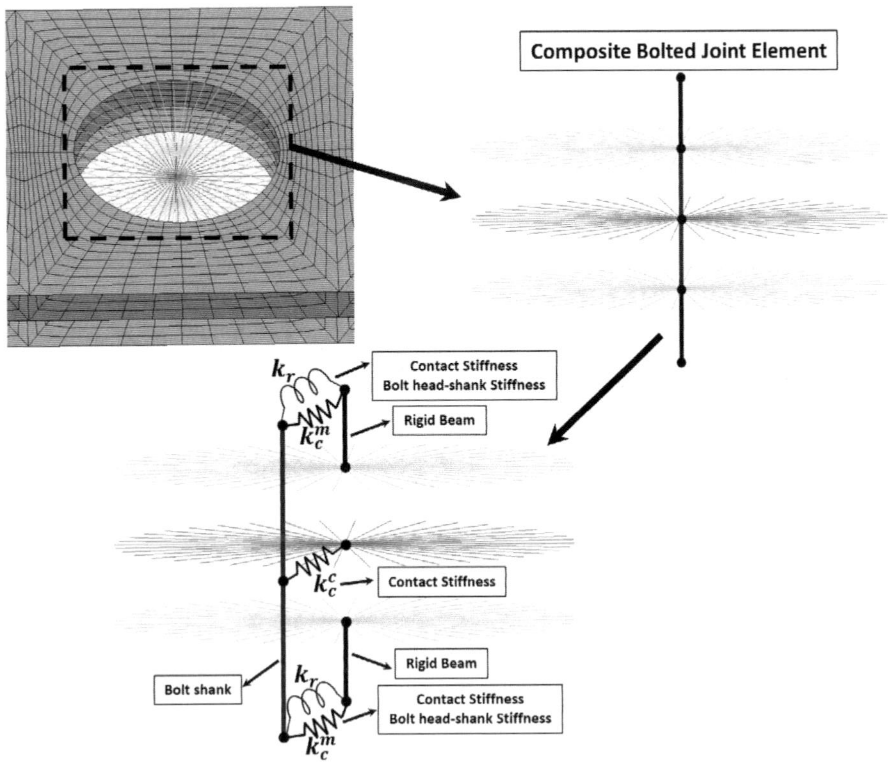

Figure 3. Configuration of the CBJE for composite-to-metal double-lap bolted joints.

3. Results

The modified CBJE for double-lap bolted joints is tested on a three-bolt hybrid specimen with aluminum outer plates and a composite material inner plate. The geometry of the specimen is depicted in Figure 4, the thickness of the plates is: $t_1 = 2.6$ mm for the outer ones and $t_2 = 5.2$ mm for the inner one. The properties of the metal materials are reported in Table 1 and those of the composite layers in Table 2; the stacking sequence of the composite plate is $[45/0/-45/90]_{5s}$.

In addition, the joint presents three titanium bolts with $d_{bolt} = 8$ mm that are preloaded with a tension stress $\sigma_{pre} = 100$ MPa and the bolt-hole radial clearance is equal to $c = 10$ μm. Considering that the friction coefficient between the aluminum and the composite plates is $\mu = 0.40$, the static friction force that sticks together the plates is $F_s = 12.06$ kN.

Moreover, the double-lap joint is tested with a traction force $F = 30$ kN. As regards the boundary conditions, the left edge of the outer plates is fully constrained and the right edge of the inner composite plate is simply supported and loaded with the traction force F.

The traction test is simulated with a shell FE model containing the CBJE for the bolted regions and with a high fidelity 3D model, that is the reference for the results verification. The shell model is composed of layered elements with 4 nodes and 6 degrees of freedom per node, whereas the 3D model presents the solid elements are 8-noded with 3 degrees of freedom per node. Regarding the 3D model, the composite plate is meshed using layered solid elements, it further contains contact elements to simulate the Hertzian contact between the bodies. Then, the two FE models are shown in Figure 5; the whole 3D depicted even though, considering the double symmetry of the double-lap joint, a quarter of the 3D model was analyzed.

Furthermore, despite the exploiting of the symmetries, the 3D model is considerably heavier

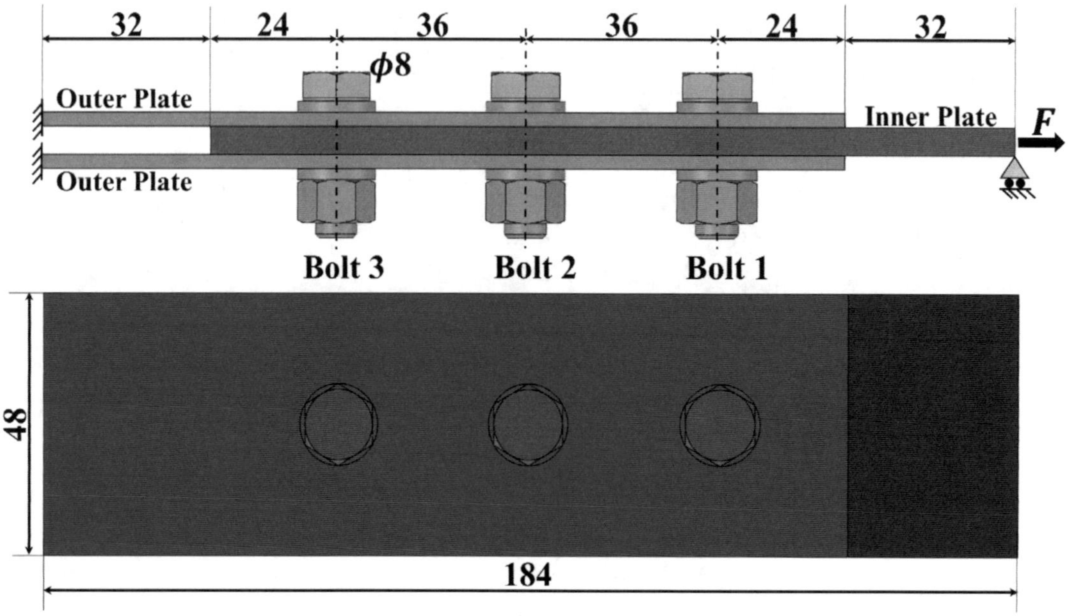

Figure 4. Geometry of the double-lap three-bolt hybrid bolted joint.

Table 1. Elastic properties of the metal materials.

	E [GPa]	$\nu[-]$
Aluminum	70	0.30
Titanium	110	0.29

Table 2. Unidirectional fiber-reinforced layer stiffness properties.

E_{11} [GPa]	E_{22} [GPa]	E_{33} [GPa]	G_{12} [GPa]	G_{13} [GPa]	G_{23} [GPa]	$\nu_{12}[-]$	$\nu_{13}[-]$	$\nu_{23}[-]$
140	10	10	5.2	5.2	3.9	0.3	0.3	0.5

from the computational standpoint. In this regard, Table 3 outlines the amount of nodes and elements composing the two FE models. The comparison between these modeling techniques reveals that the CBJE drastically reduces the simulation time that depends on the amount of degrees of freedom and, therefore, on the quantity of nodes and elements.

A first assessment of the CBJE accuracy is outlined in Figure 6, it is shown a comparison between the contour maps of the displacement u_x obtained with both the FE models. The contour map of the CBJE model reports the displacement at the mid-surface of the plates. The distribution of displacement is very similar.

The CBJE methodology is capable of accurately evaluate the global stiffness of the double-lap joint. In fact, Figure 7 compares the load-displacement curves of the joint obtained with the two FE model, and it should be noted that the degree of accordance of the curves is remarkably relevant. In addition, the curves feature two stages: the first one presents higher stiffness and the plates are stuck together by the friction forces, the second one is characterized by lower slope

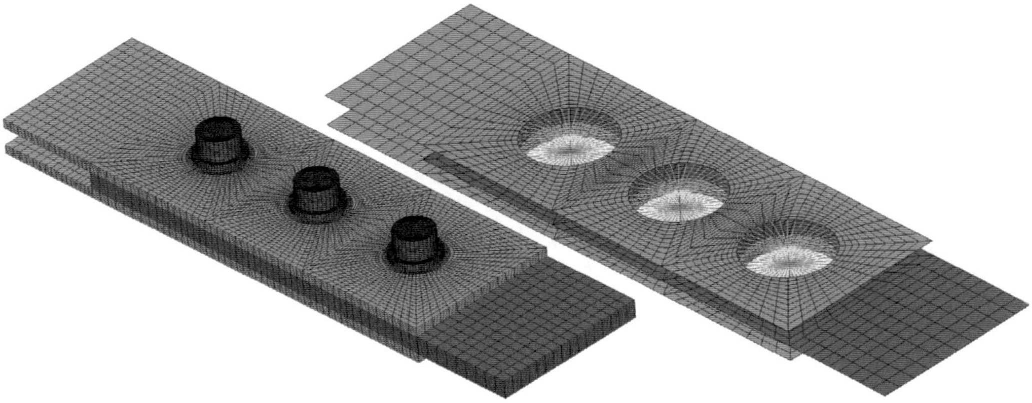

Figure 5. Double-lap three-bolt hybrid bolted joint: (left) 3D model; (right) shell model with CBJE.

Table 3. Computational burden of double-lap three-bolt hybrid bolted joint modeled with the 3D elements and the CBJE technique.

	3D FE Model	CBJE FE Model	Δ[%]
Nodes	65,976	4,248	−93.56
Elements	67,052	4,302	−93.58

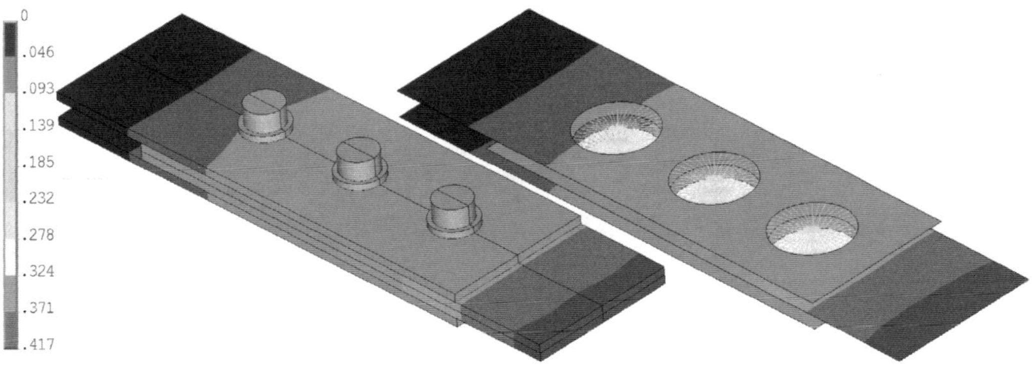

Figure 6. Contour of displacement u_x in the load direction determined by 3D model and CBJE model.

determined by the slipping of plates and the radial clearance recover.

The proposed method founded on the CBJE can also determine the load distribution between the bolts. Figure 8 reports the distribution of the external load between the bolts. It can be seen that during the initial part of the traction test the external load is reacted by the friction force acting at the mutual interface between the plates, meanwhile the bolts do not share any amount of load. Once the maximum value of the friction force F_s is equaled, the further increments of external load are reacted by the bolts and they produce the relative movement between the plates. In particular, the first bolt, i.e. Bolt 1, shares the highest portion of the applied load; it is followed by Bolt 3, whereas Bolt 2 carried the lowest amount of load. The amounts of

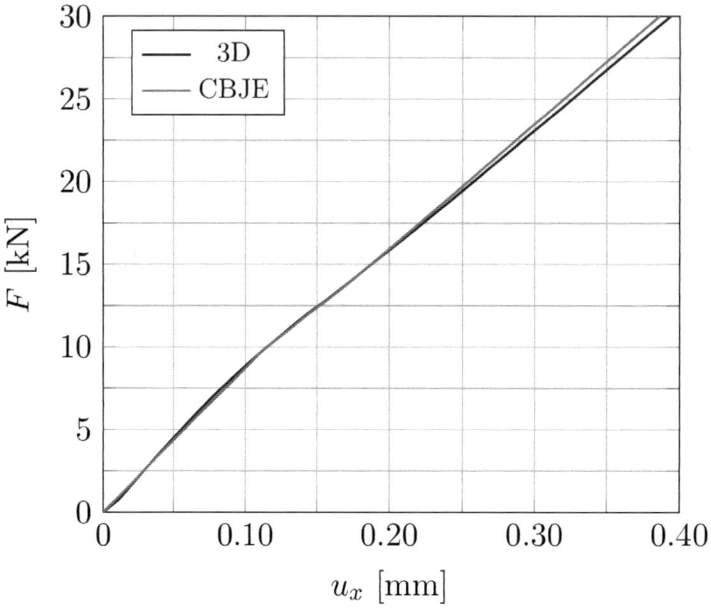

Figure 7. Diagram of load-displacement curves of the double-lap three-bolt joint determined by 3D model and CBJE model.

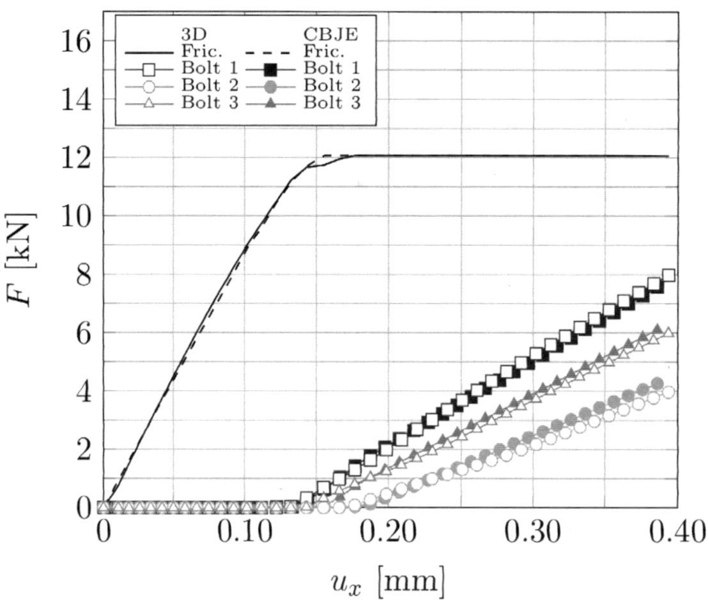

Figure 8. Distribution of bolt-load versus the joint displacement of the specimen, comparison of 3D model and CBJE model.

bolt-load are precisely assessed by the CBJE model, the differences with the 3D model are very limited. The reliable evaluation of the bolts-load allows for the accurate structural assessment of the joint.

4. Conclusions

In this paper, the extension of the CBJE methodology for hybrid composite-to-metal bolted joints is presented. The CBJE is a user-defined finite element, the characteristics of the elements derive from the theoretical reference model of bolted joint. The solution of the theoretical model is obtained in closed form solution, whereas the composite material theoretical model is solved exploiting the Ritz method.

This modified version of the CBJE is tested through the analysis of a hybrid double-lap joint, the results are compared with a 3D FE model. The main outcomes of the comparison are related to important computational savings which allows a reduction of the computational time. Then, the stiffness of the joint is accurately measured along all the load-displacement curve and the distribution of bolts-load is also precisely determined.

The combination of precision and low computational requirements is particularly advantageous during the first stage of the design process, in order to assess different configurations of the joint.

References

[1] Marannano G, Zuccarello B 2015 Numerical experimental analysis of hybrid double lap aluminum-CFRP joints *Composites Part B: Engineering* **71** 28-39

[2] Cao Y, Zuo D, Zhao Y, Cao Z, Zhi J, Zheng G, Tay T E 2021 Experimental investigation on bearing behavior and failure mechanism of double-lap thin-ply composite bolted joints *Composite Structures* **261** 113565

[3] McCarthy M A, McCarthy C T, Padhi G S 2006 A simple method for determining the effects of bolt–hole clearance on load distribution in single-column multi-bolt composite joints *Composite Structures* **73** 78-87

[4] Awadhani L V, Bewoor K 2017 Analytical and experimental investigation of Effect of geometric parameters on the failure modes in a single lap single bolted metal to GFRP composite bolted joints subjected to axial tensile loading *Materials Today: Proceedings* **4** 7345-7350

[5] McCarthy M A, McCarthy C T, Lawlor V P, Stanley W F 2005 Three-dimensional finite element analysis of single-bolt, single-lap composite bolted joints: part I — model development and validation *Composite Structures* **71** 140-158

[6] Kapidžić Z, Nilsson L, Ansell H 2014 Finite element modeling of mechanically fastened composite-aluminum joints in aircraft structures *Composite Structures* **109** 198-210

[7] Gray P J, McCarthy C T 2010 A global bolted joint model for finite element analysis of load distributions in multi-bolt composite joints *Composites Part B: Engineering* **41** 317-325

[8] Liu F, Yao W, Zhao L, Wu H, Zhang X, Zhang J 2020 An improved 2D finite element model for bolt load distribution analysis of composite multi-bolt single-lap joints *Composite Structures* **253** 112770

[9] El Masnaoui W, Daidié A, Lachaud F, Paleczny C 2020 Semi-analytical model development for preliminary study of 3D woven Composite/Metallic flange bolted assemblies *Composite Structures* **255** 112906

[10] Belardi V G, Fanelli P, Vivio F 2018 Elastic analysis of rectilinear orthotropic composite circular plates subject to transversal and in-plane load conditions using Ritz method *Composite Structures* **199** 63-75

[11] Belardi V G, Fanelli P, Vivio F 2019 First-order shear deformation analysis of rectilinear orthotropic composite circular plates undergoing transversal loads *Composites Part B: Engineering* **174** 107015

[12] Belardi V G, Fanelli P, Vivio F 2021 Ritz method analysis of rectilinear orthotropic composite circular plates undergoing in-plane bending and torsional moments *Mechanics of Advanced Materials ans Structures* **28** 963-979

[13] Belardi V G, Fanelli P, Vivio F 2021 On the radial bending of shear-deformable composite circular plates with rectilinear orthotropy *European Journal of Mechanics / A Solids* **86** 104157

[14] Belardi V G, Fanelli P, Vivio F 2018 A novel composite bolted joint element: Application to a single-bolted joint *Procedia Structural Integrity* **12** 281-295

[15] Belardi V G, Fanelli P, Vivio F 2019 FE analysis of single-bolt composite bolted joint by means of a FE analysis of single-bolt composite bolted joint by means of a simplified modeling technique *Procedia Structural Integrity* **24** 888-897

[16] Belardi V G, Fanelli P, Vivio F 2021 Theoretical definition of a new custom finite element for structural modeling of composite bolted joints *Composite Structures* **258** 113199
[17] Vivio F 2009 A new theoretical approach for structural modelling of riveted and spot welded multi-spot structures *International Journal of Solids and Structures* **46** 4006-24

Implementation of the Peak Stress Method for the automated FEM-assisted design of welded joints subjected to constant amplitude multiaxial fatigue loads

A Visentin[1], A Campagnolo[1], F Sacchet[1] and G Meneghetti[1]

[1] University of Padova, Department of Industrial Engineering, 35131 Padova, Italy

giovanni.meneghetti@unipd.it

Abstract. The Peak Stress Method (PSM) is a local approach to the fatigue strength assessment of welded structures. Starting from FE-calculated opening, in-plane shear and out-of-plane shear peak stresses at the weld toe and weld root, the PSM defines an equivalent peak stress for estimating the fatigue failure location and fatigue life of welded structures, in compliance with properly defined design curves. An interactive tool has been developed in Ansys® Mechanical to automate all calculation tasks required to apply the PSM to generic welded structures. The developed application allows to identify and analyse all weld toe and weld root lines of the structure in a full-automated way, performing fatigue life estimation on each analysed node. Finally, fatigue life results can be visualized directly on the model's geometry through dedicated contour plots. In this work, some joint geometries taken from the literature and related to steel welded joints subjected to multiaxial fatigue loads are re-analysed taking advantage of two design procedures: (i) manual application of the PSM, (ii) automated implementation of the PSM. Taking advantage of the tool developed in Ansys®, a remarkable reduction in analysis time and effort can be achieved, with respect to manual PSM analyses.

1. Introduction

According to the local fatigue approach based on the NSIF-parameters [1–3] the weld toe and weld root of an arc-welded joint can be modelled and studied as sharp V-notches having null tip radius ($\rho = 0$) and notch opening angles equal to 135° and 0° respectively, as highlighted in next figure 1. Assuming a linear-elastic behavior for the material of the welded structure, external loads produce singular stress fields in the local regions enclosing weld toes and roots. The NSIF-parameters allow to quantify the intensity of the singular stress distributions, according to equation (1).

$$K_i = \sqrt{2\pi}\cdot \lim_{r \to 0}\left[\left(\sigma_{jk}\right)_{\theta=0}\cdot r^{1-\lambda_i}\right] \quad \text{where} \quad i = 1, 2, 3 \quad \text{and} \quad \sigma_{jk} = \sigma_{\theta\theta}, \tau_{r\theta}, \tau_{\theta z} \tag{1}$$

Where the linear elastic stresses $\sigma_{\theta\theta}$, $\tau_{r\theta}$ and $\tau_{\theta z}$ are calculated close to the notch tip ($r \to 0$), along the notch bisector direction ($\theta = 0$), as shown in next figure 2. Parameters λ_i, i.e. stress singularity exponents, are function of the notch opening angle 2α of the considered V-notch site, according to [4,5].

The strain energy density (SED) averaged over a structural volume surrounding the notch tip has been adopted as a fatigue strength criterion by Lazzarin and collaborators [6–8] in order to assess weld toe and weld root fatigue failures. The closed-form expression of the averaged SED as a function of the relevant NSIF-parameters tied to mode I, mode II and mode III loadings is reported in equation (2).

$$\Delta\bar{W}=\left\{c_{w1}\frac{e_1}{E}\left[\frac{\Delta K_1}{R_0^{1-\lambda_1}}\right]^2+c_{w2}\frac{e_2}{E}\left[\frac{\Delta K_2}{R_0^{1-\lambda_2}}\right]^2+c_{w3}\frac{e_3}{E}\left[\frac{\Delta K_3}{R_0^{1-\lambda_3}}\right]^2\right\} \tag{2}$$

R_0 represents the size of the structural volume and equals 0.28 mm for welded joints made of structural steels. E represents the material Young's modulus of elasticity while e_i (i = 1, 2 and 3) are coefficients dependent on the notch opening angle 2α and on material, in terms of Poisson's ratio ν. Moreover, ΔK_1, ΔK_2 and ΔK_3 are the range values of the NSIFs tied to mode I, mode II and mode III, respectively. Finally, c_{wi} account for the mean stress effect and depend on the load ratio R_i of the i-th loading mode (i = 1, 2 or 3), according to equation (3).

$$c_{wi}(R_i) = \begin{cases} \dfrac{1+R_i^2}{(1-R_i)^2} & \text{if stress-relieved and } -1\le R_i\le 0 \\ \dfrac{1-R_i^2}{(1-R_i)^2} & \text{if stress-relieved and } 0\le R_i<1 \\ 1 & \text{if as-welded for any R value} \end{cases} \quad \text{where } i=1,2,3 \tag{3}$$

It is worth noting that c_{wi} = 1 should be adopted in the case of welded joints tested in as-welded conditions, being them almost insensitive to mean stresses, according to design standards [9].

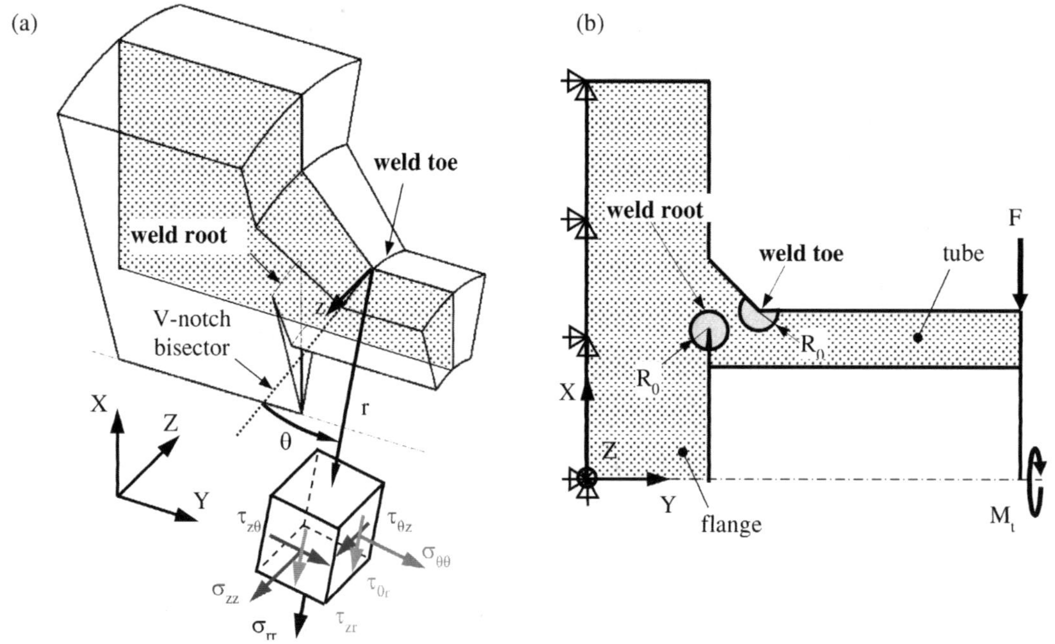

Figure 1. Details of the NSIF-based and averaged SED approaches with respect to a partial-penetration tube-to-flange welded joint under multiaxial fatigue loading. (a) Cylindrical reference system (r, θ, z) centred at the weld toe and local elastic stresses. (b) Structural volume of size R_0 centred at the weld toe and weld root, according to the averaged SED approach.

2. The Peak Stress Method (PSM)

The Peak Stress Method (PSM) is a rapid, FE-oriented tool to estimate the NSIF-terms K_1, K_2 and K_3, taking advantage of the local peak stresses, i.e. opening, in-plane shear and out-of-plane shear peak stresses, calculated from a linear elastic FE analysis [10]. One of the main advantages of the PSM is the possibility to employ coarse meshes with respect to those required by other local approaches, such as the NSIF approach. Moreover, only the linear elastic peak stresses calculated at the notch tip suffice to be considered. The estimated NSIFs values can be evaluated from the following expressions [11–13]:

$$\text{(a) } K_1 \cong K^*_{FE} \cdot \sigma_{\theta\theta,\theta=0,peak} \cdot d^{1-\lambda_1}; \quad \text{(b) } K_2 \cong K^{**}_{FE} \cdot \tau_{r\theta,\theta=0,peak} \cdot d^{1-\lambda_2}; \quad \text{(c) } K_3 \cong K^{***}_{FE} \cdot \tau_{\theta z,\theta=0,peak} \cdot d^{1-\lambda_3} \quad (4)$$

The terms $\sigma_{\theta\theta,\theta=0,peak}$, $\tau_{r\theta,\theta=0,peak}$ and $\tau_{\theta z,\theta=0,peak}$ in equation (4) refer to the peak stresses defined in a local cylindrical coordinate system centered at the node at the V-notch tip, having *z*-direction tangential to the notch tip edge, *r*-direction defined by the notch bisector line and θ as the radial direction (see next figure 3). Each peak stress is evaluated with respect to $\theta = 0$, which means that, as an example, the opening stress $\sigma_{\theta\theta,\theta=0,peak}$ acts in the normal direction with respect to the notch bisector. Parameter *d* in equation (4) refers to the average size of the finite elements to be given as input to the free mesh generation algorithm of the FE software. Parameters K^*_{FE}, K^{**}_{FE} and K^{***}_{FE} in equation (4) depend on: (i) element type and formulation; (ii) FE mesh pattern; (iii) procedure employed by the FE code to extrapolate stress values at nodes. The reader is referred to the recently published state-of-the-art review of the PSM [10] for additional detail about the method.

2.1. The Peak Stress Method with tetrahedral finite elements

The main advantage in adopting tetrahedral finite elements resides in the possibility to easily discretize either simple and complex 3D geometries without the need for sub-modeling techniques. Dealing with the application of the PSM with tetrahedral elements, an average peak stress formulation [10,14] has been introduced according to equation (5) in order to compensate the variability of peak stress values along the notch tip profile due to the intrinsic irregularity of the free tetra mesh pattern. More in detail, equation (5) performs the moving average of the peak stresses calculated on three adjacent vertex nodes, i.e. $n = k$-1, $n = k$ and $n = k$+1 and assigns the calculated average stress at the central node ($n = k$).

$$\bar{\sigma}_{ij,peak,n=k} = \left. \frac{\sigma_{ij,peak,n=k-1} + \sigma_{ij,peak,n=k} + \sigma_{ij,peak,n=k+1}}{3} \right|_{n=node} \quad (5)$$

Moreover, peak stresses evaluated at the notch tip nodes belonging to free surfaces of the analysed structure can be affected by distorted mesh patterns and are therefore to be excluded. Furthermore, only peak stresses calculated at vertex nodes located along the notch tip line must be used, while peaks stress values related to mid-side nodes are to be neglected. PSM parameters K^*_{FE}, K^{**}_{FE} and K^{***}_{FE} have previously been calibrated using either 10-node [14] and 4-node [15] tetra elements (see table 1). The calibration of parameters K^*_{FE}, K^{**}_{FE} and K^{***}_{FE} for tetra element types has been performed under the following conditions, which are more widely discussed in [10]:

a. A range of notch opening angle 2α values is supported. As reported in table 1, in the case of mode I and mode III loadings, the range $2\alpha = 0° - 135°$ has been investigated, being $2\alpha = 0°$ and $2\alpha = 135°$ the typical notch opening angles at weld root and weld toe, respectively. However, weld toe profiles having opening angles up to 150° have been analysed adopting the same calibration constants [16]. In the case of mode II loading, only $2\alpha = 0°$ [14,15] and $2\alpha = 90°$ [17] have been investigated, being the typical cases of the weld root without or with a gap.
b. A minimum mesh density ratio $(a/d)_{min}$ must be fulfilled when defining the free mesh pattern at the notch tip according to the PSM, *a* being the characteristic size of the considered notch, defined in

[10]. As reported in table 1, the value of $(a/d)_{min}$ depends on: (i) adopted FE type, (ii) considered loading mode and (iii) opening angle 2α of the analysed notch.

Table 1. Summary of parameters K^*_{FE}, K^{**}_{FE} and K^{***}_{FE} and mesh density *a/d* requirements to apply the Peak Stress Method with 10-node SOLID 187 tetra elements in Ansys® FE software [10].

Loading	PSM parameters	2α = 0°	2α = 90°	2α = 120°	2α = 135°	*a* root side	*a* toe side
Mode I	K^*_{FE}	1.05±15%	1.05±15%	1.05±15%	1.21±10%	min{*l*, *z*}	*t*
Eq. (1a)	$(a/d)_{min}$	3	3	3	1		
Mode II	K^{**}_{FE}	1.63±20%	2.65±10%	n.a.	n.a.	min{*l*, *z*}	*t*
Eq. (1b)	$(a/d)_{min}$	1	1	n.a.	n.a.		
Mode III	K^{***}_{FE}	1.37±15%	1.37±15%	1.70±10%	1.70±10%	min{*l*, *z*}	*t*
Eq. (1c)	$(a/d)_{min}$	3	3	3	3		

[a] 'Full graphics' option must be activated when calculating peak stresses according to PSM 3D

2.2. Fatigue design according to the PSM

The averaged SED can be expressed as a function of the relevant NSIF-parameters K_1, K_2 and K_3, according to equation (2). Moreover, the PSM allows to rapidly estimate the NSIFs-terms as a function of the relevant FE-calculated peak stresses, according to equations (4a)-(4c). As a result, the averaged SED itself can be re-written as a function of the relevant peak stresses. By introducing the SED expression for an equivalent uniaxial strain state (i.e. $W=(1-\nu^2)\cdot\sigma^2_{eq,peak}/2E$), an equivalent peak stress term can be defined according to equation (6) [10] and can be referred as a design stress quantity for the fatigue strength assessment of welded joints.

$$\Delta\sigma_{eq,peak} = \sqrt{c_{w1}\cdot f^2_{w1}\cdot\Delta\bar{\sigma}^2_{\theta\theta,\theta=0,peak} + c_{w2}\cdot f^2_{w2}\cdot\Delta\bar{\tau}^2_{r\theta,\theta=0,peak} + c_{w3}\cdot f^2_{w3}\cdot\Delta\bar{\tau}^2_{\theta z,\theta=0,peak}} \quad (6)$$

Equation (6) is valid for 4-node as well as 10-node tetra elements and the averaged peak stress terms $\bar{\sigma}_{\theta\theta,\theta=0,peak}$, $\bar{\tau}_{r\theta,\theta=0,peak}$ and $\bar{\tau}_{\theta z,\theta=0,peak}$ are defined according to equation (5). Parameters f_{wi} (i = 1, 2 and 3, according to the loading mode) account for peak stress averaging inside the material structural volume having size R_0 (see figure 1) [10] and are defined as in following equation (7).

$$f_{wi} = K_{FE}\cdot\sqrt{\frac{2e_i}{1-\nu^2}}\cdot\left(\frac{d}{R_0}\right)^{1-\lambda_i} \quad \text{where } i = 1, 2, 3 \quad (7)$$

The FE-calculated peak stresses involved in equations (5)-(6), as well as f_{wi} parameters in equation (7), are functions of the average FE size d adopted in the FE model (see next figure 3); however, the equivalent peak stress defined in equation (6) is ultimately independent from the FE size d owing to the multiplication of the peak stresses by the relevant f_{wi} parameters [10].

Moreover, dealing with the fatigue assessment of welded joints subjected to multiaxial loads, an approximate approach originally proposed in [18] has been adopted in the current work. The strain energy contribution tied to each separate load case j must be accounted and evaluated individually. Afterwards, all averaged SED contributions are to be combined in order to determine the total averaged SED value, according to equation (8).

$$\Delta\bar{W} = \sum_j \Delta\bar{W}_j \tag{8}$$

Introducing the SED expression for an equivalent uniaxial strain state (i.e. $W=(1-\nu^2)\cdot\sigma_{eq,peak}^2/2E$) as before, the equivalent peak stress can be re-formulated according to the following equation (9).

$$\Delta\sigma_{eq,peak} = \sqrt{\sum_j c_{w,j}\left\{f_{w1}^2\cdot\Delta\bar{\sigma}_{\theta\theta,\theta=0,peak}^2 + f_{w2}^2\cdot\Delta\bar{\tau}_{r\theta,\theta=0,peak}^2 + f_{w3}^2\cdot\Delta\bar{\tau}_{\theta z,\theta=0,peak}^2\right\}_j} \tag{9}$$

Assuming for instance the case of a partial-penetration tube-to-flange welded joint under combined bending-torsion loading (see next figure 2), the averaged SED contributions tied to the bending load (M_b) and to the torsion load (M_t) are to be accounted separately and furtherly combined according to equation (8). In order to do so, each separate load case must be considered and solved individually and. Finally, the equivalent peak stress tied to the combined bending-torsion loading condition must be calculated according to equation (9), where j identifies the single accounted load case, i.e. $j = M_b, M_t$.

The equivalent peak stress can be adopted to estimate the fatigue life of welded joints, in compliance with a proper PSM-based design curve. A local biaxiality ratio λ has been formulated as the ratio between the SED contributions due to mode II/III shear and mode I stresses, in order to guide the selection of the correct design curve to address. λ can be expressed as a function of the relevant averaged peak stresses according to equation (10), which is valid or 4-node as well as 10-node tetra elements.

$$\lambda = \frac{c_{w2}\cdot f_{w2}^2\cdot\Delta\bar{\tau}_{r\theta,\theta=0,peak}^2 + c_{w3}\cdot f_{w3}^2\cdot\Delta\bar{\tau}_{\theta z,\theta=0,peak}^2}{c_{w1}\cdot f_{w1}^2\cdot\Delta\bar{\sigma}_{\theta\theta,\theta=0,peak}^2} \tag{10}$$

The proper PSM-based design curve can be selected according to the criteria proposed in [10] and reported in table 2 for the case of joints made of structural steels, where $\Delta\sigma_{eq,peak,A,50\%}$ is the fatigue class at $2\cdot10^6$ cycles (for a survival probability of 50%) and k is the slope of the fatigue design curve. The case $\lambda = 0$ corresponds to a pure local mode I stress state, while $\lambda > 0$ implies a mixed mode opening-shear stress condition [10].

Finally, fatigue life can be estimated with respect to the desired survival probability by comparing the equivalent peak stress with the proper PSM-based fatigue design curve (see next figure 4).

Table 2. Criterion of selection for the PSM-based fatigue design curve for steel arc-welded joints [10].

Thickness, T [mm]	λ Eq. (10)	$\Delta\sigma_{eq,peak,A,50\%}$ [MPa]	k	T_σ
$T \geq 2$ mm	$\lambda = 0$	214	3.0	1.90
$T \geq 2$ mm	$\lambda > 0$	354	5.0	1.90

2.3. Step-by-step manual application of the PSM for the fatigue assessment of welded joints

The current paragraph reports all the elaboration steps required to manually apply the PSM for the fatigue assessment of a generic arc-welded structure. The reader is referred to the recently published state-of-the-art review of the PSM [10] and to [17] for a comprehensive insight on the application of the method.

a) Identify weld toes and weld roots in the structure. For each considered notch site, define the notch opening angle 2α and the characteristic notch size a, as indicated in next figure 2. Moreover, identify the thickness T characterizing the welded detail under exam.

b) Adopt a proper global element size d in order to satisfy the $(a/d)_{\min}$ PSM requirements (as a function of the opening angle 2α and loading mode, as reported in table 1) and generate a free FE mesh. Solve the FE model afterwards.

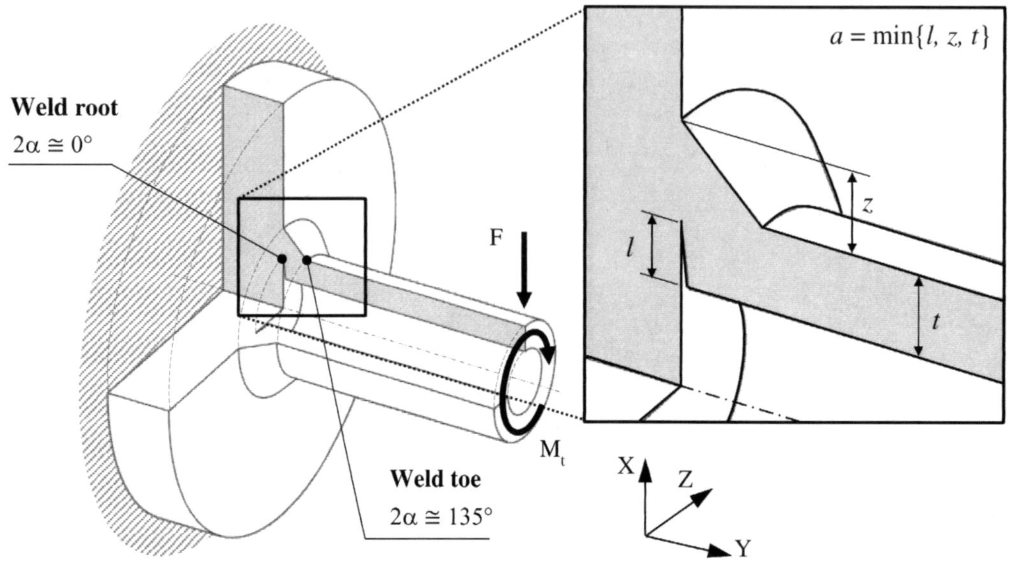

Figure 2. Tube-to-flange welded joint under combined bending-torsion loading. Criterion for the selection of the characteristic notch size a, according to the PSM.

Once solution is done, for each node on the considered notch tip line (weld toe or weld root), iterate the following steps (c-h):

c) Define a proper local coordinate system (r, θ, z) centred on the considered node, having directions as indicated in next figure 3. Retrieve the relevant peak stress components with respect to the local coordinate system (see figure 3).

d) Perform the moving average calculation on adjacent vertex nodes, in order to calculate the average peak stress components according to equation (5) and figure 3.

e) Evaluate the stress singularity exponents λ_i and the averaged SED coefficients e_i. Recently, some polynomial expressions were proposed in [17,19] to achieve rapid evaluation of λ_i exponents as a function of the notch opening angle 2α. Accordingly, two-variables polynomial expressions were also proposed in [17,19] to achieve rapid evaluation of e_i coefficients as a function of either the notch opening angle 2α and the considered material, in terms of Poisson's ratio ν.

f) Evaluate the mean stress correction factors c_{wi} according to equation (3), which account for the mean stress effect and depend on the load ratio R_i of the i-th loading mode (i = 1, 2 and 3). Evaluate the PSM parameters f_{wi} (i = 1, 2 and 3) according to equation (7), which account for peak stress averaging inside the material structural volume having size R_0 (see figure 1).

g) Calculate the equivalent peak stress, according to equation (6). Moreover, dealing with the fatigue assessment of the welded joints subjected to multiaxial loads, each different load case must be simulated separately and the equivalent peak stress tied to the combined load case must be evaluated using equation (9), according to the approximate approach described in paragraph 2.2 and originally proposed in [18].

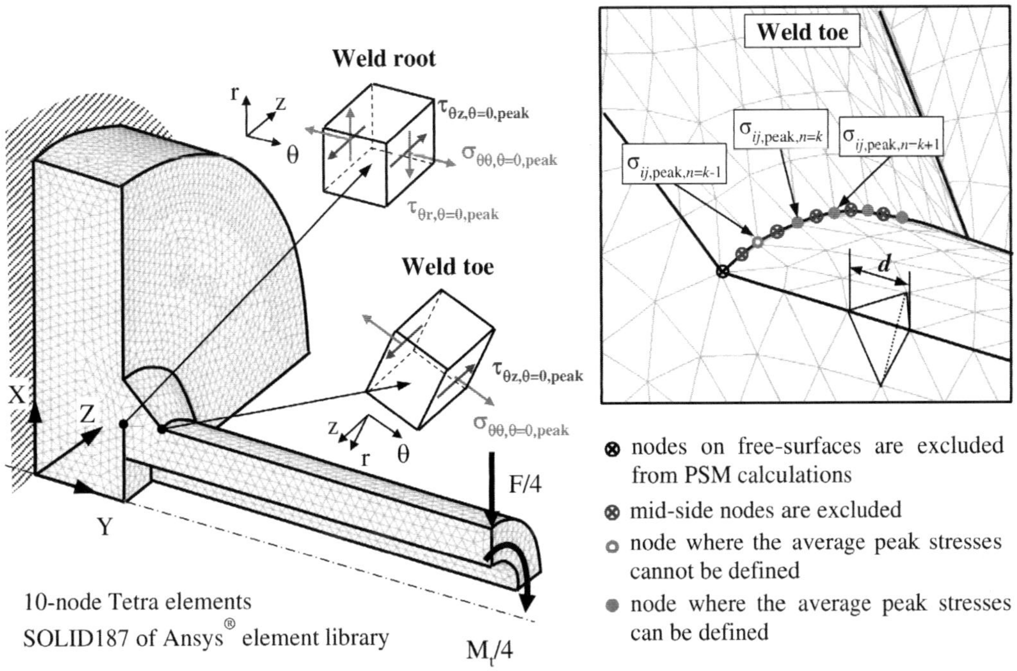

Figure 3. FE model of a tube-to-flange welded joint under combined bending-torsion loading. Definition of local coordinate systems and extraction of the peak stresses according to the PSM.

h) Calculate the local biaxiality ratio λ according to equation (10), in order to address the proper PSM-based design curve. Finally, perform fatigue life estimations with respect to the desired survival probability by comparing the equivalent peak stress calculated at (g) with the proper PSM-based fatigue design curve, according to figure 4.

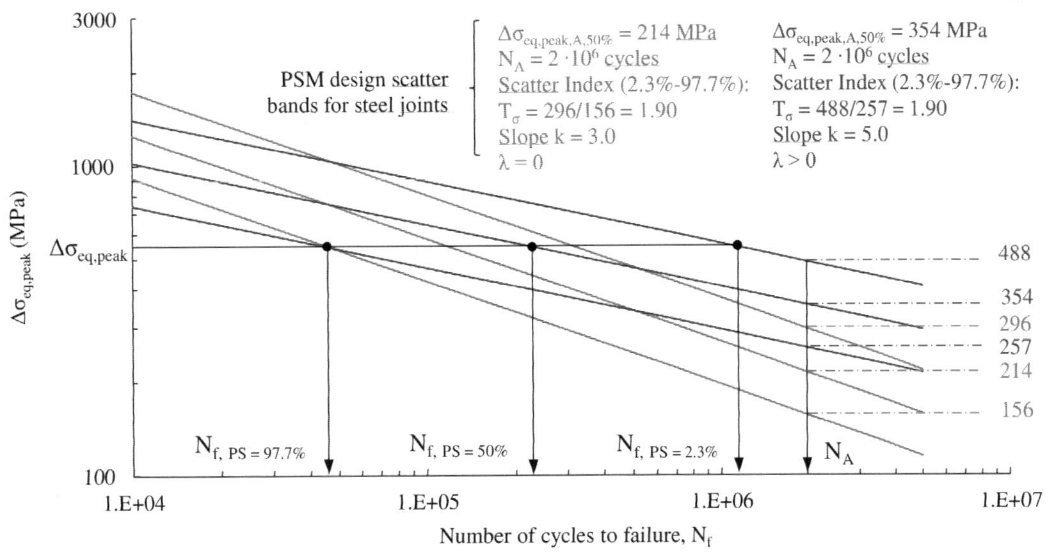

Figure 4. Fatigue strength assessment of weld toe and weld root failures according to the PSM in the case of structural steel arc-welded joints having thickness T ≥ 2 mm and λ ≥ 0 (see table 2).

3. Automated implementation of the PSM for the fatigue design of welded joints

An interactive application has been developed in Ansys® Mechanical APDL [17,19], in order to automate the fatigue strength assessment of welded structures subjected to either uniaxial and multiaxial loading conditions, according to the PSM. In this perspective, a more powerful and efficient tool has been recently developed for Ansys® Mechanical, the FE environment of Ansys® Workbench, taking advantage of Ansys® Customization Toolkit (Ansys® ACT). ACT is a programming interface, designed to grant support and integration between Ansys® FE code and modern programming languages (i.e. IronPython, Microsoft® C#, Microsoft® Jscript, XML, HTML) allowing the user to develop applications and integrate them within the different CAD/FE environments available in Ansys® Workbench, in order to automate different modeling, simulation and post-processing tasks. The application presented in this work is designed to provide the analyst a user-friendly set of tools tailored to automate each implementation task required to apply the PSM to a generic FE model (see paragraph 2.2). Therefore, two different graphical interfaces have been developed: (i) an ACT Wizard, namely an interactive application designed to guide the analyst throughout the automated PSM analysis and (ii) a dedicated toolbar, which provides different post-processing features that allow visualization and consultation of analysis results through contour plots, charts and numerical reports.

More in detail, the developed tool is designed to automate the following PSM application tasks:

a. Perform compatibility checks on the FE model according to PSM requirements, in terms of mesh density a/d with respect to $(a/d)_{\min}$ for the adopted element type (see table 1), depending on user-inputs on a, d and T_{min} (i.e. the minimum welded thickness in the structure).
b. Automatically identify weld toes and weld roots among all the edges of the structure, by locally evaluating the notch opening angle 2α along model edges (see also figure 1 and figure 2).
c. Calculate PSM-related parameters, i.e. stress singularity exponents and SED coefficients. Stress singularity exponents λ_i and SED coefficients e_i are computed taking advantage of the polynomial expressions proposed in [17,19], as function of the opening angle 2α and material Poisson's ratio ν.
d. Define and orientate a local coordinate system on each analysed peak node in order to retrieve the relevant peak stresses tied to loading mode I, mode II and mode III, according to figure 3.
e. Perform the moving average operation on adjacent vertex nodes, according to figure 3.
f. Calculate PSM parameters c_{wi} and f_{wi} (i = 1, 2 and 3) according to equations (3, 7) and combine the calculated average peak stresses into the equivalent peak stress on each analysed node, according to equation (6).
g. Calculate the local biaxiality ratio λ on each analysed node, according to equation (10).
h. Estimate fatigue life on each analysed node, referencing to the proper PSM-based design curve among those proposed in [10] according to material type, local biaxiality ratio and survival probability for fatigue life estimations (see figure 4).

The developed tool also provides post-processing features that allow the analyst to access and visualize the analysis results with different grades of detail.

a. Through contour plots. Fatigue life, equivalent peak stress and local biaxiality ratio results can be plotted directly along analysed edges on a wireframe view of the geometry, as shown in next figure 5 and figure 9. Coloured contour plots allow to rapidly single out and compare at a glance the different competing crack initiation sites of the analysed structure.
b. Through results charts. Fatigue life, equivalent peak stress and local biaxiality ratio results can be consulted taking advantage of dedicated comparison charts, which can be generated directly within Ansys® Mechanical. Comparison charts allow the analyst to simultaneously plot and compare results distributions on one or more analysed edges.
c. Through numerical reports. A comprehensive results report is generated during the automated analysis (see figure 5). All numerical values related to analysed edges and nodes, i.e. edge id, nodal coordinates, relevant peak stresses, average peak stresses, local biaxiality ratio as well as equivalent peak stress and fatigue life values for each analysed node are collected and available in a *log* file, which can be directly opened and consulted in Microsoft® Excel from the dedicated shortcut.

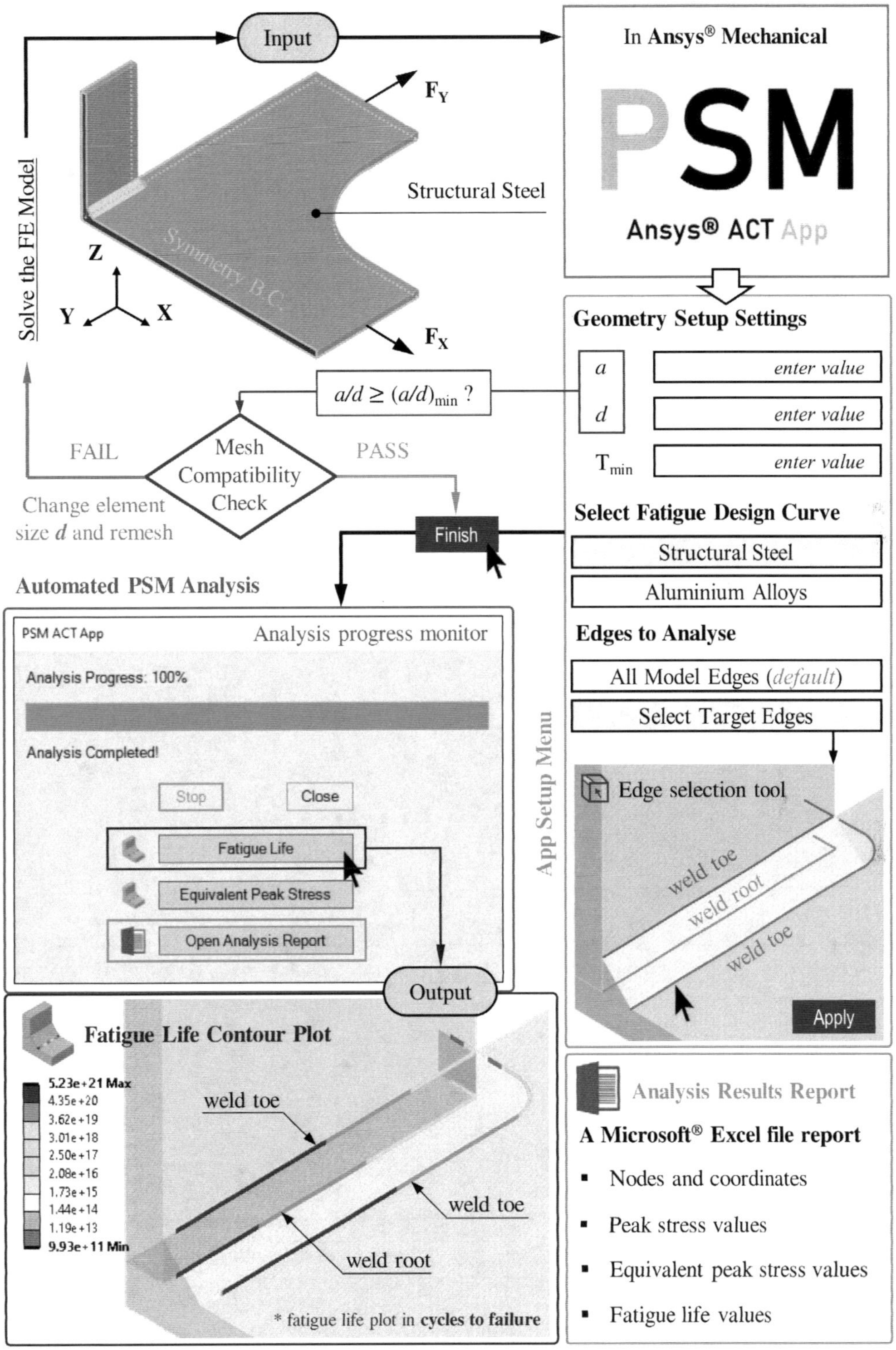

Figure 5. The automated PSM tool developed in Ansys® Mechanical. Illustration and description of the automated analysis workflow for the FE analyst.

4. Steel joints under multiaxial loads: geometries and FE analyses according to the PSM

Some experimental data available in literature, related to fatigue failures at the weld toe and weld root in steel welded joints subjected to either uniaxial and multiaxial fatigue loading have been considered and re-analysed according to the PSM [20]. The analysed joint geometries and loading conditions are reported in table 3. Previous PSM analyses were conducted in [21] on the considered geometries, taking advantage of either free meshes generated using 2D 4-node plane harmonic elements and mapped meshes generated using 3D 8-node brick elements (PLANE 25 and SOLID 185 of the Ansys® element library, respectively). The joints characterized by an axis-symmetric geometry were simulated using PLANE 25 plane harmonic elements, while the remaining geometries were simulated using SOLID 185 brick elements (K-option 2 set to 3), taking advantage of the sub-modeling technique available in Ansys® Mechanical APDL to isolate and analyze the critical areas of the joint at the weld toe and weld root.

Table 3. Arc-welded joint geometries and load cases re-analysed according to the PSM [20].

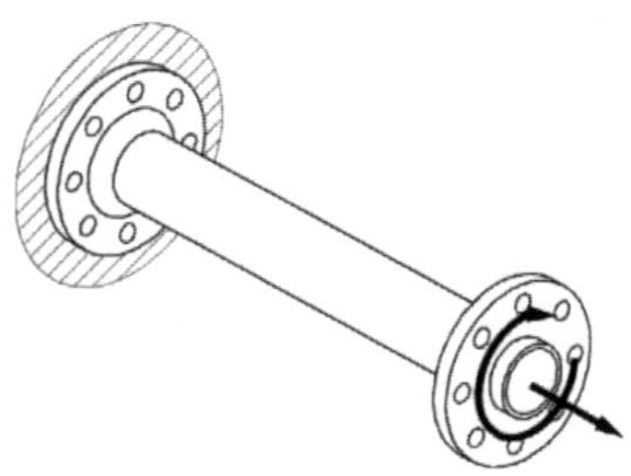	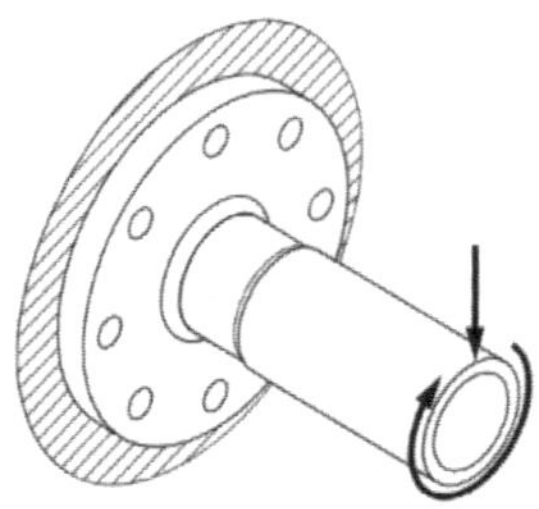	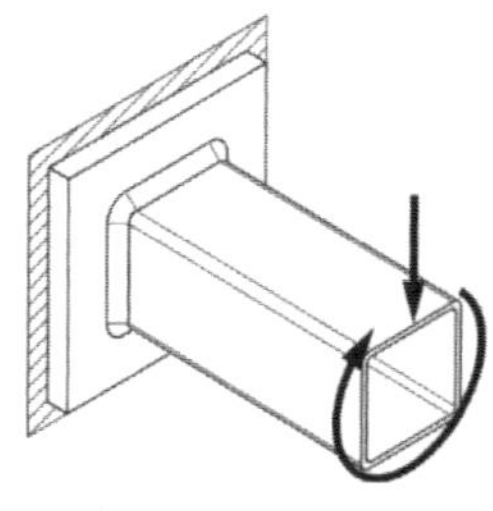
Razmjoo *et al.* [22]	Yousefi *et al.* [23]	Bäckström *et al.* [24]
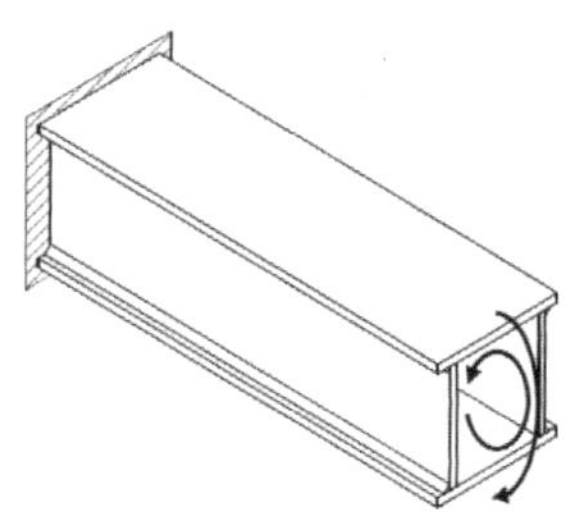	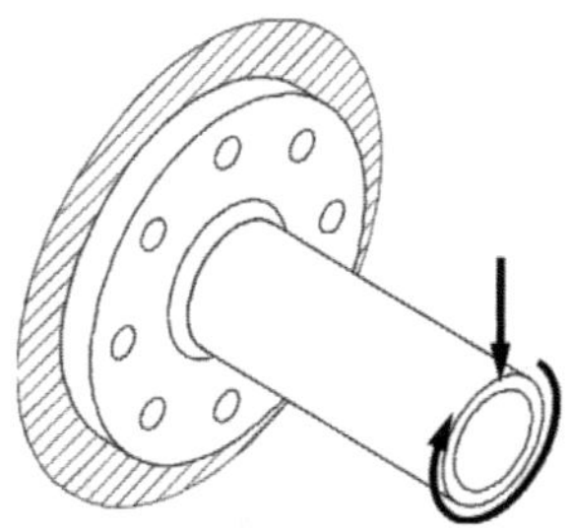	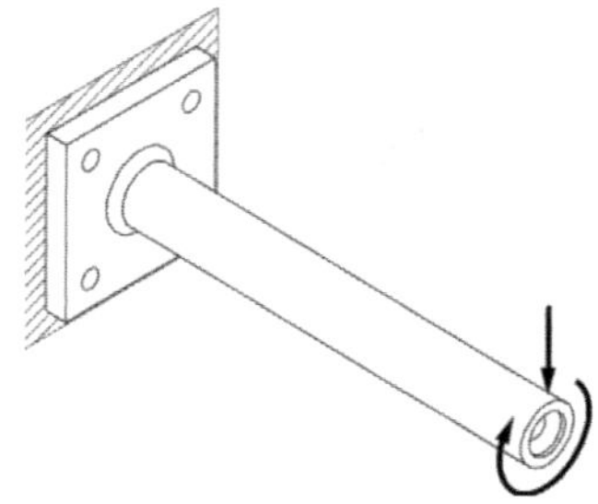
Dahle *et al.* [25]	Sonsino [26,27]	Bertini *et al.* [28]
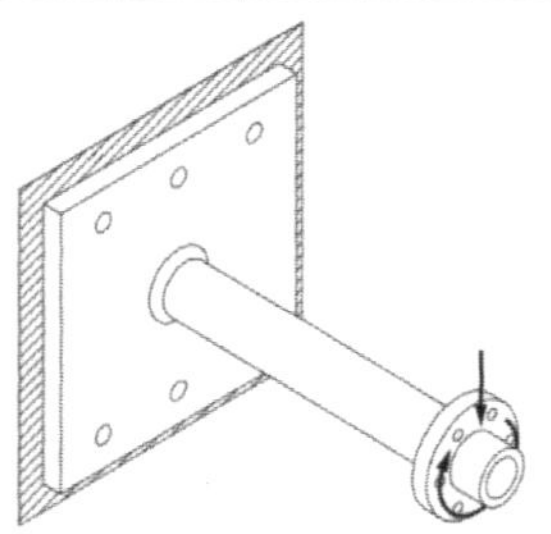	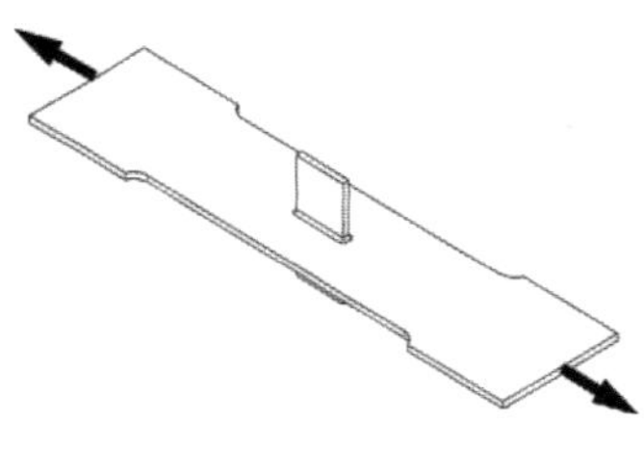	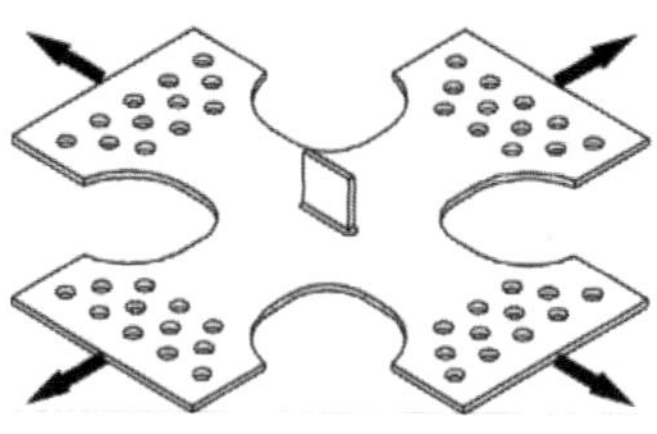
Frendo and Bertini [29]	Takahashi *et al.* [30,31]	Takahashi *et al.* [30,31]

All geometries reported in table 3 have been re-analysed by means of the following assessment approaches: (i) manual PSM application in Ansys® Mechanical APDL, as described in paragraph 2.2, and (ii) automated PSM implementation, taking advantage of the developed tool available in Ansys® Mechanical (see paragraph 3). A free mesh pattern of 3D 10-node tetrahedral elements (SOLID187 of the Ansys® element library) has been generated for each considered geometry, taking advantage of the free meshing algorithms provided either by Ansys® Mechanical APDL and Ansys® Mechanical, in the case of the manual and automated PSM analyses, respectively. Loads and constraints, as well as the adopted global element size *d*, have been applied likewise to the FE models, either in Ansys® Mechanical APDL and Ansys® Mechanical. Moreover, differently from the analyses conducted in [21], no sub-modeling techniques were required due to the higher 3D discretization capabilities offered by 10-node tetra elements with respect to plane and brick elements.

In this paper, a set of welded joints has been considered among all analysed geometries of table 3 in order to provide a deeper insight on the analysis procedures and achieved results. Information related to material properties, as well as involved welding process and testing conditions for the considered geometries is reported in table 4, while next figures 6-8 report more details on the FE analyses performed according to the PSM.

Table 4. Material properties, welding process and testing conditions of the considered steel joints [21].

Reference	Material	Yield strength [MPa]	Ultimate strength [MPa]	Welding process	Testing conditions
Yousefi *et al.* [23]	P 460	520	670	MIG	Stress-relieved
Frendo and Bertini [29]	S355 JR	360	520	-	As-welded
Takahashi *et al.* [30,31]	JIS SM400B	283	432	MAG	As-welded

All joints have un-machined welds. Joint geometries and loading conditions are reported in figures 6-8.

4.1. Yousefi et al. [23], tube-to-flange joints

The tube-to-flange joints under exam are characterized by a weld toe and a weld root with partial penetration and have been fatigue-tested under reversed (R = -1) as well as pulsating (R = 0) pure bending, pure torsion and combined bending-torsion loadings [23], as reported in figure 6 and table 5. Multiaxial loadings were applied both in-phase ($\phi = 0°$) and out-of-phase ($\phi = 90°$) by adopting a nominal biaxiality ratio $\Lambda = \sigma_{nom}/\tau_{nom} = 1$ [21]. The original experimental results are expressed in terms of nominal stress range (maximum minus minimum value) with respect to the tube and the accounted number of cycles refers to the generation of a through-the-thickness fatigue crack.

The geometry has been modelled as 1/4 of the entire welded joint, taking advantage of the axis-symmetry, and has been simulated taking advantage of a 10-node tetra free-generated 3D mesh either in Ansys® Mechanical APDL and Ansys® Mechanical, as shown in figure 6. The element size is prescribed by the mesh density requirement $a/d \geq 3$. The tube thickness is assumed as characteristic size at weld toe, namely $a = 8$ mm. Consequently, d must be at least equal to 8/3 = 2.67 mm according to the requirements in table 1, in order to achieve compatibility with respect to either mode I and mode III loadings at weld toe. A global element size of 2 mm has been finally adopted to generate the mesh pattern. Moreover, a mesh refinement was performed at the weld root, in order to achieve a local element size of ~0.3 mm (see figure 6) and guarantee PSM applicability. In this case indeed $a = 1$ mm, namely equal to the weld root length, and d must be at least equal to 1/3 = 0.33 mm, in order to achieve compatibility with respect to mode I, mode II and mode III loadings at the weld root.

All welded joints were tested under stress-relieved conditions; therefore, the nominal load ratio *R* has been taken into account by means of equation (3) and the equivalent peak stress has been calculated on each analysed node, either in the manual PSM analysis and by the automated one. As reported in table 5, both the manual and automated PSM analyses allow to correctly estimate the crack initiation site, which always occurred at the weld toe [23].

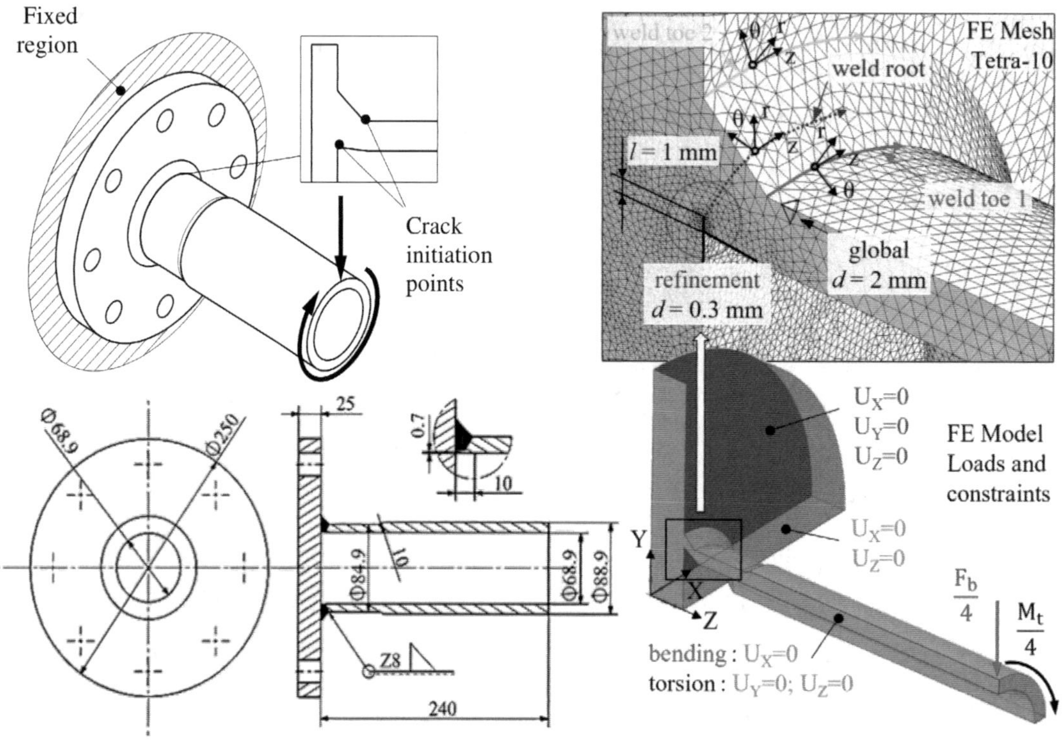

Figure 6. Tube-to-flange joint geometry by Yousefi *et al.* [23]. Loading conditions and details of the FE analyses performed according to the PSM. FE mesh generated in Ansys® Mechanical FE software.

4.2. Frendo and Bertini [29], tube-to-flange joints

The tube-to-flange fillet-welded joints under exam (see figure 7) were fatigue-tested under reversed (R = -1), as well as pulsating (R = 0), in-phase ($\phi = 0°$) and out-of-phase ($\phi = 90°$) combined bending-torsion loading. Nominal biaxiality ratios $\Lambda = \sigma_{nom}/\tau_{nom} = 0.88$ and $\Lambda = 3.25$ were adopted during the tests. The number of cycles to break-through reported in the original contribution [29] were expressed in terms of bending and torsion nominal stresses, according to Navier's formula and Bredt's formula respectively.

The geometry has been modelled as 1/4 of the entire welded joint, taking advantage of the axis-symmetry, and has been simulated by means of a 10-node tetra free-generated 3D mesh either in Ansys® Mechanical APDL and Ansys® Mechanical, as reported in figure 7. In order to guarantee PSM applicability at either weld toe and weld root with respect to mode I, mode II (at the weld root only) and mode III loadings, the mesh density requirement $a/d \geq 3$ needs to be enforced. The characteristic notch size equals the tube thickness at weld toes, namely a = 10 mm, which also corresponds to weld roots length and to weld legs length. Consequently, d must be equal to or lower than 10/3 = 3.33 mm. Finally, a 10-node tetra mesh of global element size 3 mm has been generated (see figure 7), achieving PSM applicability at both weld toes and weld roots simultaneously.

All welded joints were tested in as-welded conditions and the equivalent peak stress has been calculated on each analysed node either in the manual PSM analysis and by the automated one. As reported in table 5, both the manual and automated PSM analyses allow to correctly estimate the crack initiation site, which always occurred at the weld toe of the outer weld bead [29].

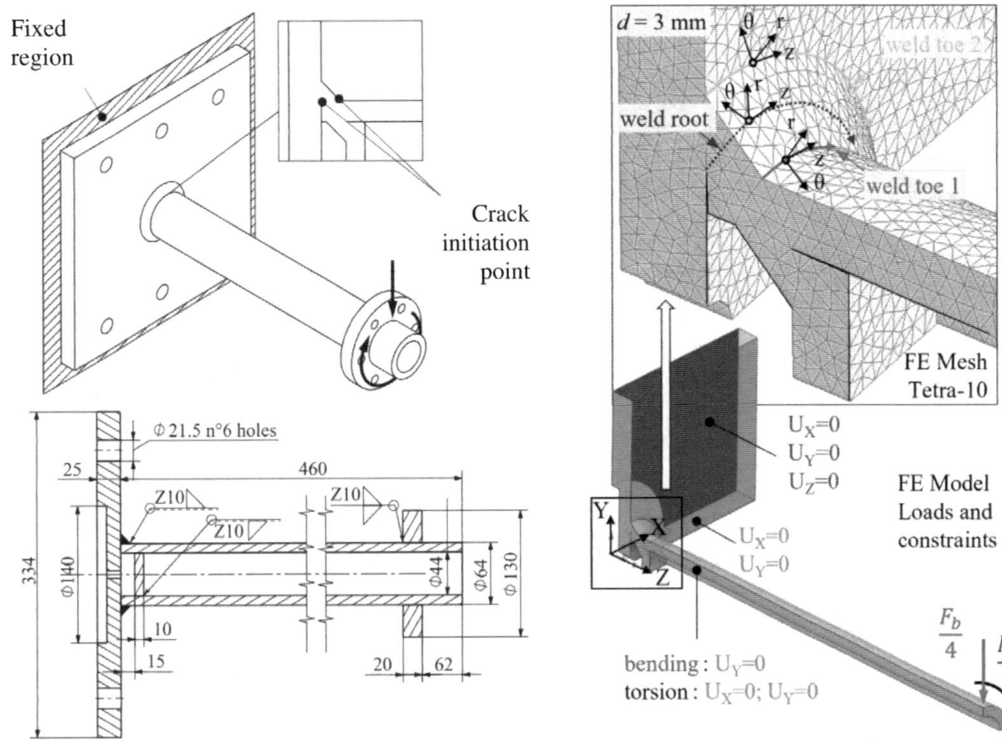

Figure 7. Tube-to-flange joint geometry by Frendo and Bertini [29]. Loading conditions and details of the FE analyses performed according to the PSM. FE mesh generated in Ansys® Mechanical FE software.

4.3. Takahashi et al. [30,31], plates with box-fillet-welded (wrap-around) joint

The geometry under exam is composed by two steel plates having a box-fillet-welded (wrap-around) joint (see figure 8). The joints were tested under pulsating biaxial fatigue loadings, either in-phase and out-of-phase, by adopting the nominal biaxiality ratios Λ reported in table 5. The number of cycles provided by the original contributions [30,31] refer to: (i) the first technically detectable crack, corresponding to a crack depth of 1÷2 mm; (ii) the number of cycles to complete failure of the welded joint, which will be considered by the re-analysis proposed in the current work.

The geometry has been modelled as 1/8 of the entire welded joint, taking advantage of the three symmetry planes characterizing the model, and has been simulated by means of a 10-node tetra free-generated 3D mesh either in Ansys® Mechanical APDL and Ansys® Mechanical, as reported in figure 8. In order to guarantee PSM applicability at either weld toe and weld root with respect to mode I, mode II (at the weld root only) and mode III loadings, the mesh density requirement $a/d \geq 3$ needs to be enforced. The characteristic notch size equals half the plates thickness at weld toes, namely a = 6 mm, which also corresponds to the weld leg length. Consequently, d must be at least equal to 6/3 = 2 mm. Finally, a 10-node tetra mesh of global element size 1 mm has been generated in order to provide a larger number of nodes for the calculations, achieving PSM applicability at both the weld toe and weld root simultaneously.

All welded joints were tested in as-welded conditions and the equivalent peak stress has been calculated on each analysed node either in the manual PSM analysis and by the automated one. Dealing with the experimental data, Takahashi *et al* [30,31] reported that fatigue failures always occurred at the weld toe with respect to the critical point A, with one exception where the fatigue crack initiated at the critical point B (see figure 8 and table 5).

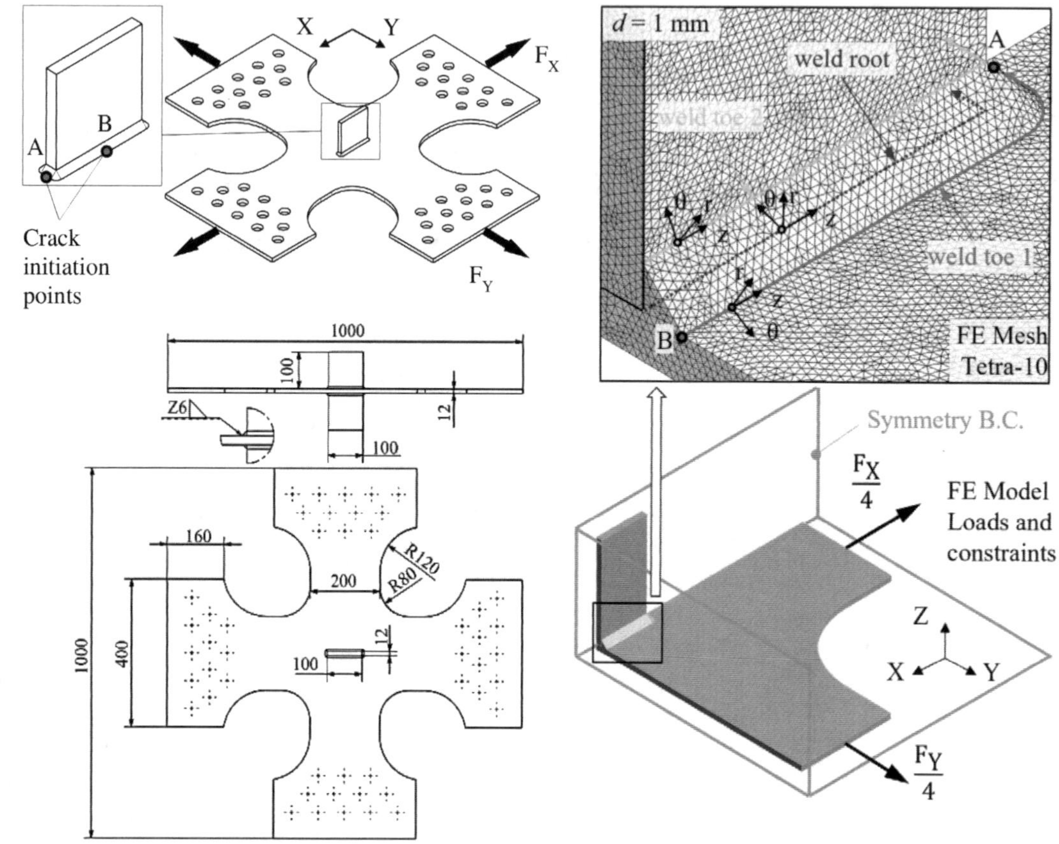

Figure 8. Plates with box-fillet-welded (wrap-around) joint geometry by Takahashi *et al.* [30,31]. Loading conditions and details of the FE analyses performed according to the PSM. FE mesh generated in Ansys® Mechanical FE software.

5. Manual and automated assessment of weld toe and weld root fatigue failures

The original fatigue results, provided in terms of nominal stress range and corresponding number of cycles to failure for each one of the considered geometries, have been re-evaluated in terms of equivalent peak stress range at the crack initiation site by means of both manual and automated PSM analyses for each geometry.

Dealing with welded joints subjected to multiaxial loading conditions, each load case has been simulated separately by means of a FE analysis, according to the approximate approach described in paragraphs 2.2-2.3. σ_{nom} = 1 MPa has been enforced for axial and bending load cases; τ_{nom} = 1 MPa has been enforced for torsional load cases. Afterwards, the equivalent peak stress range tied to each combined load case has been evaluated according to equation (9), taking account of the corresponding nominal load ratio $\Lambda = \sigma_{nom}/\tau_{nom}$ (see table 5). In order to convert the experimental data from nominal stress range to equivalent peak stress range, each nominal stress range given in the experimental fatigue tests has been multiplied by the equivalent peak stress calculated at the estimated crack initiation point from the FE model.

Table 5. Fatigue test details and PSM analysis results for the joint geometries of figures 6-8.

Reference	Load case[a]	R_σ (R_X)	R_τ (R_Y)	Phase ϕ [°]	Λ[b]	Failure location (N° failures)	$\Delta\sigma_{eq,peak}$ [MPa][c] Manual PSM	Automated PSM	Nominal stresses $\Delta\sigma_{nom}$ [MPa]	$\Delta\tau_{nom}$ [MPa]
Yousefi *et al.* [23]	B	-1	-	-	∞	Toe 1 (16)	1.703	1.771	207÷825	-
	B	0	-	-	∞	Toe 1 (9)	2.408	2.505	187÷261	-
	T	-	-1	-	0	Toe 1 (9)	2.019	2.121	-	242÷401
	T	-	0	-	0	Toe 1 (9)	2.855	2.999	-	209÷304
	B+T	-1	-1	0	1	Toe 1 (7)	2.644	2.744	145÷312	145÷312
	B+T	0	0	0	1	Toe 1 (7)	3.740	3.880	144÷358	144÷358
	B+T	-1	-1	90	1	Toe 1 (8)	2.644	2.744	140÷220	140÷220
	B+T	0	0	90	1	Toe 1 (9)	3.740	3.880	145÷312	145÷312
Frendo and Bertini [29]	B+T	-1	-1	90	3.25	Toe 1 (5)	16.028	16.724	98÷187	30÷57
	B+T	0	0	90	3.25	Toe 1 (5)	16.028	16.724	78÷178	24÷55
	B+T	-1	-1	90	0.88	Root (5)	6.242	6.282	56÷100	64÷114
	B+T	0	0	90	0.88	Root (5)	6.242	6.282	50÷90	57÷103
Takahashi *et al.* [30,31]	A	0	0	180	1.02	Toe 1 A (5)	2.994	2.916	79÷132	78÷130
	A	0	0	180	0.85	Toe 1 A (1)	2.549	2.482	111	130
	A	0	0	180	0.68	Toe 1 A (1)	2.116	2.061	87	130
	A	0	0	180	0.34	Toe 1 B (1)	1.599	1.695	45	131
	A	0	∞	0	1.52	Toe 1 A (1)	4.341	4.227	130	-85
	A	0	∞	0	1.25	Toe 1 A (1)	3.611	3.517	106	-84
	A	0	∞	0	1.02	Toe 1 A (1)	2.998	2.920	87	-84
	A	0	∞	0	0.85	Toe 1 A (1)	2.554	2.488	73	-85

[a] A = axial, B = bending, T = torsion
[b] Nominal biaxiality ratio $\Lambda = \sigma_{nom}/\tau_{nom}$
[c] Calculated with $\Delta\sigma_{nom}$ = 1 MPa and/or $\Delta\tau_{nom}$ = 1 MPa, Λ as indicated. Reported values are at the crack initiation point.

Next figures 10-11 show good agreement between the experimental fatigue results and the fatigue life estimations provided by the PSM, dealing with fatigue failures at the weld toe and weld root of the considered joints subjected to either uniaxial and multiaxial loading conditions. Moreover, figures 10-11 allow to compare the results coming from the manual PSM application with the ones generated by the automated PSM analysis. It has been noted that percent variations between the equivalent peak stress values obtained from the two approaches are always below 10% and account for: (i) the different mesh patterns generated by the two FE programs, i.e. Ansys® Mechanical APDL and Ansys® Mechanical, given the joint geometry; (ii) the different procedures adopted to define the local coordinate systems and extrapolate the relevant peak stresses on the analysed nodes. The automated algorithm performs indeed a numerical estimation of the notch opening angle 2α at each peak node and refers to nodal coordinates in order to define local coordinate systems and retrieve the peak stresses. Therefore, the automated procedure is affected by a degree of approximation which is tied to FE mesh density.

Frendo and Bertini [29]; load case **M**

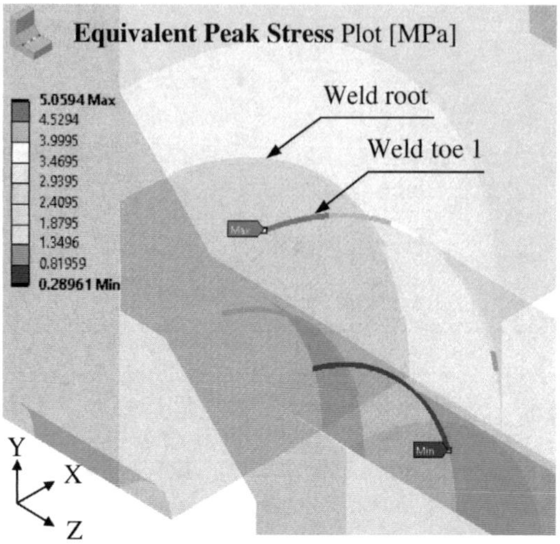

Frendo and Bertini [29]; load case $\mathbf{M_{ZZ}}$ (torsion)

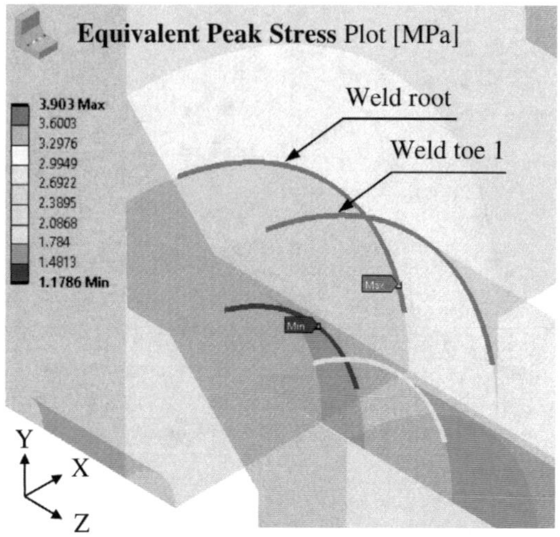

Takahashi *et al.* [30,31]; load case $\mathbf{F_X}$

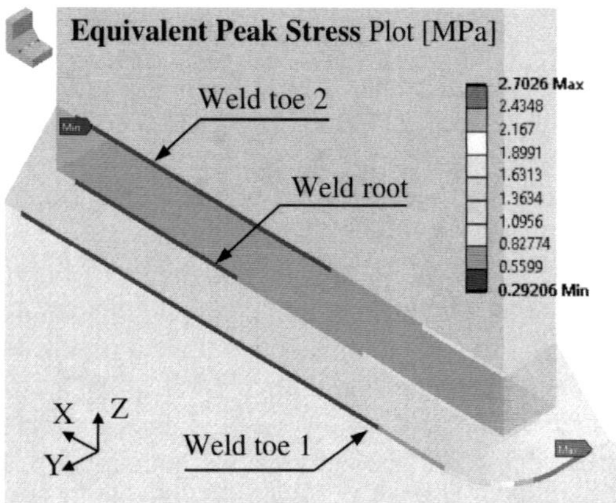

Takahashi *et al.* [30,31]; load case $\mathbf{F_Y}$

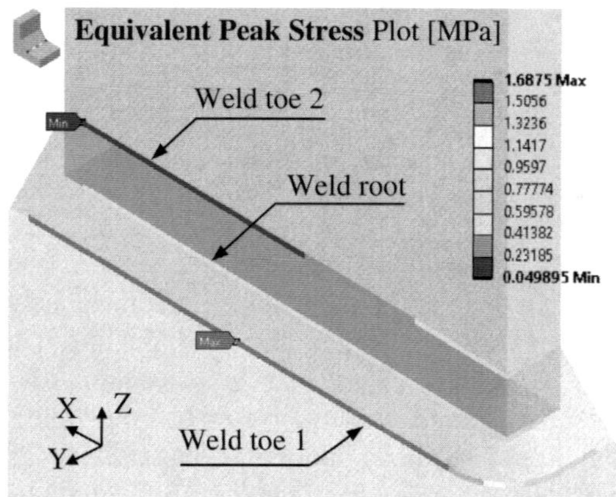

Figure 9. Examples of equivalent peak stress contour plots generated by the developed PSM tool in Ansys® Mechanical for the welded joint geometries tested in [29] and [30,31] (see figures 7-8) and simulated for each separate load case, according to the approach described in paragraphs 2.2-2.3.

6. Discussion and conclusions

The Peak Stress Method (PSM) grants the possibility to assess fatigue failures in welded joints subjected to either uniaxial and multiaxial, in-phase and out-of-phase loading conditions. In compliance with a proper PSM-based design curve, the equivalent peak stress provides a reliable design quantity able to single out the fatigue failure location in presence of multiple competing crack initiation sites (see also figure 9).

In this paper, three of the nine welded joint geometries proposed in table 3 have been considered for a deeper insight (see figures 6-8); however, the complete re-analysis work [20] considered all joint geometries proposed in table 3, for a total of 306 re-analysed experimental data.

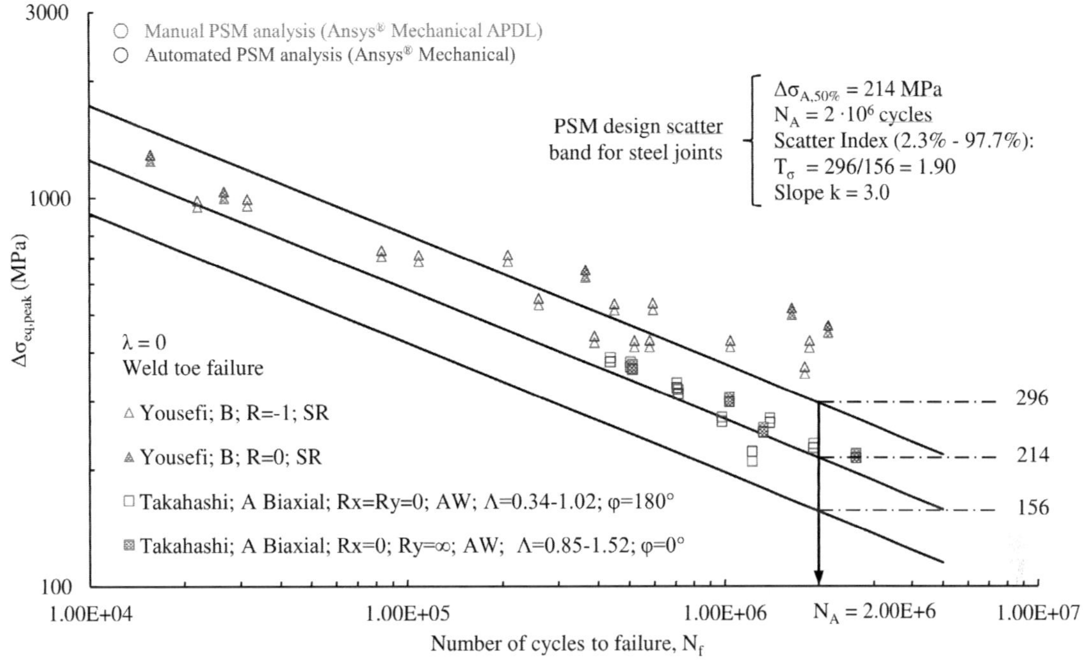

Figure 10. Fatigue assessment of weld toe failures according to the PSM. Comparison between the fatigue design scatter band ($\lambda = 0$) proposed in [10] and experimental fatigue data by [23] and [30,31]. A = axial load; B = bending load; T = torsion load; AW = as-welded; SR = stress-relieved.

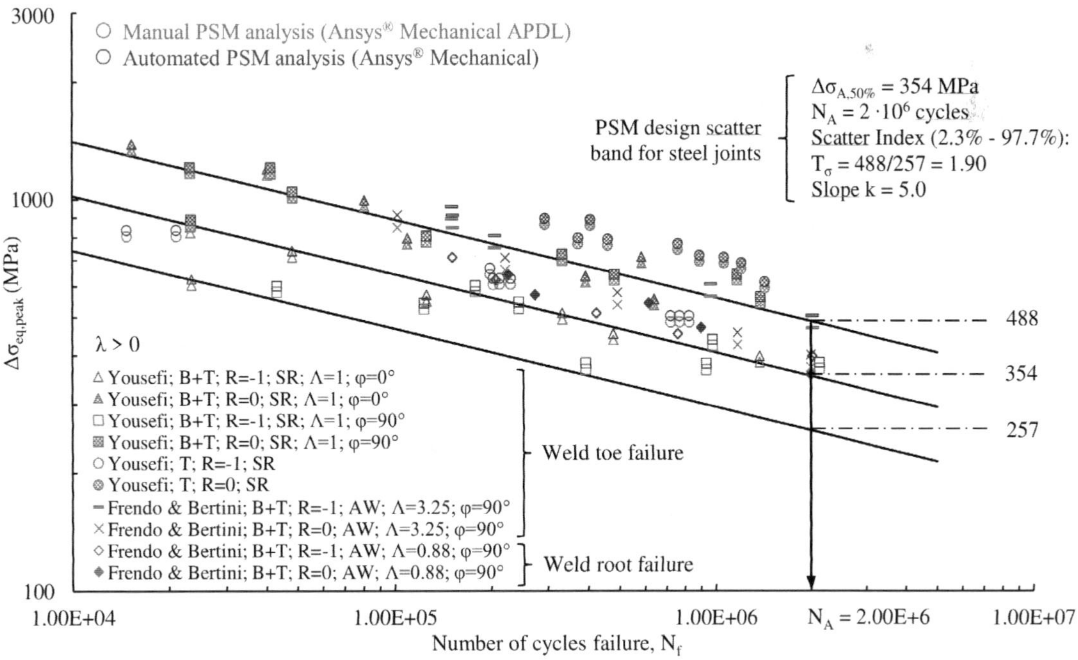

Figure 11. Fatigue assessment of weld toe and weld root failures according to the PSM. Comparison between the fatigue design scatter band ($\lambda > 0$) proposed in [10] and experimental fatigue data by [23] and [29]. B = bending load; T = torsion load; AW = as-welded; SR = stress-relieved.

The PSM allowed to successfully assess the fatigue crack initiation site in approximately 270 fatigue failure cases occurring at the weld toe or weld root, i.e. ~90% of the considered cases. Moreover, about 180 of the total 306 experimental data, i.e. ~60%, fall within the 2.3%-97.7% PSM-based design scatter bands, proving a good agreement with the proposed fatigue design curves, while the remaining data, i.e. ~40%, are on the safe side. Finally, the possibility to take advantage of the new tool developed in Ansys® Mechanical to automate PSM analyses on generic welded joint geometries allowed to achieve a remarkable reduction in calculation effort and time, with respect to the ones required to manually apply the method and elaborate the results of the FE analyses. More in detail, table 6 shares some estimations on "manual versus automated" analysis time, required to apply the PSM on each competing weld toe and weld root characterizing the geometries proposed in figures 6-8. Thanks to the coarse meshes required by the PSM to perform compatible FE analyses and the full-automated implementation in Ansys® Mechanical, the newly developed PSM application might reveal itself a useful tool in every-day design practice when dealing with the fatigue assessment of welded joints and structures.

Table 6. Analysis time comparison between manual and automated PSM approach.

Reference	Analysed nodes	**Manual** PSM analysis[a] estimated time [s]	**Automated** PSM analysis[b] estimated time [s]
Yousefi *et al.* [23]	226	~ 9000 (~ 2 h 30 minutes)	~ 40 seconds[c]
Frendo and Bertini [29]	62	~ 2400 (~ 40 minutes)	~ 20 seconds
Takahashi *et al.* [30,31]	124	~ 3600 (~ 60 minutes)	~ 40 seconds

All time estimations are given for a single PSM analysis on all weld toes and weld roots of the geometry
[a] Time required to: (i) define and orientate local coordinate systems; (ii) retrieve relevant peak stresses
[b] Time required to perform a complete automated PSM analysis, up to fatigue life estimation
[c] Separate analyses required at the weld toe and weld root due to the adopted element sizes (see paragraph 4.1)

References

[1] Lazzarin P and Tovo R 1998 A notch intensity factor approach to the stress analysis of welds *Fatigue Fract. Eng. Mater. Struct.* **21** 1089–103
[2] Radaj D, Sonsino C M and Fricke W 2006 *Fatigue assessment of welded joints by local approaches* (Woodhead Publishing Limited)
[3] Lazzarin P, Sonsino C M and Zambardi R 2004 A notch stress intensity approach to assess the multiaxial fatigue strength of welded tube-to-flange joints subjected to combined loadings *Fatigue Fract. Eng. Mater. Struct.* **27** 127–40
[4] Williams M L 1952 Stress Singularities Resulting From Various Boundary Conditions in Angular Corners of Plates in Extension *J. Appl. Mech.* **19** 526–8
[5] Qian J and Hasebe N 1997 Property of eigenvalues and eigenfunctions for an interface V-notch in antiplane elasticity *Eng. Fract. Mech.* **56** 729–34
[6] Lazzarin P and Zambardi R 2001 A finite-volume-energy based approach to predict the static and fatigue behavior of components with sharp V-shaped notches *Int. J. Fract.* **112** 275–98
[7] Livieri P and Lazzarin P 2005 Fatigue strength of steel and aluminium welded joints based on generalised stress intensity factors and local strain energy values *Int. J. Fract.* **133** 247–76
[8] Lazzarin P, Livieri P, Berto F and Zappalorto M 2008 Local strain energy density and fatigue strength of welded joints under uniaxial and multiaxial loading *Eng. Fract. Mech.* **75** 1875–89
[9] Hobbacher A F 2016 *Recommendations for Fatigue Design of Welded Joints and Components* (Cham: Springer International Publishing)
[10] Meneghetti G and Campagnolo A 2020 State-of-the-art review of peak stress method for fatigue strength assessment of welded joints *Int. J. Fatigue* **139** 105705
[11] Meneghetti G and Lazzarin P 2007 Significance of the elastic peak stress evaluated by FE analyses at the point of singularity of sharp V-notched components *Fatigue Fract. Eng. Mater. Struct.* **30** 95–106

[12] Meneghetti G 2012 The use of peak stresses for fatigue strength assessments of welded lap joints and cover plates with toe and root failures *Eng. Fract. Mech.* **89** 40–51
[13] Meneghetti G 2013 The peak stress method for fatigue strength assessment of tube-to-flange welded joints under torsion loading *Weld. World* **57** 265–75
[14] Campagnolo A and Meneghetti G 2018 Rapid estimation of notch stress intensity factors in 3D large-scale welded structures using the peak stress method ed G Hénaff *MATEC Web Conf.* **165** 17004
[15] Campagnolo A, Roveda I and Meneghetti G 2019 The Peak Stress Method combined with 3D finite element models to assess the fatigue strength of complex welded structures *Procedia Struct. Integr.* **19** 617–26
[16] Meneghetti G, Campagnolo A and Berto F 2015 Fatigue strength assessment of partial and full-penetration steel and aluminium butt-welded joints according to the peak stress method *Fatigue Fract. Eng. Mater. Struct.* **38** 1419–31
[17] Meneghetti G, Campagnolo A and Visentin A 2020 Automated fatigue strength assessment of arc-welded structures according to the Peak Stress Method *Procedia Struct. Integr.* **28** 1062–83
[18] Meneghetti G, Campagnolo A and Rigon D 2017 Multiaxial fatigue strength assessment of welded joints using the Peak Stress Method – Part I: Approach and application to aluminium joints *Int. J. Fatigue* **101** 328–42
[19] Visentin A 2020 Automated Fatigue Strength Assessment of V - Notch Structures according to the Peak Stress Method (Department of Industrial Engineering, University of Padova)
[20] Sacchet F 2021 Applicazione del metodo della tensione di picco alla previsione di vita a fatica di giunti saldati soggetti a stati tensionali multi-assiali (Department of Industrial Engineering, University of Padova)
[21] Meneghetti G, Campagnolo A and Rigon D 2017 Multiaxial fatigue strength assessment of welded joints using the Peak Stress Method – Part II: Application to structural steel joints *Int. J. Fatigue* **101** 343–62
[22] Razmjoo G 1996 *Fatigue of load-carrying fillet welded joints under multiaxial loadings* ed Abington (Cambridge, UK: TWI REF. 7309.02/96/909)
[23] Yousefi F, Witt M and Zenner H 2001 Fatigue strength of welded joints under multiaxial loading: experiments and calculations *Fatigue Fract. Eng. Mater. Struct.* **24** 339–55
[24] Bäckström M, Siljander A, Kuitunen R and Ilvonen R 1997 Multiaxial Fatigue Experiment of Square Hollow Section Tube-to-Plate Welded Joints *Proceedings of the First North European Engineering and Science Conference* ed A F Blom (EMAS Ltd.) pp 167–77
[25] Dahle T, Olsson K-E, Jonsson B, Bäckström M, Siljander A and Kuitunen R 1997 Multiaxial fatigue experiments of square hollow section tube-to plate welded joints *Proceedings of the First North European Engineering and Science Conference* ed A F Blom (EMAS Ltd.) pp 163–77
[26] Sonsino C 1995 Multiaxial fatigue of welded joints under in-phase and out-of-phase local strains and stresses *Int. J. Fatigue* **17** 55–70
[27] Sonsino C 2004 Assessment of multiaxial fatigue behaviour of welded joints under combined bending and torsion by application of a fictitious notch radius *Int. J. Fatigue* **26** 265–79
[28] Bertini L, Cera A and Frendo F 2014 Experimental investigation of the fatigue resistance of pipe-to-plate welded connections under bending, torsion and mixed mode loading *Int. J. Fatigue* **68** 178–85
[29] Frendo F and Bertini L 2015 Fatigue resistance of pipe-to-plate welded joint under in-phase and out-of-phase combined bending and torsion *Int. J. Fatigue* **79** 46–53
[30] Takahashi I, Ushijima M, Takada A, Akiyama S and Maenaka H 1999 Fatigue behaviour of a box-welded joint under biaxial cyclic loads *Fatigue Fract. Eng. Mater. Struct.* **22** 869–77
[31] Takahashi I, Takada A, Ushijima M and Akiyama S 2003 Fatigue behaviour of a box-welded joint under biaxial cyclic loading: effects of biaxial load range ratio and cyclic compressive loads in the lateral direction *Fatigue Fract. Eng. Mater. Struct.* **26** 439–48

A semi-analytical method for the calculation of double-ellipsoidal heat source parameters in welding simulation

S Trupiano[1], V G Belardi[1], P Fanelli[2], L Gaetani[3] and F Vivio[1]

[1]Department of Enterprise Engineering, University of Rome Tor Vergata, Via del Politecnico, 1, Rome, 00133, Italy.
[2]Department of Economics, Engineering, Society and Business Organization (DEIM), University of Tuscia, Largo dell'Università, Viterbo, 01100, Italy.
[3]Eleo2 Engineering S.r.l., Via Benedetto Stay, 69, 00143 Rome, Italy.

E-mail: simone.trupiano@uniroma2.it

Abstract. Structural integrity assessment of welded components requires the determination of residual stress and deformations induced by the welding process. The implementation of both a numerical static mechanical analysis and a thermal transient analysis is needed in order to accurately determine stress and strain fields. The most faithful representation of the heat flow provided by the welding torch in thermal simulations is realized with Goldak's double ellipsoid, especially in case of MIG/MAG and combined TIG-MIG/MAG welding. The correct determination of the temperature trends during the welding process simulation is essential for the subsequent mechanical FEA. In order to obtain the above-mentioned temperature trends, some parameters must be determined ex ante, e.g., the geometrical characteristics of Goldak's double ellipsoid. Generally, a large number of trial simulations must be performed in order to define these parameters. However, manual fitting is often complicated. To overcome such difficulties, in this work, it is reported a combined analytical and numerical method to determine the parameters using the genetic algorithm NSGA-II. First-attempt parameters are obtained through the solution of the Fachinotti's analytical formulation. This first stage requires a restricted set of experimental data, such as weld pool extension and one temperature trend in the proximity of the seam. Afterwards, first-attempt parameters and experimental data are used to determine the final parameters by means of a set of numerical thermal simulations guided through the genetic algorithm optimization. To confirm the good-functioning of the method, a numerical analysis for the simulation of the experimental data and another FE simulation performed with the final results of the method are presented.

1. Introduction

Following the welding process, residual stresses and distorsions generally occur in the joint. Residual stress can be added to those acting on the joint during the normal operation, resulting in unforeseen stress conditions. In addition, in structures fatigue cracks often occur in welded joints. The correct determination of residual stresses is therefore essential for static and fatigue assessments. The implementation of advanced structural simulations of welded joints requires adequate thermal simulations. These must be able to faithfully replicate the temperature trends that occur in the welding process. Rate temperature changes have a significant influence on the metallurgical phase transformations that have a pivotal role in thermal dilatation and in material behaviour, especially for ferritic steels [1, 2, 3]. Goldak's double ellipsoid [4] is the state-of-art model for thermal simulations of a moving source of heat flux. The TIG and MIG/MAG welding can be simulated with outstanding results using the double ellipsoid modelling. This model has been used for the simulation of many types of welded joints, such as butt-welded plates, girth joints, fillet and cruciform welds.

The calibration of Goldak's model is achieved by varying the parameters of the double ellipsoid in the transient FE simulations. Correct calibration is obtained by replicating the the weld pool extension of the joint and the temperature trends provided by the thermocouples in the experimental tests. This procedure is rarely straightforward and is generally carried out manually by trial and error. In the case of multi-pass welds, it is difficult to determine the parameters. Welding parameters such as voltage, current and welding speed can vary with each pass, requiring as many calibrations as passes [5]. Manual calibration is challenging even in the case of special weld pool shapes, that can be achieved by combining several heat flux models.

This work has been carried out in order to facilitate the implementation of numerical thermal simulations by eliminating the manual calibration of the heat flux. The method outlined below employs the genetic algorithm to guide the process of Goldak's parameter determination through 2D numerical models. When parametrically configured these models allow many automated simulations to be performed with a low computational burden. These simulations can be used with negligible errors when conduction in the longitudinal direction can be neglected. Such modelling may lead to increasing errors and deviations from the experimental results, where longitudinal conduction becomes important, due to the material properties or low welding speed.

An analytical approach is employed before the parameter definition by using the genetic algorithm and 2D parametric simulations. This preliminary approach enables to limit the population of 2D analyses to be performed reducing the range of research of the heat flux parameters.

2. Description of the semi-analytical method

Goldak's double-ellipsoidal heat source is defined by nine parameters [4], as shown in Eq. (1) for the front ellipsoid and in Eq. (2) for the rear ellipsoid. Generally, welding process parameters are known a priori. Welding Procedure Specification (WPS) collects data including: Voltage Vo, current I and welding speed v. The efficiency η depends on the welding process used. Geometrically, the double ellipsoid is defined by four parameters: a, b, c_f and c_r. As shown in Figure 1, a parameter is the width semi-axis, b is the depth of the double-ellipsoid while c_f and c_r parameters are the front and the rear semi-axis of the ellipsoid. The last two factors represent the fraction of the total heat deposited in the front and rear ellipsoid. These factors are related to c_f and c_r geometrical parameters through the Eqs. 3 and 4. In these relations, it is also present the α parameter that is equal to two to guarantee continuity of the q functions at $z = 0$ (Figure 1) [6].

$$q_f(x, y, z, t) = \frac{6\sqrt{3}\, f_f\, \eta\, Vo\, I}{\pi\, \sqrt{\pi}\, a\, b\, c_f}\, e^{-3\,(\frac{x^2}{a^2}+\frac{y^2}{b^2}+\frac{(z-v\,t)^2}{c_f^2})} \tag{1}$$

$$q_r(x, y, z, t) = \frac{6\sqrt{3}\, f_r\, \eta\, Vo\, I}{\pi\, \sqrt{\pi}\, a\, b\, c_r}\, e^{-3\,(\frac{x^2}{a^2}+\frac{y^2}{b^2}+\frac{(z-v\,t)^2}{c_r^2})} \tag{2}$$

$$f_f = \frac{\alpha}{1+\frac{c_r}{c_f}} \tag{3}$$

$$f_r = \frac{\alpha}{1+\frac{c_f}{c_r}} \tag{4}$$

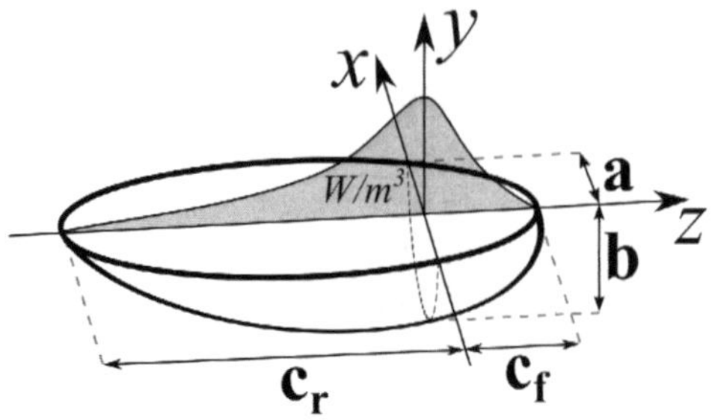

Figure 1. Double-ellipsoidal heat source shape and relative geometrical parameters.

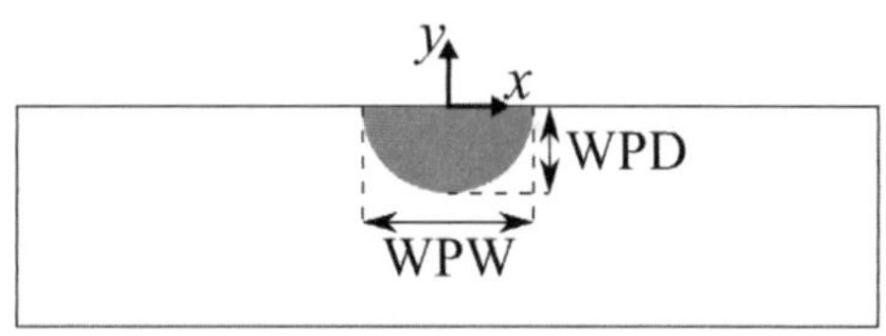

Figure 2. Depth (WPD) and width (WPW) of the molten weld pool.

Ultimately, the geometrical parameters are unknown in the simulation of a weld using the double ellipsoid schematisation of the heat flux (a, b, c_f and c_r). Often these parameters can be related to each other, as reported in [7, 8], with $c_r = 4c_f$ and $c_f = b$. In this way, the only unknowns to determine are a and b.

The method presented in this work makes it possible to determine the two unknown geometric parameters by knowing WPS parameters, weld pool extension and peak temperature T_{peak} reached in the welded joint in a point as close as possible to the seam. The extension of the molten pool is characterised by measuring the depth of penetration (WPD) and the width (WPW) of the weld, as shown in Figure 2.

The method presented is based on the use of the NSGA-II genetic algorithm [9] applied to a 2D parametric transient thermal model. In order to determine the geometric parameters, these factors are varied by the NSGA-II algorithm. For each value pair, a and b, a 2D analysis is performed and the results in terms of T_{peak} and weld pool extension (WPW, WPD) are analysed. Subsequent analyses are guided by the genetic algorithm according to the results and their deviation from the experimental data. At the end of the analyses, it is possible to determine the optimal parameters a and b by imposing a criterion for choosing between those solutions satisfying the above constraints, i.e. on Pareto front.

Through the use of Fachinotti's analytical method [6], have been obtained the starting values of the geometric variables to be set in the genetic algorithm. This analytical model solves the thermal field induced by a moving heat source in a semi-infinite body. However, this method neglects convective and radiative heat exchanges. Furthermore the thermal properties of the material are considered constant as the temperature changes. In any case, the preliminary determination of the geometrical parameters of the double ellipsoid makes it possible to reduce the optimisation burden in the subsequent phase by limiting the range of research of the a and b semi-axis. A MATLAB [10] code has been developed to determine the best set of geometric starting values solving the Fachinotti's method. In input, only the WPS data, the weld pool extension and the T_{peak} in a point near the seam are needed, even at this early stage.

3. Reference case study

In order to test the novel semi-analytical method presented in this work, a numerical reference case study has been realised. The reference case (REF) is the thermal and structural simulation of a bead-on-plate weld. Geometry and position of the seam and the dimensions of the plate are reported in Figure 3. The plate is 10 mm thick.

Transient thermal analysis of the bead on plate welding has been realized with 3D linear brick elements with 8 nodes. In order to streamline the simulations, only half the plate was simulated using symmetry. In the thermally altered zone the mesh is very dense (Figure 4) to better handle the high thermal gradients and to adequately fit Goldak's double ellipsoid equation. The elements in the center line of the seam have a side size of 0.5 mm, and the whole model is composed of 140,400 elements and 155,952 nodes. Convection and radiation are imposed on all external faces of the plate, except the symmetry face, that is considered adiabatic. The convective exchange coefficient is 15W $/m^{2\circ}C$ and the emissivity coefficient is 0.8, while the ambient temperature, which coincides with the initial temperature of the plate, has been set to 20°C. Table 1 shows the temperature-dependent thermal and structural properties of 304 stainless steel [11]. The solidus temperature of the 304 steel is 1,340°C while the liquidus temperature is 1,390°C. The welding parameters are resumed in Table. In the Table 2 are reported the double ellipsoid parameters (a, b, c_f and c_r), that have been chosen in order to obtain thermal results with physical significance, using typical values that can be found in literature for this type of welding and material. In the same Table are shown: the speed of welding v, the voltage Vo, the current I and the efficiency η.

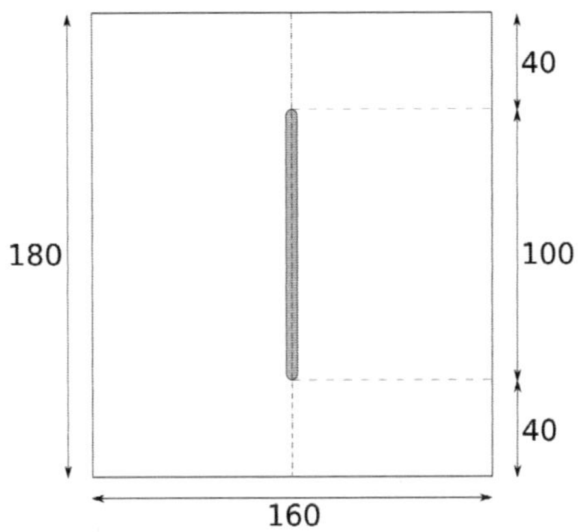

Figure 3. Bead on plate geometry and dimensions.

Table 1. Material properties of stainless steel [11].

T [°C]	c_p [J/(kg°C)]	λ [W/(m°C)]	E [GPa]	ν [−]	σ_y [MPa]	E_H [−]	α [10^{-6}/°C]
20	442	15.0	200	0.278	230	2.80	19
200	515	17.5	185	0.288	184	2.59	19
400	563	20.0	170	0.298	132	2.38	19
600	581	22.5	153	0.313	105	2.14	19
800	609	25.5	135	0.327	77	1.89	19
1000	631	28.3	96	0.342	50	0.01	19
1200	654	31.1	50	0.350	10	0.005	19
1340	669	33.1	10	0.351	10	0.001	19
1390	675	66.2	10	0.353	10	0.001	19
2000	675	66.2	10	0.357	10	0.001	19

IOP Conf. Series: Materials Science and Engineering 1214 (2022) 012023 doi:10.1088/1757-899X/1214/1/012023

Figure 4. Reference 3D FE model mesh.

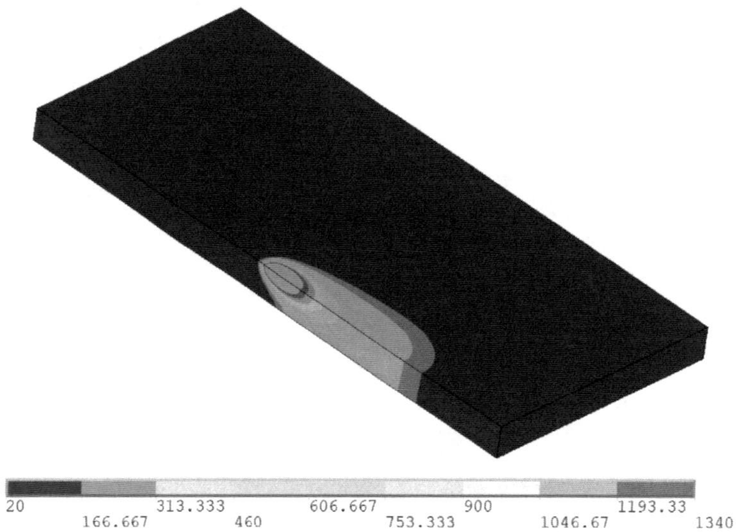

Figure 5. Transient thermal simulation of the reference FE model.

Simulation time step has been imposed very short, 0.05 seconds, to best simulate the movement of the welding torch. The reference model during the transient thermal simulation is shown in Figure 5. The grey zone in the figure represents the molten weld pool. The resulting extension of the pool and the peak temperature reached in the center line of the welding are reported in Table 3. These values represent the experimental parameters that has been used to test the method in object.

IOP Conf. Series: Materials Science and Engineering 1214 (2022) 012023 doi:10.1088/1757-899X/1214/1/012023

Table 2. Welding and double-ellipsoid parameters of the reference model (REF).

a	b	c_f	c_r	v	Vo	I	η
[mm]	[mm]	[mm]	[mm]	[mm/min]	[V]	[A]	[–]
5.9	2.8	2.8	11.2	300	18	200	0.8

Table 3. Extension of the welding pool and peak temperature of REF model.

WPW	WPD	T_{peak}
[mm]	[mm]	[°C]
8.92	3.44	2110

4. Analytical determination of the preliminary double ellipsoid parameters

As stated previously, the first step of the semi-analytical method is the determination by a MATLAB code of the starting value of the geometrical parameters a and b. This code solve numerically the Fachinotti's integral (Eq. 5), that does not have a closed-form solution. The index i in the Eqs. 6 and 7 can be referred to front f or rear r.

The solution of the integral is implemented on a numerical mesh that has a limited extension around the centre line of the seam to limit the calculation time.

$$T(x,y,z,t) = T_0 + \frac{3\sqrt{3}}{\pi\sqrt{\pi}}\frac{\eta\, Vo\, I}{\rho\, c_p}\int_0^t \frac{exp[-3\frac{x^2}{12\kappa(t-t')+a^2} - 3\frac{y^2}{12\kappa(t-t')+b^2}]}{\sqrt{12\kappa(t-t')+a^2}\sqrt{12\kappa(t-t')+b^2}} \quad (5)$$
$$[f_r\, A_r\,(1-B_r) + f_f\, A_f\,(1+B_f)]\, dt'$$

$$A_i = \frac{exp\left[-3\frac{(z-vt')^2}{12\kappa(t-t')+c_i^2}\right]}{\sqrt{12\kappa(t-t')+c_i^2}} \quad (6)$$

$$B_i = erf\left[\frac{c_i}{2}\frac{z-vt'}{\sqrt{\kappa(t-t')}\sqrt{12\kappa(t-t')+c_i^2}}\right] \quad (7)$$

Implemented code solves the integral over the domain, by discretely varying the parameters a and b. For each set of a and b parameters, the peak temperature T_{peak} and the $T_{liquidus}$ isotherm are calculated. These isotherms allow the determination of the WPD and WPW. Finally, these values are compared with the experimental ones. The pair of parameters a and b that allows the smallest deviation of peak temperature T_{peak} and weld pool extension from the experimental data are automatically selected and become the starting parameters for the next step. The maximum and minimum values of the parameters a and b used in the calculation code can be related to the experimental dimensions of the molten pool. In this application: a_{min} = 0.1 WPW, b_{min} = 0.1 WPD, a_{max} = 1.2 WPW, b_{max} = 1.2 WPD.

The preliminary parameters obtained through the analytical method above described are given in the Table 4.

5. Final geometric parameters determination using the NSGA-II algorithm with 2D parametric analysis

The last phase of geometric parameters determination employs a 2D parametric model to perform transient thermal analyses guided by the NSGA-II genetic algorithm. The 2D model

Table 4. Preliminary geometric double ellipsoid parameters.

pa [mm]	pb [mm]	pc_f [mm]	pc_r [mm]
9	5	5	20

thermally simulates the welding process by testing several sets of the a and b parameters. The FE model simulates only half plate by using the symmetry with quadrilateral 8-noded quadratic elements. The model is composed of 800 elements and 2,521 nodes, convection and radiation have been imposed in all external lines excluding the symmetry line, which was considered adiabatic. The mesh of the 2D model is shown in Figure 6. The area near the bead has square elements with 0.5 mm sides. Preliminary parameters enable to set the mean values of the variables ($a_{mean} = pa$ and $b_{mean} = pb$) and the relative ranges of variation in the NSGA-II algorithm ($a_{max,min} = \pm 60\%\, pa$ and $b_{max,min} = \pm 60\%\, pb$).

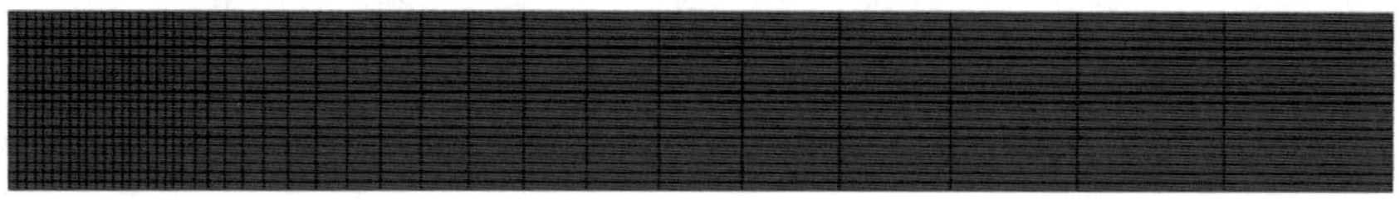

Figure 6. Mesh of the 2D parametric model.

The genetic algorithm through 200 2D simulations varies the values of a and b by comparing the results in terms of T_{peak} and melting pool extension (grey zone in Figure 7) with the reference ones. During the simulations, this comparison determines the new pairs of geometric parameters to be tested in the following one. The solution group is identified by defining the deviations

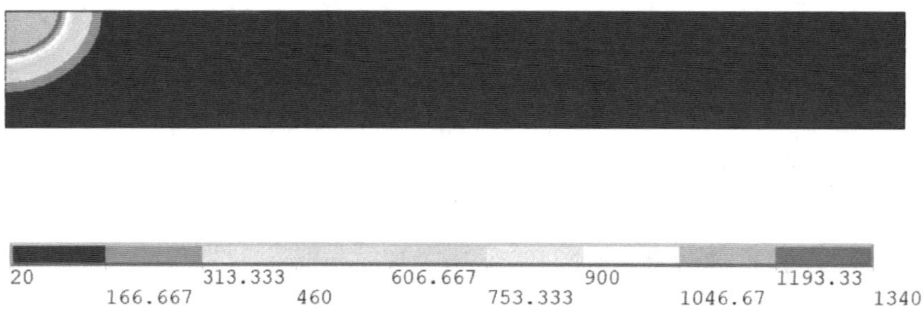

Figure 7. Temperature contour of the 2D parametric model.

deemed acceptable between the simulated and experimental results (in this case the REF model results). By means of a selection criterion, the best solution will be identified within the group to which the final pair of geometric parameters a and b is associated. In this application, the solution that minimised the sum of the deviations of the T_{peak}, WPW and WPD results was chosen, i.e. a linear choice method was set up with an equal weight function for the three results.

The final Goldak's parameters, deduced through the method presented, are reported in Table 5.

Table 5. Final Goldak's parameters.

a [mm]	b [mm]	c_f [mm]	c_r [mm]
6	3.1	3.1	12.4

6. Final thermal and structural results

In this section, the Goldak's parameters obtained by the method in object have been used to perform a 3D thermal simulation, in order to replicate the REF welding process. The results of the structural simulation employing the thermal simulation temperatures as input are also reported. The thermal and structural results of the model using the calculated double ellipsoid geometric parameters will be referred to as FEM. These results will be compared with those of the REF model, which is used as an experimental reference. The 3D FEM model uses the same boundary conditions as the REF case in terms of ambient temperatures, convection and radiation.

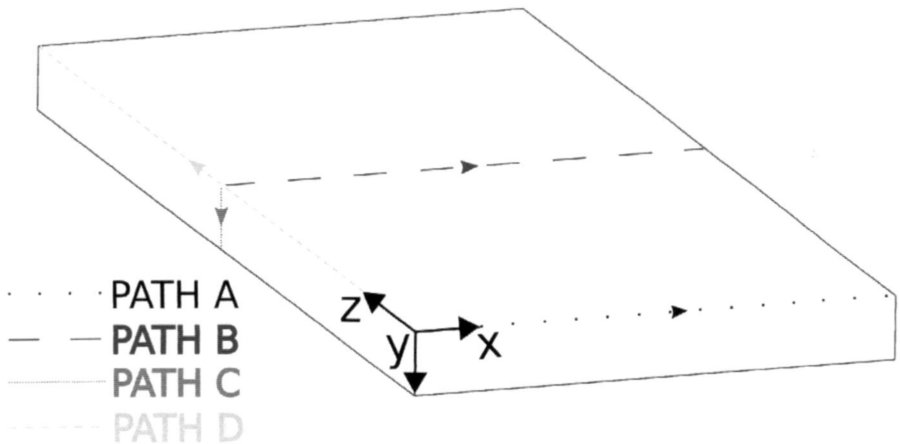

Figure 8. Cartesian axes and structural result paths A, B, C, D.

Table 6. Weld pool comparison.

		WPW	WPD
REF	[mm]	8.92	3.44
FEM	[mm]	8.39	3.39
$\Delta_{FEM-REF}$	[%]	−5.94	−1.60

The weld pool extension results and the comparison with the REF ones are reported in Table 6. WPW and WPD resulting from the thermal simulation with the final Goldak's parameters are in good agreement with the reference ones.

In Figure 8, it is showed the origin of Cartesian axes used to present the thermal results. The temperature trends are reported in Figure 9, evaluated at the center line of the seam, at middle length of plate, for both REF and FEM models. In this figure the experimental and the FEM T_{peak} values can be observed. The latter has a percent deviation of -5.50% from the REF ones. In Figures 10 and 11 are reported the temperature trends at 5 mm and 10 mm from the the center line of the seam at middle length of plate. The results are in accord with the REF ones showing a good functioning of the presented method.

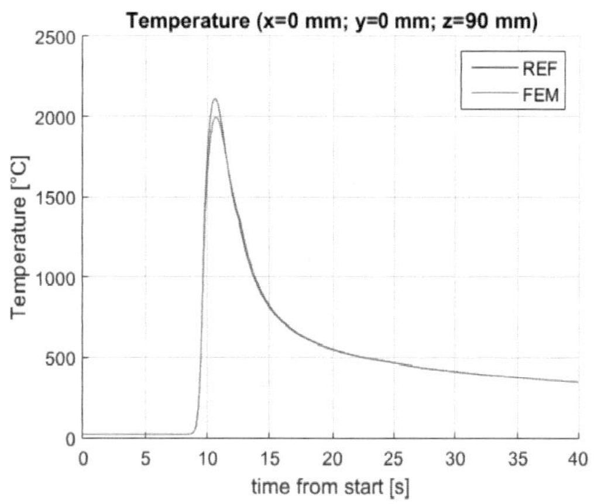

Figure 9. Temperature trends in the center line of the seam.

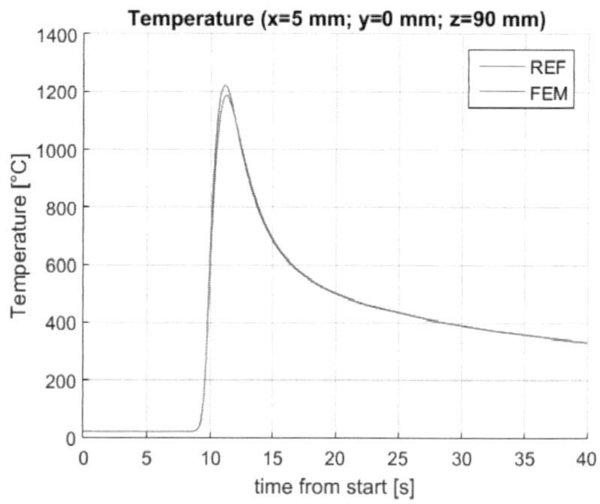

Figure 10. Temperature trends at $x = 5$ mm, $y = 0$ mm, $z = 90$ mm.

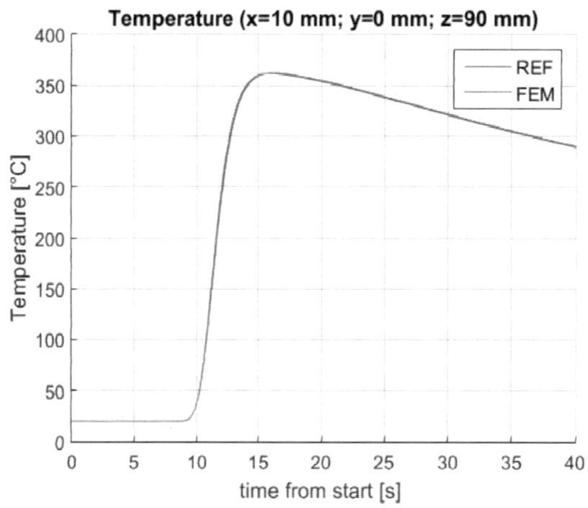

Figure 11. Temperature trends at $x = 10$ mm, $y = 0$ mm, $z = 90$ mm.

In order to further confirm the well functioning of the method, the structural results in terms of residual stress are also reported. Again, the results of the FEM model were compared with those of the REF model. In Figure 8 are reported the paths where the residual stress have been evaluated. In Figures 12 and 13 are reported the residual stress in the first 20 mm of the path A and B. Residual stress along thickness at middle length of plate are reported in Figure 14. The structural results along the seam of the bead on plate welding are shown in Figure 15. All the structural trends reported are almost superimposed with the REF ones confirming the good functioning of the proposed method.

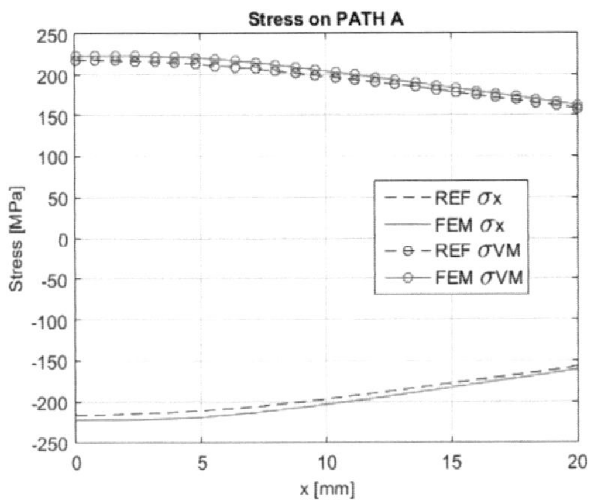

Figure 12. Stress on path A.

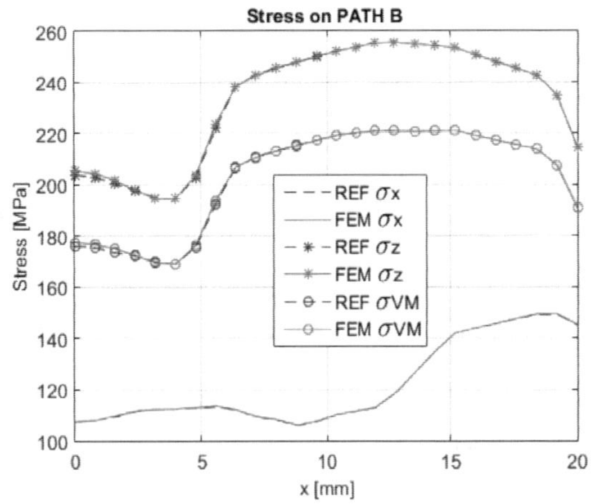

Figure 13. Stress on path B.

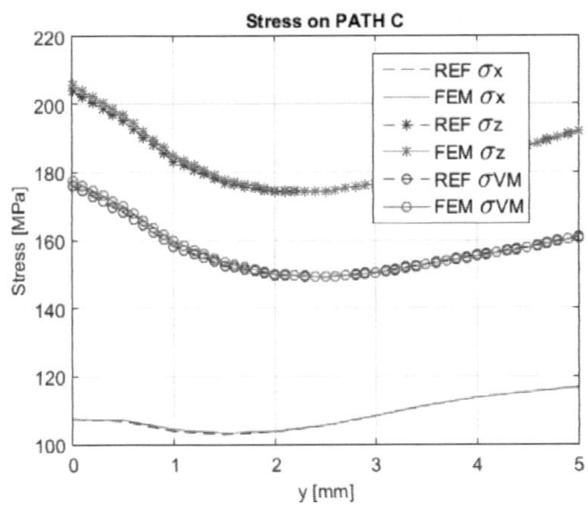

Figure 14. Stress on path C.

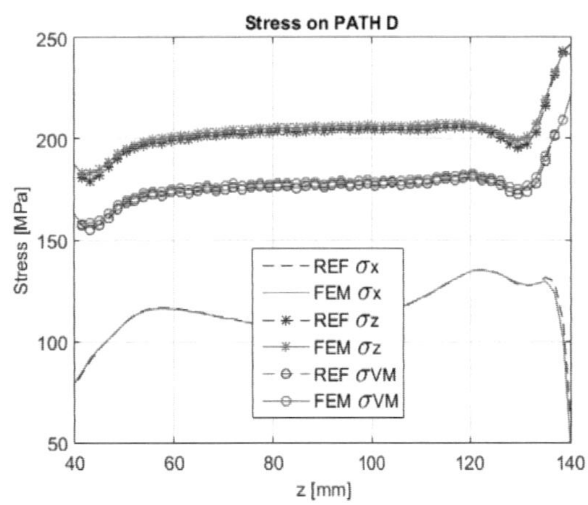

Figure 15. Stress on path D.

7. Conclusions

In conclusion, a semi-analytical method was developed for the determination of the geometrical parameters of the Goldak's ellipsoid knowing only a peak temperature near the welded seam and through the extension of the molten pool. The resolution of the Fachinotti's analytical model reduces the search range of the double ellipsoid parameters in the following step, which is implemented through the NSGA-II genetic algorithm that guides the 2D numerical simulations. It is through such simulations that the final Goldak's ellipsoid parameters are determined.

The thermal and structural results are encouraging even if obtained on preliminary numerical models of a bead on plate welding. It was decided to test the method presented on a numerical reference model and not on an experimental test in order to exclude any unknown variables that might influence the preliminary tests.

The method can be easily automated through the use of fully parametric lists that can be adapted to the simulation of butt welds or fillet welds, including those realised with equivalent or hybrid models [12, 13]. Moreover, this semi-analytical approach is well suited to determine the geometric parameters of multi-pass welds.

References

[1] Goldak J A, Akhlaghi M 2005 *Computational Welding Mechanics.* (Boston: Springer)
[2] Lindgren L E 2007 *Computational Welding Mechanics.* (Sawston: Woodhead Publishing)
[3] Deng D 2008 FEM prediction of welding residual stress and distortion in carbon steel considering phase transformation effects *Mater. Des.* **30** 359-366
[4] Goldak J, Chakravarti A, Bibby M 1984 A new finite element model for welding heat sources *Metall Mater Trans B* **15** 299–305
[5] Pu X, Zhang C, Li S, Deng D 2017 Simulating welding residual stress and deformation in a multi-pass butt-welded joint considering balance between computing time and prediction accuracy *Int J Adv Manuf Technol* **93** 2215–2226
[6] Fachinotti V, Anca A, Cardona A 2011 Analytical solutions of the thermal field induced by moving double-ellipsoidal and double-elliptical heat sources in a semi-infinite body *Int. J. Numer. Meth. Biomed. Engng.* **27** 595-607
[7] Chujutalli J, Lourenço M, Estefen S 2020 Experimental-based methodology for the double ellipsoidal heat source parameters in welding simulations *Mar Syst Ocean Technol* **15** 110-123
[8] Chujutalli J, Lourenço M, Estefen S 2016 *Methodology for the determination of heat source parameters for a FCAW process Proc. PRADS 2016* (Copenhagen)
[9] Deb K, Pratap A, Agarwal S, Meyarivan T 2002 A fast and elitist multiobjective genetic algorithm: NSGA-II *IEEE TEVC* **6** 182-197
[10] MATLAB 2019 *MATLAB version 9.7.0 (R2019b)* (Natick, Massachusetts: The MathWorks Inc.)
[11] Brickstad B, Josefson B L 1998 A parametric study of residual stresses in multi-pass butt-welded stainless steel pipes *Int. J. Press. Vessels Pip.* **75** 11-25
[12] Trupiano S, Belardi V G, Fanelli P, Gaetani L, Vivio F 2021 A novel modeling approach for multi-passes butt-welded plates *J. Therm. Stresses.* **44** 829-849
[13] Perić M, Tonković Z, Rodić A, Surjak M, Garašić I, Boras I, Švaić S 2014 Numerical analysis and experimental investigation of welding residual stresses and distortions in a T-joint fillet weld *Mater. Des.* **53** 1052-1063

IOP Conf. Series: Materials Science and Engineering 1214 (2022) 012024 doi:10.1088/1757-899X/1214/1/012024

Preliminary design of an anthropomorphic cutting and potting system for automated agriculture

M Delogu*, D Uggiosi, R Barbieri and M Pierini

Università degli Studi di Firenze, Department of Industrial Engineering (DIEF), Via di S. Marta, 50139, Firenze – Italy.

*e-mail: massimo.delogu@unifi.it

Abstract. The study presented is part of the Rural Development Programme of the Tuscany Region and the European Union: the objectives are to limit the number of manual operations, to enhance technological transfer operations, to share best practices and to reduce the carbon footprint of the Scindapsus cuttings production chain. In this context, the work presents the approach to a preliminary design of an anthropomorphic system for automated cutting and potting of cuttings. For a better understanding of the current procedures and solutions, a critical analysis of the State of the Art of agricultural automation processes available in the literature was carried out. This study was developed in parallel with market research to identify the custom components to be produced to make a choice consistent with the technical specifications. The design has dealt with the system for handling and preparing the cuttings through panels and conveyor belts, starting from the needs defined in the initial phase. The analysis considered, both at a mechanical and functional level (evaluation of times and methods), the layout of production space with a high degree of automation for cutting and potting cuttings, with particular attention to the issues of the workplace safety and the maintainability of the elements: the best configuration of resources, personnel and equipment were designed through a *what-if* scenario analysis populated by deterministic and stochastic events.

1. Introduction

The Rural Development Programme of the Tuscany Region and the European Union is a project aimed to grant an optimal management of natural resources and climate action. This is possible by stimulating a balanced rural, social, and economic development, in order to create and maintain work opportunities, as stated in regulation 1305/2013 of the European Parliament and of the Council [1]. European Union has scheduled six main goals for an appropriate rural development, such as: to promote technological transfer and know-how to agricultural activities, enhance productiveness of existing rural companies, to preserve, promote and resume ecosystems related to agriculture and forestry.

This work is framed in this legislative, economic, and environmental context and is aimed to enhance technological transfer to reduce manual operations, to share best practices between companies and to reduce the carbon footprint of the Scindapsus cuttings production chain. Nowadays this process is totally provided manually by specialized workers. This work is intended to achieve a preliminary design and optimisation of an anthropomorphic cutting and potting system for automated agriculture to reduce processing times, increase productivity and reduce the cost of the final nursery product.

It is necessary to briefly analyse the State of Art of automation in agricultural processes. In the last decades and years robots and artificial intelligence are becoming more and more used in various application fields, as they reduce risks related to dangerous operations. Originally, robots have been developed to replace humans by doing dangerous or dull operations [2], [3], but modern robot arms are widely used in every field: from heavy mechanical industries to agricultural applications. Agriculture, in fact, is regarded to be one of the most dangerous workplaces and it is mandatory to reduce the risk for workers: plenty of health problems have been identified, such as hearing loss, respiratory illnesses and different types of cancers [4], [5]. Manual operations such as pruning, harvesting, weeding, handling and digging include several risk factors such as awkward postures, iterative and prolonged trunk and knee bending and the carrying and lifting of heavy loads [[4], [6]]. Mechanisation in several operations has mitigated the aforementioned health problems, but it is necessary to develop new technologies in order to reduce environmental impact of agricultural processes and related mechanical risks [7].

Nowadays, in the era of Industry 4.0 [8], defined as "*the set of technologies, devices and processes [...] capable of operating in an integrated way along the several phases of the production process and along the several levels of the supply chain [...] that allow for self-sufficient production, integrated operations, decentralized decisions, minimum human intervention*" [9], [10], advanced robotics has been widely used and agricultural robots may represent a solution to this issue, but they should be in line with several technical requirements, such as: lightweight structure, small size and possibility to adapt to working area [11]. A mandatory technology for the application in agriculture field, characterized by a natural diversity that could not be eliminated from the final products, is the artificial vision. This field of research has grown very fast in last years especially with the application of Generative Adversarial Networks in recognize real life objects [12], in machine learning approaches [13] and generally speaking Artificial Intelligence [14]. Before introducing working methodology and the specific case study based on Scindapsus cuttings potting system, it is useful to focus on "sustainable development" definition, namely: "*development that meets the needs of the present without compromising the ability of future generations to meet their own needs*" [11], [15]; this formulation has to be considered as a goal to achieve in every design application. Industry 4.0 can contribute positively to reduction of environmental impact of processes [9], but it can increase production of waste from electric and electronic equipment (WEEE [16]) and social problems [17] such as unemployment [18], even if the correlation between these two factors has to be studied deeply, due to higher production rate [19]. In parallel with applications, also sensors and automatic data collection for the Agriculture 4.0 is a promising field of research because supporting systems are mandatory to develop integrated technological solutions [20], [21], [22]. Authors from [23], [24] in addition propose a literature review to collect Agriculture 4.0 specific tools.

This work is intended to correlate every factor mentioned above to provide a preliminary layout design of an autonomous station for cutting, processing and potting Scindapsus cuttings in a high-automation degree process, nevertheless without forgetting social, logistic, economic, and environmental aspects related to this choice.

2. Case study: Scindapsus cutting production and potting

The FLO.ROBOT project is an opportunity to study and design a high-automation station for moving, cutting, and potting Scindapsus cuttings. A similar work but with a different plant application has been proposed by authors of [25], [26], [27].

As mentioned in the previous section, Scindapsus' production chain is now totally composed by specialized workers, thus the plant (see **Figure 1**) is moved and cut by hand. According to the manufacturer, this operation is characterized by a high level of repetitive operations and it is quite slow and time-consuming.

a) b)

Figure 1: Scindapsus a) Scindapsus potted and ready to be sold b) Scindapsus potted, manually by the worker.

Scindapsus is a flowering plant and it represents the core business of the manufacturer mentioned above. Its cultivation time varies from 12 to 36 weeks. This time includes four main phases, such as:

1. Growing phase
2. Picking up the branch
3. Cuttings production
    - Every branch is cut into proper length
4. Potting phase

Cuttings' production (see **Figure 2**) is intended to be converted from a series of manual operation to a sequence of automatic ones. According to available data, idoneous branches have to be pick up manually and are characterized by widely variegated leaves and strong stem.

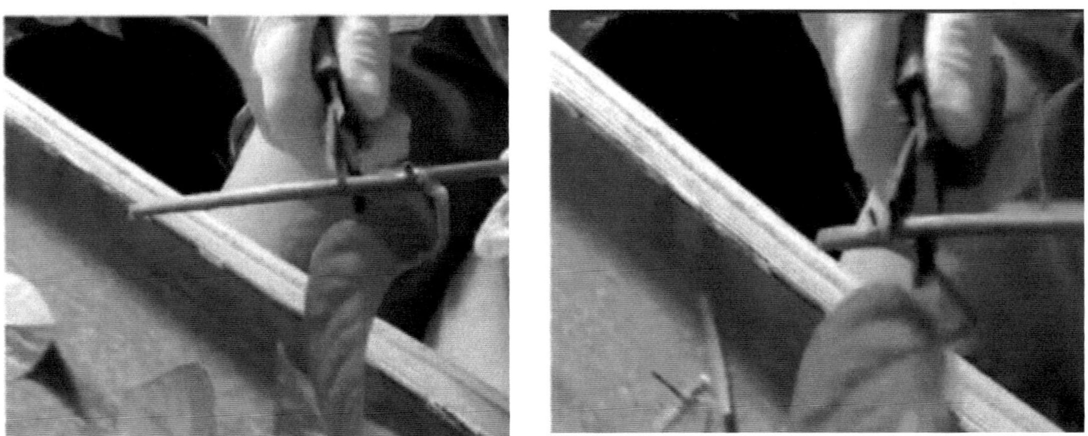

a) b)

c) d)

Figure 2. Operations needed to produce Scindapsus cuttings. a) A portion of stem is cut b) the branch is moved forward c) the first side cutting is obtained d) end of cuttings' production.

3. Design approach method

The motivation of the study relies on the relevance of such production for the manufacturing and nursery district including various companies in the area of Mount Amiata, near Siena, in the region of Tuscany (Italy), exporting worldwide. According to data and nursery prediction, the annual requirement of Scindapsus cuttings is 5,5 millions and this amount is estimated to become higher and higher.

The approach followed in the proposed activity is summarized in **Figure 3**. After a first phase of critical analysis of the State of the Art (see the the Introduction paragraph), a brainstorming is carried out to define the Customer Requirements (CRs) for the system. The CRs represented the input data for a QfD analsys aimed to the definition of the Design Requirements (DRs) in terms of different alternatives from which to choose the best one according to the CRs and on which an environmental and economic assessment was also performed.

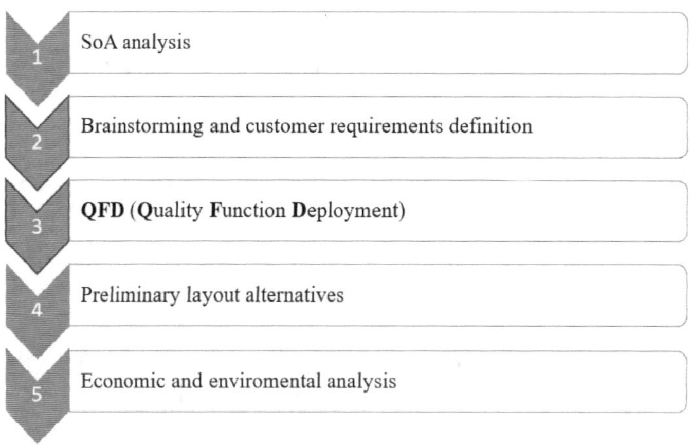

Figure 3. Design approach method.

3.1. Quality Function Deployment

As mentioned above, authors follow a structured approach Quality Function Deployment (QFD) based on to determine importance and correlation of the design parameters. In literature, this framework is largely used because of application easiness and of synthesis skills [28], [29], also when talking about environmental analysis [30]. An extract of the FLO.ROBOT QFD is shown in **Figure 4** and it has been produced by the project stakeholders with the authors' support:

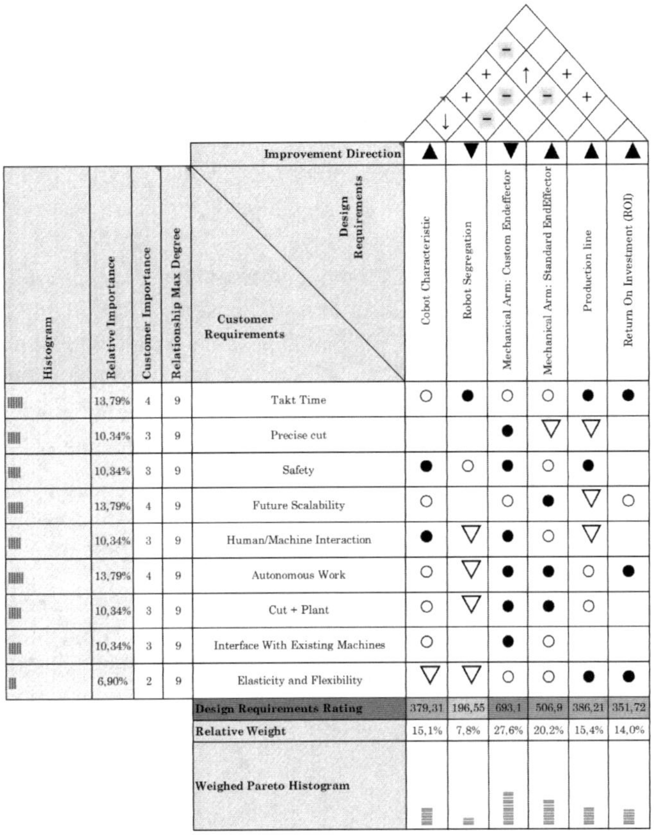

Figure 4. FLO.ROBOT QFD Application.

On the rows, the following "Customer Requirements" (CRs) have been identified:

1. Takt time: the system ability to not increase the total process takt time.
2. Precise cut: the system ability to precisely operate the cut.
3. Safety: mandatory to protect operator's safety.
4. Future scalability: the possibility to future upgrade the system.
5. Human/machine interaction: the machine operational easiness for the operators.
6. Autonomous work: the system ability to be fully operative without the operator supervision (excluding loading and unloading phases).
7. Cut and plant: the plant cutting and positioning in the pot (final product) machine ability.
8. Interface with the existing machines: system ability to be implemented in the existing job floor.

9. Elasticity and flexibility: system ability to positively react to plant changes and volume changes respectively.

On the basis of the CRs above, the following Critical to Quality parameters (CTQs) and "Design requirements" (DRs) have been defined:

1. Cobot characteristics: this means the possibility to develop a part of the machine close to a cobot, so able to share the same working areas with human operators and without barriers that could slow down the working activity.
2. Robot segregation: some parts could be dangerous for the operators and so must be not accessible to them. For this reason, a precise definition of the system layout has to be developed according to essential health and safety requirements of 2006/42/EC on Machines safety.
3. Mechanical arm_custom end-effector: a possible macro-solution to be evaluated is the design of an anthropomorphic arm equipped with an end-effector characterized by a larger work flexibility despite a larger initial cost.
4. Mechanical arm_standard end-effector: the exact opposite of the previous solution, this one provides a minor work flexibility with a minor initial cost. In addition, the hardware/software interface between all the system components could be harder to develop with a non-controllable factor as a standard end-effector.
5. Production line: this characteristic refers to the possibility to integrate the final product within the existing production line in order not to increase the total takt time or to create additional process bottlenecks.
6. Return on Investment (ROI): finally, this parameter refers to the customer need to implement a machine able to decrease the total cost of ownership of the entire process.

The solution of an anthropomorphic arm equipped with a custom end-effector results the most interesting one between the other possibilities because provides the highest rating among the DRs. This value comes from the "Coupling factor" between DRs and CRs multiplied with the "Customer Importance" vector. The symbols in fact mean a strong correlation with a dark point, a moderate correlation with a light point and a weak correlation with the light triangle: in numbers, a value of 9, 3 and 1 has been respectively assigned.

A second winning point is the high score reached in "Interface with Existing Machines": in fact, the job floor is already equipped with many existing production automation tools (a machine to put earth in the pot, semi-automatic material handling, final product packaging, …) and a project constraint was to reuse most of the existing tools.

In the QFD "roof", the cross-correlation matrix between the DRs, authors determine that the "Machine Segregation" could help improve the productivity, but the loss in flexibility and so in future development of this application let the designer to choose a more "cobot-like" solution. The "QFD roof" in fact is filled with a upward arrow in case of strong positive correlation, a downward arrow in case of strong negative correlation, a "+" symbol in case of positive correlation and a "-" symbol in case of negative correlation.

There is also some graphical information in the histogram columns defined in the left and in the matrix bottom part: those graphical approach simply resume the numerical results in order to make the study more affordable for everyone working on this project.

In conclusion, the QFD framework provides a structured approach for the designers to determine not only the best possible solution among all the available but also provides a different "problem glance" to increase the system features.

The outcomes of QFD analysis allow defining the following technical specifications functional to the identification of the different layouts and related system's components.

In **Table 1** the data related to the technical specification of the robot are summarized.

Table 1. Cobot's technical specification.

Technical specification	Range
Range of operation	700-2000 mm
Manipulation time	< 2 s
Cost	< 50000 €
After-sale service	Available
Get custom end-effector	Available

3.2. Preliminary layout alternatives

According to QFD results and technical specification three layouts have been analysed to evaluate integrability with the existing production line. In this paragraph authors will expose and discuss solutions according to project stakeholders and technical requirements.

Two mechanical layouts require a panel where the operator has to hang up Scindapsus branches and a properly designed u-shaped duct, shown in **Figure 5** in order to let the cobot move the branch.

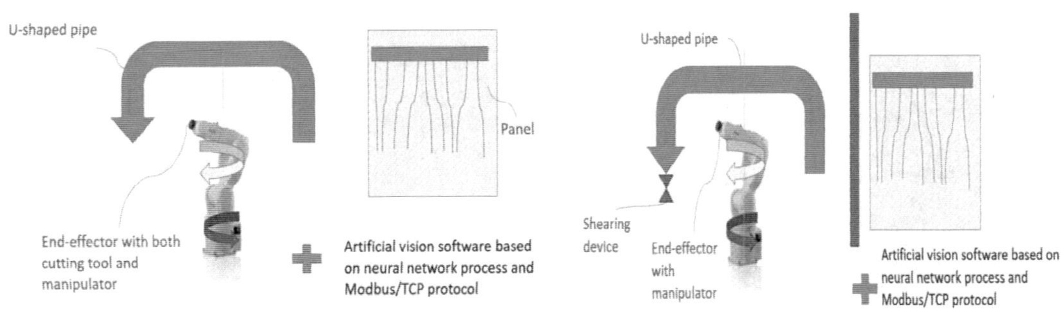

Figure 5. Schematic mechanical layouts with two solutions: 1) end-effector with both cutting tool and manipulator, 2) shearing device at the end of the pipe.

These solutions differ from each other as in the first one the cobot end-effector has to drag and cut the cuttings, while in the second layout the cutting operation is demanded to a shearing device, placed at the end of the duct. Due to the production volume, this kind of solution is not functional and may not grant working volumes required from the project stakeholders. The u-shaped duct is realized using a pipe with a radial access to enable the end-effector grab and move the Scindapsus leaves and make them assume a proper position to the stem; but this would require two or more cobots to keep present working volume:

IOP Conf. Series: Materials Science and Engineering 1214 (2022) 012024 doi:10.1088/1757-899X/1214/1/012024

in fact, using only one mechanical arm, transporting operation would unacceptably reduce productivity of the station.
Another solution is shown in **Figure 7**. Branches are properly fixed by the worker to a custom belt conveyor with elastic band sewn on it and moved by a traction pulley and a stepper motor; cutting and grabbing functions are demanded the cobot's end-effector.

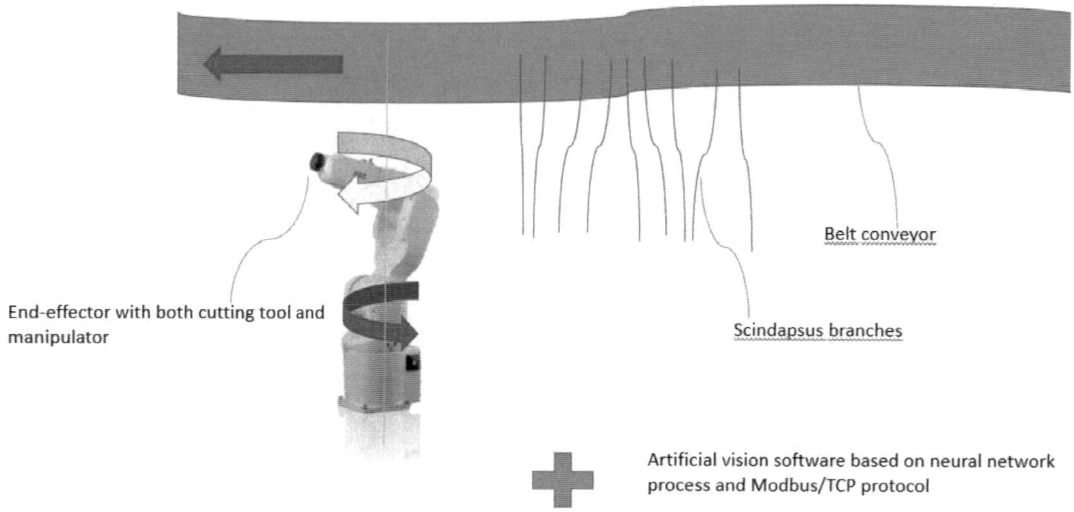

Figure 6. Schematic mechanical layout with belt conveyor.

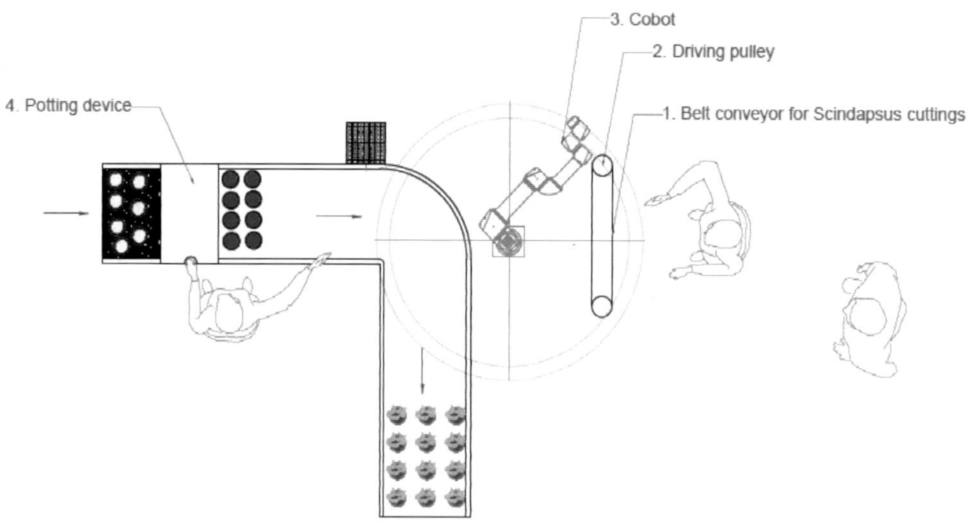

Figure 7. Mechanical layout with belt conveyor and potting device.

This solution has been analysed in further details as it respects following requirements:

- *Serviceability*. Properly formed workers can maintenance the system;
- *Future scalability*. The system presented in **Figure 7** can be modified using an higher number of pulleys in order to hang more branches
- *Good integration with the existing production plant*

3.3. Economic/Environmental impact analysis

One of the "Customer Needs" is the ROI generation of the system. It means that the machine must be economically sustainable for the company and so the revenues coming from implementation must be greater than the sum of implementation and operational costs, the total cost of ownership in other term. That is the reason why authors developed a process model to investigate the mutual interdependencies between the entire system cost drivers. The used environment is a "Discrete Events Process Simulator", a kind of sandbox where all the atom operations and resources are putted together with their mutual interaction and statistical characterization to develop Montecarlo simulation to study the system long term behaviour.

Discrete event simulation is a well-known approach to the process design to experiment different solutions in a numerical environment, so advantages are mostly in rapidity and cheapness of the attempts. In agricultural, instead, is commonly used in logistic and transportation problems [31], [32] where production issues are not very well investigated. The application presented in this paper is referred to a greenhouse production that is closer to an industrial process, this is the reason why the authors decided to implement the discrete event simulation.

Next step is to determine the statistical characterization of time and cost of each atom activity.

The BPMN standard [33] has been chosen because of its easiness of process representation and because it is a largely used standard in process analysis application. The standard models the "ring of application" as a "swimming pool" where the "lanes" are the single process owners. In **Figure 8**, the BPMN-like model is shown:

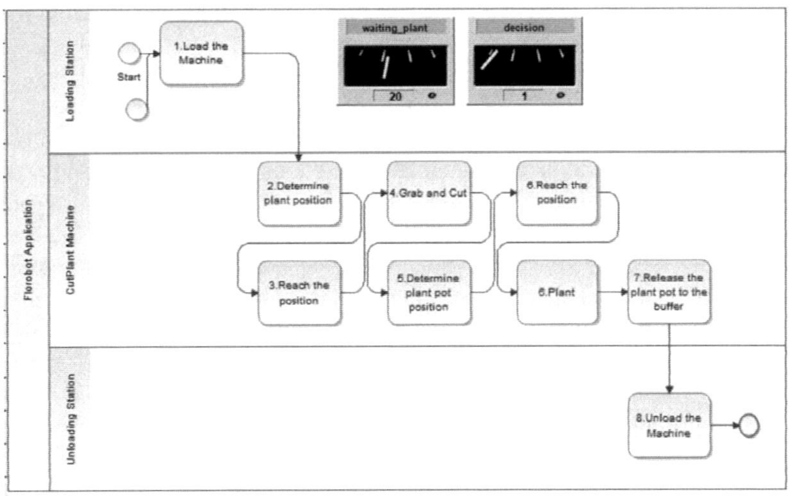

Figure 8. Process BPMN diagram.

In this model the swimming pool has been chosen as the "FLO.ROBOT Application", where the stakeholders are the mechanical system ("CutPlant Machine" in the central lane), the place where the operator loads the machine with the Scindapsus to be cut ("Loading Station" in the upper lane) and the place where the final pots are unloaded to be sent to following activities ("Unloading Station" in the lower lane). Each lane is equipped with a resource, one operator in the loading station, one operator in the unloading station and the FLO.ROBOT machine in the central lane. In **Figure 9** the FLO.ROBOT resource characterization interface is given as an example:

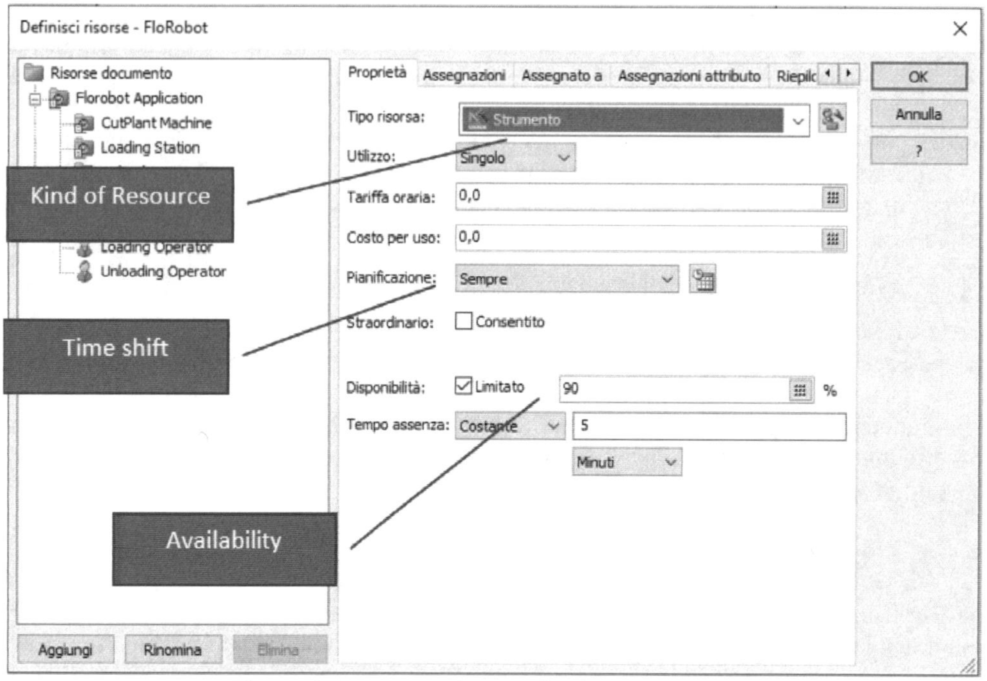

Figure 9. Resource characterization interface.

The graphical interface shows that it is possible to set up the hour cost for utilization and the cost per use of the resource. They are set at "0" because this parameter will be calculated in the final balance. In the central window part, the software allows defining the resource time planning, in other words the working days shifts where it is available: FLO.ROBOT machine is of course always available in any day or night time. Lastly, in the lower part of the windows, the Process Simulator interface allows determining the system availability such as total working time net of faults. The example given shows that the FLO.ROBOT application is 90% of time available and 10% occupied with set-ups, repairs, regulations, etc. The characterization has been developed for the "Operator" resources, too. It is important to highlight that time and costs parameters have been inferred by the data sheet information and by expert opinions collected during technical meetings. In fact, it will be possible to be more efficient with the statistical characterization when a system prototype will be realized and will be available for measurements. However, the aim of this preliminary design is to develop a design framework and to investigate different process design alternatives.

Each box in the "swimming pool" represents an atom activity in which the process has been divided, and more in details:

1. *Load the machine*: in this activity an operator set up the plants in the transportation belt to make them available for cutting process.
2. *Determine plant position*: in this activity the artificial vision system determines the position of the Scindapsus knot between the leave and the stem and communicate coordinates to the anthropomorphic arm controller.
3. *Reach the position*: this activity models the behaviour of the anthropomorphic arm when it moves from the zero position to the cutting area.
4. *Grab and cut*: in this activity the anthropomorphic arm grab the leaves under the cutting knot and then makes the cut.
5. *Determine plant pot position*: this is the activity where the artificial vision system find out the position of the plant pot and communicates the coordinates to the anthropomorphic arm controller.

6. *Reach the position*: the anthropomorphic arm moving from cutting position to the planting position.
7. *Plant*: the activity of planting a Scindapsus.
8. *Unload the machine*: in this step a second operator takes the filled pots away from the output conveyor belt and addresses them to the following activities, out of the scope of this project.

The model has been also provided with two counters that help the process analyst to debug the model before to use it. They measure the value of a supporting variable and the total number of waiting plants within the flow.

Once created the process map, the statistical characterization of each activity must be done. The software application allows adding many parameters and statistical distribution to synthesize the activity behavior. As it is possible to see from **Figure 10**, each activity has a graphical interface form to add these values:

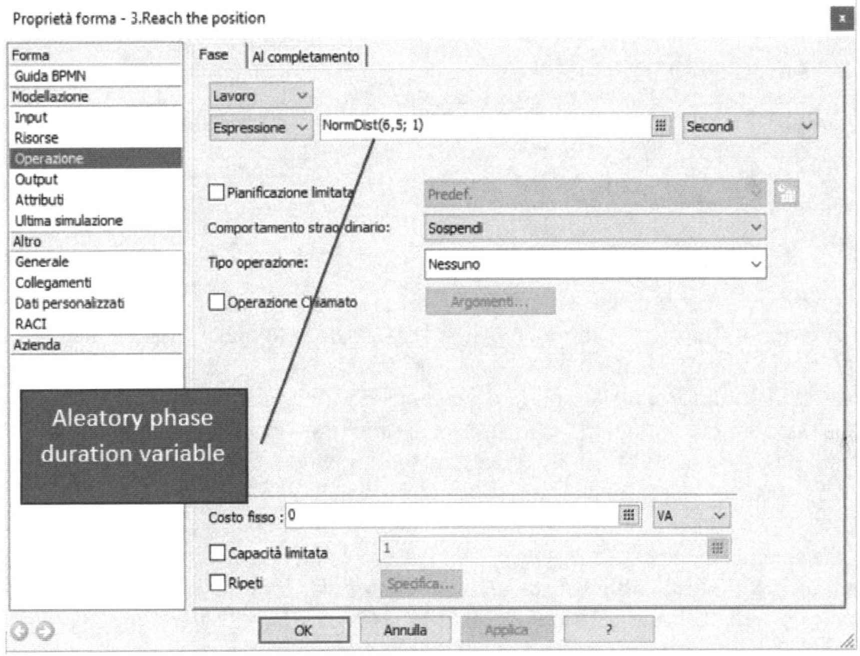

Figure 10. Activity characterization form.

As an example, the "3. Reach the position" form is shown. It is possible to see in the first text box the "NormDist(6,5;1)" that represent the statistical distribution the software will use in the Montecarlo simulation shown next in this paper. More in details, the software will use a Normal distribution with average 6,5 and standard deviation 1 seconds to determine the time needed by the anthropomorphic arm to reach the position to cut the Scindapsus. This form has been filled up for all the activity represented in the model with the data reported in **Table 2**:

Table 2. Statistical characterization parameters

Activity	**Statistical distribution (s)**
1. Load the machine	Triangle(5;15;8)
2. Determine plant position	Norm(2;0,1)
3. Reach the position	Norm(5;1)
4. Grab and cut	Norm(17;4)
5. Determine plant, pot position	Norm(2;0,1)
6. Reach the position	Norm(5;1)
7. Plant	Norm(13;1)
8. Release the plant pot to the buffer	3
9. Unload the machine	Triangle(5;15;8)

There are two further considerations to do:

1. As previously described, these numbers come from a data sheet analysis and from interviews with the work group. To be more precise it is mandatory to make a data collection campaign on the prototype to validate the model.
2. The used distributions are the normal one and the triangular. Normal has been used for tools activities where the triangular has been use for activities that involve human operator. It is only a commonly used heuristic in the process modeling when data are not enough to determine the correct statistical distribution.

Once all the parameters have been set up, the Montecarlo simulation reports the average performance of the system as reported in **Table 3** here below. The simulation time has been set up for 5 working weeks with a one shift working period (08.00 AM - 05.00 PM), 5 working days per week for the operators. Results are reported in minutes:

Table 3. Simulation results

Lane	**Number of transactions**	**Cycle Avg**	**Work Avg**	**Waiting Avg**
CutPlant Machine	2234	101,79	0,91	77,52
Loading Station	2234	290,76	0,16	1,68
Unloading Station	2234	169,64	0,15	0,42

The columns represent:
1. *Lane*: the system department.
2. *Number of transactions*: it represents the total amount of pots the system produced in the simulation period.
3. *Cycle Avg*: the average time a single plant pot needs to be completely produced. This time includes also waiting for available resources and when outside the working period.
4. *Work Avg*: the average time the pots spent being worked. This time does not include any waiting time.

5. *Waiting time*: the time each pot spends in waiting an available resource. This time does not include the time spent waiting in not working shifts.

This configuration has been considered as a base line to determine how sensible the final result is to any variation in these data. For this reason, authors developed a series of Montecarlo simulations changing the average values in the activities data input form. Results are presented in **Table 4** here below and in **Figure 11** the data interpolation:

Table 4. Sensitivity analysis results

Variation	Number of transactions	Delta production
-20% Baseline	2980	+746
-10% Baseline	2488	+274
Baseline	2234	0
+10% Baseline	1993	-241
+20% Baseline	1990	-244
+30 % Baseline	1982	-256

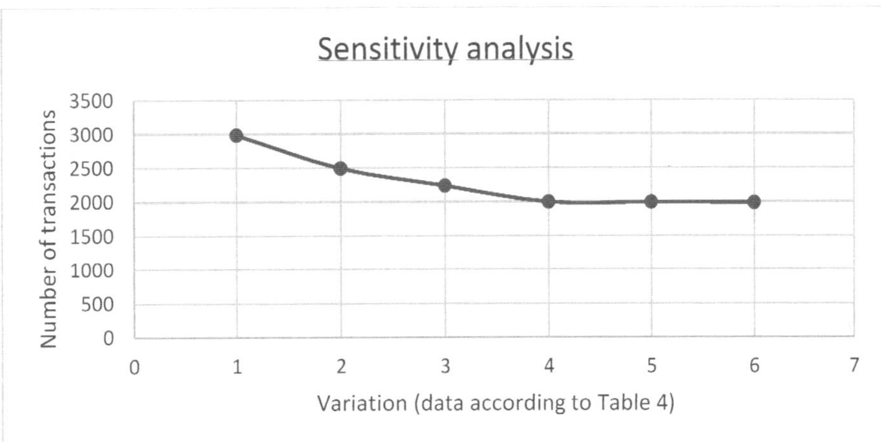

Figure 11. Sensitivity analysis results

It is clearly visible that the production converges to a value in this range of variation. It means that it is not necessary in this domain to increase the machine performance: in fact the system will give similar performances with an extra expenditure increase.

In addition, from **Table 3** it is possible to see that the Waiting Average is a low parameter value for the Loading Station and for the Unloading Station. This means that the two operators are poorly saturated and so their efficiency is low. For this reason, authors proposal is to use only one operator to load and to unload the machine to keep lower the total cost of ownership. Simulation results with only one operator are reported in **Table 5**.

Table 5. Single operator sensitivity analysis

	Variation	Number of transactions	Delta production
1	-20% Baseline	2961	+727
2	-10% Baseline	2488	+254
3	Baseline	2234	0
4	+10% Baseline	1993	-241
5	+20% Baseline	1990	-244
6	+30 % Baseline	1982	-252

Results prove that only one operator is enough to work with the FLO.ROBOT system. Savings of this results are quantifiable in one "Full Time Equivalent" (FTA) cost for the organization.

For this proposed solution a cost analysis has provided, both both from economic and environmental point of view. The data needed for this analysis are:

1. Hour cost for electric energy: authors used the average cost for a kWh in Italy that is quantifiable in 0,10691 €/kWh.
2. Power absorbed by the robotic arm: authors determined this data starting from data sheets values. The used number is 0,39 kW.
3. Carbon intensity: the average equivalent carbon dioxide released in the atmosphere. This value has been set at 255 gCO2eq/kWh as proposed by EEA, European Environmental Agency [34].

The three scenarios simulation and analysis are reported in **Table 6:**

Table 6. Environmental and economic analysis simulation results

Variation	Delta Production	Cost (€)	Delta Cost	Equivalent Carbon dioxide (gCO_2eq)	Delta Equivalent Carbon Dioxide
-10% Baseline	+254	1,22€	+0,30€	2598,03	+239,31
Baseline	0	0,92€	0	2358,71	0
+10% Baseline	-241	0,82€	-0,1€	2082,78	-275,93

The results show that the best solution with chosen layout and initial assumptions is the -10% Baseline with only one operator working on the system. Again, this result comes from the time and cost assumptions made at the beginning of this paper: once the project will be finished and the prototype ready, the presented framework will be used again with more realistic data.

4. Conclusion and final remarks

In conclusion, in this paper the authors propose a structured approach to a preliminary design of an anthropomorphic cutting and potting system for automated agriculture. The Scindapsus cutting and planting operation represent a case study characterized by a product natural variability and by a limited literature in agriculture automation field. This application needs a neural network for visual recognition and an anthropomorphic arm for handling and therefore the support of a structured approach will help the system design. The authors firstly studied Customer Requirements and Design Requirements through the Quality Function Deployment approach to prioritize the solution to be implemented: this point has been extremely important because the working team was opened to many persons and so this tool helped to focus ideas and proposal. As a result, a set of alternatives has been identified and a concept design of each one has been made to investigate pros and cons. The most viable solution resulted an anthropomorphic arm with a custom end-effector. The proposed activity includes also an economic and environmental analysis based on a discrete event simulation approach to determine the best process parameter level. In the case study, the authors have evaluated the minimal number of operators to be involved in the process and the machine speed that reduces the carbon footprint despite the lowest reduction of productivity.

Acknowledgments

The activity here presented is part of the FLO.ROBOT project funded under the call for proposals 'Integrated food chain projects (PIF) - 2017' issued by the Tuscany Region within the framework of the 2014-2020 Rural Development Programme (RDP). This publication reflects only the author's view, exempting the Community from any liability.

References

[1] "*Regulation no. 1305/2013 of the European Parliament and of the Council*" of December 2013 on support for rural development by the European Agricultural Fund for Rural Development (EAFRD).

[2] Baohua Zhang, Yuanxin Xie, Jun Zhou, Kai Wang, Zhen Zhang, State-of-the-art robotic grippers, grasping and control strategies, as well as their applications in agricultural robots: A review, Computers and Electronics in Agriculture, Volume 177, 2020, 105694, ISSN 0168-1699, https://doi.org/10.1016/j.compag.2020.105694.

[3] Tai, K., El-Sayed, A. R., Shahriari, M., Biglarbegian, M., & Mahmud, S. (2016). State of the art robotic grippers and applications. *Robotics*, *5*(2), 11.

[4] Lefteris Benos, Avital Bechar, Dionysis Bochtis, Safety and ergonomics in human-robot interactive agricultural operations, *Biosystems Engineering*, Volume 200, 2020, Pages 55-72, ISSN 1537-5110, https://doi.org/10.1016/j.biosystemseng.2020.09.009.

[5] Kirkhorn, S. R., & Schenker, M. B. (2002). Current health effects of agricultural work: respiratory disease, cancer, reproductive effects, musculoskeletal injuries, and pesticide–related illnesses. *Journal of Agricultural Safety and Health*, *8*(2), 199.

[6] Benos, L., Tsaopoulos, D., & Bochtis, D. (2020). A Review on Ergonomics in Agriculture. Part II: Mechanized Operations. *Applied Sciences*, *10*(10), 3484.

[7] Bochtis, D. D., Sørensen, C. G., Busato, P., & Berruto, R. (2013). Benefits from optimal route planning based on B-patterns. *Biosystems Engineering*, *115*(4), 389-395.

[8] Braña, F. J. (2019). A fourth industrial revolution? Digital transformation, labor and work organization: a view from Spain. *Journal of Industrial and Business Economics*, *46*(3), 415-430.

[9] Beltrami, M., Orzes, G., Sarkis, J., & Sartor, M. (2021). Industry 4.0 and sustainability: Towards conceptualization and theory. *Journal of Cleaner Production*, 127733.

[10] Castelo-Branco, I., Cruz-Jesus, F., & Oliveira, T. (2019). Assessing Industry 4.0 readiness in manufacturing: Evidence for the European Union. *Computers in Industry*, *107*, 22-32.

[11] Moysiadis, V., Tsolakis, N., Katikaridis, D., Sørensen, C. G., Pearson, S., & Bochtis, D. (2020). Mobile Robotics in Agricultural Operations: A Narrative Review on Planning Aspects. *Applied Sciences*, *10*(10), 3453.

[12] Leonardo Galteri, Claudio Ferrari, Giuseppe Lisanti, Stefano Berretti, Alberto Del Bimbo, Deep 3D morphable model refinement via progressive growing of conditional Generative Adversarial Networks, Computer Vision and Image Understanding, Volume 185, 2019, Pages 31-42, ISSN 1077-3142, https://doi.org/10.1016/j.cviu.2019.05.002.

[13] Swain, M., Singh, R., Thakur, A. K., & Gehlot, A. (2020). A machine learning approach of data mining in agriculture 4.0. *Int. J. Emerg. Technol*, *11*, 257-262.

[14] Jha, K., Doshi, A., Patel, P., & Shah, M. (2019). A comprehensive review on automation in agriculture using artificial intelligence. *Artificial Intelligence in Agriculture*, *2*, 1-12.

[15] Imperatives, S. (1987). Report of the World Commission on Environment and Development: Our common future. *Accessed Feb*, *10*, 1-300.

[16] EC, Directive 2012/19/EU of the European Parliament and of the Council of 4 July 2012 on waste electrical and electronic equipment (WEEE) Text with EEA relevance, vol. OJ L. 2012.

[17] Rose, D. C., & Chilvers, J. (2018). Agriculture 4.0: Broadening responsible innovation in an era of smart farming. *Frontiers in Sustainable Food Systems*, *2*, 87.

[18] Kurt, R. (2019). Industry 4.0 in terms of industrial relations and its impacts on labour life. *Procedia computer science*, *158*, 590-601.

[19] WEF (2016). (Future of Jobs Report: Employment, Skills and Workforce Strategy for the Fourth Industrial Revolution, Global Challenge Insight Report), http://www3.weforum.org/docs/WEF_Future_of_Jobs.pdf.

[20] Symeonaki, E., Arvanitis, K., & Piromalis, D. (2020). A context-aware middleware cloud approach for integrating precision farming facilities into the IoT toward agriculture 4.0. *Applied Sciences*, *10*(3), 813.

[21] Puranik, V., Ranjan, A., & Kumari, A. (2019, April). Automation in agriculture and iot. In *2019 4th International Conference on Internet of Things: Smart Innovation and Usages (IoT-SIU)* (pp. 1-6). IEEE.

[22] Pisanu, T., Garau, S., Ortu, P., Schirru, L., & Maccio, C. (2020). Prototype of a low-cost electronic platform for real time greenhouse environment monitoring: An agriculture 4.0 perspective. *Electronics*, *9*(5), 726.

[23] Zhai, Z., Martínez, J. F., Beltran, V., & Martínez, N. L. (2020). Decision support systems for agriculture 4.0: Survey and challenges. *Computers and Electronics in Agriculture*, *170*, 105256.

[24] Klerkx, L., Jakku, E., & Labarthe, P. (2019). A review of social science on digital agriculture, smart farming and agriculture 4.0: New contributions and a future research agenda. *NJAS-Wageningen Journal of Life Sciences*, *90*, 100315.

[25] Huang, Y. J., & Lee, F. F. (2010). An automatic machine vision-guided grasping system for Phalaenopsis tissue culture plantlets. *Computers and Electronics in Agriculture*, *70*(1), 42-51.

[26] Jin, X., Li, D., Ma, H., Ji, J., Zhao, K., & Pang, J. (2018). Development of single row automatic transplanting device for potted vegetable seedlings. *International Journal of Agricultural and Biological Engineering*, *11*(3), 67-75.

[27] Choudhary, R., Rajarathnam, M., Alekya, M., Hari, K. P., & Kumar, S. S. (2020, December). Agriculture automation system. In *Journal of Physics: Conference Series* (Vol. 1706, No. 1, p. 012089). IOP Publishing.

[28] M. Leary, C. Burvill, "Enhancing the quality function deployment conceptual design tool", 2007, Journal of Mechanical design vol. 129.

[29] Vanegas, L. V., e A. W. Labib. «A Fuzzy Quality Function Deployment (FQFD) model for deriving optimum targets». International Journal of Production Research 39, n. 1 (2001): 99–120. https://doi.org/10.1080/00207540010005079.

[30] Yüksel, Hilmi. «Design of automobile engines for remanufacture with quality function deployment». International Journal of Sustainable Engineering 3, n. 3 (2010): 170–80, https://doi.org/10.1080/19397038.2010.486046.

[31] Daniel Nilsson, Discrete Event Simulation as a Performance Analysis Tool in Agricultural Logistics Systems, IFAC Proceedings Volumes, Volume 34, Issue 26, 2001, Pages 19-24, ISSN 1474-6670, https://doi.org/10.1016/S1474-6670(17)33626-1.

[32] Aaron P. Turner, Michael P. Sama, L. Samuel G. McNeill, Joseph S. Dvorak, Tyler Mark, Michael D. Montross, A discrete event simulation model for analysis of farm scale grain transportation systems, [32]Computers and Electronics in Agriculture, Volume 167, 2019, 105040, ISSN 0168-1699, https://doi.org/10.1016/j.compag.2019.105040.

[33] Business Process Notation and Modelling, http://www.omg.org/spec/BPMN/2.0.

[34] https://www.eea.europa.eu/data-and-maps/daviz/co2-emission-intensity-8#tab-googlechartid_googlechartid_googlechartid_googlechartid_chart_11111.

IOP Conf. Series: Materials Science and Engineering 1214 (2022) 012025 doi:10.1088/1757-899X/1214/1/012025

Discrete model for 3D motion of helical springs

F De Crescenzo and P Salvini

University of Rome Tor Vergata, Department of Enterprise Engineering
via del Politecnico 1, 00133 Rome Italy

E-mail: francesco.de.crescenzo@uniroma2.it

Abstract. Large displacements of coiled springs are affected by both contact between coils and geometric non-linearities. The two phenomena also interact with each other and the only technology that efficiently accounts for both is Finite Elements Method (FEM). Such analysis can become very expensive, especially when a great number of elements and contacts are involved, as it is often the case. A discrete 3D model for fast and efficient simulation of non-linear springs is presented. The discrete model is based on equivalent beam theory and each node is associated with a contact region placed on the helix. Thus contact is accounted on its real occurrence and fast non-linear solution procedure is implemented. A study case is presented and comparison with FEM is discussed.

1. Introduction

Contact between coils is often used as a mean to increase spring stiffness and provide additional damping. One of the most common applications of such strategy are valve springs in internal combustion engines and dual-rate or progressive springs in vehicle suspensions [1]. When coils come in contact, they almost stop deforming elastically, causing a reduction in the number of active coils and an increase in overall stiffness. Simulation of systems where non-linear springs are involved requires an appropriate modelling strategy. A full simulation of the coiled wire, e.g. using Finite Elements Method (FEM), would be too expensive to run and the correct managing of contact algorithm is non-trivial. Therefore, most commercial packages provide several tools to account for non-linear springs at low computational cost. A common approach is the multi mass model, where the spring is represented by several masses connected with a sequence of stiffness and damping elements [2]. The parameters are computed from coils properties, and non-linearity arises as the gaps between consecutive masses close. It is also possible to directly provide the non-linear spring characteristics, obtained from previous computations (analytically or via finite elements) or direct experiments. However, these methods only account for vertical motion and it seems to the authors that there is a lack in literature about the influence of coil contacts on flexural and lateral behaviour, while clash between coils not only affect the stiffness, but also changes the direction of spring motion and the volume occupied during operation. Starting from Haringx work on instability [3], several researchers have studied the lateral motion of helical springs. Analytical solutions have been obtained using equivalent beams under large displacements, [4] [5], but these models cannot manage coil contacts. On the other end, the dynamics of the wire has been addressed by many authors using numerical methods: finite element [6], transfer matrix method [7], finite difference [8], dynamic stiffness matrix method [9]. However, also in those cases it is still hard, and in some

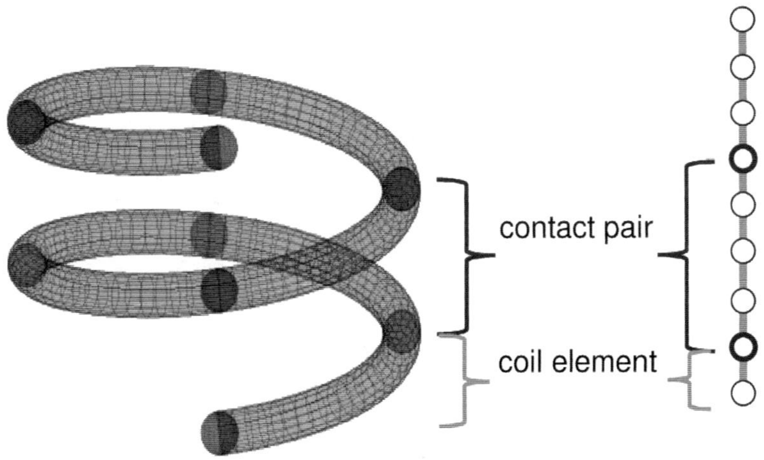

Figure 1. 3D helix and equivalent beam when four elements per coil are used.

cases unpracticable, to account for coil contacts and large deflections. A modelling strategy for two dimensional motion of non-linear springs with coil contact has been presented in [10]. The proposed model takes advantage of the equivalent beam formulation and represents the spring using several non-linear beam-like elements placed along the axis of the helix. Meanwhile, contact surfaces are related to the nodes and it is possible to introduce contact relations between adjacent coils. Such a model proved to be a good tool for quick and accurate prediction of the component behaviour and it is now extended to the general 3D case, where orientation of coils is defined by three consecutive finite rotations. The model is described in in section 2, focusing on rotation parametrization, kinematics of the elements and computation of nonlinear restoring forces. Contact is described in section 3 and the solution procedure is explained in 4. A numeric example is provided, considering FEM as benchmark in section 5 and finally conclusions are drawn.

2. 3D equivalent beam

2.1. Outline of the model

The model is based on Haringx equivalent beam [3], [11], [12], a beam-column structure with elastic properties derived from the characteristics of the spring. Nevertheless, it is worth to mention two main assumptions in Haring's work that differ from classical theory of beam-column instability:

(i) beam stiffness is kept constant, thus cross-section properties update accordingly with deformed length
(ii) only bending rotations are considered for orientation of axial and shearing forces

Thus it is possible to discretize the equivalent beam into a series of beam-like elements, that reflect previous assumptions, each node has six degrees of freedom (three translations and three rotations). The full helix in 3D can be reconstructed from nodal positions and orientations, in this manner it is possible to associate a contact zone on the helix to each node and verify the volume occupied by the spring in working conditions. A picture of the modelling strategy is given in figure 1.

2.2. Rotation parametrization

In 3D motion it is important to adopt a consistent set of rotation parameters that define the orientation of nodal frames of reference. For the present work it is found convenient to express the orientation through three consecutive rotations, as shown in figure 2:

(i) Rotation ζ about $\mathbf{e}_{3,J}$

(ii) Rotation η about $\mathbf{e}_{2,J}$

(iii) Rotation ξ about $\mathbf{e}_{1,J}$

where $\mathbf{e}_{1,J}$, $\mathbf{e}_{2,J}$, $\mathbf{e}_{3,J}$ are the unit vectors of the moving frame of reference on node J when node I is assumed to be fixed. In the undeformed configuration all moving frames coincide with the fixed global one thus:

$$\mathbf{e}_{1,J} = \mathbf{e}_{1,0} = \begin{pmatrix} 1 \\ 0 \\ 0 \end{pmatrix}, \quad \mathbf{e}_{2,J} = \mathbf{e}_{2,0} = \begin{pmatrix} 0 \\ 1 \\ 0 \end{pmatrix}, \quad \mathbf{e}_{3,J} = \mathbf{e}_{3,0} = \begin{pmatrix} 0 \\ 0 \\ 1 \end{pmatrix} \tag{1}$$

Unit vectors in the deformed configuration are change as:

$$\mathbf{e}_{1,J} = \mathbf{R}(\zeta, \eta, \xi)\mathbf{e}_{1,0} \tag{2}$$

$$\mathbf{e}_{2,J} = \mathbf{R}(\zeta, \eta, \xi)\mathbf{e}_{2,0} \tag{3}$$

$$\mathbf{e}_{3,J} = \mathbf{R}(\zeta, \eta, \xi)\mathbf{e}_{3,0} \tag{4}$$

where $\mathbf{R}(\eta, \zeta, \xi)$ is the rotation matrix, given by [13]:

$$[\mathbf{R}(\zeta, \eta, \xi)] = \begin{bmatrix} \cos\zeta & -\sin\zeta & 0 \\ \sin\zeta & \cos\zeta & 0 \\ 0 & 0 & 1 \end{bmatrix} \begin{bmatrix} \cos\eta & 0 & \sin\eta \\ 0 & 1 & 0 \\ -\sin\eta & 0 & \cos\eta \end{bmatrix} \begin{bmatrix} 1 & 0 & 0 \\ 0 & \cos\xi & -\sin\xi \\ 0 & \sin\xi & \cos\xi \end{bmatrix} \tag{5}$$

2.3. Coil kinematics and internal forces

The aim of this section is to show the procedure that allows to express generalized nodal forces as a function of generalized nodal displacements. According to Haringx assumptions, the deformation of the coils is expressed in terms of bending, twisting, extension/compression and shearing. Since it is assumed that the elastic properties of the equivalent beam are axisymmetric, bending can be expressed as a finite rotation about a generic axis $\mathbf{R}_b(\mathbf{u}, \phi)$, that orient unit vector $\mathbf{e}_{1,I}$ towards $\mathbf{e}_{1,J}$. Calling $\mathbf{e}_{1,b}$ the unit vector after a bending rotation is applied, it turns that:

$$\mathbf{e}_{1,J} = \mathbf{e}_{1,b} \tag{6}$$

where

$$\mathbf{e}_{1,b} = \mathbf{R}_b(\mathbf{u}, \phi)\mathbf{e}_{1,I} \tag{7}$$

Bending angle ϕ is computed as:

$$\phi = \arccos\left(\frac{\mathbf{e}_{1,I} \cdot \mathbf{e}_{1,J}}{||\mathbf{e}_{1,I}|| \; ||\mathbf{e}_{1,J}||}\right) \tag{8}$$

and bending axis $\mathbf{u}$ is:

$$\mathbf{u} = \begin{pmatrix} u_x \\ u_y \\ u_z \end{pmatrix} = \frac{\mathbf{e}_{1,I} \times \mathbf{e}_{1,J}}{||\mathbf{e}_{1,I}|| \; ||\mathbf{e}_{1,J}||} \tag{9}$$

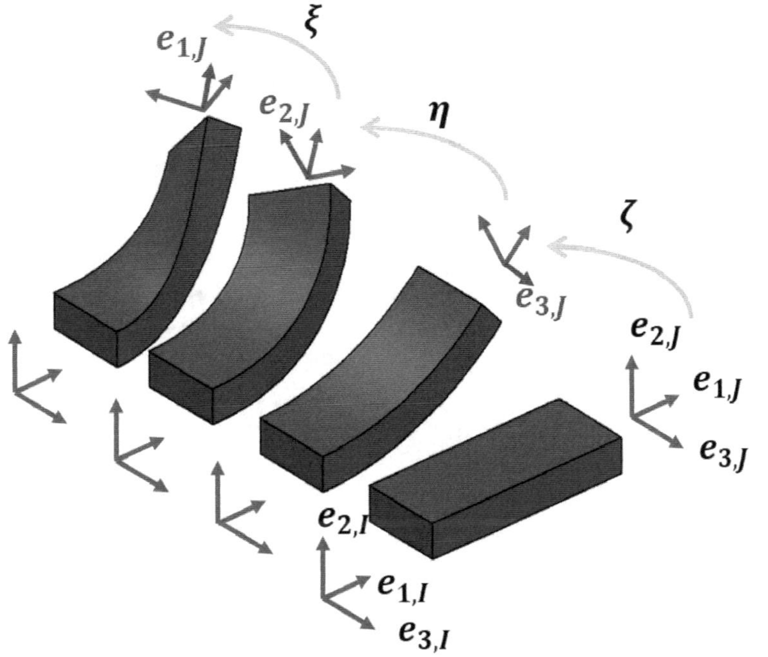

Figure 2. Finite rotations sequence

Bending rotation matrix is then [13]:

$$\mathbf{R}_b(\mathbf{u}, \phi) = \begin{bmatrix} u_x^2 V_\phi + \cos(\phi) & u_x u_y V_\phi - u_z \sin(\phi) & u_x u_z V_\phi + u_y \sin(\phi) \\ u_x u_y V_\phi + u_z \sin(\phi) & u_y^2 V_\phi + \cos(\phi) & u_y u_z V_\phi - u_x \sin(\phi) \\ u_x u_z V_\phi - u_y \sin(\phi) & u_y u_z V_\phi + u_x \sin(\phi) & u_z^2 V_\phi + \cos(\phi) \end{bmatrix} \tag{10}$$

where $V_\phi = 1 - \cos(\phi)$.
Usually (eq. 6) will not be satisfied for unit vectors $\mathbf{e}_2$ and $\mathbf{e}_3$, thus a further rotation about $\mathbf{e}_{1,J}$ is required. The required angle ψ is the twisting angle, computed as either:

$$\psi = \arccos\left(\frac{\mathbf{e}_{2,b} \cdot \mathbf{e}_{2,J}}{||\mathbf{e}_{2,b}|| \;\; ||\mathbf{e}_{2,J}||}\right) \quad \text{or} \quad \psi = \arccos\left(\frac{\mathbf{e}_{3,b} \cdot \mathbf{e}_{3,J}}{||\mathbf{e}_{3,b}|| \;\; ||\mathbf{e}_{3,J}||}\right) \tag{11}$$

The applied twisting rotation $\mathbf{R}_t(\mathbf{e}_{1,J}, \psi)$ is given again by (eq. 10) with appropriate substitutions. Of course, it is verified that:

$$\mathbf{e}_{1,J} = \mathbf{R}_b \mathbf{e}_{1,I} \tag{12}$$

$$\mathbf{e}_{2,J} = \mathbf{R}_t \mathbf{R}_b \mathbf{e}_{2,I} \tag{13}$$

$$\mathbf{e}_{3,J} = \mathbf{R}_t \mathbf{R}_b \mathbf{e}_{3,I} \tag{14}$$

According to Haringx assumptions, the elongation and shearing directions are defined by bending rotations. For every element the elongation direction $\mathbf{e}_b$ is computed considering the half-way bending rotation:

$$\mathbf{e}_b = \mathbf{R}(\mathbf{u}, \phi/2) \tag{15}$$

The length $\mathbf{l}$ of the element is given by the component of nodal distance $\mathbf{a}$ along $\mathbf{e}_b$, and shearing $\mathbf{v}$ is the other component.

$$\mathbf{a} = \begin{pmatrix} x_J - x_I \\ y_J - y_I \\ z_J - z_I \end{pmatrix} \tag{16}$$

$$\mathbf{l} = (\mathbf{a} \cdot \mathbf{e}_b)\mathbf{e}_b \tag{17}$$

$$\mathbf{v} = \mathbf{a} - \mathbf{l} \tag{18}$$

Obviously elongation Δl of the element is given by:

$$\Delta l = ||\mathbf{l}|| - l_0 \tag{19}$$

Having completely defined coil kinematics, it is possible to compute the elastic forces arising as a consequence of deformations. There are an axial force, $\mathbf{f}_{ax}$ directed along $\mathbf{e}_b$, a shearing force $\mathbf{f}_{sh}$ parallel to $\mathbf{v}$, a bending moment $\mathbf{m}_b$ applied on $\mathbf{u}$ and a twisting moment $\mathbf{m}_t$ applied on $\mathbf{e}_b$:

$$\mathbf{f}_{ax} = K_{ax}\Delta l\mathbf{e}_b \tag{20}$$

$$\mathbf{f}_{sh} = K_{sh}\mathbf{v} \tag{21}$$

$$\mathbf{m}_b = K_b\phi\mathbf{u} \tag{22}$$

$$\mathbf{m}_t = K_t\psi\mathbf{e}_b \tag{23}$$

Note that the magnitude of each force is directly proportional to the corresponding deformation through a stiffness constant that only depends on spring geometry and material properties. Their values are computed taking into account the most important component of stress in the wire [12] when an ideal flat coil is loaded:

$$K_{ax} = n\frac{Gd^4}{8D^3}\cos\alpha \tag{24}$$

$$K_{sh} = n\frac{Ed^4}{8D^3}\cos\alpha \tag{25}$$

$$K_b = n\frac{4EGd^4}{64D(E+G)}\cos\alpha \tag{26}$$

$$K_t = n\frac{Ed^4}{64D}\cos\alpha \tag{27}$$

where $\cos\alpha$ is a correction factor that accounts for the effect of the helix angle and n is the number of elements per each coil. For non-uniform coils average values are used.
Finally, nodal forces on nodes I, and J are obtained by equilibrium:

$$\mathbf{f}_I = -\mathbf{f}_l - \mathbf{f}_v \tag{28}$$

$$\mathbf{f}_J = \mathbf{f}_l + \mathbf{f}_v \tag{29}$$

$$\mathbf{m}_I = -\mathbf{m}_b - \mathbf{m}_t + \frac{1}{2}\mathbf{a} \times \mathbf{f}_2 \tag{30}$$

$$\mathbf{m}_J = \mathbf{m}_b + \mathbf{m}_t + \frac{1}{2}\mathbf{a} \times \mathbf{f}_2 \tag{31}$$

3. Contact between coils

3.1. Definition of contact

The algorithm implemented is based on penalty function. In fact, especially when several coils are involved in contact, which is often the case, penalty based methods are much easier to converge than Lagrangian multipliers. Moreover, local small compenetrations allow to represent distributed contacts.

Contact regions are spheres of the nominal wire diameter, placed at the end of each coil element, see figure 1. When deformations occur, the position of the spheres is updated according to nodal displacements and orientations. Thus contact regions are placed along the helix, while their positions are computed from the kinematics of the equivalent beams. If $\mathbf{p}_I$ and $\mathbf{p}_K$ are the position vectors of nodes I and K, which define a contact pair, the position vectors of the corresponding contact spheres are given by:

$$\mathbf{r}_A = \mathbf{p}_I + \mathbf{R}_I\mathbf{q_0} \tag{32}$$

$$\mathbf{r}_B = \mathbf{p}_K + \mathbf{R}_K\mathbf{q_0} \tag{33}$$

where $\mathbf{R}$ is the rotation matrix defined by rotation parameters as given in (eq. 5), and $\mathbf{q_0}$ is the relative position of the sphere with respect to its node.If, for example, four elements per coil are being used, $\mathbf{q_0}$ writes:

$$\mathbf{q_0} = \begin{pmatrix} 0 \\ \cos n\frac{\pi}{2} \\ \sin n\frac{\pi}{2} \end{pmatrix} \qquad \text{with} \quad n = 0, 1, 2, 3 \tag{34}$$

However, spherical contact surfaces may eventually lead to unrealistic events, with the spheres slipping on each other along the tangent to the helix. For this reason the distance vector between spheres centres is projected on a plane perpendicular to tangent unit vector $\mathbf{t}$, which is defined as the average between the tangent (helix angle effect is neglected) at both locations:

$$\mathbf{h} = (\mathbf{r}_B - \mathbf{r}_A) - (\mathbf{t} \cdot (\mathbf{r}_B - \mathbf{r}_A))\mathbf{t} \tag{35}$$

being

$$\mathbf{t} = \mathbf{R}_{AB}\mathbf{t_0} \tag{36}$$

where $\mathbf{R}_{AB}$ is the orientation of the contact pair, taken as:

$$\mathbf{R}_{AB} = \mathbf{R}(\frac{\zeta_I + \zeta_K}{2}, \frac{\eta_I + \eta_K}{2}, \frac{\xi_I + \xi_K}{2}) \tag{37}$$

and $\mathbf{t_0}$ is the tangent direction in the undeformed configuration, perpendicular to $\mathbf{q_0}$:

$$\mathbf{t_0} = \mathbf{q_0} \times \mathbf{e}_{1,0} \tag{38}$$

Contact force $\mathbf{f_C}$ is directed as the distance vector $\mathbf{h}$ and its magnitude is a power function of compenetration:

$$\mathbf{f_c} = \begin{cases} \mathbf{0} & \text{if} \quad ||\mathbf{h}|| > d^* \\ K_c(d^* - ||\mathbf{h}||)^k \frac{\mathbf{h}}{||\mathbf{h}||} & \text{if} \quad ||\mathbf{h}|| \leq d^* \end{cases} \tag{39}$$

where K_c is an appropriate stiffness constant, k is an exponent slightly greater than 1 and d^* is the equivalent wire diameter. The application point of the force is the midpoint between sphere centres:

$$\mathbf{r_C} = \frac{1}{2}(\mathbf{r_A} + \mathbf{r_B}) \tag{40}$$

Knowing contact force and its application point, it is trivial to compute corresponding forces and moments on nodes I and K.

4. Nonlinear solution

When accounting for the contribution of all elements, a system of equation is obtained, which gives internal forces as function of nodal coordinates and rotation parameters:

$$\mathbf{f}_{elastic} + \mathbf{f}_{contact} = \mathbf{f}(x, y, z, \xi, \eta, \zeta) \tag{41}$$

When external loads and boundary conditions are applied, the generic static problem writes:

$$\begin{cases} \mathbf{f}_{loads} + \mathbf{f}_{elastic} + \mathbf{f}_{contact} = \mathbf{0} \\ \mathbf{x}_{\Sigma} = \mathbf{s} \end{cases} \tag{42}$$

where Σ represents the set of degrees of freedom where boundary conditions $\mathbf{s}$ are applied. To find the solution Newton-Raphson procedure is implemented, where Jacobians of elastic and contact forces are efficiently computed applying derivation rules to the equations described in the previous sections. Stiffness matrix $\mathbf{K}$ is obtained from Jacobians, then incremental displacements are computed from residual loads $\mathbf{df}$ and assumed displacements $\mathbf{ds}$ given by boundary contitions:

$$\mathbf{dx} = \mathbf{K}_{ff}^{-1}(\mathbf{df} - \mathbf{K}_{fc}\mathbf{ds}) \tag{43}$$

where subscript $_{ff}$ refers to rows and columns corresponding to free dofs and subscript $_{fc}$ refers to free rows and constrained columns.

5. Numerical example

A numerical example is given and the results of the proposed discrete model are compared to those obtained by finite elements (FE). The study case is a conical spring, under compression, bending and a lateral force perpendicular to the bending plane. Spring ends are close coiled, so that one inactive coil is considered at each end.

Table 1. Features of the spring used in the example

Feature	Value
Wire diameter [mm]	1.5
Coil diameter at bottom [mm]	16.5
Coil diameter at top [mm]	7.5
Type of ends	close coiled
Number of total coils	8
Number of active coils	6
Constant pith [mm]	3
Total height [mm]	21
Compression Load [N]	20
Lateral load [N]	2
Bending moment [N mm]	20

The FE model is made of 3D beam elements placed along the helix. Top and bottom coils are rigidly connected to a central node, located on spring axis at the beginning and the end of active coils. Loads and boundary conditions are applied to these central nodes.
The deformed configuration is shown in figure 3, where the 3D rendering of the discrete model solution with four elements per coil (yellow coils with blue and red contact spheres), is compared to FEM results (light blue coils).

Table 2. 3D discrete and finite elements models

Feature	Discrete model	FE
Element type	Custom	3D beam
Elements per turn	3, 4, 6, 8, 12	40
Type of contact	Custom	External spherical gap, no sliding.
Contact elements per turn	3, 4, 6, 8, 12	4
BC at bottom end	all dof fixed	all dof fixed
BC at top end	all dof free	all dof free

The absolute displacements of the top centre node, where loads are applied, are plotted in figure ??. FEM and discrete models agree very well in the prediction of axial displacement. Some spread is observed when monitoring lateral displacements, which show a stronger dependence on the number of elements per coil. Solution given by the discrete model clearly changes when switching from three to four elements per coils and it seems not worthy to use more than six. A similar analysis on FE could not be run because of convergence problems. The best compromise sees four contact elements per coil. It would have been possible to use other contact types, but each one has its drawbacks, either in terms of accuracy or computational time.

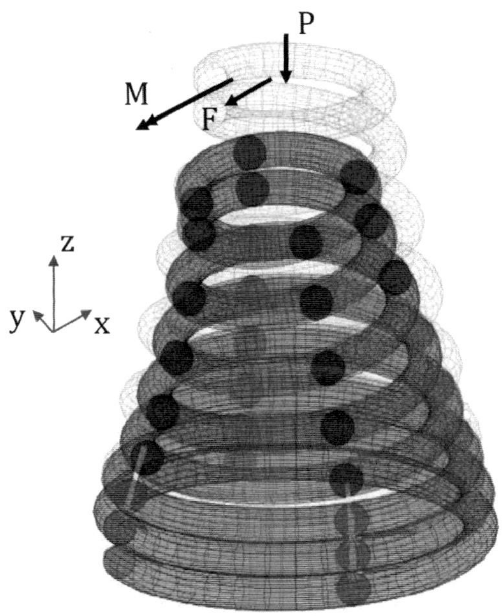

Figure 3. Deformed spring: Discrete model is yellow, FE is light blue. Blue spheres are non-active contact regions in the discrete model and red spheres are active ones. Wireframe helix is the undeformed configuration.

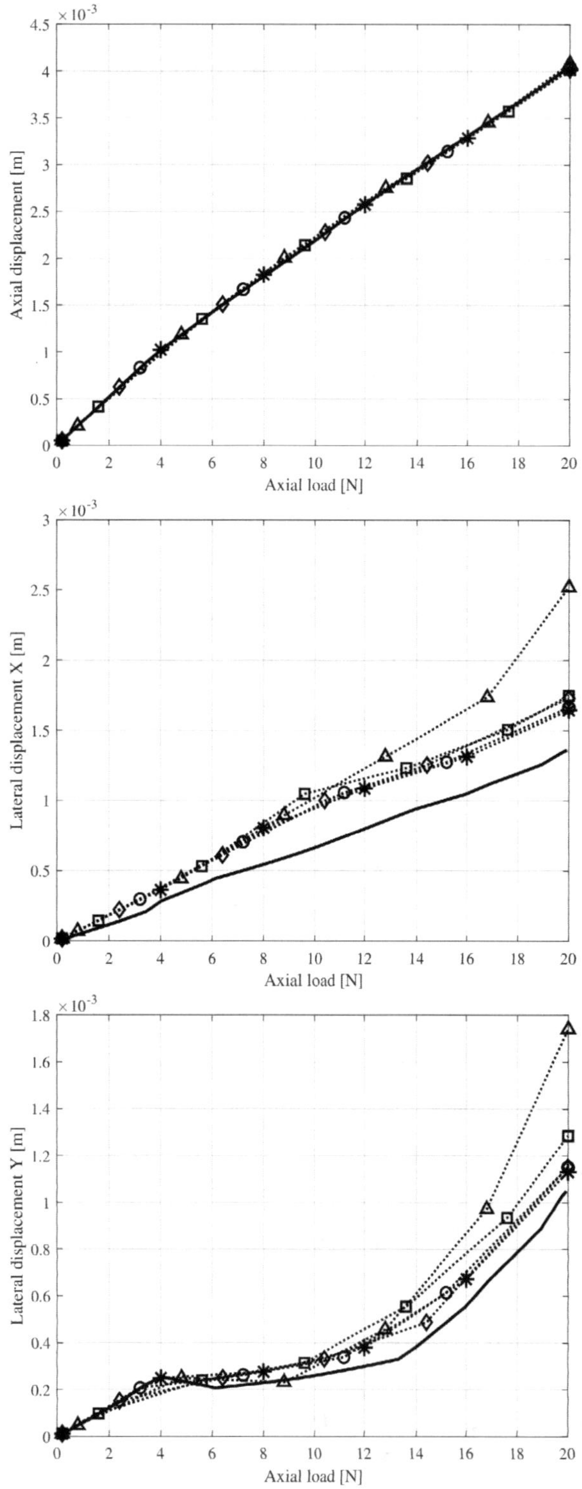

Figure 4. Displacements of top central node.
Finite elements ——
3D Discrete with 3 elements per coil △······
3D Discrete with 4 elements per coil □······
3D Discrete with 6 elements per coil ◇······
3D Discrete with 8 elements per coil ○······
3D Discrete with 12 elements per coil * ······

6. Conclusions

A 3D discrete model for non-linear springs is described and successfully applied to a case study. The model is based on Haringx equivalent beam theory, but take advantage of discretization in order to efficiently detect and simulate contact between spring coils. The peculiarity of the preposed modelling technique is the search and definition of contact on a projected geometry following deformation of the spring helix and the application of these effects on the equivalent beam nodes. The model is compared with FEM results; despite the challenges in contact managing it is shown that both modelling strategies, based on very different approaches, gave similar results, with an acceptable spread for practical purposes. The proposed 3D discrete model present several advantages in terms of computational time and model update procedures. The proposed modelling technique it is then considered a valuable basis for further developments aiming at the simulation of dynamic events.

References

[1] Blair G P, McCartan C and Cahoon W 2010 Measurement and Computation of the Characteristics of -Progressive Valve Springs *SAE International* 2010-01-1056
[2] Schamel A, Hammacher J and Utsch D 1993 Modeling and Measurement Techniques for Valve Spring Dynamics in High Revving Internal Combustion Engines *SAE Technical Paper* 930615
[3] Haringx J A 1948 On highly compressible helical springs and rubber rods, and their application fo vibration-free mountings, part I *Philips Research Reports* vol 3 (Eindhoven: Philps research Laboratories) pp 401-49
[4] Mizuno M 1960 Problem of large deflection of coiled springs, *Bulletin of JSME* **3** 95-103
[5] Michalczyk K 2015 Analysis of lateral Vibrations of the Axially Loaded Helical Spring *J. Theor. App. Mech-Pol* **53** 745-55
[6] Mottershead J E 1982 The large displacements and dynamic stability of springs using helical finite elements *Int. J. Mech. Sci.* **24** 547-58
[7] Becker L E, Chassie G G and Cleghorn W L 2001 On the natural frequencies of helical compression springs *Int. J. Mech. Sci.* **44** 825-41
[8] Lin Y Y and Pisano A P 1990 Three-dimensional dynamic simulation of helical compression springs *J Mech Des* **112** 529-37
[9] Lee J and Thompson D J 2001 Dynamic stiffness formulation, free vibration and wave motion of helical springs *J Sound Vib* **239** 297-320
[10] F De Crescenzo and P Salvini 2021 *IOP Conf. Ser.: Mater. Sci. Eng.* 1038 012064
[11] Timoshenko S P and Gere J M 1985 *Theory of elastic stability*, (New York: McGraw-Hill Book)
[12] Wahl A 1963 *Mechanical Springs* (New York: McGraw-Hill Book)
[13] Cheli F and Pennestrì E 2006 *Cinematica e Dinamica dei Sistemi Multibody* (Milano: Casa Editrice Ambrosiana)

IOP Conf. Series: Materials Science and Engineering 1214 (2022) 012026 doi:10.1088/1757-899X/1214/1/012026

On the Formulation of an Elastic-Plastic Beam Model: the Pre-Integration Idea

C Iandiorio[1] and P Salvini[1]

[1] University of Rome "Tor Vergata", Department of Enterprise Engineering,
Via del Politecnico 1 – 00133 Rome, EU, Italy
christian.iandiorio@uniroma2.it

Abstract. This paper provides a formulation of an elastic-plastic beam model. The constitutive behaviour of material after elasticity limit is considered perfect plastic. Although this is not a general plastic behaviour, it allows to minimize mathematical difficulties to fulfil the idea: pre-integration trough the thickness extended to material non-linearity. Three types of deformation mechanisms are found: fully elastic, plasticization over only one side or both of sides. The complete kinematic set of equations are obtained, together with the analytical stress trends. Yield stress-continuity equations are used to separate the elastic and plastic domains. These domains define a quasi-state diagram; Allowing to associate the normal force and bending moment acting on the section to the elastic-plastic state within the section. Note that in the case of a rectangular section, the solution is given in a fully analytical form, otherwise the domains are solved numerical solutions. But these last, may be obtained once for all, for any one-symmetrical section.

1. Introduction

The ultimate load-bearing capacity of structures, the processing through plastic deformation and the progressive crashing are among the interests that induced the necessity to extend the structural behaviour over the elastic-plastic response of materials. This scenario is usual in the production of components used in various fields, such as automotive, aerospace and transport industries. In mechanical engineering, the case of combined bending and traction stresses in plane is frequently encountered, particularly in the metal-forming of rods and metal wires [1,2,3,4,5]. The analytical study in the elastic-plastic field is troublesome [6], certainly due to the missing of the Principle of superimposition, due to non-linearity. For this reason, most of the models take into account only the bending stress, whatever is the non-linear constitutive law [7].

This paper considers the influence of the axial load in the behaviour of plastic deformation, i.e. the combined mechanism of axial and bending loads. This has an effect on the position of the neutral axis, making it possible that plasticity occur only in one side (upper or lower) or on both sides as in the case of pure bending. The prediction of the final shape after the tensile-bending process and release, where springback happens, is of particular interest. The results in terms of residual stresses and elastic return, moves considerably away from the solution given by bending loading alone.

In the Computer-Aided-Engineering the elastic-plasticity of metallic materials, or in general the inelastic behaviour, causes nonlinear force-displacement responses. The main trouble is that the material response, through its constitutive-relation, changes in the bulk according to the stress reached. This implies that the solution is not reached on a one-shot step, but by an incremental loading procedure. At each step of the increments, it is necessary to compute the actual stiffness deduced by the tangent material operator (moduli matrix) [8]. However, a numerical integration over the bulk (the section in a beam) is required, therefore a sampling over the thickness is carried out. There are two key points, often not sufficiently emphasized. The first is that this work-flow requires a sufficient number of sampling trough the thickness, which becomes burdensome especially for slender beams. The second is that non a-priori predictability about where plasticity will manifest and how quickly the yield front changes. This last implies that along the curvilinear abscissa a huge number of elements are required. At the best of authors knowledge, there is no way to avoid these complications, because the direct connection between loads and displacements is only possible in the elastic case.

A shortcut often considered in beam structures is the lumped plastic-hinge method [9,10] that, although unable to accurately describe the elastic-plastic state within the sections, is capable to estimate the overall collapse of a frame structure. This analysis takes the name of first order plastic hinge.

Another peculiarity inherent to elastic-plastic events, is the need to follow the load-history path, due to the partial irreversibility of the deformation process. In fact, the same final forces and moments acting on the section can be reached by countless different paths, which although differ for stress-strain field. Therefore, the final state (stress, deformation, displacements) is generally not a state function. To avoid these difficulties, proportional load-paths are always accounted, but this may be unrealistic if a more complex load history applies.

The model proposed in this paper allows to pre-compute over the thickness the behaviour of each section being beyond its elastic limit. Therefore, the procedure provides a way to built-up a beam element, pre-integrated through the thickness, whose actual elastic-plastic response is only governed by section loads. In other words, the stress state and the kinematic behaviour of the section only depends on two variables: axial load and bending moment. Three elastic-plastic cases of deformation-mechanics are possible: totally elastic, one-side plasticization (unilateral, upper or lower) and two-side plasticization (bilateral). In the case of monotonic and proportional loading, a state diagram emerges. It returns the deformation-mechanism given the values of the axial and the bending loads, i.e. one-shot calculation. Considering non-trivial path-loadings, the step history must be followed but the cited-on state diagram is still useful to solve each step.

As demonstrated hereinafter, the model solution is completely analytical for rectangular beam sections. For any mono-symmetrical cross-sections (e.g. circular or I-beams) the solution requires an a-priori numerical integration for the determination of the boundaries among the elastic-plastic domains, but the state function loads-displacements still exists.

2. The Elastic-Plastic Deformation

In this section, the governing equations of the elastoplastic beam are obtained. The material is isotropic, with elastic-perfectly plastic behaviour, same under tension or compression (figure 1). The cross-section is symmetrical along the *y-axis* (figure 2) and its shape keeps constant along the Cartesian abscissa. The kinematic mechanism of the displacements field is not influenced by elastic-plastic behaviour; therefore, in a planar bending the engineering type 2D-beam model shows up as:

$$U(x,y) = u(x) - y\,\psi(x) \tag{1}$$

$$V(x,y) = v(x) \tag{2}$$

The positive directions of the displacement components are shown in figure 2.

The beam model follows Bernoulli-Euler hypotheses: the shear strain is assumed null and the strain-displacements relations are:

$$\varepsilon(x,y) = u' - y\,\psi' \tag{3}$$

$$\gamma(x,y) = 0 \;\; (\rightarrow \;\; \psi = v') \tag{4}$$

The differential equilibrium is independent of the elastic-plastic behaviour; the normal force, the shear one and the bending moment derivatives are governed by:

$$N' = -q_x \tag{5}$$

$$T' = -q_y \tag{6}$$

$$M' = -T \tag{7}$$

With an elastic-perfectly plastic behaviour three deformation mechanism are possible (figure 3): totally elastic, one-side (upper or lower) plasticization and two-side plasticization. Each of them corresponds to a different kinematic set of equations (to be found) which characterizes the deformation mechanism.

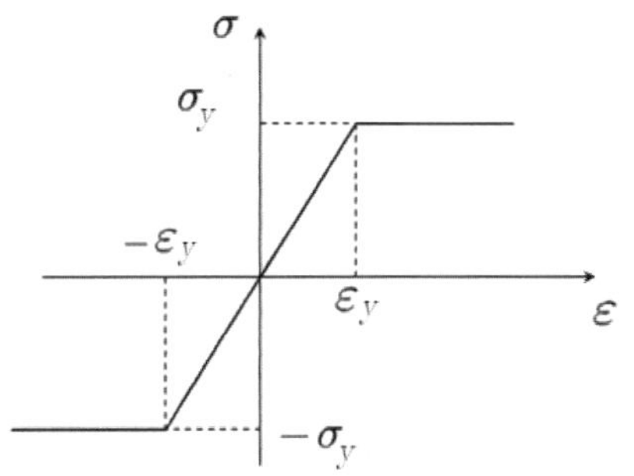

Figure 1. Elastic-perfectly plastic behaviour of material

2.1. The totally elastic case

The governing equations for the full elastic case are straightforward and, e.g., can be found in [11]. They are reported here for completeness. The normal force N, the shear force T and the bending moment M are coupled with the derivatives of displacement components as:

$$N = EAu' \tag{8}$$

$$T = -EIv''' \tag{9}$$

$$M = EIv'' \tag{10}$$

Where A and I are geometric constants, respectively, the cross-section area and the moment of inertia, that for one-symmetrical section can be computed using the knowledge of the width function $b(y)$ and the quantities h^+, h^- (figure 4):

$$A = \int_{h^-}^{h^+} b(y)\, dy \;\; ; \;\; I = \int_{h^-}^{h^+} y^2\, b(y)\, dy \tag{11}$$

Furthermore, the normal stress is expressed by the binomial Navier's formula:

$$\sigma(x,y) = \frac{N}{A} - \frac{M}{I} y \tag{12}$$

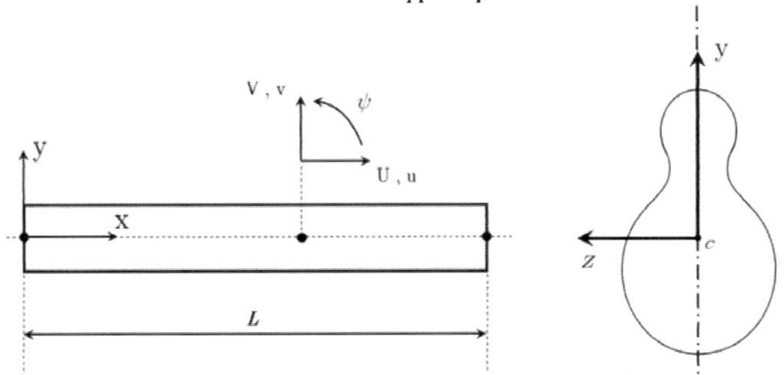

Figure 2. Positive displacement directions

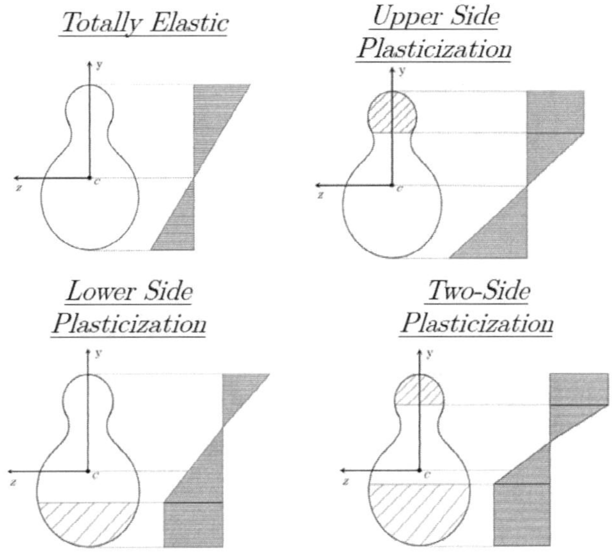

Figure 3. Possible deformation mechanisms

2.2. Upper side plasticization

In this case the section is in partial plasticization at the upper side, identified by the unknown ordinate y_{p+} (figure 5). This quantity represents the limit of the plasticization front. This case is not comprehensive of all combination of (N, M). Being the resulting forces and moments directed as the reference system, it is simple to prove that the only combinations that trigger this plasticization case requires that $(N > 0, M < 0)$ and $(N < 0, M > 0)$, the first caused by traction the second by compression yielding. Therefore, the constitutive equation can be written as:

$$\sigma = \begin{cases} sign^{+}(\sigma_y)|\sigma_y| \, , & y \geq y_{p+} \\ E(u' - y\,v'') \, , & y \leq y_{p+} \end{cases} \tag{13}$$

Where the $sign$ function is defined as:

$$sign^{+}(\sigma_y) = \begin{cases} +1\,, & (N > 0, M < 0) \\ -1\,, & (N > 0, M < 0) \end{cases} \tag{14}$$

Imposing the equivalence between the internal stress and the section force and moment resultant and using the equations (3,4,13) one obtains:

$$N(x) = sign^{+}(\sigma_y)|\sigma_y|A_{p+} + EA_e u' - ES_e v'' \tag{15}$$

$$M(x) = sign^{+}(\sigma_y)|\sigma_y|S_e - ES_e u' + EI_e v'' \tag{16}$$

By equation (7) the relation between shear force and displacement results.
In the latter equations the following geometric quantities appears: the area of the (upper) plastic region A_{p+}, the area of the elastic region A_e, the static moment of the elastic region S_e, and the moment of inertia of the elastic region I_e. These geometrical quantities are expressed by:

$$A_e(x) = \int_{h^-}^{y_{p+}(x)} b(y)\, dy \;\; ; \;\; A_{p+}(x) = \int_{y_{p+}(x)}^{h^+} b(y)\, dy \tag{17}$$

$$S_e(x) = \int_{h^-}^{y_{p+}(x)} y\, b(y)\, dy \;\; ; \;\; I_e(x) = \int_{h^-}^{y_{p+}(x)} y^2\, b(y)\, dy \tag{18}$$

It is worth noting that, differently from the full elastic case, the quantities in equations (17,18) are not constant along *x-axis*, but they change inasmuch they depend on the trend of $y_{p+}(x)$.
Extracting the expression of u' and v'' by equations (15,16), the stress trend is found using equation (13), obtaining:

$$\sigma(x,y) = \begin{cases} sign^{+}(\sigma_{sn})|\sigma_{sn}|\,, & y \geq y_{p+} \\ \left(1 + \dfrac{{S_e}^2}{I_e A_e - {S_e}^2}\right)\dfrac{N}{A_e} + \dfrac{S_e}{I_e A_e - {S_e}^2} M - sign^{+}(\sigma_{sn})\dfrac{|\sigma_{sn}|}{A_e}\left(A_{p+} + \dfrac{A\,{S_e}^2}{I_e A_e - {S_e}^2}\right) + & \\ \quad -y\,\dfrac{S_e N + A_e M - sign^{+}(\sigma_{sn})|\sigma_{sn}| A\, S_e}{I_e A_e - {S_e}^2}\,, & y \leq y_{p+} \end{cases} \tag{19}$$

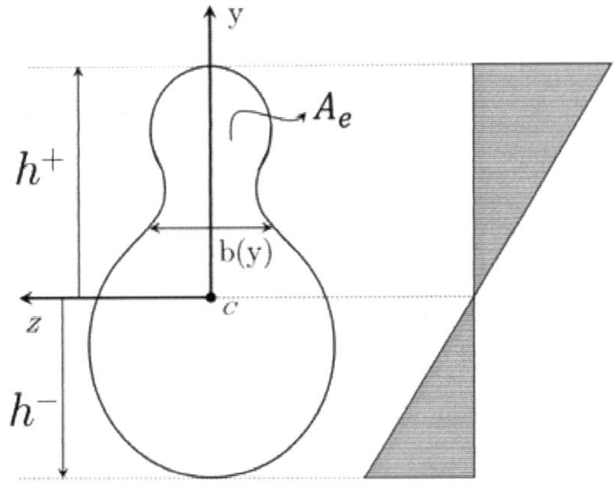

Figure 4. Totally elastic case: nomenclature

The unknow y_{p+} is solved imposing that for $y = y_{p+}$ (figure 5) the equations in (19) are equal:

$$M(S_e - A_e y_{p+}) + N(I_e - S_e y_{p+}) + sign^+(\sigma_{sn})|\sigma_{sn}|A(I_e + S_e y_{p+}) = 0 \quad (20)$$

From which it is evident that the trend of the plasticization front y_{p+} depends on the trend of $N(x)$, $M(x)$ other than the cross-section shape.

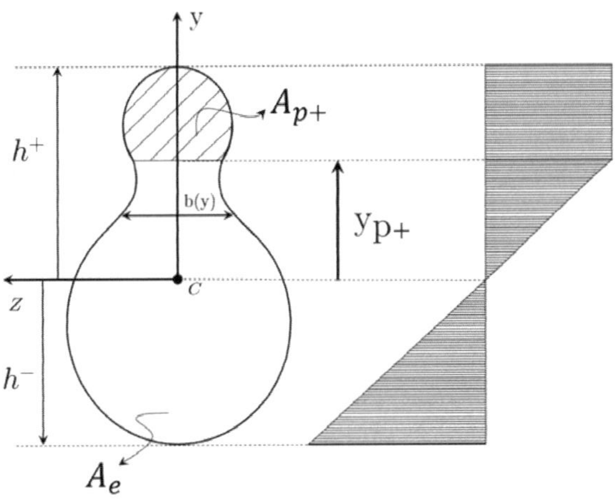

Figure 5. Upper side plasticization: nomenclature

2.3. Lower side plasticization

The equations that govern the case of lower side (figure 6) plasticization is formally identical to the previous one unless the replacement of the $sign^+$ function with:

$$sign^-(\sigma_y) = \begin{cases} +1, & (N > 0, M > 0) \\ -1, & (N < 0, M < 0) \end{cases} \quad (21)$$

and the area A_{p+} with A_{p-}.

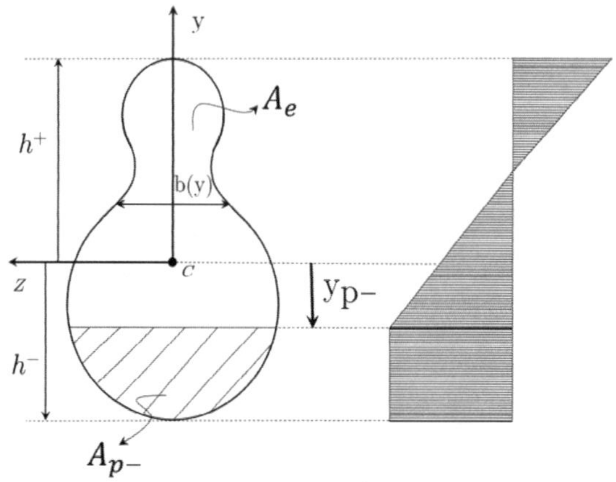

Figure 6. Lower side plasticization: nomenclature

In this case, the geometrical quantities are function of the trend of the ordinate $y_{p-}(x)$:

$$A_e(x) = \int_{y_{p-}(x)}^{h^+} b(y)\ dy\ ;\ A_{p-}(x) = \int_{h^-}^{y_{p-}(x)} b(y)\ dy \tag{22}$$

$$S_e(x) = \int_{y_{p-}(x)}^{h^+} y\, b(y)\ dy\ ;\ I_e(x) = \int_{y_{p-}(x)}^{h^+} y^2\, b(y)\ dy \tag{23}$$

2.4. Two-side plasticization

When a two-side plasticization holds, two ordinates y_{p+} , y_{p-} appear to manage the plasticization front trend (figure 7). The constitutive equation becomes:

$$\sigma = \begin{cases} sign^{\pm}(\sigma_y)|\sigma_y|\,,\ y \geq y_{p+} \\ E(u' - y\,v'')\ ,\ y_{p-} \leq y \leq y_{p+} \\ -sign^{\pm}(\sigma_y)|\sigma_y|\,,\ y \leq y_{p-} \end{cases} \tag{24}$$

Where:

$$sign^{\pm}(\sigma_y) = \begin{cases} +1\,, & (N > 0, M < 0)\ or\ (N < 0, M < 0) \\ -1\,, & (N > 0, M > 0)\ or\ (N < 0, M > 0) \end{cases} \tag{25}$$

Again, the relations between the resultants and the displacements for this case are:

$$N = sign^{\pm}(\sigma_y)|\sigma_y|(A_{p+} - A_{p-}) + EA_e u' - ES_e v'' \tag{26}$$

$$M = sign^{\pm}(\sigma_y)|\sigma_y|(S_{p+} - S_{p-}) - ES_e u' + EI_e v'' \tag{27}$$

Where:

$$A_e(x) = \int_{y_{p-}(x)}^{y_{p+}(x)} b(y)\,dy \;;\; A_{p+}(x) = \int_{y_{p+}(x)}^{h^+} b(y)\,dy \;;\; A_{p-}(x) = \int_{h^-}^{y_{p-}(x)} b(y)\,dy \qquad (28)$$

$$S_e(x) = \int_{y_{p-}(x)}^{y_{p+}(x)} yb(y)\,dy \;;\; S_{p+}(x) = \int_{y_{p+}(x)}^{h^+} yb(y)\,dy \;;\; S_{p-}(x) = \int_{h^-}^{y_{p-}(x)} yb(y)\,dy \qquad (29)$$

$$I_e(x) = \int_{y_{p-}(x)}^{y_{p+}(x)} y^2 b(y)\,dy \qquad (30)$$

Obtaining the expression of u' and v'' by equations (26,27), the stress trend is given by equation (24):

$$\sigma(x,y) = \begin{cases} sign^{\pm}(\sigma_{sn})|\sigma_{sn}|\,, \; y \geq y_{p+} \\ \left(1+\frac{S_e^{\,2}}{I_eA_e - S_e^{\,2}}\right)\frac{N}{A_e} + \frac{S_e}{I_eA_e - S_e^{\,2}} M - sign^{\pm}(\sigma_{sn})\frac{|\sigma_{sn}|}{A_e}\Big[(A_{p+} - A_{p-}) + \\ + \frac{(A_{p+} - A_{p-})S_e - A_e(S_{p+} - S_{p-})}{I_eA_e - S_e^{\,2}} S_e\Big] - y\,\frac{S_eN + A_eM}{I_eA_e - S_e^{\,2}} + \\ -y\,\frac{sign^{\pm}(\sigma_{sn})|\sigma_{sn}|\left[(A_{p+} - A_{p-})S_e - A_e(S_{p+} - S_{p-})\right]}{I_eA_e - S_e^{\,2}}\,, \; y_{p-} \leq y \leq y_{p+} \\ -\,sign^{\pm}(\sigma_{sn})|\sigma_{sn}|\,, \; y \leq y_{p-} \end{cases} \qquad (31)$$

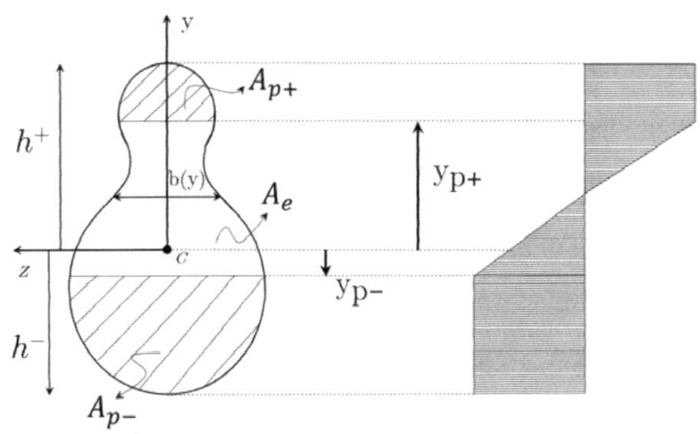

Figure 7. Two-side plasticization: nomenclature

The unknows y_{p+}, y_{p-} are found imposing that for $y = y_{p+}$ and $y = y_{p-}$ the second of equation (31) must be equal to the first or the third, respectively. Therefore, two stress-continuity equations allow to solve the two above mentioned unknowns:

$$M(S_e - y_{p+}A_e) = N[S_e\, y_{p+} - I_e] - sign^{\pm}(\sigma_{sn})|\sigma_{sn}|\,[(A_{p+} - A_{p-})(S_e\, y_{p+} - I_e) + \\ +(S_{p+} - S_{p-})(S_e - y_{p+}A_e) - I_eA_e + S_e^{\,2}]$$

$$M(S_e - y_{p-}A_e) = N[S_e\, y_{p-} - I_e] - sign^{\pm}(\sigma_{sn})|\sigma_{sn}|\,[(A_{p+} - A_{p-})(S_e\, y_{p-} - I_e) + \\ +(S_{p+} - S_{p-})(S_e - y_{p-}A_e) - I_eA_e + S_e^{\,2}] \tag{32}$$

2.5. Fully plastic state

The fully plastic condition is a special case of the previous two-side plasticization where the elastic area is null and the ordinates y_{p+} and y_{p-} coincide. Therefore, the only unknown ordinate y_p appears (figure 8). Considering null the elastic region in equation (26,27):

$$N = sign^{\pm}(\sigma_y)|\sigma_y|(A_{p+} - A_{p-}) \tag{33}$$

$$M = sign^{\pm}(\sigma_y)|\sigma_y|(S_{p+} - S_{p-}) \tag{34}$$

Where the expression of the geometrical quantities A_{p+}, A_{p-}, S_{p+}, S_{p-} are simply found imposing $y_{p+} = y_{p-} = y_p$ in the equation (28,29).

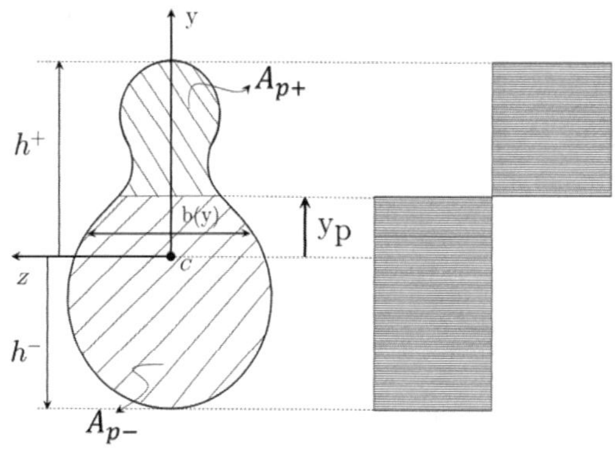

Figure 8. Fully plastic state: nomenclature

3. The quasi-State Diagram

The kinematic equations are now fully deduced for the three possible deformation-mechanics. In order to apply the pre-integration, it is necessary to deduce from the mere knowledge of (N, M) how the section plasticizes. For this purpose, it is unavoidably to solve the continuity stress equations (20,32,33,34) which imply the computation of some integrals.

The case of rectangular cross-section $b \times h$ allows a full and straightforward analytical solution that is reported in the following. The steps are not explained here because they are long, and they could mask the basic idea of this paper (only lengthy algebraic manipulations are required).

The equations (33,34) provide the limit parabola that corresponds to collapse (fully yield region):

$$m + n^2 = 1 \tag{35}$$

where a convenience notation involving dimensionless variables is adopted:

$$n = \frac{N}{N_p} = \frac{N}{bh\sigma_y} \quad ; \quad m = \frac{M}{M_p} = \frac{4M}{bh^2\sigma_y} \tag{36}$$

In which N_p, M_p are the values of the resultant force and moments that, alone, cause the entire plasticization of the cross-section (collapse).
Solving the equation (20), the trend of the plasticization front y_{p+} is found:

$$\frac{y_{p+}(x)}{h} = 2 - \frac{3m(x)}{2\big(1 - n(x)\big)} \tag{37}$$

Imposing $y_{p+} = \frac{h}{2}$ the boundary of incipient yielding of the upper-side is obtained:

$$m = \frac{2}{3}(1 - n) \tag{38}$$

Similarly, for the case of lower side plasticization the incipient lower yielding gives:

$$\frac{y_{p-}(x)}{h} = -2 + \frac{3m(x)}{2\big(1 - n(x)\big)} \tag{39}$$

$$m = \frac{2}{3}(1 + n) \tag{40}$$

In the same way, for bilateral plasticization, the equation (32) provides:

$$\frac{y_{p+}(x)}{h} = \frac{1}{2}\left[-n(x) + \sqrt{3\big(1 - n^2(x) - m(x)\big)}\right] \tag{41}$$

$$\frac{y_{p-}(x)}{h} = \frac{1}{2}\left[-n(x) - \sqrt{3\big(1 - n^2(x) - m(x)\big)}\right] \tag{42}$$

Imposing $y_{p+} = \frac{h}{2}$, $y_{p-} = -\frac{h}{2}$ the incipient two-side plasticization boundary is given:

$$m = \frac{2}{3}(1 - n)(1 + 2n) \tag{43}$$

Figure 9 shows the perfect coincidence of these analytical results when compared with the numerical calculation of the equations (20,32,33,34).

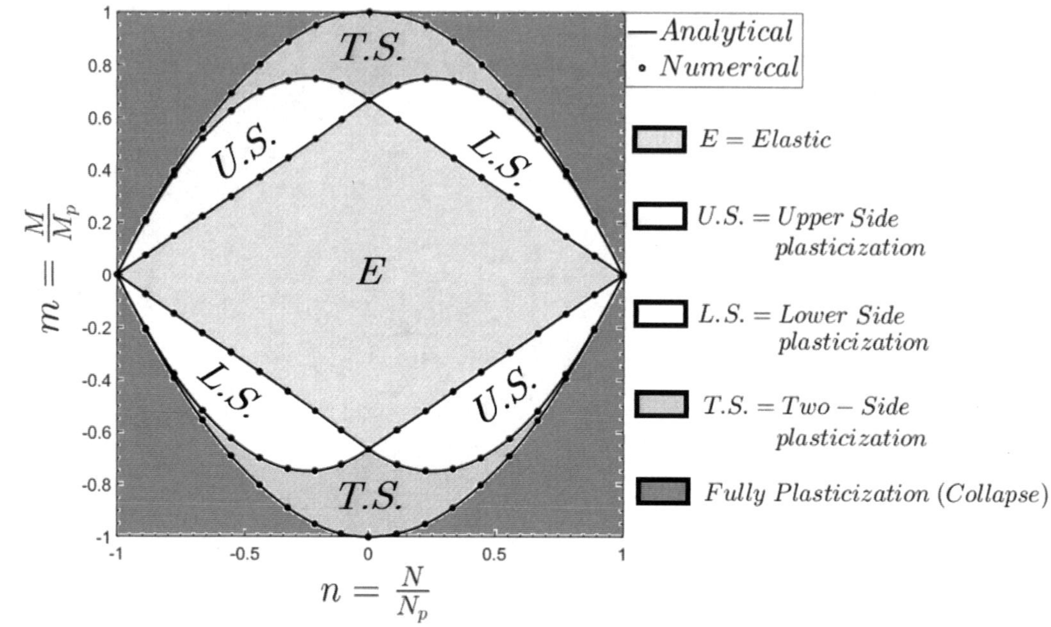

Figure 9. The quasi-state diagram

Therefore, given only the values of a pair (N, M) ,besides the elementary elastic solution, it derives the plasticization type, to which the respective kinematic equations are associated, and the trend of the plasticization front. It is worth noting that these results are lawful only for monotone and proportional path loading, this is why it is a quasi-state diagram. For complex histories a step analysis is required. At each step one should carefully account whether an additional plasticization occurs or an elastic return holds. Although, the quasit-state diagram remains useful to analyses this occurrence.

4. Conclusions

This paper addresses a complete formulation of beam model extended to elastic-perfect plastic behavior. The material behaves in same way under traction or compression. The cross-section keeps constant along the Cartesian abscissa and one-symmetry holds along the y-axis.

Three different deformation mechanics appear possible: fully elastic, one-side plasticization (upper or lower) and two-side plasticization. For all of these a different set of kinematic equations are derived, together with their analytical stress trend. For each of the elastic-plastic cases one or two stress-continuity equations appear; these equations are fundamental, since they allow to return the location of the plasticization front. This makes it possible to predict, having the end forces applied: the plasticization front trend, the type of plasticity and the kinematic response. For a monotonic and proportional path loading the diagram develops like a state function. The history-loading can be followed again using this graph, observing when a further plasticization occurs or an elastic unloading holds. For the case of a rectangular beam section the quasi-state diagram is solved analytically. For other cross-sections it is always possible to build up the diagram after the numerical solutions of the governing equations. These numerical solutions, however, can be carried out once for all, for any cross section.

References

[1] Ma J and Welo T 2021 Analytical springback assessment in flexible stretch bending of complex shapes *Int. J. of Machine Tools and Manufacture* **160**

[2] Bonfanti A, Syngellakis S and Bhaskara A 2017 Response and residual curvature of bent-stretched circular rods with applications to metal forming: closed-form solutions for elastic-perfectly plastic and hyperbolic hardening materials *Int. J. of Mech. Sciences* **123** pp 340-349
[3] Slater R 1977 Engineering and Plasticity: Theory and application to metal forming processes *Macmillan Press London*
[4] Altan T 2012 Sheet metal forming processes and applications *ASM International Ohio*
[5] Yu T and Zhang L 1996 Plastic bending: theory and applications *World Scientific Pub Co Inc*
[6] Štok B and Halilovic M 2009 Analytical solutions in elasto-plastic bending of beams with rectangular cross section *Appl. Math. Modelling* **33** pp 1749–1760
[7] Rimovskis S and Sabaliauskas 2012 Analysis of rectangular and circular cross-section power hardening elements under pure bending *Int. J. of Materials Engineering* **2** pp 84-89
[8] Zienkiewicz O and Taylor R 2000 The Finite Element Method Solid Mechanics volume 2 *Butterworth-Heinemann Oxford*
[9] Kim D, Son H, Yi Y and Hong S 2019 generalized finite element formulation for efficient first-order plastic hinge analysis *Advances in Mech. Engineering* **11** pp 1–19
[10] Ehrlich D and Armero F 2005 Finite element methods for the analysis of softening plastic hinges in beams and frames *Comp. Mech.* **35** 237–264
[11] Timoshenko S 1940 Strength of materials part I *D. Van Nostrand Company New York*

Integrated multi-objective optimisation of the support structure of heliostats in concentrated solar power plants using a genetic algorithm

L Lomazzi[1], F Cadini[1]*, M Giglio[1], C Cancro[2], G Ciniglio[2], G Graditi[2] and A Pontecorvo[2]

[1]Politecnico di Milano, Department of Mechanical Engineering, Via La Masa n.1, Milan, Italy
[2] Italian National Agency for New Technologies, Energy and Sustainable Economic Development, ENEA – Research Center, Piazza E. Fermi 1, 80055, Portici (NA), Italy

*Corresponding author: francesco.cadini@polimi.it

Abstract. The optimisation of the support structure of heliostats in concentrating solar power plants is a fundamental task aimed at attempting to reduce the high levelised cost of energy (LCOE) of current configurations. In this work, an integrated multi-objective optimisation framework is presented, which relies on the combination of a lean and fast structural model with a genetic algorithm to simultaneously minimise both the overall mass of the support structure and the mean angle of rotation of the mirror surface, which directly affects the optical efficiency of the component. A particular feature of the proposed framework is that it represents an integrated solution, i.e., it allows to simultaneously optimise the main components of the heliostat support structure, i.e., the pedestal, the truss and the back support structure, assuming they are off-the-shelf components easily available on the market. The optimisation problem is set up selecting as design variables (i) the number of elements in the back support structure and (ii) the relevant characteristics of all the components considered, i.e., section shape and dimensions, according to the components commercial datasheets. At each iteration of the optimisation process, the structural model is fed with the current design variables values and, according to some computed aerodynamic loads, it allows evaluating the displacement and rotation of the points of interest within the mirror surface. An aerodynamic model present in the literature based on experimental wind tunnel tests is used to estimate the wind forces acting on the heliostat as a function both of the mirror inclination angle with respect to the ground and of the wind direction with respect to the mirror orientation. In this work, the proposed methodology is demonstrated on a realistic case study and the results commented in detail, highlighting possible future developments and the limitations of the framework.

1. Introduction

The cost of heliostats is an impacting factor on the overall cost of concentrated solar power plants [1]. Since most of this cost is determined by the amount of material of the heliostat support structure, driving the design phase of such a structure towards weight reduction would be beneficial to reduce the high levelised cost of energy (LCOE) of current configurations [2]. Several studies have been conducted in the past years on possible cost reduction strategies, these have identified that both the heliostat size and the properties of the elements in the support structure should be optimised to achieve an effective cost reduction [3–6]. However, the selection of the structural designs should not only rely on the cost of the structure, but also on its performance. In fact, lean heliostat supports may lead to cheap structures, while allowing for deformations under operative loads, such as wind loading and gravity, that may result in unsatisfactory optical efficiency [7]. Hence, the heliostat structural design should pursue a trade-off between the cost of the system and its performance. To this purpose, it is paramount to determine the structural deformation of the reflective surface of the heliostat, which is herein referred to as *mirror*, under gravity and wind loading, since it directly affects the optical efficiency of the system [8]. The mechanical performance of the heliostat structure under the loading conditions mentioned above is typically assessed resorting to finite element analyses [9–11]. However, this type of analysis is computationally demanding and is not compatible with an efficient structural design optimisation scheme, which would benefit from faster methods to evaluate the mechanical performance of possible structures. Despite this, some works in the literature have been proposed that optimise the heliostat structure employing finite element models to evaluate the mechanical performance of the design, but the optimisation approach is limited to the evaluation of a little number of possible solutions [2,6]. Instead, an optimisation approach that evaluates a satisfactory number of possible solutions is presented in the work in [12], that employs a particle swarm optimisation algorithm to identify the best heliostat shape that simultaneously minimises the occupied field area and maximises the field efficiency. However, to the authors' best knowledge, extensive approaches of this type focused on the structural optimisation of the heliostat structure appear to be missing in the literature.

In this work, a methodological approach to the optimisation of the support structure of heliostats in concentrated solar power plants is presented. The proposed methodology relies on an optimisation framework based on a genetic algorithm, that interacts with a simplified structural model of the heliostat structure to drive the design process towards the simultaneous minimisation of the structural weight and of the mirror mean deformation. This paper is organised as follows. Section 2 presents the main features of the structural model and of the optimisation framework proposed in this work. Section 3 demonstrates the approach to the optimal design of the heliostat support structure by means of a realistic case study and discusses the main results of the optimisation process. Finally, Section 4 draws out the conclusions from this work.

2. Methodological approach

In this Section the methodological approach set up to identify the optimal configuration of the support structure of heliostats in concentrated solar power plants is presented. This approach involves the interaction between the structural model of the heliostat support structure and the optimisation method employed, which is identified in a genetic algorithm.

2.1. *Structural model of the heliostat support structure*

The heliostat structure considered in this work is shown in Figure 1.

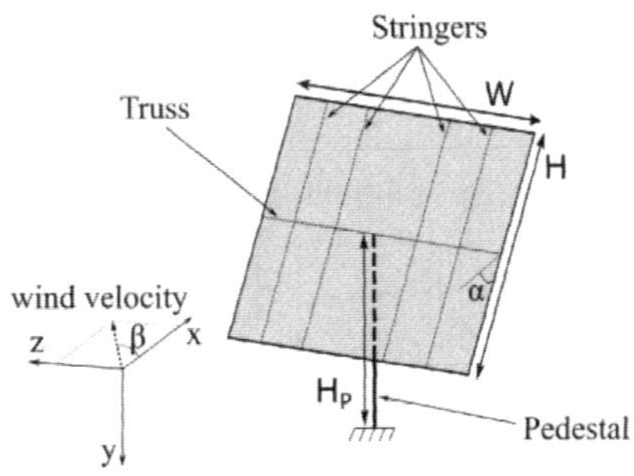

Figure 1. Heliostat considered in this work.

The heliostat structure is composed of the mirror and its support structure, that includes the stringers, the truss and the pedestal. The mirror stiffness is neglected in the computations, while all the structural components of the support structure are simplified as beams clamped at the mutual intersections. Hence, the pedestal is composed of one single element from the ground to the truss, the truss is assumed to be clamped at its geometrical centre, in correspondence of the intersection with the pedestal, while one end of each stringer is considered to be clamped at the intersection with the truss. Geometric symmetry is exploited to speed up the computations. In fact, since the structure is symmetric with respect to the planes perpendicular to the mirror surface passing through the edges mid points, structural evaluations are performed considering one quarter of the support structure only, retrieving the results at the other quarters thanks to symmetry considerations. This is shown in Figure 2, where the mirror symmetry planes are identified and the quarter selected for the computations is highlighted with black diagonal stripes.

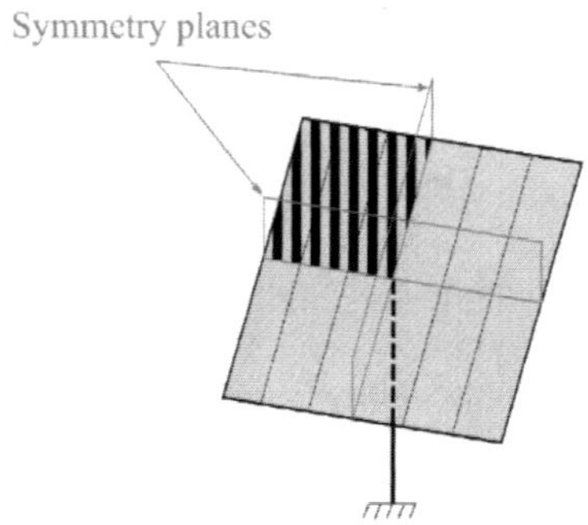

Figure 2. Symmetry planes and quarter selected for the computations (black diagonal stripes).

The structural computations are performed resorting to the Euler-Bernoulli beam theory [13]. The main assumption of the structural model is that all the loads acting on the structural elements in a position different from the beam ends are simplified as distributed equivalent loads acting over the whole length of the element. The authors verified that this assumption is reasonable and does not limit the validity of the approach, but this proof is not reported here for the sake of brevity. Moreover, all the structural elements are assumed to be described by a perfectly linear material constitutive law. This assumption allows determining the deflection and the rotation of some points of interest along the structural elements, independently considering each structural element, linearly combining these results in a later phase of the computations to get the overall displacement of the points of interest on the mirror surface.

The loading condition is represented by (i) the weight of the structural elements and of the mirror and (ii) wind forces and moments acting along the axes of the reference system shown in Figure 1. In particular, according to the considerations in the work in [14], only the aerodynamic forces along the x and the y axes and the aerodynamic moments described by vectors directed as the y and the z axes, i.e., the azimuth and elevation moments, respectively, are considered in the structural model. The magnitude of the wind loads is computed according to the equations presented in the work in [14], which are not reported here for the sake of brevity. The user-defined parameters representing the input values to the aerodynamic equations are the wind direction, which lies in the xz plane and is identified by the azimuth angle (β), i.e., the horizontal misalignment between the wind direction and the vector normal to the mirror surface, and the mirror elevation angle (α) (Figure 1), the wind peak velocity ($V_{w,p}$), the wind peak velocity-to-mean velocity ratio (r), the height of the wind velocity measuring station (H_w) and an adimensional number accounting for environmental effects (n).

No more information is reported in this Section about the structural model, since this is not the main focus of this work and since the equations employed, e.g., the Euler-Bernoulli beam theory equations, are well known to the research community.

2.2. *Optimisation method*

The optimisation method employed in this work involves a genetic algorithm. This type of algorithms is commonly used in several engineering fields to identify minima and maxima of a function of interest. In particular, with regards to structural engineering applications, this method is typically employed to optimise the structural design in terms of minimisation of the weight of the solution and maximisation of its stiffness.

In this context, embedding optimisation procedures into the design process of the heliostat structure represents a promising approach to identify the optimal structural configurations. In this work, the optimal configurations are represented by the set of structural solutions that simultaneously minimise both the weight of the support structure of the mirror and its average deformation.

With regards to the structural configuration described in Section 2.1, the design variables considered in the optimisation process are represented by (i) the number of stringers (x_1), (ii) the stringers section properties (x_2), (iii) the truss section properties (x_3) and (iv) the pedestal section properties (x_4). The design variables are all described by discrete integer numbers only. In fact, the design variables representing the section properties are identified by the corresponding index within a built-in database of off-the-shelf components easily available on the market including H, T, hollow rectangular, hollow square and hollow circular-shaped sections [15]. Two objective functions are considered in the optimisation process, i.e., the weight of the structural support structure (W) and the mean deformation of the mirror (R). The latter is included in the analysis since minimising the mirror deformation is equivalent to maximising the optical efficiency of the heliostat, which represents a paramount variable for this type of systems. Note that in this work *deformation* refers to the mirror rotation vector component lying in the mirror plane, which is computed as the mean value of the deformation measured at some points of interest within the mirror surface. Hence, identifying with α the elevation angle of the mirror (**Figure 1**) and with i the ith point of interest within the mirror surface, the objective functions considered in this work are defined by Equations (1) and (2):

$$W=W_{stringers}+W_{truss}+W_{pedestal} \tag{1}$$

$$R=\text{mean}\left(rot_{//,i}^{\alpha}\right) \quad \forall\alpha,i \tag{2}$$

where $W_{stringers}$, W_{truss} and $W_{pedestal}$ are the stringers, the truss and the pedestal weight values, respectively, while $rot_{//,i}^{\alpha}$ represents the rotation of the ith point of interest when the mirror is characterised by the elevation angle α. Note that the optimisation process considered in this work is a multi-objective optimisation process, since two objective functions are involved to evaluate the possible solutions.

More details on the specific parameters of the genetic algorithm employed in the optimisation framework are presented in the case study shown in Section 3.

3. Case Study

In this Section the proposed methodology is demonstrated by means of a case study.

The heliostat considered in this case study is shown in Figure 3. It consists of a 5mm thick mirror with surface size 6x6m^2 and density ρ_m=2.5kg/m^2 mm, backed by a support structure composed of stringers, one truss and one 3.4m high pedestal. The structural elements of the support structure are made of a generic steel with density ρ_s=7890kg/m^3, Elastic modulus E=200GPa, Poisson's ratio ν=0.33 and yield stress σ_{VM}=355MPa.

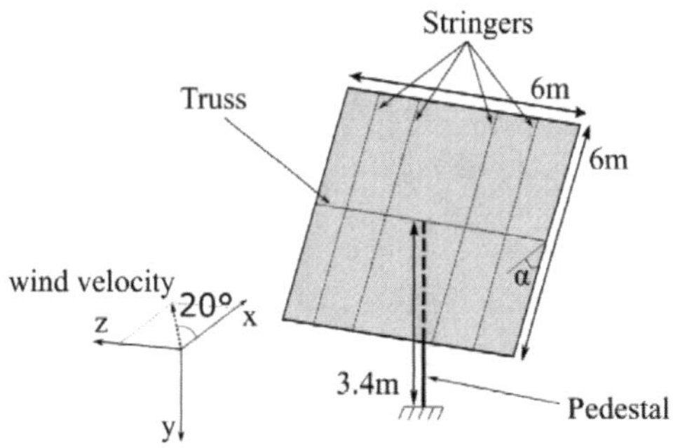

Figure 3. Heliostat structure considered in the case study.

Three elevation angles are considered in the analysis, i.e., α=[30°,50°,70°], and the wind velocity is assumed to hit the mirror with azimuth angle β=20° to consider the most critical wind loading conditions [14]. Moreover, the heliostat is supposed to be placed in free field (n=0.143 [14]) and to be hit by wind with peak velocity $V_{w,p}$=64km/h and peak velocity-to-mean velocity ratio equal to r=1.6. The wind data are assumed to have been obtained from a H_w=4m high measuring station.

As already anticipated in Section 2.2, the design variables are identified in (i) the number of stringers (x_1), (ii) the stringers section properties (x_2), (iii) the truss section properties (x_3) and (iv) the pedestal section properties (x_4). Some constraints are introduced in the analysis to limit the range of variability of the design variables. In particular, x_1 can only assume integer numbers from 2 to 8, the stringers and the truss sections can vary within the whole database, which is composed of 213 structural solutions, while the pedestal is limited to hollow rectangular, square and circular-shaped sections. For the sake of clarity, the index ranges identifying the different section shapes in the database considered in this work are shown in Table 1.

Table 1. Database composition.

Section shape	Index range
H	[1, 18]
T	[19, 29]
Hollow rectangular	[30, 116]
Hollow square	[117, 164]
Hollow circular	[165, 213]

The optimisation process is set up using the Matlab® algorithm *gamultiobj*, which is a genetic algorithm specifically implemented for multi-objective problems [16]. Note that, being this case study a multi-objective problem, a set of optimal solutions will be identified by the optimisation process, i.e., those solutions lying on the Pareto front of the problem. The initial population size is set to $N_P = 100$ individuals and is randomly generated. This number of individuals is identified as a satisfactory trade-off between computational resources required and discretisation of the Pareto front. In fact, considering a larger population would on the one hand increase the time required to evaluate the objective functions of the whole population, while on the other hand it would allow describing the numerical Pareto front with more solutions, thus increasing the accuracy of the results. Moreover, a population composed of $N_P = 100$ individuals meets the requirements of the specific genetic algorithm employed in this work [16]. The default functions governing the crossover and mutation processes need to be specialised to the case of discrete integer design variables. No further information is reported in this work regarding this coding task, since this is not considered to bring any additional insight into the proposed framework. Two stop criteria are selected, i.e., maximum number of generations $N_G = 200$ and FunctionTolerance=10^{-5} over the last $N_S = 100$ generations. The former criterion is employed to limit the required computational resources to carry out this case study, while the latter criterion is employed to stop the optimisation process whenever, simultaneously, (i) the geometric mean of the relative variation of the dispersion over the last N_S generations is lower than the FunctionTolerance value and (ii) the final dispersion value is lower than the mean dispersion value computed over the last N_S generations, where the *dispersion* is a metric measuring the movement of the set of points on the Pareto front. No more information is reported on this metrics, the interested reader is referred to [17] to go deeper into the topic.

During the optimisation process, the solver calls the structural model described in Section 2.1 to evaluate the objective functions for each individual in the current population. In particular, the genes of each individual, i.e., the design variables values, are passed to the structural model, that returns the support structure weight and the mean deformation of the mirror surface. Twelve points are selected as points of interest over which to compute the mean deformation of the mirror surface, as shown in Figure 4.

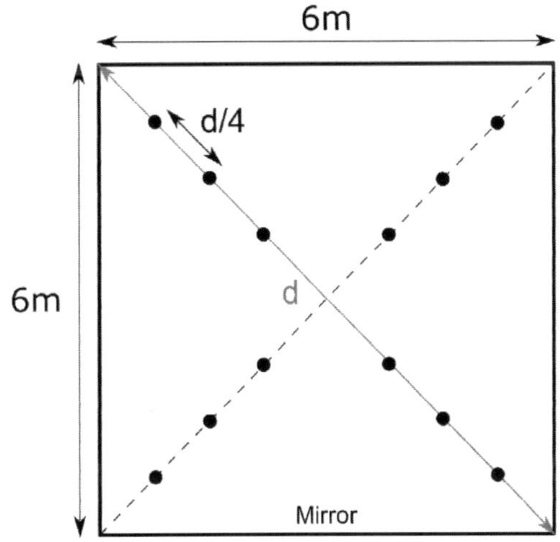

Figure 4. Points of interest on the mirror surface (black dots).

Since the structural model is implemented in the Excel software package, while the optimiser is coded in the MATLAB® environment, an interface between the two models needs to be developed. The most promising solution in terms of computational time for the analysis is identified in modelling the interface using ActiveX controls. This approach consists of creating a Microsoft server in the MATLAB® environment that allows the optimiser to directly work on the excel model, thus avoiding eventual overheads caused by the employment of the MATLAB® built-in functions *xlsread* and *xlswrite*. The adopted solution allows evaluating the objective functions of a single individual in a time $t_{ActiveX}$=2.5s, while using the built-in functions mentioned in the latter solution would require $t_{xlsread,xlswrite}$=5.6s for the same task.

With regards to this case study, the optimisation process runs for 102 generations, before stopping due to the FunctionTolerance stop criterion. The population at the last generation is reported in Figure 5.

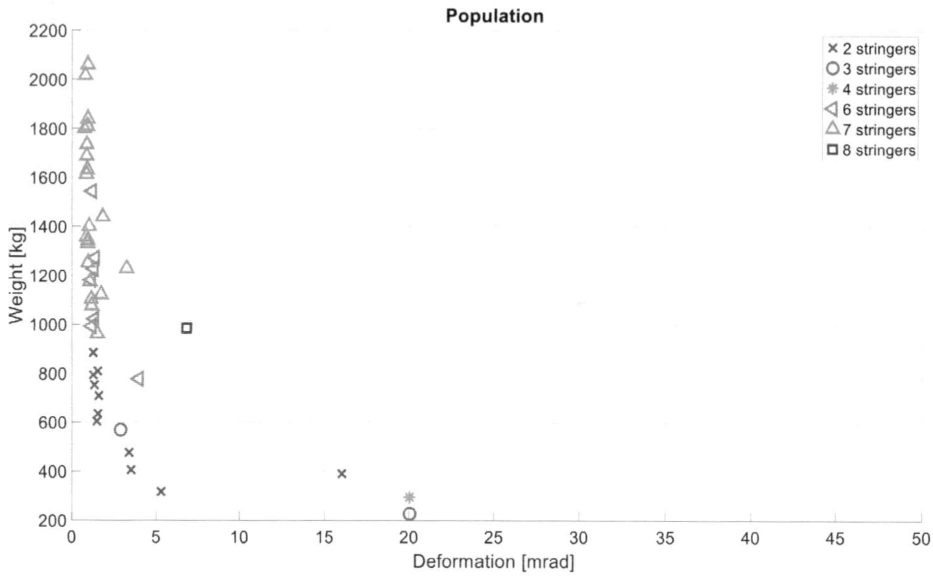

Figure 5. Population at the last generation.

The individuals shown in Figure 5 appear to be satisfactorily distributed around a hyperbolic curve limiting the optimal solutions. This hyperbolic curve represents the numerically reconstructed Pareto front and is composed of those points that represent an optimal trade-off between the objective functions, meaning that it is not possible to move from one point in the Pareto front to any other point by improving both the objective functions simultaneously. The numerical Pareto front reconstructed during the optimisation process is shown in Figure 6.

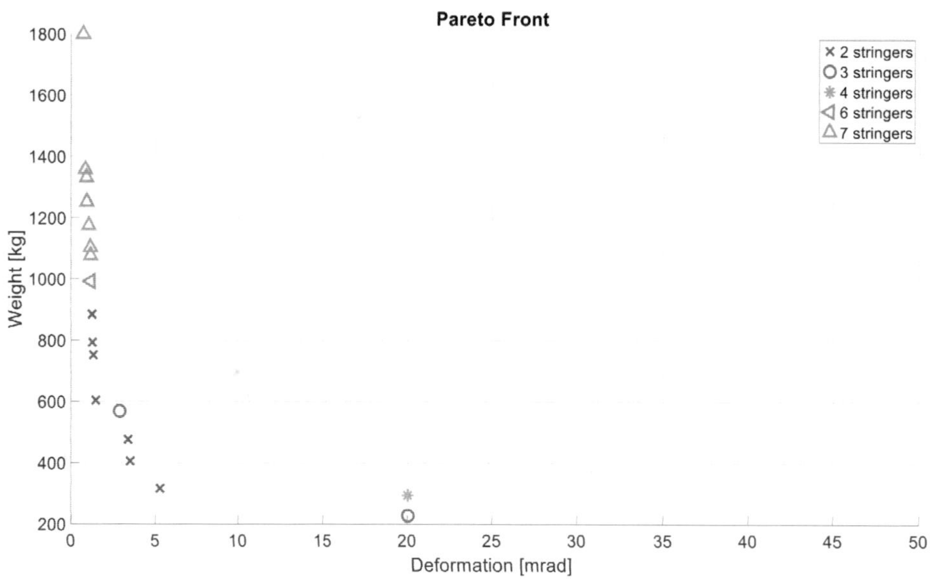

Figure 6. Numerically reconstructed Pareto front.

It is worth noting that no optimal solution with five or eight stringers, i.e., no solution with $x_1 = 5$ or $x_1 = 8$, is included in the Pareto front. This may be surprising at first sight, since the exact Pareto front would in principle be composed of solutions with any possible value of the first design variable. In fact, keeping the sections shapes for the stringers, the truss and the pedestal constant, increasing the number of stringers would reduce the mirror deformation and increase the structural weight of the solution, thus identifying a new point in the Pareto front, according to the definition reported above. However, in this case study a limited number of individuals in the population is considered, i.e., $N_P = 100$, which does not allow having a numerically reconstructed Pareto front discretised enough to include solutions with $x_1 = 5$ or $x_1 = 8$. Despite this, the composition of the Pareto front is physically sound, since the general trend appears to satisfy the intuition that increasing the number of stringers leads to moving from solutions characterised by larger deformation and smaller weight to those with smaller deformation and larger weight. Few exceptions to this rule are present in Figure 6, but no further investigation on the causes of these exceptions is reported here because out of the scope of this work.

More information can be given regarding each identified optimal solution by further post-processing the results. In particular, a safety factor (η) can be defined to identify those structural configurations that can safely carry the wind and weight loads without the plasticisation of any of the components they are composed of. The safety factor η_j of the *j*th component of the support structure is computed according to Equation (3):

$$\eta_j = \frac{\sigma_{y,j}}{\sigma_{VM,j}} \tag{3}$$

where $\sigma_{y,j}$ and $\sigma_{VM,j}$ are the yielding stress and the Von-Mises stress of the *j*th component, respectively. Note that the term "component" is generically adopted to refer to the stringers, the truss and the pedestal. The structural solution can withstand the loads without undergoing plasticisation if the safety factor of all the elements it is made of satisfies the requirement $\eta_j \geq 1$. The output of this post-processing is presented in Figure 7, where for the sake of clarity the only optimal solution with six stringers, i.e., the fittest individual with $x_1 = 6$, is shown. The individual is characterised by four numbers that identify (i) the most critical safety factor among those of the structural elements (η_{crit}), computed according to Equation (3), (ii) the index of the stringers section in the database, (iii) the index of the truss section in the database and (iv) the index of the pedestal section in the database.

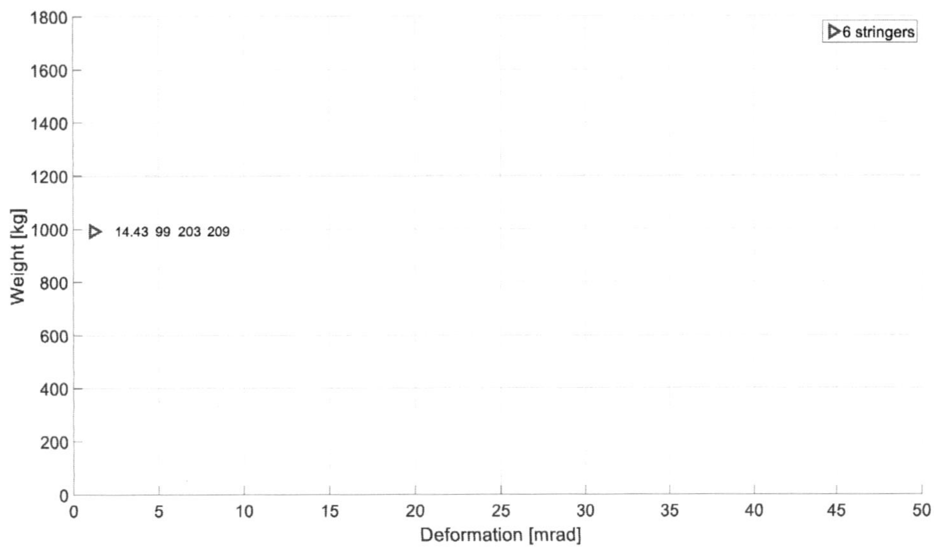

Figure 7. Detailed information of the optimal solution with six stringers.

The individual shown in Figure 7 is characterised by critical safety factor η_{crit} satisfying the requirement $\eta_{crit} \geq 1$ and the corresponding structural solution is determined by hollow rectangular-shaped stringers and hollow circular-shaped truss and pedestal. This solution is physically sound, since circular sections are assigned to those elements undergoing torque, while stringers are described by hollow rectangular sections, which are known to be effective in case no torque is applied.

Moreover, the design variables values, the objective functions values and the most critical safety factor value among those of the structural elements are reported in Table 2 for each of the individuals included in the Pareto front.

Table 2. Properties of the optimal solutions.

x_1	x_2	x_3	x_4	Weight [kg]	Deformation [mrad]	η_{crit}
2	163	116	209	885	1.26	11.55
2	162	116	209	793	1.27	12.22
2	162	115	209	753	1.34	10.14
2	155	115	209	604	1.48	11.20
2	152	115	202	476	3.39	6.48
2	192	152	205	406	3.51	5.47
2	81	108	115	317	5.29	3.59
3	191	155	206	569	2.89	6.15
3	181	142	196	228	20.04	1.74
4	60	83	153	295	20.03	1.44
6	99	203	209	992	1.18	14.43
7	112	16	211	1799	0.73	16.90
7	155	204	212	1357	0.85	14.50
7	155	204	211	1331	0.92	14.50
7	152	204	211	1252	0.94	15.10
7	193	203	211	1175	1.05	13.16
7	149	203	210	1102	1.15	13.76
7	99	203	209	1076	1.16	13.76

As expected, it turns out that the optimal sections for the stringers are mostly identified within the hollow rectangular and square sections in the database, since no torque is applied to this type of element. Moreover, most of the sections of the truss are characterised by hollow rectangular and hollow circular shapes, while the pedestal is almost always characterised by hollow circular sections, which is expected since these two structural elements are subjected to torque. Moreover, it is worth highlighting that (i) no H- and T-shaped sections are identified within the individuals in the Pareto front, (ii) none of the optimal solutions undergoes plasticisation, i.e., $\eta_{crit} \geq 1$, and (iii) some structural solutions admit deformations below the threshold 1mrad, which is a typically pursued value in this field [18].

4. Conclusions

In this work a methodological approach to the multi-objective optimisation of the support structure of heliostats in concentrated solar power plants employing a genetic algorithm has been presented. The proposed approach involves the interaction between the heliostat structural model and the optimiser. The former, which has been presented in Section 2.1, has been implemented in the Excel software package, it is based on the Euler-Bernoulli beam theory and relies on the main assumptions that (i) the mirror stiffness is negligible and (ii) the structural elements, i.e., the stringers, the truss and the pedestal, undergo linear elastic deformations. The latter, which has been described in Section 2.2, consists of the *gamultiobj* algorithm implemented in the MATLAB® environment, which has been properly modified to work with discrete integer design variables.

The proposed methodological approach has been demonstrated by means of a case study in Section 3. The optimisation has been carried out considering the wind loads from three different inclination angles of the mirror of the heliostat, keeping the wind azimuth angle constant to simulate the most critical scenario. The results of the optimisation are characterised by a Pareto front described by the hyperbolic shape typical of multi-objective optimisation problems and the optimal solutions present physically sound properties. In fact, it has been shown that the stringers are mostly assigned hollow rectangular and square sections, which is expected since no torque is exerted on this type of structural element, while solutions including hollow circular sections for the truss and the pedestal have been identified, which is likely to have been determined due to the torque applied to these two structural elements.

Despite the results of the optimisation process are realistic, the proposed methodology may be further improved to get more accurate results. The structural model employed to evaluate the deformation of the mirror surface may be refined by relaxing the assumptions reported in Section 2.1. For instance, the stiffness of the mirror may be included in the analysis and an analytical theory accounting for the loads not applied to the beam ends in a more accurate way may be implemented. With regards to the optimisation, a finer discretisation of the Pareto Front may be achieved considering a larger number of individuals, even though this would increase the required computational resources. On the other hand, the required computational time may be decreased by implementing the structural model and the optimiser in the same environment to avoid passing the data between the models. Moreover, the evaluation of the plasticisation of the structural elements may be introduced as a constraint to the optimisation process, thus allowing to directly replace those individuals described by solutions in which at least one of the components they are made of undergo plasticisation, even though this would increase the computational effort required to complete the optimisation process.

Funding

The work is part of the Research and Innovation Project "Solargrid: Sistemi sOlari termodinamici e fotovoLtaici con Accumulo peR co-GeneRazIone e flessibilità Di rete "—cod. ARS01_00532. The project has been jointly funded by the European Union and Italian Research and University Ministry (MIUR) under the Programma Operativo Nazionale "Ricerca e Innovazione" 2014–2020 (PON "R&I" 2014–2020).

Available material

The database including the sections considered in this work is available at https://polimi365-my.sharepoint.com/:b:/g/personal/10421282_polimi_it/ERIFCXbcSCdBozaCY0uKB7kB5c9h-HshXcGHmnE12fSipA?e=vq7l8z.

References

[1] Mancini T R, Gary J A, Kolb G J and Ho C K 2011 *Power Tower Technology Roadmap and cost reduction plan.* (Albuquerque, NM, and Livermore, CA (United States))

[2] Aldaz L, Burisch M, Zaversky F, Sánchez M, Villasante C and Olasolo D 2018 Heliostat structural optimization: A study of wind load effects with CFD-FEM methods *AIP Conf. Proc.* **2033** 210001

[3] Bhargav K R, Gross F and Schramek P 2014 Life Cycle Cost Optimized Heliostat Size for Power Towers *Energy Procedia* **49** 40–9

[4] Blackmon J B 2012 Heliostat size optimization for central receiver solar power plants *Concentrating Solar Power Technology* (Woodhead Publishing) pp 536–76

[5] Pfahl A 2014 Survey of Heliostat Concepts for Cost Reduction *J. Sol. Energy Eng.* **136**

[6] Von Reeken F, Weinrebe G, Keck T and Balz M 2016 Heliostat cost optimization study *AIP*

Conf. Proc. **1734** 20005

[7] Weinrebe G, Von Reeken F, Wöhrbach M, Plaz T, Göcke V and Balz M 2014 Towards Holistic Power Tower System Optimization *Energy Procedia* **49** 1573–81

[8] Todd Griffith D, Moya A C, Ho C K and Hunter P S 2015 Structural Dynamics Testing and Analysis for Design Evaluation and Monitoring of Heliostats *J. Sol. Energy Eng.* **137** 021010

[9] Zang C, Wang Z, Liang W and Wang X 2010 Structural Design and Analysis of the Toroidal Heliostat *J. Sol. Energy Eng.* **132** 041007

[10] Zang C, Wang Z and Liu X 2009 Design and analysis of a novel heliostat structure *1st International Conference on Sustainable Power Generation and Supply, SUPERGEN '09* (Nanjing: IEEE)

[11] Thalange V C, Dalvi V H, Mahajani S M, Panse S V., Joshi J B and Patil R N 2017 Design, optimization and optical performance study of tripod heliostat for solar power tower plant *Energy* **135** 610–24

[12] Belaid A, Filali A, Gama A, Bezza B, Arrif T and Bouakba M 2020 Design optimization of a solar tower power plant heliostat field by considering different heliostat shapes *Int. J. Energy Res.* **44** 11524–41

[13] Timoshenko S 1953 *History of strength of materials* (New York: McGraw-Hill)

[14] Peterka J and Derickson R 1992 *Wind load design methods for ground-based heliostats and parabolic dish collectors* (Albuquerque, NM)

[15] Oppo Srl Tabelle https://www.oppo.it/tabelle/a_elenco_tabelle.html (Accessed 2021-07-21)

[16] MathWorks Find Pareto front of multiple fitness functions using genetic algorithm https://www.mathworks.com/help/gads/gamultiobj.html (Accessed 2021-07-21)

[17] MathWorks gamultiobj Algorithm https://www.mathworks.com/help/gads/gamultiobj-algorithm.html (Accessed 2021-07-21)

[18] Sattler J C, Röger M, Schwarzbözl P, Buck R, Macke A, Raeder C and Göttsche J 2020 Review of heliostat calibration and tracking control methods *Sol. Energy* **207** 110–32

Finite element design of the connection of barriers for the protection of crowded places

E V Arcieri and S Baragetti

Department of Management, Information and Production Engineering, University of Bergamo, Viale Marconi 5, Dalmine 24044, Italy

sergio.baragetti@unibg.it

Abstract. The terrorist attacks in Europe have highlighted the problem of safety in public places. It is enough to run a vehicle at high speed towards the crowd to cause a large number of injuries and victims. In response to these phenomena, the Municipalities have perimetered sensitive areas with protection tools. The authors of this work have already designed a mobile anti-terrorism barrier capable of stopping 3500 kg and 7500 kg vehicles with a speed of 64 km/h in a few meters. In this work, the connection of two barriers by ropes is designed using Finite Element (FE) method to perimeter large areas with a small number of devices. Steel and Dyneema® ropes are studied, discussing the number of ropes and the dimensions needed.

1. Introduction

In the recent past, terrorists have wounded and killed many civilians in Nice, Barcelona, London, Stockholm and Berlin by running heavy vehicles at high speed. In response to these attacks, Municipalities have fenced off crowded places such as squares and pedestrian areas, which are sensitive targets for this type of attack.

In general, protection systems can be divided into: (i) fixed, (ii) retractable and (iii) mobile. Fixed systems are very effective although they can be dangerous if impacted by a vehicle because they can be severely damaged and some of their parts can hit people. The presence of foundations requires long installation operations that permanently modify the place where these systems are located and make them expensive. Furthermore, fixed barriers delay rescue operations because they make it difficult for any vehicle to enter the protected areas. Retractable systems (for example mobile bollards) allow the entry of authorized vehicles but they are still expensive because they are equipped with foundation and mechanisms necessary for their movement. Mobile barriers have no foundation and for this reason they can be positioned when and where they are needed and they have a low price. Among the mobile barriers, the most used are the concrete blocks, which dissipate impact energy mainly by friction and are therefore not very effective [1]. Concrete Jersey barriers are commonly adopted to prevent vehicles from entering the opposite lane being more suitable for oblique impacts than for the normal impacts, which can occur in the event of a terrorist attack [2].

The interest in the protection of certain areas is demonstrated by the number of recently designed and patented systems, which exploit different ideas to dissipate the impact energy. In the system presented in [3] part of the impact energy becomes plastic energy. Other energy absorption systems are described in [4] and in [5]. The barrier in [4] can be used as street furniture thanks to its good looks. Another patented solution presents a crash cushion that collapses when the applied axial force is greater than a threshold [6]. A kind of leverage is shown in [7]: this system raises the vehicle by changing the direction of the vehicle's momentum.

In this scenario, the authors' research group designed in collaboration with Besenzoni Defence & Protection S.r.l. and Besenzoni S.p.A. a new mobile barrier which is able to stop a 3500 kg van running at 64 km/h in less than 5 m. Thanks to this outstanding result, obtained experimentally, the barrier has been certified [1]. This system also has an aesthetic appeal, which avoids generating a sense of danger in pedestrians. The final shape of the barrier was defined thanks to mathematical and FE modelling and experimental testing [1,8-9]. FE method was also adopted to demonstrate the effectiveness of a single barrier in stopping a 7500 kg truck travelling at 64 km/h [10]. However, Municipalities should protect large areas using a limited number of barriers. In this work, conducted in collaboration with Besenzoni Defence & Protection S.r.l. and Besenzoni S.p.A., the connection of two barriers is designed using FE method. Steel and Dyneema® [11] ropes are arranged in different configurations, assessing each time the loads involved.

2. Materials and methods

This work deals with the design of the connection between two barriers of the type described in [1,8,10]. The designed barrier, shown in Figure 1, is mobile and it is good looking, so it can be used as street furniture. It consists of a cast iron base and several steel perimeter sheets. In [1] the use of aluminium was investigated as a consequence of the research group expertise in its alloys [12-13]. Any treatments such as the deposition of coatings [14-15] were not considered. The dimensions are 3000 mm x 860 mm x 1010 mm and the mass is approximately 2050 kg. In the barrier there is a bag containing 1550 kg of water, with different purposes: (i) the water allows to reach a significant mass, useful for dissipation by friction, after positioning the barrier so the transport of the system is facilitated; (ii) the release of large amounts of water following the impact takes away a lot of energy. The plastic strain of the barrier contributes to the dissipation of the impact energy [1,8-10].
The modelling techniques described in [10] were adopted for the purpose of this paper. FE analysis was conducted with Abaqus/Explicit 6.14-5, as explicit integration methods are suitable for impacts [16-18]. The behavior of the water contained in the barrier was studied using a Smoothed Particle Hydrodynamics approach [19].

Figure 1. Barrier [1,8,10].

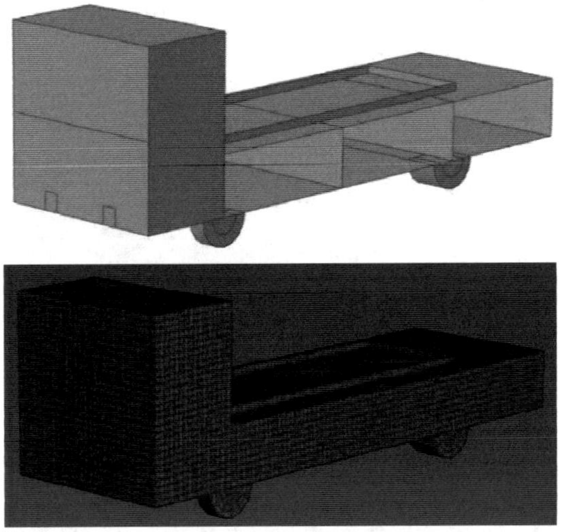

Figure 2. Model of the vehicle: geometry and mesh [10].

The normal impact of an Iveco Eurocargo ML120E24 against the center of the system consisting of two connected barriers was studied. The vehicle data are as follows: wheelbase 4185 mm; mass 7500 kg with ballast; height from the ground 358 mm at the front; position of the center of gravity in longitudinal direction 1785 mm from the front axle; vertical position of the center of gravity 970 mm above the ground [20]. The truck was modelled as two parallelepipeds of shell elements with suitable width (6 or 8 mm) and a frame of shell elements on the bodywork reproducing the ballast. The truck was stiffened to have 20% deformability and two side members and two cross members were inserted to reduce surface bending as visible in Figure 2 [10], where some surfaces are removed to show the inner part. The lack of experimental results makes the authors assume a linear elastic material behavior for the wheels. According to this hypothesis, they resist impact. Table 1 shows the materials of the FE model. For each of them it is indicated which component it is assigned to and what its properties are [8,10,21-22]. The steel properties were also assigned to the vehicle surfaces, although the density value was changed to ensure the correct position of the vehicle's center of gravity. For water, the Wilkins energy equation [23] was adopted and for this reason the reference sound speed is indicated in Table1 [8,10,21-22]. It was decided to connect the barriers with ropes. In [24] a theoretical model for the nonlinear bending of wires is proposed. For them, an indefinite elastic behavior was conservatively assumed with attention not to exceed the limit load during the simulation since exceeding this value would imply the failure of the connection in reality. The ropes were modeled with truss elements of circular section. It was assumed the use of two types of materials: steel and Dyneema®. Dyneema® is particularly suitable for traction cables and it consists of a synthetic fiber of high-strength polyethylene with very high molecular weight and extremely long molecules that allow the force to be transferred much more effectively. In terms of resistance, Dyneema® ropes are comparable to steel ropes with the advantage of not being subject to corrosion problems. They are also able to withstand the ultraviolet rays and through experimental tests this material has shown good resistance to fire [22,25]. The last property is very important because it prevents that someone burns the ropes making ineffective the system. Corrosion resistance is another important aspect in a huge variety of applications [26-27]. The ropes were positioned at different heights and connected to the vertical edges of the barrier by coupling type connections. Kinematic connections were used for the coupling of the perimeter sheets. The friction behavior in threaded connections [28] was not implemented. The ground was modelled as a surface of shell elements with high stiffness and the nodes fixed in order to avoid the deformation of the ground.

Table 1. Materials [8,10,21-22].

Material	Assigned to	Density (kg/m^3)	Young's modulus (MPa)	Poisson's ratio	Yield stress (MPa)	Reference sound speed (m/s)
S235JR (Elastic-perfectly plastic)	Perimeter sheets Vehicle (Different density)	7800	206000	0.30	235	
Cast iron (Elastic-perfectly plastic)	Base	7300	120000	0.26	250	
Water	Fluid	1000				1450
Rope (Steel - elastic)	Ropes	7800	110000	0.30		
Rope (Dyneema® – elastic)	Ropes	3000	132000	0.30		

Figure 3. Three connecting ropes and one internal rope in Dyneema®: arrangement of the ropes.

The FE analyzes performed consist of a first calculation step of 20 ms and a second one of 400 ms: in the first one, the gravitational load applied made the components come into contact; in the second, the impact between the truck and the obstacle was simulated. The pieces of the barrier and the vehicle were joined by means of constraints. In order to prevent all parts from interpenetrating, frictionless general contact was implemented throughout the model. The interactions with the friction coefficients of [10] were implemented to define the behavior between some regions of the model. In the model, the wheels cannot rotate in order to reduce the computational cost. To simulate the rolling friction, a coefficient of friction of 0.01 was implemented between the ground and the wheels.

3. Results and discussion

3.1. Three connecting ropes and one internal rope in Dyneema®

The first case analyzed refers to the impact of a 7500 kg vehicle running at 64 km/h against two barriers placed at a distance of 1 m from each other and connected by Dyneema® ropes with a diameter of 20 mm. The limit load for these ropes is 230 kN [29]. Three ropes were placed at 1 m (C1), 0.39 m (C2) and 0.13 m (C3) from the ground (Figure 3). A rope was inserted into each barrier to make the part of the barrier far from the impact zone cooperate in absorbing the impact energy. The internal ropes, I1 and I2, were inserted in the upper part, coaxial to the rope C1. In this case of connection between two barriers, the 7500 kg vehicle was stopped in 3 m. The following maximum axial loads were obtained: 183 kN for C1, 17 kN for C2, 3 kN for C3, 60 kN for I1 and 51 kN for I2. C3 is the most loaded rope and all the loads are below the limit so the ropes resist. The results are quite symmetric. Even if the model mesh was as symmetric as possible and symmetry conditions were implemented, the complexity of the problem analyzed may be responsible for this behavior. The lack of perfect symmetry overloaded a rope rather than the symmetric one. It is reasonable to suppose that in a condition of perfect symmetry the load on the two ropes would be intermediate. Since the study conducted focused on the maximum values of load obtained for the ropes, it was precautionary. Furthermore, it is difficult for a perfectly symmetric situation to occur in reality.

In addition to the analysis of the loads on the ropes, the equivalent plastic deformation (PEEQ) of the sheets after impact was also analyzed. It can be estimated that the failure of the barrier sheets occurs in those areas where the PEEQ is greater than the elongation A% of the material. In the case of S235JR steel, which is the material of the sheets, this value is 0.26 [30]. Figure 4 shows the PEEQ field for the analyzed case. By setting A% = 0.26 as the upper limit of the chromatic scale, it is possible to identify the points that exceed A%, which are the gray areas. For this first case of analysis, it can be noted in Figure 4 that these areas correspond to the impact areas and those of attachment of the ropes. The failure of these areas is useful for dissipating part of the impact energy and therefore for stopping the vehicle.

3.2. Three connecting ropes and two internal ropes in steel

Starting with the model presented in Section 3.1, one internal rope was added for each barrier, coaxial to C2. These two additional ropes are called I3 and I4 in this manuscript with I3 under I1 and I4 under I2. 16 mm steel ropes were still implemented. The stopping distance in this case is 5 m. The following loads were obtained: 168 kN for C1, 4 kN for C2, 2 kN for C3, 69 kN for I1, 46 kN for I2, 20 kN for I3 and 15 kN for I4. The C1 rope is the one that basically works among the connecting ones. The PEEQ distribution shown in Figure 5 was obtained. The position of the gray areas is similar to that of the previous cases. In summary, this configuration presents an improvement in terms of load on the connecting ropes compared to previous models, especially for the middle and lower ropes. However, in parallel with the decrease in the load on the connecting ropes, an increase in displacement of 2 m was obtained compared to the models with only one internal rope. This decrease in performance is attributable to the increased rigidity of the system due to the presence of a greater number of ropes. It was therefore decided to no longer analyze the insertion of a second internal rope in each barrier. The presence of this rope would also complicate the assembling of the single barrier as it would interfere with the water tank inside.

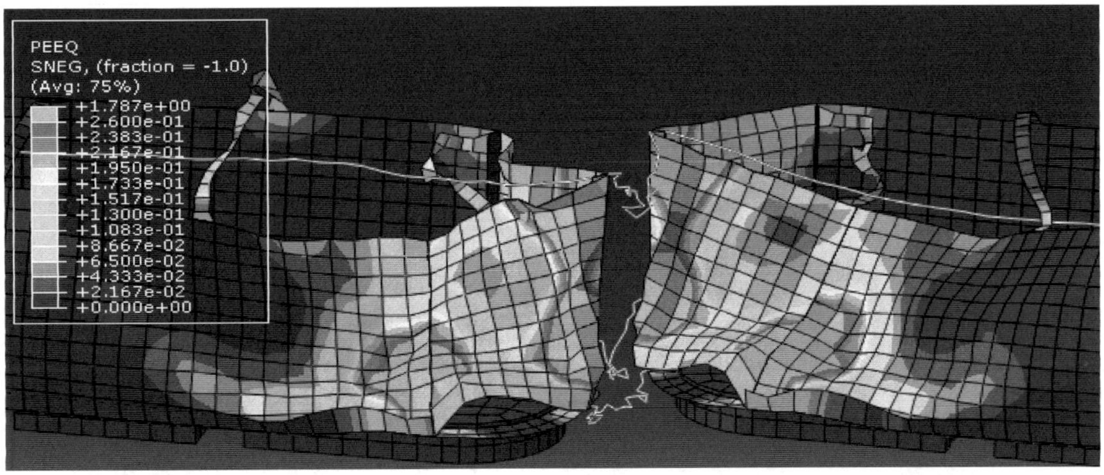

Figure 4. Three connecting ropes and one internal rope in Dyneema®: PEEQ.

IOP Conf. Series: Materials Science and Engineering 1214 (2022) 012028 doi:10.1088/1757-899X/1214/1/012028

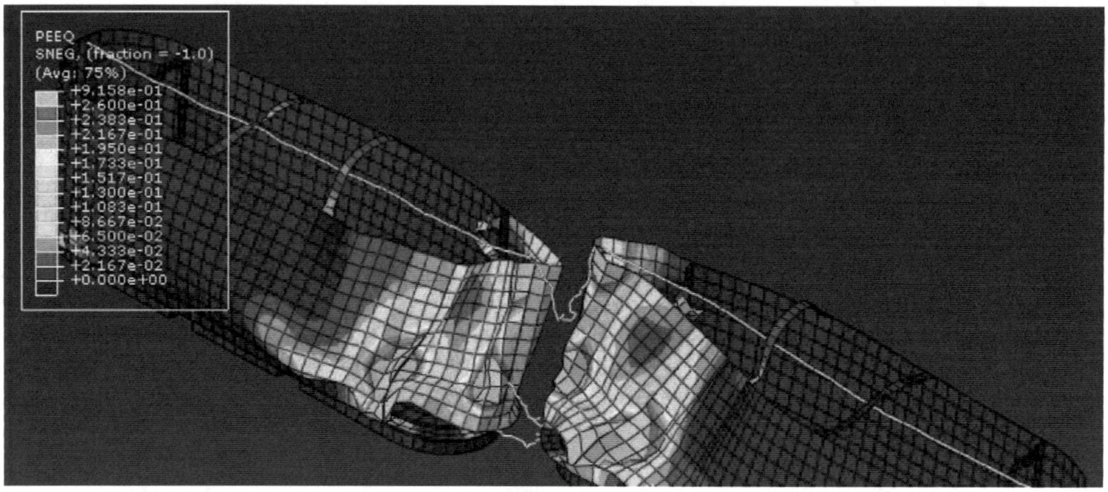

Figure 5. Three connecting ropes and two internal ropes in steel: PEEQ.

3.3. One connecting rope and one internal rope in Dyneema®

Analysing the stress state in the ropes of Sections 3.1 and 3.2 it was noted that the connecting ropes C2 and C3 were almost unloaded. It was therefore decided to keep only C1 as a connecting rope. A 60 mm diameter rope in Dyneema® was chosen. The internal ropes, on the other hand, still had a diameter of 20 mm. To identify the limiting load for the 60 mm rope, extrapolation from the data in [29] was carried out which gave a limiting load of 730 kN. Figure 6 shows the FE model of the new configuration. The following loads are obtained on the ropes: 152 kN on C1, 53 kN on I1 and 27 kN on I2. According to the results, all the ropes would resist. Analysing the PEEQ values in Figure 7, at the end of the simulation, it can be seen that the extension of the areas that exceed the percentage of elongation A% is greater. In fact, during the impact dynamics the barriers tend to get closer to each other. This behaviour is due to a greater axial stiffness of the rope, associated with the increase in the diameter of the rope. The rope, if subject to traction, attracts with greater force the barriers to which it is connected. The possible tearing of the sheets can be a favourable effect as it dissipates a part of the impact energy. In order to reduce the failure, reinforcements could be positioned in these areas. A displacement of 3 m was obtained. A configuration of this type is therefore adequate both in terms of construction and performance obtained. It is possible to replace the ropes with chains of similar properties.

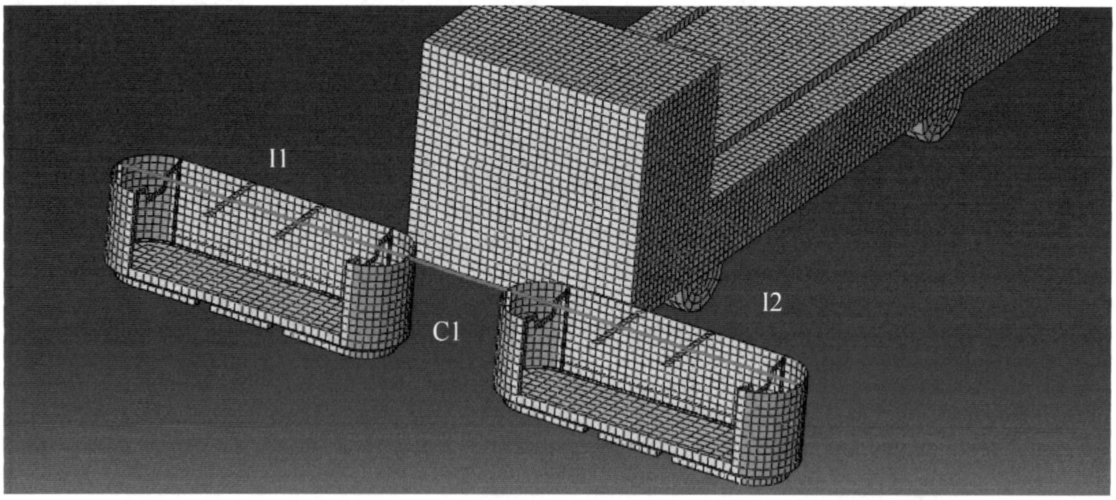

Figure 6. One connecting rope and one internal rope in Dyneema®: model.

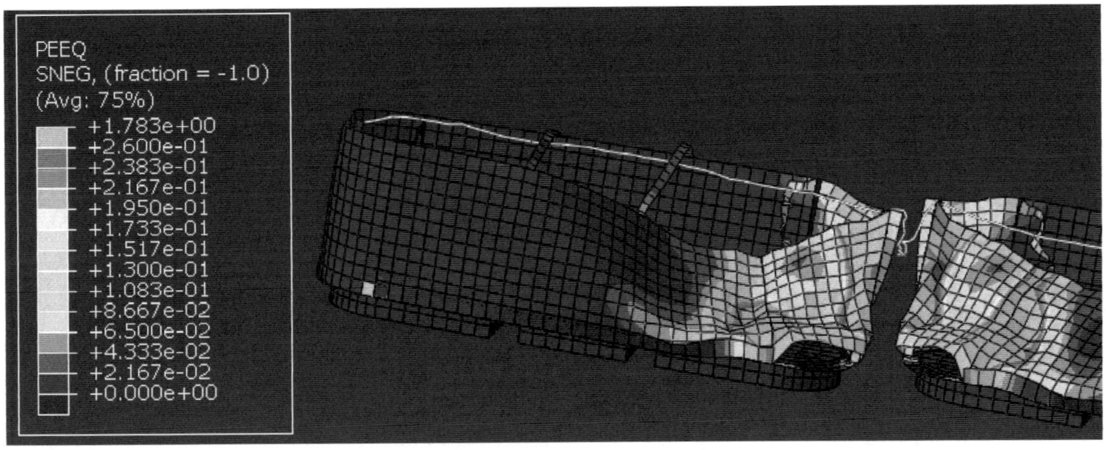

Figure 7. One connecting rope and one internal rope in steel: PEEQ.

4. Conclusions

The terrorist attacks in Europe have highlighted the issue of safety in public places. It is enough to run a vehicle at high speed towards the crowd to cause a large number of injuries and victims. The authors of this work have already designed a mobile anti-terrorism barrier capable of stopping 3500 kg and 7500 kg vehicles running at 64 km/h in a few meters. In this work, the connection of two barriers is designed using the FE method to perimeter large areas with a small number of devices. Three configurations are studied, evaluating each time the loads involved in the case of the normal collision of a truck. It was chosen a configuration consisting of the two barriers at a distance of 1 m connected by a Dyneema® rope with a diameter of 60 mm and with a coaxial internal rope with a diameter of 20 mm. As failure was not implemented in the model, the PEEQ analysis revealed which areas of the barrier could actually be damaged. They coincide with the areas where the impact occurs and where the ropes are connected. The failure of the barrier sheets could be a good effect, as it dissipates some of the impact energy.

Acknowledgments
The authors would like to thank Besenzoni Defence & Protection S.r.l., Besenzoni S.p.A., Federico Locatelli and Alessandro Sala for their support in the work.

References

[1] Baragetti S and Arcieri E V 2019 Proc. *ASME Int. Mechanical Engineering Congr. and Expo.* vol 14 (Salt Lake City) 156960
[2] Kala J, Hradil P and Salajka V 2012 *Int. J. Civ. Eng.* **6** 831
[3] Amengual Pericas A 2009 Patent EP2314772B1
[4] Shen L, Caspe M, Ji J and Wang Q 2010 *Proc. Building Design + Construction*
[5] Stevanato A 2014 Patent US2017362788
[6] Impero P 2013 Patent ZA201507239
[7] https://barriers.miframsecurity.com/products/mvb-3x/. [Accessed on 12/01/2019]
[8] Baragetti S and Arcieri E V 2019 *Procedia Struct. Integrity* **24** 91
[9] Baragetti S and Arcieri E V 2020 *Eng. Fail. Anal.* **113** 104564
[10] Arcieri E V and Baragetti S 2021 *IOP Conf. Ser.: Mater. Sci. Eng.* **1038** 012008
[11] https://www.dsm.com/dyneema/en_GB/home.html [Accessed on 03/12/2019]
[12] Baragetti S and D'Urso G 2014 *J. Mech. Sci. Technol.* **28** 867
[13] Baragetti S and Arcieri E V 2020 *Proc. Inst. Mech. Eng. Part C J. Mech. Eng. Sci.* online
[14] Baragetti S, Gelfi M, La Vecchia G M and Lecis N 2005 *Fatigue Fract. Eng. Mater. Struct.* **28** 615
[15] Baragetti S and Tordini F 2007 *Int. J. Fatigue* **29** 1832

[16] Arcieri E V, Baragetti S, Fustinoni M, Lanzini S and Papalia R 2018 *Procedia Struct. Integrity* **8** 212
[17] Baragetti S, Guagliano M and Vergani L 2000 *J. Mech. Sci. Product. Technol.* **15** 91
[18] Arcieri E V, Baragetti S and Lavella M 2021 *IOP Conf. Ser.: Mater. Sci. Eng.* **1038** 012012
[19] Monaghan J J 2005 *Rep. Prog. Phys.* **68** 1703-1759
[20] https://private.iveco.com/italy/collections/technical_sheets/Documents/usato/medi/medium%202/ML120E24FP.pdf [Accessed on 01/10/2018]
[21] http://www.teci.it/funi/standard-a-trefoli/s10-zn-s11-zn/ [Accessed on 03/12/2019]
[22] https://www.xvertical.it/dyneema-e-kevlar/ [Accessed on 03/12/2019]
[23] Wilkins M L 1999 Computer Simulation of Dynamic Phenomena (Berlin: Springer-Verlag Berlin Heidelberg)
[24] Baragetti S 2006 *Meccanica* **41** 443
[25] Locatelli F and Sala A 2019 *Modellazione numerica esplicita e implicita di strutture innovative per la sicurezza e il settore nautico: modelli teorici, numerici e riscontri sperimentali* master thesis
[26] Baragetti S 2013 *Surf. Interface Anal.* **45** 1654
[27] Baragetti S and Arcieri E V 2018 *Int. J. Fatigue* **112** 301
[28] Baragetti S, Terranova A and Vimercati M 2009 *Int. J. Mech. Sci.* **51** 790
[29] https://www.armare.it/it/. [Accessed on 03/12/2019]
[30] https://irp-cdn.multiscreensite.com/057cab6f/docs/1.3716043..pdf [Accessed on 03/12/2019]

Experimental investigation on crack propagation paths in spur gears

F Curà[1], A Mura[1] and L Corsaro[1]

[1] Dimeas Politecnico di Torino, corso Duca degli Abruzzi 24, 10129 Torino, Italy

francesca.cura@polito.it

Abstract. Spur gears subjected to bending fatigue may nucleate cracks at the tooth root fillet. In thin rim gears these cracks may propagate in a safe way (through the tooth) or in catastrophic way (through the rim). Crack propagation direction is mainly influenced by both wheel geometry parameters and crack initiation point, as already pointed out by theoretical and numerical results available in literature. Aim of this work is to set up an experimental activity in order to verify the onset of the bending crack and its propagation path in spur gears with different geometries. In particular, a special device connected to a standard fatigue machine was realized to perform bending tests for both standard and thin rim gears. During bending tests, an IR thermocamera was utilized to monitor the surface thermal profile in the tooth root fillet zone.

1. Introduction

Crack propagation in gears is a problem related not only to the life of the components, but also to the concept of failsafe design. Fail safe design means to design a component in order that, if a failure occurs, this may cause a "safe failure". This aspect is very important above all in aerospace industry. As a matter of fact, in aerospace applications the need of reducing weight brings to produce gears with very thick rim and web. Considering thin rim gears, when a crack is nucleated near the tooth root, it may propagate through the tooth (causing the loss of the entire tooth or a portion of it) or the propagation may follow a path across the wheel diameter (causing the projection of big parts of the gear that may break the gearbox and may cause serious damage to the aircraft). The first failure mode is defined as "*failsafe failure*" and the second one as "*catastrophic failure*" and, of course, has to be avoided.

Designers need to have robust design criteria in order to predict crack propagation paths and to avoid catastrophic failures. An increasing number of variables of interest, under the hypothesis of isotropic material, have been taken into account during the years and utilised in numerical simulations to achieve that goal. Traditional 2D and 3D finite elements (FEM) and extended finite elements (XFEM) models have been run, focusing on both crack initiation point and corresponding propagation direction.

On the basis of previous simulations results [1-3], it was noted that the crack propagation direction was mainly influenced by both wheel geometry parameters and crack initiation point. Results related to gear models with different web and rim thicknesses were also interpreted in ISO Standard environment [4], relating the crack path to the so called gear blank factor C_R, useful in cases of mating gears consisting of rims and webs. Results showed that the interaction between web and rim thickness may influence the crack propagation and the corresponding safe or catastrophic failure mode [1-3], [5-6].

IOP Conf. Series: Materials Science and Engineering 1214 (2022) 012029 doi:10.1088/1757-899X/1214/1/012029

For specific geometry configurations, crack propagation paths may be affected also by other parameters such as the centrifugal load. For this reason, the effect of the centrifugal load (proportional to wheel speed), related to the bending one, was investigated by means of dedicated simulations [2].

Many studies were carried on during the years, above all from the numerical point of view, but there is a lack in literature for as concerns experimental tests to verify the obtained results in term of crack path typologies. Some evidence of experimental activities can be found in [7], where test were run on wheel models.

Aim of this work is to set up an experimental activity in order to verify the onset of the bending crack and its propagation path in spur gears with different geometries. In particular, a special device connected to a standard fatigue machine was realized to perform bending tests for both standard and thin rim gears. Three gears with different geometry configurations (back ratio and web ratio values) were tested. During the bending tests, an IR thermocamera was utilized to monitor the surface thermal profile and to emphasize the stress variation in the tooth root fillet zone. Both crack initiation point and propagation path direction were pointed out, and this information was compared to the corresponding available in literature from numerical simulations.

2. Materials and Methods

Three types of spur gears (one *standard* and two *thin-rim* gears) were designed to be tested in this activity under bending fatigue conditions.

Geometrical parameters of each gear are resumed in Table 1. In particular are shown: pitch diameter (*dp*), face width (*b*), number of teeth (*z*), modulus (*m*), backup ratio (rim dimension respect to the tooth height) (*mb*), web ratio (web thickness respect to the face width) (*mw*) and gear blank factor (C_R) [4]. As an example, technical drawings of Gear 1 and Gear 2 are shown in Figure 1.

Table 1. Gears geometry.

	dp [mm]	*b* [mm]	*z*	*m* [mm]	*mb*	*mw*	C_R
Gears 1	96	20	32	3	1	1	1
Gears 2	96	20	32	3	0,3	0,1	0.63
Gears 3	96	20	32	3	0,5	0,1	0.59

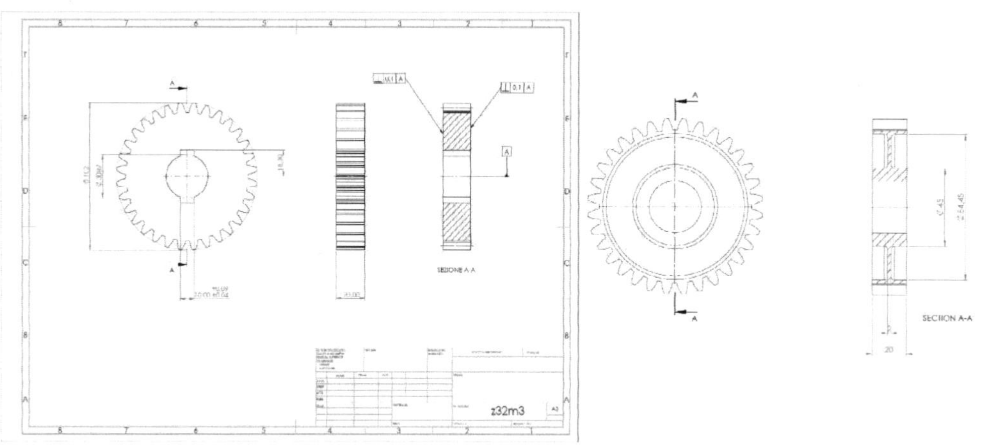

Figure 1. Technical drawings of Gear 1 and Gear 2.

Each gear is made of C45 steel without thermal treatments.

A special device connected to a standard fatigue machine was realized to perform bending tests for both standard and thin rim gears (see Figure 2, left side). A punch integrated into the device was able to apply the pulse loading to the gear tooth. In particular this punch was designed in order to load the tooth along its involute profile in both position and direction referred to the single contact point [8].

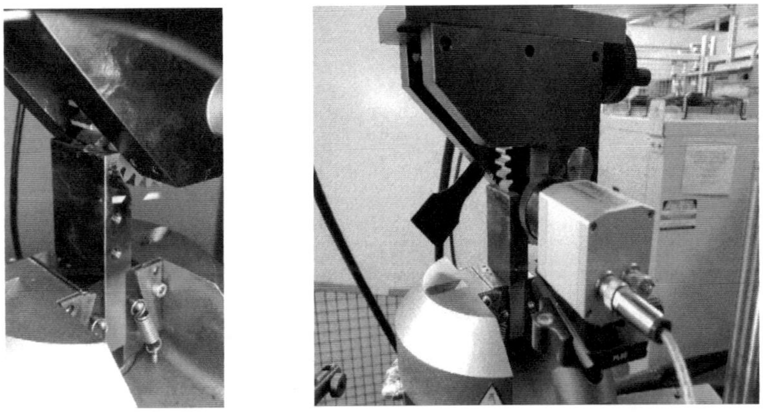

Figure 2. Test apparatus.

Fatigue tests were run with 20 Hz excitation frequency. Maximum and mean force were respectively 12200 N and 6200 N, a small static load was always applied to guarantee the contact between punch and tooth.

During bending tests, an IR thermocamera (*IR tech Timage Radiamatic "XT"*) was utilized to monitor the surface thermal profile in the tooth root fillet zone (see Figure 2, right side). The thermal detector was located at a distance of 60 mm from the gear. A reference specimen (see Figure 2, right side) allowed to acquire the environmental temperature during each test. Gears and reference specimen were both covered by black spray paint to generate a surface with an emissivity close to 0,95. Two different thermal profiles were extracted from each thermogram near the root tooth fillet, respectively an *aera* and a *point* of 2x2 pixel (located inside the area close to the crack initiation point). The relative temperature ΔT (difference between the absolute temperature and the temperature of the reference specimen) was computed for Gears. Figure 3 illustrates an example of the area position near the tooth root fillet.

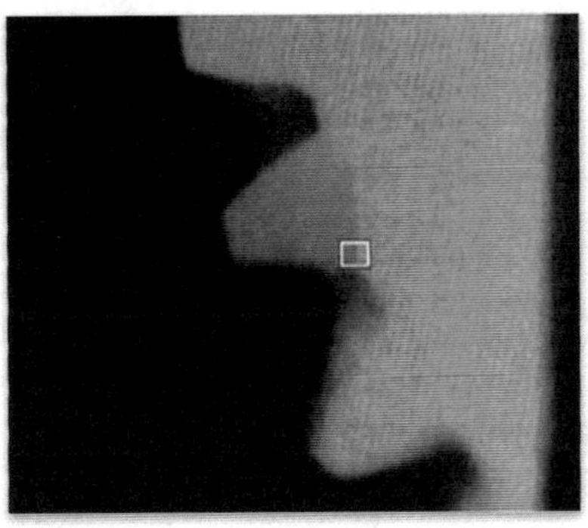

Figure 3. Position of the area.

3. Results

Results obtained in this work concern crack propagation paths caused by bending fatigue cycles. For sake of brevity, only results referred to Gear 1 and Gear 2 are reported in this paper.

Figure 4 shows, as an example, Gear 1 and Gear 2 and their failure typologies. As it can be observed, Gear 1 crack path (on the left) corresponds to a classical fail safe condition, while in Gear 2 (on the right) a crack path directed through the rim can be pointed out.

Figure 4. Gear 1 (left) and Gear 2 (right) crack paths.

Figure 5 shows an example of thermal images recorded during fatigue tests. More in detail, the left image is related to Gear 1 and, as in Figure 4, a crack propagating through the tooth can be appreciated. The right image shows the crack propagating through the rim, as predicted for Gear 2 already in the design phase.

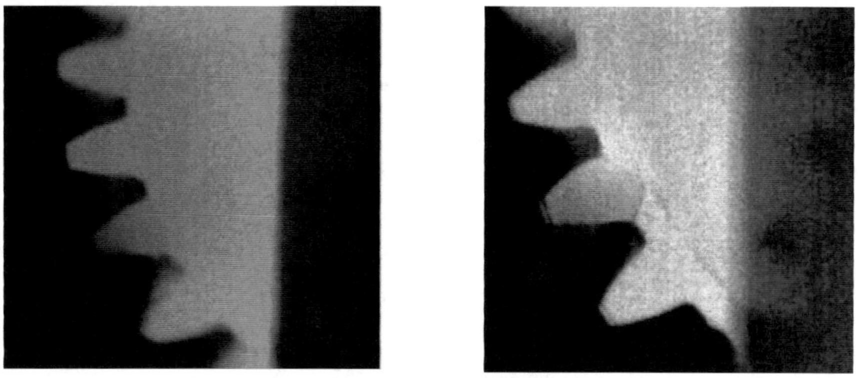

Figure 5. Thermal images of Gear 1 (left) and Gear 2 (right).

As a matter of fact, geometrical parameters of Gear 2, in particular its backup and web ratio values, were chosen to verify the corresponding catastrophic failure already emphasised by numerical simulations, and now demonstrated by fatigue tests. Rim and web geometries may be related to the well known C_R parameter [4] (see Table 1), as a function of both backup ratio and web ratio values. As already indicated in [2], the position of this parameter respectively in the tooth failure zone, in the uncertainty zone or in the rim failure zone of its graph (see Figure 6 [2]) allows us to draw some hypothesis about the possibility of safety or unsafety fracture basing only on the gear geometry.

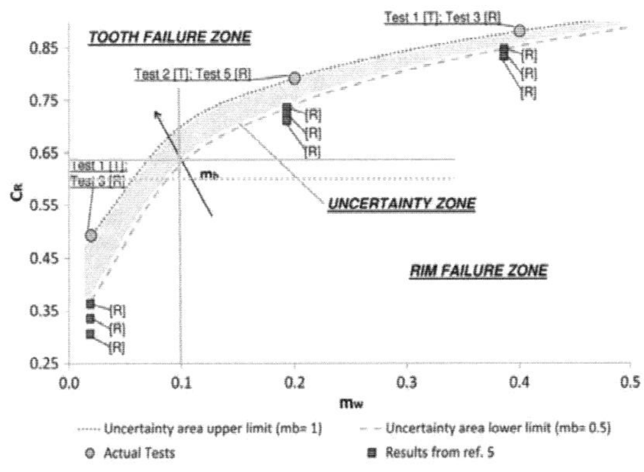

Figure 6. CR factor [2].

Figure 6 points out CR parameter values for Gears 2 and 3, characterised by different back up ratio (m_b) values and with the same web ratio (m_w), driving respectively to fail safe and catastrophic failures. Finally, Figures 7 and 8 show the thermal profiles monitored during the fatigue tests respectively for Gear 1 and Gear 2. In particular, in each Figure is reported the difference temperature ΔT (respect to the ambient temperature or reference specimen) as a function of the number of cycles, referred to both crack nucleation point and to an area around the nucleation point (nucleation area, mean value of temperature).

From the analysis of these Figures, it can be noted the increase of thermal profiles in the first phase of the test, then a stabilised condition of temperature and, finally, an abrupt variation corresponding to the failure. Nucleation point and area profiles show in both cases identical trends.

The comparison between Figures 7 and 8 shows that the fatigue life of Gear 2 is lower than that of Gear 1 in term of number of cycles (118000 cycles for Gear 1 and 104300 cycles for Gear 2), as expected. Moreover, the relative temperature ΔT level is higher for Gear 1 than for Gear 2. This unexpected result may be interpreted as a strong dissipation flow in the case of Gear 2 within the whole wheel body, due to the high displacements involved in fatigue tests.

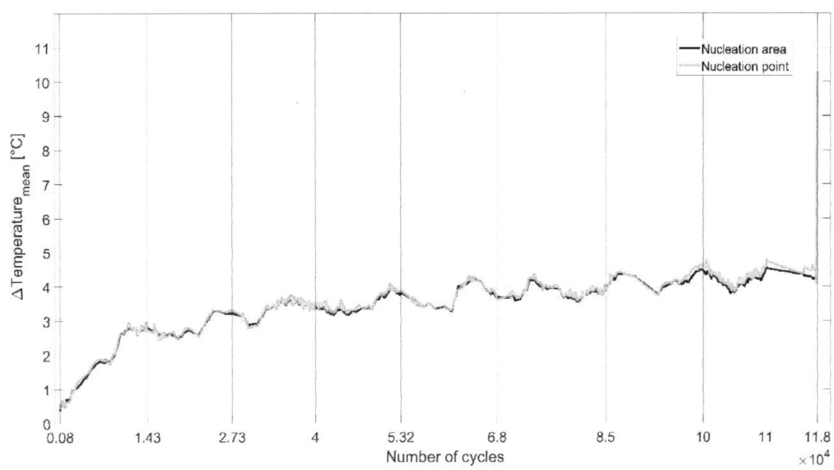

Figure 7. Gear 1 thermal profiles.

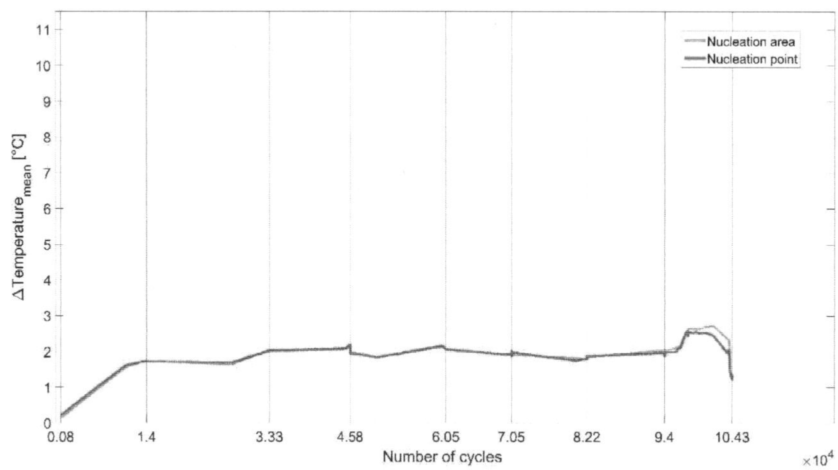

Figure 8. Gear 2 thermal profiles.

4. Conclusions

The present work represents a first step in the experimental verification of crack propagation paths in light weight gears.

To this aim, an experimental activity was set up in order to verify the onset of the bending crack and its propagation path in spur gears with different geometries. To perform bending tests, a special device connected to a standard fatigue machine was realised, consisting in a system to clamp the wheel to the machine and in a punch useful to apply the pulse loading. This loading entity was intended to be applied on a tooth involute profile in point and direction corresponding to the so called single contact point. This way, the effect of the sharing condition for the gear subjected to bending test was neglected.

During bending tests, an IR thermocamera was utilized to monitor the surface thermal profile in the tooth root fillet zone. The monitoring of surface thermal maps made possible the identification of both crack initiation point and its propagation in the first phase of the phenomenon, and not only in the last phase when fracture occurs.

On the basis of previously obtained results, based on numerical simulations, the geometry of gears was designed in order to predict the corresponding crack propagation path. Experimental results confirmed that geometrical parameters and corresponding blank factor values may represent an useful point of view providing a realistic design map for thin rimmed gears and for aerospace gears with particular web geometries.

References

[1] Curà F, Mura A, Rosso C 2015 *Fatigue & fracture of engineering materials & structure* Effect of rim and web interaction on crack propagation paths in gears by means of XFEM technique p. 1237-1245 ISSN: 1460-2695 - doi: 10.1111/ffe.12308

[2] Curà F, Mura A, Rosso C 2015 *Frattura e integrità strutturale* Effect of centrifugal load on

crack path in thin-rimmed and webbed gears vol. 9 p. 447-455 ISSN: 1971-8993 - doi: 10.3221/IGF-ESIS.34.50
[3] Curà F, Mura A, Rosso C 2016 *Fatigue & fracture of engineering materials & structure* Influence of high speed on crack propagation path in thin rim gears ISSN: 1460-2695 - doi: 10.1111/ffe.12481
[4] ISO 6336 - Calculation of load capacity of spur and helical gears, Part 1 (2019)
[5] Curà F, Mura A, Rosso C 2016 *Procedia structural integrity* Crack propagation behaviour in planet gears Vol. 2 p. 3610-3616 ISSN: 2452-3216 - doi: 10.1016/j.prostr.2016.06.450
[6] Curà F, Mura A, Rosso C 2017 *Procedia structural integrity* Investigation of crack propagation path in tube gears vol. 7 476-483 ISSN: 2452-3216 - doi:10.1016/j.prostr.2017.11.115
[7] Lewicki David G., Gear Crack Propagation Path Studies, Guidelines for Ultra-Safe Design-NASA/TM-2001-211073
[8] ISO 6336 - Calculation of load capacity of spur and helical gears, Part 3 (2019)

Design and construction of an offshore diffuser augmented wind turbine with a high efficiency alternator

G P Demelio[a*], SM Camporeale[a], A Castellano[a], F Cupertino[b] and M Torresi[a]

[a]Polytechnic University of Bari - Department of Mechanical Engineering, Mathematics and Management (DMMM), Via Orabona 4, 70126 Bari, Italy
[b]Polytechnic University of Bari - Department of Electrical Engineering and Information Technology (DEI) , Via Orabona 4, 70126 Bari, Italy
*e-mail: giuseppepompeo.demelio@poliba.it

Abstract. This paper describes the design and the assembly of a new type of a ducted wind turbine with an electric power generator adopting permanent high coercive magnets embedded in the peripheral ring of the rotor. The nominal power is 20 kW, and the maximum diameter of the external duct is 9 meters. The project has been carried out within the Marine Energy Laboratory (MEL) funded by the Italian Ministry for Education, University and Research (MIUR) and collects the most advanced technologies of naval maritime engineering and combines them with energy and turbomachinery technologies. The presence of a divergent duct enables the interception of a greater air mass flow rate, allowing the reduction of the rotor diameter and the deflections of the blades; hence, at constant tip speed ratio, a higher rotational speed compared to conventional turbines and a better efficiency of the permanent magnet power generator.

1. Introduction

The proposed work describes the design and realization of an innovative ducted wind turbine characterized by a double shrouds architecture and a generator with permanent magnets embedded within a peripheral ring realized on the rotor, thought for offshore applications

The overall reference architecture foreseen consists of a self-displacing floating complex supporting three wind turbines (Figure 1).

The activity is part of the MEL (Marine Energy Lab) project, funded by MIUR, which involves the construction of an offshore laboratory, whose original design is shown in Figure 2. The offshore lab is connected to an onshore lab by a series of cables for energy transfer and data transmission. The offshore laboratory houses the prototype turbine and the control and monitoring room.

The aim of the project is to create a flexible power generation system oriented to the needs of small, isolated marine communities with a system that allows for relatively simple operation and management.

It is foreseen that the plant will work in open sea, therefore in presence of undisturbed wind with a relatively high average value, that the efficiency of the turbine, with reference to both the fluid-dynamic and the electrical components, will be maximum, and that a system of energy transfer to the land will be realized with an optimized line of cables.

Figure 1. Conceptual project of a self-displacement wind power plant

Figure 2. Original drawing of the marine laboratory for carrying out tests on the single wind turbine

2. Ducted Wind Turbines

As is well known, on the basis of the classical theory of the Betz actuator disk [1], the maximum theoretical power that can be extracted from a wind turbine with free flow on the rotor is limited to a fraction $C_{p,max}$ equal to 16/27 of that contained in the undisturbed flow, a value very difficult to approach in actual achievements especially in small turbines. In the realization of large turbines a C_p = 0.5 could be approached. This limit is not valid in the case of ducted flow [2].

Wanting to maximize the obtainable energy with the same footprint with the adoption of a diffuser, theoretically one could reach a power coefficient C_p close to unity.

For this type of wind turbine the design of the rotoric and statoric components and of the safety and orientation devices is more complex.

In the application developed, the objectives pursued were the maximization of the extractable power in relation to the area invested by the wind and the optimization of the efficiency in the production of electrical energy, incorporating the stator part of the generator in the diffuser by means of windings made on multiple layers on printed circuit boards and including an array of permanent magnets with high coercivity in an annular rotor supported by the ends of the blades. In this way, taking advantage of the high peripheral speed of the rotor, it is avoided to introduce a mechanical transmission with function of multiplier of revolutions, necessary, in conventional applications, to have an acceptable efficiency of the alternator.

The realized turbine, shown in Figure 3, has a nominal power of 20 kW, an overall diameter of 9 meters and consists of a double diffuser whose internal structure supports the central steel hub on which the rotor is splined. The rotor has five blades and a peripheral ring on which the permanent magnets of the electric generator are mounted.

The main problem faced in the design of the rotor concerns the need to ensure adequate stiffness to limit to 12 mm the maximum gap between the permanent magnets installed on the peripheral ring and the stator in any operating condition, since it is not possible to use conductive materials in the vicinity of areas subject to high variations in magnetic flux. The peripheral ring has been realized by constructing 5 identical sectors connected to the ends of the blades and arranged with housings necessary to hold the high coercivity magnets in position with the necessary safety for the operators during assembly. A unidirectional fiberglass belt was wrapped around the ring to ensure strength and rigidity requirements. The stator has been realized with a steel supporting structure on which 16 composite sectors have been mounted, eight for each diffuser. On the internal diffuser the alternator windings are mounted on a printed circuit board.

For the execution of the preliminary tests, a support pole with relative orientation devices has been realized, which places the axis of rotation at 15 meters of height, as shown in Figure 4.

Figure 3. Turbine on the ground in the assembly phase

Figure 4. Turbine mounted on a pole for carrying out preliminary tests

3. Fluid Dynamics Design

For the design of a ducted wind turbine, the conventional blade element momentum theory, developed for free stream wind turbines, cannot be applied; hence, for this specific case, an original numerical model has been developed [2], and computational fluid dynamic (CFD) simulations were carried out for validation.

The diffuser consists of two distinct concentric annular wing shaped diffusers. This solution has been chosen in order to reduce the axial length of the diffuser. In fact, with a double diffuser, it is possible to obtain a greater diffusion without incurring in the risk of a flow separation, since the external air flow, which is channeled into the annular gap between the two elements (acting as a nozzle), accelerates energizing the boundary layer making it less keen to separate.

The aerodynamic profile of each annular wing has got a camber line designed according to a cubic Bezier curve with blade thickness defined according to the one of a NACA0004 profile. The use of the Bezier curves for the cumber line and the NACA 4 digits thickness distribution guarantees the continuity of the function that describes the profile and those of the first and second derived functions. This allows to have a more regular profile description with a lower tendency to flow separation and therefore a considerable improvement in the aerodynamic efficiency of the annular wing.

For the aerodynamic design of rotor blades, the design value of the tip speed ratio was set to$\lambda = \omega R/V_0 = 5$. The geometry of the blades was built starting from three basic profiles of the NACA 4 digits series, assigned at hub, midspan, and tip. In particular, the airfoil profiles used for the design of the blade are:

- NACA 6618 hub;
- NACA 5418 mid span;
- NACA 1318 tip.

Starting from these 3 reference profiles, parabolic interpolations were used to obtain 8 non-standardized intermediate profiles (always of the NACA 4 digits series), 4 between hub and midspan and 4 between midspan and tip.

The rotor tip radius is 3.2 m. The number of blades chosen for the rotor is 5.

The preliminary numerical model used to define the chord and twist distributions along the blade span is based on the resolution of the steady, incompressible, axisymmetric 2D, Reynolds Averaged Navier-Stokes (RANS) equations together with the mass conservation equation. Due to the low Mach numbers, it is possible to neglect the air compressibility. A finite volume approach is used to discretize the system of equations. The velocity field is obtained by solving the momentum balance equations, while the pressure field is extracted by solving a pressure correction equation in order to impose the mass conservation according to the SIMPLE algorithm (Semi Method Implicit for Pressure-Linked Equations) developed by Patankar [3]. For the turbulence closure, the standard k-ε model was used [4], which suppose an isotropic turbulence (Boussinesq hypothesis). This turbulence model was chosen being robust, computationally cost-effective, and reasonably accurate in a wide range of turbulent flows [5]. All viscous and convective terms were discretized using a first-order accurate upwind scheme.

In this preliminary phase, to avoid any excessive computational cost, otherwise required by a detailed blade-resolved simulation to evaluate the aerodynamic interaction between the blade airfoils and the flow, a hybrid approach was chosen. In the area swept by the rotor, the spatial discretization is similar to that of the rest of the domain, and, within these cells, source terms have been introduced that simulate the momentum exchange between blades and the flow. These source terms are introduced through appropriate User Defined Functions (UDFs), written in C language, and implemented into the CFD code. This approach was first proposed by Rajagopalan and Fanucci [6], who developed a 2D CFD code for Darrieus rotors and was applied in a 3D case by Torresi et al. [7]. The aerodynamic forces on each blade depend on the local velocity field, i.e. on the velocity inside each cell swept by the rotor. Therefore, an iterative approach is required to achieve solution convergence. Once both the local Reynolds number and the angle of attack have been evaluated, in each cell swept by the rotor, it is possible to compute the lift and drag coefficients, based on the polars of the chosen airfoils, and then recalculate the source terms. To obtain all the polars of the airfoils, the QBlade [8] software was used.

By drastically reducing the number of cells within the computational domain, it was possible to carry out many simulations by modifying each time the rotor geometry to evaluate the one capable of providing the best performance under the design flow conditions (see figure 5).

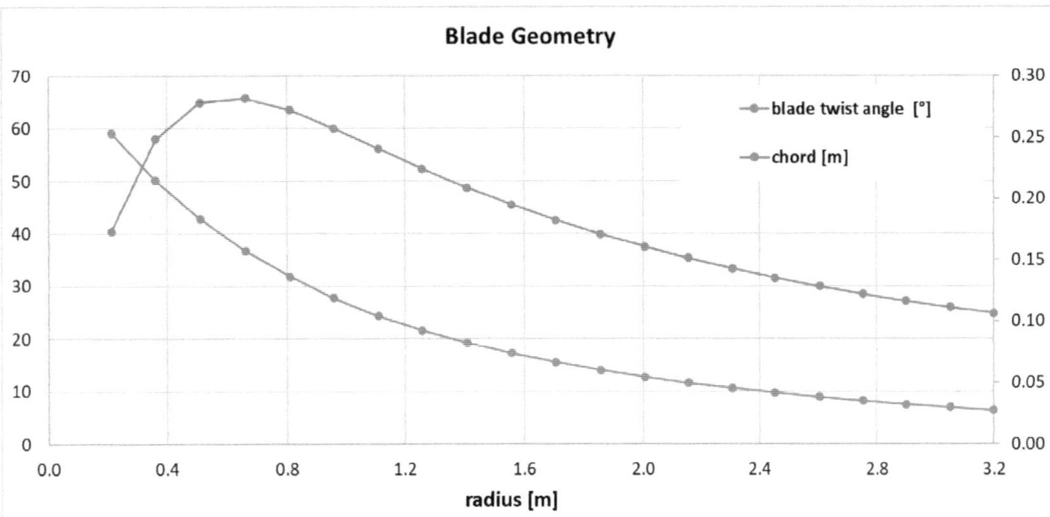

Figure 5 - Chord and twist distributions along the blade span

4. Permanent Magnets Generator

The designed generator can also work as a motor, to speed up the start-up transient when the wind speed reaches the turbine start-up threshold. It consists of a rotor on which are installed permanent magnets with high coercivity with dimensions equal to 60 x 30 x 15 mm arranged with a pitch equal to 50.6 mm. On the stator are placed the windings obtained on multilayer printed circuit boards. For weight containment and easier integration with the wind turbine structure, the stator of the electrical machine is devoid of ferromagnetic circuitry. The radial thickness of the stator of the electrical machine is therefore reduced to a few millimeters. The use of printed circuits facilitates the industrialization of the machine and contributes to the containment of losses in the electrical circuits that would compromise the efficiency of the electrical machine at higher speeds in the presence of windings made with traditional conductors and in the absence of ferromagnetic material.

The realization of the alternator was preceded by a preliminary study and by the realization of a small scale test prototype necessary for the choice of the magnets, for the definition of the type of coils for the stator and for the design of the control electronics.

The test prototype has also allowed the verification of the results of the performance calculation made with finite element software.

Once known the typical speed regimes of operation of the wind turbine and the maximum permissible gap for an acceptable performance, the expected performances have been verified realizing a test prototype in reduced scale having the same peripheral speed, same magnets, same pitch and same gap.

The small scale prototype shown in Figure 6, with a rotor diameter of 1 m, was designed to allow a relatively easy replacement of the multilayer coils on a printed circuit board. Two possible types of coils configurations were tested, the first with two coils per layer on four layers, the second with four coils per layer on two layers.

The first configuration has a lower electrical resistance and therefore potentially higher efficiency. The second configuration has the advantage of being thinner and of having a limited axial encumbrance, with relative lower production costs; moreover, not having internal layers, although it possesses a higher electrical resistance, heat dispersion is easier. After the operating tests performed on the test prototype the configuration with two coils per layer has been chosen, leading to the definition of the electrical machine and its control electronics. Figure 7 shows the printed circuits made for the coils and their final arrangement on the stator of the wind turbine.

Figure 6. Prototype for alternator testing

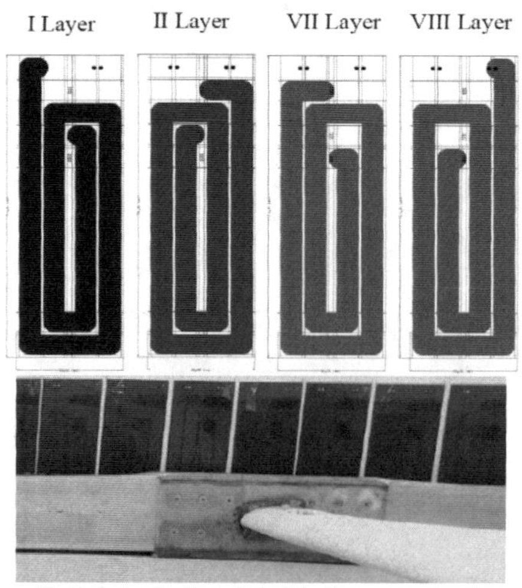

Figure 7. Printed circuit with 2 turns per layer on four layers for winding

5. Mechanical Design

5.1. Rotor

The problems of strength and reliability of this type of turbine have not yet been the object of consolidated studies since the number of realizations is very limited. In the case of double diffuser architecture and with the realization of a rotor that has a peripheral ring supported by the blades and having the function of rotating part of the electrical machine, to the knowledge of the writers, we limit ourselves to the case illustrated here.

The maximum design loads of the blades, as in the classic case of horizontal axis turbines, are those related to abnormal operating conditions, capable of determining extreme stresses much higher than those found in the presence of maximum loads in ordinary operating conditions. In the specific case the blades, having to support the ring on which the magnets are mounted, have additional design requirements related to the relatively high moment of inertia, which introduces higher moments in the case of emergency braking and gyroscopic effects, however limited by means of yaw speed control.

On this basis, in accordance with the design requirements, a lay-up consisting of unidirectional and four-axial fiberglass mats was set up and verified. In particular, the lay-up was designed as a combination of two types of layers, differentiated with respect to the orientation of the fibers:

- Quadriaxial (QA) mats: they have fibers oriented at 0°, 45°, 90°, 135° of 300 g/m2 for a total thickness of 1.1 mm.
- Unidirectional (UD): parallel fibers of 1000 g/m2 with a total thickness of 0.9 mm.

The mechanical properties of the adopted stratifications have been determined through conventional destructive experimental tests and through advanced non-destructive experimental tests. The tests were conducted at the Official Materials Testing Laboratory "M. Salvati" of the Polytechnic University of Bari. In particular, some material samples representative of the used stratifications have been subjected to tensile tests in order to obtain the mechanical parameters in terms of stiffness (technical moduli) and strength, and to ultrasonic goniometric immersion tests by means of an innovative setup that allowed to determine all the independent components of the elastic tensor that characterizes the mechanical response of each composite sample QA and UD [9].

The composite layering was defined with reference to the radial position along the blade and is shown in Table 1 below.

Table1. Layering

Radial interval[mm]	Layering	Total thickness[mm]
250-360	6QA+4UD+QA+4UD+6QA	21.5
360-510	4QA+4UD+QA+4UD+4QA	17.1
510-810	2QA+4UD+QA+4UD+2QA	12.7
810-1110	QA+4UD+QA+4UD+QA	10.5
1110-2750	6QA+4UD+QA+4UD+6QA	21.5
2750-3050	QA+2UD+QA+2UD+QA	6.9

A series of reference specimens, made with QA+2UD+QA layup, were subjected to static and fatigue tests with stress ratio R=0.1. The average ultimate stress obtained is 425 MPa (minimum value 396 MPa), while the fatigue strength curve has m = 7.9 with max = 127.5 MPa at 2-106 cycles.

A further critical element of the rotor is the peripheral ring in which the permanent magnets are embedded, as it is necessary to ensure adequate flexural rigidity to avoid interference with the stator under the action of radial elastic deformations. The radial clearance introduced at rest is 12 mm and is very limited compared to the external radius of the rotor of 3200 mm.

The ring connects circumferentially the end of the five blades and has the fundamental task of supporting the 400 permanent magnets each weighing 0.205 kg, positioned at a pitch of 50.6 mm

IOP Conf. Series: Materials Science and Engineering 1214 (2022) 012030 doi:10.1088/1757-899X/1214/1/012030

measured at the base of the magnet. In addition, to ensure adequate efficiency of the magnetic circuit, it was necessary to insert a peripheral winding of iron-magnetic plate.

The design and sizing, verified with a finite element analysis, have been realized considering the main stresses to which the ring is subjected and consisting of:

- Own weight of the structure;
- Centrifugal thrusts of the magnets and the ferromagnetic plate;
- Deflection induced by differential deformation of the blades (e.g. as a result of misalignment with respect to the wind and due to gyroscopic effects).

Figure 8 shows schematically the front and cross sections of the peripherical ring. The objective pursued was to reduce the thickness of the magnet retaining structure in the radial direction that occupies part of the air gap. For this reason, underneath the magnets, a unidirectional fiberglass belt wrapped circumferentially and contained in a U-shaped cradle made of quadriaxial mat has been realized in order to provide the necessary stiffness in the radial and circumferential directions.

The cradle is overall made of five equal arcs that were placed on a template, to allow the subsequent winding of the unidirectional. The positioning of a further cradle destined, to house the magnets, allowed the lateral winding of the unidirectional fiber to be carried out. A light filler was introduced to position the ferromagnetic plate at the intended radius, to distance the magnets during assembly and to dampen any vibrations under operating conditions. The crown was completed by winding a four-axial mat capable of transmitting to its structure the centrifugal force exerted by the magnets. The rotor assembled in all its components is shown in Figure 9. Auxiliary structures and components, such as the hub with its connections to the blades and the emergency brake, have obviously been made.

The blades engage on the hub by means of the so-called "Ikea" joint consisting of cylindrical pawls, transversely threaded and radially inserted in the base annular section of the blades, on which are mounted threaded steel shanks that engage the flanges present on the hub.

The emergency brake, suitably sized, is of the disc type and is operated by an electromagnetic actuator.

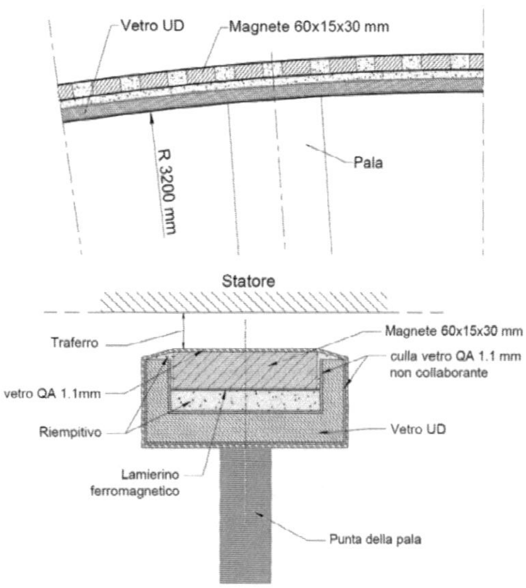

Figure 8. Sections of the crown

Figure 9. Rotor

5.2. Structure of stator and double diffuser

The function of the stator structure is to support the diffusers, the rotor hub with associated braking system, and to ensure adequate connection to the fifth wheel for yaw control for orientation.

In the offshore conceptual design, orientation is accomplished by means of the rotation of the polyhull vessel on which the turbines are placed.

For the realization of the prototype and the performance of functional tests, it was necessary to introduce the motorized rotary table for yaw control. Obviously, the flange made for the fifth wheel can be exploited to make a fixed assembly.

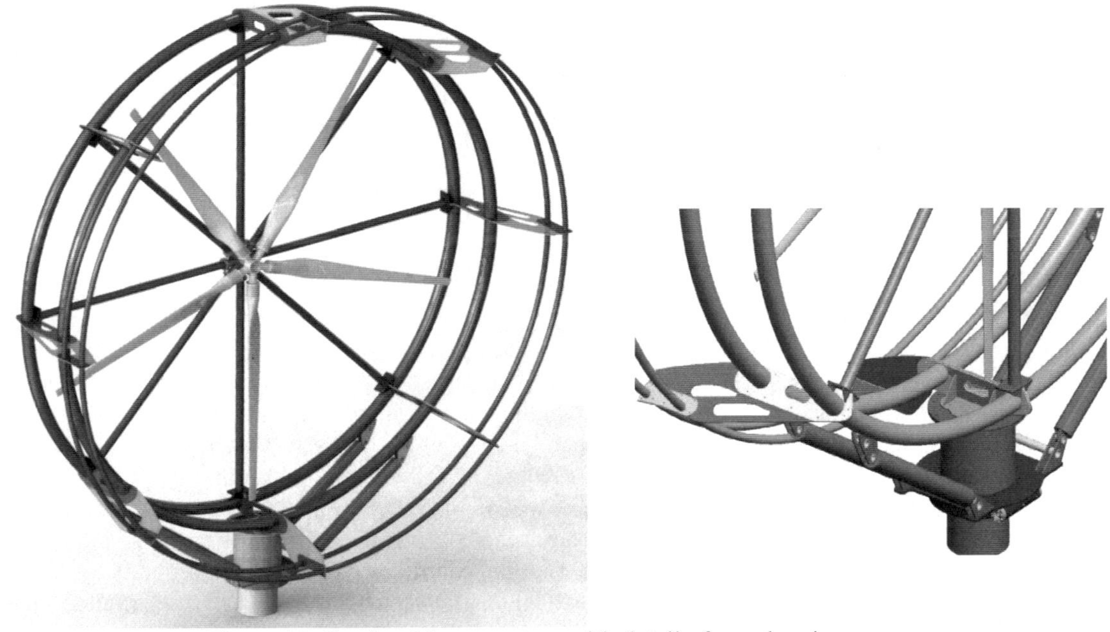

Figure 10. Steel turbine structure with detail of yaw bearing area

For road transport needs it has been necessary to realize a modular assembly that has directed the design choice on a steel structure (the project also provides an alternative in aluminum alloy) consisting of sixteen sectors, eight for the internal diffuser, eight for the external one. Each sector consists of two curved profiles with a hollow circular section that carry at the ends the connection flanges to suitably shaped flat profiles, as shown in Figure 10.

The structures of the aerodynamic profiles of the diffusers, made of fiberglass composite, are grafted onto the auxiliary transverse connections between the curved beams with circular section. There are a total of sixteen modules, eight for the inner and eight for the outer diffuser. Figure 11 shows the metal structure of the stator during the assembly tests of the modular elements of the diffusers. The structure was verified by adopting a FEM model.

Figure 11 shows the steel structure of the stator during the assembly tests of the modular elements of the diffusers.

Figure 11. Steel structure of the stator during the assembly tests of the diffuser modules

5.3. Realization of the prototype and preliminary tests

The rotor, the diffusers on the stator and the floating support structure of the offshore laboratory were made by the company Cartflow in Castelvolturno (Naples – Italy), a leading company in the production of wind turbines with the ability to make molds with robotic management processes for a length of 50 meters and with a proprietary technique of construction of the blades defined “One-Shot”, which allows the direct introduction of shear webs in the blade structure. The prototype was assembled in the laboratories of the same company to carry out the functioning tests (see Figure 4).

The steel structure of the stator was made by the company Faver (Bari - Italy).

Tests have been carried out to verify the absence of interference in the rotational motion, the efficiency of the braking system, the starting of the turbine, the yaw control system. Tests related to the verification of the performance and management of the control electronics are in progress.

An unexpected test condition of resilience to exceptional weather conditions was determined by a storm with gusts of wind that exceeded 140 km/h, which resulted in damage to the roof of the Cartflow shed, but left the turbine unscathed.

6. Conclusions

The MEL project is innovative and has ambitious goals aiming at the development of the first offshore application of ducted wind turbines on a floating hull.

The presence of a divergent duct (the double diffuser) allows the interception of a higher air mass flow rate, allowing the reduction of rotor diameters at the same rated power, increasing rotor stiffness, reducing blade deformations, and increasing rotation speed compared to conventional turbines.

The electric generator with permanent magnets embedded in the rotor ring takes advantage of the high peripheral speed improving efficiency.

The size of the target turbine is smaller than currently installed multi-megawatt (MW) offshore turbines, but it can be mounted on the hull in a dry dock, greatly reducing installation costs, making it possible to periodically transport the platform to dry dock for maintenance, and thus ensuring a very long life. Also noteworthy are the lower costs for decommissioning, at the end of life, of these floating mobile turbines compared to traditional offshore wind turbines fixed on the seabed.

The research topic developed is in accordance with the themes proposed within the Great Research and Innovation Scope 4 “Digital, Industry, Aerospace” and 5 “Climate, Energy, Sustainable Mobility” of the National Research Program PNR 2021-27.

References

[1] Torresi M, Postiglione N, Filianoti PF, Fortunato B and Camporeale SM 2016 Design of a ducted wind turbine for offshore floating platforms *Wind Engineering* Vol. 40(5) 468–474 DOI: 10.1177/0309524X16660226

[2] Palmieri M, Bozzella S, Cascella GL, Bronzini M, Torresi M and Cupertino F 2018 Wind Micro-Turbine Networks for Urban Areas: Optimal Design and Power Scalability of Permanent Magnet Generators *Energies* Vol. 11(10) DOI: 10.3390/en11102759

[3] Patankar SV 1980 *Numerical Heat Transfer and Fluid Flow*. Washington, DC and New York: Hemisphere Publishing Corporation

[4] Launder BE and Spalding DB 1972 *Lectures in Mathematical Models of Turbulence*. London and New York: Academic Press

[5] ANSYS, Inc. 2016 ANSYS Fluent User's Guide, Release 17.2

[6] Rajagopalan RG and Fanucci JB 1985 Finite difference model for vertical axis wind turbines. *Journal of Propulsion and Power* 1432–436

[7] Torresi M, Fortunato B and Camporeale SM 2013 An efficient 3D CFD model for the analysis of the flow field around Darrieus rotors. In: *Proceedings of the ASME Turbo Expo* 8 San Antonio TX 3–7 June art. no. GT2013-95564 DOI: 10.1115/GT2013-95564

[8] Marten D, Wendler J, Pechlivanoglou G, et al. 2013 QBlade: An open source tool for design and simulation of horizontal and vertical axis wind turbines *International Journal of Emerging Technology and Advanced Engineering* 3: 264–269

[9] Castellano A, Fraddosio A and Piccioni MD 2017 Ultrasonic goniometric immersion tests for the characterization of fatigue post-LVI damage induced anisotropy superimposed to the constitutive anisotropy of polymer composites *Composites Part B: Engineering* 116 122-136 DOI: 10.1016/j.compositesb.2017.02.025

Assessment of the deformation behavior of a tensegrity floor through photogrammetry

G Chiappini[1], M Coccia[2], M Rossi[2], F Marchione[3], P Munafò[3], C Scoccia[2] and C Luca[2]

[1] eCampus telematic University, Novedrate, 22060 – Italy
[2] Department of Industrial Engineering and Mathematical Sciences, Università Politecnica delle Marche, Ancona, 60131 – Italy
[3] Department of Construction, Civil Engineering and Architecture, Università Politecnica delle Marche, Ancona, 60131 – Italy

gianluca.chiappini@uniecampus.it

Abstract. Tensegrity structures are interesting elements that can be used to create innovative building components. In this work, the deformation behavior of a new type of tensegrity floor (patent n. 0001426973) was investigated through image analysis and photogrammetry, using stereoscopic algorithms. The main innovation of the proposed tensegrity floor is the structural cooperation between glass tiles and steel components, obtained using suitable adhesive joints. In order to evaluate the deformation of the floor under different loading conditions a non-contact measurement system was developed and validated on a scale prototype of the floor. The used method is a stereoscopic optical technique based on the grid method, the validation was performed comparing the measured displacement with the one measured in the same floor with linear displacement sensors. The stereoscopic images were processed with Matlab®, furthermore, a combination of interpolation algorithms were employed to generate the full-field map of the displacement. Two configurations of the floor (with and without the adhesive joints) and three loading conditions were investigated. The analysis demonstrates that the adhesive joints produce a reduction of the overall deflection of the floor of about 60%, falling within the admissible values established by Italian building code. The measurement system is reasonably accurate and can be used in future applications where linear transducers cannot be installed.

1. Introduction

Tensegrity is a structural principle based on a series of components subjected either to tension or compression arranged in such a way that the compressed members (usually bars or struts) do not mutually interact while the pre-stressed tensioned members (usually cables or tendons) describe the geometry of the system. No structural member experiences a bending moment and there are no shear stresses within the system. This can produce exceptionally strong and stiff structures for their mass and for the cross section of the components. Renè Motro describes the tensegrity structures as "systems in a stable selfstress state, they include a discontinuous set of compressed components inside a continuum of tensioned components" [1].

Figure 1 shows a tensegrity floor, developed and patented by one of the authors [2, 3], which is a hybrid glass-metal structure characterized by a light-weight metallic sub-frame sustaining a glass deck. Due to its high transparency, the floor is particularly suitable for applications in case of coverage of archaeological sites and the revitalizing of historical buildings, where it is necessary to have a view underneath the floor level.

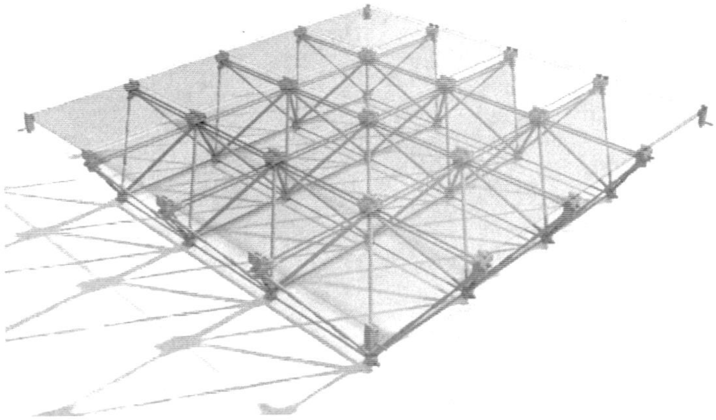

Figure 1. Figure with short caption (caption centred).

The idea of using tensegrity structures with a metallic subframe was already applied by architects for their aesthetic values, see for instance refs. [4, 5]. However, the main novelty of this tensegrity structure is the introduction of the glass plates as structural elements which contribute to the global stiffness of the system. Usually, the glass panels do not produce an increase of stiffness of the whole building and they do not contribute to the underlying metallic supporting structure, even in the glazing facades that are commonly named *structural*. In the patented solution [6, 7], which is the object of this study, instead, the glass is no longer in the simply supported configuration but it is connected to the structure through a specifically designed adhesive joint, and it contributes actively to the deformation reduction of the whole system, allowing it to satisfy the limits imposed by building codes.

In order to evaluate the effectiveness of such approach and assess the contribution given to the stiffness of the floor by the adhesive joint, load tests were carried out on a floor 1:2 scale prototype. The aim of this work was to develop a measurement technique that exploits photogrammetry to evaluate the displacements and deflection of the floor during load tests. The development of a reliable non-contact measurement method is particularly important for future applications such as the *in-situ* monitoring of the floor deformation and for the performance analysis of the proposed solution.

In this work, the measurements were carried out on the prototype under two conditions: (i) with the glass tiles adhesively bonded to the steel nodes and (ii) with the glass tiles simply supported to the metallic frame. The results of the optical measurements were compared with an independent displacement value measured using mechanical LVDT sensors.

2. Material and methods

2.1. Tensegrity floor prototype

The prototype of the floor is made of a 4×4 grid of glass tiles, measuring 30×30 cm^2 each, the overall size of the prototype is 1.2×1.2 m^2 (see figure. 2). As illustrated in the schematic of figure. 2, the glass tiles are supported by a series of metal cubes, named *upper cubes*, that constitutes the first level of the tensegrity structure, the structure is then composed by a second level of metal cubes, named the *lower cubes*. In each level, the cubes are connected to each other with a set of orthogonal strands subjected to axial tension, named *upper* and *lower strands*, respectively. Finally, the two levels are connected through a set of lateral elements, named *lateral strands,* which are the compressed part of the tensegrity structure and are made of a series of rigid beams.

The cubes have holes for the passage of the cables while small connecting forks, realized on the external surface, are used to connect the beams of the lateral strands. Glass panels are bonded with the upper cubes with the epoxy adhesive 3M ™ Scotch-Weld ™ DP7240B/A.

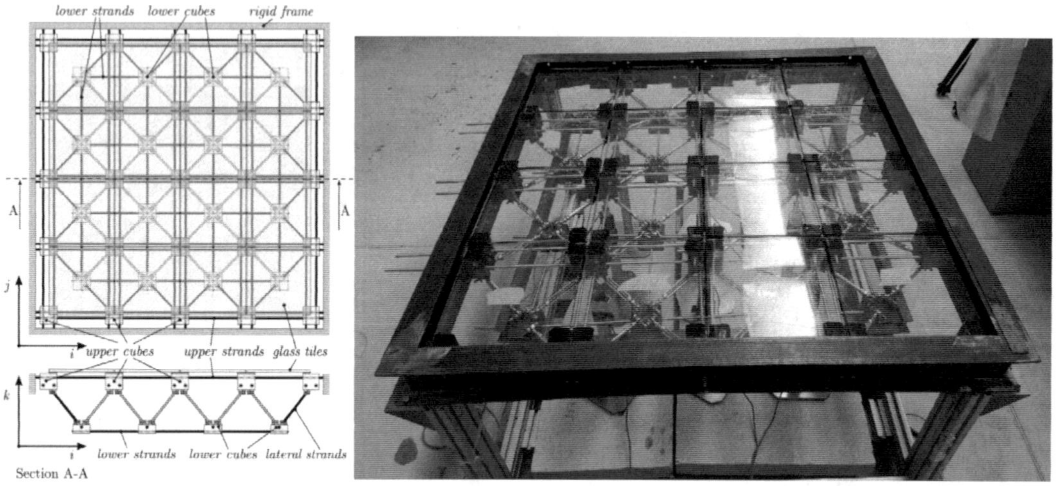

Figure 2. Floor prototype.

2.2. *Experimental test*

The prototype was tested with three different loading conditions, as shown schematically in figure 3. The loading conditions were determined in accordance with the Italian building code NTC2018, i.e.:

- a uniformly distributed load of 400 kg/m^2;
- a load of 200 kg/m^{2}applied to the centre of the floor;
- a load of 200 kg/m^{2}applied to one side of the floor.

Each load case was repeated three times and the results presented and discussed in the following sections are the average of the measurements obtained in the three repetitions.

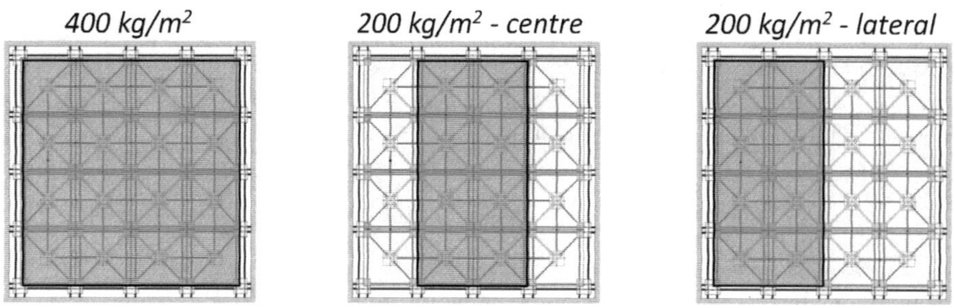

Figure 3. Load conditions.

2.3. *Experimental setup*

The displacement and deflection of the floor was measured with two different techniques: one based on a stereoscopic optical technique applied to the upper glass surface and the other based on the use of displacement sensors applied to the lower part of the floor. In figure 4 the used equipment and the experimental setup are shown.

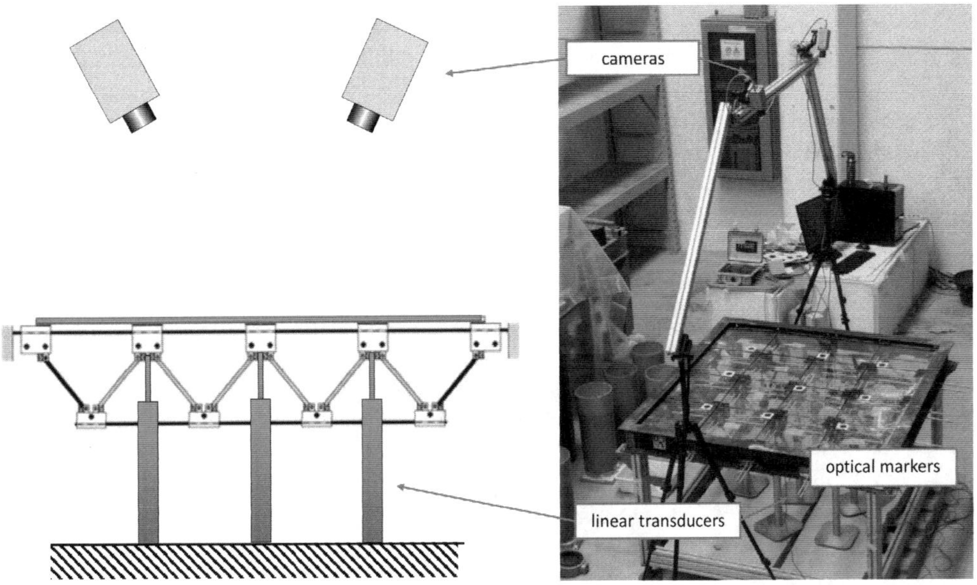

Figure 4. Experimental setup.

The displacement sensors used to measure the lower displacement of the floor, are the M.A.E. PT50T, i.e. spring loaded linear displacement transducers with a stroke of 50 mm.

Table 1. Features of the cameras and calibration parameters.

		CAMERA 1	CAMERA 2
Sensor	[Type]	CMOS	CMOS
Sensor pixel size	[μm]	6.8 × 6.8	6.8 × 6.8
Sensor resolution	[pixel]	1280 × 1024	1280 × 1024
Frame rate	[fps]	1	1
Lens	[Type, mm]	C mount, 12	C mount, 12
Working distance	[mm]	1500	1500
Sensor noise	[gray level, dB]	0.94, −19	0.94, −19
Displ. accuracy (st. dev)	[pixel, mm]	±[0.02, 0.09]	±[0.02, 0.09]
CALIBRATION DATA			
fx – fy	[pixel]	1779 - 1775	1792 - 1785
cx – cy	[pixel]	621 - 495	636 - 519
Tx – Ty – Tz	[mm]	784 – -36 – 212	
αx – αy – αz	[rad]	0.0053 – 0.7118 – 0.0285	

The adopted optical technique [8, 9, 10] is a grid method that allows to measure the shape of the surface subjected to the testing loads. Operatively, a grid of circular markers was glued on the surface, as depicted in figure 4, then a stereoscopic system made of two digital cameras (model Pixelink®B371F, CMOS sensor with 1280x1024 pixel resolution) was used to obtain the picture of the floor at two different angle in order to apply stereography, after a suitable camera calibration. Table 1 shows the characteristics of the cameras and the obtained calibration parameters.

Starting from the undeformed configuration, images were saved for each loading condition. The images were then post-processed by image binarization and bubble analysis, obtaining the 3D coordinates of the markers in a global reference frame by stereoscopic algorithms. Finally, the 3D coordinates of the markers in the deformed and undeformed configuration were used to calculate the vertical displacement of the floor. The whole process is described in figure 6; in the deformed images, the weights used to load the floor are visible, the position of cameras and markers was indeed optimized so that the weights do not interfere in the stereoscopic analysis. The last plot of figure 6 shows how each point of the undeformed (blue) and deformed (red) configuration can be expressed in the same coordinate system to derive the deformation of the floor.

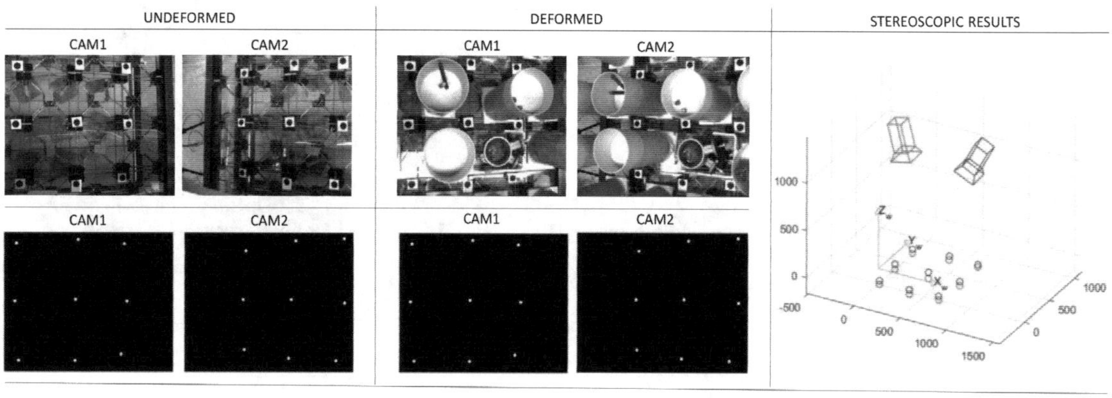

Figure 5. Image analysis and stereoscopic results.

2.4. Data processing

Figure 6 shows the results obtained for the first load case, where a uniformly distributed load of 400 kgf/m^{2}was applied to the floor. The developed optical technique was used to measure the vertical displacement close to the optical markers (black dots in figure) while the linear transducers were applied in the lower surface close to the lower cubes (red dots in figure). Since the position of the optical markers and the linear transducer is not coincident, a difference is observed between the two measurements that, for certain points is rather high, see for instance in points 4, 6, 7 and 8. To overcome this issues and have a fair comparison, the measurement of the displacement was computed at the exact position of the linear transducer using an interpolation procedure which starts from a discrete grid of 9×9 points.

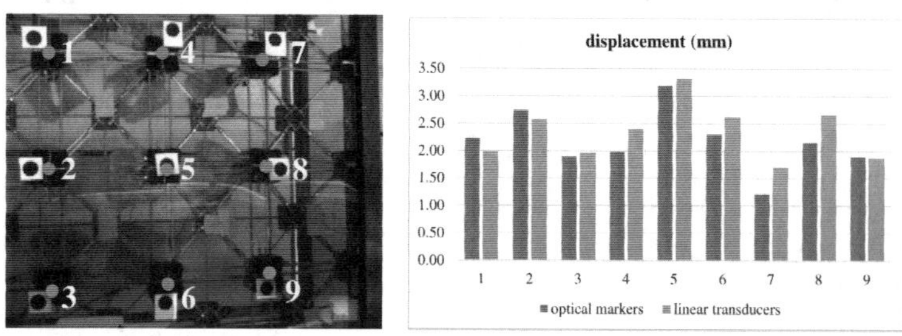

Figure 6. Resulting displacements with the two different measurement techniques. The values corresponding to the optical technique were calculated close to the optical markers (black dots) while the linear transducers measurement were performed close to the lower cubes (red dots)

The interpolation was performed combining three different algorithms available in Matlab®, in particular, the final displacement was obtained as the average of the values obtained from the different algorithms. The used algorithms are:

- *cubicinterp* –bicubic spline interpolation;
- *biharmonicinterp* - biharmonic spline interpolation (Matlab® griddata method);
- *thinplateinterp* - Thin-plate spline interpolation;

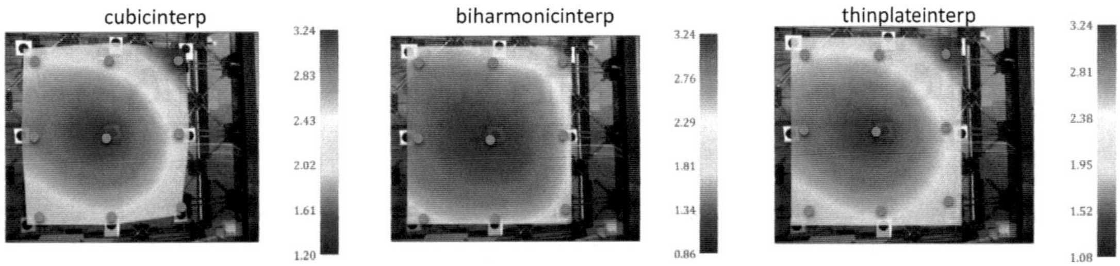

Figure 7. Extrapolation of optical results

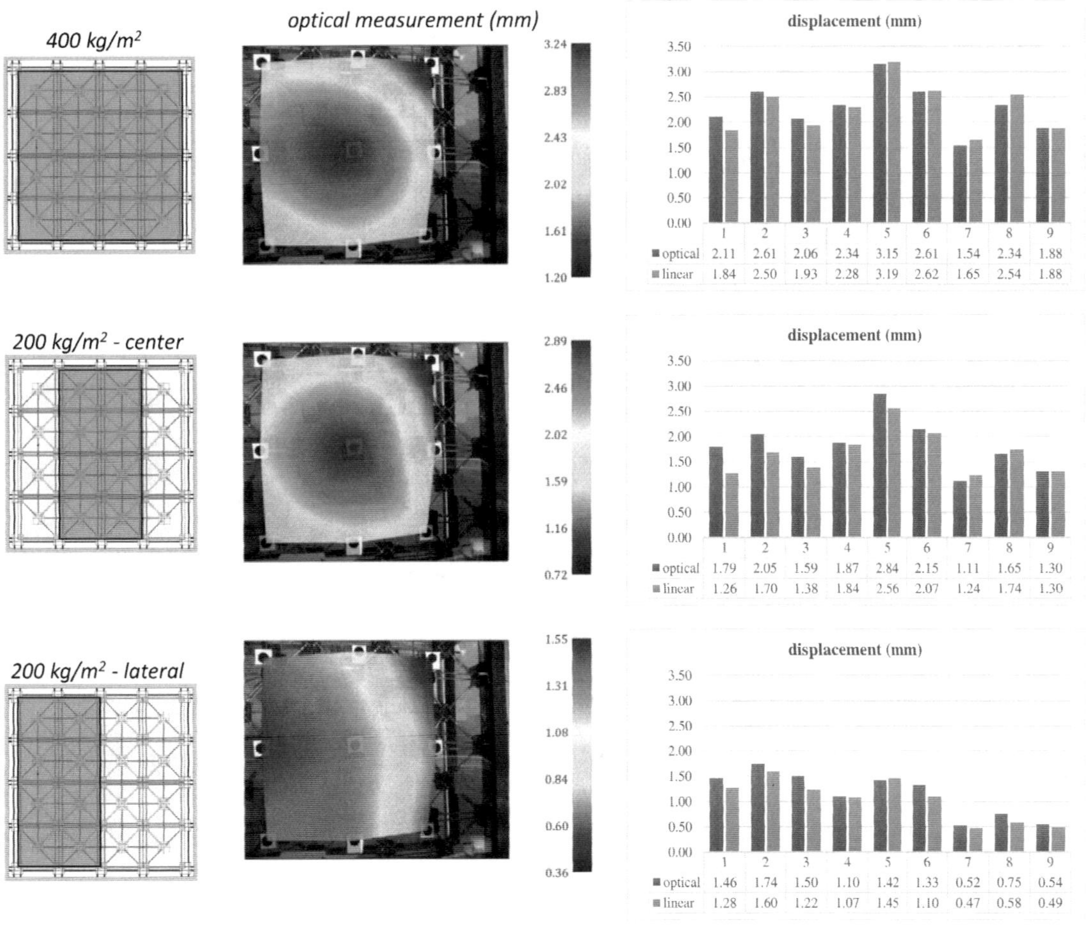

Figure 8. Results for the floor with glued tiles

IOP Conf. Series: Materials Science and Engineering 1214 (2022) 012031 doi:10.1088/1757-899X/1214/1/012031

From this interpolation, a continuous map of the floor displacement can also be derived. Figure 7 shows the displacement maps obtained with the three different algorithms for the first load case, the three methods provide similar results. From here on, therefore, only the maps obtained combining the three methods will be shown. The comparison between optical method and linear transducer will be conducted at the exact position of the linear transducer (red dots), obtained through interpolation.

3. Results and discussion

This section presents the results of all measurements made with the three different loading conditions for the prototype with adhesively bonded glass panels (figure. 8) and with simply supported glass panels (figure. 9). For each condition, a scheme of the applied load, a map of the vertical displacement and a comparison between the optical measurement and the linear transducer at the 9 control points is provided. The first row of figure 8 illustrates the results obtained for the first loading condition, which is the same discussed in figure 7. In this case, a better agreement is found between the measurement of the optical method and the one obtained with the linear transducer, the large difference observed in points 4, 6, 7 and 8 is consistently reduced. A residual error is still present, which can be ascribed to the fact that the linear transducer is applied in the lower surface of the floor while the optical measurement is performed on the top, however, the obtained accuracy is reasonably good for the sake of the intended application.

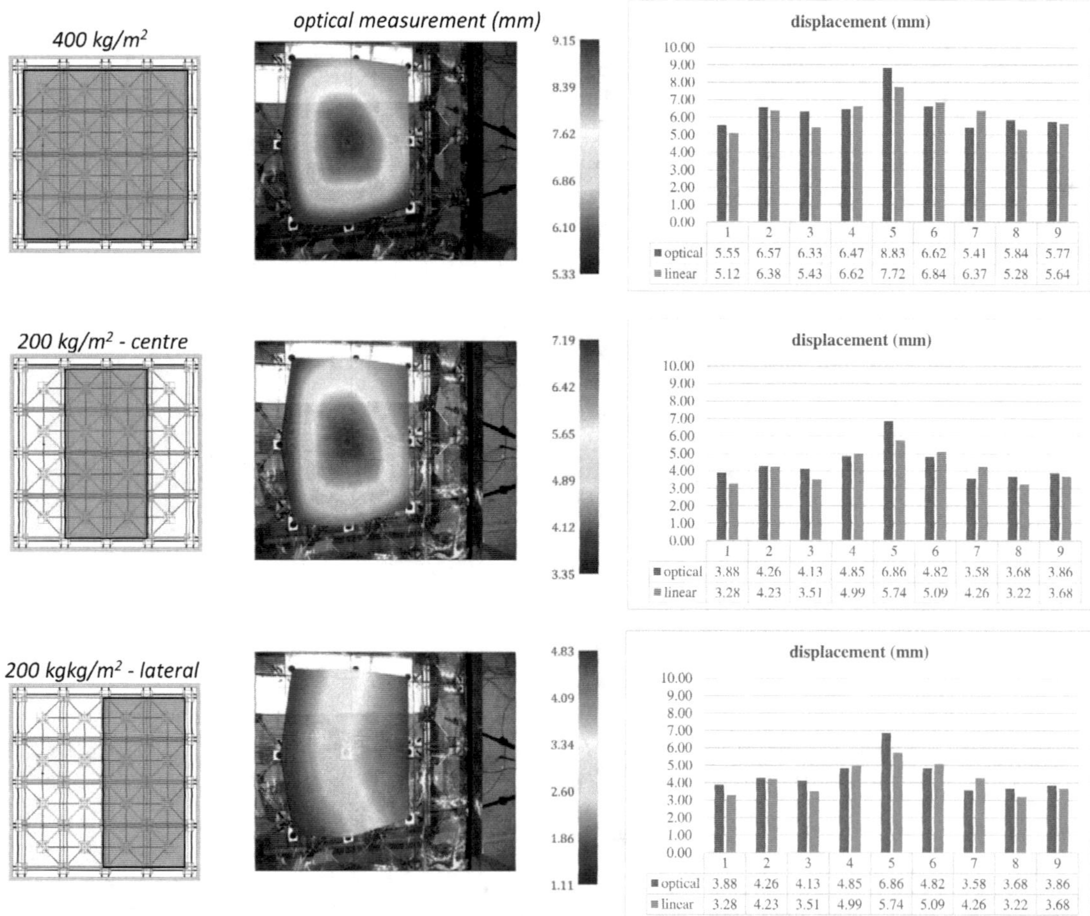

Figure 9. Results for the floor with simply placed glass tiles

To show the potentiality of the method, figure 9 illustrates the displacement of the floor obtained when the tiles are simply supported in the metallic frame without the adhesive joint. A consistent increase of the displacement is observed in all cases, which is around the 60%. Figure 10 shows the comparison with respect to the first loading condition. The optical measurement allows to capture the stiffness reduction efficaciously and can be considered as a valid alternative to the standard measurement performed with mechanical transducer.

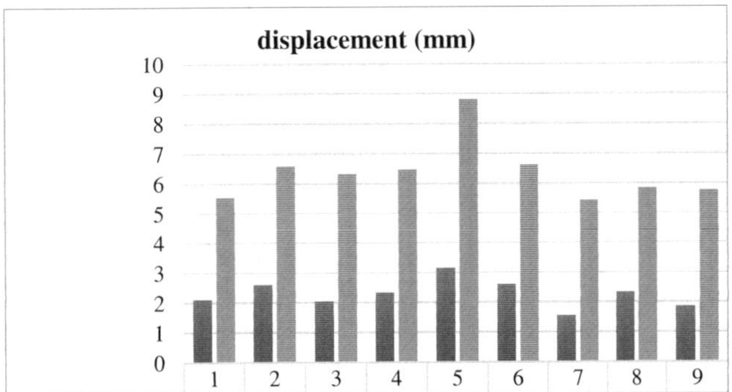

	1	2	3	4	5	6	7	8	9
■ glued tiles	2.11	2.61	2.06	2.3352	3.1505	2.6096	1.54	2.34	1.88
■ not glued tiles	5.55	6.57	6.33	6.47	8.83	6.62	5.41	5.84	5.77

Figure 10. Comparison of the displacements measured between the floor with glued tiles and simply placed

4. Conclusion

In this work, the result of a series of experiments performed on a tensegrity floor scale prototype was presented and discussed. The prototype, patented by one of the authors, is a hybrid glass-metal structure characterized by a light-weight metallic sub-frame sustaining a glass floor decking. The peculiarity of the developed floor is that the glass is not simply supported but it is connected to the structure through a specifically designed adhesive joint. The tests were conducted to evaluate the contribution given to the global stiffness of the floor by the adhesive joints.

The aim of this work was to develop a measurement technique based on photogrammetry to evaluate the displacements and deflection of the floor during this type of load tests. The optical measurement was compared with the one obtained in the lower part of the floor with linear transducers: the comparison was made exploiting multiple interpolation functions of Matlab® to evaluate the displacement at the exact position of the sensor.

The tests and measurements showed a consistent contribution of the adhesive joint to the overall stiffness of the floor: with glued glass tiles there is a stiffness reduction of about 60%. The non-contact measurement technique is enough accurate and reliable to perform this type of test. In the future, the developed measurement technique will be used to evaluate different types of structural adhesives and to optimize the floor and the adhesive joint.

References

[1] Motro, R., 2003. Tensegrity: structural systems for the future. Elsevier.
[2] Munafò, P., 2017. Tensegrity floor, May 3. Patent N.0001426973.
[3] Alderucci, T., Terlizzi, V., Urso, S., Borsellino, C., and Munafò, P., 2018. "Experimental study of the adhesive glass-steel joint behavior in a tensegrity floor". International Journal of Adhesion and Adhesives, 85, pp. 293–302.
[4] Quirant, J., Kazi-Aoual, M., and Motro, R., 2003. "Designing tensegrity systems: the case of a double layer grid". Engineering structures, 25(9), pp. 1121–1130.

[5] Cimmino, M., Miranda, R., Sicignano, E., Ferreira, A., Skelton, R., and Fraternali, F., 2017. "Composite solar facades and wind generators with tensegrity architecture". Composites Part B: Engineering, 115, pp. 275– 281.

[6] Scoccia, C., Palmieri, G., Callegari, M., Rossi, M., Carbonari, L., Munafò, P., Marchione, F., Chiappini, G., Multibody Analysis Of A Tensegral Servo-Actuated Structure For Civil Applications, Proceedings of the ASME 2021 International Design Engineering Technical Conferences & Computers and Information in Engineering Conference, August 17-20, 2021, Online, Virtual

[7] Scoccia, C., Carbonari, L., Palmieri, G., Callegari, M., Rossi, M., Munafò, P., Marchione, F., Chiappini, G., Design of a tensegrity servo-actuated structure for civil applications, ASME Journal of Mechanical Design.

[8] Simoncini, M., Forcellese, A., Mancini, E. et al. Experimental and numerical investigation on forming limit curves of AA6082 aluminum alloy at high strain rates. Int J Adv Manuf Technol 112, 1973–1991 (2021).

[9] Sasso, M., Mancini, E., Chiappini, G., Simoncini, M., Forcellese, A., Adapted Nakazima test to evaluate dynamic effect on strain distribution and dome height in balanced biaxial stretching condition (2018) International Journal of Mechanical Sciences, 148, pp. 50-63.

[10] Eusebi, A.L., Bellezze, T., Chiappini, G., Sasso, M., Battistoni, P., Influence of aeration cycles on mechanical characteristics of elastomeric diffusers in biological intermittent processes: Accelerated tests in real environment (2017) Water Research, 117, pp. 143-156.

IOP Conf. Series: Materials Science and Engineering 1214 (2022) 012032 doi:10.1088/1757-899X/1214/1/012032

Dynamic behaviour of a battery pack for agricultural applications.

F Mocera[1], S Martelli[1] and M Costamagna[1]

[1] Politecnico di Torino, Department of Mechanical and Aereospace Engineering, Corso Duca degli Abruzzi 24, 10129 Torino, Italy

E-mail: francesco.mocera@polito.it

Abstract. Fight and contrast against climate change and global warming can be considered the most important challenge for the next decades. One of the most involved sectors is the one related to the carriage of passengers and goods, and, recently, also the work vehicles too. The agricultural machines and tractors are no exception to this. Indeed, agricultural industry is the second contributor in terms of pollutant emissions. So, research about agricultural machinery is focusing itself about the development of more sustainable propulsion systems such as hybrid or full-electric solutions. One of the most important components of a hybrid or full electric vehicle is the battery pack. The lack of adequate vibration isolation is the main cause of battery pack failure during operation. In the field of the agricultural vehicles, since the maximum speed is quite low and the weight of the various subsystems are high, combined with heavy working cycles, the dynamic analysis of the battery pack focuses on its low frequency behaviour. In this context, this paper focuses the attention on the dynamic behaviour of a battery pack, thought for a plug-in hybrid electric orchard tractor, through simulations obtained taking advantage of FEM and multibody software. In particular, it will be illustrated the dynamic behaviour of the battery pack while the tractor is moving on a bumpy road and when the tractor chassis is subjected to an impulsive load.

1. Introduction

Recently, the agricultural machines field has begun to show attention to the development of hybrid-electric solutions [1]. The main problem related to the design of hybrid or electric tractors is the high overweight with respect to the traditional ICE vehicles. This may cause some problems, such as high pressure on the soil, reduction of tractor's maximum speed and high energy consumption [2].

Considering agricultural machines, since vehicles speed range is quite low (the maximum vehicle speed is usually set to 40 km/h) and their weight is very high, due to heavy duty cycles during operation, the dynamic analysis of the battery pack should focus the attention on low frequencies vibrations. Usually, battery packs for electric vehicle are the result of sheet metal components assembled by bolts or welds, whose prestress due to joining may influence the battery pack dynamic response [3]. The correct design of the battery pack is very important: indeed, the main cause of battery packs long-term failure is represented by inappropriate vibration isolation. While the tractor is moving, low-frequency vertical vibrations are transmitted from the road to the vehicle. Furthermore, the presence of road irregularities such as potholes and bumps may trigger resonance phenomena [4]. For these reasons, is fundamental that battery pack natural frequencies are not included between 5 – 30 Hz. Indeed, resonance may compromise Li-Ion cells life [5] or trigger a series of exothermic reactions that may cause gas leak, fire or explosion. Thus, from the dynamic point of view, the main design goals should pursue high mechanical strength, maximum natural frequencies and optimization of the battery pack weight [6].

2. Case Study

The battery pack, described in this paper, is intended to be installed in a PHEV orchard tractor, according to a parallel hybrid-electric architecture [7,8]. According to this, the cells, which are contained in the accumulator, may be considered as an impulsive power source of energy [9] that takes action to help ICE engine in certain situations. The main features of an orchard tractor are the compact shape, the operation in tight space and on high slope roads. Other important characteristics of the orchard tractor are the reduced wheelbase (< 2m) and track width (< 1,6m), height less than 2,5 m and a maximum steering angle of approximately 50°. The battery pack of this case study is characterized by a nominal voltage of 640 VDC and a capacity of 16 kWh. The battery pack is constituted by 200 Li-Ion cells (electrical configuration 200s1p) subdivided in 8 modules connected in series. Each module (Figure 1.a) presents 25 Li-Ion cells (electrical configuration 25s1p). The module shape is similar to a box, whose size features are 233mm x 230mm x 270mm (height x width x depth). In addition to this, the battery pack also presents an electronic converter (EC) (Figure 1.b) which contains all the sub-systems required to guarantee battery safety and to monitor and manage the cells (BMS, relays, fuses etc.). The EC shape is substantially a box, whose size features are 500mm x 230mm x 260mm (width, hight and depth). Both battery module and EC case are made of sheet metal.

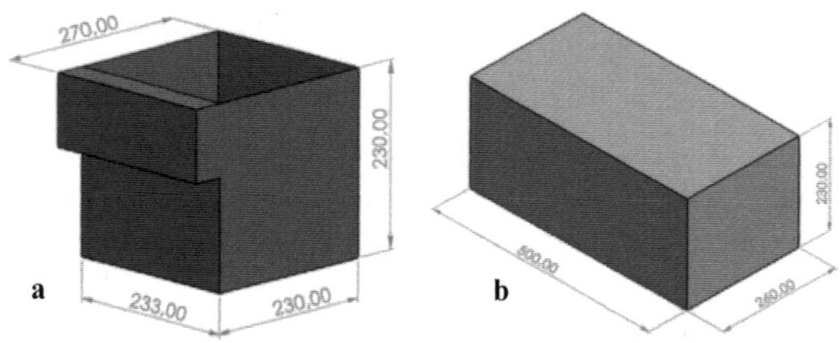

Figure 1. a) Battery module; b) Electronic converter.

Considering the battery pack location in the vehicle, while for automotive field it is well-known [10], for agricultural vehicles the battery pack location is one of the most important challenges. The main aim was to preserve the traditional tractor shape in such a way that the vehicle functionality and visibility could remain unchanged. Another goal was to achieve a high compactness level of electrical and electronic components. For these reasons, the battery pack has been placed in the area included between the cab and ICE bonnet (Figure 2).

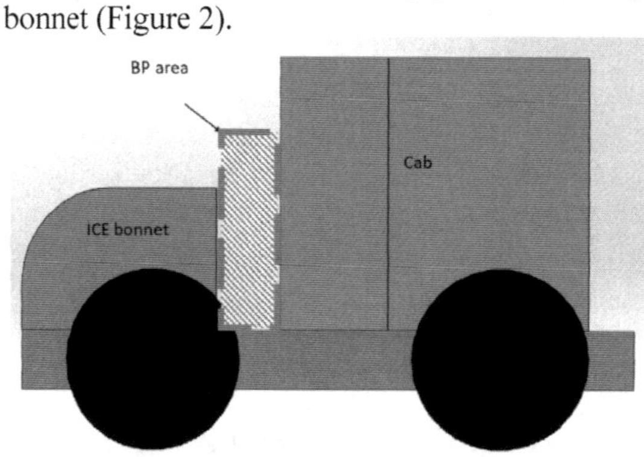

Figure 2. Battery pack location.

As consequence, the battery case presents a horseshoe shape, in which all the element mentioned before are contained (Figure 3.a). The battery case is fixed to the vehicle chassis with 10 anti-vibration supports (Figure 3.b): 6 supports fix the battery pack bases to the chassis, while the others fix the battery pack back wall.

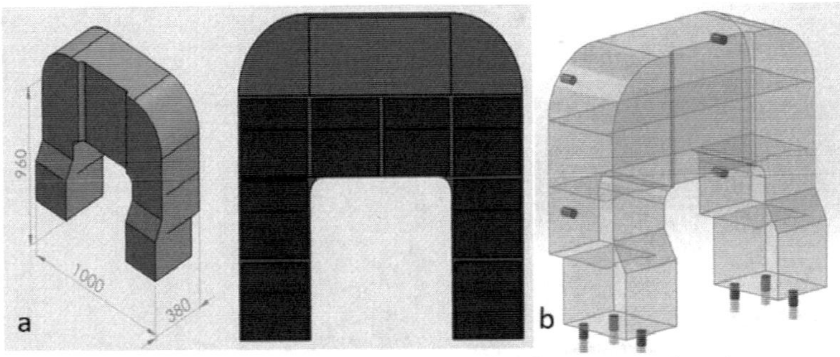

Figure 3. a) battery case; b) Anti-vibration supports location.

2.1 FEM model

First step to model the system is the definition of its geometry. From a geometrical point of view, since all the structures are made of sheet metal, each component was realized with surface modelling technique to better represent their plane stress field while allowing for the component thickness parameterization. To complete the geometric model, there are 2 surfaces under the battery pack bottom plates, representing the chassis plates, on which the accumulator has been installed. Secondly, through the pre-processing environment, each metal sheet of every component was related with a "contact-bonded" constraint to simulate welds. The same kind of constraint was used also to simulate modules and electronic converter fastening: indeed, their bottom plate are bounded to battery pack case (Figure 4).

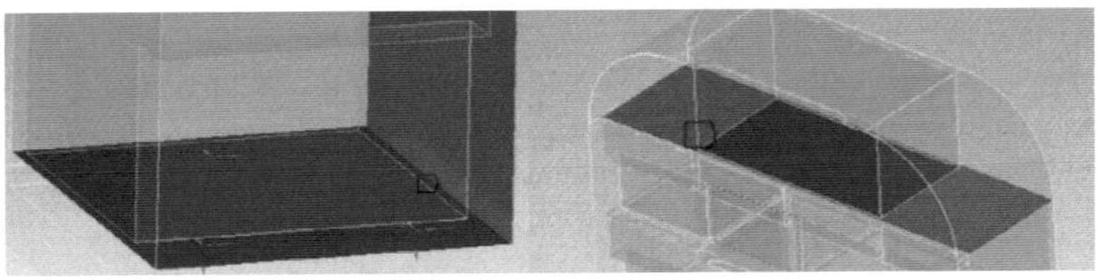

Figure 4. Contact constraints examples.

To simulate Li-Ion cells and safety components weight, a distributed mass of 30 kg has been applied on the bottom plates of the modules and of the EC. In this way, it was possible to well approximate the inertial proprieties of the system without straining computing power. The anti-vibrating supports are fixing elements and, at the same time, vibrations isolation elements. To represent this condition each anti-vibrating support was characterized by 3 orthogonal COMBIN14 spring elements (Figure 5). Each group of springs presents a longitudinal-stiffness (corresponding to the value in the catalogue) and a traverse-stiffness (corresponding to the half catalogue value).

Figure 5. Anti-vibrating supports FEM modelling.

Concerning the meshing process, all the system components were discretized with SHELL281 elements (6 d.o.f. for each node), using "Curvature" size function. The maximum element dimension, after convergence analysis procedure, has been set to 30 mm. The result has been a homogeneous mesh grid composed by 28037 nodes and 8078 elements (Figure 6) without excessive mesh distortions.

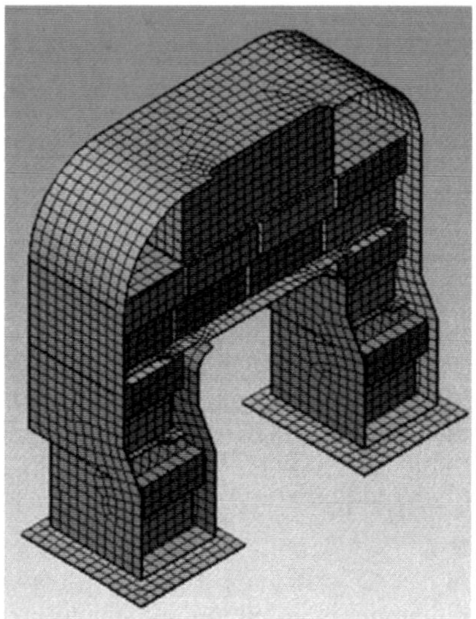

Figure 6. System discretization.

3. Modal Analysis and overload test

The first simulations executed to evaluate the dynamic behaviour of the system were the modal analysis and the overload test. The observation of the system natural frequencies is fundamental to avoid resonance phenomena. Focusing on the battery pack system, it is sufficient to have the lowest natural frequencies higher than 20-30 Hz. In this paper, the first 7 vibrating modes will be considered, as the lowest number of modes to obtain a reliable result. The overload test is important because, the tractor, during operation, may overcome an obstacle, such as a bump, at high-speed. It consists of a transient analysis, structured as shown in figure 7.

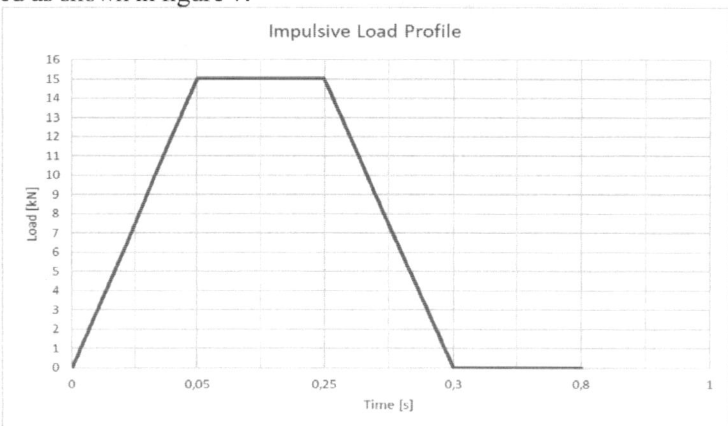

Figure 7. Impulsive load profile for overload analysis.

The system main features are shown in the following table:

Table 1. Battery pack main features.

DATA	VALUE
Battery case walls thickness	4 mm
Modules/EC walls thickness	2,5 mm
Chassis plates thickness	5 mm
Damping Factor	3 %

Concerning the materials composing the different structures, 3 macro-cases has been evaluated:

Table 2. Macro-cases description.

Macro-case	Battery case material	Module/EC material
1	S235JR (steel)	S235JR (steel)
2	S235JR (steel)	EN-AW-3003 (Al alloy)
3	EN-AW-3003 (Al alloy)	EN-AW-3003 (Al alloy)

For each macro-case 3 different kind of anti-vibrating support have been tested:

Table 3. Anti-vibrating supports main features.

Thread	Stiffness (N/mm)	Max. deformation (mm)	Length (mm)	Diameter (mm)
M12	1500	5	25	75
M12	2200	4	25	75
M12	3300	3	25	75

The main aim of this simulation campaign is the research of the best anti-vibrating support and materials combination.

The designed goals set for the modal analysis and overload test are:

- Natural frequencies >20 Hz.
- Maximum allowable stress equal to 1/3 of the corresponding material yield stress.
- Maximum vertical deformation equal to 5 mm.
- No interferences.
- Amplitude of oscillation of EC support plate must be < 1 mm in absolute value.

The amplitude of oscillation of the EC support, considered 1 side node and 1 central node of the EC support plate, has been evaluated in this way:

$$\text{Amplitude}(t)=v_{centralnode}(t) - v_{sidenode}(t) \quad (1)$$

Where:

- Amplitude: amplitude of oscillation as a function of time expressed in mm.
- $u_{bdnsq`kmcd}$: vertical deformation of the central node of the EC support plate (Figure 8).
- $u_{rhcdmcd}$: vertical deformation of the side node of the EC support plate (Figure 8).

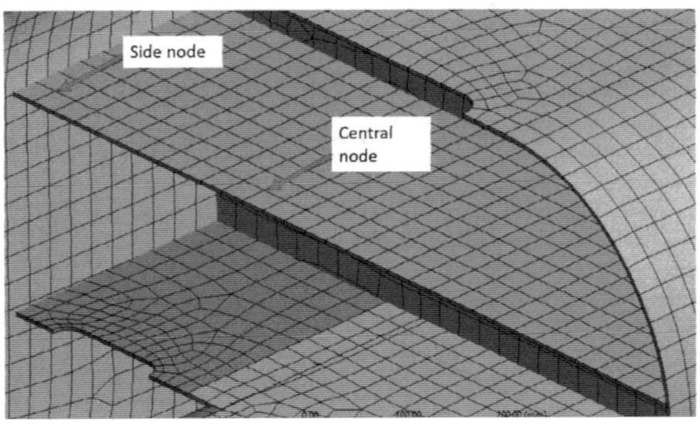

Figure 8. Node selected for amplitude of oscillation study.

All the simulations did not consider the damping coefficient of the anti-vibrating support, because the support damping factor is extremely complex to model (elastomeric material, non-linear behaviour, lack of a strong theoretical basis). Furthermore, this choice fitted well in a design logic according to the worst-case principle.

3.1 Result discussion

The results obtained from the modal analysis (Figure 9) showed that:

- Increasing anti-vibrating support stiffness, the natural frequencies grow.
- Using same-stiffness anti-vibrating supports, if the density and stiffness of the material used decrease, the natural frequencies of the modes that describe a global behaviour of the system grow.
- Using same-stiffness anti-vibrating supports, if the density and stiffness of the material used decreases, the natural frequencies reduce themselves, when the corresponding vibration mode describes a local phenomenon.

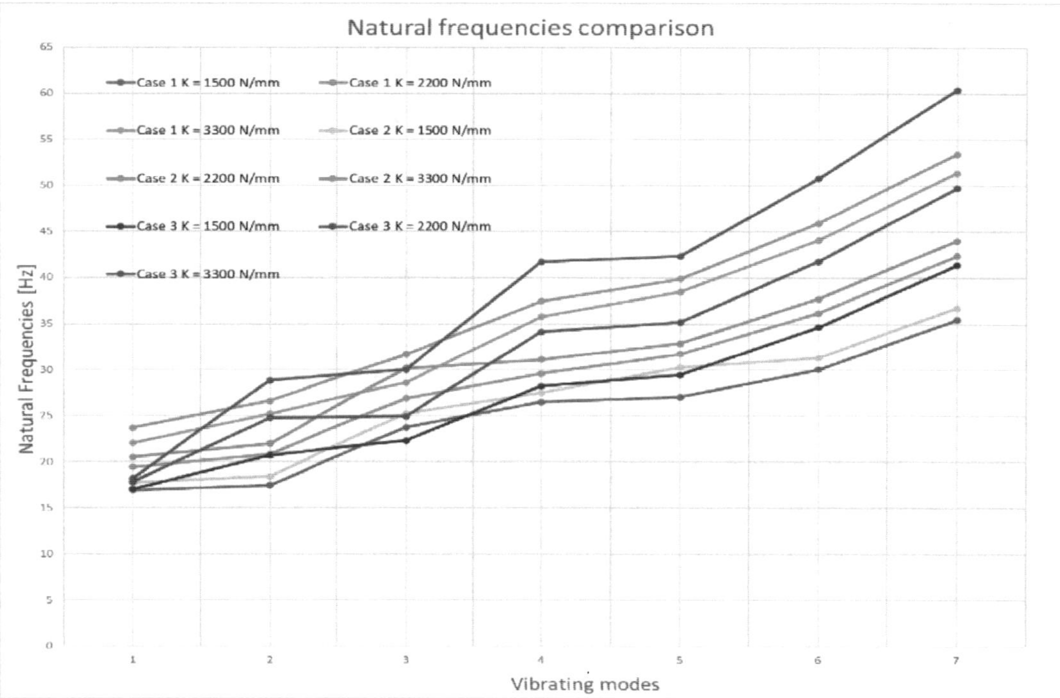

Figure 9. Natural frequencies comparison.

The results of the overload test are shown in the following table:

Table 4. Overload test results.

				Macro-Case 1				
Support stiffness (N/mm)	σ_Case (MPa)	σ_module (MPa)	σ_EC (MPa)	SF Case	SF module/EC	V_Case (mm)	V_module (mm)	V_EC (mm)
1500	80,88	28,99	28,91	2,91	8,13	5,84	5,45	5,79
2200	85,34	27,88	31,24	2,75	7,52	4,17	3,82	4,12
3300	89,88	29,33	29,03	2,61	8,01	3,2	2,87	3,16
				Macro-Case 2				

Support stiffness (N/mm)	σ_Case (MPa)	σ_module (MPa)	σ_EC (MPa)	SF Case	SF module/EC	V_Case (mm)	V_module (mm)	V_EC (mm)
1500	70,01	19,964	12,99	3,36	8,26	5,60	5,34	5,56
2200	67,83	19,416	12,438	3,46	8,50	3,93	3,89	3,75
3300	99,35	20,7	20,483	2,37	7,97	3,5	2,88	3,45
Macro-Case 3								
Support stiffness (N/mm)	σ_Case (MPa)	σ_module (MPa)	σ_EC (MPa)	SF Case	SF module/EC	V_Case (mm)	V_module (mm)	V_EC (mm)
1500	81,59	29,32	19,83	2,02	5,63	6,88	5,43	6,77
2200	87,79	31,86	15,72	1,88	5,18	4,86	4,35	4,78
3300	92,58	36,64	13,4	1,78	4,50	3,35	3,48	3,26

The transient analysis showed that:

- Increasing the mass of the system, using same stiffness anti-vibrating supports, the maximum vertical deformation reduces itself.
- In every single sub-case, the average stress is very low (less than 10 MPa), while the peak values are recorded during the load application phase.
- The modules and the electronic converter are characterized by very low stress value.
- The maximum stressed areas are the EC support plate and the internal connection zone (Figure 10).
- In general, the EC support plate oscillation is similar for each case, and it is less than 1 mm almost for each sub-case (Figure 11).

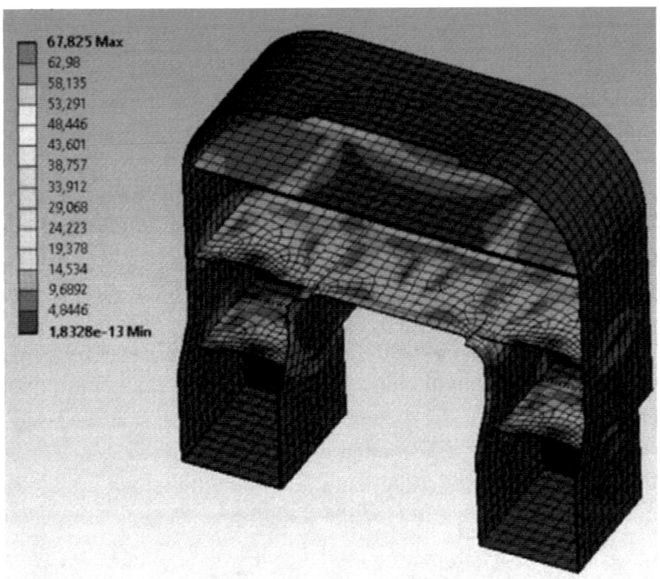

Figure 10. Macro-case 2, K = 2200 N/mm, stress distribution.

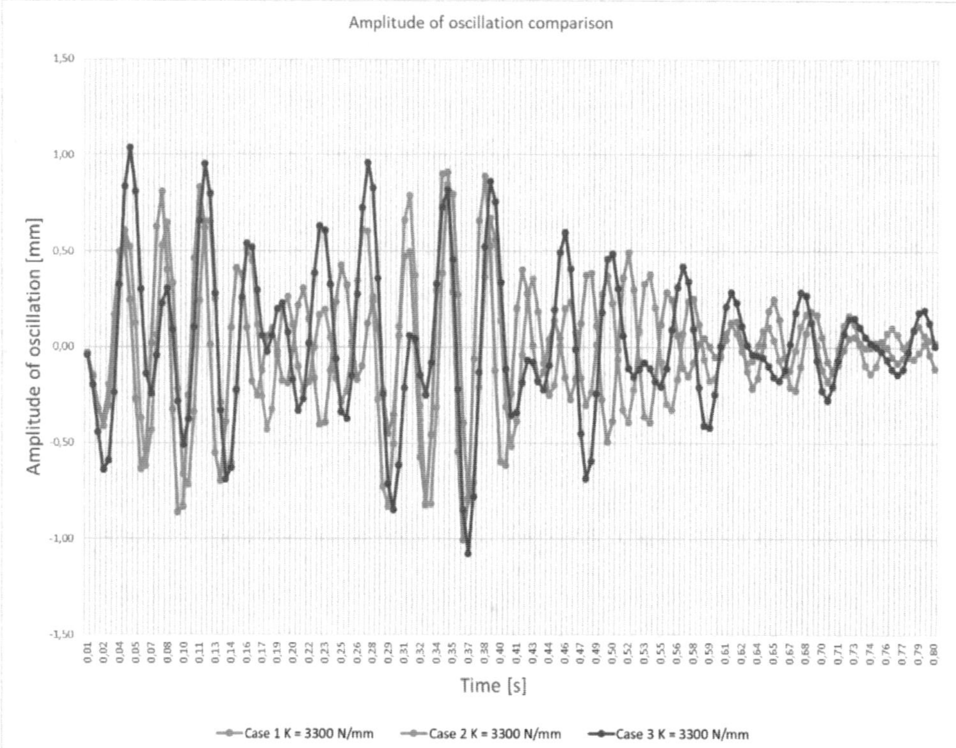

Figure 11. EC support plate Amplitude of oscillation comparison.

Most of the impulsive load is absorbed by the battery case structure, securing the modules and the EC. As said before, the main aim of these simulations has been the research of the best combination structures materials – anti-vibrating support. Only one combination achieved all the design goals: Macro-case 2 (battery case made of S235JR steel and modules and the electronic converter made of EN-AW 3003) using anti-vibrating supports with stiffness K = 2200 N/mm. However, also macro-case 2 and anti-vibrating supports with stiffness K = 3300 N/mm showed very interesting results, especially some improvements of the system geometry. In conclusion, following the results obtained in this simulation campaign, the battery pack was characterized by a good dynamic response to an impulsive load.

3.2 Improved model

What has been described until now represents the basic model of the battery pack. To improve the dynamic behaviour of the system, some changes has been implemented. In particular, the updates consist of:

- Addition of 2 ribs on the EC support plate (Figure 12).
- Battery case cover thickness reduced from 4 to 3 mm.
- Modules and EC walls thickness reduced from 2,5 to 2 mm.

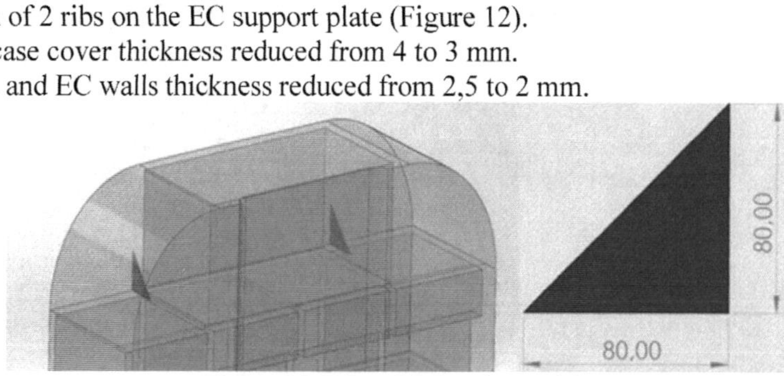

Figure 12. Ribs added to the model.

So, the improved model has been subjected to the same tests done for the basic model. In this case, the combination anti-vibrating support – structural materials used are the macro-case 2 and support stiffness K = 2200 N/mm and K = 3300 N/mm. The results of the modal analysis are illustrated in the following figure:

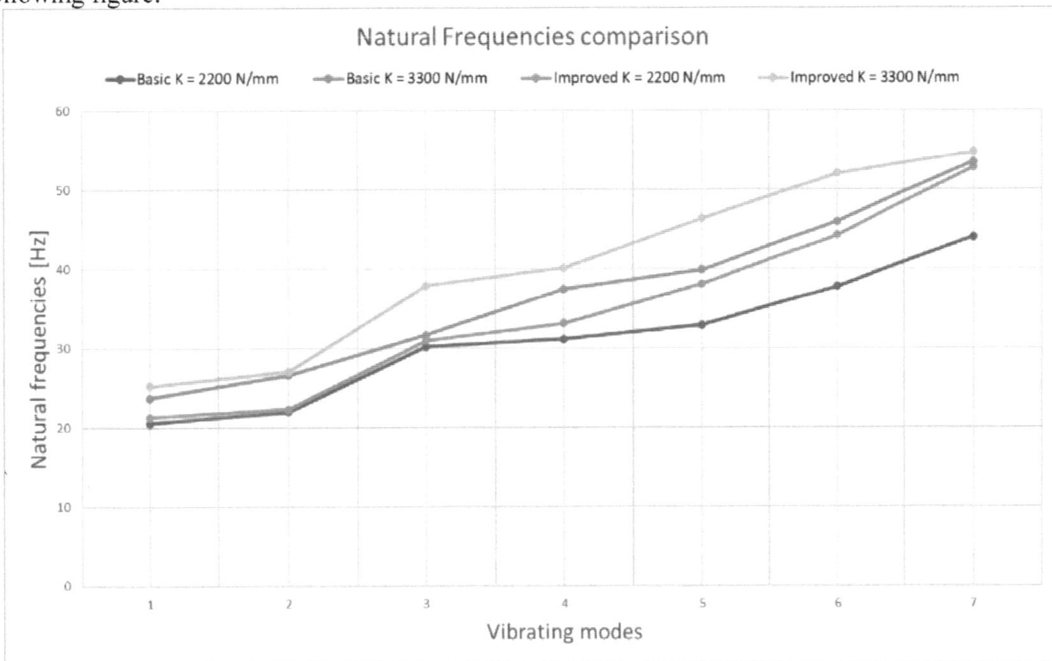

Figure 13. Natural Frequencies comparison between basic and improved model.

Observing fig. 13, going from the basic model to the improved one, the natural frequencies of the system grow. Considering the vibrating modes with the highest modal participation factor (Figure 14), there is an increment of 3,6% (K = 2200 N/mm) and 6,4% (K = 3300 N/mm) for the mode "a", and an increment of 46% (K = 2200 N/mm) and 64% (K=3300 N/mm) for the mode "b".

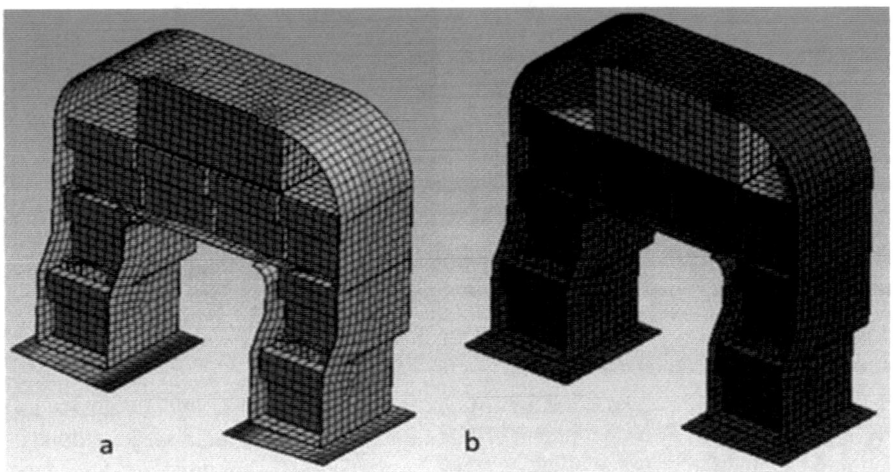

Figure 14. Vibrating modes with the highest participation factor.

The reasons that can explain this improvement are the system weight reduction (- 10%) and the presence of the ribs on the EC support plate. The results of the overload test are shown in the following table:

Table 5. Overload test result and comparison between basic and improved model.

Support stiffness (N/mm)	σ_Case (MPa)	σ_module (MPa)	σ_EC (MPa)	SF Case	SF module/ EC	V_Case (mm)	V_module (mm)	V_EC (mm)
2200 Basic	67,83	19,42	12,44	3,46	8,50	3,93	3,89	3,75
3300 Basic	99,35	20,7	20,48	2,37	7,97	3,5	2,88	3,45
2200 Improved	65,26	23,5	5,64	3,6	7,02	3,68	3,7	3,55
3300 Improved	74,4	26,05	4,99	3,16	6,33	2,91	2,99	2,87

Observing *table 5* it is evident the reduction of the maximum stress on the battery case structure: indeed, considering anti-vibrating support with stiffness K = 3300 N/mm, there is a reduction about 25 % respect to the basic model. Furthermore, the entire system is not subject to an intense field of stress with an average stress recorded less than 5 MPa. The most stressed area is the internal connection zone (Figure 15), while the EC support plate, thanks to the presence of the ribs, is not much stressed anymore (maximum stress recorded 40 MPa). Also, for the improved model, almost all the impulsive load is absorbed by the battery case structure. Concerning the maximum vertical deformation, the result obtained for the improved models are better than the corresponding of the basic one. Furthermore, the values obtained for battery case, modules and electronic converter are very similar to each other, confirming the absence of excessive oscillating phenomena.

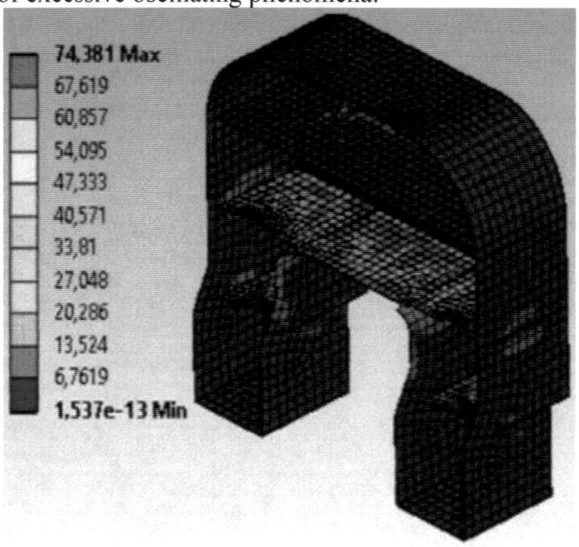

Figure 15. Stress distribution improved model K = 3300 N/mm.

Concerning about the EC support plate amplitude of oscillation, observing figure 16, it is clear the advantage brought by the ribs: indeed, the reduction of the amplitude of oscillation is about 76% respect to the basic model. Considering these results, each sub-case shows a good dynamic behaviour, and all the design goals are achieved. However, the best performance is obtained with support stiffness K = 3300 N/mm, so these kinds of supports will be used for the next step: the Virtual Field Test.

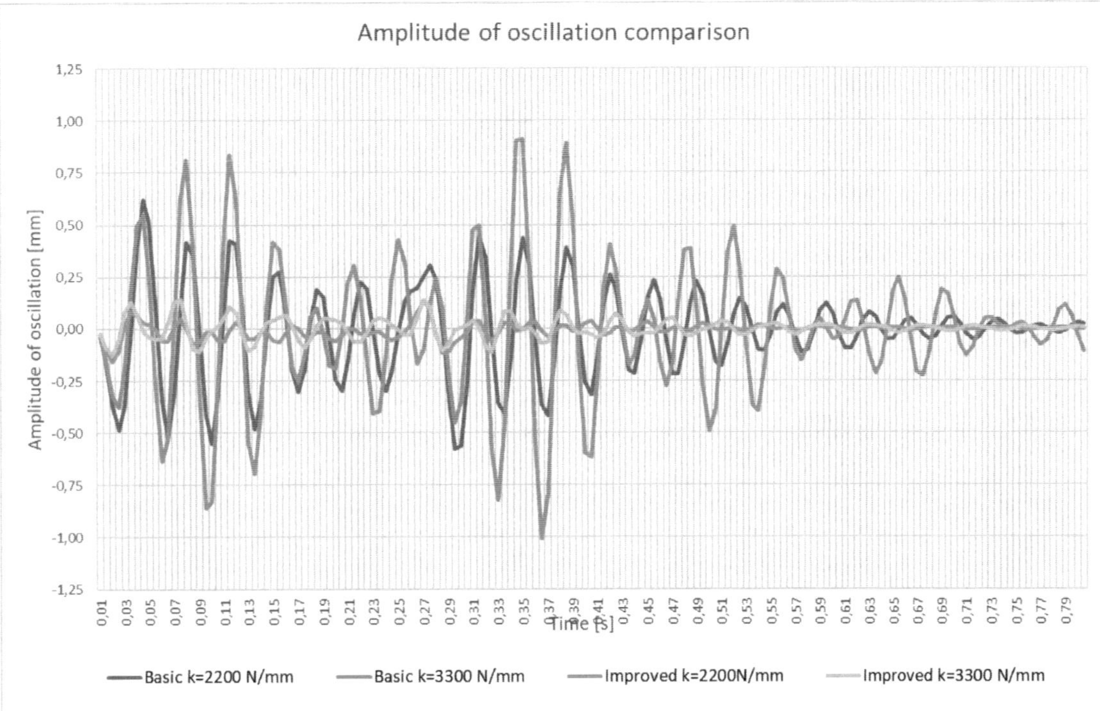

Figure 16. Amplitude of oscillation comparison between basic and improved model.

4. Virtual Field Test

The purpose of the virtual field test is to evaluate the dynamic behaviour of the battery pack, with respect to the entire vehicle, during hard work condition situations. The aim is to replicate what happens during a real Field Test [11]. These tests are based on experience for lack of tests standardization and because they must take into account the available computing power. The virtual field tests are born combining Multibody analysis and FEM simulations. In particular, the multibody model was used to obtain a series of load profiles, applied to the battery pack's supports. Then, the load profiles constitute the input loads of a transient analysis, similar to the ones done in the overload tests, to evaluate the dynamic behaviour of the battery pack. The design goals set for these virtual field tests were:

- Load frequency application must not be close to system natural frequencies.
- Maximum allowable stress equal to 1/2 of the corresponding material yield stress.
- Absence of interferences.
- Relative vertical deformation between BP and chassis < 2 mm.
- EC support plate amplitude of oscillation < 1 mm.

4.1 Virtual prototype

The virtual prototype consists of a series of rigid bodies, connected to each other with different kind of joint to limit their degrees of freedom. As said before, the tractor, described in this paper, is an orchard tractor. Dimensional and mass data are illustrated in table 6 and table 7.

Table 6. Dimensional data orchard tractor.

Data	Value
Wheelbase	2265 mm
Length	3740 mm

Width	1536 mm
Track width	1216 mm
Maximum height	2100 mm
Ground clearance	326,3 mm
Height at seat	1300 mm
ICE Dimensions	759 mm x 571 mm x 806 mm
Tyre	320/70 R20
Transmission	4WD

Table 7. Weight data orchard tractor.

Component	Mass (kg)
Chassis	255
Front and rear axle	180
Gearbox	150
Transmission	250
ICE and bonnet	745
Tyres	200
Driver and on-board instruments	200
Battery Pack	400
Electric motor	30
Chassis internal masses	200
Axles internal masses	150
Total MASS	2760

Concerning the torque regulation, it is based on vehicle speed target, and it is characterized by asymptotic behaviour. The asymptote corresponds to the maximum transmissible torque by the wheels to the ground and this limit is set to:

- 1000 Nm for each tyre in case of high-speed transmission mode.
- 2500 Nm for each tyre in case of slow-speed transmission mode.

Diesel combustion energy and battery pack are constrained to the chassis with *fixed joints*, while cab and wheels are bounded to the chassis with *bushing joints*. So, in this way the model can take into account the deformability of the tyres and of the ground too [12,13]. Finally, two markers have been positioned in correspondence to the chassis plates on which the battery pack is installed. In this way the value of the load transmitted to the battery pack from the chassis can be obtained and used to the FEM transient analysis.

4.2 Virtual Field Test 1 & 2

In the first two virtual field test, the tractor is moving through a bumpy road formed by 10 speed bumps, distant 2 meters from each other. Each speed bump presents an arc of circumference shape. The main features of a single speed bump are showed in Figure 17:

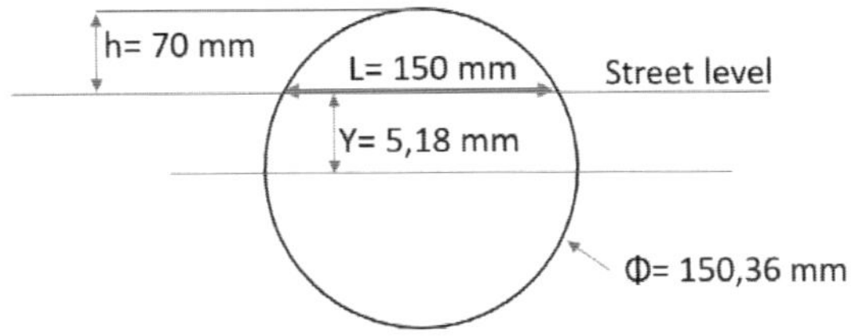

Figure 17. Virtual Field Test 1 & 2 Bump description.

The tractor is moving at 10 km/h, in slow-speed transmission configuration. Furthermore, since the aim of these tests is the simulation of an working condition, at the vehicle back end, a lumped mass of 1000 kg, simulating the rear attached implement, was positioned. The difference between virtual field test 1 & 2 (Figure 18) is that the second test presents each speed bump cut in two and every single half-part is distanced of 1 metre to the other one.

Figure 18. Virtual field test 1 (on the left) and 2 (on the right).

The results of the multibody simulation, that constitutes the input for FEM transient analysis are shown in the following figures:

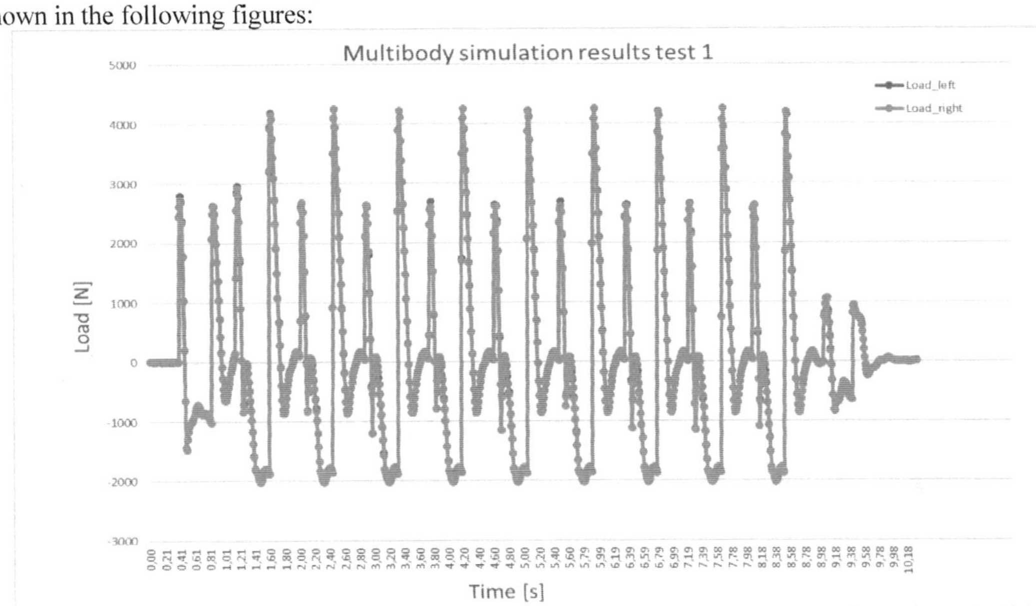

Figure 19. Results multibody simulation Test 1.

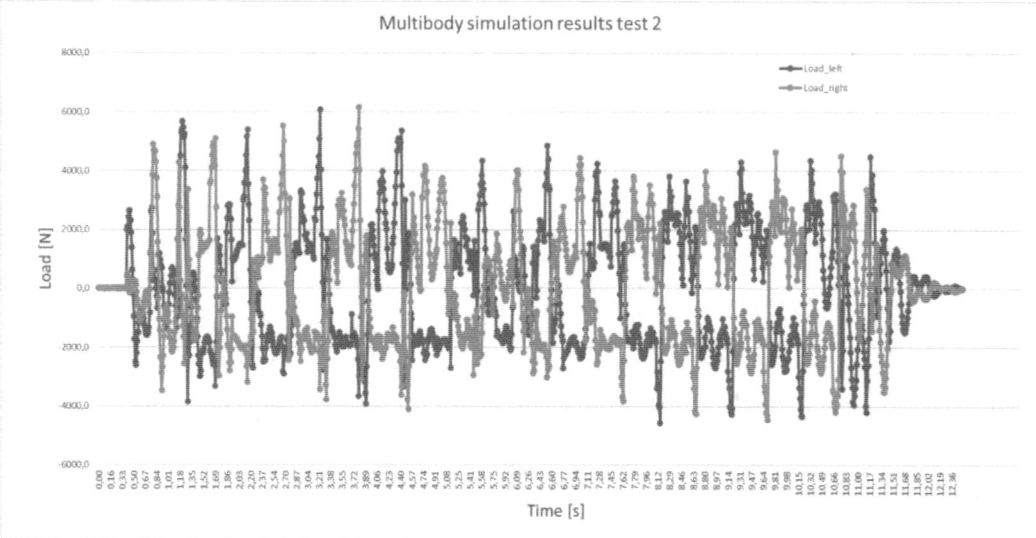

Figure 20. Results multibody simulation Test 2.

4.3 Virtual Field Test 3 & 4

In the last two field tests, the tractor is moving through a bumpy road formed by 10 speed bumps, distant 2 meters from each other. Each speed bump presents an arc of circumference shape. The main features of a single speed bump are showed in figure 21:

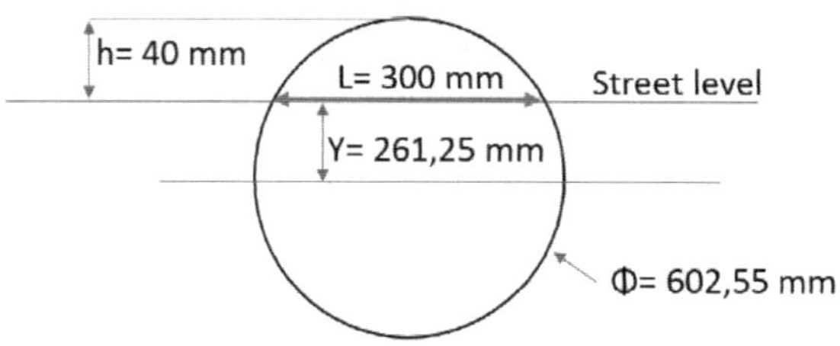

Figure 21. Virtual Field test 3 & 4 Bump description.

The tractor is moving at 30 km/h, in fast-speed transmission configuration. The difference between virtual field test 3 & 4 (Figure 22) is that the second test presents each speed bump cut in two and every single half-part is distanced of 1 metre to the other one.

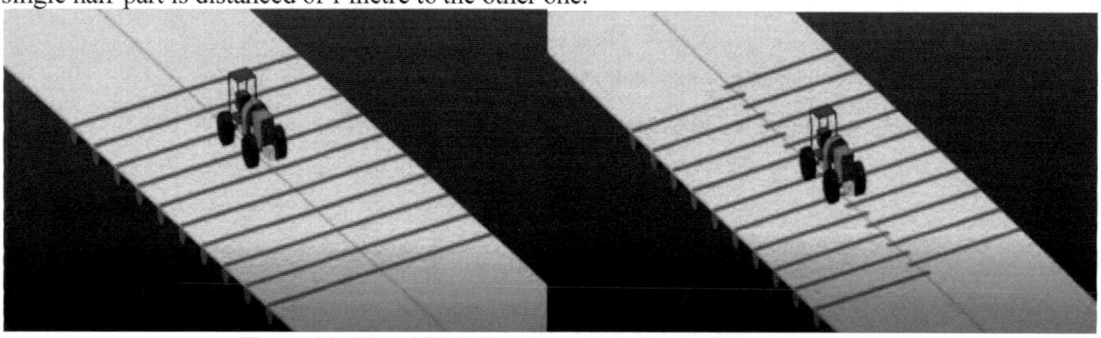

Figure 22. Virtual Field Test 3 (on the left) and 4 (on the right).

The results of the multibody simulation, that constitutes the input for FEM transient analysis are shown in the following figures:

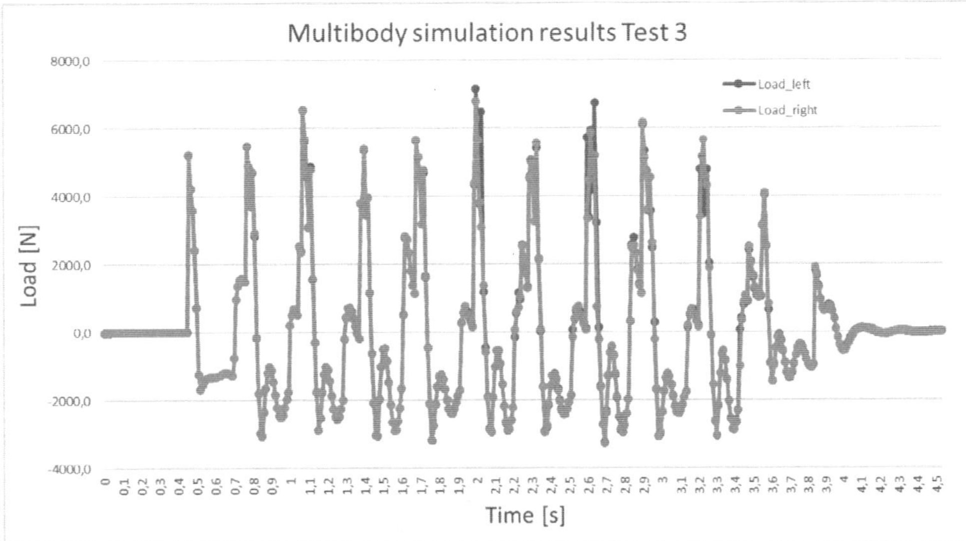

Figure 23. Results multibody simulation Test 3.

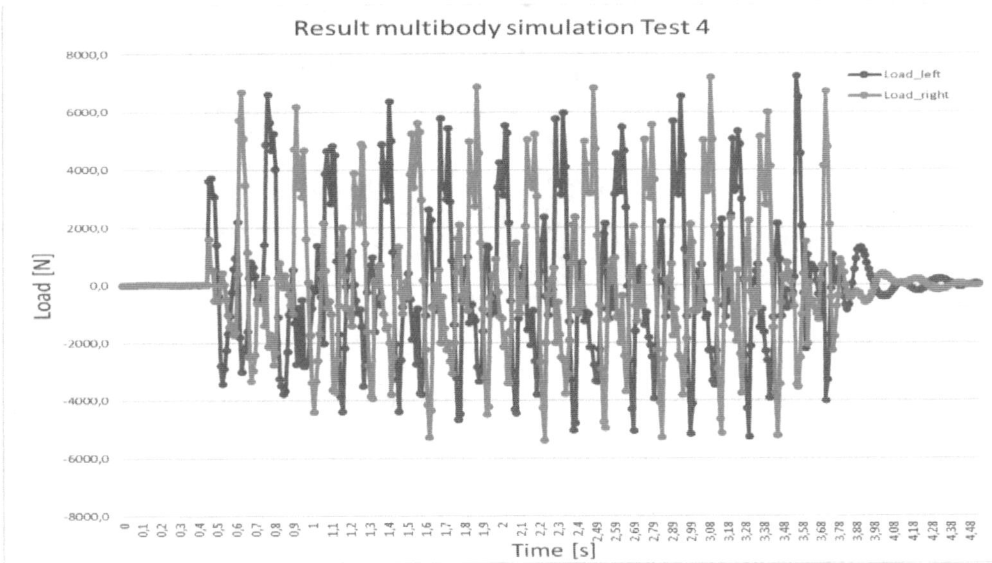

Figure 24. Results multibody simulation Test 4.

4.4 Virtual Field Tests results discussion

Table 8. Overall results Virtual Field Tests.

Field Test	1	2	3	4
Load application frequency	1,4 Hz	2,8 Hz	4,2 Hz	8,33 Hz
Minimum SF encountered	3,6	3	2,5	2,1
Relative vertical deformation between BP and chassis	1,2 mm	0,6 mm	1,6 mm	0,9 mm
Amplitude of oscillation EC support plate	0,4 mm	0,5 mm	0, 55 mm	0,8 mm
Interference	No	No	No	No

Table 8 shows that all the design goals are achieved, so the battery pack thought for this application has a quite positive response to heavy working conditions. Considering load application frequency, it is clear that the orchard tractor battery pack is subjected to low frequency working conditions, confirming that the frequency working range of the tractor are lower than 30 Hz. Another remarkable point is the correlation between load frequency application and stress level developed: indeed, if the frequency grows, the safety factor decreases. The highest stress level occurs during virtual field test 3 and 4. So, the most critical conditions are ruled by vehicle speed (consequently the frequency) rather than speed bump shape and dimension. The minimum Safety factor encountered is related to the battery case structure, while modules and electric converter shows a minimum safety factor upper to 10. So, even in this case almost all the input load is absorbed by the external structure of the battery case. In almost all cases, the most stressed component (Figure 25) is the EC support plate, but its amplitude of oscillation is very low thanks to the ribs presence.

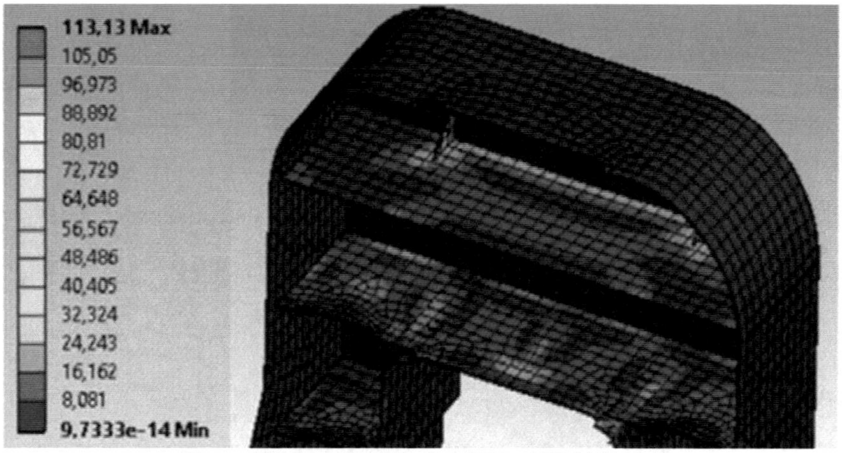

Figure 25. Stress distribution Virtual Field Test 4.

Considering relative vertical deformation between battery pack and chassis, it appears that the anti-vibrating support (quale in definitiva) are appropriate for this kind of application because the maximum allowable value is never overcome.

5 Conclusions

This paper shows how, even for agricultural applications, a battery pack influences dynamic behaviour, weight, and architecture of an orchard tractor. Indeed, the battery pack is an important component, not only in terms of performance, but also in terms of safety and preservation of the driver and the entire vehicle too. For these reasons, an appropriate dynamic design of the accumulator is crucial. The most important result obtained from this paper is that is possible to install, from a dynamic and mechanical point of view, a big size battery pack, in a small size tractor, as the orchard tractor, preserving its traditional shape and functionality. The most critical design goals to achieve are:

- Low natural frequencies.
- High mechanical stresses, near the yield value.

So, the main design aspects, to pay attention to, are:

- System mass.
- Structure stiffness.
- Mechanical strength.

Concerning the structural material choice, it must be the result of the best compromise in terms of:

- Mechanical strength.
- Easy manufacture.

- Low cost.

The structure stiffness must be as much high as possible to guarantee higher natural frequencies and avoid resonance phenomena. The most important contribute to stiffness structure is given by the anti-vibrating supports. Anyway, stiffening local intervention are recommended (as mechanical ribs) in strategic point of the structure too (for example in presence of cantilevered masses).
The mass of the system is fundamental, and it must be as low as possible to improve vehicle performance and dynamic behaviour because the system mass has a strong impact on natural frequencies and inertial proprieties of the structure.

This work is open to some possible improvements such as considering the non-linear spring characteristic of the anti-vibrating support to better approximate its behaviour. Another possible improvement may be the use of input data which come from real field test. Finally, this paper can be considered as the starting point to configure a complete system fatigue study.

References

[1] Somà A 2017 Trends and Hybridization Factor for Heavy-Duty Working Vehicles *Hybrid Electric Vehicles* (InTech)

[2] Sitompul J, Zhang H, Noguchi R and Ahamed T 2019 Optimization study on the design of utility tractor powered by electric battery *IOP Conference Series: Earth and Environmental Science* **355**

[3] Hong S-K, Epureanu B I and Castanier M P 2014 Parametric reduced-order models of battery pack vibration including structural variation and prestress effects *Journal of Power Sources* **261**

[4] Arora S, Shen W and Kapoor A 2016 Review of mechanical design and strategic placement technique of a robust battery pack for electric vehicles *Renewable and Sustainable Energy Reviews* **60**

[5] Mocera F, Soma A and Clerici. D 2020 Study of aging mechanisms in lithium-ion batteries for working vehicle applications *2020 Fifteenth International Conference on Ecological Vehicles and Renewable Energies (EVER)* (IEEE)

[6] Shui L, Chen F, Garg A, Peng X, Bao N and Zhang J 2018 Design optimization of battery pack enclosure for electric vehicle *Structural and Multidisciplinary Optimization* **58**

[7] Mocera F and Somà A 2021 A Review of Hybrid Electric Architectures in Construction, Handling and Agriculture Machines *New Perspectives on Electric Vehicles [Working Title]* (IntechOpen)

[8] Mocera F and Somà A 2020 Analysis of a Parallel Hybrid Electric Tractor for Agricultural Applications *Energies* **13**

[9] Warner J 2015 Battery Pack Design Criteria and Selection *The Handbook of Lithium-Ion Battery Pack Design* (Elsevier)

[10] Sankaran G and Venkatesan S 2021 Standardization of electric vehicle battery pack geometry form factors for passenger car segments in India *Journal of Power Sources* **502**

[11] Renius K T 2020 Tractor Tests *Fundamentals of Tractor Design* (Cham: Springer International Publishing)

[12] Nicolini A, Mocera F and Somà A 2019 Multibody simulation of a tracked vehicle with deformable ground contact model *Proceedings of the Institution of Mechanical Engineers, Part K: Journal of Multi-body Dynamics* **233**

[13] Mocera F, Somà A and Nicolini A 2020 Grousers Effect in Tracked Vehicle Multibody Dynamics with Deformable Terrain Contact Model *Applied Sciences* **10**

The contact problem in FEM analysis of filiform structure for large deployable reflectors

R Guglielmini, E Marotta and P Salvini

Department of Enterprise Engineering, University of Rome "Tor Vergata", Via del Politecnico 1, 00133 Rome, Italy

salvini@uniroma2.it

Abstract. Space reflectors need a continuous increase of size to improve the quality of the communication. Considering the small space available in launchers, the antenna must be assembled within the vector and deployed in the space environment. The antenna requires a thin metallic mesh sustained by an appropriate net structure. The best reflector efficiency is reached when all the metallic wires keep in stable contact, allowing the electrical field to be homogeneous. The number of wires involved is enormous, so that, the modelling requires a simple structural approach connected to a reliable contact tool. Single wires can be modelled by beams, but this opens a non-trivial contact managing. As a matter of fact, the wires describe a thin surface, but the contact is realized by 3D crossing. This point is the main objective of the present work which suggests a way to overcome the crossing while maintaining the structure modelling in a 2D field. The contact efficiency needs that the wires do not exceed the elastic strength, causing a lowering of loading contacts due to irreversible deformations. Taking advantage of the present approach, an investigation over the percentage of active contacts in a full mesh is performed, so that optimally applied tensile loads are identified.

1. Introduction

The capability to communicate using a satellite connection is strongly linked to the quality of the signal. Satellites receive and send messages overtaking any obstacles thanks to their high altitude in the atmosphere. They operate in a specific orbit that has itself an area of service on the earth called footprint, generally variable in time. For receiving a message, the satellite must reflect it on a single point, the fulcrum of the parabolical structure, by means of a reflector, also called LDR (Large Deployable Reflector). To optimize the number of bits correctly read by the satellite, the reflector requires particular attention to the metal mesh regularity. LDR contains several knitted thin wires in order to minimize the cost of the entire mission reducing the weight of materials carried in orbit.

Research on reducing costs in the aerospace field has shifted to the development of structures with customizable functions that can provide enhanced functions for smart products [1]. The technological process of textile manufacturing offers the potential to produce a new generation of material systems that extends traditional textile applications [1], [3], [4]. Due to the particularity of fabrics and the complex relationship between stress and deformation, the mechanics of textiles makes the modelling of these structures quite tricky [5]. There are three main types of textile fabrics: woven, knitted and non-woven fabrics [4], [6]. The most commonly used configuration sets are described in detail in [7]. The stitches of the knitted fabric are arranged in rows and columns, roughly equivalent to the warp and weft directions of woven structures. Knitted fabrics have some advantages over woven and non-woven fabrics: they have higher elongation, so they allow the production of "flexible fabrics" with a wider range of available deformations [8]. This statement is particularly meaningful and focuses attention on knitted mesh structures, the characteristics of which can be planned and calibrated in advance to achieve the desired effects.

The first attempts to gain a mechanical model and analysis of textile structures began about a hundred years ago. The first publication appeared in a report by R. Haas in the National Advisory Committee for the US Air Force in 1918 [9]. This article discusses theoretical and practical aspects, as well as possible tests and applications. However, Haas's work has been unknown for a long time, and Pierce's work [10] is considered the starting point for mechanical modelling of textile structures. Since then, researchers have begun to work on the application of existing analysis methods, used in other engineering fields. In aerospace applications, knitted fabrics have become essential for constructing reflectors for deployable antennas used on communication satellites [11], [12].

(a)
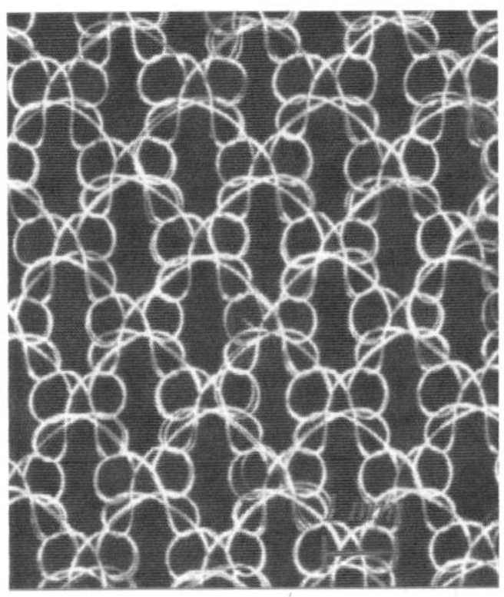
(b)

Figure 1. **(a)** A satellite orbit around the Earth; **(b)** Knitted metallic mesh

Different aspects must be considered very carefully when designing the LDR; electromagnetics, structural, kinematics, dynamics and thermals, and technological limits must not be ignored. Most reflectors on orbit are made of strictly parabolic structures. The introduction of LDR supports the need to greatly increase the size of the reflection area and so the signal-to-noise ratio, while considering the reduction in the weight and cost of the transmitter. At present, the size of the reflector used is as large as 15 meters, but a larger diameter, such as 30 meters, is currently being developed [13], [14]. To achieve this dimensioning, the antennas must be equipped with a deployable mechanism to open in orbit. The parabolic surface reflects electromagnetic waves, and it is made of metal knitted fabric. The knitted fabric is sustained and fixed in position by a net that does not have any radio capability. Using these devices, satellites can benefit of a low weight structure [15]. The geometric accuracy and the compliance of the LDR must be accurately studied. Therefore, a structural analysis of the whole must consider the sustaining structure, the net where the metallic mesh is applied, and the metallic mesh itself.

The main difficulty in mesh characterization comes from the low overall stiffness associated with large displacements and slight loads: these features induce special attentions. For example, a basic aspect is the grasping system. If it is not optimized, it will introduce a local hardening effect, thereby changing the overall elasticity assessment [16]. Testing for structural characterization requires dedicated equipment. In [17], the experimental test was carried out using a customized two-axis electromechanical device. The testing machine is equipped with a clamping system to reduce the adverse effects just mentioned [18].

The mesh could be weaved in different ways, but it influences the SNR (Signal Noise Ratio). The best solution is to have a warp knitted reflector with Atlas-Atlas stitches. Voids make structure lighter but can negatively affect the reception of the signal: if the void dimension is bigger than ten times the wavelength of the signal, the antenna cannot reflect the signal. A good match of the different wires is realized if the mutual contact keeps stable and this condition makes the almost isotropic movement of electric charges all around the net. The number of the active contacts is although governed by the boundary conditions applied at the mesh, i.e., the tensile loads applied while connecting the mesh to the net during the building phase on earth. An optimal service configuration that guarantees the maximum number of contacts is performed, considering the stiffness of all components forming the reflector and assembling a model to study by Finite Element Analysis. The critical point is that the contact must remain active also when the satellite travels on a shadow area, and the structure is subjected to contraction, or when it is totally exposed to the sun radiation, suffering dilatation [19]. These combined effects cause to withstand a large temperature variation, from *-190°C* to *+140°C* [20]. The metal wire is usually made of molybdenum or tungsten, covered with gold, and its diameter is very thin (approximately *15 microns*), compared to the diameter of the loops (few millimetres).

The enormous computing power available nowadays allows the development of commercial computation codes with advanced mechanical analysis capabilities and it promotes the adoption of numerical methods by textile designers. Finite Element software provides different strategies to be used to correctly model knitted fabrics. Explicit FEM can be used to simulate small or large deformed meshes in static or dynamic analysis [21]. Implicit FEM is more effective for small deformations. The software allows the adoption of different element types to model different geometric shapes: generally, beam elements are used for yarn modelling, even if solid and shell element discretization is also recurrent. [22], [23]. In one-dimensional modelling, each wire is discretized with a simple beam, rod, or spring element. The knitting produces a basic pattern that repeats periodically to form the entire mesh [24], [25], [26]. Difficulties arise because in Finite Element codes, the identification of contact between objects is usually one of the most time-consuming operations [27], and the number of wires in contact is in the order of millions per square meter, very hard to manage. For this reason, only a very limited fabric portion has been modelled so far [28], [29]. Contact problems introduce strong nonlinearity that affects convergence [30]. The whole reflector, knitted by several wires, can be easily modelled in a 3D representation using CAD software; but the attempt to solve the mesh structure by modelling all curved wires through beams introduces a huge amount of elements and elements-in-contact, that makes the computation lengthy and convergence is hard to achieve.

The authors [31] are developing a new 2D technique to model the wires with a limited number of *DoFs* and a simplified contact algorithm; but there is the exigence to validate the approach through a Finite Element Analysis, that is performed in the present paper.

In the simulations reported below, a 2D modelling is preferred, that, however, induces some problems to manage the contact. Hereinafter some different approaches are taken in exam to identify the most profitable technique for contact.

2. Method

Using the classical features and the element in the library of Finite Element codes, a 3D modelling of the reflectors should be the best choice. However, this approach requires too much *DoFs* and the solution provides a long computation time and convergence difficulties.

As it is clear in figure 2, the crossing among wires realizes out of the mesh plane. Although, the diameter of the wires is so small, compared to the loops, that it seems very reasonable to condense the overlapping 3D contact into a 2D action largely reducing the contact complexity.

In such 2D representation, a couple of wires crosses in two points, if these keep distance the contact is non-active, when they tend to approach, so close as *1.25* times the diameter, the contact activates.
Another aspect regards the considerable reciprocal sliding, the contact region may move while loading so that the a-priori paring of the elements in possible contact, necessary for the modelling, turns troublesome. A good idea to represent the 3D contact through a 2D modelling is to interpose a spot element within the interlaced loops, which is free to move in plane, but is potential contact with both wires. However, as the previous definition suggests, it has free rigid motions, and the solution is

therefore impossible using static analysis (if some part of the model is free to move the stiffness matrix $[K]$ becomes singular and the elastic problem $[K]\{x\} = \{F\}$ has no solution). To provide a static solution, one must apply some constraints at all contact spots, but this may affect the overall behaviour of the mesh.

Keeping within static analysis, the problem with this element could be solved through three different strategies. All of them introduces additional elements providing contact that are restrained in some way to prevent the lability of the model. Generally, this is obtained by introducing dummy springs that require energy to move or by adding dummy friction on the plane where movement is possible.

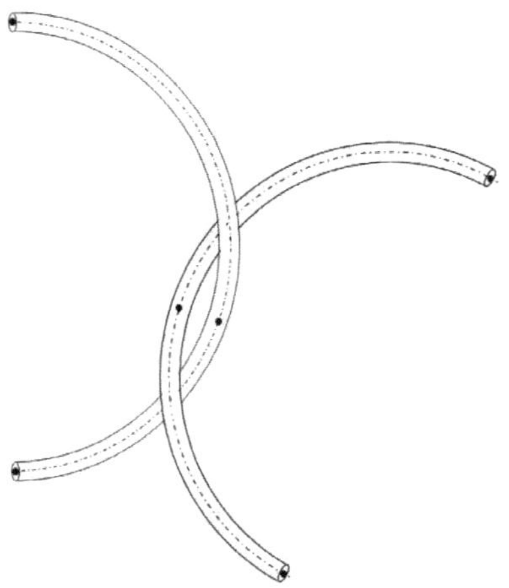

Figure 2. 3D crossing between wires.

An interesting alternative is made possible if a dynamic analysis is conceived, in fact, any lability is overcome since inertia loads are always connected to movements. This option is also investigated, considering that the structure requires the modelling of two features: stiffness and inertia matrices.

2.1. Single contact point

For this research, we referred to a commercial Finite Element Code, ANSYS®, that provides several algorithms for non-linear contact features. The procedure introduces a relationship between *DoFs* of elements not structurally connected before contact. The technique follows two alternative approaches (or mixes them) known as Lagrange multiplier or Penalty mode. Whatever is the method, the different algorithm is to be chosen in a group of elements that share all the information to define contact features. These elements are classified in two families: contact and target elements. Generally, contact and targets are defined over an extended region that considers both parts that might come into contact. This is difficult to implement when 2D analysis is performed. In fact, the two overlaps (see figure 2) are misinterpreted as contact even if they are not close. The way to overcome this difficulty is to limit the contact region, of only one of the wires, to a simple spot. However, this position only works if one is aware of the effective location of the contact, even when some sliding occurs. This is generally not known a-priori. An example is represented in figure 3 where two wires knot themselves in the middle with particular boundary conditions: inferior ends are totally constrained, while superior edges are respectively free to move in vertical direction and in both vertical and horizontal one, with an imposed displacement in the last. Wires have a circular section of *15* μm and a curvature of *1 mm*. It is interesting to highlight how it is difficult to locate the contact point in the undeformed configuration. This is troublesome for two wires, for an assemblage of several wires this prediction is almost impossible. Moreover, the contact spot may also skip the target region and the contact is missed, eliminating the physical crossing action.

The Analysis considers three nodes beam element BEAM189 [32], one contact element CONTA175 in the left wire, and target elements TARGE170 with cylinder shape, all over the right wire. Hereinafter the material considered is tungsten (W) with a density of *19250 kg/m³* and Young's modulus assumed *400 GPa*.

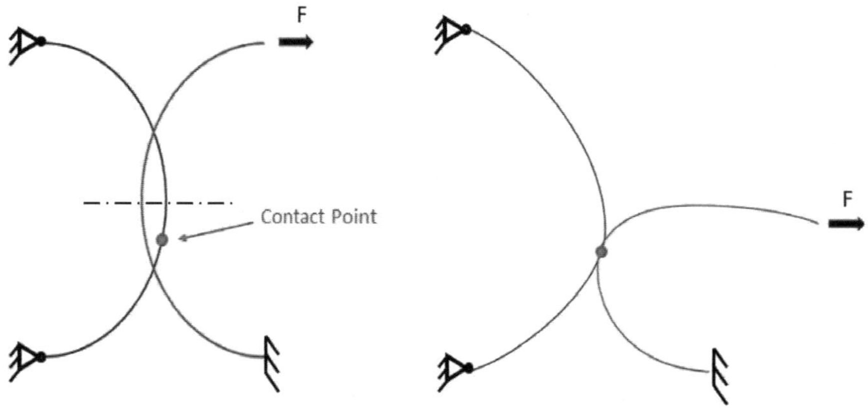

Figure 3. Contact of two 2D wires with a single contact point vs target.

In order to consider the thickness of the beam in contact is possible to account for it using a contact offset (CNOF command [33]) accounting of the wire diameters. As an alternative, the shape of the elements going into contact can be defined through the command TSHAP that provides the shape of the target elements that remain rigid during the interaction. In this case, TSHAP is set as CYLI (cylindrical) fixing its value with the radius of the wire. As already said, in the evolution of the simulation the contact could slip away and nobody else contributes to keeping the two wires bonded. Furthermore, the contact spot remains fixed on one wire, but this could be far from reality because the sliding interests only one of the two wires.

2.2. Interposing a rigid body restrained by springs

An idea to afford this problem provides an external element connecting the two wires. This is free to move thanks to additional constraints. This element could be a single point or a cylinder inside the common region, or a ring (modelled through beams) embracing both wires (see figure 4).

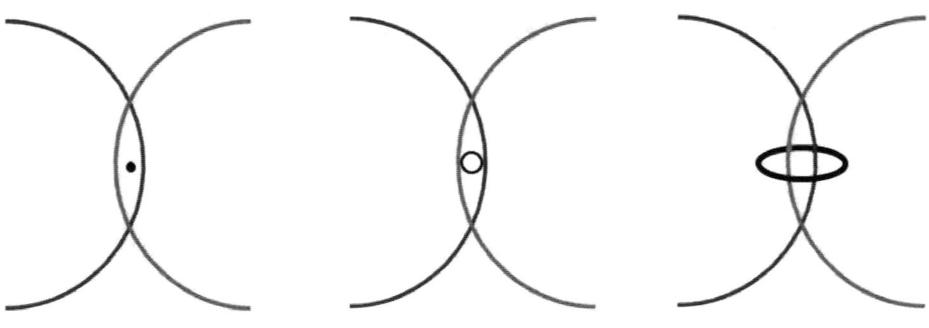

(a) **(b)** **(c)**

Figure 4. Three different external elements to model a 2D wires contact in 2D: (a) single point; (b) cylinder; (c) ring.

To avoid the free movement of this interlacing element springs to the ground are added. It is not easy to find an optimal value for these external springs. Too high values cause the deformation to assume non-physical shapes, too low values cause difficulties in the convergence during large deformation analysis. For the analysis here considered the optimal stiffness value is close to *0.001 Nm*. But this value is tuned for the two wires considered. For a larger wired structure, this value could not be suitable. The other solution regards a rigid ring embracing the two loops. External springs are also necessary to avoid lability. In this last case, the contact is no more a spot and is extended to the whole internal region of the ring.

All the proposed solution considered in this section allows the full sliding of both wires and the contact may be not fixed a-priori that was the pitfall of previous Single Contact Point approach.

Comparing the three solutions for contact here provided, it is clear that the simplest one, to obtain the sliding of the contact region, is to generate a single body in the between the two wires. However, the effect of the added external stiffness to the single point unpredictably affects the results.

2.3. Plane whit friction contact
As an alternative to the added springs, friction can be added to the single bodies introduced. A friction plane is introduced. This plane is bigger than the structures and is made from four-point linked in a quadrilateral element TARGE170 with TSHAPE, QUAD. Over that single point between the wires is then modelled another contact CONTA175 element. These two elements are then paired together thanks to a new real constant set dedicated to the couple. Although possible, this solution introduces additional difficulties in terms of non-linear effects. For the interaction of two wires it works, but for more complete knitted structures the difficulties increase exponentially.

2.4. Dynamic Analysis
Structure analyses within static conditions need a well-constrained and numerically stable model to achieve reasonable results. The wires are very thin with a low stiffness so that it is easy to incur bad-conditioning problems. In addition, the external springs or the friction constant used to stabilize the modelling must be appropriately chosen. If a dynamic simulation is assumed, there is the need to introduce masses at all *DoFs*, but lability is prevented by inertia loads always present during displacements. The mass values of the single bodies may be set in the order of less than half a loop of the wire. The evolution time chosen for the analysis assumes a significant value and, in connection with the masses is chosen so that almost no vibrations occur while applying loads. From the point of view of computing time, using lumped mass evaluation does not significantly change the overall values, but also smooths the convergence difficulties. Comparing to the added stiffness or friction plane approach for the static analysis, it is much easier to tune the single body masses so that their influence is minimal.

2.5. Modelling the LDR
Moving from a simple modelling consisting in two wires to the complex metallic mesh involves an exponential growth of the problem in term of computing time and contact analysis. In fact, the mesh is composed of a huge number of loops that knot together. Let us consider figure 5 representing a portion of fabric. This structure develops with 2 types of loops: a circular and an oval one that is mirrored and repeated (evidenced in red). Observing the mesh, these two loops along the longitudinal direction of weaving generate an essential structure that could be called “primary portion” (blue in figure 5). Adding a primary portion with a mirrored one, (violet in figure 5) the “elementary wire path” is completed and similarly, it repeats towards the warp direction.

From the point of view of FE analysis of the primary portion model is crucial to forming the entire structure. Using straight beams, no less than *400* nodes are necessary to get a reliable structural behaviour of the primary portion. This is mainly because of the contact identification and sliding. When several primary portions are present, a full mesh requires a massive number of *DoFs* that grow together with the size of the mesh. In addition, each circular loop can activate three contacts that, as described in paragraph *2.4*, require the insertion of contact bodies within the crossing region.

In order to reduce the computational cost a representative part is found observing the mesh, optimizing the power of the simulation and searching results in it: primary portion could be associated to a single edge, or element, that subsequently generate a structure made by rhombus. The rhombus could be defined as "complete" when in every edge, all the contact points are potentially activated and they are enough far from external bound, so not affected by boundary conditions.

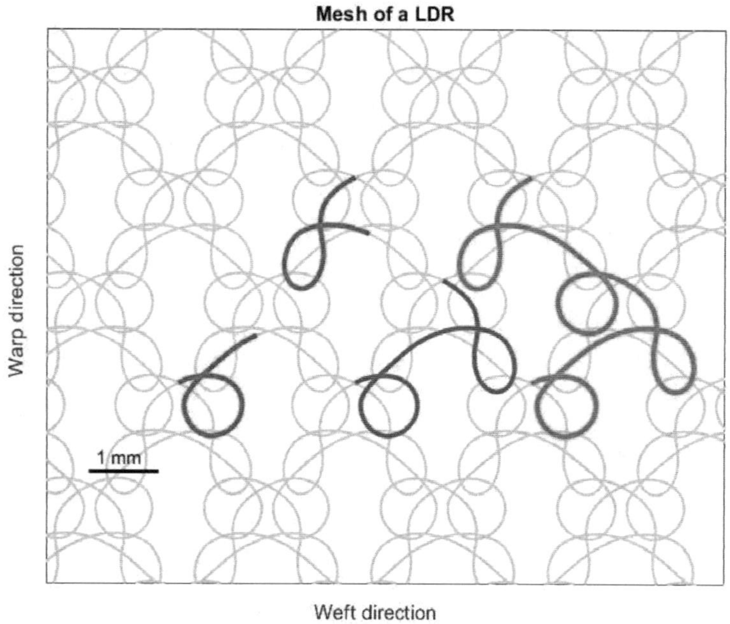

Figure 5. Description of the different portion in the mesh. From left to right: circular loop and oval one; 'primary part' and 'elementary wire path'.

The need to model a consistent portion of fabric to account of its structural response is given by the difficulty in applying boundary conditions. These are applied on the edges of each wire, but they introduce local disturbance that makes difficult this characterization. Therefore, the characterization is completed on the central rhombus (see figure 6) that is less affected by boundary loading.

Figure 6. Location of the contact bodies and the repeated rhombus pattern.

3. Results

The primary goal is to investigate the number of active contacts under differing boundary conditions. As written before, a homogeneous distribution of electric charge in every load condition reduces SNR and improve the quality of data exchange. The percentage of active contacts is strongly related to initial boundary conditions and should be maximal to enhance the performance of the reflector. The model is so submitted to different load conditions and percentage of active contact is taken into account. One of the modelled structure deformations is represented in figure 7 that refers to an isotropic displacement of *1.2 mm*.

The meaningful structure suggests investigating the status of contact bodies in the complete rhombus placed in the middle of the model. In the rhombus, there are 12 contact bodies for a total of 24 elements of contact (everyone describes the interference between two wires and needs two different contact pairing).

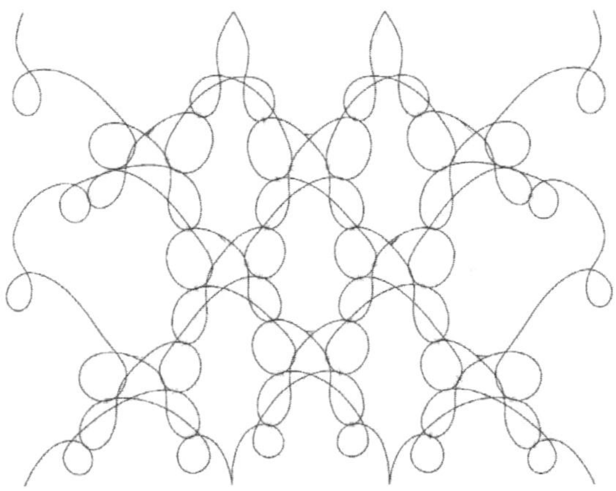

Figure 7. Deformed configuration of the mesh submitted to an imposed displacement at all boundaries both in warp and weft direction. Imposed displacement is of *1.2 mm*.

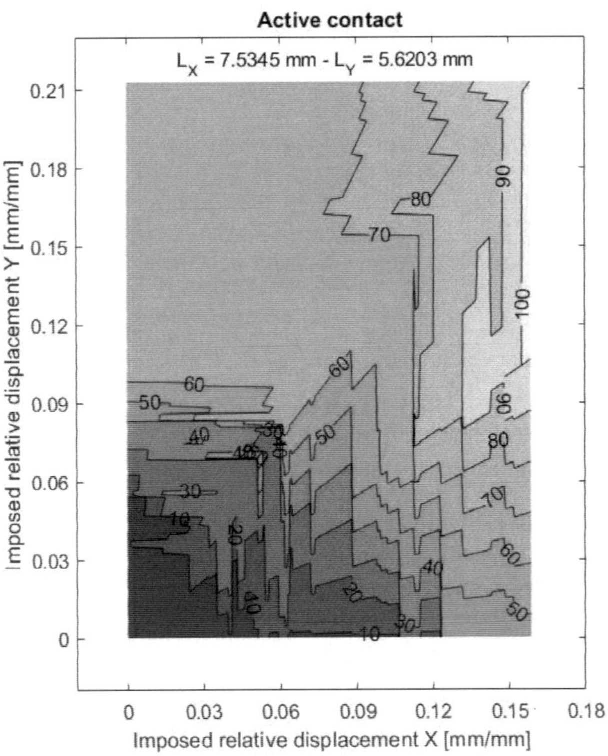

Figure 8. Contour map of percentage active contacts. The axes give the imposed deformations.

IOP Conf. Series: Materials Science and Engineering 1214 (2022) 012033 doi:10.1088/1757-899X/1214/1/012033

External loads are supplied by means of imposed displacements at the end of every wire. Figure 8 condense the results of many computations. It is evident that poor strains cause a low number of activated contacts. However, it is interesting to highlight the non-isotropic behaviour of contacts in the warp and weft directions. The warp direction (vertical in figure 8) is stiffer than the warp one. However, from the point of view of contacts, they need to activate a sufficient strain in the weft direction.

Deformations in the weft direction are, for this reason, crucial to optimize the behaviour, but a little change of them causes the loss of a huge number of contacts; on the opposite, the mesh in the weft direction is more stable from the point of view of activated contacts. Around a homogeneous 10% of deformation is necessary to activate a significant number of contacts for this Atlas-Atlas configuration. This result concerns this particular mesh and cannot be generalized. What is general is the method that applies to any other metal knitted mesh.

In correspondence of the 10% deformation, it was carried on a thermal verification of the contact stability. The results show that at the temperature changes considered (*-190°C* to *140 °C*) the percentage of contact keeps stable, so this is not a cause of malfunctioning of the reflecting mesh.

4. Conclusions

The present paper concerns the analysis of contacts of a metal mesh subjected to imposed deformations. The analysis is conducted on a specific knitted mesh, but the method has a general meaning. In the paper, FE analysis has been adopted to gain a reference for a more specific method developed by the authors. In fact, convergence difficulties make it difficult to analyse a full antenna mesh with commercial finite elements. Nevertheless, the finite element can be used when considering a small portion to achieve its characteristics in terms of stiffness and contact diffusion.

Among all possible techniques to manage the contact in a 2D frame, many of them have been reported in the paper. The resulting best way considers the insertion of a massive contact body and conducting the analysis in a dynamic contest. This allows to represent the contact, which is a physical 3D crossing of wires, as the coupling of two loops in a 2D reference. This method also allows the sliding of contact points which is determinant in the analysis of knitted metal meshes.

References

[1] Roddy D. 2001 *Satellite Communications* London McGraw-Hill.

[2] Horrocks A R and Anand S C 2000 *Handbook of Technical Textiles* UK: CRC Press

[3] Hu J L 2004 *Structure and mechanics of woven fabrics* New York CRC Press

[4] Liu D et al. 2017 On the role of material architecture in the mechanical behavioir of knitted textiles *International Journal of Solids and Structures* **109** 101

[5] Parson E M, Weerasooriya T, Sarva S and Socrate S 2010 Impact of woven fabric: Experiments and mesostructure-based continuum-level simulations *Journal of the Mechanics and Physics of Solids* **58** 1995

[6] Misnon M I, Islam M, Epaarachchi J and Lau K 2014 Potentiality of utilising natural textile materials for engineering composites applications *Materials and Design* **59** 359

[7] Robinson A T C and Marks R 1973 *Woven cloth construction* Manchester: The Textile Institute

[8] Ruan X and Chou T W 1996 Experimental and theoretical studies of the elastic behavior of knitted-fabric composites *Composites Science and Technology* **56** 1391

[9] Haas R and Dietzius A 1918 The stretching of the fabric and the shape of the envelope in nonrigid ballons *Annual Report, National Advisory Committee for Aeronautics* **16** 149

[10] Peirce F T 1937 The geometry of cloth structure *Journal of the Textile Institute* **28** 43

[11] Li T and Su J 2011 Electrical properties analysis of wire mesh for mesh reflectors *Acta Astronautica* **69** 109

[12] Thomson M W 2000 The astromesh deployable reflector *Antennas and Propagation Society International Symposium* **80** 435

[13] Ozawa S 2014 Design Concept of Large Deployable Reflector for Next Generation L-Band S.A.R. Satellite *Proceedings of the 2nd International Conference on Advanced Lightweight Structures and Reflector Antennas* Tbilisi, Georgia **2** 43

[14] Datashvili L, Maghaldadze N, Pauw J, Endler S, Baier H and Gigineishvili L. 2014 Advanced architectures of large deployable mesh reflectors: from medium to very large size *Proceedings of the 2nd International Conference on Advanced Lightweight Structures and Reflector Antennas* Tbilisi, Georgia **2** 7

[15] Shi H, Yuan S and Yang B 2018 New Methodology of Surface Mesh Geometry Design *Journal of Spacecraft and Rockets* **55** 266

[16] Hanabusa Y, Takizawa H and Kuwabara T 2013 Numerical verifcation of a biaxial tensile test method using a cruciform specimen *Journal of Materials Processing Technology* **213** 961

[17] De Salvador W, Marotta E, Pennestrì E, Salvini P and Valentini P P 2016 Some problems arising during the experimental characterization of compliant knitted mesh *ESA Antenna Workshop* Noordwijk, The Netherlands **37** 1

[18] Marotta E 2018 Structural Modelling of metallic knitted mesh *PH.D. Thesys University of Rome Tor Vergata- Italy*

[19] Bedukadze G and Datashvili L 2014 Some problems in ground testing of large space lightweight structures *Proceedings of the 2nd International Conference Advanced Lightweight Structures and Reflector Antennas* Tbilisi, Georgia **2** 123

[20] Scialino L, Salvini P, Migliorelli M, Pennestri E, van 't Klooster K, Santiago-Prowald J, Valentini P P, Rodriguez G and Gloy Y 2014 Structural Characterization and Modeling of Metallic Mesh Material for Large Deployable Reflectors *Proceedings of the 2nd International Conference on Advanced Lightweight Structures and Reflector Antennas* Tbilisi, Georgia

[21] Kyosev Y 2006 Computational model of loops of a weft knitted fabric *3rd International Textile, Clothing and Design Conference* Dubrovnik, Croatia **3** 1

[22] Nilakantan G, Keefe M, Bogetti T A, Adkinson R and Gillespie Jr J W 2010 On the finite element analysis of woven fabric impact using multiscale *International Journal of Solids and Structures* **47** 2300

[23] Vassiliadis S, Kallivretaki A and Provatidis C 2010 Mechanical Modelling of Multifilament Twisted Yarns *Fibers and Polymers* **11** 89

[24] Duhovic M and Bhattacharyya D 2006 Simulating the deformation mechanisms of knitted fabric composites *Composites* **37** 1897

[25] Ben Boukaber B, Haussy G and Ganghoffer J F 2007 Discrete models of woven structures *Composites* **38** 498

[26] Valentini P P, Falcone M, Marotta E, Pennestrì E and Salvini P 2016 Theoretical and experimental characterization of a FEM element assembly for the simulation of very compliant knitted mesh *Int. J. Numer. Meth. Engng* **107** 419

[27] Wriggers P 1995 Finite Element Algorithms for Contact Problems *Computational Methods in Engineering* **2** 1

[28] Zhou G, Sun X and Wang Y 2004 Multi-chain digital element analysis in textile mechanics *Composites Science and Technology* **64** 239

[29] Miao Y et al 2008 Mechanics of textile composites: Micro-geometry *Composites Science and Technology* **68** 1671

[30] Vassiliadis S 2011 *Advances in Modern Woven Fabrics Technology: The State of the Art* IntechOpen UK

[31] Marotta E, Massimi L and Salvini P 2020 Modelling of structures made of filiform beams: Development of a curved finite element for wires *Finite Elements in Analysis and Design* **170**

[32] ANSYS Inc. 2011 *ANSYS Mechanical APDL Reference*

[33] ANSYS Inc. 2004 *ANSYS Contact Technology Guide*

Active thermography technique for barrier coatings characterization

F Curà[1], R Sesana[1], L Corsaro[1] and R Mantoan[2]

[1] DIMEAS Politecnico di Torino, corso Duca degli Abruzzi 24, 10129 Torino, Italy
[2] ATLA, Via Secondo Caselle 10, 10023 Chieri (Turin), Italy

francesca.cura@polito.it

Abstract. Aim of this paper is to define and set up an experimental procedure, based on active thermography, for the characterization of thermal barrier coatings for industrial applications, above all in aerospace field. The developed procedure is intended to be a fast and reliable method, alternative to the consolidated one described in International Standards as ISO18755 and ISO18555. In particular, this approach consists of a pulsed active thermography, in transmission configuration, obtained by means of a laser excitation. Temperature data processing, according to and adapting Standard procedures, allowed us to determine thermal parameters as conductivity and diffusivity. Obtained results were compared in terms of thermal properties variation with respect to base and coated materials, and in term of different coating procedures. These results were also compared to those available in literature.

1. Introduction

In industrial aerospace field, coatings are widely applied and a typical application is as thermal barrier coating (TBC) for underneath metallic layers. In particular, Ceramic Thermal Barrier Coatings (TBCs) on superalloy components are used successfully in land-based gas turbine and aircraft engines.

These coatings are generally made by either air plasma spraying (APS) or electron beam physical vapour deposition (EB-PVD) [1]. In general, EB-PVD TBCs show a superior durability due to the columnar structure, but they are very expensive compared to APS TBCs. Nowadays, Suspension Plasma Spraying (SPS) technology provides a promising methodology to produce advanced thermal barrier coatings using nanopowders [2]. This process is capable of producing coatings with different types of structure, from very dense and compact ones to columnar with high volume of vertical intercolumnar voids. As a matter of fact, SPS coatings represent a challenging field of research, both for academic interest and commercial applications. Increased coating characteristics, such as low thermal conductivity and high level of thermal cycling resistance, may be actually reached by SPS technology; SPS coatings can generate columnar microstructures typical of electron-beam physical vapor deposition (EB-PVD) TBC, but at a much lower cost [3].

To this aim, the characterization of thermal diffusivity and conductivity properties is critical for process and material selection in industrial manufacturing of turbines and other components. Recently, thermographic techniques have been developed to characterize mechanical and thermal properties of components with coatings.

Aim of this paper is to define and set up an experimental procedure, based on active thermography, for the characterization of coatings for industrial applications. This procedure is intended to be a fast and reliable method, alternative to the consolidated one described in International Standards. In this research activity, it was applied to different industrial coating processes to investigate the corresponding thermal properties. In particular, in the present paper two coating processes (Atmospheric and Suspension Plasma Spray, APS and SPS) applied to the same base material and the same coating material were investigated.

The active thermography analysis provided the measurement of surface temperature of specimens undergoing a thermal excitation, applied by means of a laser pulse (Pulsed Technique). Temperature data processing, according to and adapting Standard procedures, allowed obtaining thermal conductivity and diffusivity parameters. These results were compared in terms of thermal properties variation with respect to base and coated materials, and in term of different coating procedures.

2. Materials and Methods

Thermal barrier coatings investigated in this study are made of *Yttria stabilized Zirconia* powder. Two types of deposition process were utilised, respectively *Suspension Plasma Spray* (SPS) and *Atmospheric Plasma Spray* (APS). The base material (substrate) was an *Inconel alloy 601*.

Figure 1 shows the specimens adopted in this experimental activity.

Figure 1. Specimens.

Shape and dimension of specimens were chosen according to Standards [4] and [5]. A SEM analysis was performed to evaluate both structure and thickness of coatings, generated by SPS and APS deposition processes. Thickness values were necessary to calibrate experimental set up parameters. Figure 2 shows SEM results of SPS and APS specimens, emphasising both coatings structure and their thickness. A microscope analysis was also performed to evaluate both thickness of Inconel 601 specimen and of metallic layer (Inconel 601) in coated specimens.

Physical characteristics values (density and specific heat) adopted for the estimation of thermal conductivity may be found in [6] for substrate (Inconel 601) and in [7] for coatings (see Table 1).

The experimental equipment was composed by a thermal camera, a laser excitation source and a PC control unit. The IR thermo camera is a FLIR A6751sc with sensitivity lower than 20 mK and 3-5 μm spectral range, while the laser source can generate a maximum power of 50 W concentrated in a small surface.

The experimental configuration was in "transmission mode". Figure 3 shows both equipment and configuration. More in detail, Inconel 601 was excited with an heating period of 50ms at the maximum power, while specimens coated were excited with an heating period of 5ms at the maximum power.

Table 1. Physical characteristics of materials.

Material	ρ [kg/m^3]	c [J/kg °C]
Inconel 601 (substrate)	8110	448
Yttria Stabilized Zirconia (coating)	5200	467

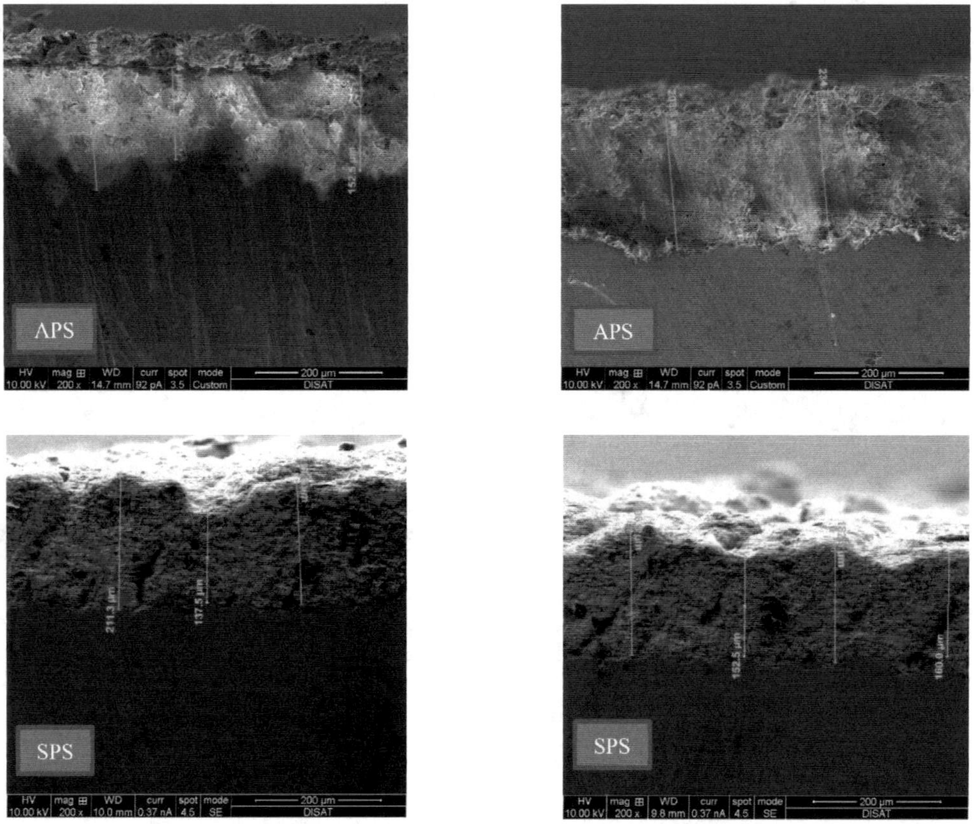

Figure 2. SEM results of SPS and APS specimens.

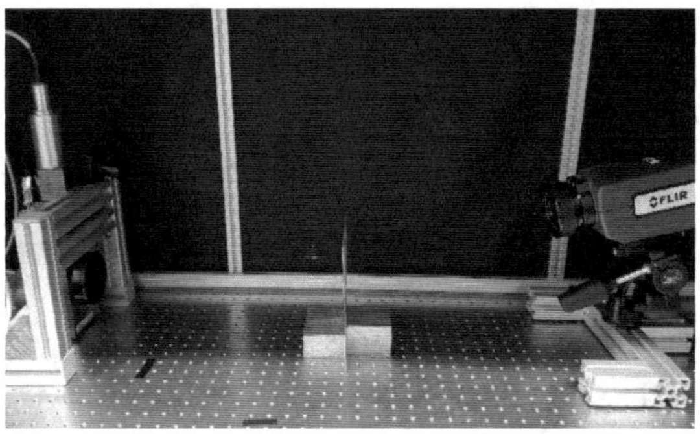

Figure 3. Experimental equipment.

Thermal characterization methods for barrier coatings follows Standards [4] and [5]. ISO18755 Standard [5] allows to determine the thermal parameters (thermal diffusivity and thermal conductivity) of a single material by using the *half rise time method*. ISO18555 Standard [4] may be used for multilayer materials (as metallic coatings), allowing to determine the thermal parameters (thermal diffusivity and thermal conductivity) of each layer by using the *areal heat diffusion time method*.

Standards [4] and [5] are generally also used to compute the apparent thermal diffusivity of multilayer materials.

Figure 4 shows an example of "Temperature rise curve" and the related parameters useful for the computation of thermal diffusivity values from experimental data (*areal heat diffusion time* "A" and *half rise time* "$t_{1/2}$").

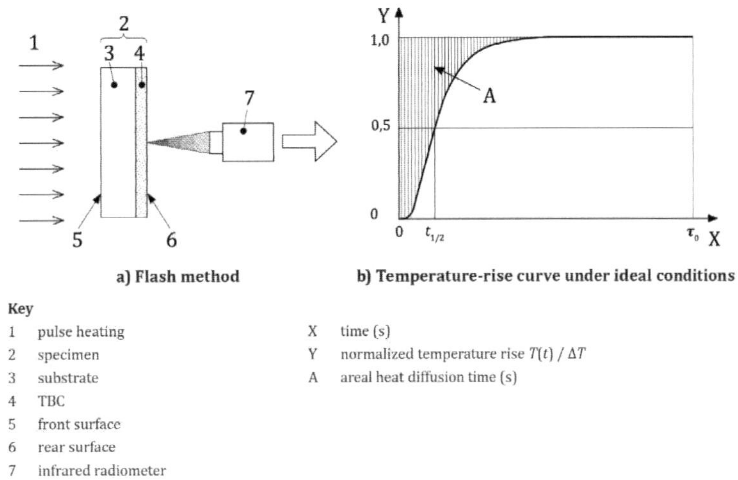

Figure 4. Temperature rise curve [4].

In this work, the temperature rise curve was obtained from the temperature profile acquired by the IR thermo camera.

Standards [4] and [5] also propose different corrections to consider particular phenomena occurring during the experimental test. *Centroid method* (to correct the finite pulse period of the excitation source) and the *Cowen method* (to correct the heat losses) were adopted.

3. Results

Thermograms of all specimens were acquired during the tests. Temperature profiles utilised in the procedure were related to a box located in the centre of each specimen.

A dedicated filtering process (Gaussian filter) was performed to reduce the noise due to the high frame rate acquisition (786 Hz for the Inconel 601 and 600 Hz for the specimens with coating). In particular, 50 points were used to filter Inconel 601 thermal profiles and 20 points were used to filter coatings thermal profiles. This procedure allowed us to better identify the starting point of the heating profile.

Figure 5 shows the relative temperature ΔT (difference between measured temperature, T, and ambient temperature, Tamb) of each specimen. A specific thermal excitation was tuned for the two different types of samples (substrate and substrate+coating), according to their different thermal behaviour (thermal conductivities). Five test replications were considered for Inconel 601 specimen (substrate material) (Test 1, 2,....5), to better compute the heat diffusion time, useful for coatings thermal characterization. Three test replications were taken into account for APS and SPS specimens (Test 1, 2, 3).

Figure 6 shows a comparison between the relative temperatures of APS and SPS, measured with the same thermal excitation.

By analysing Figures 5 and 6, it may be pointed out that Inconel 601 base material and Inconel 601 with coatings show a different thermal behaviour. As expected, the base material emphasises a substantially higher relative temperature respect to the corresponding of coated specimens (see Figure 5). On the other hand, specimens with TBC, regardless of the production process, show a similar relative temperature trend during the tests (see Figure 6).

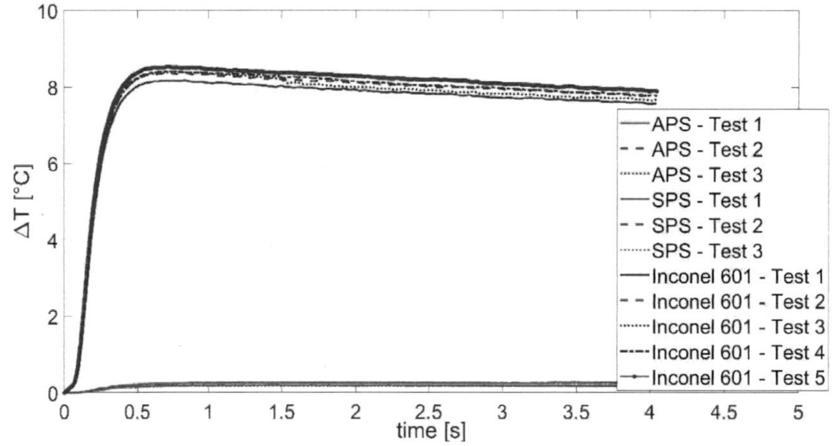

Figure 5. Relative temperature ΔT of each specimen.

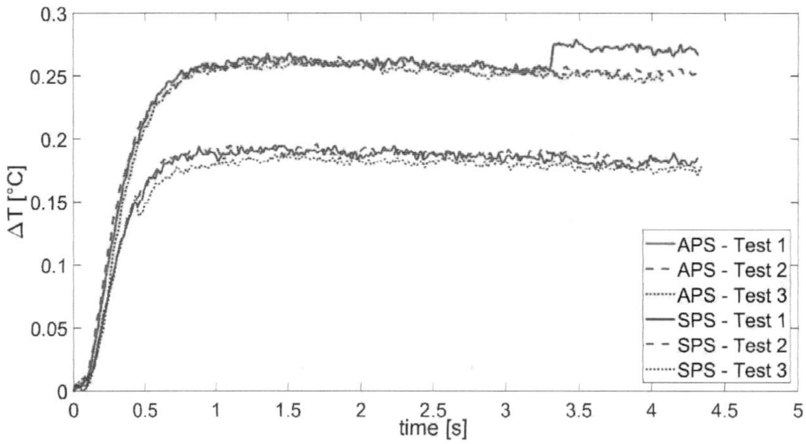

Figure 6. SPS and APS: relative temperature ΔT.

All temperature profiles were processed, according to Standards [1] and [2], by considering the temperature increment up to the maximum temperature value. Figure 7 shows the normalized relative temperature profile of each test with respect to its maximum value. This elaboration allowed us to determine the parameters of interest (*areal heat diffusion time* "A" and *half rise time* "$t_{1/2}$") and to realize the thermal characterization of both substrate and coatings.

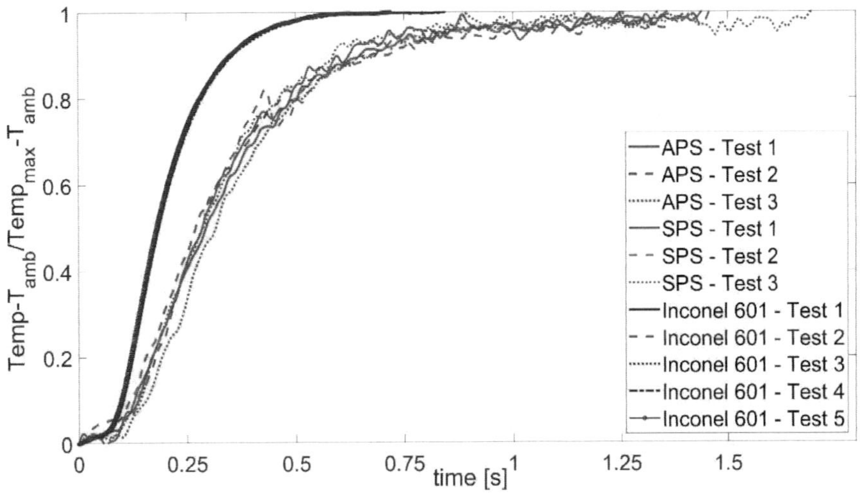

Figure 7. Normalized relative temperature ΔT of each specimen.

Table 2 illustrates the results concerning the thermal characterization of the substrate (averaged values of all replications) (Inconel 601) by applying Standards [4] and [5] formula. Ideal thermal diffusivity and thermal conductivity values calculated by ISO18755 Standard [5] (monolayer ceramic material model) are presented in the left side of the table (α_{ideal} and k_{ideal}), while the corresponding values obtained by ISO18555 Standard [4] (multi layer material model) are shown in the right side one (α_{1-i} and $k_{1-ideal}$). Ideal thermal parameters [5] were also corrected by both Centroid Method (α_{tc} and k_{tc}) and Cowen Method (α_{Cowen5}, k_{Cowen5}, $\alpha_{Cowen10}$ and $k_{Cowen10}$) and represented in the same table (Table 2, left side). Finally, Table 2 reports the calculated heat diffusion time of the substrate (τ_1).

Table 2. Thermal parameters for Inconel 601.

Thickness: 2 mm	ISO 18755 [5]								ISO 18555 [4]		
	α_{ideal} [m²/s]	k_{ideal} [W/m°C]	α_{tc} [m²/s]	k_{tc} [W/m°C]	α_{Cowen5} [m²/s]	k_{Cowen5} [W/m°C]	$\alpha_{Cowen10}$ [m²/s]	$k_{Cowen10}$ [W/m°C]	$\alpha_{1-ideal}$ [m²/s]	$k_{1-ideal}$ [W/m°C]	τ_1 [s]
Inconel 601	3.31E-06	12.02	3.82E-06	13.90	3.84E-06	13.96	3.83E-06	13.90	3.53E-06	12.82	1.2504

Table 3 illustrates the corresponding results related to coated specimens (substrate and TBCs), as for base material the results are averaged on all replications. Tests performed on coated specimens allowed us to characterize both SPS and APS coatings, once the heat diffusion time (τ_1) of the substrate was calculated (see Table 2). The layout of Table 3 (SPS and APS coatings) is the same already utilised for Table 2, the left side refers to the ISO18755 Standard [5] approach and the right side to the ISO18555 Standard [4] one. In particular, respectively for APS and SPS coated specimens, the following results are shown in the left side of the table: ideal apparent thermal diffusivity values ($\alpha_{app-idea}$), apparent thermal diffusivity values corrected by both Centroid Method (α_{app-t}) and Cowen Method ($\alpha_{app-Cowen5}$ and $\alpha_{app-Cowen10}$, respectively 5 and 10 times the half rise time "$t_{1/2}$"). The right side of the table shows: ideal apparent thermal diffusivity values ($\alpha_{app-idea}$) of substrate+coatings, ideal thermal diffusivity and conductivity values of coatings (α_{2-ide} and k_{2-ide}).

Table 3. Thermal parameters for SPS and APS coatings.

		ISO 18755 [5]				ISO 18555 [4]		
	Mean thickness [mm]	$\alpha_{app-ideal}$ [m²/s]	α_{app-tc} [m²/s]	$\alpha_{app-Cowen5}$ [m²/s]	$\alpha_{app-Cowen10}$ [m²/s]	$\alpha_{app-ideal}$ [m²/s]	$\alpha_{2-ideal}$ [m²/s]	$k_{2-ideal}$ [W/m°C]
APS	0.1868	3.18E-06	3.20E-06	3.09E-06	3.17E-06	3.11E-06	1.31E-07	0.32
SPS	0.1759	2.64E-06	2.67E-06	2.63E-06	2.66E-06	2.58E-06	1.24E-07	0.30

From the analysis of Table 2, it can be observed that for Inconel 601 both Standard analytical formula ([4] and [5]) provide similar diffusivity and conductivity values. These values are in a good agreement with the corresponding provided by the material producer [6] (k =11,2 to 12,7 W/m°C).

A similar consideration can be stated referring to TBCs results, basing on the analysis of Table 3. More in detail, it can be concluded that obtained thermal parameters values for SPS and APS coatings are comparable to those available in literature [7] as order of magnitude. A great tuning of data is difficult to be reached due to the variability of TBC thickness that strongly influences the obtained values.

4. Conclusions

The results obtained in the present work allow us to draw the following conclusions.

The experimental procedure defined and set up for thermal characterisation of industrial TBCs, based on active thermography, provided very good results comparable to those available in literature. As wished and hypothesised, this methodology can be thought as a fast and reliable method, alternative to the consolidated one described in International Standards ISO18755 and ISO18555, for which it is required a dedicated equipment.

Furthermore, this developed technique, together with its processing algorithms, provides a robust method to be used in industrial environment. As a matter of fact, from coatings point of view, it can be conclude that, despite of the difficulty in define specimens with quite constant thickness, obtained thermal diffusivity and conductivity values are representative of the actual behaviour of TBCs obtained by different production processes.

Finally, particularly for as concerns the present industrial application, it can be concluded that SPS and APS coatings show a substantial identical thermal behavior and the choice to employ one of these can be left to cost considerations.

References

[1] Z. Tang, H. Kim, I. Yaroslavski, G. Masindo, Z. Celler, D. Ellsworth 2011 *Proc. of International Thermal Spray Conference and Exposition* Novel Thermal Barrier Coatings Produced by Axial Suspension Plasma Spray (Hamburg, German)

[2] S. S. Dautov et al 2019 J. *Phys.: Conf. Ser. 1281 012008* Increasing thermal and mechanical properties of thermal barrier coatings by suspension plasma spraying technology

[3] Zhaolin Tang December 2015 *Thermal Spray of Suspensions & Solutions Symposium (TS4)* Axial Suspension Plasma Spraying (ASPS). Its Development and Application (Montreal, Canada)

[4] International Standard ISO 18555 (2016). Metallic and other inorganic coatings — Determination

IOP Conf. Series: Materials Science and Engineering 1214 (2022) 012034 doi:10.1088/1757-899X/1214/1/012034

of thermal conductivity of thermal barrier coatings
[5] International Standard ISO 18755 (2005). Fine ceramics [advanced ceramics, advanced technical ceramics] — Determination of thermal diffusivity of monolithic ceramics by laser flash method
[6] Special Metals – 2005, www.specialmetals.com
[7] Megumi Akoshima et al 2011 *J. Appl. Phys. 50 11RE01* Thermal Diffusivity Measurement for Thermal Spray Coating Attached to Substrate Using Laser Flash Method (Japan)

IOP Conf. Series: Materials Science and Engineering 1214 (2022) 012035 doi:10.1088/1757-899X/1214/1/012035

Multiscale stress analysis of porosity clusters as mechanical models of degenerated graphite defects in cast iron

R Rizzoni*, **P Livieri and R Tovo**

Department of Engineering, University of Ferrara, Via Saragat 1, 44122 Ferrara, Italy

*Email: raffaella.rizzoni@unife.it

Abstract. We extend previous results for the stress analysis of a cluster of degenerated graphite in spheroidal cast iron, modelled as a multiscale three-dimensional composite. At the microscale, precipitates of degenerated graphite are modelled as spheroidal voids in a linear elastic matrix. At the mesoscale, clusters of degenerated graphite are aggregates of spheroidal voids. These aggregates are studied by means of a numerical analysis based on finite element simulations. As a second approach, the clusters are seen as homogeneous inclusions made of an equivalent porous elastic material. The average elastic properties of the porous material are calculated using an approach proposed by Tandon and Weng and based on Eshelby's equivalent principle and Mori-Tanaka's estimate. Comparison of the two approaches gives encouraging results in terms of average properties.

1. Introduction

Cast iron is an engineering material with a wide range of applications, such as turbines, pipes, machines, and automotive industry parts. In nodular cast iron, graphite spheroids or nodules provide large ductility and fatigue strength. Deviations from the spherical shape and arrangement of graphite particles are strongly related to temperature evolution and distribution during casting procedures and subsequent heat treatments. In thick-walled casted components, degenerated graphite microstructures frequently emerge resulting in deterioration of the mechanical properties [1-5].

In literature, ductile cast iron is often modelled as a composite material characterized by two phases: graphite nodules, acting as reinforcement, and the metal matrix. In this context, micromechanical modelling approaches have been proposed to understand the role played by graphite nodules during damage and failure [6-9]. In these works, cells models based on spherical or spheroidal graphite particles arranged regularly in a metallic matrix have been investigated. Graphite particles are frequently modelled as voids or cavities, because they are typically seen to undergo decohesion from the matrix, cf. [7,10,11].

For degenerated graphite, an approach treating graphite particles as porosity has been explored by the authors in [12-14]. In particular, at the microscopic scale, graphite precipitates are modelled as spheroidal voids, randomly oriented and orderly or disorderly distributed in a linear elastic isotropic matrix. The average elastic properties of the porous equivalent material can be evaluated numerically by the finite element (FE) method or analytically by means of Mori-Tanaka's (MT) estimate, as proposed by Tandon and Weng [15]. At a larger scale (mesoscale), a cluster of degenerated graphite is viewed as a single spheroidal inclusion embedded into the elastic isotropic matrix and made of equivalent porous material. Eshelby's fundamental solution [16-18] is then applied to calculate the stress distribution internal and external to the inclusion modelling the cluster. Specifically, the proposed two-scale model allows characterizing average stress distribution inside the cluster depending on the following microstructure parameters: the density of graphite particles, their aspect ratio, the orientation and the aspect ratio of the cluster of degenerated graphite [13,14].

In the present paper, we extend our previous work in two ways. First, we compare the average elastic properties, calculated both numerically and analytically, with some experimental data available in the literature for porous materials. Next, we compare the results of external stress distribution calculated analytically (TW approach and Eshelby's solution) with those estimated numerically via FE simulations.

The plan of the paper is the following. Section 2 reviews the proposed multiscale model for cast iron with degenerated graphite. In Section 3, the experimentally calculated elastic properties of some porous materials are compared with the corresponding elastic properties calculated both numerically and analytically. Section 4 is devoted to the comparison between the results for stress analysis obtained via FE simulations and the average stress distribution calculated using the multiscale model. Some concluding remarks and future directions of the work are illustrated in Section 5.

2. Review of multiscale model for cast iron with degenerated graphite

Spheroidal cast iron is modelled as a composite material in which spheroidal voids with aspect ratio t (the ratio of length to semi-diameter) are randomly dispersed in a three-dimensional isotropic matrix.

At the microscale level of the composite, the average elastic properties of the porous equivalent material have been evaluated by using two different approaches: analytically, as proposed in [15,19], and numerically by the finite element (FE) method.

At the mesoscale level, for the analytical approach, a clusters of voids is described as a spheroidal inclusion having the average elastic properties calculated at the microscale. Numerically, clusters containing ordered and disordered distribution of randomly oriented spheroidal voids have been analyzed.

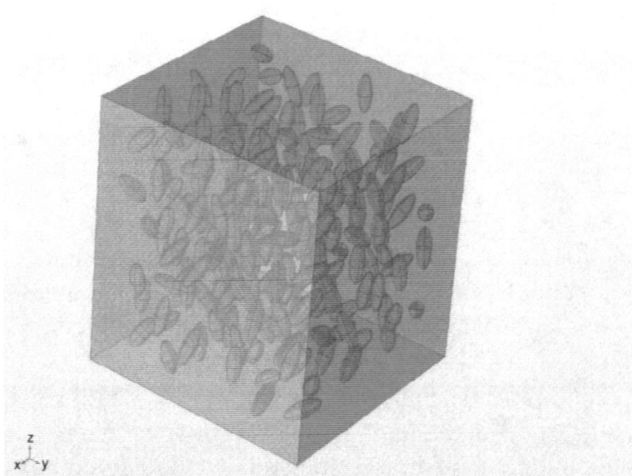

Figure 1. Example of microscale configuration numerically analyzed: disordered distribution of randomly oriented spheroids in a cube of L×L×L mm (L=6 mm).

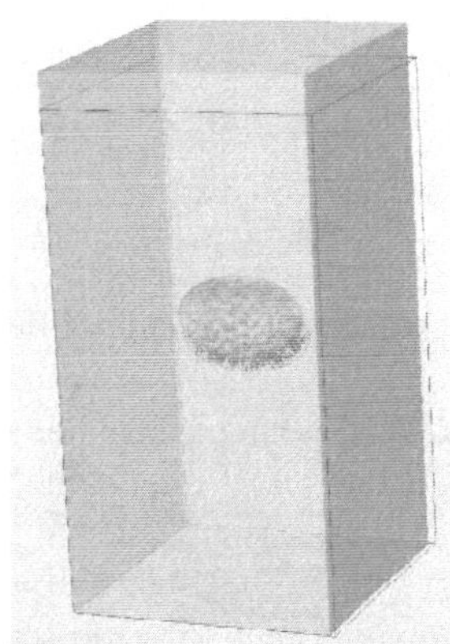

Figure 2. Example of mesoscale structure: cluster of spheroidal voids disorderly distribution inside a spheroidal volume.

2.1. Microscale level: porous material

Applying Eshelby's equivalent principle and Mori-Tanaka's hypothesis of the average stress in the matrix, Tandon and Weng have determined the average bulk and shear moduli of the equivalent porous material in terms of the porosity, λ, their aspect ratio, t, and the elasticity properties of the matrix [15,19]. From the estimated average bulk and shear moduli, the average Young modulus and Poisson ratio can be calculated under the assumption of macroscopically isotropic porous material. The latter assumption follows from the randomness of voids' orientations and it has been verified numerically.

In addition to the previous homogenization procedure, a three-dimensional FE analysis has been carried out using commercially available software (COMSOL Multiphysics). Cubes of linear elastic material with Young modulus and Poisson ratio E=206 GPa and ν=0.3, respectively, and ordered and disordered void distributions have been studied. The configuration shown in figure 1 represents a typical example. For the ordered distribution, randomly oriented voids are placed at the centers of a uniform grid with cube elementary cells of 1×1×1 mm. For the disordered distribution, the voids have centers placed randomly inside the volume and they do not overlap. For both types of distributions, the voids are spheroids with t ranging between 0.1 and 0.8.

2.2. Mesoscale level: a cluster of voids

A cluster of degenerated graphite is modelled as an inclusion perfectly bonded to an elastic three-dimensional matrix subjected to a given remote uniform load σ_0. The stress components inside and at the boundary of the inclusion have been calculated by applying Eshelby's solution [16-18].

Numerically, clusters of spheroidal voids have been tested in COMSOL under uniaxial tensile loading, as shown in figure 2. Inside the cluster, ordered and disordered distributions of voids have been taken into account.

3. Comparison with experimental data available in literature

The analytical and numerical estimates of average elastic properties calculated as summarized in Section 2 have been compared with the experimental results obtained in [20, 21]. In [20], 20×20×150 mm beams made of high strength hemihydrate plaster containing randomly distributed polystyrene agglomerates have been tested in four-point bending conditions. The content of polystyrene has been varied to realize different levels of porosity, which have been characterized by X-ray computed tomography and optical imaging on bidimensional sections. Typical load-displacement curves show a quasi-brittle behavior.

The elastic modulus $\bar{E}$ is found to reduce with increase porosity and the following result (in GPa) was obtained in the regression analysis:

$$\bar{E} = 14.28 - 0.28\,\lambda \tag{1}$$

with adjusted determination coefficient $R^2 = 0.90$. Figure 3 shows the graph of the relation (1), together with the estimate obtained by applying the Tandon and Weng approximation [15,19] and the numerical estimate obtained via the FE method. For both estimates, the porosity is modelled with disordered distributions of equal spherical voids.

Figure 3 indicates that the approach by Tandon and Weng and the numerical analysis of microstructures made of defects distributions are in good agreement with experimental data, with percent errors lower than 15% for porosity levels below 10%. For larger porosity values, the relative errors are found to increase. A possible reason may be the effect of the real pore microstructure, differing from the idealized distributions of spherical voids considered in calculating the estimates.

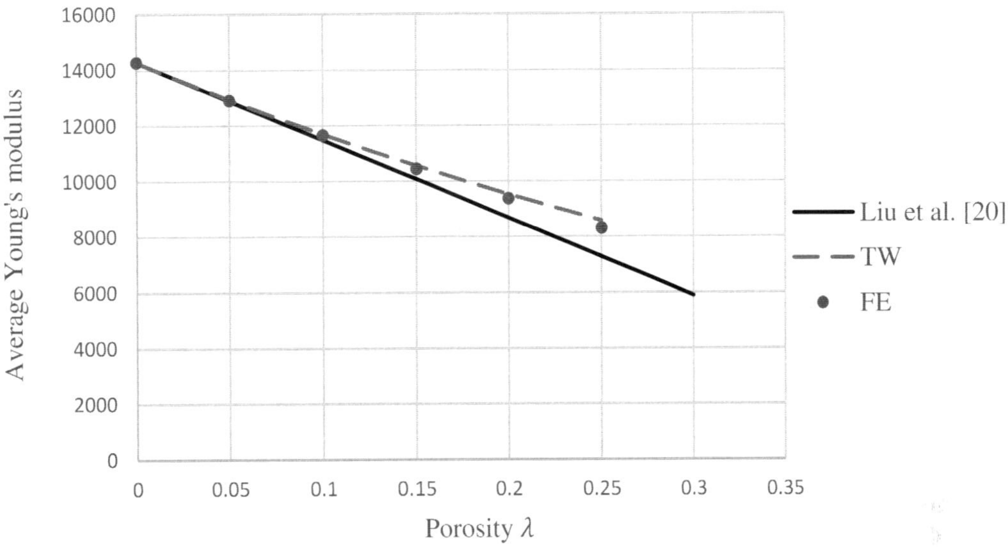

Figure 3. Average elastic modulus vs. porosity in a quasi-brittle porous material (see text for description). The black line corresponds to the linear regression calculated by Liu et al. with experimental data on beams in four point bending tests [20]. The dashed curve is the estimate obtained by applying the approach proposed by Tandon and Weng [15,19]. The dots correspond to the numerical estimate calculated analyzing disordered distributions of spherical voids.

In [21], Young's modulus of nodular cast iron $\bar{E}_{ci}$ was measured by a pulse-echo elastic-wave technique. The volume fraction of the graphite nodules was determined by quantitative metallographic techniques. Several specimens were examined in three different conditions: as-cast (AC), annealed (A) and normalized-and-tempered (NT). Table 1 shows the experimentally measured values together with the estimates provided by the approximated method of Tandon and Weng (TW) and by the numerical analysis (FE). The two estimates were obtained by considering levels of porosity equal to the measured values of graphite volume fraction. For the considered values of the latter, the data listed in Table 1 indicate an excellent agreement between the experimental data and the analytical and numerical results.

Table 1. Average elastic modulus vs. porosity in nodular cast iron: comparison of model predictions with experimental data obtained from literature for cast iron in as-cast (AC), annealed (A) and normalized-and-tempered (NT) conditions [20].

Type	Graphite (%)	$\bar{E}_{ci}$ [20] (GPa)	TW (GPa)	FE (GPa)	Error TW (%)	Error FE (%)
AC	9.2	176.0	174.0	171.0	1.1	2.9
A	10.5	169.0	170.0	166.0	0.6	1.7
NT	9.2	169.0	174.0	171.0	2.9	1.2

4. Validation of multiscale model via average stress distribution

To validate the multiscale model, the average stress distribution calculated analytically has been compared with the stress distribution calculated numerically in a porous spheroidal cluster under uniaxial tensile loading, as shown in Figure 2.

A rectangular box of dimensions 25×25×40 mm, with a cluster having semiaxes equal to 6 and 3 mm, has been numerically studied in COMSOL. Inside the cluster, the material is characterized by an ordered distribution of small spherical voids with a radius of 0.4 mm. The elastic constants of the matrix are as before, E = 206 GPa and ν = 0.3. Uniaxial traction was applied by imposing vanishing vertical displacement at the bottom surface of the box and a uniform distributed traction surface load σ at the top surface.

Table 2 shows the values of the stress components, averaged over the equatorial line of the spheroidal surface of the cluster and normalized with respect to the applied load σ. Table 2 shows also the normalized average values of the same stress components calculated by using the multiscale model, i.e. via Eshelby's solution at the external surface of the inclusion made of the equivalent porous material. The data listed in the Table indicate an overall good agreement between the two sets of average stress components. However, stress peaks strongly depending on relative position of neighbouring cavities cannot be detected by the multiscale model, in which defects are smeared out into equivalent porosity.

Table 2. Normalized average stress components calculated via a FE analysis performed over the porous cluster (label "FE") and via the multiscale model (label "MM"). The stress components are calculated over the equatorial line of the cluster's surface (for the FE model) and at the external surface of the inclusion (for the MM model), and they are averaged over the equatorial line and normalized with respect to the applied surface load σ.

Model	σ_x/σ	σ_y/σ	σ_z/σ	σ_{xy}/σ	σ_{xz}/σ	σ_{yz}/σ
FE	0.057	0.022	1.287	-0.002	-0.005	-0.010
MM	0.048	0.015	1.326	0.000	0.000	0.000

5. Conclusion

In this paper, we have extended previous research on stress analysis in clusters of degenerated graphite in spheroidal cast iron, modelled as a multiscale three-dimensional composite. Clusters are viewed as spheroidal aggregates of spheroidal voids embedded in a three-dimensional linear elastic matrix. Two different approaches have been introduced: a numerical analysis based on finite element simulations and an analytical approach based on homogenization and Eshelby's solution. The two approaches are validated by comparing the predicted average elastic properties of the porous material inside the clusters with experimental data obtained from the literature [20]. A good agreement is observed, indicating that homogenization procedures are effective in estimating the elastic properties of spheroidal cast iron.

As a second validation, the stress components averaged over the equatorial line of the spheroidal surface of the cluster and normalized with respect to the applied load have been determined by either of two approaches. The comparison between the two sets of data indicates a good agreement. However, stress distribution calculated via FE analysis shows local stress peaks at the boundary of the cavities, strongly depending on their relative positions. These stress peaks cannot be detected by the analytical approach based on homogenization.

As a future development of the present model, extended approaches improving defects interactions should be explored, such as models based on Eshelby's formalism with polynomial eigenstrains [23]. Other issues to be explored could be the bonding and the damage of the graphite particles at the interface with the matrix, cf. [24-27].

References

[1] Hütter G, Zybell L and Kuna M 2015 *Eng. Fract. Mech.* **144** 118
[2] Rausch T, Beiss P, Broeckmann C, Lindlohr S and Weber R 2010 *Procedia Eng.* **2** 1283
[3] Costa N, Machado N and Silva F 2010 *Int. J. Fatigue* **32** 988
[4] Lacaze J, Asenjo I, Mendéz S, Sertucha J, Larranaga P and Suárez R 2012 *Int. J. Metalcasting/Winter* 35
[5] Kaczorowski J, Jozwiak K and Innocenti M 2013 *J. Fail. Anal. Prev.* **13** 445
[6] Brocks W, Hao S and Steglich D 1996 *J. Phys. VI* **6** 43
[7] Bonora N and Ruggiero A 2005 *Int J Solids Struct* **42** 1401
[8] Bonora N, Gentile D, Ruggiero A and Iacoviello F 2003 *Proc. 9th Int. Conf. on the Mechanical Behaviour of Materials*, Genève, Swiss
[9] Collini L and Nicoletto G 2005 J. Strain. Anal. Eng. **40** 107
[10] Andriollo T, Thorborg J and Hattel J 2016 *Int. J. Solids Struct.* **100** 523
[11] Dong M, Prioul C and François D 1997 *Metall. Mater. Trans. A* **28** 2245
[12] Cova M, Livieri P, Rizzoni R and Tovo R 2017 *Procedia Struct. Integrity* **7** 446
[13] Rizzoni R, Livieri P, Tovo R and Cova M 2020 *Theor. Appl. Fract. Mech.* **109** 102731
[14] Rizzoni R, Livieri P and Tovo R 2021 *IOP Conf. Ser. Mater. Sci. Eng.* **1038** 012047
[15] Tandon G P and Weng G J 1986 *Compos. Sci. Tech.* **27** 111
[16] Eshelby J D 1957 *P. Roy. Soc. A-Math. Phy.* **241** 376
[17] Eshelby J D 1959 *P. Roy. Soc. A-Math. Phy.* **252** 561
[18] Eshelby J D 1961 in *Progress in Solid Mechanics*, eds. N Sneddon and R Hill vol. **2**, 89
[19] Pan H H, Weng G J 1995 *Acta Mech* **110** 73
[20] Liu D, Šavija B, Smith G E et al. 2017 *Int J Fract* **205** 57
[21] Speich G R, Schwoeble A J and Kapadia B M 1980 ASME. *J. Appl. Mech.* **47** 821
[22] Mura T 1987 *Micromechanics of Defects in Solids* (Springer Netherlands)
[23] Fond C, Riccardi A, Schirrer R et al. 2001, *Eur. J. Mech. A/Solids* **20** 59
[24] Lebon F, Dumont S, Rizzoni R et al 2016 *Compos. Part B-Eng.* **90** 58
[25] Bonetti E, Bonfanti G, Lebon F and Rizzoni R 2017 *Meccanica* **52** 1911
[26] Guinovart-Sanjuán D, Rizzoni R, Rodríguez-Ramos R et al 2017 *Compos. Struct.* **176** 539
[27] Livieri P and Tovo R 2017 *Int. J. Fatigue* **101** 363

IOP Conf. Series: Materials Science and Engineering 1214 (2022) 012036 doi:10.1088/1757-899X/1214/1/012036

Evaluation of the thermomechanical Shape memory polymers in equi-biaxial condition by hydraulic bulge test

M Coccia[1], E Farotti[1] and A Lattanzi[1]

[1] Department of Industrial Engineering and Matematical Science, Via Brecce Bianche, Ancona, 60131, Italy

E-mail: `m.coccia@pm.univpm.it`

Abstract. Shape Memory Polymers (SMPs) are materials capable of changing their primary shape to a secondary shape thanks to the so called Shape Memory Effect (SME) phenomenon. The shape-shifting is achieved through the action of an external stimulus, such as heat, electricity, pH, etc. In this paper, experiments on a thermally actuated thin film of a Shape Memory Thermoplastic Polyurethane (SMPU) were performed to calibrate the parameters of a constitutive model which accounts the rubbery/glassy phase transition mechanism behind the shape memory behaviour. In particular, thermomechanical uniaxial tensile tests have been carried out in order to the Young modulus and Poisson's ratio above/under glass transition temperature and the fixity/recovery ratio. Additionally, the hydraulic bulge test (HBT) in a thermally controlled loading/unloading cycle was used to study the behaviour of the SME at large strains under equi-biaxial stress state. The corresponding outcomes were, therefore, employed to validate the results of the initial calibration by means of a Finite Element (FE) simulation of the HBT.

1. Introduction

Shape memory polymers (SMPs) are mechanically functional smart materials, that are particularly interesting for biomedical applications [1, 2], aerospace applications [3, 4] and actuators [5, 6]. The SMPs own a shape memory effect (SME), i.e. under certain external stimulus such as the magnetism [7], heat [8, 9, 10], electricity [11], light [12, 13, 14] and some specific chemicals [15], it is possible to deform the material from his original shape to a temporary one and vice versa. The transition temperature which allows the shape-shifting is the glass transition temperature (T_g); basically, when the material is above this transition temperature, the polymer change his mechanical properties, activating the SME. The increasingly use of smart materials in different application fields requires indeed a specific characterization based on their chemical and mechanical properties.

In this paper, in order to characterize the SME behaviour, the typical thermo-mechanical cycle on uniaxial tensile tests was performed. The outcomes allows to identify the two main parameters that represent the nature of the SME: the efficiency of the material in maintaining the temporary shape, and the ability to recover an intermediate shape. A schematic of the thermomechanical cycle is depicted in Figure 1, according to[16].

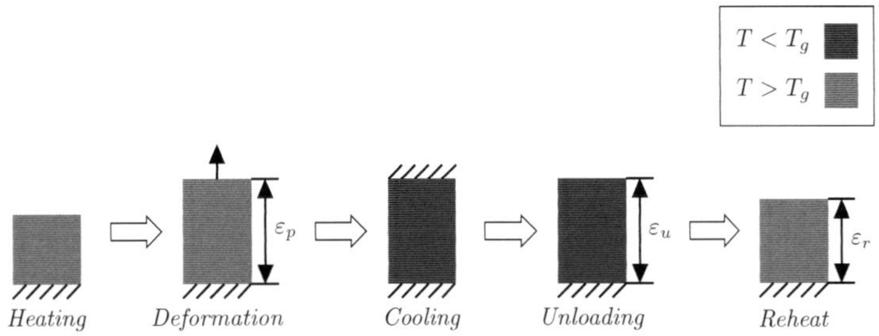

Figure 1. Schematic of the typical thermomechanical cycle for the activation of the shape Memory Effect.

Typically, the cycle is composed of five steps (i.e. heating, deformation, cooling, unloading and reheating) which are obtained by controlling the temperature and the deformation during the test [17]. In particular, in the first and the second step, the specimen is heated up to temperature above the glass transition temperature ($T > T_g$) and deformed to a prescribed value (ε_p), respectively. Thus, in the third step, the specimen is rapidly cooled to a temperature below the glass transition temperature ($T < T_g$) while maintaining ε_p constant. In these three steps a new shape was impressed and learned to the specimen. Then, maintaining the temperature under T_g, the specimen is unloaded until the stress is reduced to zero; note that strain on the specimen can decrease at the end of the step to ε_u. This allows us to compute a measure of the SME's capability to keep the prescribed deformation, i.e. the fixity ratio:

$$R_f = \frac{\varepsilon_u}{\varepsilon_p} \times 100 \tag{1}$$

In the last step, the specimen is heated again, exceeding the T_g with no load applied; in this phase the change of shape occurs and the specimen tends to recover its initial shape. However, at the end of the heating phase, not all of the prescribed deformation is recovered, leading to a residual deformation ε_r. Thereby, the so called recovery ratio can be computed as:

$$R_r = \frac{\varepsilon_p - \varepsilon_r}{\varepsilon_p} \times 100 \tag{2}$$

In this work, a comparison between a equi-biaxial thermomechanical test and a simulation is performed. In particular, a description of biaxial test by means a HBT to determine the behaviour of the SMPs was reported. Through an optical equipment, the specimen shape was recorded during the test, and the DIC technique used to measure the deformation of the specimen, while a transducer detects the inflation pressure. Then, a numerical model on the HBT test and the material used to describe the SMPs mechanical behaviour was described. In order to simulate a SMP the determination of some parameters are necessary to describe the SME of the material. More specifically, uniaxial standard test was conducted to determine Young's modulus and Poisson's ratio above and below the transition temperature. Otherwise a thermomechanical uniaxial test was performed to known the principal properties for describe the SME.

2. Materials and specimens

The material used in this work is a Shape Memory Thermoplastic Polyurethane (SMPU), supplied by MAIP GROUP® in the shape of laminated foils. Both the standard uniaxial specimens (ISO 527-2) [18] and the circular bulge test specimens (with a diameter of 140 mm) were cut from the same 0.4 mm thick foil.

The external stimulus used to activate the SME of the material is the temperature. In order to determine the glass transition temperature (T_g) of the polymer a differential scanning calorimetry (DSC) analysis was performed. According to the procedure described in ASTM D 3418 standard [19], the sample was tested in the range from -20°C to 250°C. The analyses was carried out by means of a Seiko® EXSTAR6000 differential scan calorimeter on a 12 mg of sample cut from the foil. The scanning was conducted by using dry nitrogen having a flow rate of 50 mL min^{-1} during the whole measurement session. The outcomes from the differential scanning calorimetry are reported in Figure 2.

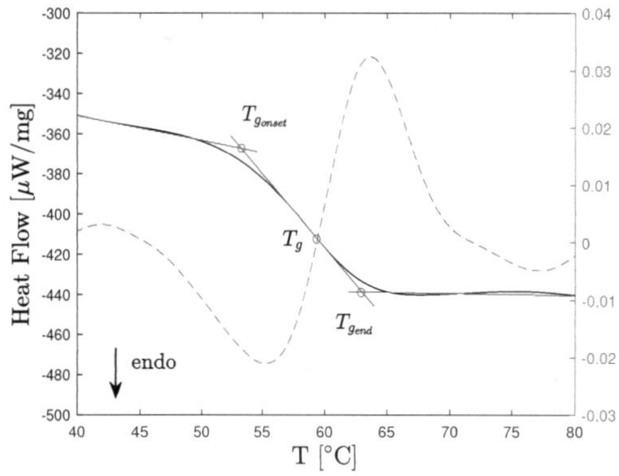

Figure 2. DSC of thermoplastic shape memory polyurethane. Black curve represents the heat flow depending on temperature, dashed curve represents the first derivative.

It can be observed that the transition starts at the onset temperature $T_{g_{onset}}$ equal to 53.30°C and ends at the transition $T_{g_{end}}$ equal to 63.48°C. The glass transition temperature, T_g, has been determined as the inflection point, equivalent to 59.45°C [20].

3. Experimental setup and measurement technique

In order to activate the shape memory nature of the material, the mechanical tests, either uniaxial and biaxial, were conducted by controlling the temperature inside a climatic chamber. In both cases, the average temperature was monitored and recorded by using 3 K-type thermocouples suspended on the specimen. The temperature signals were acquired through National Instruments SCXI 1000DC acquisition system, equipped with Ni 1102C multi-channel signal conditioner module. The monoaxial and thermomechanical tests were performed by using an electromechanical testing machine (Zwick/Roell®Z050), equipped with a 500 N load cell with an accuracy of 0.01 N. On the other hand, the thermomechanical tests in balanced biaxal stress state were performed using a hydraulic bulge test machine. A schematic view of the set-up is reported in Figure 3. To ensure a homogeneous heating on the specimen during the HBT, also the water used for forming the specimen was maintained at temperature above T_g, while the fluid pressure was measured with a pressure transducer with an accuracy of 0.01 bar. The whole deformation history of specimens was recorded during all the tests employing two Pixelink® BU371F cameras (1280 × 1024 pixel^2 8-bit sensor); in this way, the displacement fields – and the derived quantities – on the specimens' surfaces were measured through 3D Digital Image Correlation (DIC) technique. Here, the image analysis has performed by commercial software MatchID® 2021, using a subset size of 29 pixel and step size of 4 pixel for the correlation. Following the ISO/DIS 16808, the equi-biaxial tensile stress has been calculated according to

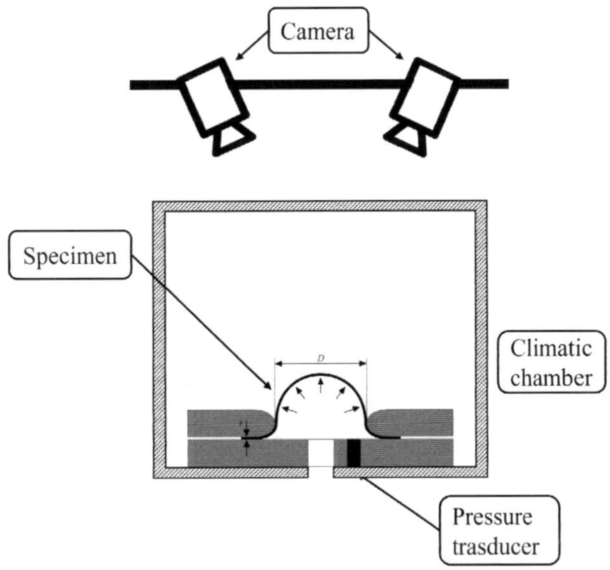

Figure 3. Bulge test set-up

the equation:

$$\begin{cases} \sigma_1 = \sigma_2 = \frac{PR}{2t}, \quad \sigma_3 = 0 \\ R = \frac{1}{k}, \quad k = \frac{1}{2}(k_{xx} + k_{yy}) \\ t = t_0 e^{\delta_3} \\ \delta_3 = ln(\varepsilon_3 + 1) = -2\delta_1 = -2\delta_2 \end{cases} \tag{3}$$

Where P was the pressure of the water, measured by a pressure transducer, D the diameter of the upper flange hole and t_0 the initial thickness of the specimen.

4. Numerical model

In order to simulate the thermomechanical cycle conducted in biaxial stress state, a numerical model of a HBT has builted, by using the finite element simulation analysis software ABAQUS/Standard®. The numerical model of the HBT is composed by two main parts: the specimen and the circular upper die, the latter modelled as analytical rigid surface. The external circumference of both parts is fixed, while pressure loads and temperature have been applied according to the sequence measured during the test illustrated in Figure 4.

The frictional contact between the upper die and the specimen surface is modelled assuming a static frictional coefficient of $\mu_s = 0.35$, typical of polymer-steel interfaces.

The material model used to describe the SME is proposed by Boatti et al. in [21], and implemented in the FE code by means of a material user subroutine (UMAT). This model is based on the temperature-dependent response of SMPs, identifying different regions corresponding to the glassy and rubbery region, respectively above and below the T_g (as depicted in Section 5.1). In particular, the model describes the five steps in Figure 1, and consider allows to employ the parameters describing both the imperfect shape-fixing and incomplete shape-recovery. The HBT information about the geometry and the characteristic of the used FEM model are listed in Table 1.

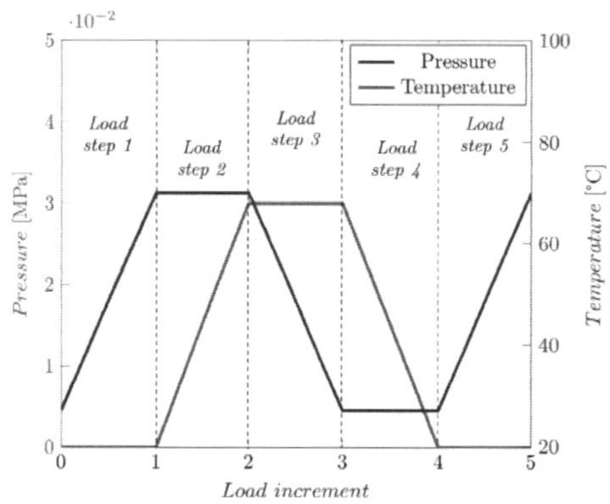

Figure 4. Step thermomechanical simulation

Table 1. Characteristics of the FE model

Geometry of HBT		FEM characteristics	
Blank size	140 mm	Type of element	C3D8 (8-nodes, full-integration)
Die diameter	90 mm	Number of elements	9987
Thickness	0.4 mm		

5. Results and discussion

5.1. Uniaxial tensile test under warm condition

The calibration of the SMP constitutive model requires the elastic properties associated with the glassy and rubbery phases of the SMPU; these data were retrieved from uniaxial tensile tests under controlled temperature. In Figure 5 we report the true stress-strain curves below and above the transition temperature up to 15% of strain.

Here it is possible to observe that at 25°C the material exhibits the typical mechanical response of PU characterized by an elastic phase reaching the upper yield point, followed by a softening behaviour and a constant plateau region; here the DIC measurement highlights the formation on a necking zone, as shown in Figure 5A. Differently, above the glass transition temperature of the material, namely 70°C, the material shows an almost liner behaviour, in accordance with the rubbery nature, without the formation of necking (Figure 5B). The Young's modulus and the Poisson's ratio determined from the tests are reported in Table 2.

Table 2. Region properties

	Glassy Region	**Rubbery Region**
E [MPa]	2460	13
ν	0.40	0.48

5.2. Uniaxial thermomechanical test

The efficiency of the SME of the SMPs in terms of fixity and recovery ratios, has been estimated by performing a uniaxial thermomechanical cycle. In Figure 6 is reported the cycle performed at 70°C and 20% of prescribed strain. In particular, according to (1) and (2), from the curve of the cycle is possible to determine the fixity ratio $R_f = 99.14\%$ and the recovery ratio $R_r = 83.67\%$.

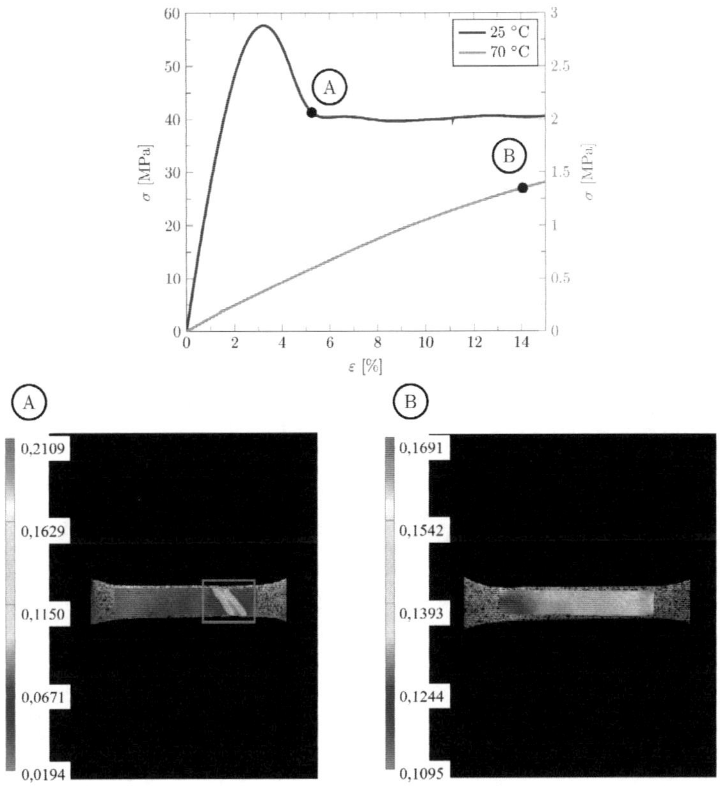

Figure 5. Experimental Stress and Strain curves at two different temperatures with DIC measured strain fields.

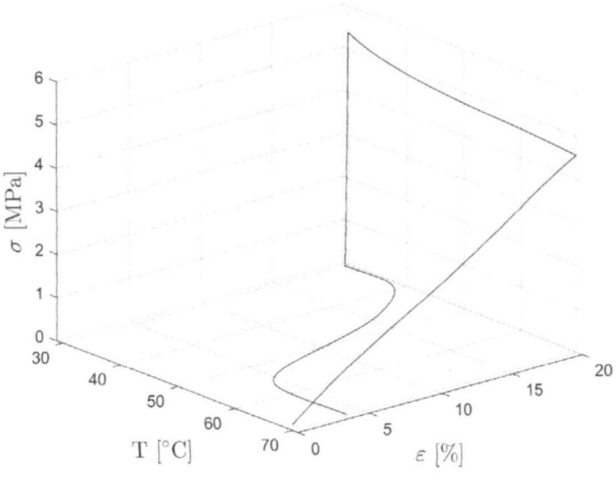

Figure 6. Thermomechanical monoaxial cycle

These data, together with the elastic properties of the glassy and rubbery phases of SMPU retrieved from the pure uniaxial tensile tests, have been used to calibrate the HBT simulation.

5.3. *Comparison between biaxial thermomechanical test and simulation*

Figure 7 shows the comparison between the experimental data and the numerical model of the HBT. The analysis shows that the ultimate strain is significantly lower in the simulated case, indicating that the SME is different from a uniaxial to a biaxial stress condition. In particular, the experimenta thermomechanical cycle in equi-biaxial condition points out a reduction of the Fixity and Recovery ratios, whose values are $R_f = 92.61\%$ and $R_r = 76.35\%$ respectively.

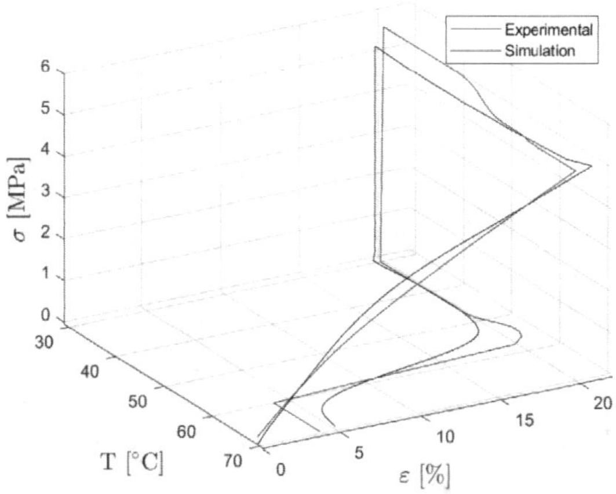

Figure 7. Comparison of thermomechanical biaxial cycle between experimental data and numerical simulation of HBT.

6. Conclusions

In this work, a HBT on a thin membrane of SMP with glass transition temperature at 60°C have been presented, exploiting the DIC technique to the deformation evolution on the specimen surface. The efficiency of the shape memory effect is evaluated by comparing the HBT experimental data with a numerical simulation of the test, whose material coefficients regulating the SMP constitutive model were calibrated from uniaxial tensile thermomechanical tests. The FE simulation provides a consistent replication of the bulge test; however, a reduction of the Fixity and Recovery ratios is observed, suggesting a small influence of the stress state on the SME of the polymer.

References

[1] Lendlein A and Langer R 2002 Biodegradable, elastic shape-memory polymers for potential biomedical applications *Science*, 296(5573):1673–1676
[2] Zhao W, Zhang F, Leng J and Yanju Liu 2019 Personalized 4d printingof bioinspired tracheal scaffold concept based on magnetic stimulated shape memorycomposites *Composites Science and Technology* 184:107866
[3] Liu Y, Du H, Liu L and Leng J 2014 Shape memory polymers and their composites in aerospace applications: a review *Smart Materials and Structures* 23(2):023001
[4] Lan X, Liu Y, Lv H, Wang X, Leng J and Du S 2009 Fiber reinforced shape-memory polymer composite and its application in a deployable hinge *Smart Materials and Structures* 18(2):024002
[5] Chen T, Bilal O R, Lang R, Daraio C and Shea K 2019 Autonomous deployment of a solar panel using elastic origami and distributed shape-memory-polymer actuators *Phys. Rev. Applied* 11:064069
[6] Peng Q, Wei H, Qin Y, Lin Z, Zhao X, Xu F, Leng J, He X, Cao A and Li Y 2016 Shape-memory polymer nanocomposites with a 3d conductive network for bidirectional actuation and locomotion application *Nanoscale* 8:18042–18049
[7] Li W, Liu Y and Leng J 2015 Selectively actuated multi-shape memory effect of a polymer multicomposite *J. Mater. Chem. A* 3:24532–24539
[8] Du H, Liu L, Zhang F, Leng J and Liu Y 2019 Triple-shape memory effect in a styrene-based shape memory polymer: Characterization, theory and application *Composites Part B: Engineering* 173:106905
[9] Wang K, Jia Y G and Zhu X X 2015 Biocompound-based multiple shape memory polymers reinforced by photo-cross-linking *ACS Biomaterials Science & Engineering* 1(9):855–863
[10] Gautrot J E and Zhu X X 2009 Shape memory polymers based on naturally-occurringbile acids *Macromolecules* 42(19):7324–7331
[11] Qi X, Dong P, Liu Z, Liu T and Fu Q 2016 Selective localization of multi-walled carbon nanotubes in bi-component biodegradable polyester blend for rapid electroactive shape memory performance *Composites Science and Technology* 125:38–46
[12] Zhang H, Xia H and Zhao Y 2014 Light-controlled complex deformation and motion of shape-memory polymers using a temperature gradient *ACS Macro Letters* 3(9):940–943
[13] Xu X, Fan P, Ren J, Cheng Y, Ren J, Zhao J and Song R 2018 Self-healing thermoplastic polyurethane (tpu)/polycaprolactone (pcl)/multi-wallcarbon nanotubes (mwcnts) blend as shape-memory composites *Composites Science and Technology* 168:255–262
[14] Du W, Jin Y, Lai S, Shi L, Shen Y and Yang H 2020 Multifunctional light-responsive graphene-based polyurethane composites with shape memory, self-healing, and flame retardancy properties *Composites Part A: Applied Science and Manufacturing* 128:105686
[15] Le X, Lu W, Xiao H, Wang L, Ma C, Zhang J, Huang Y and Chen T 2017 Fe3+-, ph-, thermoresponsive supramolecular hydrogel with multishape memory effect *ACS Applied Materials & Interfaces* 9(10):9038–9044
[16] Tobushi H, Hara H, Yamada E and Hayashi S 1996 Thermomechanical properties in a thin film of shape memory polymer of polyurethane series *Smart Materials and Structures* 5(4):483–491
[17] Atli B, Gandhi F and Karst G 2009 Thermomechanical characterization of shape memory polymers *Journal of Intelligent Material Systems and Structures* 20(1):87–95
[18] ISO 527-2:2012 - plastics — determination of tensile properties
[19] ASTM D3418 - standard test method for transition temperatures of polymers by differential scanning calorimetry
[20] Liang C, Rogers C A and Malafeew E 1997 Investigation of shape memory polymers and their hybrid composites *Journal of Intelligent Material Systems and Structures* 8(4):380–386
[21] Boatti E, Scalet G and Auricchio F 2016 A three-dimensional finite-strain phenomenological model for shape-memory polymers: Formulation, numerical simulations and comparison with experimental data *International Journal of Plasticity* 83:153–177

Corrosion Resistance Behaviour of recycled AlSi10Mg alloy: Surface Morphology and Acoustic Emission Investigation

C Barile[a], C Casavola[a], PK Vimalathithan[a] and G Renna[a]

[a] Dipartimento di Meccanica, Matematica e Management, Politecnico di Bari, Via Orabona 4, 70126 – Bari, Italy

claudia.barile@poliba.it

Abstract. The corrosion resistance behaviour of recycled AlSi10Mg alloy prepared using Selective Laser Melting (SLM) process is investigated. The specimens are exposed to salt solution attack (5% NaCl) atomized in uniform droplets, inside a controlled Salt Spray Test (SST) chamber for 1000 h. The surface morphology of the specimens exposed to different predefined exposure times (0 h, 6 h, 48 h, 168 h, 480 h and 1000 h) are investigated under Scanning Electron Microscope (SEM). The SEM micrographs shows the salient features of the corrosion attack such as the formation both of pits and corrosion products on samples surface in different exposure times in the SST chamber. Similarly, the Acoustic Emission (AE) signals generated during the corrosion process are recorded for the different exposure times. The AE waveforms are studied using advanced waveform processing techniques. The waveforms, in their time-frequency domain, provide detailed information on the characteristic features of the acoustic source. The different AE sources have been characterized from the time-frequency analysis of the waveforms.

1. Introduction

Additive Manufacturing (AM) techniques theoretically eliminates the necessity of chip formation or the requirement of lubrication during the manufacturing process. This enables the conservation of stock material and avoids any toxic chemical wastage as a result of the lubrication. However, the unused and un-melted stock materials during these processes are rarely recycled or reused. Particularly, in laser-based AM processes such as Selective Laser Sintering (SLS) or Selective Laser Melting (SLM), some of the feed stocks are deposited over the heated bed and exposed to the regions of proximity to the high-energy laser source [1-4]. This results in the change in the particle size, density, and morphology, which may affect the reusage of these feed materials [5, 6]. Nonetheless, some research works in the past have shown the successful reuse of the recycled feed materials in the SLM process [7, 8].

In this research work, the corrosion resistance behaviour of the recycled AlSi10Mg powder in the test specimens manufactured using the SLM process is studied. The test specimens prepared from the recycled feed material are exposed to SST chamber for simulating the environment-corrosion process for up to 1000 h. Acoustic Emission (AE) technique is used for monitoring the corrosion progressing the material exposed to the salt solution attack. The stress waves generated from a material due to the irreversible deformation occurring within the material propagates at an ultrasonic velocity. Recording and analysing these stress waves using a piezoelectric transducer forms the basic principle of AE technique [9].

The acoustic waves generated from different sources of corrosion progression such as spallation, formation of corrosion products, and pit formation may differ from each other in terms of their amplitude and frequency content. This study aims in understanding the corrosion behaviour of the AlSi10Mg recycled powder manufactured using SLM process in terms of the time-frequency characteristics of the acoustic waves generated from the test specimens.

2. Experimental Procedure

2.1. Materials and Methods

Recycled AlSi10Mg powder is used as the feed material for preparing dog-bone shaped test specimens using SLM process in RenAM 500 M system. A high-power Nd:YAG laser source of 400 W is traced over the deposited AlSi10Mg powder on the preheated bed to prepare the specimens. The recoater is traced along the Y-axis laser scanned along the X-axis to produce a layer thickness of 20 μm. The test specimens are stress relieved by keeping them in an ambient temperature of 300 °C for about 2 h before air-cooling [10]. The schematic of the test specimen is presented in Figure 1.

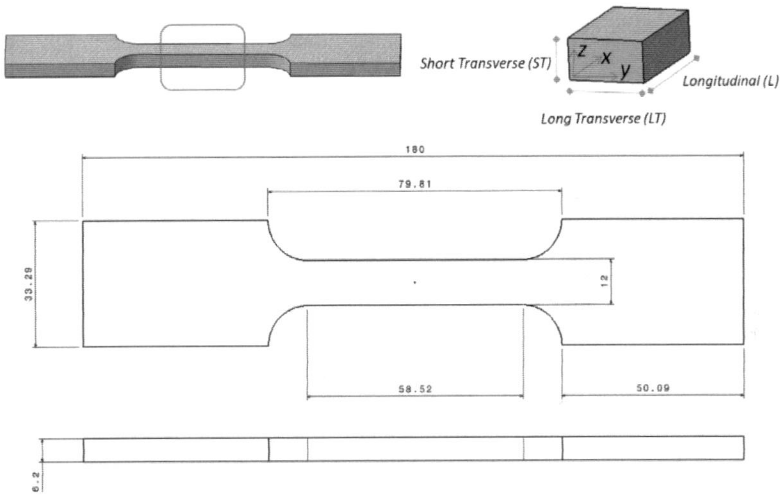

Figure 1. Schematic of test specimen exposed to the SST chamber.

The prepared specimens are placed in a SST chamber (Weiss Technik SC-KWT 450) to simulate the corrosion environment, at temperature of 35 °C. The test is carried out according to the ASTM B117 standard by exposing the test specimens to a 5% NaCl micrometric salt spray with a humidifier temperature and pressure of 48 °C and 0.98 bar, respectively. The exposure times are chosen according to DIN EN ISO 9227 standards: 2, 6, 24, 48, 96, 168, 240, 480, 720 and 1000 hours. After each exposure time, the specimen is taken out, cleaned and then air dried for further analysis. Particularly, the loss mass, density, and surface roughness were evaluated. The mass loss was determined by reweighing and subtracting the panel mass after exposure from its original mass, analogously to the procedure of [11]. The sample weights were evaluated by an analytical balance with 0.1 mg of sensitivity, while the density was calculated using the water displacement method (via the Archimedes method).

The surface roughness of SLS samples post corrosion treatment was measured using a Taylor Hobson Surtronic 25 portable profilometer with a cut-off length of 0.8 mm and a measurement length of 8 mm. Particularly, Ra and Rt roughness parameters were evaluated. Ra is the Roughness Average value that represents the arithmetic average of the absolute values of the roughness profile ordinates, while Rt is Roughness Total height value that indicates the maximum peak to valley height within the evaluation length. The roughness reported was the average of five values scanned from different areas on the coating surface.

Surface morphologies analysis is performed on the Long Transverse-Longitudinal section (LT-L plane) of the test specimens under Zeiss EVO-MA10 scanning electron microscope (SEM).

For recording the AE generation during the exposure of the specimens to the SST chamber, a piezoelectric sensor with a ceramic core, specially designed for harsh environment is used. The sensor ISR 15 is mounted on the specimens inside the SST chamber. A thin layer of silicone grease is smeared to the transducing surface of the sensor to improve the acoustic coupling. The acoustic waves above a detection threshold of 25 dB are recorded, and pre-amplified by 40 dB using a 2/4/6 AE pre-amplifier. The waves are recorded at a sampling rate of 1 MHz.

2.2. Processing of Acoustic Emission Waves

The recorded acoustic waves are studied in their time-frequency domain using Continuous Wavelet Transform (CWT). Analytical Morlet wavelet with a symmetry parameter of 3, with 3 octaves and 32 voices per octave is used. This wavelet is from a family of analytical wavelets, whose Fourier transform vanishes at negative frequencies. The analytical wavelets are most suitable for CWT, particularly for the group of nonstationary waves such as acoustic waves.

3. Results and Discussions

3.1. Corrosion Behaviour

Salt spray tests were performed in order to understand the corrosion behaviour of the SLS specimen made from recycled AlSi10Mg powder. Corrosion progression was assessed by mass loss and surface roughness measurements, microscopical examination and Acoustic Emissions (AE).

Figure 2 illustrates the ratio between the mass loss due to the corrosive attack and the sample surface exposed to the aggressive environment (mg/cm^2 in Figure 2a), and density values of the specimen during salt spray tests (Figure 2b). It should be observed that the mass loss of the sample was low and almost constant up to 24 h, after this exposure time the trend of mass loss was linearly-decreasing with increasing exposure time (Figure 2a). This decreasing mass loss is related with a weight gains of sample due to an important progressive formation of corrosion products from this time on. It is interesting to note that sign inversion of mass loss (from loss to mass gains) occurs between 24 and 48h of exposure time. As can be observed, following 48 h the mass gain was about 78% after 168 h and 98% after 1000 h.

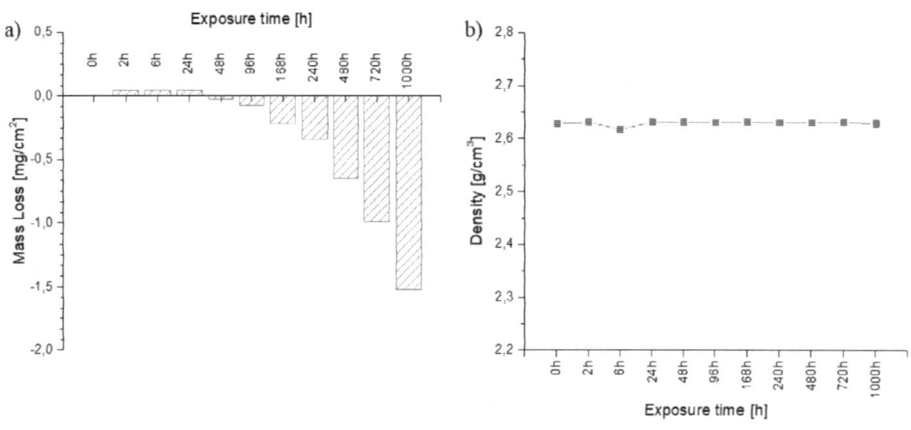

Figure 2. Analysis of the effect of exposure to a controlled corrosive environment of the SLS specimen made from recycled AlSi10Mg powder: a) mass loss (mg/cm^2) and b) density value (g/cm^3) as function of exposure time (h)

Therefore, SLM sample fabricated using recycled powder is characterized by a low surface defectiveness during the first 24 h, exhibiting a corrosion rate almost constant; however, following this exposure time the sample displayed a decrease in corrosion rate as exposure time increased. Besides, by looking at Figure 2b it is worthwhile to note that density values of the samples remained constant with increasing exposure time in SST chamber, indicating that no occurrences are significantly interconnected to the formations and/or changes of open and closed porosity, which could lead to a reduction in physical properties.

Figure 3 shows the Ra and Rt surface roughness of sample as a function of the exposure time in SST chamber. The graph revealed that the Ra and Rt surface roughness values fluctuate slightly with the increasing exposure time. This behaviour is likely due to some simple considerations related to the corrosion phenomenon. During the accelerated aging treatment there is a concomitance of two mechanisms. One, there may be the formation of new micro-porosities (pitting) emerging on the surface as well as an increase in existing ones; on the other hand, the formation of corrosion products. In agreement with literature, the pits on the surface, for the longer exposure times, may also be partially filling by means of material products, whose morphological characteristics depend largely on the corrosive atmosphere that generated them [12, 13], leading to a slight decrease in the Rt roughness.

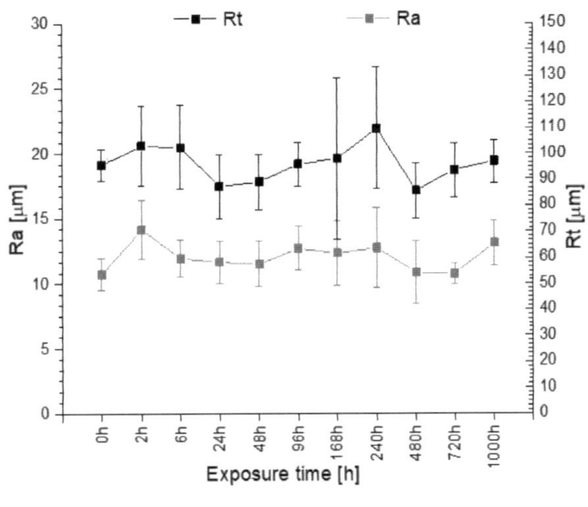

Figure 3. Ra (Roughness Average value) and b) Rt (Roughness Total height value) with respect to exposure time interval to salt spray test.

A careful analysis of the morphological features of the SLM specimen exposed to different duration of the exposure time are depicted in Figure 4. As can be observed, the surface morphology of the as-received SLM sample (before aging - 0 h) was characterized by rough texture resulted mainly from partially melted powder adhering to the surface (balling phenomena). In agreement with literature [6-14], the surface of the sample subjected to SST were characterized by local corrosion: usually manifested by random formation of pits. Specifically, randomly shaped and round pits of small-sizes and shallow could be observed on the surfaces during the first 48 h (see the high-resolution SEM images in Figure 4b and 4c). The corrosion progressed along with exposure time, and simultaneous pits formation on the surface of the samples and refilling of the same by corrosion products occurred. Particularly, it was observed that, following the 48 h of exposure time, the samples surfaces were mainly characterized by corrosion products. After 480 h the surface appears to be more covered by corrosion products with different sizes and morphology. At the end of tests (1000 h), high magnification analysis indicated that the corrosion pits were substantially enlarged and became slightly more marked, inside which it places the corrosion products (Figure 4f). Therefore, morphological analysis using SEM confirmed the remarks of the mass loss results and surface roughness measures.

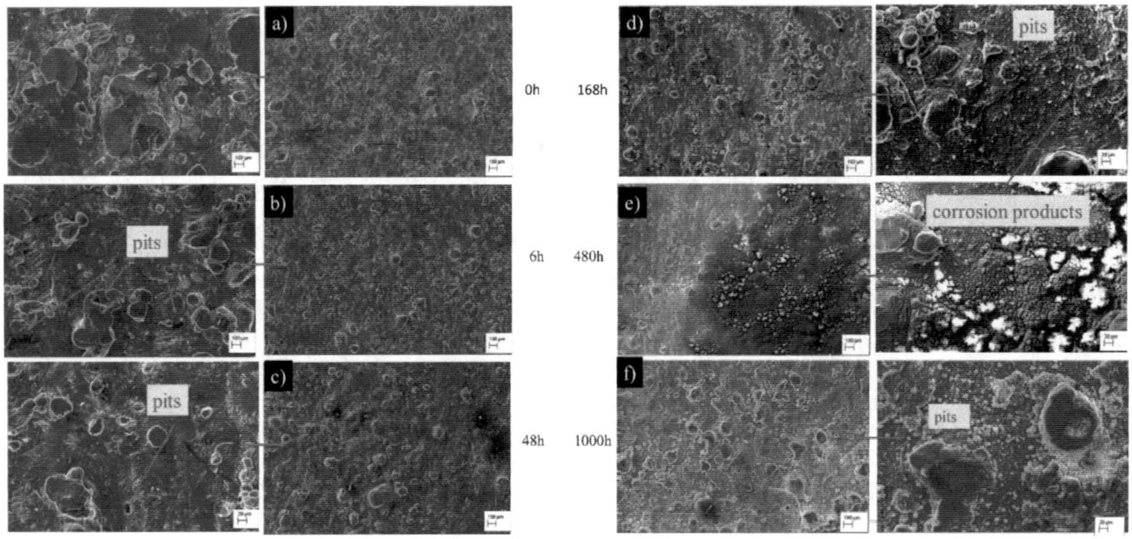

Figure 4. Secondary SEM images of sintered specimen surfaces following 0, 6, 48, 168, 480 and 1000 h of salt spray tests.

3.2. *Acoustic Emission Results*

Acoustic waves generated during the progression of the corrosion in the AlSi10Mg test specimens are discussed in this section. From the morphological analysis in Section 3.1, it is evident that the corrosion products and pits formed at different exposure times. The formation of these pits and the corrosion products lead to the emission of stress waves, which is propagated through the specimen at different frequencies. The velocities of these waves and the mode of propagation also differs based on their sources. Therefore, differences in their time-frequency characteristics can be observed.

The acoustic waves generated from the different exposure times are analysed in their time-frequency domain. First, during the initial stages of corrosion (between 0 to 6 h), two specific group of acoustic waves are being observed. The time-frequency characteristics of the representative waves from these groups are analysed using CWT results and are presented in Figures 5a and 5b. The first wave in Figure 5a has a frequency centred around 275 kHz to 300 kHz but it has a very short duration. Besides, this wave vanishes after 2800 sample duration. The second wave in Figure 5b, however, has a longer duration of about 7500 samples. Besides, the frequency at which the magnitude maximum magnitude observed is very closely localized to 300 kHz and this magnitude reduces as the reverberations can be observed over the duration before vanishing entirely at 7500 samples.

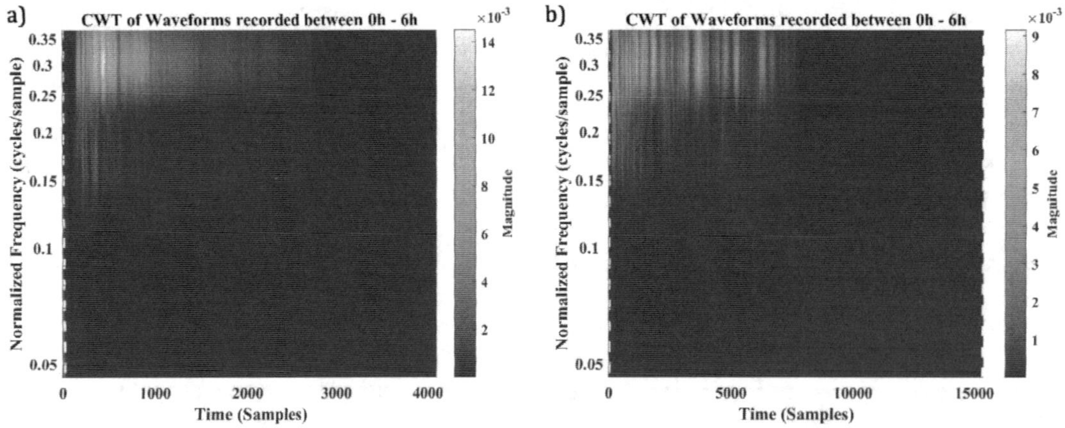

Figure 5. CWT of Waveforms recorded during 0h to 6h duration.

IOP Conf. Series: Materials Science and Engineering 1214 (2022) 012037 doi:10.1088/1757-899X/1214/1/012037

The CWT of the acoustic waves generated during the initial pit formation around 48h exposure time are presented in Figures 6a and 6b. During the pit formation, two more distinct group of acoustic waves are observed. The first acoustic wave in Figure 6a has the frequency with the maximum magnitude around 250 kHz to 300 kHz, similar to the waves from Figure 5, however, it is very short in duration, amounting for about only 700 samples before vanishing. The second group of the acoustic wave in Figure 6b also has the frequency localized around 250 kHz to 300 kHz but some reverberations can be observed up to 1500 samples. First, there is a clear difference between the waveforms during the initial stages of exposure (in Figure 5) and during the pit formation. Although the frequency localization of these waveforms is around 250 kHz to 300 kHz, they vary significantly in terms of their time localization and reverberations.

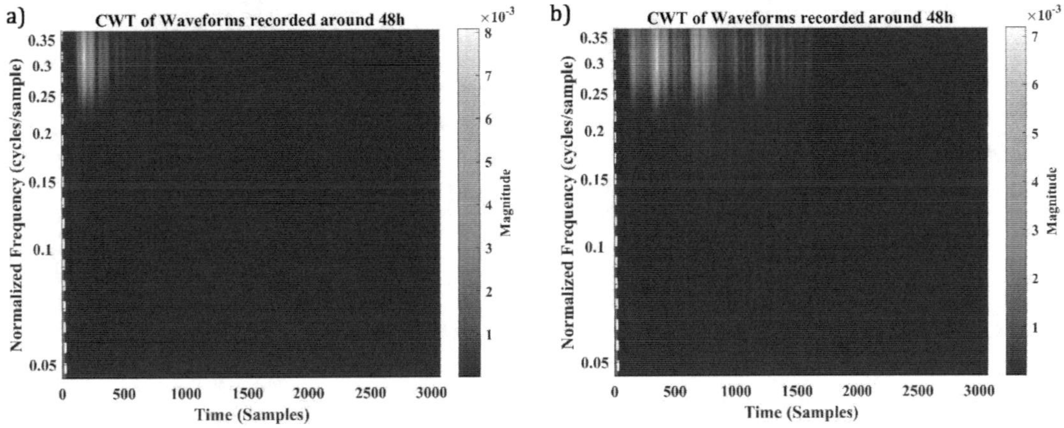

Figure 6. CWT of Waveforms recorded around 48h duration.

Corrosion products have started to form following 48 h and these grow with increasing exposure time, as mentioned in Section 3.1. The CWT of the acoustic waves recorded during 168-480 h exposure time is presented in Figure 7. Interestingly, during this stage, only one group of acoustic waves are recorded. Two examples of this group of acoustic waves are presented respectively in Figures 7a and 7b. The frequency of these acoustic waves is also localized between 250 kHz to 300 kHz but there are two significant differences between these waves and the acoustic waves during the initial stages and pit formation. The frequency localization of these acoustic waves is much shorter in duration and the magnitude of the frequency is much higher. The acoustic waves from the pit formation in Figures 6a and 6b have a maximum magnitude of 7 x 10^{-3}, but the magnitude of waves during the apparent corrosion product formation is between 2.5 x 10^{-2} and 4.5 x 10^{-2}.

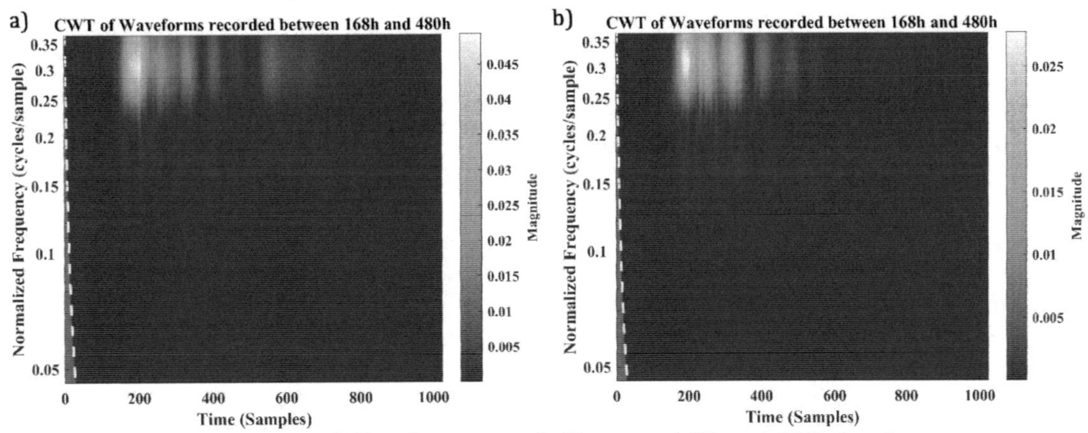

Figure 7. CWT of Waveforms recorded between 168h and 480h duration.

While analysing the acoustic waves at the end of the corrosion exposure, between 720h and 1000 h, formation of corrosion products is observed. Theoretically, the CWT of the acoustic waves observed in this duration must be similar to the acoustic waves in Figure 7. Coincidentally, the frequency localization and the maximum magnitude of the frequency of the acoustic waves generated between 168h and 480h, and 720h and 1000h are the same. The CWT results of the acoustic waves from the end of corrosion exposure is presented in Figure 8. The results show that the maximum magnitude of the frequency around 300 kHz is between 2.2×10^{-2} and 3.3×10^{-2}. This can lead to the conclusion that there is a significant similarity in the frequency localization and the magnitude of the acoustic waves from the formation of corrosion products between different duration.

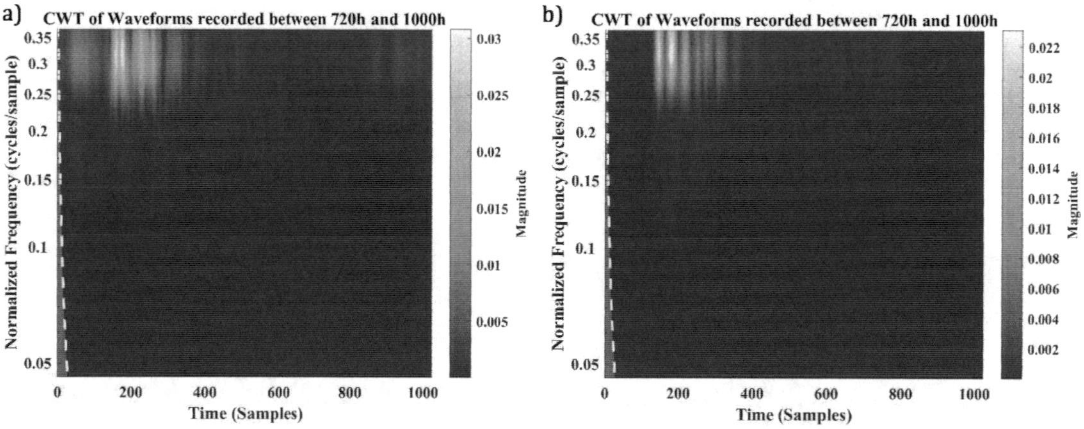

Figure 8. CWT of Waveforms recorded between 720h and 1000h duration.

A general conclusion made from these results about the time-frequency characteristics, the magnitude of the localized frequency and the different sources of the corrosion progression. Pit formation generally generates acoustic waves with low magnitude and longer duration with the frequency localized around 250 kHz and 300 kHz. There are also a lot of reverberations observed in these waves. The formation of corrosion products also generates acoustic waves in the same frequency range but having much larger amplitude and very short duration. Deeper analysis on this time-frequency characteristics can be helpful in identifying the amount of pit and corrosion product formation during the exposure of the recycled AlSi10Mg powder to the controlled corrosive environment.

4. Conclusion

The corrosion behaviour of the AlSi10Mg manufactured using SLM process from the recycled feed material is studied. In general, the corrosion tests performed on the specimens show good corrosion resistance. The surfaces of the samples subjected to SST are characterized by local corrosion. The morphological analysis on the test specimens at different exposure time shows the formation of pits and corrosion products. Particularly, samples are characterized by the low surface defectiveness during the first 24 h, exhibiting a corrosion rate almost constant, following this exposure time the samples displayed a decrease in corrosion rate as exposure time increased due to an important progressive formation of corrosion products from this time on.

Corrosion pits started to form before 48h, and the acoustic waves generated during this same duration have low magnitude, longer duration and frequency localized between 250 kHz and 300 kHz. Formation of other corrosion products occur around 168h to 480h of exposure and also during the final stages of exposure up to 1000 h. The acoustic waves generated during these periods shows a distinct feature of having larger magnitude and higher duration with the frequency around 300 kHz. Therefore, it can be concluded that the corrosion progressions such as formation of pits and corrosion products

IOP Conf. Series: Materials Science and Engineering 1214 (2022) 012037 doi:10.1088/1757-899X/1214/1/012037

can be differentiated using the acoustic emission. The corrosion behaviour, in general, of the recycled AlSi10Mg specimens can be characterized qualitatively using this technique.

References

[1] Asgari H, Baxter C, Hosseinkhani K and Mohammadi M 2017 On microstructure and mechanical properties of additively manufactured AlSi10Mg_200C using recycled powder. *Mater Sci Eng-A* **707** 148-158

[2] Kim DK, Hwang JH, Kim EY, Heo YU, Woo W and Choi SH 2017 Evaluation of the stress-strain relationship of constituent phases in AlSi10Mg alloy produced by selective laser melting using crystal plasticity FEM *J Alloy Compd* **714** 687-697.

[3] Dehoff RR, Tallman C, Duty CE, Peter WH, Yamamoto Y, Chen W and Blue CA 2013 Case study: additive manufacturing of aerospace brackets *Mater Sci Forum* **171(3)**.

[4] Yusuf SM, Hoegden M and Gao N 2020 Effect of sample orientation on the microstructure and microhardness of additively manufactured AlSi10Mg processed by high-pressure torsion *Int J Adv Manuf Tech* **106(9)** 4321-37.

[5] Tang HP, Qian M, Liu N, Zhang XZ, Yang GY and Wang J 2015 Effect of powder reuse times on additive manufacturing of Ti-6Al-4V by selective electron beam melting *Jom* **67(3)** 555-563.

[6] Rafieazad M, Chatterjee A and Nasiri AM 2019 Effects of recycled powder on solidification defects, microstructure, and corrosion properties of DMLS fabricated AlSi10Mg *Jom* **71(9)** 3241-52.

[7] Ardila LC, Garciandia F, González-Díaz JB, Álvarez P, Echeverria A, Petite MM, Deffley R and Ochoa J 2014 Effect of IN718 recycled powder reuse on properties of parts manufactured by means of selective laser melting *Phys Proc* **56** 99-107.

[8] Maamoun AH, Elbestawi M, Dosbaeva GK and Veldhuis SC 2018 Thermal post-processing of AlSi10Mg parts produced by Selective Laser Melting using recycled powder *Additive Manuf* **21** 234-247.

[9] Barile C, Casavola C, Pappalettera G and Vimalathithan PK 2020 Application of different acoustic emission descriptors in damage assessment of fiber reinforced plastics: A comprehensive review *Eng Fract Mech* ***235*** 107083.

[10] Barile C, Casavola C, Pappalettera G and Vimalathithan PK 2020 Acoustic emission descriptors for the mechanical behavior of selective laser melted samples: An innovative approach *Mech Mater* **148** 103448.

[11] ASTM International B 117, Standard Practice for Operating Salt Spray (Fog) Apparatus, (2003).

[12] Barile C, Casavola C, Campanelli SL and Renna G 2019. Analysis of corrosion on sintered stainless steel: mechanical and physical aspects *Eng Fail Anal* **95** 273-282.

[13] Liu C, Leyland A, Bi Q and Matthews A 2001 Corrosion resistance of multi-layered plasma-assisted physical vapour deposition TiN and CrN coatings *Surf Coat Tech* **141(2-3)** 164-173.

[14] Revilla RI, Liang J, Godet S and De Graeve I 2016 Local corrosion behavior of additive manufactured AlSiMg alloy assessed by SEM and SKPFM *J Electrochem Soc* **164(2)** C27.

Adhesion of viscoelastic media: an assessment of a recent JKR-like solution

G Violano [1], G Orlando [1], G P Demelio [1] and L Afferrante [1]

Polytechnic University of Bari, Department of Mechanics, Mathematics and Management, Via E. Orabona 4, Bari, 70125, Italy

E-mail: guido.violano@poliba.it

Abstract. Adhesion of soft compliant solids is irreversible and rate-dependent. As a result, two different paths are observed in loading-unloading adhesion experiments because of dissipation occurring in the unloading phase. An effective surface energy is usually introduced to take account of such dissipation.

Here, by exploiting a recent theoretical solution developed to study the detachment of a rigid sphere from a viscoelastic substrate (Violano et al., 2021), two different approaches are considered to calculate the surface energy. The first approach is based on the phenomenological equation derived by Gent & Schultz (A. N. Gent & J. Schultz, 1972), the latter exploits Persson & Brener theory for viscoelastic crack propagation (B. N. J. Persson & E. A. Brener, 2005).

In both cases, results are observed to be in good agreement with experimental data taken from the literature.

1. Introduction

"*Adhesion of viscoelastic solids is irreversible and rate dependent. It presents a difficult problem in contact mechanics and is an ongoing field of research*". This is an excerpt from a private letter - dated November 18, 2003 - written by K.L. Johnson. The recipient of the letter (V.L. Popov), recently presented it in the form of a note concerning the history of JKR theory [1].

Although eighteen years have passed since Johnson's letter, adhesion of viscoelastic solids is still an open challenge for experts in contact mechanics and tribology. This is clearly demonstrated by recent efforts in experimental [2, 3], numerical [4, 6, 5] and theoretical works [7, 8].

In the contact between elastic spheres, Johnson, Kendall & Roberts (JKR) theory [9] predicts a pull-off force (i.e., the tensile load required for detaching the spheres) $F_{\text{PO}} = 1.5\pi\Delta\gamma R$, being R the equivalent radius of curvature of the spheres and $\Delta\gamma$ the adiabatic value of the surface energy. However, experiments on soft materials (e.g., rubbers [10], elastomers [11], organic tissues [12]) have shown that the pull-off force increases by enhancing the unloading rate. As a result, the effective surface energy $\Delta\gamma_{\text{eff}}$ observed in experiments is higher than its adiabatic value. Such behavior is related to rheological properties of soft matter.

In classical theoretical solutions [13, 14, 15], the debonding of a rigid indenter from a viscoelastic half-space is studied in the framework of Linear Elastic Fracture Mechanics (LEFM). The edge of contact line is seen as the tip of a propagating crack. Gent & Schultz (GS) [16] and Maugis & Barquins [13] (MB), first introduced phenomenological expressions to relate the effective surface energy with the crack tip velocity V_c. In this class of theoretical models, two common assumptions are done: i) debonding starts from a complete relaxed state of the material; ii) viscous dissipation is localized exclusively close to the crack tip.

Such assumptions are relaxed in recent numerical solutions [4, 6, 5]. Indeed, we developed a finite-element algorithm for studying the adhesion between a rigid sphere and a viscoelastic substrate [4]. When unloading starts from a relaxed state of the material, the pull-off force (and hence $\Delta\gamma_{\rm eff}$) is found to monotonically increase with $V_{\rm c}$ up to an asymptotic value reached at high unloading velocities. On the other hand, if unloading starts from an unrelaxed state of the material, the relation between pull-off force and $V_{\rm c}$ takes the shape of a bell curve in a double-logarithmic plot.

Despite their potentiality and accuracy, numerical methods may require significant time and computational costs. For this reason, in a recent work [17], we proposed an extension of JKR theory to the contact between viscoelastic spheres. Such theoretical solution is here compared with the experimental outcomes of Ahn & Shull (AS) [11] by exploiting the phenomenological equation of Gent & Schultz to describe the dependence of $\Delta\gamma_{\rm eff}$ on $V_{\rm c}$. In addition, estimating $\Delta\gamma_{\rm eff}$ from the viscoelastic crack propagation theory of Persson & Brener [19], a second comparison with the experimental data of Lorenz et al. [18] is proposed. Results show, in both cases, the model is quite in good agreement with reference data.

2. JKR-like viscoelastic model

The problem under investigation is sketched in Fig. 1.

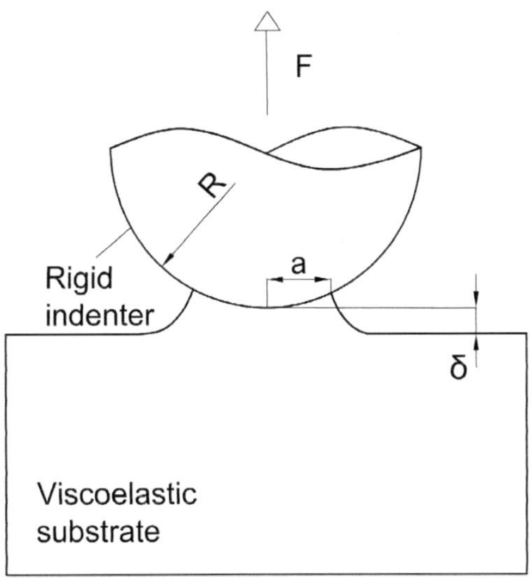

Figure 1. The problem under investigation: a rigid spherical indenter pulled apart from a viscoelastic half-space.

The model developed in Ref. [17] assumes the half-space behaves as a linear elastic medium during the loading phase. In other words, loading is described by the classical JKR theory. Such condition can be reproduced in adhesion experiments by i) performing the loading phase at small driving velocity of the indenter [20], ii) waiting a sufficient long dwell time before unloading as complete relaxation of the viscoelastic substrate must be reached [2]. JKR theory [9] returns the applied force F and the penetration δ as functions of the contact radius a:

$$F = \frac{4}{3}\frac{E^* a^3}{R} - \sqrt{8\pi E^* a^3 \Delta\gamma} \tag{1}$$

$$\delta = \frac{a^2}{R} - \sqrt{\frac{2\pi a \Delta\gamma}{E^*}}, \tag{2}$$

being E^* the plain strain elastic modulus.

In order to extend JKR theory to viscoelastic contacts, we suggested (Ref. [17]) to replace the adiabatic surface energy $\Delta\gamma$ with its effective value $\Delta\gamma_{\text{eff}}$ in eqs. (1-2).

To estimate the effective surface energy we can exploit the phenomenological equation introduced by Gent & Schultz [16] and Maugis & Barquins [13]:

$$\Delta\gamma_{\text{eff}}(a) = \Delta\gamma[1 + (V_c(a)/V^*)^n], \tag{3}$$

where the viscoelastic parameters V^* and n can be measured experimentally [13, 2].

Alternatively, when the dependence of the viscoelastic modulus on the frequency of excitation is known, we can use Persson & Brener theory for viscoelastic cracks [19], according to which

$$\Delta\gamma_{\text{eff}}(a) = \Delta\gamma \left[1 - \frac{2E_0}{\pi}\int_0^{2\pi V_c(a)/s} d\omega \frac{F(\omega)}{\omega} Im\left(\frac{1}{E(\omega)}\right)\right]^{-1} \tag{4}$$

where $F(\omega) = [1 - (\omega s/(2\pi V_c))^2]^{1/2}$. In eq. (4), $s(V_c)$ is the crack tip radius and $E(\omega)$ is the viscoelastic modulus, being $E_0 = E(0)$ its value at vanishing excitation frequencies ω. Moreover, PB showed that $s(V_c)/s_0 = \Delta\gamma_{\text{eff}}/\Delta\gamma$, where s_0 is the crack tip radius at vanishing contact line velocity and takes values of the order of 1 nm.

The contact line velocity is related to the contact radius a by $V_c(a) = -da/dt$. If we assume that unloading occurs under fixed pulling velocity $V = -d\delta/dt$, then V_c can be written as $V_c = V \cdot da/d\delta$.

To estimate $da/d\delta$, we initially assume constant $\Delta\gamma$ in eq. (2), so that the contact line velocity can be derived as

$$V_c(a) = V\left(\frac{2a}{R} - \sqrt{\frac{\pi\Delta\gamma}{2aE^*}}\right)^{-1}. \tag{5}$$

Substituting the above relation, in eq. (3) or (4) one can calculate the effective surface energy $\Delta\gamma_{\text{eff}}$, which in turn can be used in JKR eqs. (1-2) to obtain force and penetration in the viscoelastic case.

Experiments on soft materials [2, 21, 22] show the existence of a *stick zone* during the initial phase of debonding. Therefore, according to Ref. [17], in the initial stick phase, we assume the spherical indenter behaves as a circular flat punch of radius equal to the value $a_{\max}$ reached at the end of loading. As a result, we have

$$\delta = \delta_{\max} - \frac{F_{\max} - F}{2E^* a_{\max}}. \tag{6}$$

being, $F_{\max}$ and $\delta_{\max}$ the force and the penetration at the beginning of unloading, respectively.

Eq. (6) holds for $\delta_c \leq \delta \leq \delta_{\max}$, being δ_c the critical penetration over which a starts to decrease. On the contrary, for $\delta < \delta_c$, contact is described by eqs. (1-2) with $\Delta\gamma$ replaced by $\Delta\gamma_{\text{eff}}$.

3. Results

A first comparison of the predictions of the described model is proposed with the experimental data collected by Ahn & Shull (AS) [11], which carried out JKR adhesion tests between a model acrylic elastomer and a glassy polymeric substrate. Specifically, hemispherical lenses of lightly cross-linked PolyN-ButylAcrylate (PNBA) with radius of curvature $R = 1$ mm were subjected to loading/unloading cycles on flat PolyMethylMethAcrylate (PMMA) substrates.

In AS experiments, loading phase is performed by increasing the load step by step and waiting a fixed time at the end of each step. Moreover, before unloading, the maximum load $F_{\max}$ is maintained constant for a dwell time of 10 min. In such case, quasi-static conditions are ensured (see Ref. [23] for explanation) and JKR solution can be used to estimate the elastic modulus and the adiabatic surface energy (for AS experiments we find $E^* = 0.17$ MPa and $\Delta\gamma = 22$ mJ/m^2). During unloading, the contact line velocity $V_c = -da/dt$ is obtained by capturing pictures of the contact area at fixed frame rates.

Figure 2 shows the trend of V_c with the contact radius a, for different pulling velocities $V = -d\delta/dt$. Markers refer to experimental data, while solid lines are the approximate solutions returned by eq. (5). Unfortunately, the values of V are not reported in Ref. [11], so we have used V as a fitting parameter in eq. (5). Moreover, we have assumed that unloading occurs at constant V.

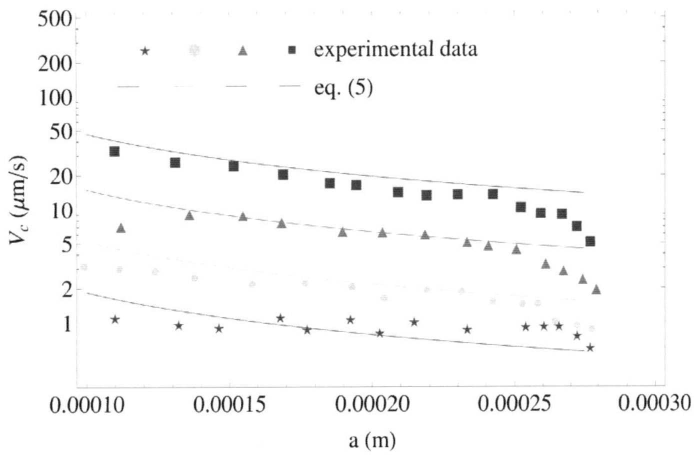

Figure 2. The contact line velocity V_c as a function of the contact radius a. Markers refer to experimental data from Ref. [11], while solid lines denote theoretical predictions of eq. (5). Results are shown for different pulling velocities ($V = 0.2$, 0.8, 2.3, 7.2 μm/s).

Following Maugis & Barquins (MB) [13], AS calculated the energy release rate G with

$$G = \frac{(F_H - F)^2}{6\pi R F_H}, \tag{7}$$

where $F_H = 4E^*a^3/(3R)$ is the Hertzian load and F is the applied load (known from experiments).

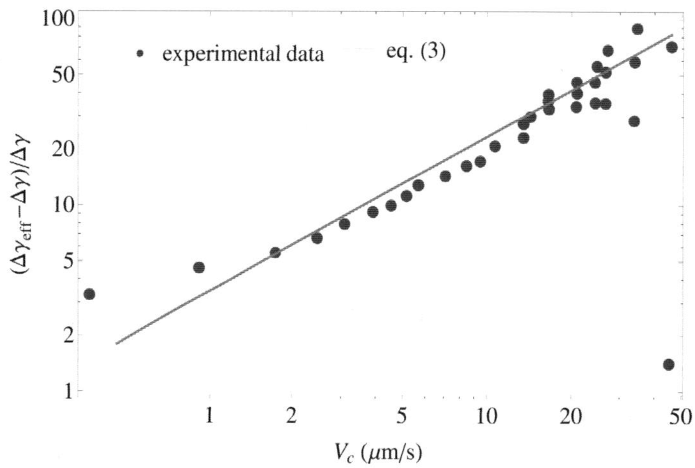

Figure 3. The relative increase in the effective surface energy $(\Delta\gamma_{\text{eff}} - \Delta\gamma)/\Delta\gamma$ in terms of the contact line velocity V_c. Markers refer to experimental data from Ref. [11], while solid line is the fit obtained by means of eq. (3) with $V^* = 2.24 \times 10^{-7}$ m/s and $n = 0.83$.

In the framework of LEFM [13], at equilibrium $G = \Delta\gamma_{\text{eff}}$. Therefore, experimental data can be used to fit the values of the parameters V^* and n in eq. (3), as shown in Fig. 3, where the relative increase in the effective surface energy $(\Delta\gamma_{\text{eff}} - \Delta\gamma)/\Delta\gamma$ is given in terms of the contact line velocity V_c. Again, markers refer to experimental data, while solid line is the fit obtained by eq. (3), which returns $V^* = 2.24 \times 10^{-7}$ m/s and $n = 0.83$.

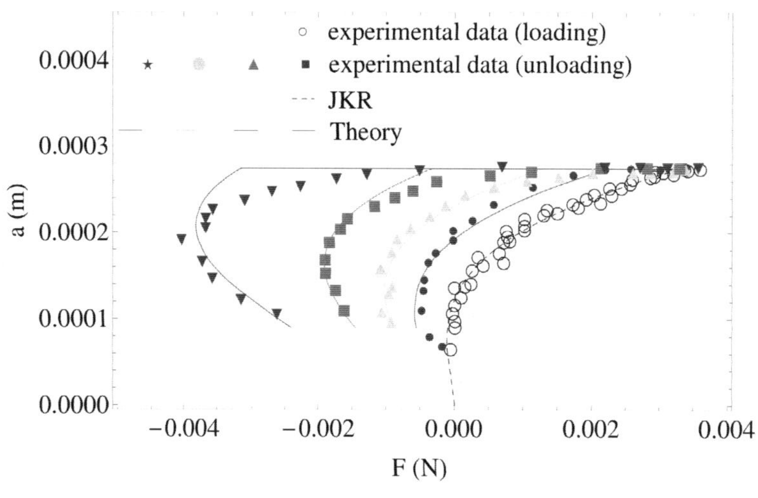

Figure 4. The contact radius a in terms of the applied load F. Markers refer to experimental data from Ref. [11], while lines denote theoretical predictions. Results are shown for different pulling velocities ($V = 0.28$, 0.8, 2.3, 7.2 μm/s).

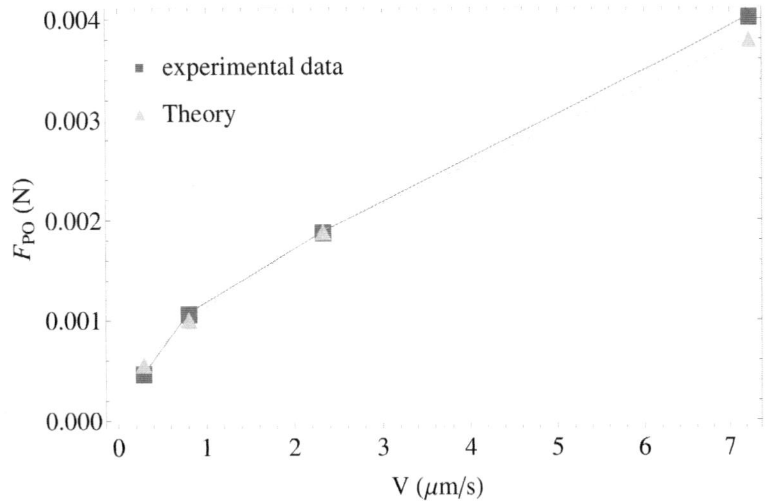

Figure 5. The pull-off force F_{PO} as a function of the pulling velocity V. Red squares refer to experimental data from Ref. [11], while green triangles are theoretical predictions.

Figure 4 shows the curves relating the contact radius a with the contact force F for increasing pulling velocities V. Black empty circles are the experimental data collected in the loading phase, for which viscoelastic effects are negligible. JKR elastic theory (black dashed line) is clearly working in this case. Colored markers refer instead to the experimental data collected during unloading at different V. Adhesion hysteresis occurs as loading and unloading paths are different and the stick zone is observed to increase with the pulling velocity. Results of the theoretical model (colored solid lines) match quite well the experiments, even if at high pulling rates the stick time is slightly overestimated.

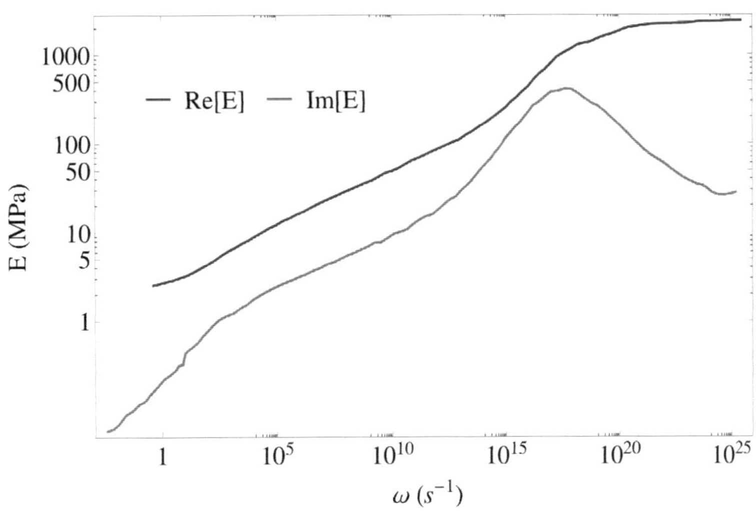

Figure 6. The real and imaginary part of the viscoelastic modulus $E(\omega)$ of the PDMS used in the experiments of Ref. [18].

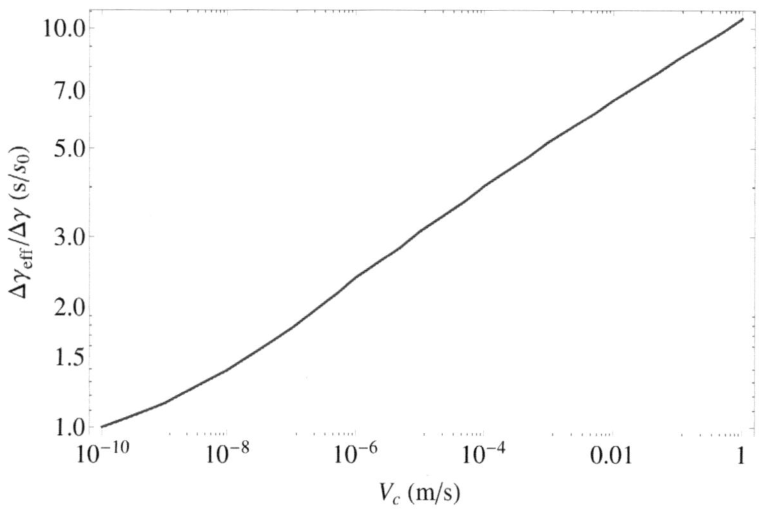

Figure 7. The effective surface energy $\Delta\gamma_{\rm eff}/\Delta\gamma$ in terms of the contact line velocity $V_{\rm c}$ for the PDMS sample used in Ref. [18]. Equation (4) has been used for the calculation of $\Delta\gamma_{\rm eff}$.

The pull-off force, i.e., the maximum tensile (negative) load measured during unloading, can be calculated by solving the equation $dF/da = 0$. In figure 5 we compare the measured pull-off force with the theoretical predictions, for increasing pulling velocity V. In such case, a very good agreement is observed also at relatively high rates.

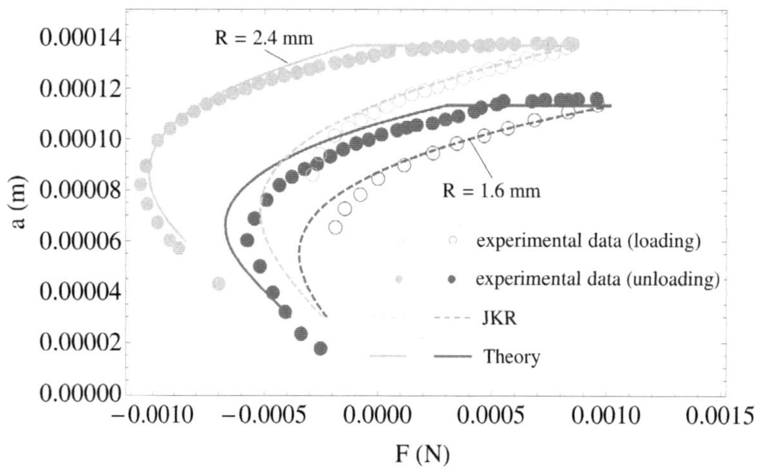

Figure 8. The contact radius a in terms of the applied load F. Markers refer to experimental data from Ref. [18], while lines denote theoretical predictions. Results are shown for different radius of curvature $R = 1.6$, 2.4 mm and pulling velocity $V = 0.1$ μm/s.

The theoretical predictions of the model are also compared with the experimental findings given in Lorenz et al. [18], where an optical microtribometer is used to perform loading-unloading experiments between polished silicon nitride spheres and a flat PolyDiMethylSiloxane (PDMS) elastomer. Lorenz et al. [18] also measured the viscoelastic modulus $E(\omega)$ of PDMS in a wide range of frequencies of excitation (see Fig. 6). So, the effective surface energy can be calculated by eq. (4), and its dependence on contact line velocity V_c is shown in Fig. 7.

Figure 8 shows the loading-unloading curves relating a and F for two values of the radius of the spheres used in the experiments (namely $R = 1.6$ mm and $R = 2.4$ mm). Loading data (empty markers) are fitted using JKR theory (dashed lines), which returns $E^* = 2.55$ MPa and $\Delta\gamma = 46$ mJ/m^2. Unloading (filled markers) is assumed to be performed at $V = 0.1$ μm/s according to Ref. [24]. Once again, the theoretical predictions (solid lines) closely match the experimental data. Also, notice an increase in the sphere radius results in an increase of the dissipated energy (wider area between the loading and unloading curves) as the region involved by viscous effects becomes larger.

4. Conclusions

In this work, JKR-like experiments performed between soft bodies are interpreted in the light of a very simple theoretical model, where viscous effects are considered by replacing in JKR equations the adiabatic surface energy $\Delta\gamma$ with the effective value $\Delta\gamma_{\text{eff}}$ which takes account of dissipation.

First, the predictions of the model are compared with the experimental findings of Ref. [11]. In such case, the effective surface energy is calculated by the phenomenological equation proposed by Gent & Schultz (GS) [16] and Maugis & Barquins [13].

A second set of experimental data, taken from Lorenz et al. [18], is then used to test the model when $\Delta\gamma_{\text{eff}}$ is estimated according to Persson & Brener theory for viscoelastic cracks [19].

In general, notwithstanding its simplicity, the proposed model correctly captures the rate-dependent behaviour occurring in the unloading phase. However, we stress the model is strictly valid when viscoelastic dissipation occurs close to the contact line edge, i.e., under the assumption that the bulk material behaves elastically.

[1] Popov V L 2021 A Note by KL Johnson on the History of the JKR Theory. Tribology Letters **69(4)** 1-3.
[2] Violano G, Chateauminois A and Afferrante L 2021 Rate-dependent adhesion of viscoelastic contacts, Part I: Contact area and contact line velocity within model randomly rough surfaces. Mechanics of Materials **160** 103926.
[3] Das D and Chasiotis I 2021 Rate dependent adhesion of nanoscale polymer contacts. Journal of the Mechanics and Physics of Solids **156** 104597.
[4] Afferrante L and Violano G 2021 On the effective surface energy in viscoelastic Hertzian contacts. arXiv preprint arXiv:2107.03796.
[5] Van Dokkum J S, Pérez-Ràfols F, Dorogin L and Nicola L 2021 On the retraction of an adhesive cylindrical indenter from a viscoelastic substrate. Tribology International, **164** 107234.
[6] Müser M H and Persson B N 2021 Crack and pull-off dynamics of adhesive, viscoelastic solids. arXiv preprint arXiv:2108.02031.
[7] Ciavarella M 2021 An upper bound for viscoelastic pull-off of a sphere with a Maugis-Dugdale model. The Journal of Adhesion, 1-14.
[8] Persson B N J 2021 A simple model for viscoelastic crack propagation. The European Physical Journal E **44(1)** 1-10.
[9] Johnson K L, Kendall K and Roberts A 1971 Surface energy and the contact of elastic solids. Proceedings of the royal society of London. A. mathematical and physical sciences **324(1558)** 301-313.
[10] Roberts A D 1979 Looking at rubber adhesion. Rubber Chemistry and Technology **52(1)** 23-42.
[11] Ahn D. and Shull K R 1996 JKR studies of acrylic elastomer adhesion to glassy polymer substrates. Macromolecules **29(12)** 4381-4390.
[12] Han G, Eriten M and Henak C R 2020 Rate-dependent adhesion of cartilage and its relation to relaxation mechanisms. Journal of the mechanical behavior of biomedical materials **102** 103493.
[13] Maugis D and Barquins M 1980 Fracture mechanics and adherence of viscoelastic solids. In Adhesion and adsorption of polymers pp. 203-277. Springer, Boston, MA.
[14] Greenwood J A and Johnson K L 1981 The mechanics of adhesion of viscoelastic solids. Philosophical Magazine A **43(3)** 697-711.
[15] Muller V M 1999 On the theory of pull-off of a viscoelastic sphere from a flat surface. Journal of Adhesion Science and Technology **13(9)** 999-1016.

[16] Gent A N and Schultz J 1972 Effect of wetting liquids on the strength of adhesion of viscoelastic material. The Journal of Adhesion **3(4)** 281-294.
[17] Violano G, Chateauminois A and Afferrante L 2021 A JKR-like solution for viscoelastic adhesive contacts. Frontiers in Mechanical Engineering **7** 25.
[18] Lorenz B, Krick B A, Mulakaluri N, Smolyakova M, Dieluweit S, Sawyer W G and Persson B N J 2013 Adhesion: role of bulk viscoelasticity and surface roughness. Journal of Physics: Condensed Matter **25(22)** 225004.
[19] Persson B N J and Brener E A 2005 Crack propagation in viscoelastic solids. Physical Review E **71(3)** 036123.
[20] Dalvi S, Gujrati A, Khanal S R, Pastewka L, Dhinojwala A and Jacobs T D 2019 Linking energy loss in soft adhesion to surface roughness. Proceedings of the National Academy of Sciences **116(51)** 25484-25490.
[21] Baek D, Hemthavy P, Saito S and Takahashi K 2017 Evaluation of energy dissipation involving adhesion hysteresis in spherical contact between a glass lens and a PDMS block. Applied Adhesion Science **5(1)** 1-11.
[22] Deruelle M, Hervet H, Jandeau G, and Léger L 1998 Some remarks on JKR experiments. Journal of adhesion science and technology **12(2)** 225-247.
[23] Acito V, Ciavarella M, Prevost A M, and Chateauminois A 2019. Adhesive contact of model randomly rough rubber surfaces. Tribology Letters **67(2)** 54.
[24] Krick B A 2021 Private communication.

Simulation of freight trains with up to three traction units in radio communication

L Cantone

University of Rome Tor Vergata, Dept. of Enterprise Engineering "Mario Lucertini", Via del Politecnico 1, 00133, Rome, Italy

E-mail: Luciano.Cantone@uniroma2.it

Abstract. Paper reports the main results of a systematic study on longitudinal train dynamics (LTD) of long freight trains, equipped with radio communication. The simulation results have been used to prepare an experimental test campaign to test the Distributed Power System (DPS) technology. The simulations refer to up/down and level track and they compare the LTD of trains with and without DPS, for different train operations and radio link conditions. The DPS technology is proved (by simulations and test) to be a very effective way to increase the efficiency of future freight trains.

1. Introduction

Railway Undertakings (RUs) are always interested to increase the efficiency of freight trains: one way to reach this goal is to increase the hauled mass by a single train, by lengthening it. Limits to increment of train mass and length are given, among others, by the in-train forces exchanged by consecutive or adjacent vehicles. These forces are caused by the fact that the traction and braking force are not the same for each vehicle: i.e., only Traction Units (TUs) apply traction force to accelerate the train and the braking force is not the same or (at least) not applied at the same time by all vehicles. The Longitudinal Train Dynamics (LTD), i.e., the oscillation of the vehicles along the direction of motion, and the longitudinal (or in-train) forces are the consequences of this situation, which is typical of freight trains but not of passengers' trains. High compressive in-train forces (also known as Longitudinal Compressive Forces - LCF) cause derailments (i.e., safety problems) whereas high tensile in-train forces (also known as longitudinal tensile forces - LTF) cause train disruption (i.e., efficiency problems): both must be avoided during train operation. These in-train forces can penalize the freight train efficiency more than the infrastructure limitation, or, in other words, they do not allow to extract all the capacity from current railway infrastructures.

There are several initiatives in Europe to improve the efficiency of freight trains. One of them has been financed under the Shift2Rail (S2R) initiative "Moving European Railway Forward", within the Innovative Program 5 (IP5). This Horizon 2020 Research and Innovation (R&I) initiative aims to develop the necessary technology to complete the Single European Railway Area (SERA). The Open Call Project, Marathon2Operation (M2O), has cooperated with the FR8RAIL II Call for Members Project to demonstrate the feasibility of freight trains equipped with Distributed Power System (DPS), using radio communication to control the Traction Units. By implementing the DPS it is possible to reduce the in-train forces.

The concept of DPS is under investigation in Europe since many years from major RUs, e.g., during the European Union Framework Programme 7 (FP7) MARATHON Project trains up to 1500 m were tested by SNCF: DPS increases the points from which the power is applied along the train, and it also increases the points from which the braking is initiated, by improving the synchronicity of braking application and reducing the in-train forces. The most versatile way to realize such idea is the realization of long freight trains made by coupling two or more trains: these trains have more TUs and can carry more load in a safe and efficient way.

Within the cooperation framework between M2O and FR8RAIL II, the simulations of in-train forces have been performed using the UIC TrainDy software [1], originally developed by the University of Rome Tor Vergata with the financial and technical support of Faiveley Transport of Italy (a Wabtec Company). This software has been validated against more than 30 experimental tests made available by the major European RUs: DB AG, SNCF and TRENITALIA; it has been also used to replicate the results of another experimental test campaign carried out by FFL4E Project, within S2R framework, in May 2019 [2].

In general, the problem of computing the in-train forces has been extensively reviewed in [3], where there are references according to numerical solvers used, the connection models of rail vehicles, the traction and dynamic brake models and other relevant aspects of LTD codes along with several application fields of LTD codes. In [4], several worldwide codes are benchmarked with differences among them depending on the simulation scenario, however such a benchmark neglects the pneumatic issue because of the relevant differences among the LTD codes.

This paper reports the simulation of LTD performed with TrainDy to select the train configuration to test, within the experimental test campaign of FR8RAIL II, and to compute the in-train forces under various working conditions, experimentally replicated. Currently, the experimental results are not available for dissemination, but we can anticipate that the accordance with the simulations is (very) satisfactory. In order to evaluate the benefits of the new technology (i.e., the DPS), it is followed the approach of the UIC Leaflet 421 [5], which envisages the implementation of the relative approach: a new (potentially un-safe) system (i.e. type of train or a train with a new technology and so on) is compared against an already existing (potentially safe) system. This approach fulfils the Common Safety Methods (CSM) adopted by the European rules (Commission Implementing Regulation (EU) 2015/1136 of 13 July 2015).

Several results reported in this paper can be found also in M2O deliverables [6] and [7]; however, this paper adds further insights to those results, highlights and systematizes the main assumptions and conclusions of these deliverables. The paper is organized as follows: a) section 2 is dedicated to explain the features of the train and of the track on which the train is tested; b) section 3 reports the way used to determine the experimental train make up and the train operations simulated; c) section 4 shows the main results in terms of in-train forces and stopping distances, for different working conditions of DPS; d) section 5 reports the conclusions and the future work on the subject.

2. Conditions for the experimental test campaign

Within the framework of collaboration among FR8RAIL II and M2O Projects, FR8RAIL II had the responsibility to organize and run the experimental tests and M2O the responsibility to determine the conditions under which these tests are executed in safe conditions (see M2O deliverables [7] and [8]). Consequently, the wagons and the types of TUs to test have been decided by FR8RAIL II, following availability considerations and industrial interests for future applications, whereas track was selected in agreement with M2O, to determine severe conditions for the DPS system in terms of availability of radio signals and effect of up/down hill on in-train forces. The track section (between the stations of Kronach and Probstzella) with the highest variations of slope (around 27 ‰) is reported in figure 1: numerical simulations have been performed both on level and on up/down hill track in order to study the effect of track on in-train forces.

The TUs selected for the tests were two BR187 (from family TRAXX AC3) and one BR188 (from family TRAXX MS3), very similar with respect to performances for in-train forces; then four types of wagons were selected: Eanos_x-59, Res-676, Facns-124, Facns-133. The wagons use composite brake shoes of type LL and are equipped with different types of buffers and draw gears but the exact correspondence with the used wagons was not known at the moment of the simulations, e.g., for Eanos_x-59 the different force-stroke characteristics reported in table 1 were possible (for the complete variability see Appendix A of [7]). This uncertainty is considered in the simulations by randomly considering different coupling elements for each wagon.

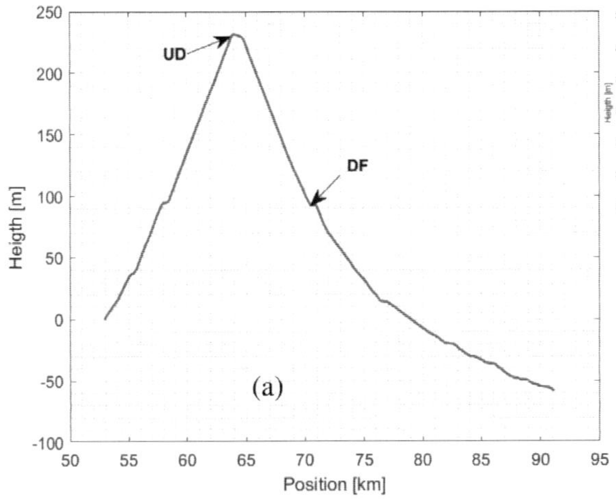

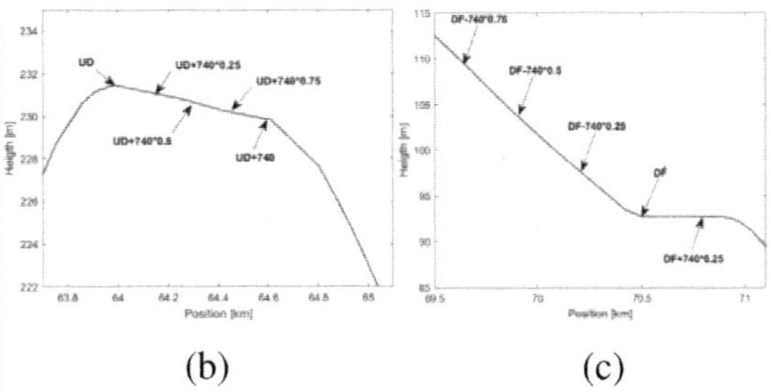

Figure 1. Section of railway track with the highest variation of slope and indications of point UD and DF (a). Starting points around UD (b) and around DF (c).

Table 1. Force-stroke characteristics for coupling elements of Eanos_x-59.

Buffing gears	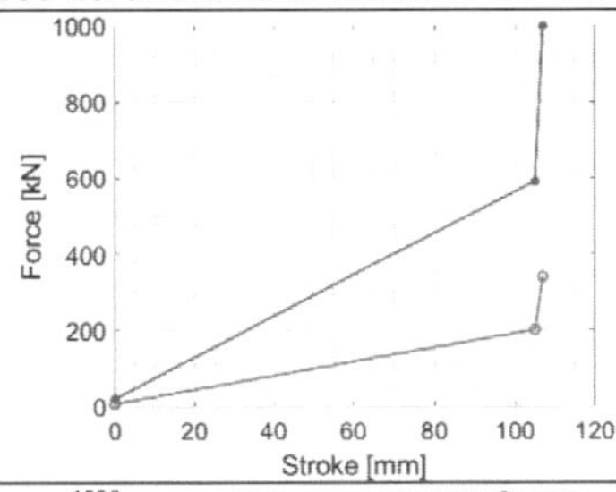	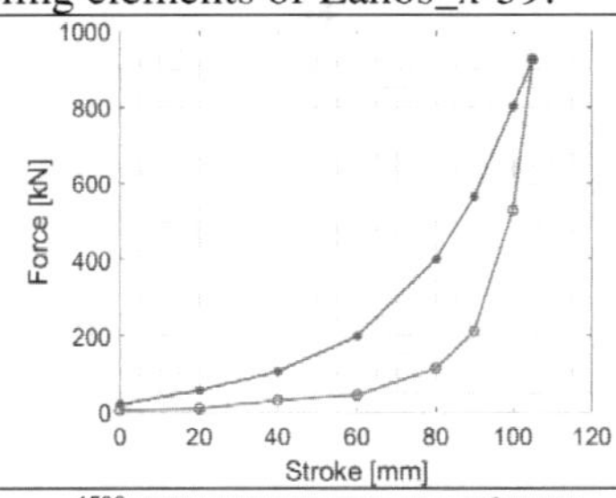
Draw gears	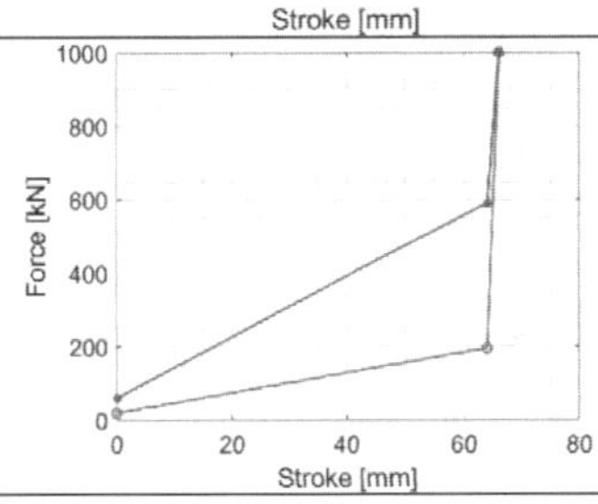	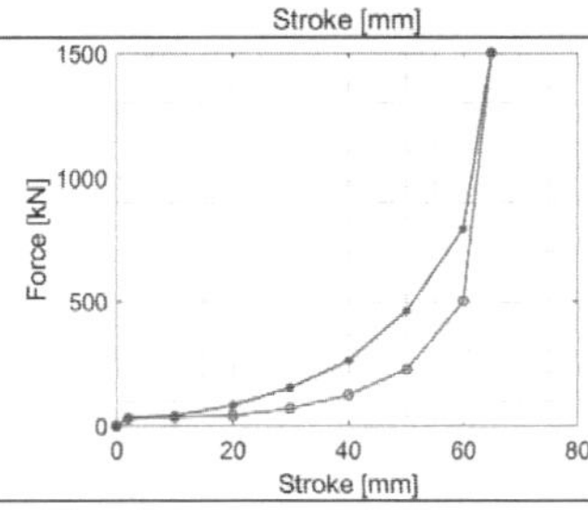

Only Eanos_x-59 wagons are fully loaded to reach the mass of 1731 ton and the length of 644 m (including the TUs). A train like this is already admitted to the traffic on the selected track and it has been tested with DPS functionalities to study the effect of this technology. Beyond the previous wagons, one measuring coach is placed close to the BR188 TU (at the beginning of the train) and it is used to store the devices needed to handle the measured quantities; the brake of measuring coach is not active.

The train make up is: BR188, Measuring Coach, Wagons of types listed before, BR187, Wagons of types listed before, BR187 (LWLWL acronym is used). It is tested moving the train in two directions (forward "FW" and backward "BW"), i.e., in FW the measuring coach, close to BR188, is at the beginning of the train; in BW it is vice-versa. Moreover, to increase the scope of experimental test campaign, the train is tested both in goods (G) and in Long Locomotive (LL) braking regimes (see [9] and [5], respectively).

3. Determination of the trainset to test

Given the vehicles, the track, the braking conditions and the running directions of the train in the previous section, the positions of the wagons are permuted between the TUs in order to find a trainset which provides low in-train forces in all different test conditions. To find the best wagon order, four types of trainsets are considered: one is without the DPS functionality and represents a train currently admitted to the traffic (reference, REF, train), the other three are DPS trains. In accordance with the position of active TU, following nomenclatures are used (when the TU is not active it behaves as a wagon, therefore it does not vent the brake pipe):

- LWL indicates a trainset in which the active TUs are at the beginning and at the end. The TU in the middle is not active. Pictogram is , for reference system (REF) and it is for DPS system.
- LWLW indicates a trainset in which the active TUs are at the beginning and in the middle. The TU at the end is not active. Pictogram is .
- LWLWL indicates a trainset in which all TUs are active. Pictogram is

Trainsets of reference train family is LWL; trainsets of DPS family are LWL, LWLW and LWLWL.

About the train manoeuvres, the emergency braking from 30 km/h after the train acceleration is used as manoeuvre since it provides high levels of in-train forces. This manoeuvre is simulated considering a radio communication between the TUs () and assuming it is lost at the same time () the emergency braking is commanded by the Driver: these conditions are the worst in agreement with the experience and the results reported in [6]. When the radio link is on, the command at the leading TU, because of DPS technology, is replicated by the remote TUs with a delay given by the technology implemented (4G, LTE radio protocol has been used in the tests): consider that for REF trains, the venting of brake pipe at the remote TU is not allowed, unless the pressure in brake pipe is 3.5 bar. When the radio link is lost, the remote TU controls the pressure in brake pipe and DPS technology vents the brake pipe when a pressure drop is locally detected (0.2 bar is used in *TrainDy* simulations); local air pressure reduction in brake pipe is performed at steps (stepwise): target pressure is 4.5 bar, when a pressure drop in brake pipe of 0.2 bar is detected with respect to 5 bar; then the next target pressure is 4 bar, when another pressure drop in brake pipe of 0.2 bar is detected with respect to 4.5 bar and so on up to a full service braking with target pressure 3.5 bar. For REF train family, in order to simulate the radio link lost, a delay of 5 s between the actions of the two Driver is used, after an Experts' evaluation.

Summarizing, the trainsets are four (one REF and three DPS), the train operations used are two: assuming the radio link between the TUs is active or it is lost. The braking regimes are two (G and LL) and the working directions are two: forward and backward. Since the measuring coach does not brake (as said above), the braking regime LL is possible only in BW direction, according to the operational German rules.

Figure 2 and figure 3 report the in-train compressive (LCF) and tensile (LTF) forces for an emergency braking commanded when the train is at 30 km/h after an acceleration: each circle refers to a train, with a specific order of wagons. The figures provide a one-to-one comparison among the same train but with DPS technology activated or not. It is worthwhile to mention that, to have a fair comparison between REF LWL and DPS LWLWL, the percentage of traction force for each TU is around 67% the maximum value, in the latter case. In this way, the effect of DPS technology on in-train forces is enhanced, since it is the only variation in the comparison; in particular, if the point is below the bisectrix it means that the DPS train is safer than the REF counterpart. In figure 2 the radio link between the TUs is working, whereas in figure 3 it is not; the figures allow the following considerations:

- The in-train compressive forces are always worst in REF trains than in DPS trains, therefore the new technology reduces the risk of derailment, whatever is the condition of radio link. Of course, when the radio link is on, the in-train compressive forces are lower. This result applies also to trains in G regime, even if they are not shown here, for sake of brevity; furthermore, the in-train compressive forces are lower in G than in LL braking regime both for REF and DPS trains.

- The in-train tensile forces are usually better for DPS than for REF trains; the trainset LWLW (i.e., the typical situation obtained when two trains are coupled) provides higher LTF with DPS in some situations: depending on wagon mass arrangement (see figure 2) and on the conditions of radio link (see figure 3). Moreover, the values of LTF are higher when the radio link is on than when it is lost, even if this is a particularity of LL regime in which the LTF are higher than those in G regime.

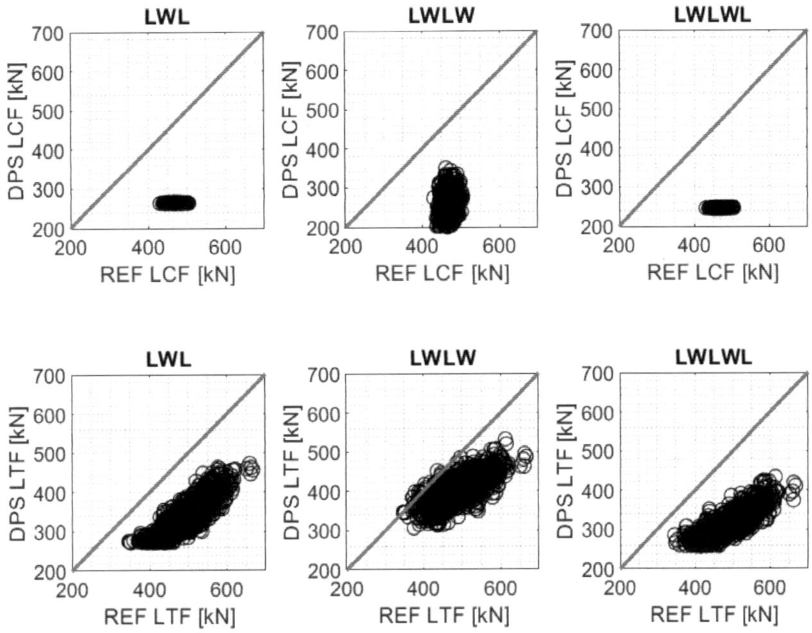

Figure 2. Acceleration up to 30 km/h followed by an emergency braking, braking regime LL and radio link working, "BW" direction.

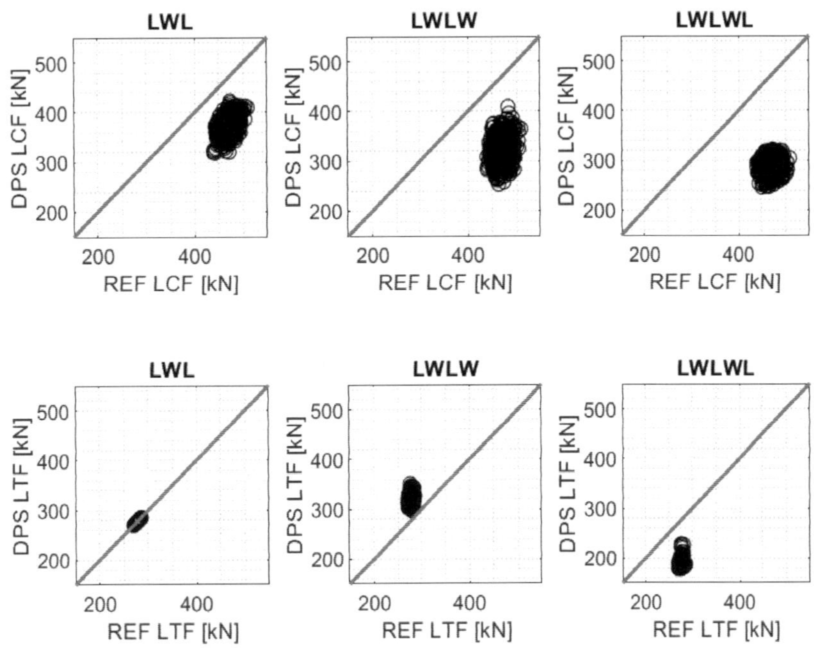

Figure 3. As in figure 2, but the radio link among TUs does not work.

- The variability of in-train forces is given by the spreading of the points: it is higher for LTF when the radio link works, and it is higher for LCF when it is lost. Anyway, these results confirm the benefits of a specific wagons ordering on the in-train forces: a suitable wagon order (when it is possible) is an effective way to safely increase the train hauled mass.
- The one-to-one comparison here reported shows the benefits of DPS technology on in-train forces without the meaningful and overlapped effect of wagon mass arrangement. In case the REF train family would have been compared directly to DPS train family, using histograms or cumulative plots, an overlap between the in-train forces would have been witnessed, giving the wrong idea that for some wagons arrangements the in-train compressive forces of DPS train could have been higher than those of REF, which is not the case.

Figure 4 reports the screenshot, from *TrainDy* software, of selected trainset, which will be further simulated in the next sections: this specific trainset minimizes the sum of in-train tensile and compressive forces in all scenarios analysed: with and without DPS (REF), in G and LL braking regime, in both directions, and with and without radio link; in this way, the selected trainset is suitable to be tested in all experimental conditions.

Position	Type	Manoeuvre	Wagon Length [m]	Train Length [m]	Brake Pipe Length [m]	GP coupl. status	Load [t]	Tare [t]
1	BR188	Man01L	18.9	18.9	37.8		0	86
2	Measuring		26.4	45.3	29.04		0	63
3	Facns-133		16	61.3	16		0	22
4	Res-676		19.9	81.2	19.9		0	24.5
5	Facns-124		19.04	100.24	19.04		0	25
6	Facns-133		16	116.24	16		0	22
7	Res-676		19.9	136.14	19.9		0	24.5
8	Facns-124		19.04	155.18	19.04		0	25
9	Eanos x-59		15.74	170.92	15.74		60	23.56
10	Facns-124		19.04	189.96	19.04		0	25
11	Facns-124		19.04	209	19.04		0	25
12	Eanos x-59		15.74	224.74	15.74		60	23.56
13	Facns-133		16	240.74	16		0	22
14	Eanos x-59		15.74	256.48	15.74		60	23.56
15	Eanos x-59		15.74	272.22	15.74		60	23.56
16	Eanos x-59		15.74	287.96	15.74		60	23.56
17	Res-676		19.9	307.86	19.9		0	24.5
18	Facns-133		16	323.86	16		0	22
19	BR187	Man01G	18.9	342.76	37.8		0	84
20	Eanos x-59		15.74	358.5	15.74		60	23.56
21	Eanos x-59		15.74	374.24	15.74		60	23.56
22	Eanos x-59		15.74	389.98	15.74		60	23.56
23	Eanos x-59		15.74	405.72	15.74		60	23.56
24	Facns-133		16	421.72	16		0	22
25	Facns-133		16	437.72	16		0	22
26	Res-676		19.9	457.62	19.9		0	24.5
27	Eanos x-59		15.74	473.36	15.74		60	23.56
28	Eanos x-59		15.74	489.1	15.74		60	23.56
29	Facns-124		19.04	508.14	19.04		0	25
30	Facns-133		16	524.14	16		0	22
31	Res-676		19.9	544.04	19.9		0	24.5
32	Res-676		19.9	563.94	19.9		0	24.5
33	Res-676		19.9	583.84	19.9		0	24.5
34	Res-676		19.9	603.74	19.9		0	24.5
35	Facns-124		19.04	622.78	19.04		0	25
36	BR187	Man01G	18.9	641.68	37.8		0	84

Figure 4. *TrainDy* software screenshot of the selected trainset.

4. Simulations of LTD for selected trainset

In this section, two types of results are shown: the results in terms of in-train forces on up/down hill track of figure 1 and the stopping distance on level track. The results in terms of in-train forces on up/down hill are reported since this type of track is capable to enhance the LTD, whereas the results on stopping distance are provided to compare the DPS and REF trains for different working conditions of radio link.

4.1. In-train forces on up/down hill, with technical parameters

In this section, two train operations are considered (for the complete list of simulations needed to check the in-train forces of DPS trains against those of REF trains refer to [6]).

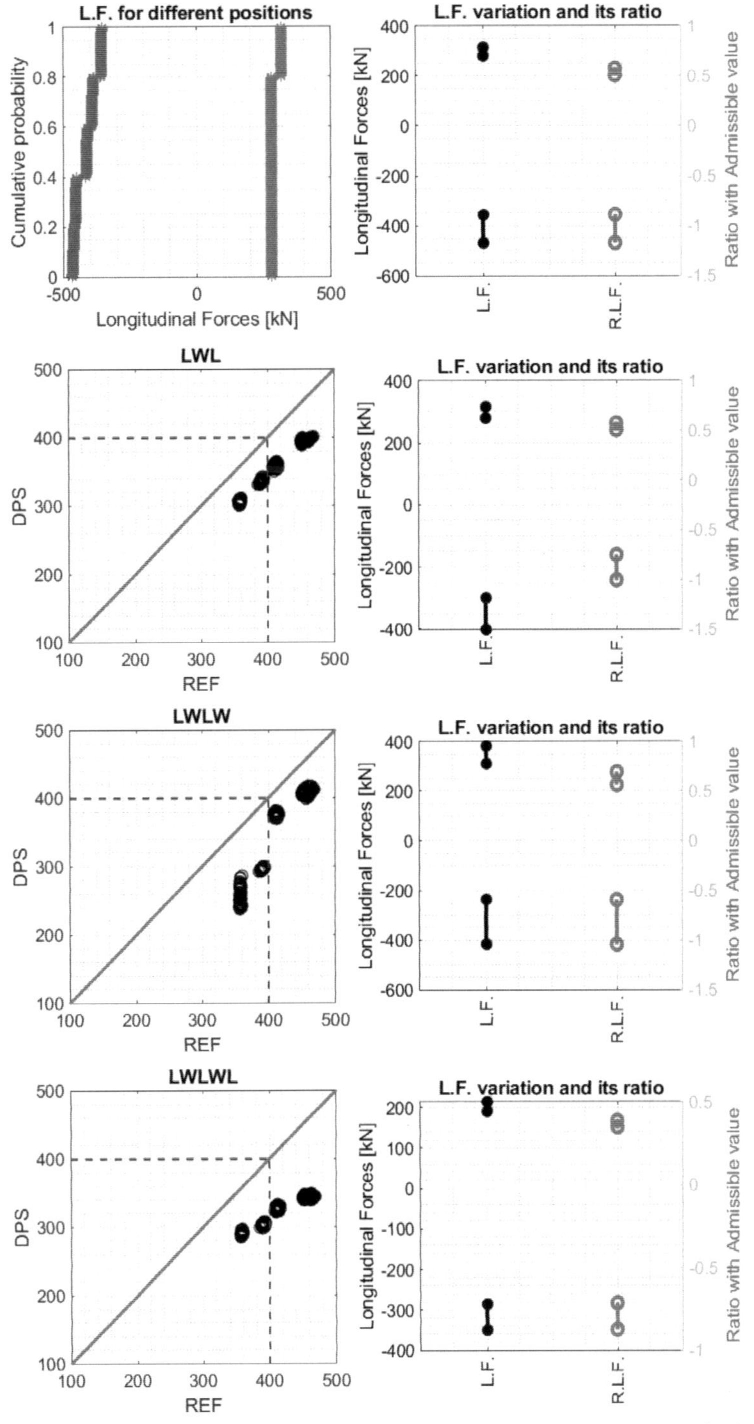

Figure 5. *Braking regime is LL, train position around point UD of figure 1. On the left the forces are reported, in [kN] for REF and for the different arrangements of DPS; on the right, the minimum and maximum Longitudinal Forces (left axis) and the ratios against their admissible values (right axis).*

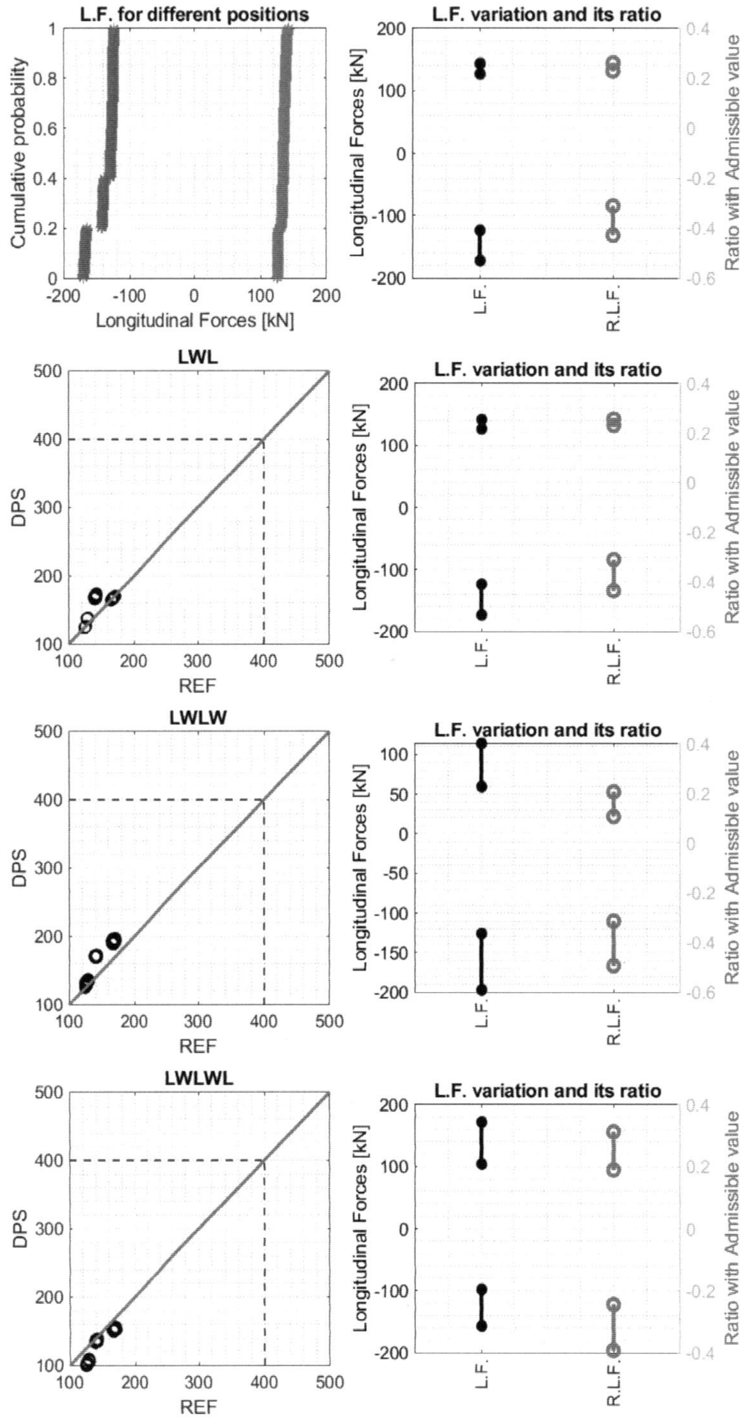

Figure 6. *Braking regime is G, direction is FW, train position around point DF of figure 1. The content is like figure 5.*

Namely:

- Traction up to 30 km/h followed by an emergency braking, when there is a radio communication loss (emergency braking is applied when the radio link is lost) and train is around point UD (see figure 1)

- Electrodynamic braking + First Application Step of braking (target pressure in brake pipe is 4.5 bar) followed by an emergency braking, when the speed is 30 km/h. In this scenario, the radio link is lost before the emergency braking is applied: for this combination of train operation and track (the train is around point DF of figure 1) this condition is the worst, according to simulations reported in [6].

Figure 5 reports the comparison of in-train forces between REF and DPS trains for different positions around point UD where the emergency braking is commanded, figure 1 (b), and for different variations of the technical parameters considered most influencing for the in-train forces, see [10]: the first row shows the cumulative frequency for REF trains on the left and the variability of LCF and LTF on the right (LTF are positive and LCF are negative, in this representation). On the right column of this figure, the ratio between in-train forces and their admissible values is also reported: when this ratio is bigger than +1, there is a risk of train disruption, whereas when it is lower than -1, there is a risk of train derailment. The other graphs of this figure report the one-to one comparison between REF and DPS trains, as in previous figures, only in terms of LCF (and not LTF, for brevity): the values of 400 kN for LCF are highlighted since these values are the maximum LCF mentioned in the UIC 421. Since the simulated trainset is always the same, the dispersion of in-train forces is given by a) the different positions on the track where the emergency braking (EB) is commanded and b) the variation of technical parameters; since the results are clustered according to the position at which the EB is commanded they show that, for this application, the position on the track where the emergency braking is commanded influences in-train forces more than the technical parameters variation. As before, the LCF of DPS trains are lower than those of REF trains.

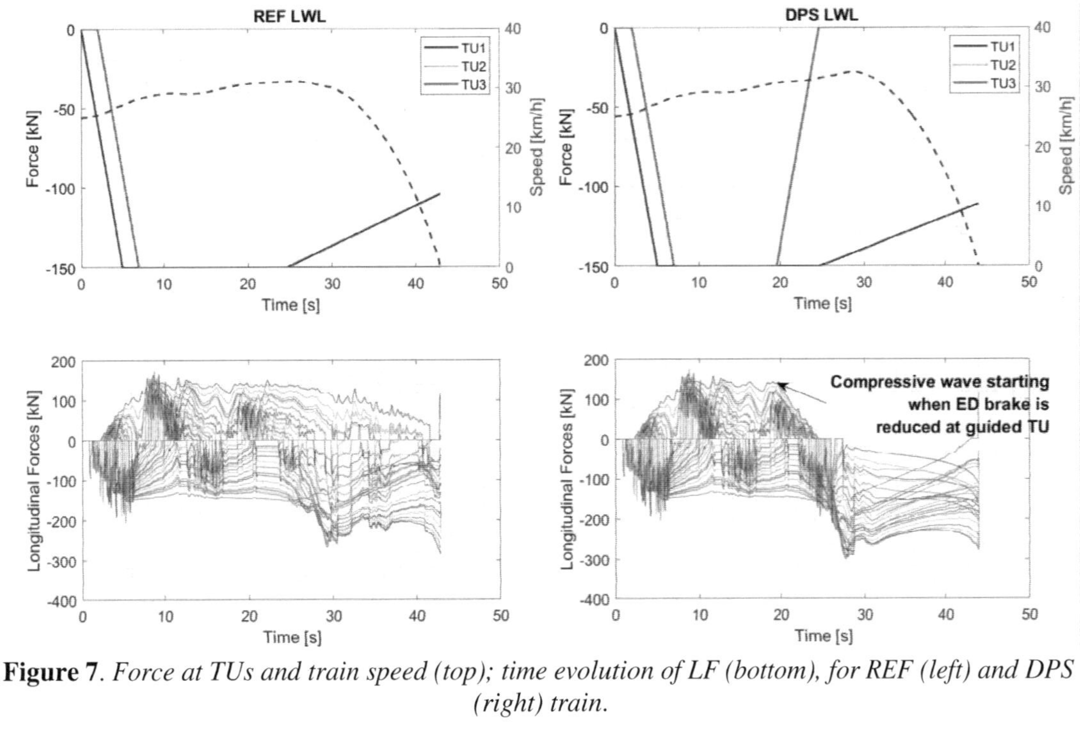

Figure 7. *Force at TUs and train speed (top); time evolution of LF (bottom), for REF (left) and DPS (right) train.*

Figure 6 is like figure 5 but the train performs the other manoeuvre listed before; moreover, it operates in G regime and in direction FW. For this manoeuvre, the DPS train is usually worse than the REF train and this is confirmed also by other results reported in [6]. The reason for this behaviour is the current implementation of DPS: when the radio link is lost, the electrodynamic force reduces automatically, and this causes a compressive wave (an example is given in figure 7), resulting in higher LCF.

The automatic removal of electrodynamic brake (and of traction force) when there is a radio link problem for some seconds (currently 2 s) is a conservative behaviour that can cause non-necessary in-train forces. This behaviour can be improved with a future development of DPS system in different ways, e.g., reducing the gradient of TU force reduction.

4.2. Considerations on stopping distance

This section analyses the performance of DPS in terms of stopping distance (table 2), considering the occurrence of DPS failure in detecting the pressure drop of 0.2 bar in brake pipe. The analysis is deterministic and only in regime G, where the highest stopping distances are expected. There is no need to employ a statistical analysis to show the benefits of DPS solution, since changing the parameters changes the numeric results but not the conclusions about the benefits of DPS with respect to reference system (REF). REF system is considered in nominal mode and in degraded mode (the TU *interlock* intervenes at 4.5 bar reducing the force at TU and braking is commanded by second Driver when pressure in brake pipe is 3.5 bar).

DPS system is considered working in three ways:

- On, radio link in "on".
- Off, radio link is "off". When DPS is activated the stepwise reduction of pressure is triggered by the detection of air pressure drops in brake pipe.
- Fail, DPS fails in detecting the pressure drop of 0.2 bar in brake pipe: the traction force is reduced by the intervention of TU interlock system.

Two train operations are considered:

- Emergency braking from (full) acceleration.
- Emergency braking from coasting.

Emergency braking is commanded when train speed is 100 km/h and it is assumed that the DPS does not perform any action on TU force unless there is an air pressure drop in brake pipe.

Table 2. Comparison of different stopping distances

Emergency braking at 100 km/h from full traction				
		Nominal	Degraded	
REF	LWL	863	864	
		On	Off	Fail
DPS	LWL	812	847	864
	LWLW	811	841	855
	LWLWL	833 (806)	875 (844)	898 (860)
Emergency braking at 100 km/h from coasting				
		Nominal	Degraded	
REF	LWL	793	794	
		On	Off	Fail
DPS	LWL	768	789	794
	LWLW	766	792	794
	LWLWL	762	790	794

The results on stopping distance (measured as difference between the final running distance and the position where the air pression drop starts in brake pipe at the first TU), show that:

- When radio is working, the stopping distances with DPS are always better than the reference train. Above consideration usually stands also when radio communication is lost; when this is not true the red colour is used. When the DPS is not able to detect the air pressure drop of 0.2 bar (see Fail column), the stopping distances are the same or lower than the REF system (except for LWLWL in emergency braking after full traction).
- For REF system and train length around 650 m there is no meaningful difference between nominal and degraded condition.

- For emergency braking from full traction, looking at DPS LWLW, when DPS fails detecting the air pressure drop, the stopping distance is lower than case LWL, since the activated TU is in the train middle, and it detects before the air pressure drop coming from the leading TU and traction force is removed before than case LWL.
- For emergency braking from full traction, looking at DPS LWLWL, there are two sets of results:
  - For full traction: the provided traction force is higher than the REF case (since there are three TUs)
  - For 67% of maximum traction at each TU: the provided traction force is roughly the same of REF case. Results are displayed within parentheses.

  Providing the same amount of power of the reference case results in lower stopping distances also for this trainset. It is important to note that also results on Longitudinal Forces have been computed considering, for traction force, the same amount of reference case, therefore, this result does not provide a bigger constraint.

  Looking at UIC 544-1 [11], for this type of train, having a percentage of braked weight around 80%, the allowed stopping distance in brake position G is around 920 m, which is higher than all simulated values.
- For emergency braking from coasting, when the radio connection in on, the stopping distances decrease from LWL to LWLWL: having closer or higher number of discharge points in brake pipe, improves the braking efficiency.
- For an emergency braking from coasting, when the radio connection in lost, the stopping distances are very similar for all positions and number of activated TU: small differences seem caused by a different internal dynamic, which has a minor effect on stopping distance.

5. Conclusions

The main conclusion of this paper is that the Distributed Power System (DPS) described here always improves the safety of freight trains currently in service when radio communication is available, with respect to Longitudinal Compression Forces (LCF). When there is a radio communication loss, the DPS train is usually better than the reference (REF) train with respect to LCF: the scenarios in which this conclusion is not valid refer to an initial application of electro dynamic brake, which can be optimized. Above conclusions do not depend on the track gradient, i.e., the DPS train is better than the REF train with respect to LCF, even if the LCF values depend on the track gradient. The section on stopping distance has shown that DPS technology is able to reduce the stopping distance and this statement is true for all the working conditions of DPS analysed. In general, an optimized mass arrangement is beneficial also with DPS technology and it can increase the hauled mass, safely. Further studies are needed to optimize the behaviour of DPS brake to better consider the trainset in which this technology is implemented: i.e., different and more optimized behaviours are possible according to different trainset layouts (LWL, LWLW or LWLWL).

Acknowledgments

The Author wants to acknowledge all the Colleagues of M2O and FR8RAIL II Projects which have contributed to the common research activity.

6. References

[1] Cantone L 2011 *TrainDy*: the new Union Internationale des Chemins de Fer software for freight train interoperability *Proc. Inst. Mech. Eng. F* **225** 57-70.
[2] Cantone L, Toubol A, et al 2020 *D3.1 – M2O LTD Simulations Report*,
[3] Cole C, Spiryagin M, Wu Q and Sun Y Q 2017 Modelling, simulation and applications of longitudinal train dynamics *Veh. Syst. Dyn.* **55** 1498-1571.

IOP Conf. Series: Materials Science and Engineering 1214 (2022) 012039 doi:10.1088/1757-899X/1214/1/012039

[4] Wu Q *et al* 2018 International benchmarking of longitudinal train dynamics simulators: results, *Veh. Syst. Dyn.* **56** 343-65.
[5] UIC 421: Rules for the consist of braking of international freight trains, 9th edition, January 2012.
[6] Cantone L, Toubol A, et al 2020 *D3.3 – Test Demonstrators with DPS, LTD simulation report*,
[7] La Rovere S, Cantone L, et al 2020 *D3.2 – Test Demonstrators with DPS, Specific application Safety case*,
[8] Gianoglio E, et al, 2021 *D4.1 – Test Demonstrators with DPS, Safety assessment report*,
[9] UIC 540: Brakes - Air brakes for freight trains and passenger trains, 7th edition, July 2016.
[10] Cantone L, Laghi D and La Rovere S 2021 A statistical study on long freight trains equipped with radio communication within Shift2Rail *IOP Conf. Ser.: Mater. Sci. Eng.* **1038** 012034
[11] UIC 544-1: Brakes – Braking performace, 5th edition, June 2013.

Vehicle localization combining non-linear state observer with artificial neural network

M Montani[1], S Amirante[1], C Annicchiarico[2] and R Capitani[1]

[1] Department of Industrial Engineering of Florence, Via Santa Marta, 3 50139 Florence
[2] Meccanica 42 srl, via Ezio Tarantelli 15, 50019 Sesto Fiorentino (Fi)

E-mail: margherita.montani@unifi.it

Abstract. The developing of autonomous drive is needed to make people life more comfortable and safer, and one of the important skills to make possible the reliability of the all control system is a good localization of the vehicle. In this study, a no-linear state observer was developed using the Unscented Kalman Filter (UKF) algorithm, to estimate the global position, global orientation, and local speeds of a car inside a known path. A characterization of the sensors input measures was made and the measures of longitudinal and lateral vehicle speed were added using an Artificial Neural Network (ANN) trained in simulated manoeuvres. In this way, it was possible to reduce the error that the observer make on the estimation of the lateral vehicle speed, and so of the side slip angle, making possible an improvement of the control activity. To assess this increase in performance, a Montecarlo analysis was made comparing the architecture proposed, ANN+UKF, with state observed, UKF, with no input measure of lateral speed. The tests were done in co-simulation environment of Vi-Grade's CarRealTime software and Matlab-Simulink.

1. Introduction

The need to reduce situations that endanger people's health and safeguard their wellbeing encourage the automotive sector to develop control system that prevent human faults. But to allow these control systems to improve their performance a proper estimation system is needed to know as accurate as possible the actual states of the vehicle. For these reasons, in this study a comparison of a no-linear state observer with and without a measure of the lateral speed of the car was made trying to reduce the side slip angle estimation errors in limit of adherence manoeuvres.

The side slip angle estimation involved many research projects because of its important in the vehicle dynamics stability controls developed in the last years [1] [2]; as well as its importance in the development of autonomous driving systems [3]. The problem is that a direct measurement of the car's speeds implies the used of expensive sensors not usable in commercial car. So, in the state of the art, it is possible to find different methodologies to overcome this problem and implement a system that doesn't involve the use of direct measurement [4] [5].

The most commonly strategies consist of: a model based approach with state observers; and a black box approach using the artificial intelligence capability of approximate any functions. The first strategy cited has been widely investigated by researchers. Some authors implemented a closed-loop state feedback observer [6], others a non-linear observer based on Lypunov asymptotic stabilization function

[7], and others a sliding mode observers [8] [9]. However, with all these methods, the estimation is strongly influenced by the uncertainties of the vehicle dynamics. To reduce the errors made, adaptive methods are developed [10], [11], even if the most common method to estimate side slip angle was the Kalman Filter (KF). It was used in many different ways: linear KF [12], and non-linear such as Extended Kalman Filter (EKF) [13] [14] [15] and Unscented Kalman Filter (UKF) [16] [17] [18]. However, the UKF is preferred over the linear KF because it reduces modelling error by representing non-linear dynamics, and over the EKF because it is only reliable for systems which are almost linear within the operating frequency range, and become unstable due to the need to calculate the Jacobians at every time step. Anyway, modelling errors cannot be completely eliminated with model-based observers.

A completely different approach is the one based on machine learning techniques and, specifically, supervised learning. The main algorithm used in the literature is Artificial Neural Networks (ANN) which have been demonstrated to be capable of approximating any function [19]. For this methodology, too, different approaches were employed with regard to both the architecture of the layers and the input quantities. Some authors preferred to use only the inherit platform sensor measures, vehicle accelerations and yaw rate [20]; and others that increase the number of input considering also the steering wheel angle and the wheels speed improving the network capability to extrapolate the vehicle model [21]. If these examples implement a deep feed-forward neural network, better results are obtained with Recurrent Neural Network (RNN) layer, being able to learn the temporal links of the various signals [22].

However, even this strategy have issues because it needs to have a big data train to cover all the car's working range. Therefore, a mixed approach [23] can ensure to have the advantages of both the strategies: training the ANN to predict the car speeds on a set of manoeuvres with different combinations of lateral and longitudinal vehicle dynamics, and using the network output as input measures of the UKF, that can consider the vehicle dynamic and correct the lack in generalization of the network.

To evaluate the beneficial that the ANN bring to the UKF developed to an autonomous car as side slip angle estimator and car locator, an in-depth statistical analysis was done through the Montecarlo analysis.

After the introduction done in the first Section, in the second Section the state observer, UKF, developed will be explained with the dynamic model used and the simulated sensors characterization; in the third Section the architecture, the training phase and the results of the ANN used will be shown; in the end Section the Montecarlo analysis will be explained, and the achieved results and the conclusions shown.

2. State observer

The state observer developed is composed of an Unscented Kalman Filter (UKF) [24] able to estimate the actual global position and orientation, and local longitudinal and lateral speed of a car inside a known path. The UKF is a non-linear state observer that uses the unscented transform to propagate the so-called 'sigma points', a group of symmetrically distributed points around the previous estimated system states which contain the information of the expected mean value and variance of the system. Representing the non-linear system in discrete time form with noise as follow:

$$x_{k+1} = f(x_k, u_k, t_k) + v_k \tag{1a}$$

$$y_k = g(x_k, t_k) + w_k \tag{1b}$$

where $x_k \in \mathbb{R}^n$ represents the state vector, $u_k \in \mathbb{R}^m$ is the input vector, $y_k \in \mathbb{R}^q$ the measurement vector, and the process noise v_k and measurement noise w_k are white Gaussian noises, i.e. zero mean and R_k^v, R_k^w as process and measurement covariance matrices, the UKF algorithm can be presented considering the state vector x_k with mean value $\tilde{x}_k$ and covariance Q_k, and the following steps:

(i) initialise mean value and variance matrix:

$$\tilde{x}_0 = \mathbb{E}[x_0] \tag{2a}$$

$$Q_0 = \mathbb{E}[(x_0 - \tilde{x}_0)(x_0 - \tilde{x}_0)^T] \tag{2b}$$

where $\tilde{x}_0$ is the a posteriori estimation of the expected mean value for k = 0 and Q_0 is the a posteriori estimation of the variance matrix for k = 0.

(ii) Calculate sigma points χ_{k-1}:

$$\chi_{k-1} = [\tilde{x}_{k-1} \quad \tilde{x}_{k-1} + A_{k-1} \quad \tilde{x}_{k-1} - A_{k-1}] \tag{3a}$$

$$A_{k-1} = \sqrt{(n+\lambda)Q_{k-1}} \tag{3b}$$

$$\lambda = \alpha^2(n+k) - n \tag{3c}$$

where λ is a scaling parameter, the constant α determines the spread of the sigma points around x_{k-1} and is usually set to a small positive value. The constant k is a secondary scaling parameter. To calculate the square root of covariance matrix Q_{k-1}, the Cholesky factorization was used for which a Hermitian positive-definite matrix B can be decomposed as $B = LL^+$ with L being a lower triangular matrix with real and positive diagonal terms.

(iii) Time update by transforming the sigma points with the non-linear functions:

$$\chi_{k|k-1} = f(\chi_{k-1}, u_{k-1}, t_k) \tag{4a}$$

$$\Upsilon_{k|k-1} = f(\chi_{k-1}, t_k) \tag{4b}$$

and computing the a priori estimation of the expected mean value $\tilde{x}_{\bar{k}}$, variance matrix $Q_{\bar{k}}$ and measurement estimation $\tilde{y}_{\bar{k}}$:

$$\tilde{x}_{\bar{k}} = \sum_{i=0}^{2n} W_i^{(m)} \chi_{i,k|k-1} \tag{5a}$$

$$Q_{\bar{k}} = \sum_{i=0}^{2n} W_i^{(c)} [\chi_{i,k|k-1} - \tilde{x}_{\bar{k}}][\chi_{i,k|k-1} - \tilde{x}_{\bar{k}}]^T + R_k^v \tag{5b}$$

$$\tilde{y}_{\bar{k}} = \sum_{i=0}^{2n} W_i^{(m)} \Upsilon_{i,k|k-1} \tag{5c}$$

where the W_i weights are:

$$W_0^{(m)} = \lambda/(n+\lambda) \tag{6a}$$

$$W_0^{(c)} = \lambda/(n+\lambda) + (1 - \alpha^2 + \gamma) \tag{6b}$$

$$W_i^{(m)} = W_i^{(c)} = 1/\{2(n+\lambda)\} \tag{6c}$$

where γ is used to incorporate prior knowledge of the distribution of the state vector (for Gaussian distributions,$\gamma = 2$ is optimal).

(iv) Measurement update by computing the measurement estimation variance Q_{ykyk} and covariance matrix Q_{xkyk} between $\tilde{x}_{\bar{k}}$ and $\tilde{y}_{\bar{k}}$:

$$Q_{ykyk} = \sum_{i=0}^{2n} W_i^{(c)} [\Upsilon_{i,k|k-1} - \hat{y}_{\bar{k}}][\Upsilon_{i,k|k-1} - \hat{y}_{\bar{k}}]^T + R_k^w \tag{7a}$$

$$Q_{xkyk} = \sum_{i=0}^{2n} W_i^{(c)} [\chi_{i,k|k-1} - \hat{x}_{\bar{k}}][\Upsilon_{i,k|k-1} - \hat{y}_{\bar{k}}]^T \tag{7b}$$

and finally, calculating the Kalman gain K_k, a posteriori estimation of expected mean value $\tilde{x}_k$ and variance matrix Q_k:

$$K_k = Q_{xkyk} Q_{ykyk}^{-1} \tag{8a}$$

$$\tilde{x}_k = \tilde{x}_{\bar{k}} + K_k (y_k - \tilde{y}_{\bar{k}}) \tag{8b}$$

$$Q_k = Q_{\bar{k}} - K_k Q_{ykyk} K_k^T \tag{8c}$$

2.1. Vehicle model

The model used by the state observer is a double-track vehicle model [25] shown in Figure 1.

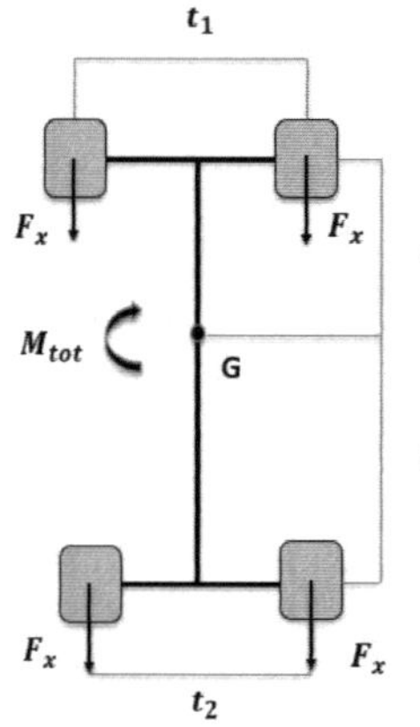

Figure 1: Double-Track vehicle model.

The state vector used is:

$$x = [X, Y, \theta, v_x, v_y, r, \alpha_{ij}] \tag{9}$$

Where X,Y are global car position on the track; θ is orientation of the car, i.e. yaw angle; vx, vy are local vehicle longitudinal and lateral speed; r is yaw rate; and α_{ij} are lateral slips of the four wheels, where ij stands for front-left, front-right, rear-left, rear-right. The evolution of these states from the previous step to the actual one is described by the following time-discrete equations:

$$X_{(k+1)} = X_k + dt(v_{x(k)}cos(\theta) + v_{y(k)}sin(\theta)) \tag{10a}$$
$$Y_{(k+1)} = Y_k + dt(v_{x(k)}sin(\theta) + v_{y(k)}cos(\theta)) \tag{10b}$$
$$\theta_{(k+1)} = \theta(k) + dt(r_{(k)}) \tag{10c}$$
$$v_{x(k+1)} = v_{x(k)} + dt(\frac{(F_{xfl} + F_{xfr})cos\delta_f + (F_{xrl} + F_{xrr}) - (F_{yfl} + F_{yfr})sin\delta_f)}{m} + + r_{(k)}v_{y(k)}) \tag{10d}$$
$$v_{y(k+1)} = v_{y(k)} + dt(\frac{(F_{xfl} + F_{xfr})sin\delta_f + (F_{yfl} + F_{yfr})cos\delta_f - (F_{yrl} + F_{yrr})}{m} + - r_{(k)}v_{x(k)}) \tag{10e}$$
$$r_{k+1} = r_k + dt(a((F_{yfl} + F_{yfr})cos\delta_f + (F_{xfl} + F_{xfr})sin\delta_f) - b(F_{yrl} + F_{yrr}) - \frac{t}{2}((F_{xfl} - F_{xfr})cos\delta_f + (F_{yfl} - F_{yfr})sin\delta_f) - \frac{t}{2}(F_{xrl} - F_{xrr}))/I_z \tag{10f}$$
$$\alpha_{ij(k+1)} = \alpha_{ij(k)} + dt(\frac{v_{y(k)}}{rel} - \frac{v_{x(k)}}{rel}tan(\alpha_{ij(k)})) \tag{10g}$$

Where $F_{x/yfl}, F_{x/yfr}, F_{x/yrl}, F_{x/yrr}$ are the longitudinal and lateral wheel forces, a, b the front and rear semi-wheelbase, t the car track, rel the lateral wheels relaxation length, δ_f the front wheel steering angle, and dt the sample time choosing to be 0.001 s.

Thanks to this model, the state observer is able to reduce the error made in the estimation of the unknown variables combining the measured signals and their noise dispersions with the model signals and their dispersions. The dispersion of the model and of the measure is found by a tuning process in co-simulations between CarRealTime dynamic vehicle model and Matlab-Simulink implementation of the state observer.

The measurement vector used by the state observer is:

$$y = [X, Y, \theta, v_x, v_y, r, a_x, a_y] \tag{11}$$

Where X,Y are global position of the car inside the path measured by the GPS sensor; θ is global orientation of the car inside the path, measured by the yaw rate integral, reset to the camera heading angle measurement and global orientation of the road lines every time that the car is straight and the side slip angle is approximately zero; r, ax, ay are vehicle yaw rate, longitudinal and lateral acceleration, measured by the IMU sensor; vx, vy are longitudinal and lateral local vehicle speed, measured by two strategies that are compared in this paper, showed in Figure 2.

The first one doesn't have the lateral speed as measure signal but as a state of the filter, and measures the longitudinal speed by the longitudinal acceleration integral, reset on the mean wheels speed at a specfic time interval during acceleration or uniform drive, as shown in the equation below:

$$Vx = \begin{cases} \int axdt & if \quad StatusBrake = 1 \\ \int axdt \quad reset \quad max(\frac{\omega_{fl}+\omega_{rr}}{2}, \frac{\omega_{fr}+\omega_{rl}}{2}) & if \quad StatusBrake = 0 \end{cases} \tag{12}$$

The second strategy involved the used of an Artificial Neural Network (ANN) able to estimate the actual value of the longitudinal and lateral speed knowing the car acceleration ax, ay, the yaw rate, r, the steering wheel angle, δ, and the wheel speeds, ω_{ij}. The ANN developed will be explain in the next chapter.

To better represent the real behaviour of the input and measure signals, a characterization of their noise was made by experimental tests, Figure 3.

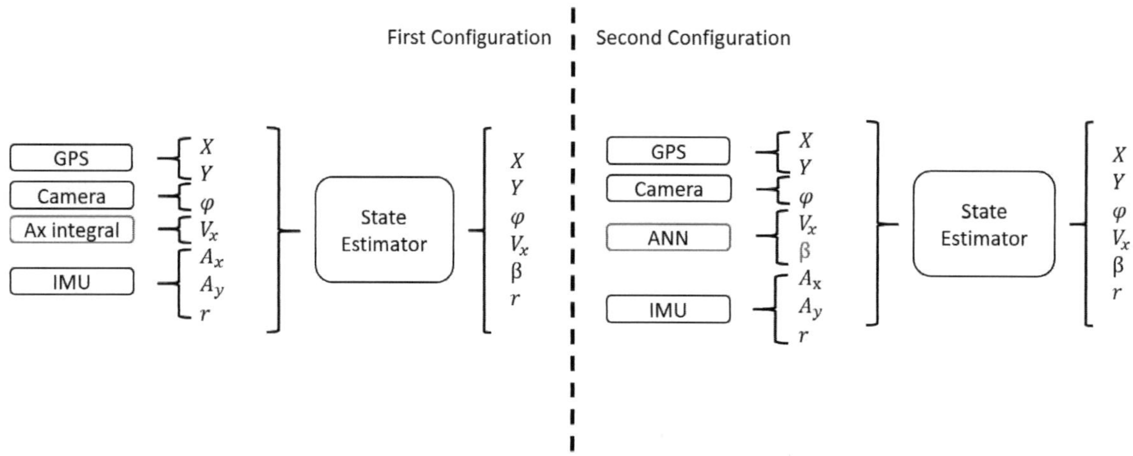

Figure 2: Representation of the two architectures tested.

Figure 3: Sensors noise characterization.

2.2. Performance of the State Observer

The UKF developed without the combination with the ANN was tested in a set of manoeuvres, performed in co-simulation environment of Matlab-Simulink and vehicle dynamic software of VI-Grade, CarRealTime.

The results obtained in the estimation of the side slip angle are shown in the Figure 4 and underlines a good performance about the mean absolute value of the side slip angle estimation, 0.29 deg, but shows the limit of the filter in the prediction of the vehicle dynamic at the limit of handling manoeuvres, where the max absolute error achieves 1.5 deg, that considering a maximum value of side slip angle admissible for safety of 6 deg represents approximately 25% of the value.

3. Artificial Neural Network to car speeds estimation

As shown in the previous Section, by the tests made, a not acceptable error on the lateral speed estimation has been seen in the filter working at limit of adhesion situations. For this reason, it was thought to develop and use as signal measure inputs an ANN [22], that trained in a varied set of car dynamics conditions can estimate the longitudinal and lateral vehicle speed reproducing the non-linearities that a simplify

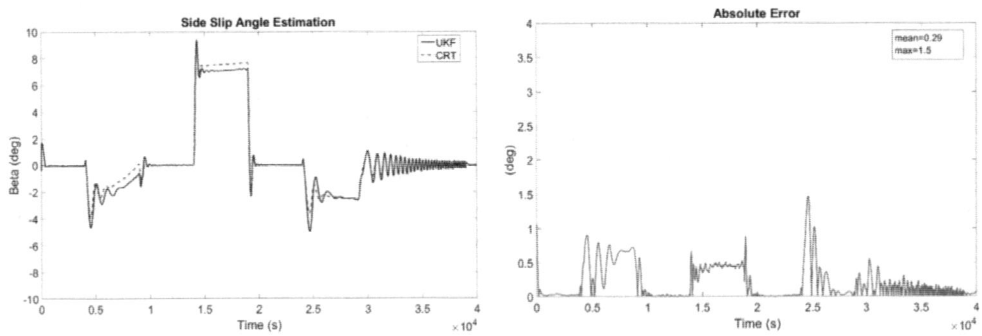

(a) **(b)**

Figure 4: Results obtained in a set of different manoeuvres by the Unscented Kalman Filter on the estimation of the side slip angle: a) Actual and estimated values of the side slip angle; b) Absolute error made in the estimation.

mathematical car model can't represent. The ANN implemented is composed by 8 input, I, and 2 output, O, that are:

$$I = [a_x, a_y, r, \delta, \omega_{ij}]; O = [v_x, v_y]; \tag{13}$$

Where ax, ay and r are the signal measured by the inertial platform, δ is the steering angle at the front wheels, and ω_{ij} are the four wheels speed.

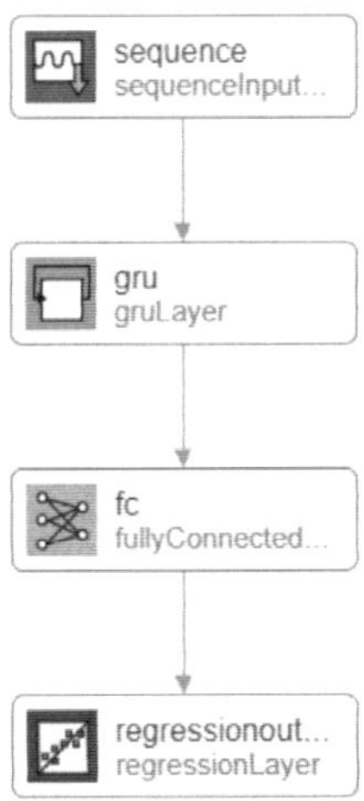

Figure 5: Architecture of the Neural Network.

The structure of the NN is shown on Figure 5, where:

- **Sequence layer**: is the Input-Layer and is responsible for feeding inputs into the network in the form of a sequence;
- **GRU layer**: is the first Hidden-Layer and is a Recurrent Neural Network (RNN) able to link signals in time by remembering past instants;
- **Fully Connected layer**: is the second Hidden-Layer and can speed up the training phase and make the network recognise similar sequences by compacting the data;

- **Regression layer**: is the Output-Layer and for typical regression problems, a regression layer must follow the final fully connected layer computing the half-mean-squared-error loss.

Then, the ANN, structured as described, was trained using a set of maneuvers representative of vehicle dynamics, on low and high friction condition, carried out using software simulations and simulations at a static driving simulator. The Data-Set used to train (70% of the data), validate (15% of the data) and test (15% of the data) the ANN is listed in the Table 1 and the ANN parameters are shown in Table 2.

Table 1: Data-Set to train, validate and test the Neural Network.

Friction	Type	Amount time (s)	Vx range (km/h)	Ay range (g)
1	Race-Track	250	10:200	0:1
	Sinus Steer	1200	5:150	0.2:0.8
	Sine Sweep	190	30:80	0.4 @1to4Hz
	Step Steer	1360	5:150	0.2:1
0.7	Step Steer	140	30:150	0.2:0.7
	Sine Sweep	160	30:80	0.4 @1to4Hz
0.5	Step Steer	100	30:150	0.2:0.5
	Sine Sweep	120	30:80	0.4 @1to4Hz

Table 2: Neural Network parameters.

Layer	Neurons	Activation function	Loss function
sequence	8	-	-
GRU	64	State: tanh Gate: sigmoid	-
FullyConnected	2	-	-
Regression	2	-	Mean Squared Error

3.1. Performance of the Artificial Neural Network

To asses the performance of the ANN, not only the manoeuvres defined by Table 1 were tested but than the ANN was implemented in Matlab-Simulink and was tested in co-simulation with the dynamic vehicle software, CarRealTime, to evaluated its real-time performance.

Table 3: Neural Network results.

	Step Steer	Sweep Steer
RMSE Vx	0.092 km/h	0.137 km/h
RMSE Vy	0.082 km/h	0.133 km/h

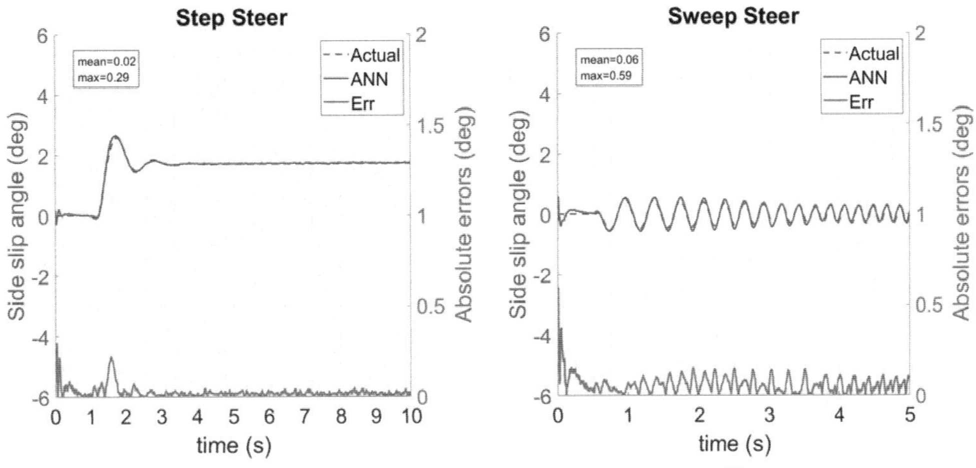

(a) (b)

Figure 6: Actual, estimated and absolute error values in terms of side slip angle estimation obtained by the ANN in Step Steer maneuver (a) and Sine Sweep (b).

The results obtained in side slip angle estimation by testing the ANN in a Step Steer manoeuvre with high lateral load, and in a Sine Sweep manoeuvre, taken from the Data-set, are shown in Figure 6, where acceptable errors are achieved in longitudinal and lateral speeds estimation, summarized in Table 3. In fact, the mean and maximum absolute errors achieved are significantly reduced compared to the UKF by up to 10% error over a maximum of 6 degrees side slip angle proving that the network has been trained in a good way and had learn the input and output link.

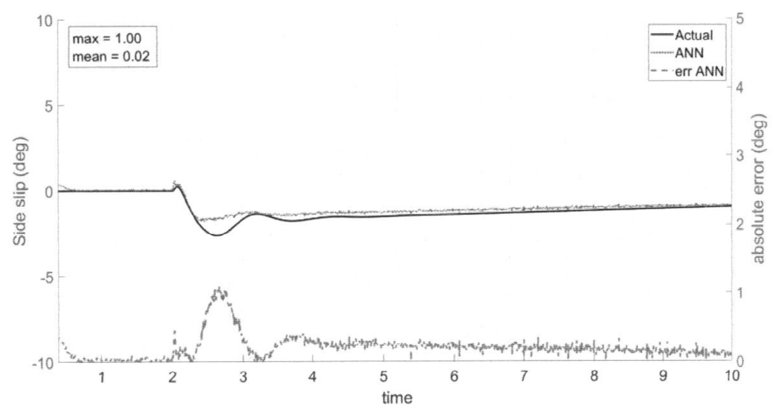

Figure 7: Results in terms of side slip angle estimation obtained by the ANN in Step Steer manoeuvre unrepresentative of the training dataset.

Testing the network with on-line run of the vehicle model, the results shown that the network generalization worsen as in the step manoeuvre of Figure 7 where the maximum error made is double the previous one.

This proved that the neural network probably has generalisation problems and that the training dataset needed to be extended to represent all dynamic driving conditions of the vehicle, thus lengthening the development time of the estimator.

4. Validation of the mixed approach

In this Section, firstly the simulation environment and then the testing methodology used to validate the robustness and accuracy of the mix-approach algorithm will be shown. The benefits of the proposed algorithm for both the longitudinal velocity and side slip angle estimation will be described.

In the results shown, the two strategies compared will be named as:

- UKF: localization algorithm composed by only the state observer without the ANN;
- UKF-ANN: localization algorithm composed by the mix between the state observer and the ANN.

4.1. Simulation environments

The tests were made in lap-top Intel(R), Core(TM) i7-7700HQ, CPU @2.80GHz and Ram 16 GB choosing a sample time of 0.001 s. The ANN and the UKF are implemented in Matlab-Simulink and to consider the vehicle dynamics, co-simulation was run between Matlab-Simulink and the Vi-Grade CarRealTime software that ensures the real-time dynamic response of the complex 14 DoF vehicle model.

4.2. Testing methodology

Since the main problem of the neural networks is to guarantee that the ANN provides an acceptable estimation error also outside the testing manoeuvres, and the problem of the UKF is to be able to describe the dynamic uncertainties linked to the tyres, it was thought to join the two systems to compensate the respective lacks of precision. In order to validate and verify the potential of this architecture under varying initial conditions and vehicle dynamics, a statistical analysis was carried out using the Montecarlo method [26]. This analysis is a statistical one and involves running the estimation system many times for the same manoeuvre by initializing the noise of the input variables in a random and always different way. During the state observer development phase, it was concluded that side slip angle estimation was most affected by variations in longitudinal speed, lateral acceleration and grip conditions.

Table 4: List of the manoeuvres used to assess the influence of different dynamic parameters in the estimators performance through the MonteCarlo analysis.

Influence parameters	Step Steer Manoeuvres			
	Steering angle (deg)	Speed (km/h)	Lateral acceleration (g)	Friction
Speed	40	50	0.25	1
	40	80	0.4	1
	40	100	0.6	1
	80	50	0.5	1
	80	80	0.9	1
	80	100	0.95	1
Friction	40	80	0.4	1
	40	80	0.45	0.7
	40	80	0.45	0.5
Lateral acceleration	40	80	0.4	1
	60	80	0.7	1
	80	80	0.9	1

Therefore, it was decided to compare the robustness and accuracy of the UKF-ANN architecture with the UKF by performing a series of step manoeuvres where vehicle speed, steering angle and grip were varied. The manoeuvres repeated 50 times for the Montacarlo analysis are shown in Table 4 and a description of the results is shown in the next subsection.

4.3. Estimation results

For each type of manoeuvre repeated by randomly initialising the noise present in the input signals, the Root-Mean-Squared-Error (RMSE) performed in the 50 tests was calculated, thus ensuring the reliability of the estimate. The RMSE obtained from the two configurations was then compared as the influence parameters varied. About longitudinal speed estimation, as shown in Figure 8, the use of the neural network as input to the UKF reduces the error especially in conditions of high lateral engagement (lateral acceleration at 0.9 g) and low adhesion conditions (friction coefficient at 0.5) reducing the RMSE by about 0.05 km/h. Only in one case a worsening of the estimate is noted (steering wheel angle at 80 deg and vehicle speed at 100 km/h), but it can be seen that the difference is only 0.001 km/h.

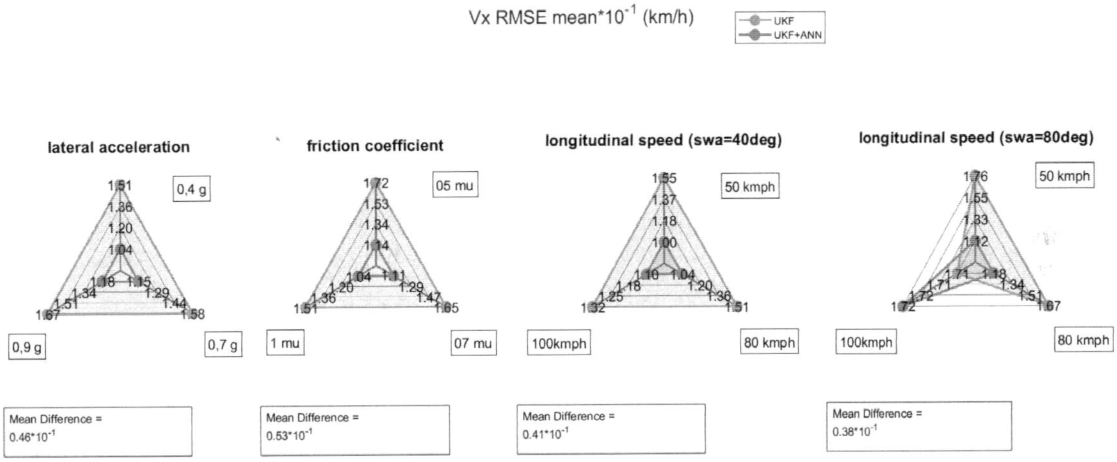

Figure 8: Statistical results of the MonteCarlo analysis for the estimation of the longitudinal speed.

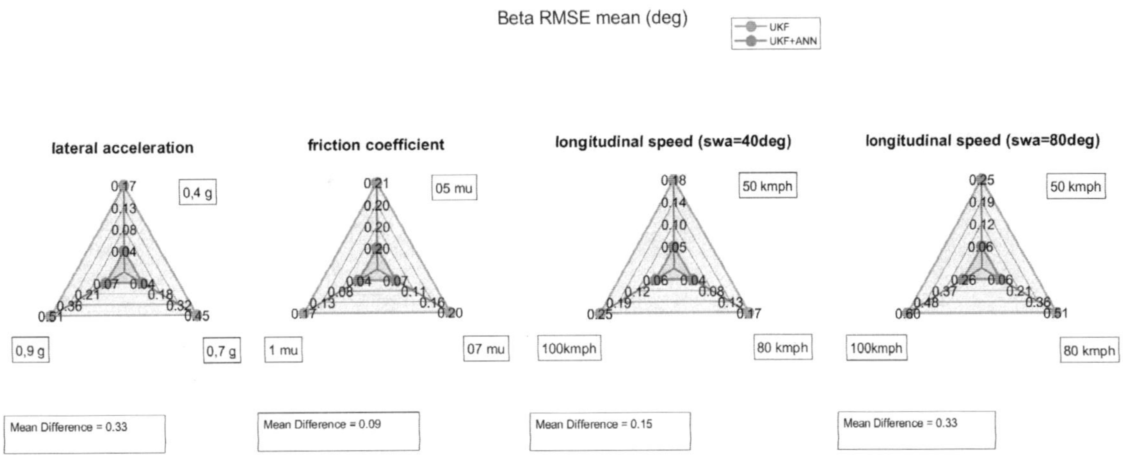

Figure 9: Statistical results of the MonteCarlo analysis for the estimation of the side slip angle.

Instead, about the side slip angle estimation, the advantages of this mix architecture are more visible. In fact, as shown in Figure 9, the UKF-ANN combination ensures in all the tests performed a RMSE

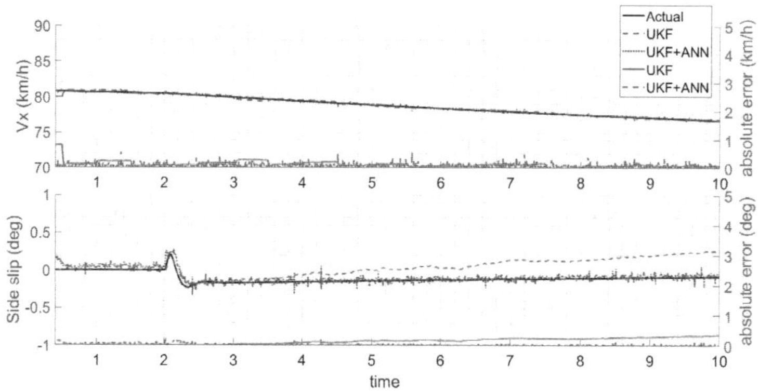

(a)

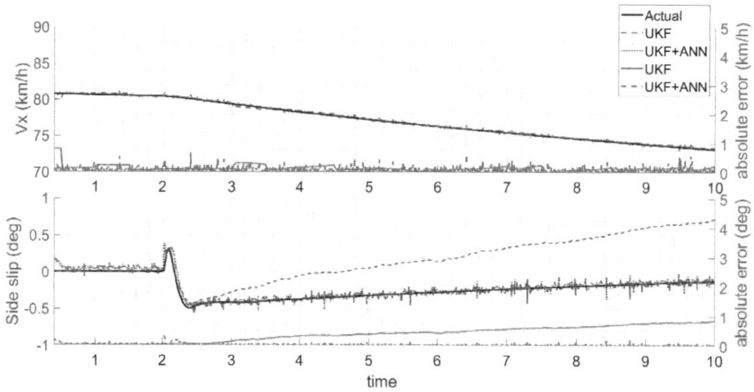

(b)

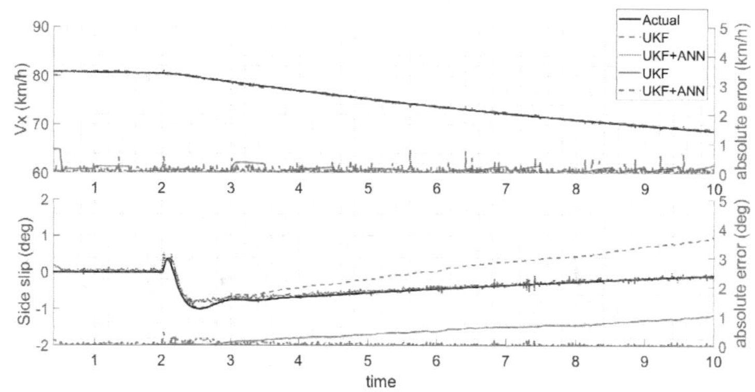

(c)

Figure 10: Results obtained in a step manoeuvre with 80 km/h and different steering angle: a) 40 deg; b) 60 deg; c) 80 deg.

reduction of almost 0.33 deg, achieving benefit especially in high lateral engagement and in high longitudinal vehicle speed. Moreover, from the results shown, it possible to see that the RMSE remains almost the same value changing the influence parameters, confirming the greater robustness achieved by UKF-ANN compared to UKF, which instead presents RMSE values varying with the dynamic conditions tested.

IOP Conf. Series: Materials Science and Engineering 1214 (2022) 012040 doi:10.1088/1757-899X/1214/1/012040

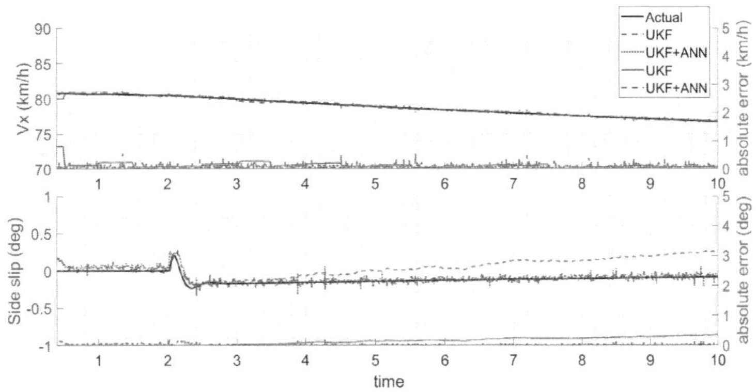

(a)

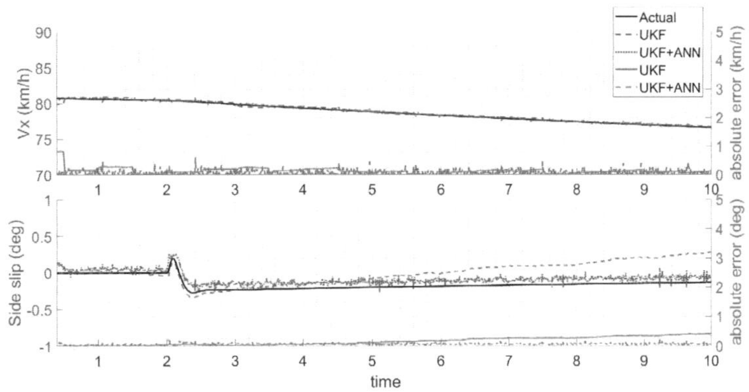

(b)

Figure 11: Results obtained in a step manoeuvre with a steering angle of 40 deg and a longitudinal speed of 80 km/h with nominal (a) and 0.7 (b) friction coefficient.

A confirmation of these results is shown in Figures 10 and 11, where the results obtained in the longitudinal speed and side slip angle estimations by changing the steering angle between 40 deg, Figure 10a, 60 deg, Figure 10b, and 80 deg Figure 10c, and by changing the friction coefficient between 1, Figure 11a and 0.7 Figure 11b are presented. In these figures, it possible to see how the absolute error made by the UKF configuration grows with the time and with increasing steering angle and thus, keeping the longitudinal speed constant, with increasing lateral acceleration. Instead, the absolute error made by the UKF-ANN configuration is almost the same in the three dynamic layout.

In addition, when changing the road path conditions both the configuration increases a bit the absolute error but the UKF-ANN configuration maintains a lower one than the UKF configuration.

5. Conclusions

In conclusion, the current study was carried out with the intention of reduce the errors made by the state of art systems in the side slip angle estimation. Therefore, a mixed approach was investigated combining the potential of the Artificial Intelligence to approximate all the function without knowing a dynamic model, and the state observer capabilities of combining measure signal and dynamic model to reduce the estimation errors.

To test and validate the proposed architecture, a Montecarlo analysis was done by exploring the influence that some dynamic parameters have on the estimation process, such as longitudinal speed, lateral acceleration and friction coefficient.

The results obtained show how the used of a ANN as sensor measure of the vehicle longitudinal and lateral velocities allows a reduction of the estimation errors in all the states tested up to a 10% error reduction considering a maximum side slip angle value of 6 deg, ensuring more robust estimation to vehicle dynamic changes. In this way, it would be possible to improved the performance of trajectory planning and tracking systems, as well as increased reliability of ESC stability systems implemented in autonomous, or standard cars.

References

[1] Van Zanten A T *et al.* 2002 Evolution of electronic control systems for improving the vehicle dynamic behavior *Proceedings of the 6th International Symposium on Advanced Vehicle Control* vol 2 p 9
[2] van Zanten A, Erhardt R, Landesfeind K and Pfaff G 2000 *IFAC Proceedings Volumes* **33** 95–102
[3] Litman T 2021 *Victoria Transport Policy Institute* 46
[4] Guo H, Cao D, Chen H, Lv C, Wang H and Yang S 2018 *IEEE/CAA J. Autom. Sinica* **5** 418–431
[5] Chindamo D, Lenzo B and Gadola M 2018 *Applied Sciences* **8** 355
[6] Weida Wang, Lijuan Yuan, Sheng Tao, Wei Zhang and Tianshu Su 2010 Estimation of vehicle side slip angle in nonlinear condition based on the state feedback observer *2010 IEEE International Conference on Automation and Logistics* (IEEE) pp 632–636
[7] Grip H, Imsland L, Johansen T A, KalKkuhl J C and Suissa A 2009 *IEEE Control Systems* **29** 36–52
[8] Shraim H, Ouladsine M, Fridman L and Romero M 2008 *International Journal of Vehicle Design* **48** 230
[9] Cadiou J C, Hadri A E and Chikhi F 2004 **218** 14
[10] Zhang H, Huang X, Wang J and Karimi H R 2015 *Mechatronics* **30** 338–347
[11] You S H, Hahn J O and Lee H 2009 *Control Engineering Practice* **17** 1367–1379
[12] Ryu J, Rossetter E J and Gerdes J C 2002 *Proceedings of the AVEC International Symposium on Advanced Vehicle Control* 8
[13] Doumiati M, Victorino A, Charara A and Lechner D 2010 A method to estimate the lateral tire force and the sideslip angle of a vehicle: Experimental validation *Proceedings of the 2010 American Control Conference* (IEEE) pp 6936–6942
[14] Huang Y, Bao C, Wu J and Ma Y 2017 *Journal of Electrical and Computer Engineering* **2017** 1–9
[15] Li L, Jia G, Ran X, Song J and Wu K 2014 *Vehicle System Dynamics* **52** 280–308
[16] Morrison G and Cebon D 2016 *Vehicle System Dynamics* **54** 1601–1628
[17] Antonov S, Fehn A and Kugi A 2011 *Vehicle System Dynamics* **49** 1497–1520
[18] Cui Q, Ding R, Zhou B and Wu X 2018 *Proceedings of the Institution of Mechanical Engineers, Part D: Journal of Automobile Engineering* **232** 1237–1252
[19] Cybenkot G 1989 *Mathematics of Control, Signals, and Systems* **2** 303–314
[20] Sasaki H and Nishimaki T 2000 A Side-Slip Angle Estimation Using Neural Network for a Wheeled Vehicle *SAE 2000 World Congress*
[21] Melzi S, Resta F and Sabbioni E 2006 Vehicle Sideslip Angle Estimation Through Neural Networks: Application to Numerical Data *Volume 2: Automotive Systems, Bioengineering and Biomedical Technology, Fluids Engineering, Maintenance Engineering and Non-Destructive Evaluation, and Nanotechnology* (ASMEDC) pp 167–172
[22] Srinivasan S, Sa I, Zyner A, Reijgwart V, Valls M I and Siegwart R 2020 *arXiv:2003.06917 [cs]*
[23] Novi T, Capitani R and Annicchiarico C 2019 *Proceedings of the Institution of Mechanical Engineers, Part D: Journal of Automobile Engineering* **233** 1864–1878
[24] Julier S, Uhlmann J and Durrant-Whyte H 1995 A new approach for filtering nonlinear systems *Proceedings of 1995 American Control Conference - ACC'95* vol 3 (American Autom Control Council) pp 1628–1632
[25] Guiggiani M 2018 *The Science of Vehicle Dynamics: Handling, Braking, and Ride of Road and Race Cars* second edition ed The Sciene of Vehicle Dynamics (Springer)
[26] Kroese D P, Brereton T, Taimre T and Botev Z 2014 *Wiley Interdisciplinary Reviews: Computational Statistics* **6** 386–392

Development of a 2D finite element model for the investigation of the tread braked railway wheels thermo-mechanical behaviour

M Magelli

Politecnico di Torino, Department of Mechanical and Aerospace Engineering, C.so Duca degli Abruzzi 24, 10129, Torino

matteo.magelli@polito.it

Abstract. The main drawback of tread braking is the rail wheel heating due to friction sliding. Thermal stresses and strains can lead to wheel surface damages while at the same time fast cooling from high temperature can cause microstructural changes, with possible local formation of brittle martensite. Nowadays, the investigation of the wheel and shoe thermomechanical interaction is typically performed using finite element (FE) codes. The paper shows the development, implementation and preliminary validation of a 2D plane FE model for the calculation of the thermal field produced in a tread braked wheel due to drag and stop braking operations. The model includes a structural contact module for the static calculation of the normal and tangential contact pressure at the wheel-shoe interface and a thermal transient module, which computes the wheel temperature considering the friction heat flux and the cooling due to air convection and rail chill.

1. Introduction

Tread braking is the braking system traditionally installed on freight wagons, however other applications in the railway field include metro vehicles and locomotives, where tread braking acts as a support to electrodynamic braking or as a back-up in case of failure of the dynamic braking system. The working principle of tread braking is based on the friction forces generated at the wheel-shoe interface when a brake block is pressed against the wheel surface. With respect to disc braking, tread braking features several benefits, mainly low costs, ease of installation, reduction of the wheelset mass and rotational inertia and the generation of a polishing effect on the wheel surface, which can improve the wheel-rail contact conditions [1]. Nevertheless, the generation of friction forces at the wheel-shoe interface necessarily leads to friction heating of the wheel itself, possibly causing local damages on the wheel surface or even local changes in the wheel steel microstructure, with formation of brittle martensite, in case a quick cooling occurs from high temperature. The friction heat partitioning between the wheel and the brake shoe is strongly dependent on the brake shoe configuration and material, as nowadays P10 (cast iron with phosphorus) shoes are being replaced by composite ones. These new compounds are specifically designed with the aim to reduce rolling noise [2,3], however, due to their composition, a higher percentage of the total friction heat flows into the wheel rather than into the shoe [4]. Moreover, composite shoes can feature a wide range of frictional characteristics according to their composition [5,6].

Due to the reason mentioned above, it is essential to gain a good understanding of the wheel thermal behavior consequent to tread braking, in order to optimize braking operations, to design new brake shoe compounds and to enhance the safety of the whole vehicle. Furthermore, the availability of a wheel-shoe thermal model is an essential tool for the development of new air brake monitoring systems [7-9]. Since it is not an easy task to perform on track tests to study the wheel-shoe thermomechanical behavior, due to issues related to high costs, interruption of the line traffic, long times, poor reproducibility of test conditions, and impossibility to analyze all the operating conditions of interest, numerical simulations seem to be the most promising strategy for the investigation of this phenomenon. Although the railway literature witnesses analytical model for the calculation of the temperature field due to friction heating at the wheel-rail and wheel-shoe contact interfaces [10-13], the most common approach nowadays is finite element (FE) codes, implemented in commercial software packages, thanks to the huge computational power of modern computers. The models developed for the investigation of the temperature field arising in tread braked wheels are typically either 3D [14-17] or 2D axisymmetric [3,4,18,19], but 3D models, despite being able to be the most accurate solution, are computationally demanding, while the main drawback of 2D axisymmetric models is that they only consider average values of the heat flows in circumferential direction, thus only computing an average value of the wheel tread. Therefore, they struggle in the computation of local hot spots which are the main source of microstructural changes and local defects. Finally, also 2D plane models [20-23] exist, which however commonly perform an a priori assumption of the wheel-shoe contact pressure distribution.

The present work shows the development of a 2D plane FE model, implemented in the commercial software package ANSYS Mechanical APDL, able to compute the temperature field of a tread braked wheel. The model includes a static structural contact module, which computes the contact pressure distribution, and a transient thermal module, which calculates the wheel thermal field due to tread braking, taking into account the friction heat, the air convection and the rail chill effect. The FE model can reproduce both 1Bg and 2Bg block configurations in drag and stop braking operations, but the present paper only deals with drag braking, which is considered to be more harmful for the wheel from a thermal point of view. The main inputs for the FE models are calculated by means of an external Matlab routine. To save the total computational time, the mechanical and thermal problems are decoupled, i.e., the contact module runs first to compute the pressure distribution, and then the outputs of this module are used to compute the friction heat flowing into the wheel, which becomes the main input of the thermal module performing the calculation of the wheel thermal field.

The paper is organized as follows: the next section describes the model structure and the modelling strategy, then focus is given to the implementation of the contact and thermal modules in ANSYS Mechanical APDL. Finally, the main outputs obtained from the preliminary simulations performed with the codes are shown, and conclusions are drawn also giving emphasis to the future upgrades needed by the code and to the future developments of the activity.

2. Model description

The tool described in the present paper includes three main modules, i.e., an external Matlab routine, a FE static contact module and a FE transient thermal module. The Matlab routine computes the main parameters related to the braking operation based on the longitudinal vehicle dynamics according to the prescription of the UIC 544-1 leaflet [24], and it was built starting from an existing longitudinal train dynamic simulator developed by the author's research group [25,26]. The static contact module calculates the normal and tangential pressure distributions which are then used in the transient thermal module to compute the evolution of the wheel temperature during the drag or stop braking operation, using a decoupled approach. Figure 1 sketches the structure of the tool described thoroughly in the present paper.

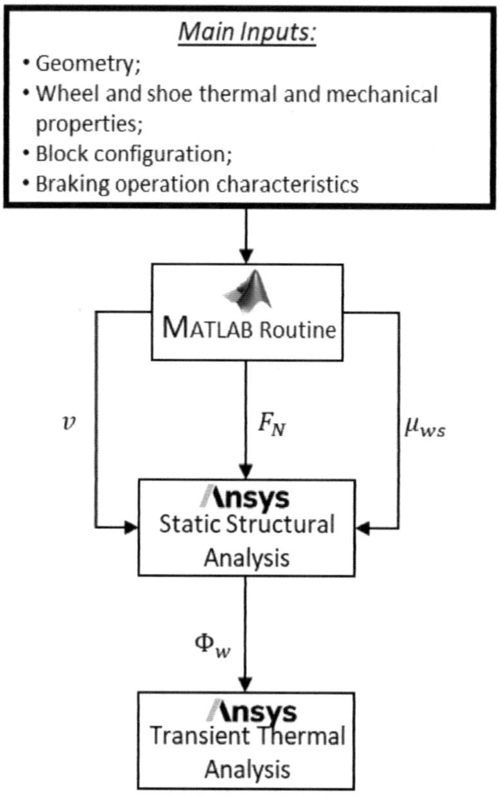

Figure 1. Scheme of the proposed tool for the simulation of the thermo-mechanical behaviour of tread braked wheels.

The Matlab routine relies on the equations of the vehicle longitudinal equilibrium and of the wheelset rotational equilibrium, written under the simplifying hypothesis that the braking effort is distributed uniformly among all wheelsets of the vehicle, see equations 1 and 2. Please note that in equations 1 and 2, M_w is the wagon weight (kg), g is gravity acceleration, i_s is the track slope (‰), $\ddot{x}$ is the vehicle deceleration, N_{ax} is the number of wheelsets of the vehicle, F_x is the longitudinal wheel-rail force acting on the wheelset, C_B is the braking torque on the wheelset, J_{yy} is the wheelset rotational inertia, $\dot{\omega}$ is the wheelset deceleration, r is the wheel radius and finally F_{Res} is the sum of the rolling and aerodynamic resistant forces. Although the last term can be calculated according to different expressions as a function of the vehicle speed $\dot{x}$ [27], the modified Davis's equation, suggested in the international benchmark of longitudinal dynamics simulators [28,29] is used, see equation 3, as it features an explicit dependency on the vehicle mass. The braking torque acting on the wheelset is related to the pressing force acting at the wheel-shoe interface F_b according to equation 4, in which N_{ws} is the number of blocks per wheel and μ_{ws} is the wheel-shoe friction coefficient. For cast iron brake shoes, the latter can be calculated as a function of the wheel-shoe pressing force F_N, and the relative speed V_s, using the Karwatzki's equation, see equation 5, in which the normal force and the relative speed must be expressed in kN and km/h, respectively. The inputs for the simulation of a drag braking operation are the vehicle running speed and the track slope. Moreover, since the vehicle is running at constant speed, the wagon and wheelset decelerations are equal to zero. Therefore, it is easy to calculate the wheel-rail longitudinal contact force from equation 1 and then the braking torque from equation 2. Then, the wheel-shoe normal contact force is obtained by solving equations 4 and 5 using the Newton-Raphson's method [30]. The Matlab routine was validated against the UIC-approved TrainDy software package [31], with a good agreement between the outputs computed by the two codes in the same simulation scenarios.

$$\frac{M_w g i_s}{1000} + M_w \ddot{x} - F_{Res} - N_{ax} F_x = 0 \tag{1}$$

$$C_B = J_{yy}\dot{\omega} + F_x r \tag{2}$$

$$F_{Res} = \frac{M_w}{1000}\left(2.943 + 89200\frac{N_{ax}}{M_w} + 0.11\dot{x} + \frac{1581\dot{x}^2}{M_w}\right) \tag{3}$$

$$C_B = 2N_{ws}\mu_{ws}F_N r \tag{4}$$

$$\mu_{wb} = 0.6\frac{V_s + 100}{5V_s + 100}\frac{16/g\,F_N + 100}{80/g\,F_N + 100} \tag{5}$$

Shifting focus to the contact module, it performs the calculation of the normal and tangential contact pressures at the wheel-shoe contact interface starting from the knowledge of the pressing force and of the wheel-shoe friction coefficient and it can deal with both 1Bg and 2Bg block configurations. The wheel-shoe contact is solved using the finite element methods, applying the augmented Lagrangian algorithm [32], with an automatic update of the normal contact stiffness at each iteration. The contact algorithm considers the effects of sliding and friction at the wheel-shoe contact interface, and a constant friction coefficient is defined, since the wheel-shoe contact is a sliding contact and therefore the transition between static and dynamic friction coefficient does not occur.

Finally, once the contact problem is solved, the normal and tangential pressure distributions are used to calculate the total friction heat generated at the wheel-shoe contact interface Φ_{tot}, while the portion of heat flowing into the wheel Φ_w, which is the main input of the thermal module, is calculated introducing a partitioning factor α, see equations 6-9, in which θ is the angular coordinate, V_s is the sliding speed, κ is the thermal diffusivity, λ is the thermal conductivity, ρ is density, c is specific heat, S is the contact area and finally b and w refer to the block and wheel, respectively. The thermal module considers the dependency of the wheel thermal properties on speed using polynomial regression from experimental data available in the literature [33], however for the sake of ease of implementation, the partitioning factor is computed only once at the beginning of the simulation, using the values calculated at 100°C.

$$\Phi_{tot}(\vartheta) = \mu_{ws}p_n(\vartheta)V_s \tag{6}$$

$$\Phi_w(\vartheta) = \alpha\Phi_{tot}(\vartheta) \tag{7}$$

$$\alpha = \left\{1 + \left(\frac{\kappa_w}{\kappa_b}\right)^{\frac{1}{2}}\frac{\lambda_b}{\lambda_w}\frac{S_b}{S_w}\right\}^{-1} \tag{8}$$

$$\kappa_{w/b} = \frac{\lambda_{w/b}}{\rho_{w/b}c_{w/b}} \tag{9}$$

The thermal module also considers the wheel cooling due to convection with ambient air and rail chill effect, as schematically shown in figure 2 for a tread braked wheel with 1Bg block configuration. The air convection coefficient depends on the wheel translational speed, and it is calculated according to the Churchill and Bernstein's correlation [34], see equations 10-13, where Nu is the Nusselt Number, Re is the Reynolds number, Pr is the Prandtl number, V_x is the wheel translational speed, r is the wheel radius, λ, ν and c are thermal conductivity, kinematic viscosity and mass specific heat, respectively, and finally subscript a refers to air properties.

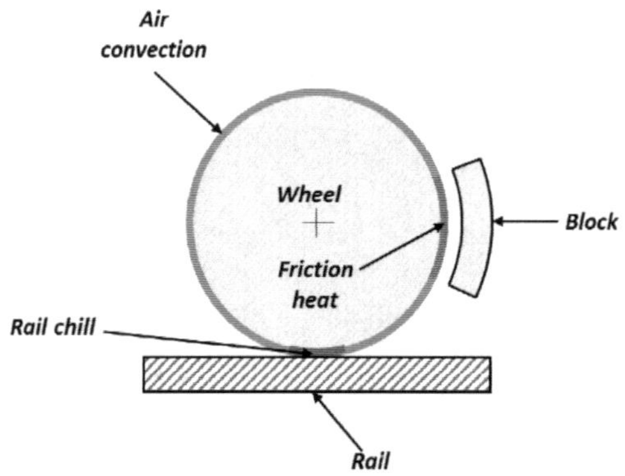

Figure 2. Sketch of the heat fluxes acting on a tread braked wheel (example for 1Bg configuration).

$$h_a = \frac{Nu\lambda_a}{2r} \tag{10}$$

$$Nu = 0.3 + \frac{0.62(Re)^{1/2}(Pr)^{1/3}}{\left[1+\left(\frac{0.4}{Pr}\right)^{2/3}\right]^{1/4}}\left[1+\left(\frac{Re}{282000}\right)^{5/8}\right]^{4/5} \tag{11}$$

$$Re = \frac{2rV_x}{\nu_a} \tag{12}$$

$$Pr = \frac{\rho_a \nu_a c_a}{\lambda_a} \tag{13}$$

Lastly, the rail chill effect, which is physically a conductive phenomenon, is treated by introducing an equivalent convection coefficient h_{rc}, only applied to the portion of the wheel which is in contact with the cooler rail, following the approach stated by Vakkalagadda et al. [4]. The convection coefficient h_{rc} is calculated as a function of the wheel translational speed V_x, assuming that the slip at the contact interface is negligible, of the longitudinal semi-axis of the contact ellipse a_x, calculated according to the Hertzian theory, and of the wheel and rail thermal effusivity β, see equations 14-17, in which σ_{wr} represents the wheel-rail thermal conductance and subscripts r and w refer to the rail and wheel, respectively.

$$\sigma_{wr} = \frac{14\gamma\beta_r(V_x)^{1/2}}{(2\pi)^{3/2}(a_x)^{1/2}} \tag{14}$$

$$\gamma = \left(1+\frac{\beta_r}{\beta_w}\right)^{-1} \tag{15}$$

$$\beta_{w/r} = \left(\lambda_{w/r}\rho_{w/r}c_{w/r}\right)^{1/2} \tag{16}$$

$$h_{rc} = \frac{a_x\sigma_{wr}}{4r} \tag{17}$$

3. Implementation of the FE modules

Both the contact and thermal FE models are implemented in the commercial software package ANSYS Mechanical APDL. The contact module is a 2D static structural model of the wheel-shoe contact, which can be used for both the 1Bg and 2Bg configurations. Figure 3 shows the wheel-shoe contact model built for a 2Bg block configuration. The wheel is represented as a hollow disc, with an external radius equal to 460 mm and an inner radius equal to 80 mm, thus not reproducing for the sake of simplicity the real wheel profile in axial direction. On the other hand, each block is represented as a circular sector with a length of 320 mm and a radial thickness equal to 60 mm, in accordance with the standard dimensions for Bg blocks [35]. Both the wheel and the block are meshed with the bidimensional, 8-node quadratic PLANE183 elements, featuring two degrees of freedom (d.o.fs) per node, assuming plane stress conditions and setting the axial width of both elements to 80 mm. The d.o.f of the wheel nodes lying at the inner radius in circumferential direction is suppressed, in order to avoid the wheel rotation, however the contact module accounts for the wheel spinning by means of the CMROTATE command. The brake block holder is not modelled in detail, so it is simply represented by means of a set of MPC184 elements, which are defined such that they behave as rigid links, transferring the load uniformly to the brake block, under the hypothesis that the brake block is far less stiff than its holder. Each rigid beam of the same block is connected to an external node, placed 607 mm away from the wheel center in radial direction, on which the pressing force calculated by means of the external Matlab routine is imposed. The external nodes are constrained so that their translation along the y-axis is prevented, while they are free to move along the x direction. The wheel-shoe conformal contact is faced introducing bi-dimensional surface-to surface contact elements, namely a layer of 3-node CONTA172 elements placed on the wheel outer periphery paired with a corresponding layer of TARGE169 elements, placed on the brake shoe inner surface. The conformal wheel-shoe contact problem is solved using the augmented Lagrangian method, adjusting the normal contact stiffness automatically at each iteration, depending on the mean stress of the underlying PLANE183 elements, so that a satisfying trade-off between computational time and good conditioning of the contact matrix can be reached. Since the contact problem is nonlinear, the solution is obtained by means of the Newton-Raphson iteration scheme, forcing the solver not to perform a stiffness matrix symmetrization to improve convergence.

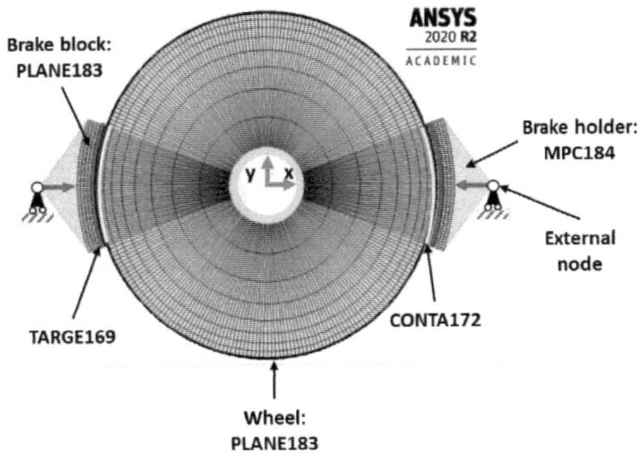

Figure 3. Wheel and block mesh in the contact module for 2Bg block configuration.

On the other hand, the thermal module is a 2D FE transient module which is able to compute the evolution of the wheel temperature during braking operations. In order to reduce the number of d.o.fs of the model and to consequently improve the computing speed, only a 45° circular sector of the wheel is modelled, see figure 4, under the assumption that the heat flux conduction inside the wheel in circumferential direction is negligible with respect to the conduction in radial direction. Simulations not presented in the paper were performed under the same scenarios with the reduced model and a model of the entire wheel, and a good agreement between the outputs of the two models was observed, thus confirming the validity of the hypothesis. The wheel is meshed by means of the 8-node quadratic PLANE77 elements, with a single d.o.f corresponding to the node temperature, setting the element behavior to plane with specification of axial thickness. The thermal boundary conditions are applied to overlying layers of 2-node SURF151 elements, specifying a plane behavior with thickness definition. As noticeable from figure 4, five sets of SURF151 elements are introduced in the model. The lateral edges and the inner radius of the wheel are covered with elements to which an adiabatic condition is superimposed, while the nodes at the outer radius are overlaid with two layers of SURF151 elements, one for the application of the friction heat flux and one for convection boundary condition, with the latter including both the convection with ambient air and the rail chill effect, which is treated as a convective phenomenon despite being a conductive heat transfer, as previously mentioned.

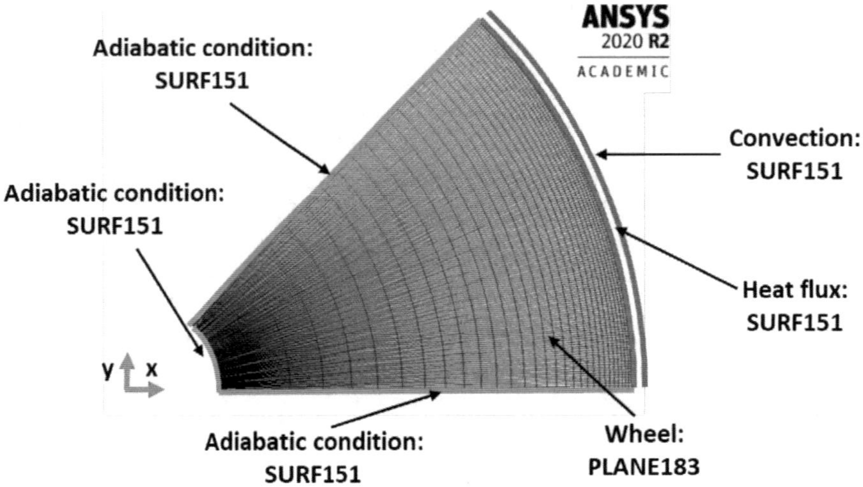

Figure 4. Mesh of the thermal module.

The application of the thermal loads accounts for the wheel rotation, however in the model the wheel is fixed and the thermal loads are rotated thanks to the definition of tables, which are a specific ANSYS built-in data type used for the definition of time and space varying loads. Tables in the models act as 3D look-up tables defining the value of the thermal load at specific time steps and angular coordinates, so that values at intermediate points are automatically calculated through simple interpolations. The thermal problem is described in ANSYS by a system of nonlinear ordinary differential equations (o.d.es), and it is solved using the Crank-Nicolson [36] scheme for numerical integration, enabling the predictive-corrector ANSYS built-in algorithm for automatic adjustment of the time step-size at each iteration and the Newton-Raphson algorithm for the solution of the nonlinear algebraic system obtained at each time step.

4. Preliminary results

This section presents the outputs obtained by the preliminary simulations performed with the contact and thermal modules. Due to the lack of experimental data, it is not easy an easy task to validate the code, however these preliminary simulations were run to assess that the outputs produced by the code are in line with the expectations. The preliminary simulations were run considering the properties of P10 brake shoes, however the code can be launched considering different shoe compounds.

Figure 5 shows the distribution of the wheel-shoe normal contact pressure computed both neglecting and considering the effect of friction and sliding at the contact interface. Two values of the normal force are considered in the plot, i.e., 28.3 and 40 kN, respectively. The former is the pressing force exerted on a reference four-axled freight wagon when the braking pressure is equal to 3.8 bar, while the latter corresponds to the maximum value considered by the UIC 544-1 leaflet for the Bg block configuration. As expected, the maximum value of the contact pressure increases if the normal force is larger. Furthermore, when friction and sliding are taken into account, the pressure distribution is asymmetric, and a peak at the leading edge is noticeable, due to a block jamming effect. The asymmetry of pressure distribution is not surprising, as the resulting contact force must always pass through the external node, to ensure the rotational equilibrium of the block.

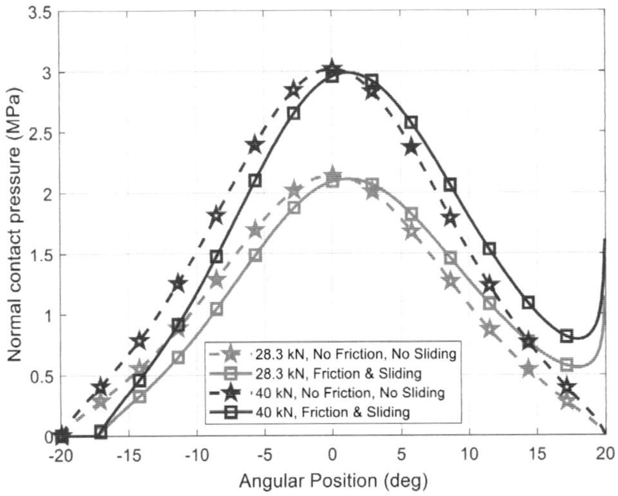

Figure 5. Effect of frictional sliding and normal force on the normal contact pressure distribution computed by the contact module.

A second simulation was run to identify the differences in the wheel temperature with 1Bg and 2Bg block configurations for the same drag braking operation on a 20-ton axle-load freight vehicle, considering the operational parameters suggested by the EN 13979 standard [37], i.e., track slope equal to 21‰ and running speed of 60 km/h. To reduce the computational times, a short-lasting manoeuvre is considered, and the simulation time is equal to 2 minutes, corresponding to a total braking distance of approximately 2 km. Obviously, in case the braking effort is exerted with a single block, the normal force is higher, and the Matlab routine computes a normal force of 13.84 kN and a friction coefficient of 0.14, while, in case the 2Bg configuration is considered, the pressing force on each block and the corresponding friction coefficient are equal to 5.24 kN and 0.18, respectively. Figure 6 shows for both cases the friction heat flux flowing into the wheel at the wheel-shoe interface. Obviously, the heat flux is higher for the 1Bg configuration, as only one contact interface, which must provide the required braking torque, is present.

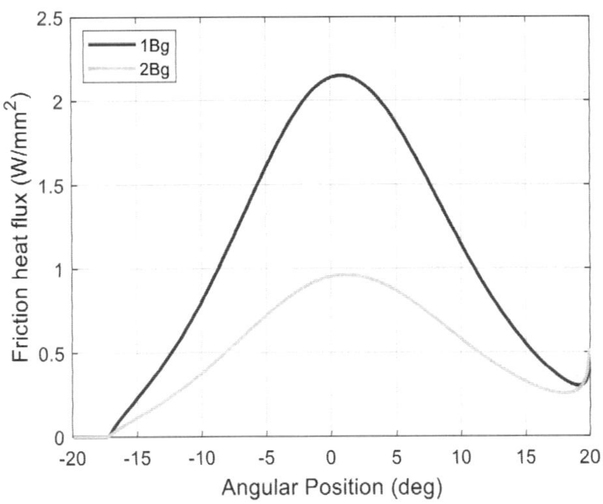

Figure 6. Heat flux distribution for 2Bg and 1Bg block configurations for the same braking operation.

Figure 7 shows the temperature evolution on a node located on the wheel surface during the drag braking operation, computed with the thermal module. As expected, the temperature for the 1Bg configuration is higher, due to the larger heat flux flowing into the wheel at the wheel-shoe contact interface. The zoomed view at the bottom in figure 7 shows that the temperature rise due to the contact with the shoe is much larger for the 1Bg configuration with respect to the 2Bg configuration. Furthermore, the difference between the maximum and minimum temperature values recorded during a single wheel revolution is far larger for the 1Bg configuration, and this means that the 1Bg configuration can be more harmful from a thermal point of view. Finally, the detailed view in the upper left corner of figure 7 highlights a steeper temperature drop when the node is not in contact with the rail due to the rail chill effect.

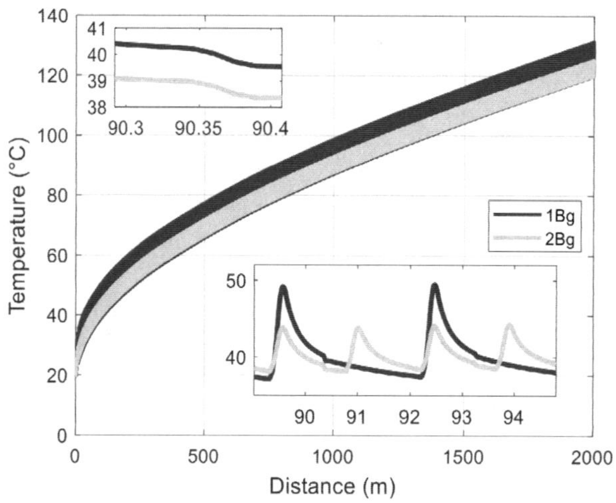

Figure 7. Wheel surface temperature during a 2 km drag braking operation at 60 km/h for 1Bg and 2Bg arrangements.

Lastly, figure 8 shows the temperature at the end of the simulation for different radial positions, and it can be noticed that the wheel is heating only in a limited range of radial coordinates where a steep temperature gradient can be observed, while the central part of the wheel is still at the initial temperature of 20°C. Therefore, since the temperature stresses and strains concern a limited peripheral portion of the wheel, it can be concluded that, as expected, local defects can arise in a limited radial sector of the wheel, and a thermal model is essential to predict the maximum wheel temperature during a braking operation, as it is not an easy task to provide the vehicle with on-board temperature sensors able to perform measurements on the external surface.

A big concern of the proposed model is the computational time, with the bottleneck of the computation obviously being the numerical integration of the o.d.es solved in the thermal module. In fact, the simulations described above, run on a computer featuring a Windows 10 Pro OS, 64-GB internal memory and an Intel 8-core i9 processor, took 2h 18 min for the 1Bg configuration and 2h 30 min for the 2Bg arrangement, thus being 69 and 75 times slower than real time, respectively. Therefore, in the author's opinion, the introduction of simplifying hypotheses, such as the development of a 2D model of a limited portion of the wheel, supported by experimental data for the model calibration, are crucial for the code to provide accurate outputs in a reasonable amount of time.

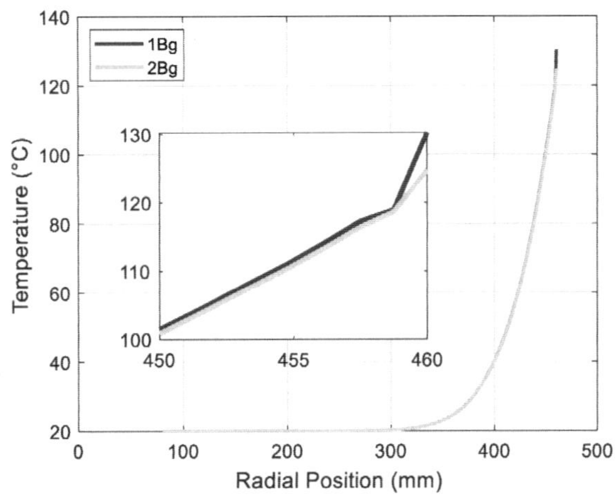

Figure 8. Temperature at the end of the drag braking simulation for different radial positions.

5. Conclusions

The paper shows the development, implementation and preliminary validation of a tool for the estimation of the temperature field produced in tread braked wheels during drag braking operations. The tool includes an external Matlab routine, for the calculation of the wheel-shoe contact force, and two FE modules implemented in ANSYS Mechanical APDL, namely a static structural contact module and a transient thermal module. The contact algorithm implemented in the structural module considers both friction and sliding at the contact interface, which lead to an asymmetry in the pressure distribution and in a peak at the leading edge, which is related to a block jamming effect. The thermal module calculates the temperature evolution due to friction heat flux, air convection and rail chill. The wheel is considered as fixed and rotating thermal loads are applied to its surface, thanks to the definition of proper ANSYS tables, which are a powerful data type for the application of time and space varying loads.

The computational effort is a primary concern in the context of this activity, and to keep the computational times as short as possible, the two FE models are decoupled, i.e., the interaction between the contact pressure and the surface temperature, as well as wear phenomena, are neglected. Furthermore, only a 45° sector of the wheel is modelled in the thermal module, with the definition of an adiabatic boundary condition at the lateral edges, under the assumption that the heat conduction in circumferential direction has a minor impact with respect to the heat transfer in radial direction. However, a good match was observed between the outputs of the reduced model and a model of the full wheel, thus confirming the validity of such hypothesis.

The preliminary simulations presented in the paper showed that the code provides results in good agreement with the expectations, and also proved that the 1Bg configuration can be more dangerous in terms of thermal stresses and strains in the wheel with respect to 2Bg configuration, under the same braking operation.

Much work is still needed to obtain a full validation of the code, which however needs to be performed against a large set of experimental data. Therefore, the author's research team is currently verifying the feasibility of designing a scaled rig for the reproduction of different braking maneuvers with all kinds of shoe material, in order to calibrate the tool parameters, keeping the computational times short. Furthermore, the code could undergo several upgrades, such as the coupling of the structural and thermal modules and the introduction of a wear module, however always paying great attention to computing speeds.

References

[1] Günay M, Korkmaz M E and Özmen R 2020 An investigation on braking systems used in railway vehicles *Eng. Sci. Technol. an Int. J.* **23** 421-31

[2] Fec M C and Sehitoglu H 1985 Thermal-mechanical damage in railroad wheels due to hot spotting *Wear* **102** 31-42

[3] Vernersson T and Lundén R 2007 Temperatures at railway tread braking. Part 3: Wheel and block

temperatures and the influence of rail chill *Proc. Inst. Mech. Eng. F: J. Rail Rapid Transit* **221** 443-54

[4] Vakkalagadda M R K, Vineesh K P and Racherla V 2015 Estimation of railway wheel running temperatures using a hybrid approach *Wear* **328-329** 537-51

[5] Cantone L and Ottati A 2018 Modelling of friction coefficient for shoes type LL by means of polynomial fitting *Open Transp. J.* **12** 114-27

[6] Vakkalagadda M R K, Srivastava D K, Mishra A and Racherla V 2015 Performance analyses of brake blocks used by indian railways *Wear* **328-329** 64-76

[7] Bosso N, Magelli M and Zampieri N 2020 Calibration and development of a multi-axis roller bench for monitoring the braking system of a railway vehicle *Ing. Ferrov.* **75** 501-23

[8] Bosso N, Magelli M and Zampieri N 2020 Validation of a brake monitoring system using a multi-axle roller-rig *WIT Transactions on The Built Environment* vol 199, ed G Passerini, J M Mera and R Takagi (Southampton, UK: WIT Press) pp. 151-62

[9] Somà A, Aimar M and Zampieri N 2021 Simulation of the thermal behavior of cast iron brake block during braking maneuvers *Appl. Sci.* **11** 5010

[10] Ertz M and Knothe K 2002 A comparison of analytical and numerical methods for the calculation of temperatures in wheel/rail contact *Wear* **253** 498-508

[11] Knothe K and Liebelt S 1995 Determination of temperatures for sliding contact with applications for wheel-rail systems *Wear* **189** 91-9

[12] Tudor A and Khonsari M M 2006 Analysis of heat partitioning in wheel/rail and wheel/brake shoe friction contact: An analytical approach *Tribol. Trans.* **49** 635-42

[13] Tudor A, Radulescu C and Petre I 2003 Thermal effect of the brake shoes friction on the wheel/rail contact *Tribol. Ind.* **25** 27-32

[14] Haidari A and Hosseini-Tehrani P 2014 Fatigue analysis of railway wheels under combined thermal and mechanical loads *J. Therm. Stresses* **37** 34-50

[15] Haidari A and Tehrani P H 2015 Thermal load effects on fatigue life of a cracked railway wheel *Lat. Am. J. Solids Struct.* **12** 1144-57

[16] Milošević M, Stamenković D, Tomić M, Milojević A and Mijajlović M 2012 Modeling thermal effects in braking systems of railway vehicles *Therm. Sci.* **16** S515-S26

[17] Pradhan S and Samantaray A K 2019 A recursive wheel wear and vehicle dynamic performance evolution computational model for rail vehicles with tread brakes *Vehicles* **1** 88-115

[18] Faccoli M, Ghidini A and Mazzù A 2018 Experimental and numerical investigation of the thermal effects on railway wheels for shoe-braked high-speed train applications *Metall. Mater. Trans. A* **49** 4544-54

[19] Vernersson T and Lundén R 2010 Wear of disc brakes and block brakes–influence of design on modelled wear for repeated brake cycles *Proceeding of the 16th International Wheelset Congress (IWC16)* (Cape Town)

[20] Petereson M 2002 Two-dimensional finite element simulation of the thermal problem at railway block braking *Proc. Inst. Mech. Eng. C: J. Mech. Eng. Sci.* **216** 259-73

[21] Suresh Babu A and Siva Prasad N 2009 Coupled field finite element analysis of railway block brakes *Proc. Inst. Mech. Eng. F: J. Rail Rapid Transit* **223** 345-52

[22] Teimourimanesh S, Vernersson T and Lundén R 2014 Modelling of temperatures during railway tread braking: Influence of contact conditions and rail cooling effect *Proc. Inst. Mech. Eng. F: J. Rail Rapid Transit* **228** 93-109

[23] Thuresson D 2004 Influence of material properties on sliding contact braking applications *Wear* **257** 451-60

[24] UIC 544-1:2014 *Brakes - braking performance*

[25] Bosso N, Magelli M and Zampieri N 2020 Long train dynamic simulation by means of a new in-house code *WIT Transactions on The Built Environment* vol 199, ed G Passerini, J M Mera and R Takagi (Southampton, UK: WIT Press) pp. 249-59

[26] Bosso N, Magelli M and Zampieri N 2021 Development and validation of a new code for longitudinal train dynamics simulation *Proc. Inst. Mech. Eng. F: J. Rail Rapid Transit* **235** 286-99

[27] Bosso N, Magelli M, Rossi Bartoli L and Zampieri N 2021 The influence of resistant force equations

and coupling system on long train dynamics simulations *Proc. Inst. Mech. Eng. F: J. Rail Rapid Transit*

[28] Spiryagin M, Wu Q and Cole C 2017 International benchmarking of longitudinal train dynamics simulators: Benchmarking questions *Veh. Syst. Dyn.* **55** 450-63

[29] Wu Q, *et al.* 2018 International benchmarking of longitudinal train dynamics simulators: Results *Veh. Syst. Dyn.* **56** 343-65

[30] Ryaben'kii V S and Tsynkov S V 2006 Numerical solution of nonlinear equations and systems *A theoretical introduction to numerical analysis* (Boca Raton, FL, USA: Chapman and Hall/CRC)

[31] Cantone L 2011 Traindy: The new union internationale des chemins de fer software for freight train interoperability *Proc. Inst. Mech. Eng. F: J. Rail Rapid Transit* **225** 57-70

[32] Simo J C and Laursen T A 1992 An augmented Lagrangian treatment of contact problems involving friction *Comput. Struct.* **42** 97-116

[33] Vernersson T 2007 Temperatures at railway tread braking. Part 2: Calibration and numerical examples *Proc. Inst. Mech. Eng. F: J. Rail Rapid Transit* **221** 429-41

[34] Churchill S W and Bernstein M 1977 A correlating equation for forced convection from gases and liquids to a circular cylinder in crossflow *J. Heat Transfer* **99** 300-6

[35] UIC 542:2010 *Brake parts - interchangeability*

[36] Crank J and Nicolson P 1947 A practical method for numerical evaluation of solutions of partial differential equations of the heat-conduction type *Math. Proc. Cambridge* **43** 50-67

[37] EN 13979:2020 *Railway applications - wheelsets and bogies - monobloc wheels - technical approval procedure* Part 1: Forged and rolled wheels

Investigating encapsulation design strategy of photovoltaic cells in the case of a solar race car

A Pavlovic [1*], V Mikhnych [2, 3], M Bertoldi [4] and C Fragassa [1]

[1] University of Bologna (UNIBO), Department of Industrial Engineering, Via Fontanelle 40, Forlì, Italy

[2] Aix-Marseille University, Campus Universitaire de Saint Jérôme, Marseille Cedex 20, 13397 Marseille, France

[3] Karlsruhe Institute of Technology (KIT), Karlsruhe School of Optics & Photonics, Schloßpl. 19, 76131 Karlsruhe, Germany

[4] University of Bologna (UNIBO), Department of Electrical, Energy and Information Engineering "Guglielmo Marconi", Viale del Risorgimento 2, Bologna, Italy

[*] ana.pavlovic@unibo.it

Abstract. The efficiency in converting solar energy into electricity is fundamental wherever photovoltaic panels are present, still more crucial in the design of racing solar vehicles. Even minimal reductions in conversion ratio, maintained for the long solar races, cause solar cars to lose race positions and competitiveness. Here we introduce a numerical-experimental study for choosing the best combination of materials to encapsulate cells in solar roofs. The tangible expectation is to improve the performance of the monocrystalline silicon cells used in our solar vehicle by maximizing heat dissipation to the environment. The operating temperature is in fact a determining factor for efficient conversion, with efficiency drops of the order of 5% every 10 °C. Different stratifications, some of which quite unusual in solar panel design, were compared by transient thermal simulations and experiments. Specifically, five alternatives were analyzed, varying in the presence and thickness of the encapsulation materials (ETFE, EVA and PET). The main scope of the work, however, was not choosing the best among several specific hypotheses, but the development of an accurate numerical model able to predict the behavior of the solar panel in conditions close to the expected ones. This model, in fact, has provided valuable help in optimizing the vehicle design by allowing to evaluate the effect of alternative materials and construction solutions in the cell's construction housing structure.

1. Introduction

For the last few decades, our world has been facing a wide variety of challenges such as climate change, depletion of natural resources and catastrophic fossil fuel impacts. All this resulted in the worldwide energy crisis [1]. This situation demands an urgent search for sustainable energy technology for the future. Nowadays, there is no doubt that solar energy is destined to be one key energy technology soon. Photovoltaics (PV) is the fastest-growing electrical energy generation source in the world [2] and has undoubtedly a "bright" future. Transition to photovoltaics contributes importantly to environmental protection. Solar energy is renewable energy and can be used in many ways, including solar streetlights, solar mobile batteries, solar buildings, and solar cars. With an increased interest in solar energy, solar cars are coming to the fore.

In order to counteract urban pollution and global warming, electric vehicles are an effective and powerful option. Solar cars refer to the electrical vehicle which is powered by solar energy gained from PV panels that are placed on the roof of the car. The PV module is an interconnected assembly of individual PV cells. Currently, many PV cell types are available in the market. However, because of the high cost, stability, availability of materials, reliability, and lifetime limitations, their widespread availability for commercial applications, such as solar cars, is limited. Besides the above factors, for racing conditions, the conversion efficiency of sunlight into electricity due to light-absorbing materials (such as crystalline silicon, amorphous silicon, gallium arsenide, and organic compounds) remains the most critical parameter [3]. Now, the crystalline silicon PV module is the best choice for onboard electrified vehicles in terms of performance, economy, stability, and availability. The typical efficiency of these cells is about 16%, while commercially this brings about 24% efficiency [4] [5].

In [6] Evans concludes that the efficiency of the PV module decreases by 0,38-0,42% with the temperature growth for silicon cells for one degree. Similar conclusions are also given in [7], which shows a decrease in the total output power of a PV system by 0.4–0.5% under standard test conditions when the module temperature rises by 1°C. In this paper, we are dealing with solar cells, and they have a fundamental effect on the PV module temperature change. Zhou and Yi [8] findings confirm this since the solar cell carried the highest temperature distribution in the PV module.

The current research essentially pursues two goals. The first is to develop a precise three-dimensional finite element (FE) thermal model of five laminated silicon solar cell layups using ANSYS analysis software. FE design helps us predict the actual condition of the PV module under operation and to determine the thermal behavior of each layer of the cell. The second aim is to validate the model by performing infrared (IR) thermography experiments for the same solar cell samples. These two steps will help to obtain the thermal characteristics of the PV module and consequently optimize its performance. For the accuracy of the model, the material, optical and thermal properties of all layers will be well defined. The numerical model with the scope to study photovoltaic modules performance was introduced in [9]. However, in that case, the research was stressed on solar radiation and ambient temperature effect. Furthermore, Chou et al. [10] and Chen et al. [11] studies included similar approaches. Nevertheless, research [10] was focused on the high-concentration PV solar package, while [11] was based on electronic packages.

Literature suggests some recent efforts in thermal modeling of solar cells and PV modules. The study [12] shows that the thermal behavior of the PV module was investigated by applying computational fluid dynamics (CFD) simulation along with the finite element method. Furthermore, to predict the effect of PV module temperature on output power and efficiency, both linear and nonlinear models were developed in [13]. Research [14] proposed a swarm optimization algorithm for calculating the temperature inside the PV module with the relation of the module and electrical efficiency. Moreover, [15] present arguments to emphasize that reducing an operating temperature contributes to a gain in cell and module efficiency. They indicate that operating temperature has a notable impact on the electrical efficiency of the PV module. This finding is supported by [16] who further point out that the temperature rise of the module is a significant issue that influences the efficiency of the solar cell. Moreover, study [17] states that PV cell temperature becomes also one of the key parameters for the long-term performance of the module.

However, none of the reported studies directly combines all specific aspects characterizing the present paper regarding the panel (i.e., monocrystalline silicon, advanced materials for encapsulation, additional layers for structural purposes, material alternatives) and methods (i.e., thermal transient analysis, absence of CFD dissipative effects, empirical validation). In the present research, several alternative layups of the PV modules were considered with the scope to investigate the effect of design changes in terms of thermal behavior and efficiency, before incorporating them on a solar vehicle [18] [19] [20]. Presented research is continuation of the previously presented paper [21].

2. Materials

2.1. Solar Cells Selection

Currently, monocrystalline silicon solar cells are considered being the best choice for onboard applications for solar vehicles because of their stability, cost, and overall performance. In this research, we use monocrystalline silicon wafer solar cells fabricated by KPE. Provided solar cells have the following main geometrical characteristics:

- Cell area: 125mm x 125mm
- Cell diameter: 150mm
- The shape of the cell: pseudo-square with rounded corners
- Average wafer thickness: 200~220, ± 30μm

At the same time, we consider the additional technical cell dimensions:

- Length of wafer edge: 125mm ± 0.5mm
- Corner to corner minimum / maximum length: 149mm / 151mm
- Flake type chips maximum width / depth: 1.5mm / 0.5mm

We also consider that the number of flake type chips should not exceed 2 per single solar cell. Moreover, monocrystalline silicon is cut from the cylindrical ingots which are grown by the Czochralski process. Because of this, the cell's corners look clipped. The performance of solar cells largely depends on their electrical efficiency. That is why electrical parameters should be also established. In our case, the chosen grade (i.e., PS1) provides a conversion efficiency of 17.25% ± 0.25 while the maximum peak power for the chosen case is 2.56 ± 0.05 Wp. All parameters are measured respecting standard testing conditions: 1,000 W/m^2, AM 1.5G, 25°C.

2.2. Sample composition

The PV module used in our paper comprises an individual solar cell. To preserve the high efficiency and durability, solar cells require corresponding protection which is achieved by a complex mix of various materials. The module is formed by laminating the following layers:

- Frontsheet: ETFE
- Encapsulant: EVA
- Solar Cell: Monocrystalline silicon cell
- Encapsulant: EVA
- Backsheet: PET

This combination was chosen since each material has a property that directly affects the cells' performance.

2.2.1. Frontsheet

The frontsheet typically gives most of the protection, rigidity, and structure for the shape of the cell. Glass is commonly used as a frontsheet in PV modules due to its durability and rather high transmittance. However, in the current layup, we introduced the ETFE material as the frontsheet. ETFE fabricates from a copolymer of ethylene and tetrafluoroethylene resin and afterward extruded into incredibly thin films.

ETFE is a fluorite-based plastic widely applied in the automotive industry that can suffer high temperatures while promoting solar efficiency. This became possible due to the ethene molecule which is extremely stable and allows to withstand high thermal stress and volatile chemicals. In general, the initial function of ETFE was to take over greenhouse glass, and that is why ETFE is lighter and stronger [22]. The major advantages of ETFE compared to glass are weight reduction, extremely high transmissivity, and flexibility of the laminated panel. Essentially, the weight of the ETFE coat is merely 2% of a glass sheet weight. Moreover, the ability of the ETFE to transmit light from ultraviolet to infrared (full light spectrum) brings to an improvement in the efficiency of the entire module. Electric failures are avoided because of the much more efficient insulating power provided by ETFE. The properties such as self-cleanness, excellent electrical resistance, stiffness, resistance to radiation, impact strength and high corrosion resistance make ETFE exceptionally attractive as a frontsheet of a solar cell [23]. Finally, advantage compared to glass is that ETFE is easily recyclable at the end of the solar panel life, and its lightness makes it possible to make a much smaller carbon footprint produced by transportation.

2.2.2. Encapsulant
An essential component of the PV module that produces optical communication, structural support, the thermal conductivity of electrical and physical insulation is an encapsulant [24]. An encapsulant is a material that melts, fully encloses the solar cell and the associated circuitry. Currently, the most used encapsulant in the PV industry is ethylene-vinyl acetate (EVA) copolymer. We chose EVA as an encapsulant since it has good bonding characteristics with our ETFE frontsheet and PET backsheet. EVA links the rest of the layers because of its adhesive nature. At the same time, a thin layer of EVA prevents contamination between the layers since it creates a clean particle-free vacuumed space.

2.2.3. Backsheet
The backsheet is the last layer of material in the PV module. The key tasks of the backsheet are protection, keeping the shape of the entire module and additional insulation of the module from humidity. In this work, polyethylene terephthalate (PET) was taken as a backsheet. PET is plastic based on a fluorine molecule that makes this material very flexible. Such mechanical parameters as exceptionally small Young's modulus and a greater elastic elongation in comparison with other materials prove that PET is one of the best flexible materials at the moment [25]. Afterward, flexibility becomes an extremely critical property for solar cell's subsequent installation into the car's roof. Furthermore, PET refers to those flexible materials that are greatly transparent and show optical transmittance of ~85% over the visible range [26].

2.3. Lamination process
A highly transparent ETFE frontsheet takes a role of a protection layer. ETFE is followed by a thin EVA sheet that spreads all over the frontsheet. We place the EVA layer on the solar cell with the photosensitive surface facing down, while another EVA layer is used to cover the solar cell's backside. Finally, the PET backsheet, which represents an insulating plastic material, is positioned on the second EVA layer.
The next step is a lamination of all layers, which was completed in an autoclave. A plate of glass (removed at the very end) was used to provide planar support for all layers during the autoclaving process. For example, in a standard laminator, the machine worktable takes the function of this glass plate. All pieces were laid on a plate of glass in the following order: PET, EVA, Silicon, EVA, EFTE (Figure 1a).
The entire lamination process takes place in a vacuum in order to overcome the effect of contaminations and penetrating gases. To obtain a vacuum, we used a vacuum bag and a vacuum pump with an applied pressure of 1 bar (Figure 1b). The absence of air helped to achieve structural stabilization of the system and uniform polymerization. For the EVA sheets to polymerize, the plate was heated to 120°C for 90 minutes. Also, the vacuum allows the EVA to be uniformly distributed over the other layers, excluding

the non-functional edges. The following step is cooling (Figure 1c), after which the solar cell is taken out from the vacuum bag. Finally, double-sided tape (cod. 468MP. by 3M) connects to the edges of the entirely laminated solar cell, which subsequently links the solar cell to the carbon fiber composite polymer (CFRP). CFRP is used for the purpose of representing actual conditions since the roof of a solar racing car is made from the same material.

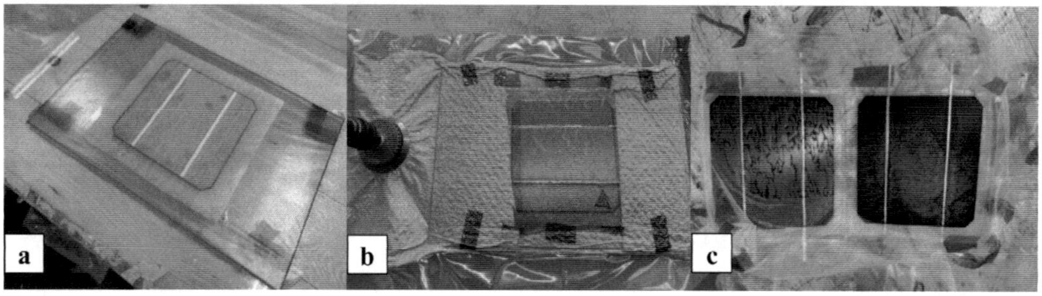

Figure 1. Solar cell lamination process: a) layers stuck (EFTA, EVA, Silicon, EVA, PET) putting on a glass sheet as support for lamination (back view); b) vacuum treatment (1 bar) with a vacuum bug and vacuum pump (back view); c) solar cells after curing in the autoclave (front view).

2.4. Layer Combinations
In this study, we analyzed 5 different layups, which were obtained by removing one or two layers and changing the thickness of the EVA encapsulant (see Table 1).
The first layup involved a complete set of all five layers. Several additional investigations were completed to examine the effect of the various layers on solar cell temperature behavior. Therefore, the second layup was missing the backsheet, while the third layup was tested without the frontsheet. It is necessary to indicate that in all three layups, we used a thick EVA. In comparison, in the fourth layup, we cut down the thickness of the EVA by half (showed as EVA Halved in Table 1). At the same time, in the fifth layup, the thickness of the EVA was diminished by 10 times from the initial thick EVA (EVA thin). In all the aforementioned compositions, double-sided tape and a CFRP plate were used.

Table 1. Variations of layups under investigation.

Layer	Function	Material	Thickness (mm)	Layups				
				1	2	3	4	5
1st	Frontsheet	ETFE	0.28	+	+			+
2nd	Encapsulant 1	EVA Thick	0.2	+	+	+		
		EVA Halved	0.1				+	
		EVA Thin	0.02					+
3rd	Solar Cell	Silicon	0.15	+	+	+	+	+
4th	Encapsulant 2	EVA Thick	0.2	+	+	+		
		EVA Halved	0.1				+	
		EVA Thin	0.02					+
5th	Backsheet	PET	0.2	+		+	+	
6th	Adhesive	Tape	0.13	+	+	+	+	+
7th	Support	CFRP	2	+	+	+	+	+

2.5. Physical and thermal materials' properties
The next essential step was to determine the thermal properties of all materials used. Table 2 illustrates the thermal properties of materials along with references t where they were collected from. With some polymers, properties differed depending on chosen reference and therefore it required a specific choice

(such as an average value). However, the variations between the values were not very significant and thus it did not affect the eventual results of the research.

Table 2. Physical and thermal properties of selected materials.

Material	Density kg/m³	Thermal Conductivity W/(m·K)	Specific Heat Capacity J/(kg·K)	References
ETFE	1730	0.24	1172	[27] [28]
EVA	945	0.35	2090	[29]
Silicon	2330	148	700	[30] [31] [32] [33]
PET	1350	0.275	1275	[34] [35]
CFRP	1490	6.83	1130	[36]
Tape	1012	0.19	2000	[37]

3. Methods

3.1. Theoretical background

Drawing an analogy between the human body and the solar panel, we state that the sun's rays affect the body in such a way that its temperature becomes greater than the ambient temperature. Parameters such as ambient temperature (T_{amb} in K), specific heat (Q_s in W/m^2), body's absorption coefficient (c_a) determine the thermal equilibrium of the body.

At the same time, if the physical system comprises several layers formed of various materials, it becomes extremely problematic to figure out the overall absorption coefficient. The physical system analyzed in this work is such a system and is illustrated in Figure 2. In this case, it makes sense to describe the thermodynamic problem. Therefore, useful relationships and considerations are given below.

Firstly, induction heat exchange with the external environment is negligible due to the fact that our physical system is closed. The system can be considered closed since the solar panel does not touch the floor, or in other words, our physical system is isolated from the bottom.

Secondly, the following expression is valid for thermal equilibrium:

$$Q_{source} = Q_{conv} + Q_{irr} \tag{1}$$

where Q_{source} is heat transferred from an energy source to the body.
Q_{irr} - heat emitted by the body through radiation.
Q_{conv} - heat emitted by the body by convection.

Both Q_{conv} and Q_{irr} depend on body temperature (T_{body}) and ambient temperature (T_{amb}).

Newton's Law describes the behavior of convection, which is proportional to the difference between body and ambient temperatures:

$$Q_{conv} = h \cdot |T_{body} - T_{amb}| \tag{2}$$

where h - the heat transfer coefficient by convection.

At the same time, the behavior of the irradiation is expressed under the Stefan-Boltzmann law:

$$Q_{irr} = A_0 \cdot \sigma_0 |T_{body}{}^4 - T_{amb}{}^4| \tag{3}$$

where $\sigma_0 = 5.67 \cdot 10^{-8}$ W/m^2·k - the *Stefan-Boltzmann constant;* A_0 – exposed area.

The area A_0 is the orthogonal projection of the area of the irradiated body to the angle of the incident rays (a). This area is also referred to the cosine of this angle of incidence defined by Lambert's cosine law. However, we deal with the entire area of the panel since in our experiment $a = 0$ and

cos(a) = 1. Moreover, such physical aspects as the body's geometry, the properties of fluid and air, as well as the conditions of their possible motion are directly related to the coefficient of convection ***h.*** The range of values suitable to the system under test is ***h*** = 2.30-2.75 W/m^2·k. They were determined by making use of parameters such as Nusselt and Grashof numbers. The exact estimation of these parameters can be directly delegated to the simulation code.

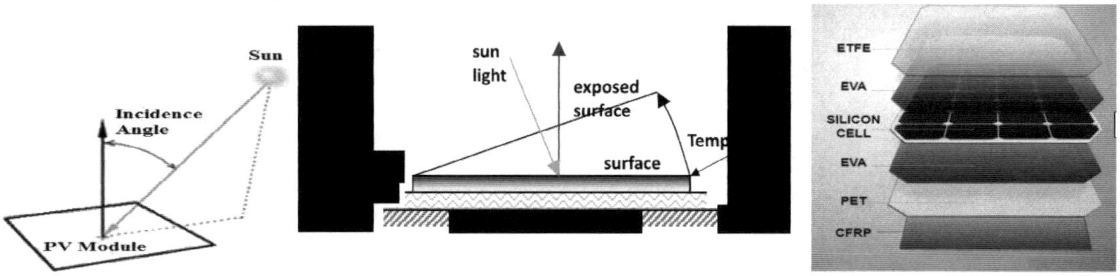

Figure 2. Heat transfer scheme where the radiation acts (with incidence angle ***a***) on a rectangular body in the environment changing its temperature. This body comprises a PV module, made by stratification of different layers (ETFE, EVA, Silicon Cell, EVA, PET), laid on a CFRP panel, before its positioning on supports.

3.1.1. Temperatures
When all additional considerations are identified (geometry, materials, etc.) it is sufficient to apply equations (1)-(3) in order to determine the body temperature ($\mathbf{T_{body}}$), since the heat transmitted by the source ($\mathbf{Q_{source}}$) and the ambient temperature ($\mathbf{T_{amb}}$) in our situation are known (1000 W/m^2 and 25 °C respectively). Furthermore, measuring temperature experimentally is another efficient way to determine $\mathbf{T_{body}}$, which was shown in this work as proof of the simulation results.
As we can observe in figure 2, the solar panel in this work consists of the stratification of various layers, some of which are even transparent. In this regard, it is necessary to clarify what the temperatures from equations (2) and (3) represent.
The surface temperature controls heat transfer, while the change in temperature inside the body is determined by thermal conductivity with a heat flux proportional to the thermal gradient (Fourier's law for thermal conductivity). This ensures the redistribution of heat, which affects the temperature in ways that directly depend on the thermodynamic system. For example, a linear trend between a higher temperature on an exposed surface and a lower temperature on a hidden surface (in the shade) is obtained when the sun's rays are exposed to a homogeneous panel. However, it is highly important to consider that in the case of a solar panel, each layer has its own thermal conductivity, which has a direct impact on the appearance of more complex temperature trends.

3.1.2. Irradiation and diffusion
Concerning the phenomenon of diffusion, the temperature of the outermost layer which is in contact with air serves as the temperature ($\mathbf{T_{body}}$) in equation 2. This wall temperature ($\mathbf{T_{wall}}$) is equivalent to the temperature of the first layer, usually represented by the frontsheet (unless there is no frontsheet in the layup). It is also essential to consider the phenomenon of irradiation. The temperature in equation 3 usually refers to the first layer. However, in our case, the temperature of the silicon is considered to be the wall temperature $\mathbf{T_{wall}}$, since the two layers above silicon (EFTE and EVA) are practically transparent. The silicon temperature relates to the one measured experimentally.

3.2. Numerical simulation

3.2.1. Software and analysis choice
In this research, we established a numerical simulation using the transient thermal analysis in ANSYS Workbench® (Ver. 2020.R2). Since an essential part of the investigation was to determine thermal

variations over time, we preferred transient thermal analysis to a steady-state test. Such thermal factors as temperature distribution, heat fluxes, heat gain or loss, and others characterize thermal analysis. The system also does not undergo fluid dynamics changes such as flowing water or air, so CFD analysis was unnecessary.

3.2.2. *Model building*

After an appropriate analysis system was chosen, the setup of the model starts with adding materials and specifying their properties. Table 1 and Table 2 show the thickness, mass, thermal conductivity, and specific heat capacity for each layer. The geometry was then created using ANSYS Design Modeler, the computer-aided design (CAD) program used by ANSYS. The geometric model was built by stacking 7 layers of 125x125 mm, and then the corresponding material with the properties already specified in the previous step was added to each of the layers. The layers were assembled with bonded contacts, which ensure the full transmission of forces and heat. In general, the geometry of the simulated sample was reproduced precisely identical to the actual sample up to the clipped corners at the edges (see section 2.1 Solar Cells Selection). In total, five different designs were generated, one per layup.
The next step in the computational analysis was to break up the geometry into pieces in a process called meshing. This discretization was completed with finite elements (FE) in SOLID186. The finite element is a higher-order solid element that exhibits quadratic displacement behavior and is determined by 20 nodes each with 3 degrees of freedom (i.e. nodal displacements along x, y, z). We carried out the discretization of the system for (up to) 198060 nodes and 27702 elements.

3.2.3. *Thermal hypotheses*

The following hypotheses and conditions were used for the numerical thermal analysis in finite element software to obtain the temperature distribution of the PV module:

- The total irradiance flux of 1000 W/m^2 acts on the panel uniformly and orthogonally.
- The first ETFE layer reflects 7% (= 70 W/m^2) and absorbs 10% (= 100 W/m^2) of the transmitted irradiance.
- The remaining energy (= 830 W/m^2) is not applied on EVA but the silicon cells in the form of constant heat flux.
- Only 3% of the heat flux (= 25 W/m^2) is reflected from the silicon and reaches the ETFE front sheet, while the rest is absorbed by the silicon layer (= 805 W/m^2).
- The transmissivity of the EVA is taken to be unity, which simplifies EVA to fully transparent matter.
- Heat is transferred between sheets by conduction.
- Heat transfer outwards gets through the ETFE and CFRP layers, respecting the phenomena of radiation and convection.
- Transmission and reflection are carried out by silicon layer in the absence of the ETFE frontsheet (3rd and 4th layups).
- For ETFE and CFRP, the emissivity (generally equal to the absorption coefficient corresponding to Kirchhoff's Law) was taken equal to 0.89.
- The convective heat transfer coefficient respectively to the ambient temperature reaches 5.67 $W/m^2\,°C$ (e.g. no wind conditions).
- Material properties are temperature independent.

3.3. *Experimental validation*

To prove the finite element analysis that can predict the temperature distribution of the solar PV module, we additionally performed Infrared Thermography (IRT) measurement. For this test, suitable equipment was used under controlled irradiation and environmental conditions to heat the PV module. At the same time, the temperature was measured over the entire surface of the module. Five different PV module layups (see Table 1) were thermally measured, while each layup measurement was repeated 3 times to achieve precise results. During the examination, all samples were placed on cardboard support on which solar lamps were directed (see Figure 3). The laminate real-time temperatures were tracked using a FLIR T440® thermal camera with a frame rate of 60 Hz, spectral range between 7.5 and 13 um, and a thermal resolution of 76.800 pixels. This equipment provides thermal and visual imagery, spot size resolution, and reliable temperature measurement even in the transition phase when the sample has not yet reached final equilibrium. The tests were conducted under the standard conditions with an irradiance of 1000W/m^2 and ≈ 24.5°C ambient temperature (varying ±1.5°C). As a result, we discovered a temperature rise over the entire surface of the PV module.

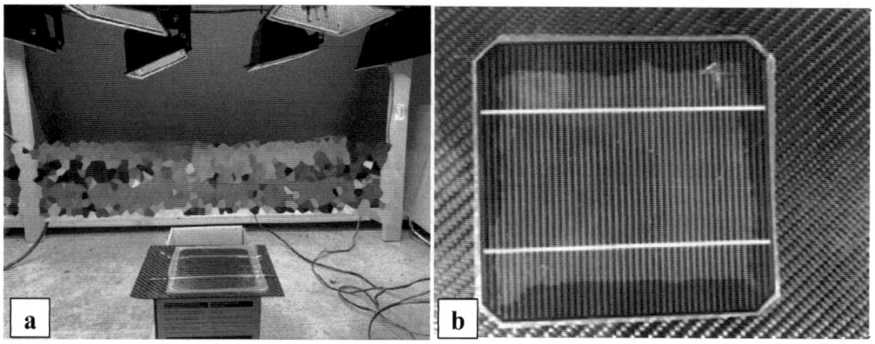

Figure 3. Entirely laminated solar cell under investigation: a) experimental mock-up; b) solar cell.

4. Results and Discussion

4.1. Simulation results

The heat distribution on layers of the laminated solar cells was collected using numerical simulation. In total, 15 tests were completed since 3 evaluations were run for 5 various specimens. In Figure 4, we can identify the temperature propagation over the silicon solar cell (3rd layer) for a 5th layup (absence of backsheet and reduced thickness of EVA). For a visual presentation of temperature propagation, six arbitrary time points were taken (60 sec, 150 sec, 450 sec, 750 sec, 900 sec, 1200 sec). Overall, we can clearly observe the temperature upward trend over time. During the 1st minute, the temperature raised from the ambient temperature (≈ 24.5°C) to 35 °C, while during the next 90 seconds the raise continued and equaled to 11°C. After 450 seconds from the starting point of the test, the temperature reached 65°C, whereas after the first 750 seconds the temperature came to 72°C. The dynamic growth continued resulted in 73.5°C during the following 150 seconds until reaching the equilibrium value (75°C obtained in 1200 seconds).

Moreover, the selected type of analysis helped to realize the heat distribution along each of the layers of the laminated sample independently. Figure 5 illustrates the differences in temperatures between 6 various layers in the case of thermal equilibrium (75°C). The temperature is uniformly distributed over the surface, which does not allow to get details by automatic color scales. This fact makes it practically inconceivable to perform a visual assessment. Due to this reason, the comparison of numeric parameters was done instead of color maps comparison.

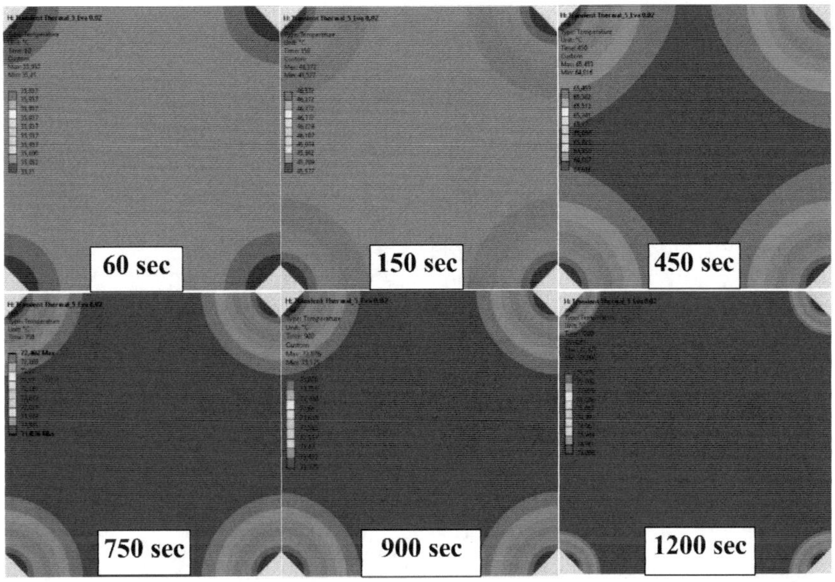

Figure 4. Simulated heat distribution over silicon solar cell layer in 5 arbitrary time points in case of 5th layup: 60 sec, 150 sec, 450 sec, 750 sec, 900 sec and 1200 sec.

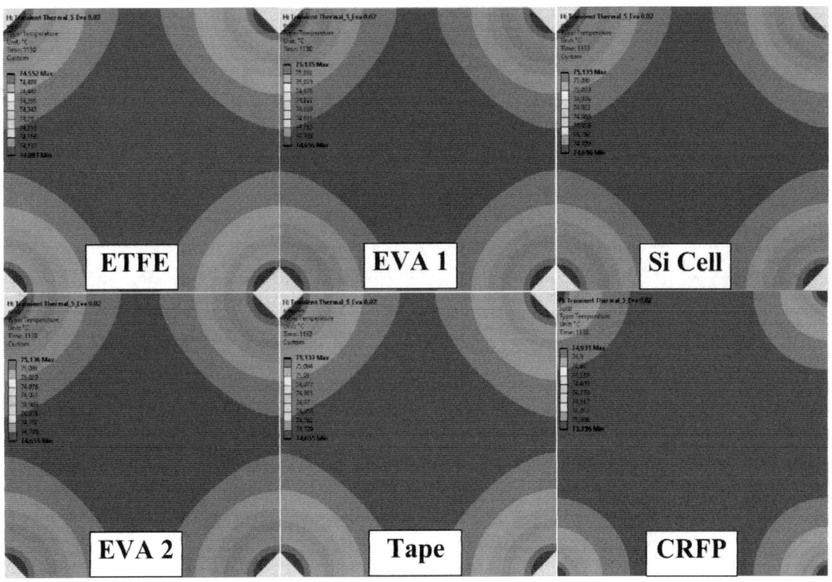

Figure 5. The temperature distribution of each layer of the laminated solar cell at the time of the thermal equilibrium (case of 5th layup, steady state).

The highest temperature values of all layers for various layup options obtained after the simulation are given in Table 3. Additionally, Figure 6 shows the temperature contrast for each layer and each component independently. As a result of the retrieved analytical data, we clearly observe that the peak temperature of the EVA encapsulant does not evolve further than 4 degrees, regardless of the chosen layup. At the same time, as predicted, because of the absence of the frontsheet (3rd and 4th layups), the temperature of the EVA layer came to the maximum values in comparison with the rest of the layups. At the same time, the peak temperature change for double-sided tape and CFRP did not exceed 4.15 degrees. Finally, for the case of the silicon solar cell, we see that the highest temperature varies within 4 degrees from one layer to another.

Table 3. The highest temperatures of all layers detected during the simulation taking into account 5 various layups.

Layer		1st	2nd	3rd	4th	5th	6th	7th
Function		Frontsheet	Encapsulant 1	Cell	Encapsulant 2	Backsheet	Adhesive	Support
Material		ETFE	EVA	Silicon	EVA	PET	Tape	CFRP
Layups	1	67.14	74.12	73.99	74.01	73.90	73.59	73.59
	2	75.40	76.01	77.26	76.25	-	76.08	76.07
	3	-	77.63	77.80	77.81	77.70	77.40	77.40
	4	-	77.87	77.96	77.96	77.96	77.74	77.73
	5	74.56	75.15	75.15	75.14	-	75.15	75.14

Examining the details of heat transitions between layers, we discovered that the maximum temperature with 3rd, 4th and 5th layups fluctuates by less than 1 °C. However, in 2nd layup, the peak temperature contrast between the ETFE frontsheet and the silicon cell reaches approximately 2 °C. Furthermore, by analyzing the peak temperatures between the ETFE sheet and the first EVA layer in the case of the 1st layup, we detected a sharp upsurge of about 7 °C.

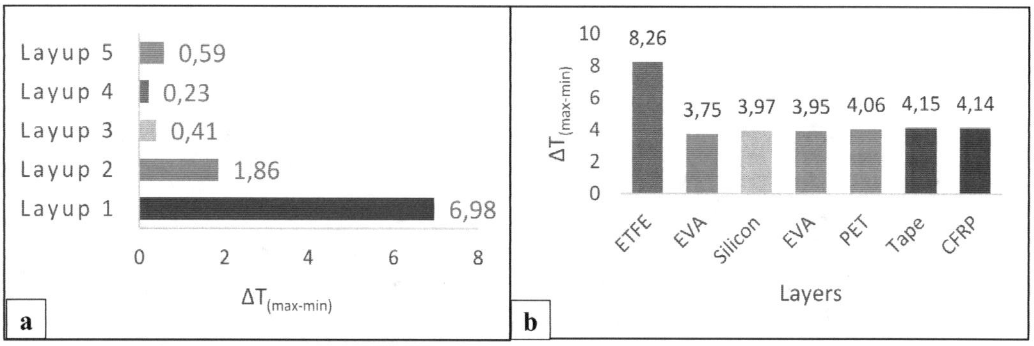

Figure 6. Temperature difference $\Delta T_{(max-min)}$: a) between 7 layers within a single layup; b) between 5 layups within a single layer of the laminated cell.

A direct graphical comparison of the highest temperatures obtained with different layups is displayed in Figure 7a. Here we can see that visually the maximum temperature of the various layers does not change much depending on the layup used. The exception is the 1st layup (or reference layup), considering all 7 layers and with the regular thickness of the EVA layer (0.2 mm), where the ETFE temperature is significantly lower than in the case of other layups (which was already mentioned in the analysis of Table 3). At the same time, Figure 7b illustrates a comparison of the maximum temperatures but in the context of layers of a laminated solar cell. First, because of this analysis, we can certainly see that removing the PET backsheet (2nd layup) has a negative impact on the thermal performance of the entire solar cell, as the temperature of each layer increases compared to a reference sample (1st layup). Second, the established model demonstrates that the layup without the presence of the frontsheet (3rd layup) also adversely affects the temperature of the entire sample. In particular, the temperature of each layer was raised sufficiently by about 4 °C related to the reference layup (except for two EVA sheets, where the maximum temperature increased by ~3.5°C). Next, a drop-off in the EVA thickness by two times (4th layup) basically did not affect the temperature characteristics of the sample, enhancing the maximum temperature of each layer by less than 0.3°C. Finally, considering the absence of the PET backsheet followed by a tenfold cut down in the thickness of the EVA (5th layup), the sample's thermal characteristics were improved (on average 1°C lower T for each layer) compared to an identical specimen with a regular thickness of the encapsulant (2nd layup). Furthermore, it should be noted that the 5th layup is the second most efficient layup after the 1st in terms of temperature characteristics. Except for the anomalous difference in the maximum temperatures of the ETFE, the temperature of all other layers increased slightly by ~1°C.

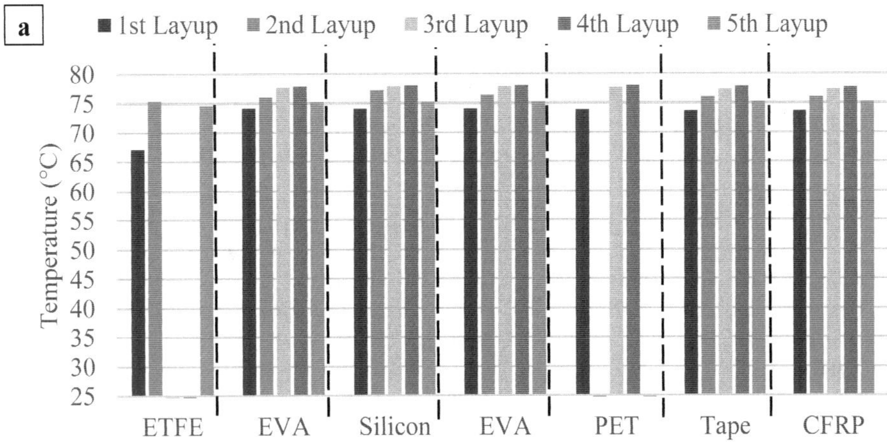

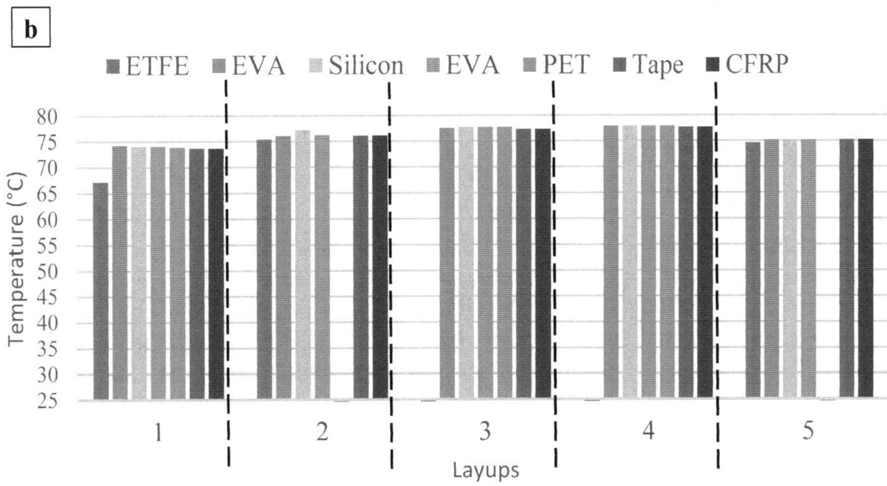

Figure 7. a) Maximum temperatures comparison in terms of different layups; b) maximum temperatures comparison in terms of different layers.

In addition to the classic thermal evaluation of the solar cell, it was essential to assess how one or another layup influences the weight of the fully laminated solar cell and consequently the solar panel weight since it is an extremely crucial parameter of a solar car. That is why the specific weight of the solar cell was determined for all 5 layups. Since the weight is directly proportional to the specific weight of the material (specified in Table 2), it was sufficient to compare the specific weights of the stacked layers of all 5 solar cell variations (see Figure 8) in order to get a full picture of lightness improvement.

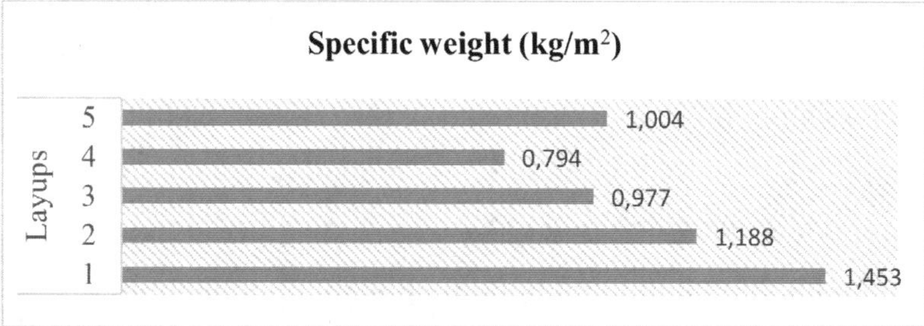

Figure 8. Specific weights of stacked layers in the case of 1st - 5th layups.

According to the results, the ETFE removal in the case of the 2nd layup was followed by a weight reduction of the entire structure by 18.2%. At the same time, a weight decreases of 18.7% was discovered in the absence of an ETFE frontsheet and an EVA thickness cut down from 200mm (3rd layup) to 100mm (4th layup). If the usefulness of running the 2nd layup is still under investigation due to the previously reported rise in the sample temperature, then a twofold loss in the EVA thickness provides a substantial decrease in weight with a minor rise in temperature. Therefore, if weight reduction turns into the primary goal, then applying the 4th layup instead of the 3rd one becomes a relevant choice. Finally, dealing with 5th and 1st layups it can be stated that PET backsheet removal can substantially reduce weight by 40%, accepting the cell's slight temperature rise of 1 °C.

4.2. Experimental results

The results of an infrared thermographic experiment were temperatures measured until the solar sample attained thermal equilibrium. As an illustration of the thermal behavior of the sample during an examination, Figure 9 displays the results of the 5th layup measurements at 5 specific points in time (t = 60 sec, 150 sec, 450 sec, 750 sec, 900 sec) using the FLIR T440 thermal camera. From Figure 9 we can observe that the cross (or probe) was concentrated on the center of the solar cell, where, in theory, the maximum temperature should be focused. This allowed to define the temperature values in the same place and afterward make a corresponding comparison. Also, on the right side of each picture, there is a temperature scale that was set before the start of the experiment. As a result, it allows to compare pictures at different times and visually determine the important heat propagation features over the sample. Further, we assumed all measurements refer to the 3rd and most significant layer of the laminated sample, i.e. the silicon solar cell. This assumption was based primarily on the two following points:

- the initial two layers of the solar cell (ETFE and EVA) are highly transparent.
- simulation results showed that temperature changes between EVA and silicon are minimal.

4.3. Experiment vs simulation comparison

First, as a result of the experiment, it was discovered that the heat flux does not spread uniformly over the surface of the cell, although the opposite was considered during the simulation. In fact, there was a slight shift in heat concentration towards the left part of the sample. This led to a minor deviation from the theoretical data because the maximum temperature moved from the center to other zones of the solar cell. In the end, the difference between the maximum registered temperature and the measurements in the center of the cell came to 5-6°C. However, by measuring and confirming the irradiation value (1000W/m2) at the center of the sample, the effect of inhomogeneity is overcome. Second, it was found that the regularity of the thermal distribution depends on the boundary conditions and local states at the boundaries. Besides this, in accordance with the simulation, the experiment clearly proved that the sample's edge regions are cooled unevenly.

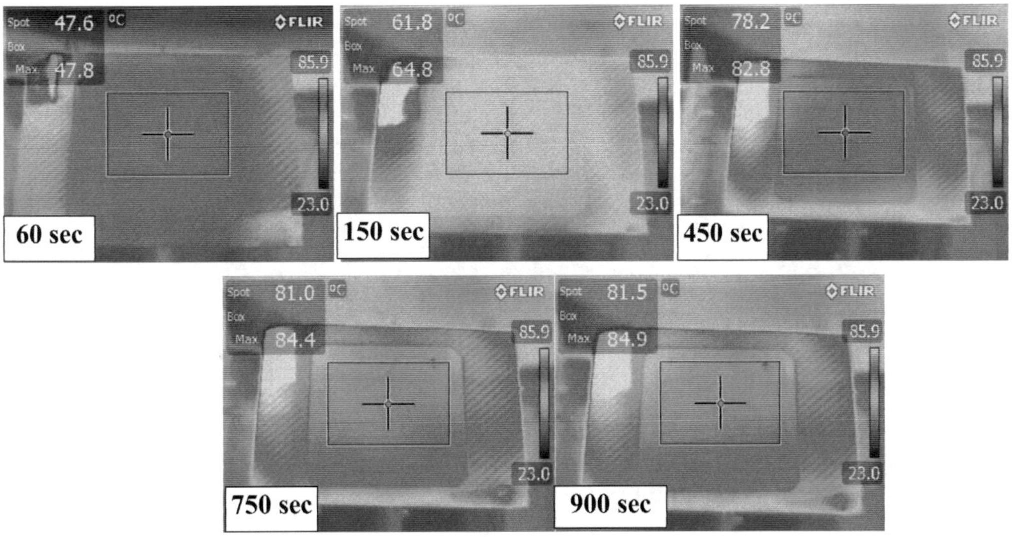

Figure 9. Experiment results: heat distribution over silicon solar cell layer in 5 arbitrary time points in case of 5th layup: 60 sec, 150 sec, 450 sec, 750 sec, 900 sec.

4.4. Thermal steady-state analysis

Figure 10 illustrates a comparison between experimental and simulation results in the case of steady-state conditions for the silicon cell layer. Specifically, 3 empirical tests were completed for each of the 5 layups (tests for a specific layup are marked with the same color, for example, 3 yellow bars in the case of the 2nd layup). In addition, green bars display the finite element simulation results. As a more accurate identification of the data, Table 4 was established, which included the average temperature value for each of the layups. In addition, Table 4 introduces the standard deviation for each layup, as well as the percentage error. It is worth noting that all experimental measures did not undergo strong differences from test to test, as evidenced by a standard deviation of ~1°C for 2nd-4th layups (for 5th layup even reached 0.3°C). However, the deviation value reached 3°C in the case of the 1st layup, which is 4%. Another interesting point is that the average temperature of the silicon layer varied within 10°C, and the greatest variation was recorded between 1st and 5th layups. Eventually, after a direct comparison of the experimental and simulation data, we identified that the simulation predicted the average experimental temperatures with an accuracy of 2.8% in 4 cases out of 5, which is an exceptionally precise result.

Table 4. Differences between empirical temperatures at the midpoint and FEM analysis temperatures at the midpoint.

Layup	Test 1	Test 2	Test 3	Average value	Standard deviation	Simulation Results (FEM)	Percentage error (%)
1st	67.6	72.7	73.4	***71.2***	*±3.2*	**73.2**	2.8
2nd	75.1	74.7	76.1	***75.3***	*±0.7*	**75.4**	0.1
3rd	76.2	77.0	78.3	***77.2***	*±1.1*	**77.8**	0.8
4th	78.7	80.3	80.6	***79.9***	*±1.0*	**78.0**	-2.3
5th	81.2	81.4	81.7	***81.4***	*±0.3*	**75.5**	-7.3

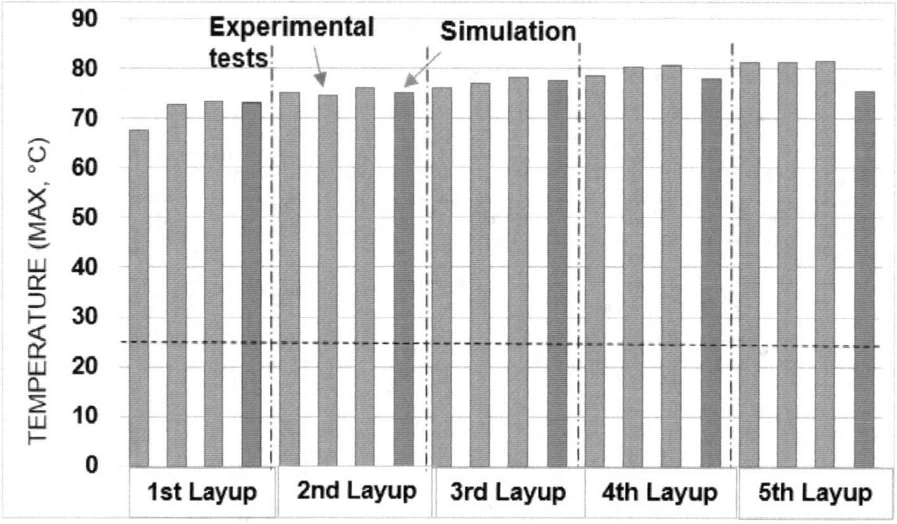

Figure 10. Comparison between experimental (3 tests per layup) and simulation results (1 per layup) in terms of the maximum temperature for each layup.

4.4.1. Thermal transient analysis

Figure 11 depicts a correlation between experimental and simulation results in the case of transient conditions for the silicon cell layer. Like the steady-state investigation, Figure 11 takes into account all 5 possible layups. Analyzing the following graphs, the first factor to note is that practically all experimental curves are superimposed on each other in the case of every layup. This is especially true for the curves of Test 2 and Test 3, which overlap almost perfectly regardless of the selected layup. For the 2nd and 5th layups, empirical curves for all 3 tests are completely overlapped. At the same time, for the 1st, 3rd and 4th layups, the Test 1 curve is located slightly lower than others, which indicates an inclination towards lower temperature values. Since the temperature variations between the tests are not significant (several degrees °C), a variation in the ambient temperature (**T_{amb}**) can be a rational explanation. The next step is to relate the experimental curves to the curve obtained from the simulation. We observe that in almost all layup options (except for the 5th layup), the analytic curves successfully predict the limiting values of the experimental curves, which was also demonstrated in the steady-state analysis. However, it is crucial to emphasize that significant differences are present in the transition zone. This attitude of the analytic curve may be due to the fact that during the experiment, the system accumulates heat much faster than predicted. Another possible reason for this behavior may be the inability to determine the thermal insulation of the layers under the silicon accurately.

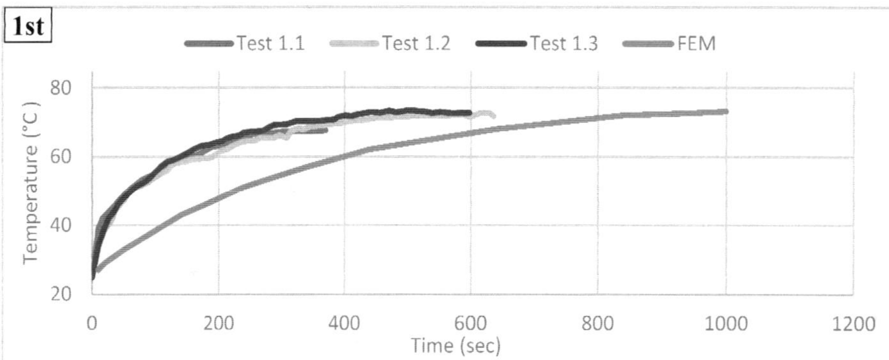

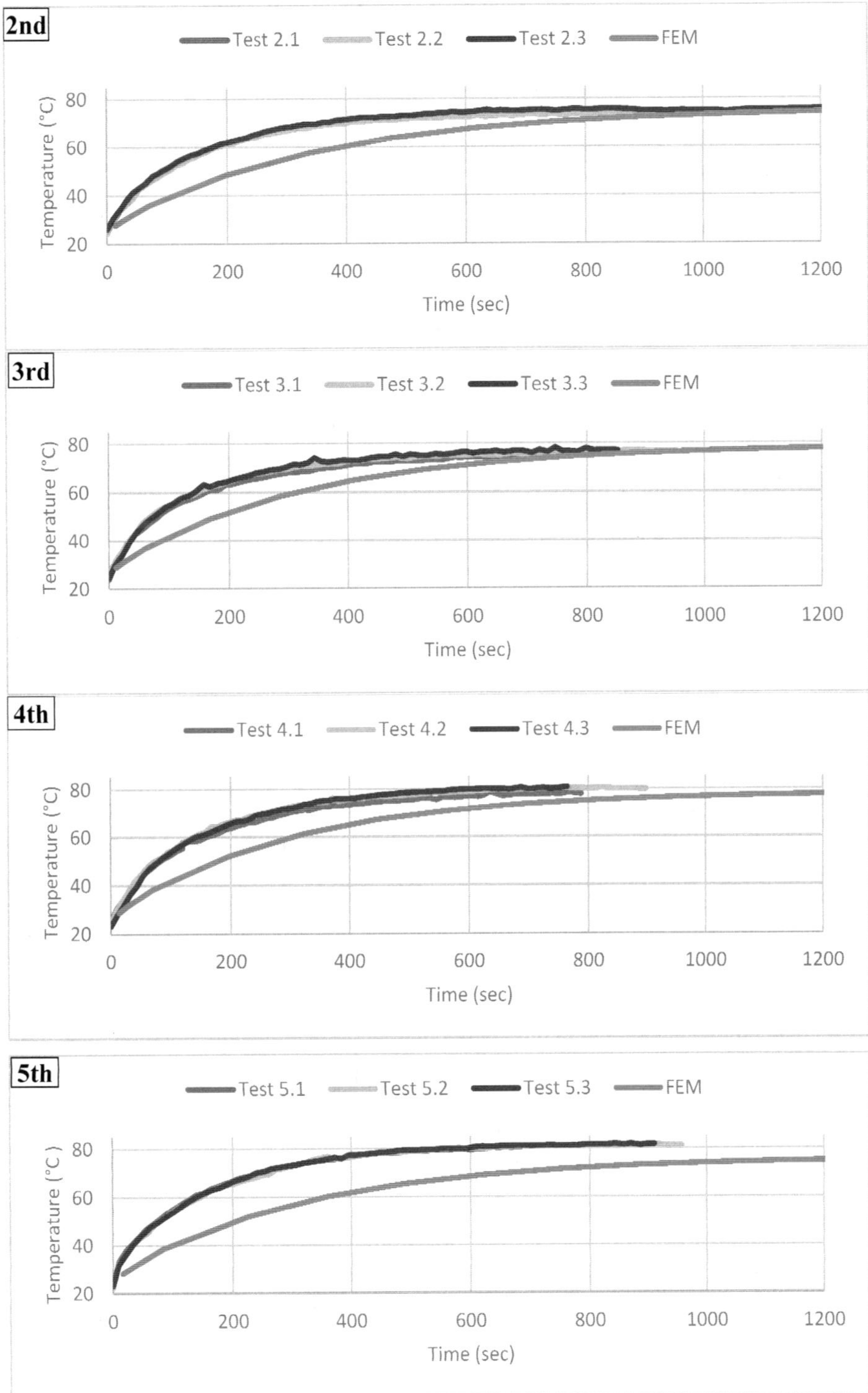

Figure 11. Comparison between experimental (3 tests per layup) and simulation FEM results (1 per layup) in the case of transient conditions.

5. Conclusions

The highest efficiency in solar energy conversion is a fundamental aspect wherever photovoltaic panels

IOP Conf. Series: Materials Science and Engineering 1214 (2022) 012042 doi:10.1088/1757-899X/1214/1/012042

are present, but it is even more critical in the case of racing solar cars. Minimal losses in the conversion efficiency, which however remain present for the many hours of solar racing, can cause the solar car to lose several positions in the standings, especially influencing top vehicles.
The nominal power of photovoltaic panels usually refers to standard test conditions, with solar radiation of 1,000 W/m^2, an air density of 1.5 kg/m^3 and a temperature of the photovoltaic cells of 25 °C. In many cases, however, operational temperature conditions should be considered since the thermal coefficients of the panels can vary considerably. In the case of silicon panels, for instance, this change reaches the highest value between all different technologies, up to 0.5% /°C. It means that the efficiency of the panel can be reduced by 5% every 10 °C of temperature over the standard 25 °C. Therefore, in the presence of cell temperatures of 60°, the drop with respect to the nominal power values can reach up to 17-18%.
Here the goal was to improve performance by maximizing thermal dissipation thanks to proper design solutions, involving both the aspects of materials selection and layer thicknesses, for encapsulating monocrystalline silicon solar cells. In this numerical-experimental study, five functional stratifications of monocrystalline silicon cells were made and compared through a transient thermal analysis, validated thanks to experimental temperature measurements.
These results are very promising with a good prediction of temperatures in stationary conditions (less so in transient ones). In some cases, an extreme accuracy was obtained with errors of the order of 0.1-0.8%. Even when lower (7%), the accuracy is enough for an appropriate model utilization.
This kind of information allows finding the right compromise in the cell layup design between the lowest temperature on the cells able to guarantee the highest conversion efficiency and the structural lightness, which cannot be reduced at will, given the functional purposes of each material layer. For instance, it was possible to conclude that a remarkable design option consists in the removal of the last layer (i.e., PET backsheet) to significantly lighten (-40%) the solar panel. This is possible only if the panel is properly inserted in the vehicle design [38] [39] considering convenient support and protective structure (Figure 12).

Figure 12. Solar cars recharging batteries in a pit-stop during the American Solar Challenger: our vehicle is on the right.

The aim of the work, however, was not so much to choose the best among the specific construction hypotheses of the photovoltaic modules investigated here, but above all to set up an efficient numerical model for solar cell design. Thanks to this new validated model it is now possible to easily investigate the behavior of the alternative solar panels in conditions closer to the ones expected from reality. In this sense, different and more complex boundary conditions are awaited in the case of the solar vehicle such as the true three-dimensional geometry of the roof, the angle of incidence of the sun in its different positions, the presence of EV modules side by side, airspeeds while the vehicle is moving. Moreover, in terms of vehicle design optimization, the new numerical model allows to verify the effect of aspects such as the use of different geometries and materials in the solar panel housing structure, as well as the actual usefulness of different passive and active cooling hypotheses proposed by the designers.

6. References

[1] Zendehdel M, Yaghoobi Nia N and Yaghoubinia M 2020 Emerging Thin Film Solar Panels. Reliability and Ecological Aspects of Photovoltaic Modules.
[2] Pvps Iea 2018 Snapshot of global photovoltaic markets. Report IEA PVPS T1-33.
[3] Minak G, Fragassa C, and de Camargo F V 2017 A brief review on determinant aspects in energy efficient solar car design and manufacturing. In International Conference on Sustainable Design *and Manufacturing*, 847-856.
[4] Abdelhamid M, Singh R, Qattawi A, Omar M, and Haque I 2014 Evaluation of on-board photovoltaic modules options for electric vehicles. IEEE Journal of Photovoltaics, 4(6): 1576-1584.
[5] Abdelhamid M 2014 Comparison of an Analytical Hierarchy Process and Fuzzy Axiomatic Design for Selecting Appropriate Photovoltaic Modules for Onboard Vehicle Design. Int. J. Mod. Eng., 15: 23–35.
[6] Evans D L 1981 Simplified method for predicting photovoltaic array output. Sol. Energy, 27: 555–60.
[7] Alami A H 2014 Effects of evaporative cooling on efficiency of photovoltaic modules. Energy Conversion and Management, 77: 668-679.
[8] Zhou J, Yi Q, Wang Y and Ye Z 2015 Temperature distribution of photovoltaic module based on finite element simulation Sol. Energy, 111: 97–103.
[9] Usama Siddiqui M, Arif A F M, Kelley L and Dubowsky S 2012 Three-dimensional thermal modeling of a photovoltaic module under varying conditions Sol. Energy, 86: 2620–31.
[10] Chou T-L, Shih Z-H, Hong H-F, Han C-N and Chiang K-N 2012 Thermal Performance Assessment and Validation of High-Concentration Photovoltaic Solar Cell Module IEEE Trans. Compon. Packag. Manuf. Technol., 2: 578–86.
[11] Chen W-H, Cheng H-C and Shen H-A 2003 An effective methodology for thermal characterization of electronic packaging IEEE Trans. Compon. Packag. Technol., 26: 222–32.
[12] Acanski M, Popovic-Gerber J and Ferreira B 2010 Thermal modeling of the module integrated DC-DC converter for thin-film PV modules Proc. 14th Int. Power Electron. Motion Control Conf. EPE-PEMC 2010.
[13] Pantic L S, Pavlović T M, Milosavljević D D, Radonjic I S, Radovic M K and Sazhko G 2016 The assessment of different models to predict solar module temperature, output power and efficiency for Nis, Serbia Energy, 109: 38–48.
[14] Sánchez Barroso J C, Barth N, Correia J P M, Ahzi S and Khaleel M A 2016 A computational analysis of coupled thermal and electrical behavior of PV panels Sol. Energy Mater. Sol. Cells.,148: 73–86.
[15] Skoplaki E and Palyvos J A 2009 On the temperature dependence of photovoltaic module electrical performance: A review of efficiency/power correlations Sol. Energy., 83: 614–24.
[16] Chou T-L, Shih Z-H, Hong H-F, Han C-N and Chiang K-N 2007 Investigation of the thermal performance of high-concentration photovoltaic solar cell package. International Conference on Electronic Materials and Packaging, 1–6.
[17] Mavromatakis F, Kavoussanaki E, Vignola F and Franghiadakis Y 2014 Measuring and estimating the temperature of photovoltaic modules Sol. Energy., 110: 656–66.
[18] Minak G, Brugo T M, Fragassa C, Pavlovic A, de Camargo F V and Zavatta N 2019 Structural Design and Manufacturing of a Cruiser Class Solar Vehicle J. Vis. Exp. JoVE.
[19] Pavlovic A, Sintoni D, Fragassa C and Minak G 2020 Multi-Objective Design Optimization of the Reinforced Composite Roof in a Solar Vehicle Appl. Sci., 10: 2665.
[20] Pavlović A, Sintoni D, Minak G and Fragassa C 2020 On the modal behaviour of ultralight composite sandwich automotive panels Compos. Struct., 248: 112523.

[21] Pavlovic A, Fragassa C, Bertoldi M and Mikhnych V 2021 Thermal Behavior of Monocrystalline Silicon Solar Cells: A Numerical and Experimental Investigation on the Module Encapsulation Materials J. Appl. Comput. Mech., 7: 1847–55.
[22] Llorens J 2015 Fabric Structures in Architecture.
[23] Burton W S and Noor A K 1997 Structural analysis of the adhesive bond in a honeycomb core sandwich panel Finite Elem. Anal. Des., 26: 213–27.
[24] Czanderna A W and Pern F J 1996 Encapsulation of PV modules using ethylene vinyl acetate copolymer as a pottant. A critical review Sol. Energy Mater. Sol. Cells 43.
[25] Li X, Li P, Wu Z, Luo D, Yu H-Y and Lu Z-H 2021 Review and perspective of materials for flexible solar cells Mater. Rep. Energy, 1: 100001.
[26] Lee C, Park A, Cho Y, Park M, Lee W I and Kim H W 2008 Influence of ZnO buffer layer thickness on the electrical and optical properties of indium zinc oxide thin films deposited on PET substrates Ceram. Int., 34: 1093–6.
[27] Ethylene Tetrafluoroethylene (ETFE) : MakeItFrom.com.
[28] Tefzel-properties-handbook.pdf.
[29] Ethylene Vinyl Acetate (EVA) : MakeItFrom.com.
[30] Properties of Silicon - El-Cat.com.
[31] France-Lanord A, Merabia S, Albaret T, Lacroix D and Termentzidis K 2014 Thermal properties of amorphous/crystalline silicon superlattices J. Phys. Condens. Matter, 26: 355801.
[32] Abe H, Kato H and Baba T 2011 Specific Heat Capacity Measurement of Single-Crystalline Silicon as New Reference Material Jpn. J. Appl. Phys., 50: 11RG01.
[33] Lee B, Liu J Z, Sun B, Shen C Y and Dai G C 2008 Thermally conductive and electrically insulating EVA composite encapsulant for solar photovoltaic (PV) cell Express Polym. Lett., 2: 357–63.
[34] Polyethylene terephthalate - online catalogue source - supplier of research materials in small quantities - Goodfellow.
[35] 2003 Polyethylene Terephthalate Polyester (PET, PETP) - Properties and Applications - Supplier Data by Goodfellow AZoM.com.
[36] Zheng X, Kim S and Park C W 2019 Enhancement of thermal conductivity of carbon fiber-reinforced polymer composite with copper and boron nitride particles Compos. Part Appl. Sci. Manuf., 121: 449–56.
[37] 3MTM Adhesive Transfer Tape 468MP Technical Datasheet.
[38] Fragassa C, Pavlovic A and Minak G 2020 On the structural behaviour of a CFRP safety cage in a solar powered electric vehicle Compos. Struct., 252: 112698.
[39] Ali H T, Akrami R, Fotouhi S, Pashmforoush F, Fragassa C and Fotouhi M 2020 Effect of the stacking sequence on the impact response of carbon-glass/epoxy hybrid composites Facta Univ. Ser. Mech. Eng., 18: 69–77.

Acknowledgments
This research was carried out within the international collaboration project 'Two Seats for a Solar Car', an intervention funded by the Italian Ministry of Foreign Affairs and International Cooperation (MAECI) with the scope to convert a racing solar car into a solar road vehicle. Methodological and technological results are exploited inside the Central European Initiative framework as part of the 'Composite for All' and 'ATC.EVO' transfer of technology actions.

IOP Conf. Series: Materials Science and Engineering 1214 (2022) 012043 doi:10.1088/1757-899X/1214/1/012043

Comfort assessment for electric kick scooter decks

Michelangelo-Santo Gulino, Giulio Vichi, Giovanni Zonfrillo, Dario Vangi

Department of Industrial Engineering, Università degli Studi di Firenze, Via di Santa Marta, 50139, Firenze, Italy

E-mail: giovanni.zonfrillo@unifi.it

Abstract.

The LEONARDO project, funded under the Horizon 2020 framework, is aimed at developing a new concept of micromobility electric vehicle combining the benefits of an e-kick scooter (rapid learning curve, stability) and those of a monowheel (greater agility). The goal is the development of a vehicle able to increase the consensus towards electric micromobility for enabling a relevant spread of this green transport solution, thanks to specific expedients like: (a) the possibility of exploiting an auxiliary battery pack to increase the path traveled by the vehicle; (b) a mass reduced to 10 kg, allowing to easily lift and move the vehicle to routinely encountered upper-level spaces (steps, the trunk of a vehicle, etc.). The weight reduction compared to similar vehicles necessarily involves the replacement or elimination of bulky components like suspension elements; for this reason, the deck must have intrinsic shock absorbing characteristics to maximize the rider comfort.

The objective of the present work is the identification of a class of decks for e-kick scooters that best meets the user comfort needs in an urban environment. To this end, experiments have been performed replacing the aluminium deck of a commercially available electric scooter with decks of different materials; the comfort offered by each deck has been measured compatibly with the ISO 2631:1997 standard on a closed test circuit, in correspondence of six different obstacles that the rider may tackle in a typical urban scenario. The results of the study, read in a statistical key by regressive methods and analysis of variance, evidence highest compatibility with maximum comfort criteria for the 11.5 mm thick bamboo deck; namely, the test campaign evidences how the maximum comfort for the rider can be obtained with a deck having a stiffness similar to that of bamboo. In this context, the bamboo deck lacking of shock absorbers demonstrates superior performance compared to aluminium decks equipped with suspensions for the same application, with a mass saving of approximately 65%. Special care should however be paid to the forces the deck could sustain: moving from an aluminium to a bamboo deck which is 20 times less stiff, forces decrease of only 20%. Overall, these highlights represent a comprehensive set of elements to pilot the design of micromobility vehicles, from both standpoints of comfort and resistance.

1. Introduction

Electric micromobility is currently seen as one of the major strategies to quickly achieve significant decreases of pollutants in urban environments within a limited time window, as the one dividing the European Commission from a reduction of 55% in greenhouse gases (2030 [1]). A mono-user, light transportation mean is the most appropriate traveling solution compared not only to alternatives like a combustion engine vehicle, but also to an electric passenger car: taking the latter as a reference, a micromobility vehicle significantly decreases the electric energy demand for transportation, an especially crucial point considering that non-sustainable sources accounted for more than 80% of the whole European production in 2020. In particular, the use of electric kick scooters (e-scooters) drastically increased in the last years, both in private and sharing contexts; it is additionally expected that the request for e-scooters will constantly increase by 25% each year until 2025 [2]. Most of travels by e-scooters however typically substitute path otherwise covered by other sustainable transportation modes as walking [3], implying that a limited reduction on the number of circulating heavy, multi-user vehicles is currently observed. For these reasons, increasing the consensus towards micromobility solutions by users of other traditional vehicles is a priority to achieve a more sustainable transport system.

From the user perspective, e-scooter inadequacy as a transportation mean depends on two fundamental causes: the high mass and the reduced autonomy. A high mass (typical e-scooters on the market are down to 13 kg) discourages the user to carry the vehicle in its routine activities, being for instance less prone to lift and carry it to upper-level spaces (steps, trunk of a vehicle, etc.). Analogously, 85% of e-scooters are employed for less than one hour per day because of the limited energy stored in battery packs [4]. To this end, the aim of the LEONARDO project (MicrovehicLE fOr staNd-Alone and shAReD mOblity[1]) is twofold: to develop a microvehicle with a total mass lower than 10 kg and whose battery packs could be shared among the users in a battery sharing scheme; the vehicle implements a fixed battery pack, while an auxiliary battery pack can be picked up by the users at a recharging station to increase the traveled paths [5]. The vehicle is conceptualized so as to feature hybrid characteristics between an e-scooter and a monowheel, i.e., combining the rapid learning curve and the stability of an e-scooter with agility and capability to easily overcome road asperities of a monowheel (higher wheel diameter [6]).

Decreasing the mass of the vehicle requires renouncing to well-established concepts as shock absorbers, typically mounted on the front and/or rear wheels of an e-scooter. A deck with suitable properties in terms of vibration attenuation hence appears as the most feasible solution to decrease the vehicle mass without affecting the riding comfort for the user. From such standpoint, decks for e-scooters available on the market are mainly represented by aluminium structures that, without a shock absorber, are extremely stiff. A research by Koontz et al. [7] demonstrated that, in general, lighter and less stiff vehicle structures feature higher maneuverability and comfort for the user. Specific investigations on e-scooter motion have also been performed by Cano-Moreno et al. [2, 8] through simulations, determining that the use of less stiff shock absorbers and wheels could ameliorate the user experience from both health and comfort standpoints (the speed being the same). Nevertheless, the trip was simulated by random vibrations as prescribed within the ISO 8608:2016[2] standard, and did not investigated which types of obstacles in an urban environment could produce such oscillations; analogously, no information is available regarding the comfort experienced by the user when shock absorbers are excluded from the vehicle system. Nonetheless, $in-situ$ (on road) measurements could significantly differ from the results of simulation or lab tests, because conditions like the rider position and weight distribution can be fully accounted for only in real road contexts [9].

Based on such assumptions, the present study aims at analyzing the comfort provided on road by different decks in terms of material, while subjected to obstacles that can be typically encountered in urban contexts. Since these tests are required to pilot the design of an innovative vehicle concept, the forces discharged on the deck additionally require special attention. Vibrations and forces on the decks are hence monitored by an accelerometer and a load cell, respectively. The main reference for comfort assessment in the study is represented by the ISO 2631:1997[3] standard, while tests are performed on a closed circuit featuring several types of obstacle.

2. Materials and methods

The present Section introduces the fundamental elements enabling determination of comfort characteristics for different types of decks in diverse conditions. These elements are represented by comfort assessment methodologies, the experimental campaigns, and the procedures employed to analyze comfort data resulting from such experiments.

[1] `https://leonardoproject.eu/`
[2] `https://www.iso.org/standard/71202.html`
[3] `https://www.iso.org/standard/7612.html`

2.1. *Comfort assessment*

Comfort assessment for a vehicle finds a valid reference in the ISO 2631:1997 regulation. The standard reports that diverse frequencies in the rider oscillations are associated with a different perceived comfort, from 0.4 Hz to 100 Hz. Hence, the acceleration must be filtered with a 0.4 Hz-100 Hz band-pass filter and appropriately weighted and expressed by the Root Mean Square (RMS) as follows:

$$a_{v_{ij}} = \sqrt{\frac{1}{B_j}\int_0^{B_j} a_{ij}^2(f)df} \tag{1}$$

$$a_{w_i} = \sqrt{\sum_j \left(W_j a_{v_{ij}}\right)^2} \tag{2}$$

Based also on the visualization of Figure 1, i is the considered direction among x, y, and z, $a_{ij}(f)$ is the i-th acceleration component in the j-th one-third octave band, B_j is the j-th one-third octave band, $a_{v_{ij}}$ the RMS acceleration value for the i-th direction in the j-th one-third octave band, W_i the weighting factor for the i-th one-third octave band given in specified tables and a_{w_i} the weighted acceleration value in the i-th direction. Composing $a_{\mathrm{w_i}}$ values in the three directions and considering a standing rider, the following equation is derived from the standard:

$$a_{TOT} = \sqrt{a_{w_x}^2 + a_{wy}^2 + a_{w_z}^2} \tag{3}$$

The standard indicates a_{TOT} as the main parameter to be considered for comfort assessment.

Figure 1. Visualization of the reference system according to ISO 2631:1997, considering the user in a standing position as in the case of an e-kick scooter rider.

2.2. *Experimental campaign*

2.2.1. Deck categories and evaluation of related stiffness Experimental campaigns have been performed employing a commercial e-kick scooter that, in its original form, relies on an aluminum deck equipped with a shock absorber (ICEWHEEL E9sD); such deck has been identified as deck A_1. To pilot the choice of the deck category that best performs in terms of comfort provided to the user, diverse decks have been produced and replaced with the original on the e-kick scooter frame:

- The original aluminium deck, after removal of the shock absorber (**deck A_2**);

- A deck constituted of two layers of bamboo, one with the fibers disposed along x and the other along y (Figure 1); a fiber glass fabric is interposed between the layers and the deck has a total thickness of 11.5 mm (**deck B**);
- A deck made of eight maple sheets with an additional non-slip grip, with a thickness of 12.5 mm (**deck C**);
- A deck from three fir sheets alternated with plywood, with a thickness of 12.5 mm (**deck D**);
- A deck made of six fir sheets alternated with plywood, with a thickness of 25 mm (**deck E**);
- A deck made of three layers of curved bamboo obtained by vacuum resin bonding, with a thickness of 15 mm (**deck F**).

To characterize the various decks (some of which are visible in Figure 2), a three-point bending test was performed on a MTS 810 universal machine in displacement control, determining stiffness and modulus of elasticity. After preliminary three-point bending qualitative tests, decks D, E, and F have been discarded because of excessive fragility, very similar stiffness compared to deck A_2, and very similar stiffness to deck C, respectively. It was then assumed that the remaining decks represent three significantly different types of objects in terms of stiffness on which comfort analyses could be based.

Figure 2. Decks A_2, E, B, C (from left to right); deck E has been discarded from subsequent tests, because of high similarity with deck A_2 in terms of stiffness.

For the aluminium (A_2), bamboo (B), and maple (C) decks, maximum loads of 880 N, 870 N, and 1320 N were respectively applied without reaching the ultimate tensile strength. In each test, a length of support span of 400 mm was kept. In Table 1, the results of the three-point bending tests are reported for the three decks. Thickness of deck A_2 has not been reported because of its complex geometry.

Table 1. Values of the main considered variables for the three-point bending tests on the considered decks.

Deck	Thickness	Width	Maximum load	Maximum deflection
A_2 - Aluminium	/	145mm	1320N	0.43mm
B - Bamboo	11.5mm	200mm	880N	5.52mm
C - Maple	12.5	200mm	870N	4.72mm

The modulus of elasticity E is obtained from the classical relation that apply to beams subjected to bending moments:

$$E = \frac{L^3 (F_2 - F_1)}{4bt^3 (a_2 - a_1)} \tag{4}$$

where L is the length of support span, b the width of the specimen, t the specimen thickness, F_2-F_1 the load increase in the straight section of the load-deformation curve and a_2-a_1 the increase in deflection in the center line of the specimen (corresponding to F_2-F_1). The stiffness S is obtained from the following relation:

$$S = \frac{48EI}{L^3} \tag{5}$$

where I is the moment of inertia of the beam. Equations 4-5 assumes isotropic behaviour of the beam, condition that might be unsuitable for specific class of materials. Therefore, assuming F_2 and F_1 respectively corresponding to 80% and 10% of the maximum load, the following results are obtained:

- Deck A_2: Stiffness 3121.9 N/mm
- Deck B: Stiffness 161.8 N/mm – Modulus of elasticity 8531 MPa
- Deck C: Stiffness 184.1 N/mm – Modulus of elasticity 7762 MPa

Because of the complex shape of the aluminium deck A_2, the modulus of elasticity for the latter could not be calculated. Its stiffness has been consequently obtained based on the slope of the load-deformation curve from the bending test. The three stiffness values confirm that such decks cover a reasonably wide range of stiffness.

2.2.2. Test circuit Based on the stiffness of each deck, differences in terms of comfort to the user are expected to emerge while riding on an urban context. Definition of a test circuit was hence required for comfort assessment of each deck. The circuit, closed to the traffic, should guarantee the presence of obstacles typically encountered in urban road environments. Several sections have been consequently considered including different types of obstacle, indicated in Figure 3 as follows:

1 Rough asphalt – $\mathbf{O_1}$;
2 Manhole without asperity – $\mathbf{O_2}$;
3 Manhole with asperity (slight gutter) – $\mathbf{O_3}$;
4 Ramp – $\mathbf{O_4}$;
5 Slight bump – $\mathbf{O_5}$;
6 Smooth surface – $\mathbf{O_6}$.

Between O_3 and O_4, a straight section with a homogeneous road surface of 45 m is present, in which maximum speed for the vehicle (25 km/h) could be achieved.

2.2.3. Instrumented vehicle After stiffness characterization of the decks, their behavior in real road environments had to be tested in the closed circuit of Figure 3. Considering that comfort can be expressed by a superposition of accelerations on diverse directions based on the ISO 2631:1997 standard, the vehicle needed to be instrumented to acquire acceleration values. To this end, accelerations to which the decks are subjected while the rider is using the kick scooter have been acquired by a three-axis accelerometer mounted in the center of the deck, with vertical z axis facing upwards (opposite to gravity); the y axis of the accelerometer corresponds to the longitudinal direction of the vehicle, coherently with Figure 1. The accelerometer had a resolution of 80 mV/g and the sampling frequency was set to 1 kS/s. At each test,

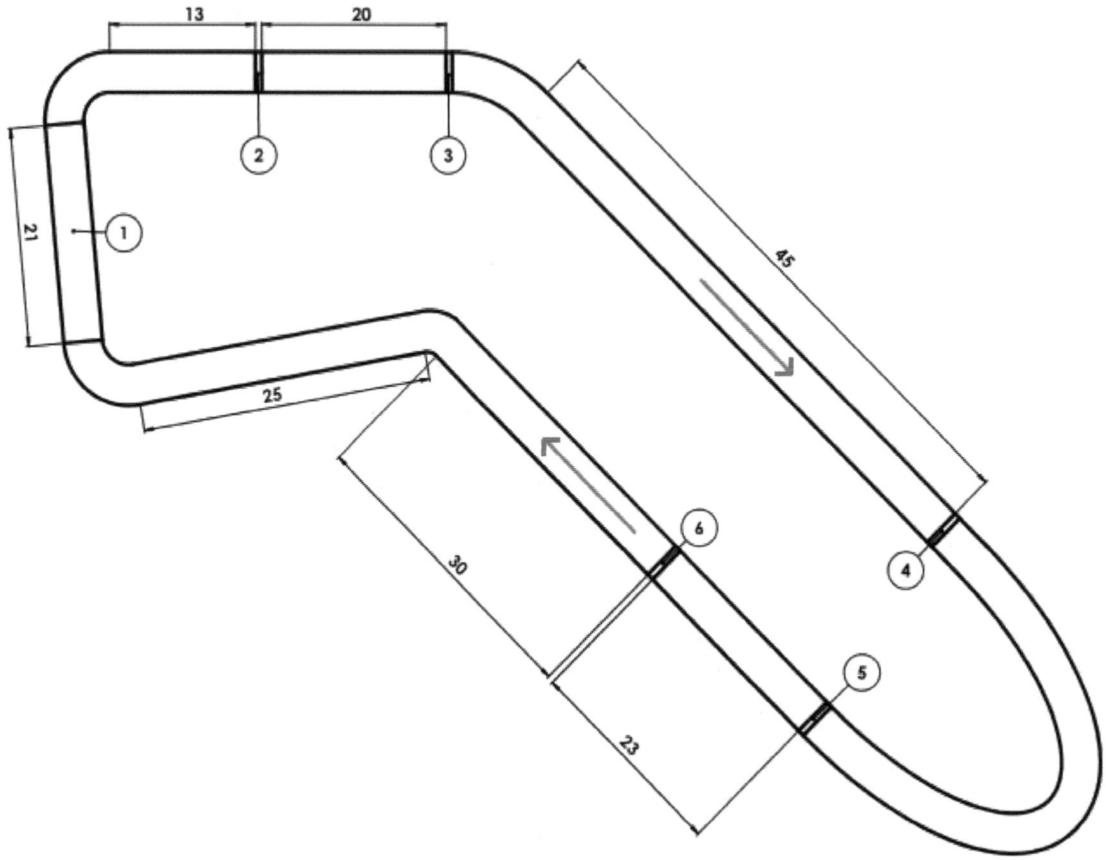

Figure 3. Considered circuit for the experimental campaigns; different sections in terms of road obstacles and distance among them are reported: rough asphalt (**1**), manhole without asperity (**2**), manhole with asperity (**3**), ramp (**4**), gutter (**5**), smooth surface (**6**).

a different deck was mounted on the e-kick scooter frame, i.e., A_1, A_2, B, C. The tests were carried out under the total mass (comprising user and overall mass of acquisition instruments) of 80 kg.

In addition, since forces discharging on the deck are fundamental in its design process (static and dynamic resistance), their monitoring is associated with fundamental highlights. Solely the forces to which the two most diverse decks in terms of stiffness are subjected were acquired along the circuit (i.e., A_2 and B), to evidence only major differences among the decks; accelerometer and load evaluations were performed in different tests. A load cell was hence mounted on a rigid panel, and interposed between the deck and the panel on which the rider stands. The load cell averaged the signals from four different strain gauges located at the vertices of the panel, with a sampling rate of 100 S/s. In Figure 4a, the load cell mounted between the panel and the deck is evidenced, with indication of the four strain gauges it is constituted of. Figure 4b depicts the vehicle ready for the acceleration test, in case of aluminium deck on which the accelerometer is mounted. Figure 4c evidences the vehicle ready for the load test, in which the mounted panel and the National Instruments© acquisition board (NI mod. 9291) are also shown.

2.3. Data analysis

Accounting for the indications included in the ISO 2631:1997 standard, the perceived comfort increases as a_{TOT} decreases, because the vibration energy transferred from the road to the rider reduces. It was hence first required to obtain the value of a_{TOT} for each test. A significance analysis was subsequently

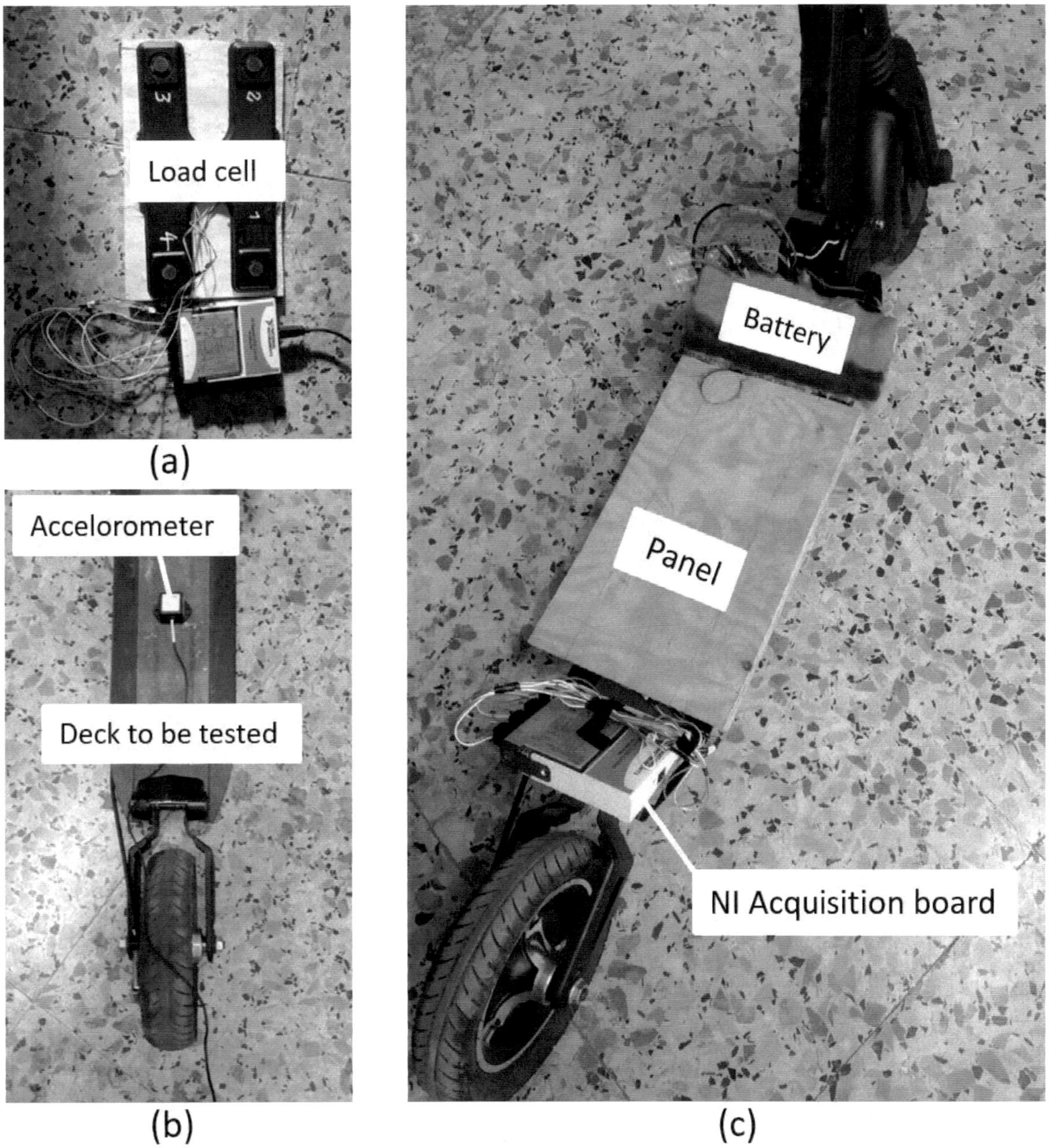

Figure 4. (a) Instrumented panel; (b) deck A_2 with the three-axis accelerometer; (c) deck A_2 with the instrumented plate.

necessary to determine the influence of the various decks on the provided comfort, the encountered obstacles being the same.

A program written in LabVIEW© was used for data acquisition. A post-processing program was also written in LabVIEW© to digitally filter acceleration data between 0.4 Hz and 100 Hz and obtain a_{TOT} in a specific time interval. The time intervals corresponding to each obstacle were isolated from the acquisition data. A systematic criterion was therefore established for a proper identification of the time intervals: for O_1, an interval of 7 s was considered starting from the beginning of the circuit (progressive obstacle), while for all other obstacles an interval of 1 s was established, with a lower limit of 0.2 s before the absolute peak (or valley) corresponding to the obstacle and an upper limit of 0.8 s after. The time intervals for the load cells acquisition were analogously defined, and the corresponding load signals similarly processed.

Once obtained all the necessary data and processed as described above, a significance study was carried out by means of regression analysis by the Minitab© statistical software, for the determination of the parameters significantly influencing the comfort provided by the diverse decks. This was obtained by referring to an Analysis Of Variance (ANOVA) process.

3. Results and discussion

Table 2 summarizes the results of the ANOVA applied to the processed signals from the accelerometer; decks have been categorized based on their material, rather than on their value of stiffness: this is a requirement to compare the different decks (lacking a shock absorber) with deck A_1 (on which a shock absorber was implemented). A p-value lower than 0.05 for a variable is considered as an indicator of significant influence of such variable on a_{TOT}. In the analysis, deck A_1 and obstacle O_1 are used as the baseline deck and obstacle, respectively. It can be highlighted that, in terms of a_{TOT}, decks A_2 and C do not provide statistically significant differences compared to deck A_1. Instead, the use of the bamboo deck B has a statistically significant influence on the comfort perceived by the rider, independently on the type of encountered obstacle. Transversally, all types of obstacle significantly influence the comfort if compared to obstacle O_1. Nevertheless, a regression (odds ratio analysis) is also required to gain quantitative information regarding such an influence.

Table 2. Results of the ANOVA related to the parameter a_{TOT} (significant variables are highlighted by *).

Parameters	P-value
Constant	0.000*
Deck A1	-
Deck A2	0.119
Deck B	0.020*
Deck C	0.220
Obstacle O1	-
Obstacle O2	0.001*
Obstacle O3	0.000*
Obstacle O4	0.000*
Obstacle O5	0.000*
Obstacle O6	0.001*

The linear regression applied to the specific problem provides the following expression for a_{TOT}:

$$\begin{aligned} a_{TOT} = 7.10 + 0.0A_1 + 2.40A_2 - 3.73B - 1.86C + 0.00O_1 + 8.04O_2 + 8.85O_3 + \\ + 12.96O_4 + 12.88O_5 + 7.62O_6 \end{aligned} \tag{6}$$

where the categorical variables A_1, A_2, B, and C represent the different decks, while the categorical variables from O_1 to O_6 the obstacles of the test circuit. In the equation, these categorical variables gain a value of 1 if the variable is taken into account, 0 otherwise. Since the coefficient related to deck B is the lowest among all decks, it can be derived that the parameter a_{TOT} reaches a minimum when deck B is employed in the experiment, i.e., maximum comfort. Deck C also provides a low value of a_{TOT}, but its statistical non-significance (Table 2) does not allow to assess its superiority compared to the aluminium deck with shock absorber. The same applies to the aluminum deck without shock absorber (A_2), that provides the highest values of a_{TOT} on average but influences a_{TOT} in a statistically non-significant way. In addition, obstacles O_2 and O_3 similarly affect the values of a_{TOT}, as well as O_4 and O_5. Nonetheless, Equation 6 features a high goodness-of-fit, with an adjusted value of R^2 equal to 84.9%.

Figure 5 depicts various graphs that highlight the difference in terms of comfort between the worst and the best deck based on the coefficients of Equation 6, i.e., respectively A_2 and B. It is derived that the bamboo deck provides the highest comfort in whichever condition, in particular in case of the progressive obstacle (O_1). Influence of the deck category on additional parameters has been considered in the analysis, e.g., value and number of the acceleration peaks in different directions; nevertheless, the main effect of the deck reflects on the a_{TOT} parameter. No further information is hence provided on such parameters.

For what regards the forces discharged on the deck, graphs are presented in Figure 6 regarding the same decks of Figure 5. It can be seen that the values of both maximum force and standard deviation force for the bamboo deck are lower than those for the aluminium one, if averaged among all cells. This is independent of the considered obstacle; however, the maximum force can reach values of 1400 N in a single cell, despite the 80 kg mass of the rider considered in the experiments. This obviously depends on the position of the rider centre of mass with respect to the deck centre. Even if stiffness for the bamboo deck is almost 20 times lower than that of aluminium, the maximum force to which the deck is subjected is only 20% lower on average. These results provide fundamental highlights for the design phase of a bamboo deck (or a deck with analogous stiffness), in terms of forces that can be admitted in transient regimes. Considering that the bamboo deck approximately accounts for 0.6 kg instead of 1.7 kg for the assembly constituted of the aluminium deck and suspension, this solution entails a mass decrease of 1.1 kg (65%).

4. Conclusions

The present work assessed the comfort provided by different types of deck to the rider for the design of an innovative electric kick scooter solution, in which mass is limited by the absence of shock absorbers. The primary reference for comfort assessment was the ISO 2631:1997, that considers the root mean square of accelerations along three axis as the main comfort indicator. Each frequency component of the acceleration signal is differently perceived by the rider, so that a frequency weighting on the signals was performed following the standard. The forces discharged on the deck were also acquired by a load cell, fundamental to define the maximum stress the deck will sustain in transient regimes.

The results demonstrated that the bamboo deck outperforms other types of considered decks, independently of the obstacles which can be encountered by the vehicle in a typical road environment (rough asphalt, manholes, etc.). More in general, based also on the results from a linear best-fitting process, the highest comfort is provided by decks with low stiffness; a linear fit is sufficient to more than adequately describe the comfort provided by the deck (R^2=84.9%). Special care must be however paid towards the forces acting on the deck: experiments demonstrated that, even if the stiffness lowers of 20 times, the maximum forces on the deck would only decrease of about 20%.

The results provide suggestions to design a deck which can be light while preserving its comfort characteristics compared to decks implementing shock absorbers. Several types of obstacles have been considered which are typically found in urban roads; therefore, the obtained indications represent a comprehensive set of elements which can be employed to pilot the design phase of future, lighter micromobility solutions.

IOP Conf. Series: Materials Science and Engineering 1214 (2022) 012043 doi:10.1088/1757-899X/1214/1/012043

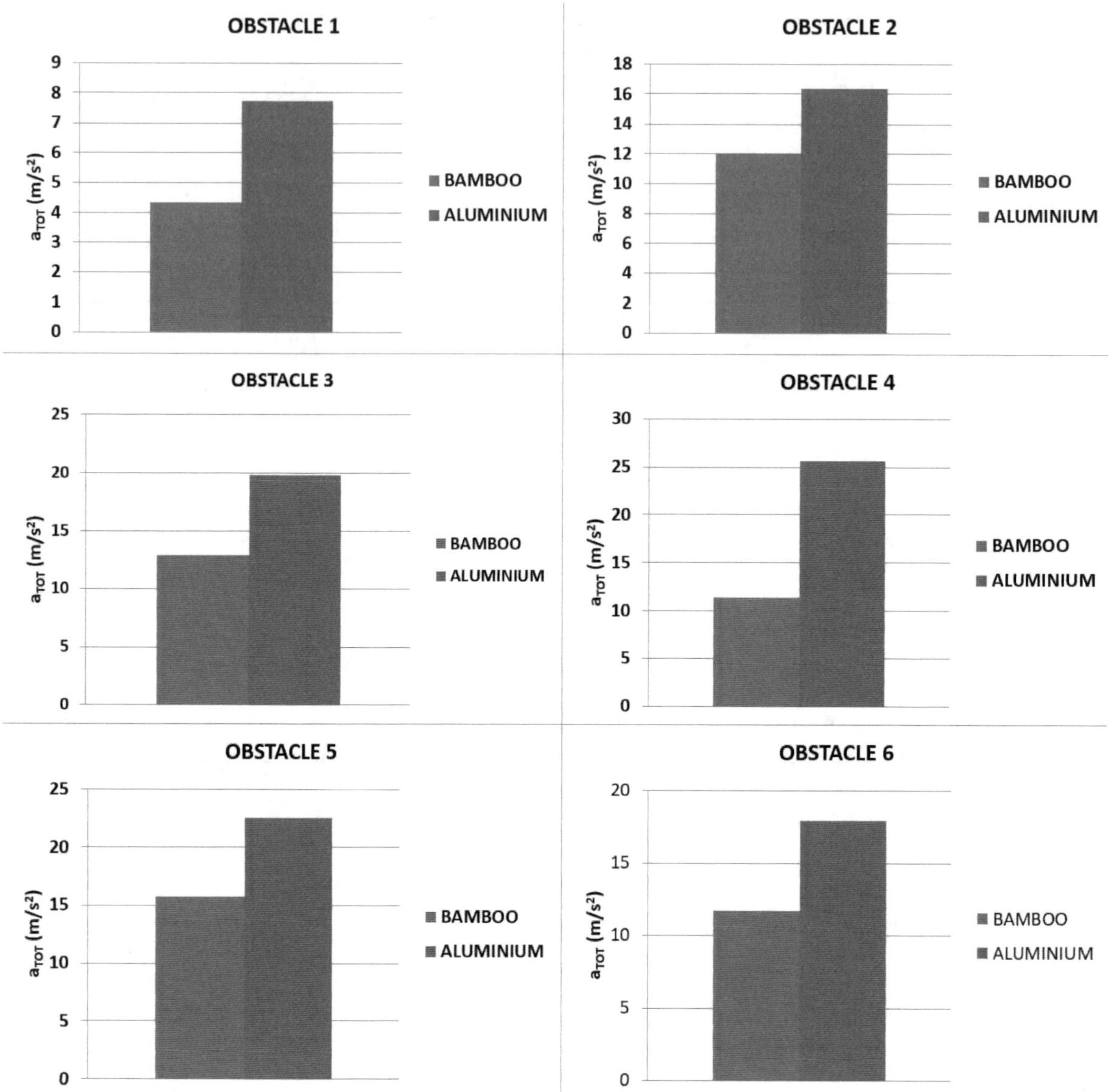

Figure 5. Comparison between the parameter a_{TOT} associated with deck A_2 (aluminium without shock absorber) and B (bamboo).

Acknowledgments

The work has been performed in the context of the LEONARDO project (MicrovehicLE fOr staNd-Alone and shAReD mOblity), funded by the European Commission under Grant Agreement n°101006687 in the Horizon 2020 framework.

References

[1] European Commission. Stepping up Europe's 2030 climate ambition: Investing in a climate-neutral future for the benefit of our people. *COM (2020) 562*, 2020.

[2] Juan D Cano-Moreno, Manuel E Islán, Fernando Blaya, Roberto D'Amato, Juan A Juanes, and Enrique Soriano. E-scooter vibration impact on driver comfort and health. *Journal of Vibration Engineering & Technologies*, pages 1–15, 2021.

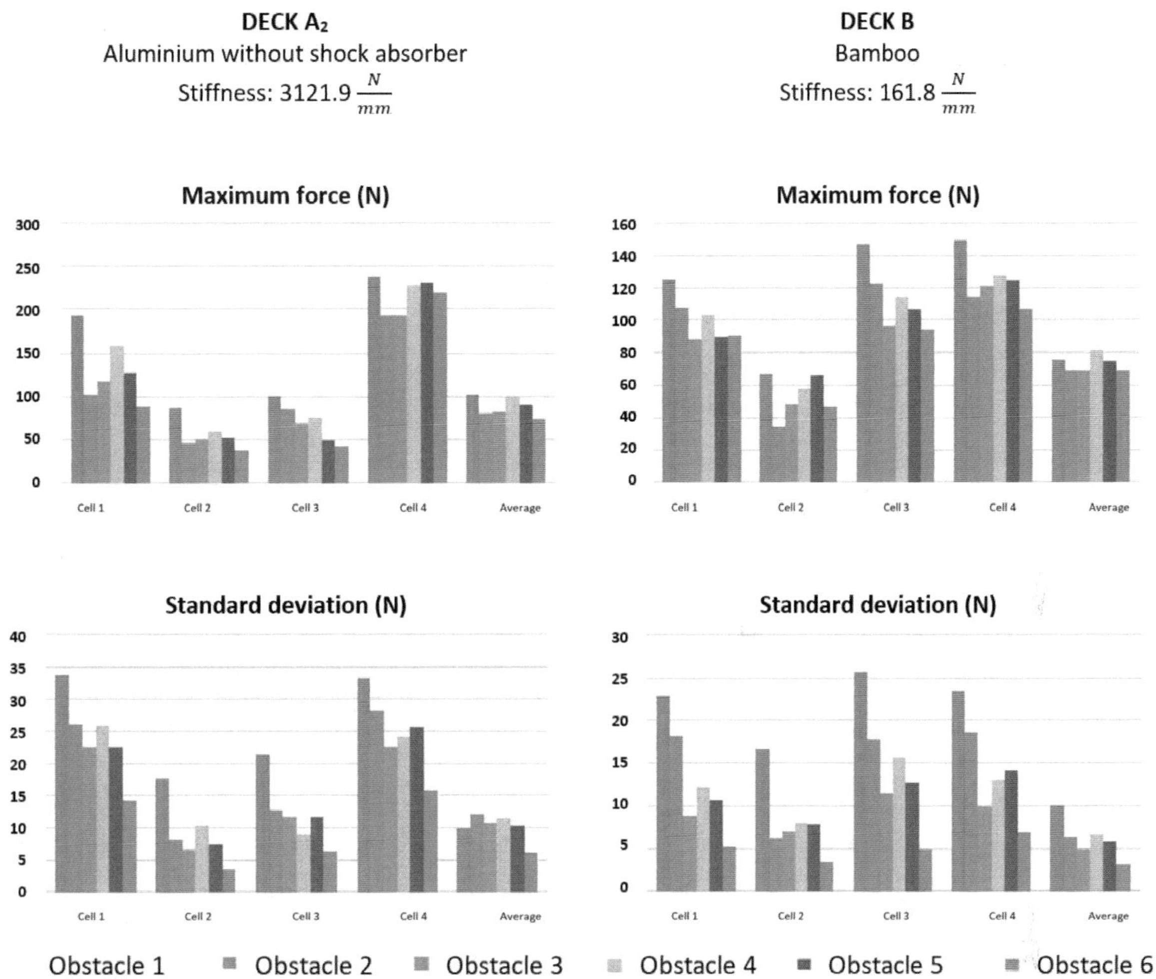

Figure 6. Comparisons between the two most distant categories of deck in terms of stiffness, as a function of maximum force and standard deviation of the force acting on each of the four cells (and corresponding average).

[3] Barbara Laa and Ulrich Leth. Survey of e-scooter users in Vienna: Who they are and how they ride. *Journal of transport geography*, 89:102874, 2020.

[4] Qingyu Ma, Hong Yang, Alan Mayhue, Yunlong Sun, Zhitong Huang, and Yifang Ma. E-scooter safety: The riding risk analysis based on mobile sensing data. *Accident Analysis & Prevention*, 151:105954, 2021.

[5] Maurizio De Lucia, Carlo Cialdai, Matteo Messeri, and Dario Vangi. Shared system of rechargeable batteries.

[6] Jie Gao, Aimin Sha, Yue Huang, Liqun Hu, Zheng Tong, and Wei Jiang. Evaluating the cycling comfort on urban roads based on cyclists' perception of vibration. *Journal of Cleaner Production*, 192:531–541, 2018.

[7] Alicia M Koontz, Eric D Brindle, Padmaja Kankipati, David Feathers, and Rory A Cooper. Design features that affect the maneuverability of wheelchairs and scooters. *Archives of physical medicine and rehabilitation*, 91(5):759–764, 2010.

[8] Juan David Cano-Moreno, Manuel Islán Marcos, Fernando Blaya Haro, Roberto D'Amato, Juan A. Juanes, and Enrique Soriano Heras. Methodology for the study of the influence of e-scooter vibrations on human health and comfort. In *Proceedings of the Seventh International Conference on Technological Ecosystems for Enhancing Multiculturality*, pages 445–451, 2019.

[9] Julien Lépine, Yvan Champoux, and Jean-Marc Drouet. Test protocol for in-situ bicycle wheel dynamic comfort comparison. *Procedia engineering*, 147:568–572, 2016.

IOP Conf. Series: Materials Science and Engineering 1214 (2022) 012044 doi:10.1088/1757-899X/1214/1/012044

Vehicle stability controller HiL validation on static simulator

F Alfatti[1,2], C Annicchiarico[2] and R Capitani[1]

[1] DIEF, Università degli Studi di Firenze, Via Santa Marta 3, 50139 Firenze (FI), Italy
[2] Meccanica 42 srl, via Ezio Tarantelli 15, 50019 Sesto Fiorentino (FI), Italy

E-mail: `federico.alfatti@unifi.it`

Abstract. In the automotive world, Hardware in the Loop (HiL) methodologies are used to speed up the design and calibration process of the various mechatronics components installed on the vehicle. Installing production parts on advanced driving simulator enables the performance assessment in realistic scenarios, including a test driver earlier in the design process. The scope of this work is to evaluate the effect of ABS and ESC system, replicating their influence on the steering wheel. A brake by wire production system has been installed on a static simulator, equipped with a EPSiL steering bench, capable of accurately replicate tie-rod forces on a production steering unit. Full brake, sine steer maneuvers have been carried out both on real vehicle and on virtual environment. The system showed its capabilities of replicating the same functionalities of the real vehicle, extending the static simulator potentialities to support the activities on calibration and test of vehicle systems.

1. Introduction

Tentative in order to achieve the objective of reducing the number and fatality of road accidents [1] has made vehicle control and safety functionalities more complicated, making the urge for a faster yet reliable method of testing such components a current trends in the industry. For example, the Autonomous Emergency Braking (AEB) functionality has made the brake unit an active component, requiring a brake action even if the driver is not applying any force to the brake pedal [2].

Moreover, the search for higher vehicle efficiency is steering the market to more controllable and energy efficient brake by wire units, where the brake pedal is hydraulically disconnected from the calipers, and the oil is pressurized by the intervention of an active actuator [3].

The current development process of this type of units presents its challenges, given the intrinsic multi-domain specifications required [2, 4]. To achieve the final design, of automotive mechatronic components and their functionalities usually requires that every step must be concluded with a test phase, defining that the design and its validation must be carried out at the same time and always before going further in the development. We refer to the graphical representation of this process as a "v-shape" [5]. Various validation methodologies are available, depending on the design phase. Usually, a Model in the Loop (MiL) is firstly implemented. Given a basic but functionally-equivalent virtual model of the system in development, designers can evaluate high-level specification and algorithms. Going further, with Software in the Loop the system in testing is the software of the unit, evaluating its programming and efficiency.

After this, a first physical prototype is available, and it's used for Hardware in the Loop testing [6]. For example, a vehicle brake system, composed by actuator unit, brake pedal and calipers, is assembled outside the vehicle, and undergoes validation cycles in order to evaluate the functionality of both the physical part and the electronic control [7]. ABS, ESC logic are roughly calibrated during these tests [8, 9, 10], but a fine tuning phase is only assigned to real-vehicle testing with an human driver.

In order to include the driver earlier in the design loop, the use of driving simulators is increasing importance [11]. The aim of this work is to propose a methodology for brake by wire unit performance assessment that can serve as an intermediate validation phase between HiL test and real vehicle tests [12].

A commercial brake by wire unit has been installed on a static driving simulator, on which a steering bench is installed. This bench acts on the real steering system, which is totally installed on the sim, and it is able to accurately reproduce forces on tierods and consequently the steering torque; the system has proved itself capable in [13] to achieve the calibration of a prototype EPS, tuning the steering feeling with expert test drivers and subjective methodologies used in road vehicles. The brake system installation in this static simulator aims to reproduce the effect of ABS and ESC intervention on steering wheel torque, extending its accuracy.

A brief description of the static simulator and brake loop is introduced. Then, the validation methodology is presented: in order to assess if the unit has same behavior between real and simulator installation, several braking maneuvers are carried out, comparing the behavior during ABS and ESC intervention. Then, a full-brake maneuver is carried out on the simulator, comparing the performances of a commercial Model in the Loop and brake unit ABS functionality in terms of driver feedback on the steering wheel torque.

2. Real-time driving simulator setup

The experimental activity has been carried out on the static simulator, supplied by Meccanica 42, shown in figure 1. It is composed by three main parts; the real time simulator, the EPSiL steering bench, and the Braking unit, as in figure 2. The real time simulator comprehends a Concurrent-RT machine, which manages the simulation environment. This machine acquires the data coming from by the Human-Machine interface, interfaces it with the vehicle model, and sends a response to the force actuators guaranteeing the max delay of a millisecond (typical timestep for the simulation). This type of machine is necessary to use the Hardware in the Loop benches installed on the simulator.

2.1. EPSiL steering bench

The EPSiL steering bench integrates a complete steering system in a human in the loop driving simulator [14]. Two torque motors and transmission act directly on the tie-rods of a production steering system by the means of a rocker, accurately reproducing the kinematic movement of the suspension. With this architecture the rig can reproduce other components of the force beyond the forces on the xy plane, improving tracking performances between vehicle model and actual force. The tie-rods, equipped with loadcells, rack, EPS, steering column, and steering wheel are integrated in the system.

Figure 1. Static simulator setup.

The described steering system layout replicates exactly the one of an actual vehicle model. In this way, steering bench is put into simulation loop, enhancing driving simulator experience and providing the perfect tool for steering feedback virtual development. In fact, through CAN connection, it is possible to change EPS tuning parameters and calibrate the system, while driver can assess the effect of each modification on multiple proving grounds. The unit proved itself capable of accurately reproduce real-vehicle tierod forces, resulting in the correct reproduction of the steering feeling [13].

2.2. Brake unit

A commercially available brake-by-wire system has been integrated on the simulator. The installation is described in figure 2. The stock brake plant is mounted on the simulator: the brake tubes, with the correct diameter and length, connects the electromechanical unit to the 4 stock brake calipers, which acts on the original brake discs. The brake pedal is directly integrated on the unit, and it acts as the main driver-machine interface.

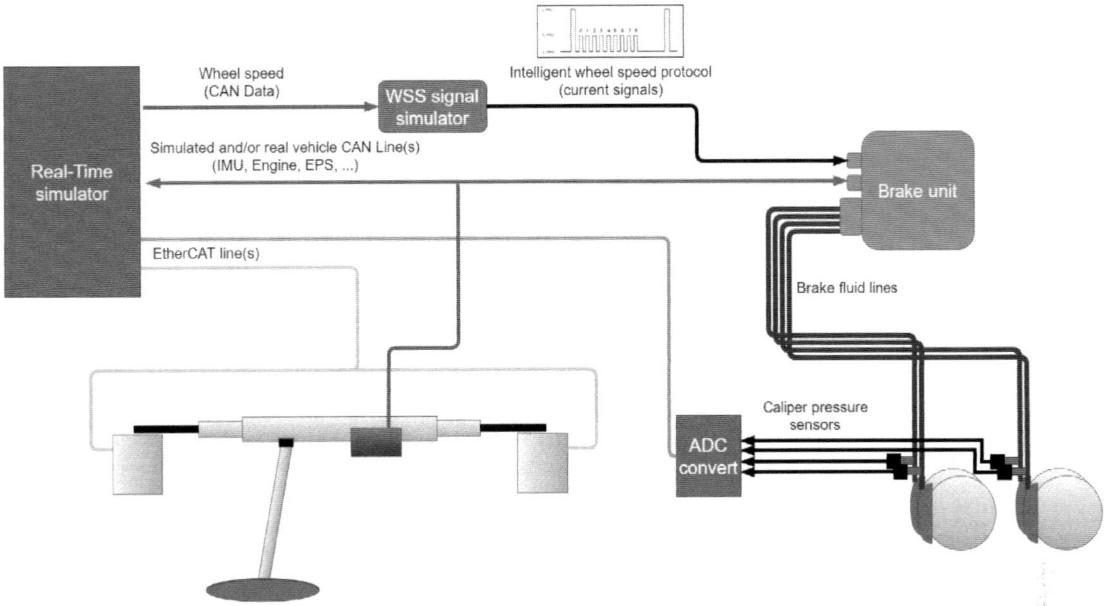

Figure 2. Simulator scheme.

The brake unit receives through the CAN communication line all the IMU signals that it usually receives on the car, simulated by the vehicle model on the real time computer. However, this unit needs the intelligent wheel speed sensors signal via a direct communication line [15, 16]. To overcome this problem, a wheel speed simulator has been implemented, which converts the signal of the wheel speed (sent via CAN from the RT machine) to the typical signal of the intelligent wheel speed sensors.

Lastly, 4 pressure sensors have been installed on the calipers, and their signals are sent back to the vehicle model via an EtherCAT measurement line, closing the loop, controlling the vehicle deceleration after a brake application by the driver. Brake unit and steering system belong to the same vehicle.

3. Methodology

The validation methodology is now presented. The goal is to have a simulator that can reproduce the vehicle behavior and driver feeling of the real scenario.

A previously validated vehicle model has been installed on the simulator, and adapted to the new control loop, given the brake unit presence. Then, the following set of maneuvers has been carried out by a driver on the simulator:

- Full brake
- Pulse brake
- Sine steer
- Brake in turn

To demonstrate the efficacy of the brake bench, the installed brake unit functionality must be coherent with the one of a vehicle during a real drive test. The same set of maneuvers has been carried out on an instrumented vehicle in a testing track. A tuning phase of the road parameters has been necessary to guarantee that the virtual and real results are comparable.

After this, a steering torque sensor has been installed on the simulator, and a maneuver with ABS intervention has been carried out, comparing a model ABS with the HiL ABS, to evaluate the effect of the brake unit on steering feeling.

3.1. Tuning
One of the main advantages to use a driving simulator is the possibility to consider the influence of various parameters on the driving test. For example, it is possible to modify the track adherence very easily, resulting in a faster evaluation of the control system in various different conditions, which is not possible or at least much more difficult to investigate on the real track.

However, track condition information is needed to correlate the behavior of the unit; the braking distance and the ABS intervention greatly depends on the track adherence. We decided to treat this value as our tuning parameter: without accurate information on the track condition, we compare the results of a full-brake maneuver from high speed to standstill in terms of braking distance. If this parameter is higher (or lower) in the simulated maneuver than the one in real-scenario, the adherence value is decreased (or increased), and the simulation re-run.

The resulting value of 0,85 is reasonable, given the fact that tests were carried out after a rainy day on a dry asphalt, reducing the available adherence. Once that the braking distance is comparable, we proceed with the evaluation of the remaining maneuvers.

4. Results
4.1. Full Brake
The first test is a full brake maneuver, from high speed to standstill. This is useful to evaluate the ABS control intervention. Starting from high speed, the driver applies full brake pedal. In figure 3 the time results are presented, in terms of longitudinal deceleration, wheel speed, brake corner and master cylinder pressure.

It is possible to spot a qualitative correlation of the latter signals between the two test environment. The major difference however emerges from the evaluation of the wheel speed: at lower speed, the signal oscillation has some noticeable differences. This behavior is caused by the tire model, which is unreliable when the braking moment causes large values of slip at lower speed. However, the results at speeds higher than 30km/h can be considered as valuable.

The KPI analysis that follows takes into account this limitation, excluding the results below this speed. Four indexes, which are maximum pressure, mean pressure, braking distance and deceleration on brake pressure ratio have been considered, and are shown in figure 4. The values have been scaled in respect to the real vehicle performance to easily compare in the same scale all the indexes; for this reason, the real test KPI has value one in all the following graphs. For all the four indexes, the error is always lower than 10% of the real test value.

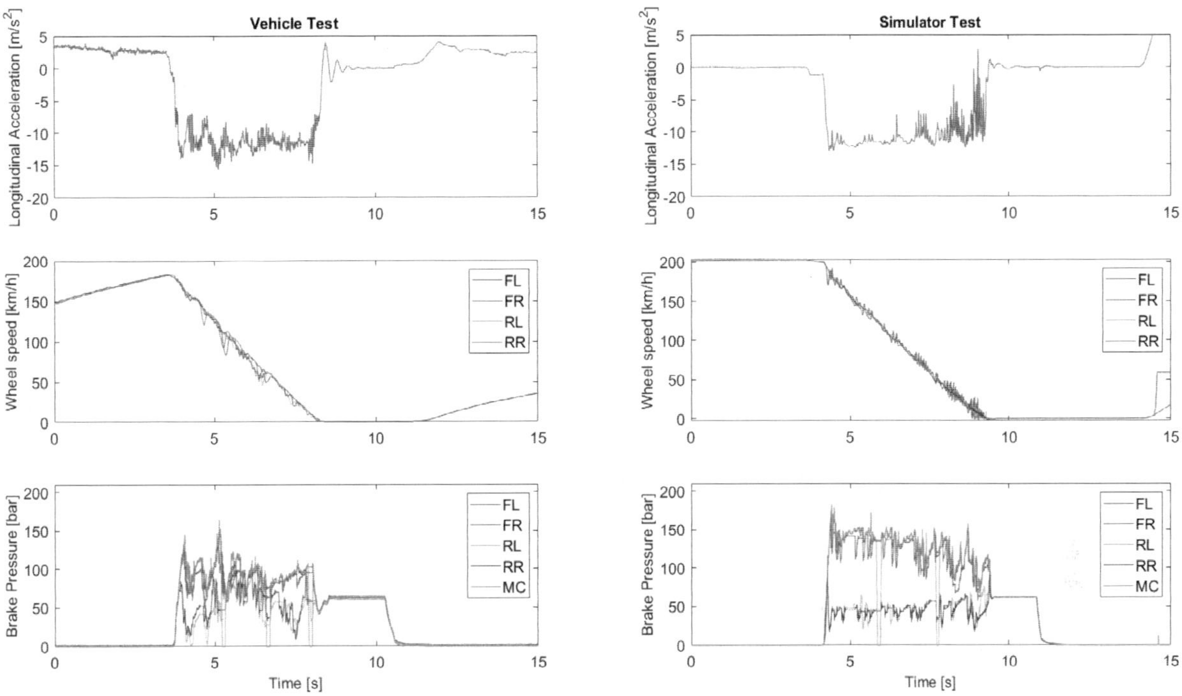

Figure 3. Full brake maneuver result.

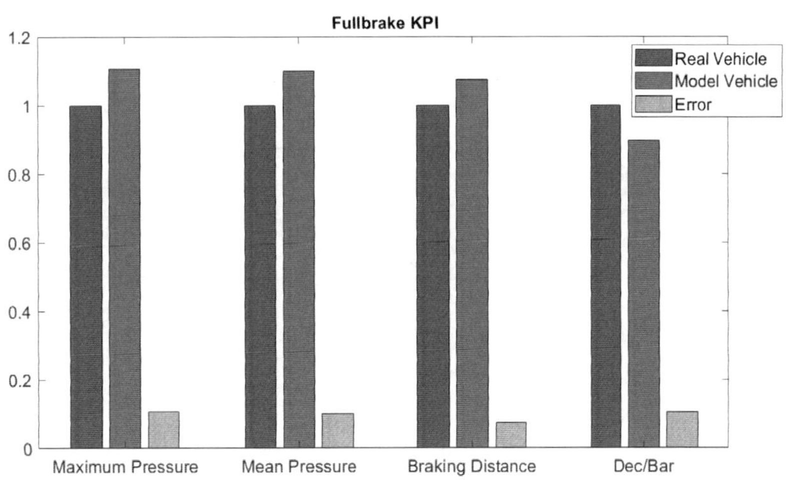

Figure 4. Full brake KPI comparison.

4.2. Pulse Brake

The second test is a pulse brake, which consists in a repeated brake application of 3 seconds, followed by a brake release. The results are presented in figure 6. This is a dynamical maneuver, and we can see that the correlation is higher than the one derived by the first maneuver. This is due to the higher speeds at which the brake momentum is applied, avoiding the numerical noise due to the inefficacy of the tire model, as already discussed.

In figure 5 the KPI comparison is presented, and again the error is less than 10 percent. The indexes for this maneuver are the same as the previously described, given the similarity of the vehicle dynamics involved. The deceleration-pressure ratio has been excluded from the analysis because already assessed during the previous test.

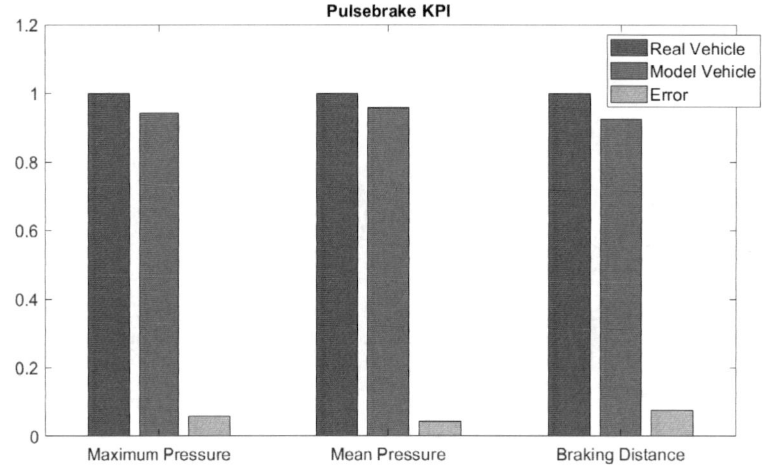

Figure 5. Pulse brake maneuver KPI comparison.

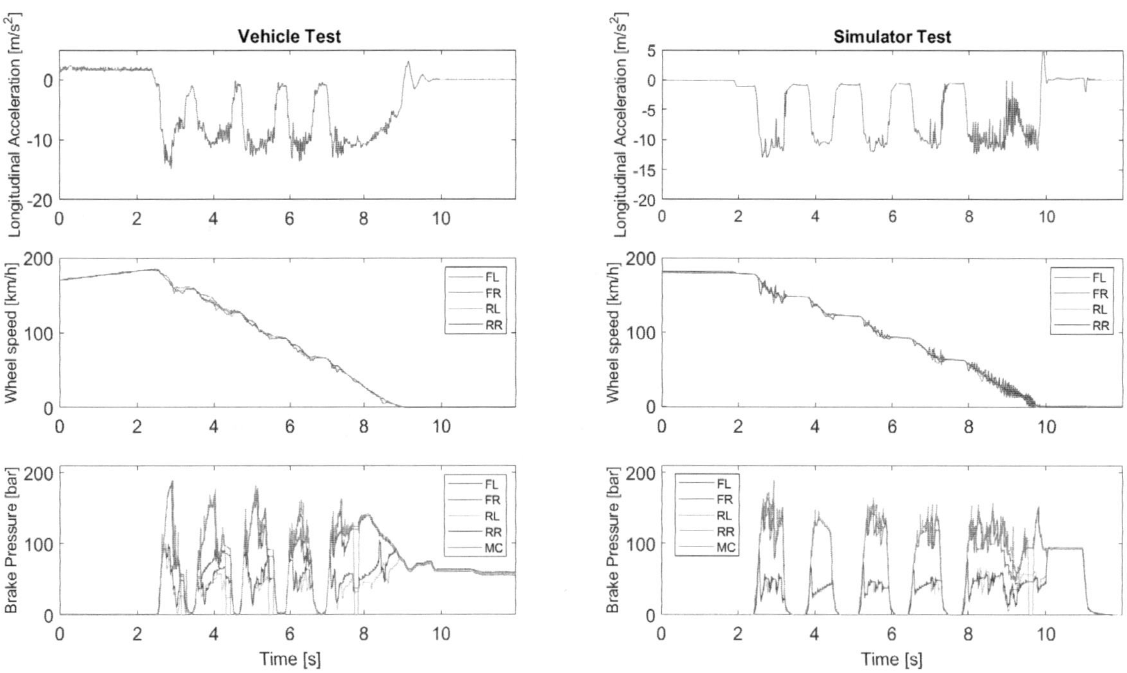

Figure 6. Pulse brake maneuver result.

4.3. Sine Steer

The third test is a sine steer in order to evaluate ESC intervention. A sine steer with frequency of $1Hz$ and amplitude 45 degrees has been carried out at 100 km/h. The results are presented in figure 7 and the chosen key performance indexes are shown in figure 8. These indexes are maximum brake pressure, sideslip angle, and yaw rate. Still, the error for the three KPI is less than 8%.

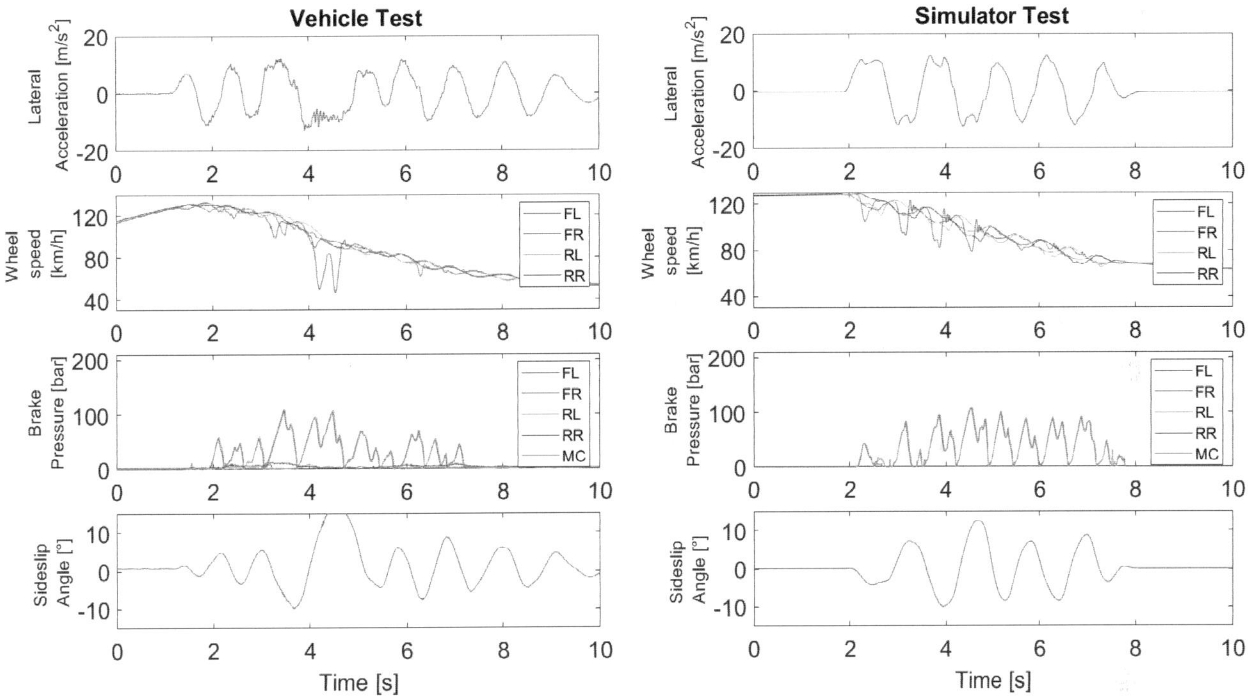

Figure 7. Sine steer maneuver result.

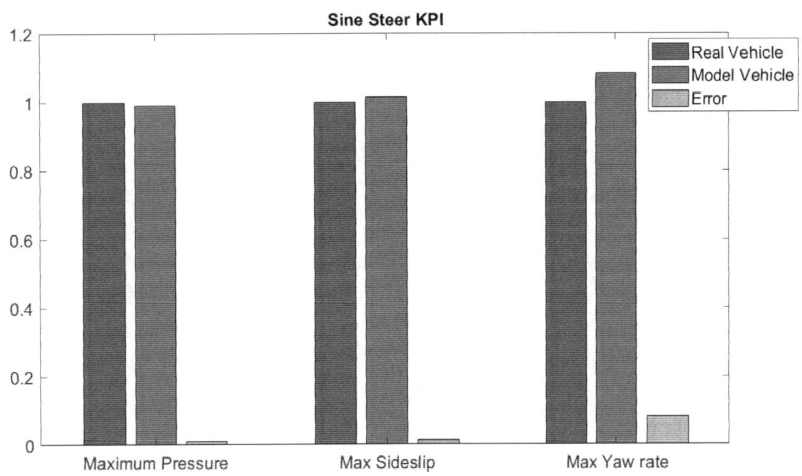

Figure 8. Sine steer maneuver KPI comparison.

IOP Conf. Series: Materials Science and Engineering 1214 (2022) 012044 doi:10.1088/1757-899X/1214/1/012044

4.4. Brake in Turn

The fourth and last test is a constant steer maneuver with the application of a full brake intervention with non-zero steering angle. The results and the KPI comparison are shown in figure 10 and 9.

The result of this maneuver are highly dependent on the tire model limitations, which introduces high wheel speed oscillation, excluding large quantities of data from the KPI analysis. The indexes for this maneuver are maximum brake pressure and sideslip angle, and the ratio between maximum yaw rate and steer angle. Again, the relative error is less than 10%.

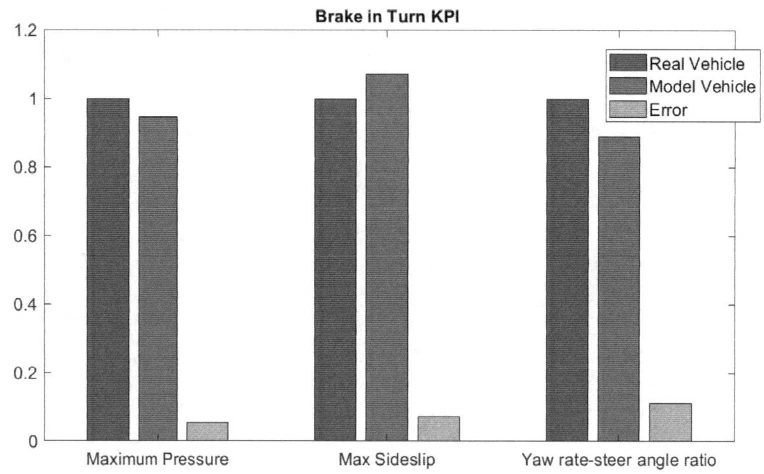

Figure 9. Brake in turn maneuver KPI comparison.

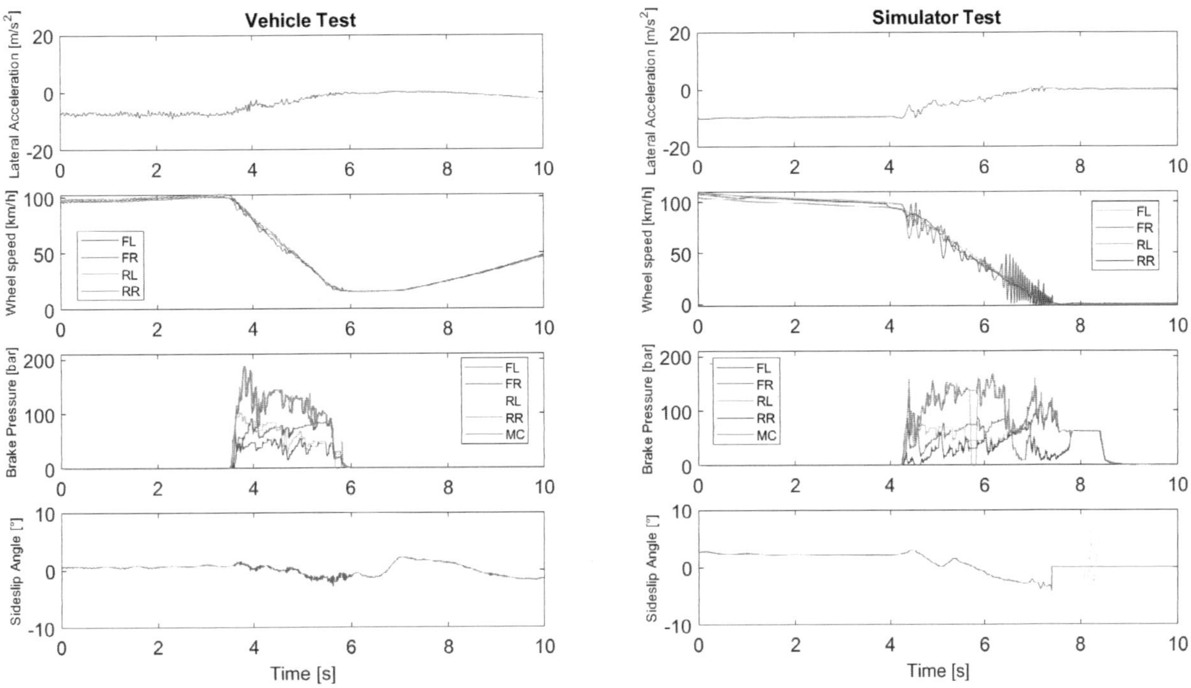

Figure 10. Brake in turn maneuver result.

4.5. Driving feeling analysis

Given the good correlation attained during the previous maneuvers, our goal is to demonstrate the impact that the installation of the brake unit can have on the driver feeling.

A subjective comparison between driver feedback with and without the brake unit during the simulation has been carried out. The possible interfaces that the driver can have with the simulator are via steering wheel torque and pedal vibration. For this reason, a steering wheel sensor has been installed.

The same maneuver, a full brake event from the speed of 200km/h, is carried out on two different simulations; the first one has a model ABS system pre-installed on the vehicle model, which follows a longitudinal slip target, and the brake unit serves only as a brake pedal input for the simulation.

In the second, the brake unit is active, and the brake loop is closed on the caliper measured pressure. The time results are presented in figure 11.

The unit functioning is highlighted by a more realistic behavior and driver feeling: thanks to the high bandwidth of the steering bench, it is possible to observe higher frequencies of the steering wheel torque; this fidelity is demonstrated also with an higher frequency content of the steering wheel torque, as presented in figure 12.

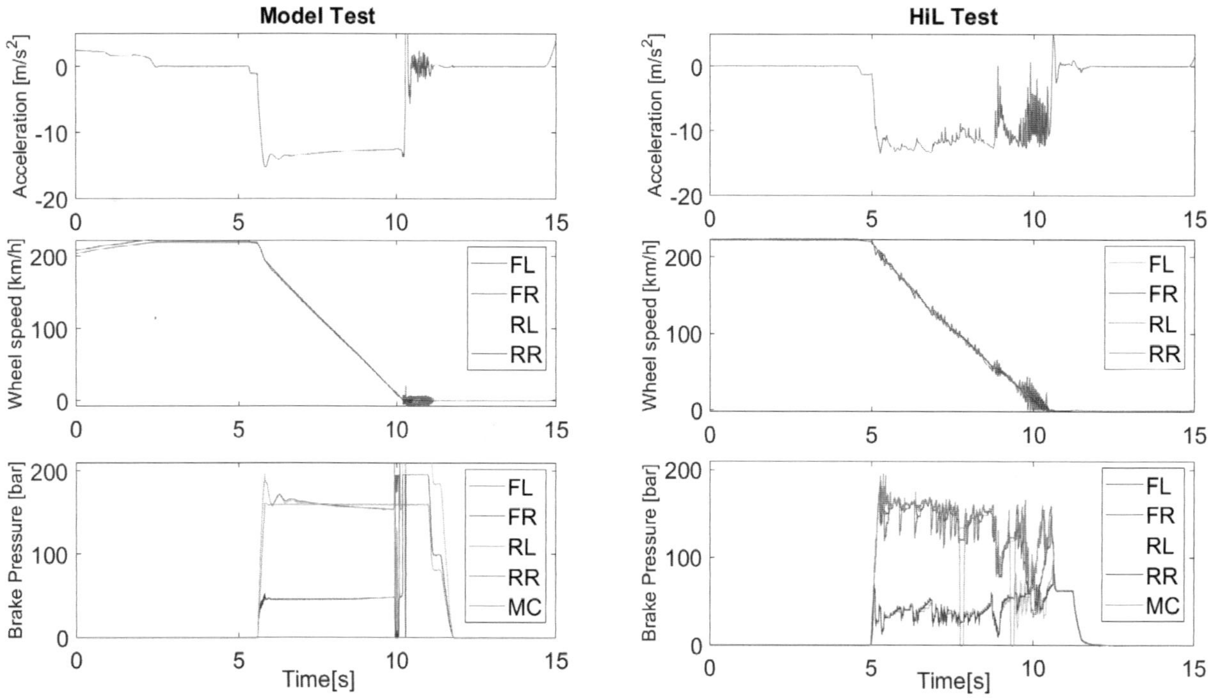

Figure 11. Model and HiL ABS comparison.

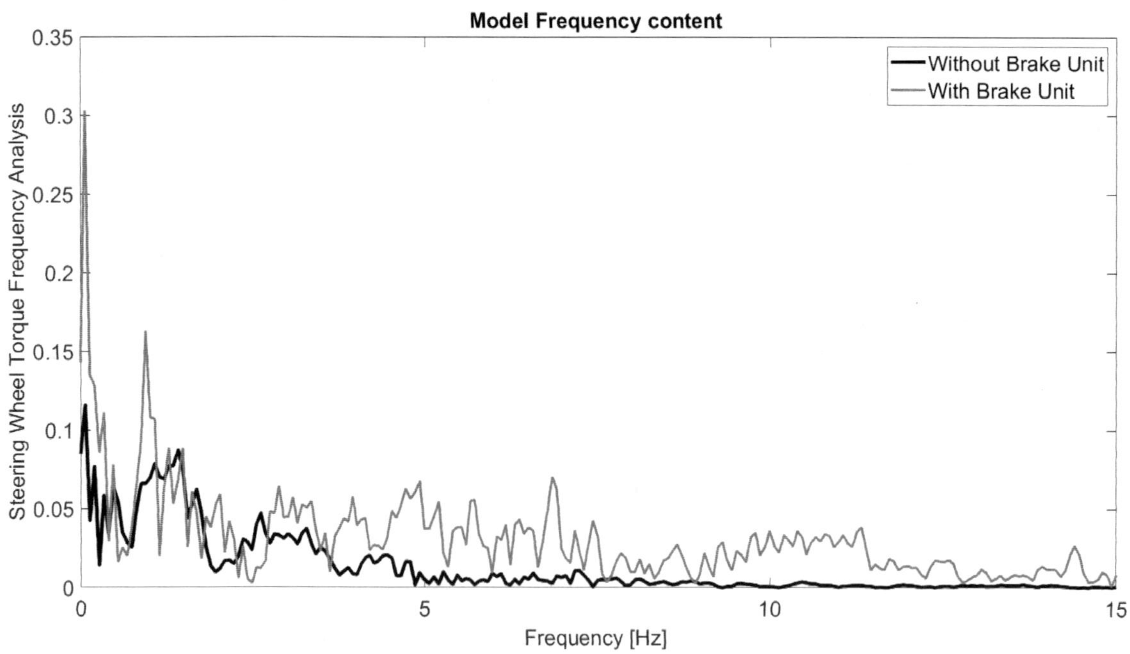

Figure 12. Steering torque frequency content comparison.

5. Conclusions

The validation process and the simulator setup has been presented. The test campaign results, summed up with the Key performances indexes, show that a good correlation between the unit behavior in the simulated and real scenario. Thanks to low error value between the KPI's, typically lower than 10%, we can affirm that this type of installation accurately reproduces the brake unit influence on the vehicle dynamics, thus allowing for reduced development time and cost of a commercial brake by wire unit; an higher number of driving scenarios can be investigated without the constraints typical for a real-world test campaign.

Moreover, this installation allows for a more immersive simulation environment, reproducing the correct driving feeling on this type of simulator where both the complete production steering and brake system are installed. Lastly, it is possible to compare the stock stability control performances with custom ones in a complete virtual environment.

It will also be possible to investigate the Advanced Driver Assistance System (ADAS) unit functionalities on the simulator, continuing the extension of driving scenarios that can be covered, as seen in [17]. Thanks to this, all the brake unit functionalities will be reproduced on the simulator, allowing for a complete and integrated testing of the product in a virtual environment, speeding up the development process.

Abbreviations

The following abbreviations are used in this manuscript:

ABS	Anti-lock Braking System
ADAS	Advanced Driver Assistance System
AEB	Autonomous Emergency Braking
ASIL	Automotive Safety Integrity Level
BW	By-Wire
BBW	Brake-By-Wire
CAN	Controller Area Network
ESP	Electronic Stability Program
EPS	Electric Power Steering
EPSiL	EPS in the Loop
ESC	Electronic Stability Control
HiL	Hardware in the Loop
IMU	Inertial Measurement Unit
KPI	Key Performance Index
RT	Real Time

References

[1] Maurer M, Gerdes J C, Lenz B and Winner H (eds) 2016 *Autonomous Driving* (Berlin, Heidelberg: Springer Berlin Heidelberg) ISBN 978-3-662-48845-4 978-3-662-48847-8

[2] Winner H, Hakuli S, Lotz F and Singer C (eds) 2016 *Handbook of Driver Assistance Systems* (Cham: Springer International Publishing) ISBN 978-3-319-12351-6 978-3-319-12352-3

[3] Yu L, Liu X, Xie Z and Chen Y 2016 (American Society of Mechanical Engineers Digital Collection)

[4] Soltani A and Assadian F 2016 *IFAC-PapersOnLine* **49** 32–38 ISSN 24058963

[5] Johanson M 2011 *New Trends and Developments in Automotive System Engineering* ed Chiaberge M (InTech) ISBN 978-953-307-517-4

[6] Gong X, Ge W, Yan J, Zhang Y and Gongye X 2020 *Actuators* **9** 15 number: 1 Publisher: Multidisciplinary Digital Publishing Institute

[7] Montani M, Capitani R and Annicchiarico C 2019 *Procedia Structural Integrity* **24** 137–154 ISSN 24523216

[8] Sabbioni E, Cheli F and d'alessandro v 2011 pp 2011–01–0032

[9] Sorniotti A and Velardocchia M 2004 pp 2004–01–2770

[10] Montani M, Vitaliti D, Capitani R and Annicchiarico C 2020 *Energies* **13** 6183 ISSN 1996-1073

[11] Guan H, Duan C G and Lu P P 2014 *Applied Mechanics and Materials* **644-650** 76–80 ISSN 1662-7482

[12] Bruschetta M and Minen D 2019 *9th International Munich Chassis Symposium 2018* Proceedings ed Pfeffer P (Wiesbaden: Springer Fachmedien) pp 147–162 ISBN 978-3-658-22050-1

[13] Talarico E M, Raimondi G, Alfatti F, Vitaliti D and Annicchiarico C 2021 *12th International Munich Chassis Symposium 2020: chassis.tech plus* Proceedings ed Pfeffer P (Springer Vieweg) p 16

[14] Vinattieri F, Wright T, Capitani R, Annicchiarico C and Danisi G 2016 Target setting and structural design of an EPS-in-the-Loop test bench for steering feeling simulation SAE Technical Paper 2016-01-1559 SAE International Warrendale, PA iSSN: 0148-7191, 2688-3627

[15] Pančík J and Beneš V 2019 *Industry 4.0: Trends in Management of Intelligent Manufacturing Systems* ed Knapčíková L and Balog M (Cham: Springer International Publishing) pp 111–120 ISBN 978-3-030-14010-6 978-3-030-14011-3 series Title: EAI/Springer Innovations in Communication and Computing

[16] Pusch S 2000 pp 2000–01–0082

[17] Ye Y, Zhao J, Wu J, Zhu B, Deng W and Chen X 2018 pp 2018–01–1450

IOP Conf. Series: Materials Science and Engineering 1214 (2022) 012045 doi:10.1088/1757-899X/1214/1/012045

Holistic eco-design tool within automotive field

A Antonacci[1], F Del Pero[1], N Baldanzini[1] and M Delogu[1]

[1] Department of Industrial Engineering, University of Florence, Italy

andrea.antonacci@unifi.it

Abstract. In the last decades, sustainability has become a critical issue for the vehicle industry. Today automotive manufacturers are subjected to several government regulations aimed at making the industry more environmentally sustainable. In this context, many research and industrial activities are focused on the eco-design field by developing innovative materials, novel manufacturing technologies, and new methodology; the main objective is to combine the best design performance with increased environmental responsibility. In this paper, a holistic design and sustainability analysis method based on a multi-criteria approach is presented to support the designer in the early conceptual phase finding the best trade-off in terms of performance and sustainability. The new methodology generates a single-score index (SSI) based on product design/sustainability metrics; data normalization, weighting, and aggregation processes are also included in the evaluation method. This method is integrated into an eco-design framework developed through HyperWorks and MATLAB software tools. A case study, based on a reference geometry, a load case, and a list of materials and processes chosen by the designer, is described to show the potential of the proposed methodology. Finally, a critical review and concluding remarks are described following the SSI results; suggestions for further improvements are made.

1. Introduction

Over the past few decades, there has been a growing interest in sustainable development [1], with environmental and cost sustainability aspects being the main pillars for business and societal decision-making. In the industrial field, the eco-design of new products represents a key element to help policymakers, stakeholders, and customers to make decisions towards a more sustainable society [2]. In this context, the Life Cycle (LC) sustainability assessment becomes essential for decision support and sustainable development indicators are widely used to assess the environmental profile of new products.

Road transport contributes heavily to the overall environmental impact of anthropogenic activities (about 20% of GHG emissions are caused by road transportation [3][4][5]). For this reason, vehicle manufacturers are facing strong regulatory pressure pushing for sustainability in several aspects, such as reducing resource consumption in production, improving the recyclability of materials, and reducing tailpipe emissions during operation. Sustainable design, investment in clean technologies, and value creation for local and global communities [5][6] are some of the efforts made by the automotive companies to achieve these challenging goals.

Regarding the design aspect, lightweighting, manufacturing processes/technologies optimization, vehicle assemblies/components redesign, and more efficient materials usage are the most common strategies used by Original Equipment Manufacturers (OEMs). Innovative design solutions certainly offer improvements in driving and technical features (e.g., in terms of performance, road safety, and drivability) and energy consumption [7]; however, the production and manufacturing of new materials and advanced components are usually associated with higher GHG emissions compared to conventional design solutions. For instance, Carbon Fiber Reinforced Plastics (CFRP) cause significantly greater impacts than conventional steels both in raw material acquisition and in semi-finished product processing [8]. Moreover, the innovative materials usage is often associated with greater problems in the End-of-Life (EoL) management: lower recovery rates and greater amounts of waste sent to landfill (due to the unavailability of separation processes and recycling technologies) are the main causes of an increased environmental burden. Thus, the innovative design solutions, even if remarkable technical advantages are obtained, may involve increased environmental impacts in the production and EoL phases, with an overall negative effect in the LC perspective [9-10].

For these reasons, the development of advanced solutions requires the conception of eco-design methodologies capable of simultaneously addressing design performance and sustainability aspects in the whole vehicle LC perspective. Life Cycle Sustainability Assessment (LCSA) is considered a suitable approach to assess sustainability in the early design phase of products. Considering the whole Life Cycle (LC), this methodology compares different product concepts, providing information on possible improvements (e.g., alternative materials or processes) and clearly describing the possibility of trade-off. Multi-Criteria Decision Analysis (MCDA) methods are used in the implementation of LCSA, to help designers and product developers solve decision problems to determine the best option when evaluating a set of alternatives based on multiple and conflicting criteria.

The automotive eco-design literature presents a huge number of articles, where type, objective, the complexity of the application process, and available inventory data vary significantly [11-12]. In this context, the sustainability analysis is considered a supporting, functional tool for validating lightweight designs. Delogu et al. [13], Sun et al. [14], Sun et al. [15], and Cecchel [16] are some of the many papers underlining this statement. These studies develop LCA of innovative automotive solutions compared to a baseline design, performing a detailed assessment of the car assets considered, including all the LC phases (production, use, and EoL), and assuming different impact categories to express the results. However, the performance/design requirements are not considered, because the competing concepts are evaluated assuming that they are strictly equivalent from a functional point of view. Fassi et al. [17] and Deng et al. [18], instead, go beyond the above studies, performing sustainability assessment on design equivalence basis. Other works focus on the development of eco-design tools rather than the pure assessment of specific case studies. Examples for this are represented by Ghadimi et al. [19] and Delogu et al. [20], which define holistic assessment methods integrating environmental, economic, and social aspects. The studies presented above incorporate the sustainability issue into the concept development phase, providing valuable support for designers and decision makers by measuring and making transparent sustainable information. However, these researches consider the environmental analysis downstream of the design process and they use it purely for validation purposes.

The only studies that integrate sustainability aspect into the design process using a systematic computer-aided design procedure are Reimer et al. [21], Russo and Matina [22] and Russo and Rizzi [23]. Reimer et al. proposes a methodology to support the conceptual design of lightweight body panels for Electric Vehicles (EVs) with a combined perspective on structural integrity and Green-House-Gas (GHG) emissions. The refined tool analytically combines numerical design and LC GHG estimation, thus providing a systematic impact assessment of design decisions on mechanical performance, lightweighting and sustainability potential. Russo and Matina and Russo and Rizzi present an eco-design approach implemented in a software framework that supports the designer in optimizing design solutions for automobiles from an environmental perspective. The integration of CAE, Life Cycle Modeling (LCM) and LCA tools represents the core of the method, as a supporting tool for the designer in the conceptualization design stage. The above studies represent very interesting attempts to integrate design and sustainability issues. That said, the conceived methods are suitable to automatically generate the optimized solutions, but the data processing requires a direct involvement of the user.

The review of automotive eco-design literature highlights the need for holistic eco-design methods able to assess both performance, design, and sustainability aspects, to evaluate at the same time implications of different nature over the entire product LC.

This paper describes a holistic eco-design approach to support designers in the early design stage, simultaneously considering design requirements and sustainability aspects. An accessible and easy-to-use tool, functional to generate and compare different design options in the automotive field, is provided. The combination of design requirements, concept physical-technical characteristics and LC inventory data allows the framework to automatically generate different design alternatives, providing a set of indicators that quantify both design performance and sustainability profile. Finally, a tailored single-score index developed through MCDA methods is used to rank and select the most promising design options.

2. Materials and Method

The presented work defines a method functional to compare alternative design solutions for mono-material automotive components considering both design and sustainability aspects. An integrated single score indicator (Product Sustainability Level, PSL) is developed on the multi-criteria decision methods; the overall methodology is implemented within the Design and Sustainability Analysis (DeSA) tool, targeted at providing a ready-to-use tool for engineers and designers. The DeSA allows broadening the range of considered design options and at the same time easing the comparative assessment process.

The following paragraphs present the design method according to two main sections:

- selection and sustainability analysis phase, where design and sustainability pillars are described, and related mid-point indicators are defined.
- classification phase, where PSL is assessed using the mid-point indicators calculated for each solution and the ranking of design alternatives is performed.

A schematic overview of the DeSA framework is presented in Figure 1.

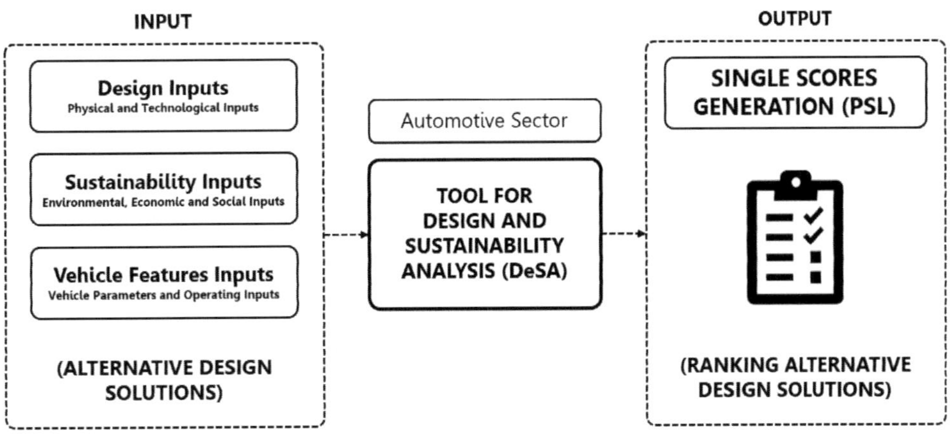

Figure 1. DeSA framework.

2.1. Selection and sustainability analysis

The DeSA tool performs the design and sustainability analysis based on a series of input data belonging to three main areas (in Figure 1):

- Design inputs, in terms of the main physical and technological features of the component in the automotive asset.
- Sustainability inputs, including Life Cycle Inventory (LCI) data functional to assess the environmental profile of the entire component LC.
- Vehicle features inputs, in terms of parameters and operating conditions that characterize the automotive component LC.

The list of parameters required by the methodology, for each of the three main areas indicated above, is reported in Table 1.

Table 1. Methodology input data.

Input Data		
Design Inputs	**Sustainability Inputs**	**Vehicle Features Inputs**
Primary Shape Design limitations Load case Boundary conditions	Materials Data Processes Data	Vehicle Lifetime Reference Vehicle Powertrain Characteristics Driving Cycle

The PSL indicator represents the final output, and it represents the sustainability level of different design solutions. The calculation of PSL includes four major steps (Figure 2): selection, sustainability analysis, and classification. The following sections describe in detail each stage of the PSL assessment.

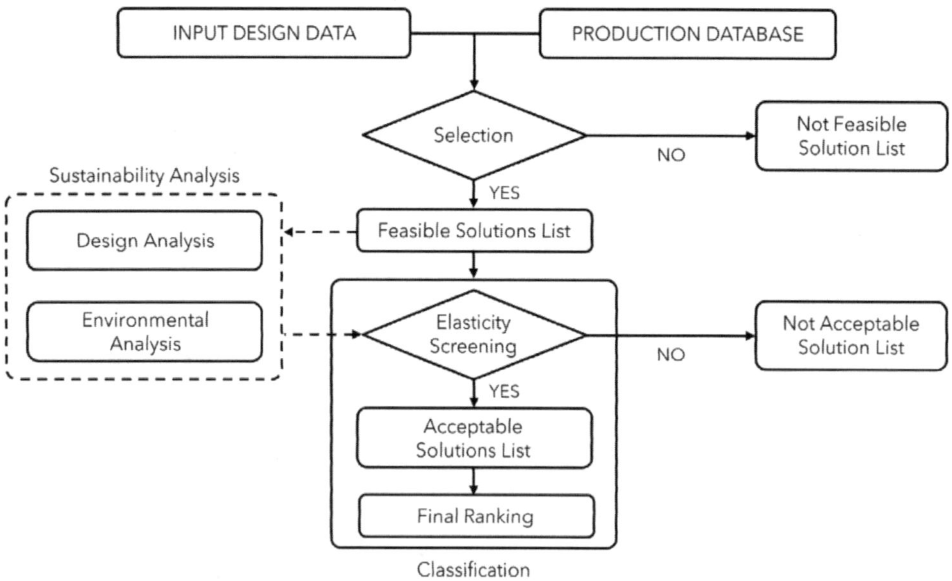

Figure 2. DeSA methodology main phases.

2.1.1 Selection. This phase determines the list of all design solutions that are feasible from a technical point of view and that satisfy the design requirements at the same time. The screening applies design constraints and design choices in terms of shape, material, and process. The methodology uses two main types of design constraints: the requirements on physical and technological features which characterize the different design options (see Table 2) are the first type and they are imposed by the tool's user. Instead, the technical compatibility between the material, manufacturing process, and component shape (for instance, the materials that can be processed through a specific manufacturing process) represents the second type of design constraints. Thus, the simultaneous application of physical, technological, and compatibility constraints allows discarding all the "shape-material-process" combinations that do not meet the design constraints, obtaining the list of feasible options.

Table 2. Design requirements imposed by the designer.

Design Requirements
Mass
Section thickness
Batch size
Tolerance
Surface roughness

The first step of the selection phase is the choice by the designer of the component's primary shape, which is selected based on the following available options:

- Prismatic (1D): circular, noncircular.
- Sheet (2D): flat, dished.
- Three-dimensional (3D): solid, hollow.

The definition of a production database (see Figure 2) is the selection's second step, where the entirety of materials and manufacturing processes available to the designer is stored. The database's materials section characterizes materials in terms of all properties used to establish if a material can be used for a specific application. The database classifies materials according to the following scheme:

- Metallic materials and alloys.
- Polymeric materials.
- Ceramic materials and glasses.
- Hybrid materials (foams, composites, and natural materials).

Table 3. Material properties in production database.

Material Properties
Density (ρ)
Young modulus (E)
Poisson ratio (η)
Yield strength (σ)
Global Warming Potential (GWP_{mat})

Table 3 reports the technical and environmental properties that characterize the materials in the database. The database's manufacturing section collects all the applicable manufacturing processes organized into a hierarchical order (see Figure 3) and defined in terms of the following features, reported in Table 4:

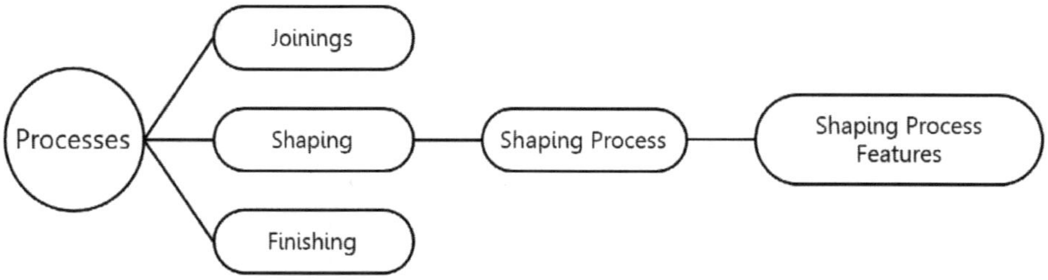

Figure 3. Scheme of the manufacturing processes hierarchy.

- Material features: the subset of materials that the process is able to use and transform.
- Shape features: list of shapes that can be made through the process (in this work, only shaping process family, reported in Figure 3, is considered).
- Physical features: physical properties related to the size (mass, section thickness), tolerance, roughness, and batch size.
- Environmental features: Global Warming Potential (GWP), caused by the process applied to the different database's materials.

Table 4. Process features in production database.

Process Features			
Material Features	**Shape Features**	**Physical Features**	**Environmental Features**
Materials List	Shapes List	Size Range Batch size Tolerance Surface roughness	Global Warming Potential (GWP_{man})

The third and final step of the selection is the application of constraints to all the available design options. The methodology applies the constraints using

- Numerical limitations referring to intrinsic physical and technological features of both materials and manufacturing processes (see Table 2)
- Shape-process compatibility (see Table 4)
- Material-process compatibility (see Table 4)

and they are analytically modelled through bar charts and binary matrices. Technical requirements and shape-material-process compatibility are combined to perform the selection, thus obtaining all the feasible solutions (Figure 2) which are characterized through the following features:

- shape (fixed by the designer);
- design constraints (provided by the case study);
- material and primary shaping process (obtained from the selection phase).

The final list of all the feasible design options represents the input for the sustainability analysis.

2.1.2 Sustainability analysis. This phase evaluates the different design alternatives obtained in the selection step through the Product Sustainability Level (PSL), an aggregated single-score indicator that considers both the design performance and the environmental profile. The sustainability analysis is structured in two main sections, dealing with design and environmental aspects, respectively.

Design analysis. In this section, the Finite Element Method (FEM) simulation modeling of all the feasible solutions provided by the selection step is performed. The design performance is assessed through a Performance Index (PI), calculated basing on performance level in terms of structural integrity provided by the design solution. In this work, PI will be defined as a safety factor provided by the ratio between the material's yield stress and the stress level on the component obtained through FEM simulations. For the FEM simulations setting, the analysis needs the following design input data:

- Fixed design input data (DI): geometric features (G) and boundary conditions (BC), which remain unaltered when passing from one feasible design solution to another (see "design inputs" of Table 1).
- Variable design input data (VI): material (MAT) and manufacturing process (MAN) obtained from the selection phase, variables when passing from one design solution to another (see "sustainability inputs" of Table 1).

The variable design input data are provided as ranges instead of single values. For this reason, the generic design solution S_i provided by the selection phase presents several sub-solutions. The number of sub-solutions (n_i) changes in terms of material and process features and depends on the variation range of the variable data. Considering the generic data range, $I = [I_{min}\ I_{max}]$, the range size φ_I is calculated as follows in equation (1):

$$\varphi_I = \frac{I_{max} - I_{min}}{I_{avg}} \tag{1}$$

Where:
I_{min} = the minimum value of generic data range
I_{max} = the maximum value of generic data range
I_{avg} = the average value of generic data range (i.e., $(I_{max} + I_{min})/2$)

The number of ranges is directly related to the materials properties and the processes features inserted in the production database (see Table 3 and Table 4). Therefore, the solution domain space Ω_{sol} is defined in equation (2):

$$\Omega_{sol} = \prod_{k}^{M} \varphi_{I_k} \tag{2}$$

Where:
M = the number of data ranges used for solution domain calculation
φ_{Ik} = the size of k-generic data range

Each value of Ω_{sol} will be divided by the minimum value obtained from several alternatives S_i (i.e., $\min(\Omega_{sol})$) and finally converted, via linear regression into simulation runs between values determined by the designer. An example of the domain space is presented in Figure 4.

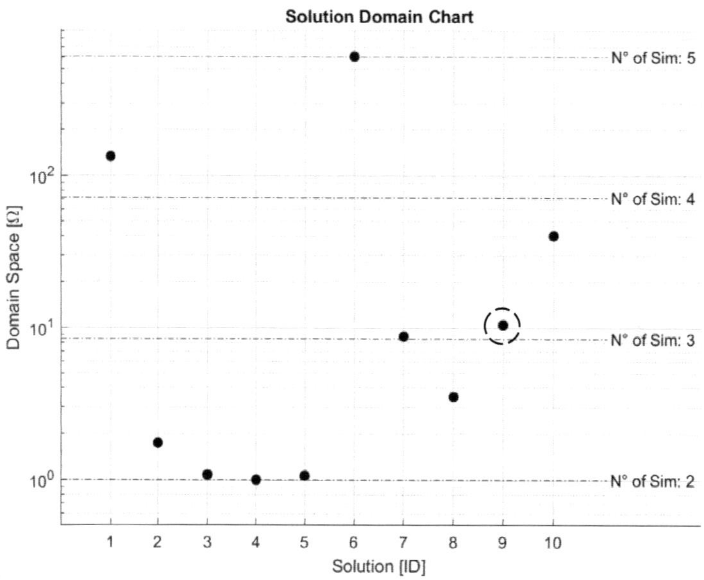

Figure 4. Example of domain space of design solutions.

In this example, the designer defines a minimum number of simulations for the generic solution (n_{min} = 2), and a maximum value (n_{max} = 5); for instance, the solution identified with ID = 9 will have a number of sub-solutions ($n_i \rightarrow n_9$) equal to 3.
Assuming S_{ij} as the generic sub-solution *j* of S_i, the PI of S_{ij} is function of both fixed and variable design inputs, as provided by the following relation (equation (3)):

$$S_{ij}=f(DI,VI_{ij})=f(G,BC,MAT_{ij},MAN_{ij}) \tag{3}$$

The sample of variable data V_{ij} to be considered for the generic solution S_{ij} is obtained through the Latin Hypercube Sampling (LHS) [24][25].

The calculation of parameter elasticity (EL) for each sub-solution S_{ij} completes the design analysis; EL is defined as the potential improvement of a sub-solution without compromising the design performance. The elasticity will be defined according to equation (4):

$$EL=PI\text{-}1=\frac{\sigma_y}{\sigma_{FEM}}\text{-}1 \tag{4}$$

Where:
σ_y = yield stress of the material component.
σ_{FEM} = stress level on the component calculated through FEM simulations.

The methodology provides the calculation of *PI* (function of design inputs) and *EL* (function of *PI*) through the combined use of MATLAB and Altair HyperWorks simulation software for each design sub-solution S_{ij} (see equation (5)):

$$S_{ij} \rightarrow \begin{cases} PI_{ij}=f(DI,VI_{ij}) \\ EL_{ij}=f(PI_{ij}) \end{cases} \tag{5}$$

IOP Conf. Series: Materials Science and Engineering 1214 (2022) 012045 doi:10.1088/1757-899X/1214/1/012045

As a consequence, each sub-solutions S_{ij} is characterized by a combination of PI-EL indexes. The methodology provides that:

- PI is used to calculate the PSL (see paragraph 2.2. – Classification);
- EL is used in the classification phase (see the following steps).

Environmental analysis. This step is aimed at assessing the environmental profile of the feasible sub-solutions S_{ij} provided by the design analysis section. The environmental assessment is carried out by means of the Life Cycle Assessment (LCA) impact category Global Warming Potential (GWP) according to the ILCD 1.09 impact assessment method [26]. The GWP indicator is determined considering the environmental impacts related to the entire Life-Cycle (LC) of the component, subdivided into the following main stages:

- materials acquisition;
- manufacturing;
- use;
- End-of-Life (EoL).

The GWP assessment is carried out according to the following calculation framework (see equations (6-19) reported in Table 5).

Table 5. List of Life Cycle Equations in DeSA methodology.

	Life Cycle Equations		
Production	$GWP_{prod}=GWP_{mat}+GWP_{man}$		(6)
	$U=\frac{mil_{comp}}{mil_{veh}}$		(7)
	$GWP_{raw}=\frac{m_{ij}\cdot gwp_{ij}^{mat}}{U}$		(8)
	$GWP_{man}=\frac{m_{ij}\cdot gwp_{ij}^{man}}{U}$		(9)
Use	$GWP_{use}=GWP_{WTT}+GWP_{TTW}$		(10)
	ICE Vehicles (ICEV)	Electric Vehicles (EV)	
	$GWP_{WTT}=gwp_{fuel}\cdot EC_{comp}$	$GWP_{WTT}=gwp_{mix}\cdot EC_{comp}$	(11)
	$GWP_{TTW}=CO_{2veh}\cdot mil_{veh}\cdot\frac{EC_{comp}}{EC_{veh}}$	$GWP_{TTW}=0$	(12)

End of Life

$$GWP_{EoL}=GWP_{shr}+GWP_{sep}+GWP_{rec}+GWP_{enr}+GWP_{disp} \quad (13)$$

$$GWP_{shr}=\frac{m_{ij}\cdot gwp^{shr}}{U} \quad (14)$$

$$GWP_{sep}=\frac{m_{ij}\cdot gwp_{ij}^{sep}}{U} \quad (15)$$

$$GWP_{rec}=\frac{m_{ij}\cdot \chi_{ij}^{mat}\cdot gwp_{ij}^{mat}}{U} \quad (16)$$

$$GWP_{enr}=\frac{m_{ij}\cdot gwp_{ij}^{enr}\cdot (1-\vartheta_{ij}^{enr})}{U} \quad (17)$$

$$GWP_{disp}=\frac{m_{ij}\cdot gwp_{ij}^{disp}\cdot \vartheta_{ij}^{enr}}{U} \quad (18)$$

LC

$$GWP_{LC}=GWP_{Mat}+GWP_{Man}+GWP_{Use}+GWP_{EoL} \quad (19)$$

Where:
GWP_{prod} = environmental impact of the production phase [kg CO_{2eq}].
U = the number of potential utilizations of the component (based on its durability) [-]; it is the ratio between the predictable duration of component (mil_{comp}) and the vehicle's lifetime (mil_{veh}).
mil_{comp} = component mileage [km].
mil_{veh} = vehicle mileage [km].
GWP_{raw} = environmental impact of the material acquisition to produce the component [kg CO_{2eq}].
gwp^{mat}_{ij} = environmental impact per mass unit of the material of generic sub-solution S_{ij} [kg CO_{2eq}/kg].
m_{ij} = component mass of generic sub-solution S_{ij} [kg].
GWP_{man} = environmental impact of the manufacturing process [kg CO_{2eq}].
gwp^{man}_{ij} = environmental impact per mass unit of the manufacturing process related to the generic sub-solution S_{ij} [kg CO_{2eq}/kg].
GWP_{use} = environmental impact of the use phase [kg CO_{2eq}].
GWP_{WTT} = environmental impact of well-to-tank use stage phase [kg CO_{2eq}/kg].
gwp_{fuel} = environmental impact of fuel production (for internal combustion engine vehicles) [kg CO2eq/kg].
gwp_{mix} = environmental impact of electrical mix grid production (for electric vehicle) [kg CO_{2eq}/kWh].
EC_{comp} = amount of Energy Consumption during operation of component; [kg] for ICEVs, [kWh] for EVs.
GWP_{TTW} = environmental impact of tank-to-wheel use stage phase [kg CO_{2eq}].
EC_{veh} = amount of Energy Consumption during operation of vehicle; [kg] for ICEVs, [kWh] for EVs.
CO_{2veh} = vehicle CO_2 emissions [kg CO_{2eq}/km].
GWP_{EoL} = environmental impact of the end-of-life (e.g., recycling, recovery, disposal) [kg CO_{2eq}].
GWP_{dis} = environmental impact of component disassembly [kg CO_{2eq}].
GWP_{shr} = environmental impact of component shredding [kg CO_{2eq}].
gwp^{shr} = environmental impact per mass unit of shredding of generic sub-solution S_{ij} (constant average value for all materials) [kg CO_{2eq}/kg].
GWP_{sep} = environmental impact of shredded material component separation [kg CO_{2eq}].

gwp^{sep}_{ij} = environmental impact per mass unit of separation of generic sub-solution S_{ij} [kg CO_{2eq}/kg].
GWP_{rec} = environmental impact of separated material component recycling [kg CO_{2eq}].
X^{mat}_{ij} = material substitution factor of generic sub-solution S_{ij} (quota of avoided primary production impact due to recycling) [-].
GWP_{enr} = environmental impact of post-separation material component energy recovery [kg CO_{2eq}].
gwp^{enr}_{ij} = mass-specific GWP of incineration with energy recovery for sub-solution Sij (referred to plastic fraction) [kg CO_{2eq}/kg].
ϑ^{enr}_{ij} = share of fibers for generic sub-solution S_{ij} (composite materials) [-].
GWP_{disp} = environmental impact of residual component disposal [kg CO_{2eq}].
gwp^{disp}_{ij} = environmental impact per mass unit of disposal of generic sub-solution S_{ij} [kg CO_{2eq}/kg].
GWP_{LC} = environmental impact of LC component [kg CO_{2eq}].

As a consequence, each sub-solution S_{ij} is characterized by a GWP_{LC}, which is used to calculate the *PSL* together with *PI* (see paragraph 2.2. – Classification).

To perform the sustainability analysis, the methodology provides the definition of a sustainability database which provides the GWP of all LC processes for the competing design sub-solutions. Additionally, the calculation framework requires the setting of user-define operating conditions that characterize the considered case study. Table 3 and Table 4 report the sustainability database GWP parameters for material-process combination, while the user-defined operating conditions of the case study are reported in Table 1 (see vehicle features inputs).

2.2. Classification

Once sustainability analysis is carried out for each of the sub-solutions S_{ij}, classification phase is performed in the following sections:

- PSL calculation of alternative solutions.
- Elasticity screening, where new constraints and design choices are applied in terms of component elasticity.
- Final ranking of acceptable solutions.

PSL Calculation. The Multi-Criteria Decision Analysis (MCDA) is the methodological principle of the PSL calculation, which integrates complex criteria to evaluate alternative decisions or solutions [27]. In this study, the values of indicators described above are used to obtain a tailored overall efficiency score, i.e., PSL. The PSL is evaluated from methodology proposed via three operations: normalization, weighting, and score aggregation; this score is performed using a weighted sum formula to combine the relative importance weights of criteria with performance scores for each product alternative.

Physical data collected for each aspect cannot be summed up together directly, due to the inconsistency and difference of units of measurements [28]. Thus, the values of these indicators need to be scaled into dimensionless values such that they can be analyzed and compared. This process is called normalization; in general, there are different statistical methods that can be used for this phase, but in this work, the "Max method" is used for the normalization of indicators [29]. According to this method, each indicator with an impact (positive or negative) on the product's sustainability ($I_k^{\mp}$) is normalized using the equation:

$$I_{Nk}^{\mp} = \frac{I_k^{\mp}}{I_k^{\mp max}} \tag{20}$$

Where:

I_{Nk} = normalized indicator for the dimension of generic sustainability k with an impact on overall sustainability.
I_k = indicator for the dimension of sustainability k with an impact on sustainability.
I^{max}_k = the highest value of the indicator for the dimension of sustainability k with an impact on sustainability (e.g., $I_k^{\mp max} = max(I_k^{\mp})$).

Regarding the weighting, several methods are commonly used, which are: equal weighting, subjective weighting (such as Analytic Hierarchy Process (AHP)), and weighting from analytical approaches. To compare the sustainability performance of several product solutions by mean of the proposed framework, the equal weighting will be applied in this methodology. In this case, weights w_i are defined by the equation (21) (i = 1, …, E is index of sustainability elements):

$$\sum_{k=1}^{E} w_k = 1 \tag{21}$$

In the aggregation phase, the normalized data are systematically aggregated into the single score based on the weighting factors assigned before and a conclusive sustainability index is finally derived. The Product Sustainability Level (PSL) is expressed by the following relation (equation (22)):

$$\text{PSL} = \sum_{k=1}^{E} w_k \cdot I_{N_k} \tag{22}$$

Where:
w_i = the weighting factor for the i_{th} aspect (i.e., design, environmental, economic, and social aspect).
I_{Ni} = the normalized score of i_{th} aspect.

In this case, considering only the design and environmental aspects (see sustainability analysis), the following formula (equation (23)) will be defined for DeSA methodology:

$$\text{PSL}_i = \text{K} \cdot \left(w_1 \cdot \frac{1}{cor + PI_{N_i}} + w_2 \cdot \frac{1}{cor + EI_{N_i}} \right) \tag{23}$$

Where:
K = PSL scale factor chosen by designer
cor = Correction coefficient (equal to 0.1).
w_j = the weighting factor for the j_{th} aspect (design and environmental aspects)
PI_{Ni} = the normalized score for the design aspect of i_{th} alternative solution
EI_{Ni} = the normalized score for the environmental aspect of i_{th} alternative solution

With the presence of the correction coefficient (cor), the value of PSL varies between the following limit values (see equation (24)):

$$\begin{cases} max(PSL_i) = 10K \\ min(PSL_i) = 0.91K \end{cases} \tag{24}$$

Elasticity screening. Once each index has been calculated for each sub-solutions S_{ij}, PSL is calculated (according to the above paragraph). The sub-alternatives results are merged since variable design input data (VI) were stored as ranges; a PSL range (PSL^{range}), as well as elasticity range (EL^{range}), is defined. If i represents the index of the generic solution, while n_i (in this case $n_i = m$) indicates the number of sub-solutions, the PSL^{range} of solution S_i is defined in the following way, through the minimum and maximum values of the sub-alternatives.

$$S_i \rightarrow \begin{cases} \{PSL_{i1}, PSL_{i2}, \ldots, PSL_{im}\} \rightarrow PSL^{range} = \left[min(PSL_{ij}), max(PSL_{ij})\right] \\ \{EL_{i1}, EL_{i2}, \ldots, EL_{im}\} \rightarrow EL^{range} = \left[min(EL_{ij}), max(EL_{ij})\right] \end{cases} \quad (25)$$

All solutions analyzed in the sustainability phase can be represented through ellipses, in a PSL-EL "bubble chart" (see an example in Figure 5). Only solutions with positive elasticity (i.e., EL>0) are considered, since they can be improved from the sustainability point of view without affecting the performance of the product itself.

A new screening is assessed through an Elasticity Line (EL_L) and Flexibility Filter (FF), whose values are defined through the designer experience. Two distinct zones in the bubble chart are defined by EL_L (as shown in Figure 5):

- A (Acceptable Zone): solutions with high EL improvement margins.
- B (Non-acceptable Zone): solutions with low EL improvement margins.

The acceptable solutions (S^{acc}_i) are those that present an elasticity value (EL) greater than the limit imposed by the designer (i.e., EL_L) and fall into the A zone; for these reasons, they will be considered in the final ranking. Instead, the design alternatives not acceptable (i.e., depending on the choice of the designer, it will not be able to fulfil the performance), fall into the B zone and they are discarded.

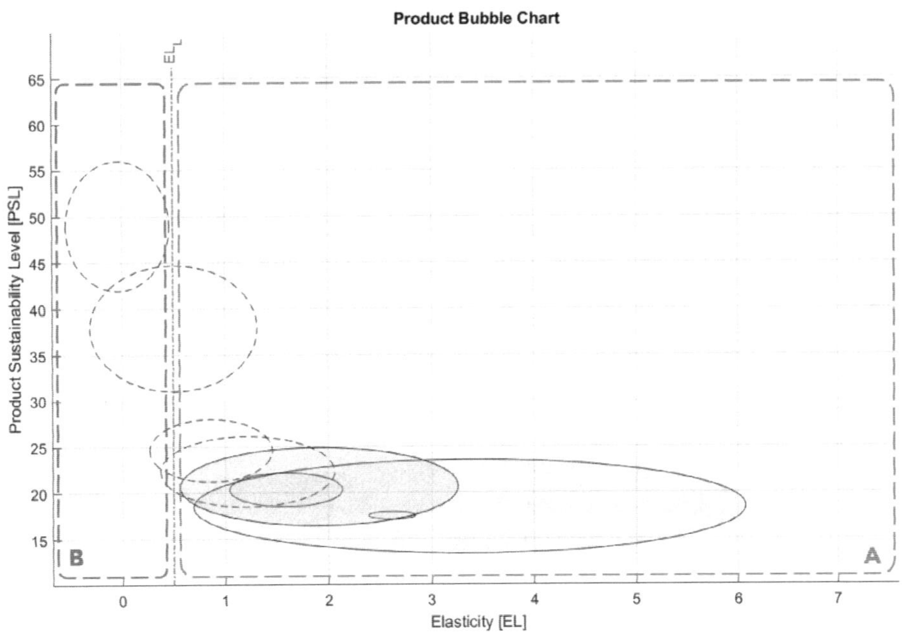

Figure 5. Bubble chart of design solutions with FF=100%.

If the generic solution S_i is present in both zones, the solution will be accepted or discarded depending on the filtering level (given by FF) set by the designer, according to equations (26-29):

$$\%EL_i^{right} \lesseqgtr FF \tag{26}$$

$$\%EL_i^{right} = \frac{diff_i^{range}}{EL_i^{range}} \tag{27}$$

$$diff_i^{range} = max(EL_{ij}) - EL_L \tag{28}$$

$$EL_i^{range} = max(EL_{ij}) - min(EL_{ij}) \tag{29}$$

Where:
$\%EL^{right}_i$ = elasticity range percentage to the right of elasticity line (E_L).
EL^{range}_i = elasticity range of solution S_i.
$diff^{range}_i$ = elasticity range of solution S_i to the right of elasticity line (E_L).
FF = flexibility filter (fixed by the designer).

- if EL_{right} < FF, the generic solutions that touch EL_L will be discarded and not considered in the final ranking phase (see Figure 5).
- if EL_{right} >= FF, the generic solution that touch EL_L will be accepted and considered in the final ranking phase (see Figure 6).

For instance, EL_L is equal to 0.5 in Figure 5 and Figure 6 (i.e., safety factor is equal to 1.5). Considering the first figure, the FF value is equal to 1 (or 100%); in this case all solutions that intersect EL_L are not considered; instead, in the other figure, the FF value is equal to 0.5 (or 50%) and the solutions having 50% of the EL range to the right of EL_L are considered in the next step of classification.

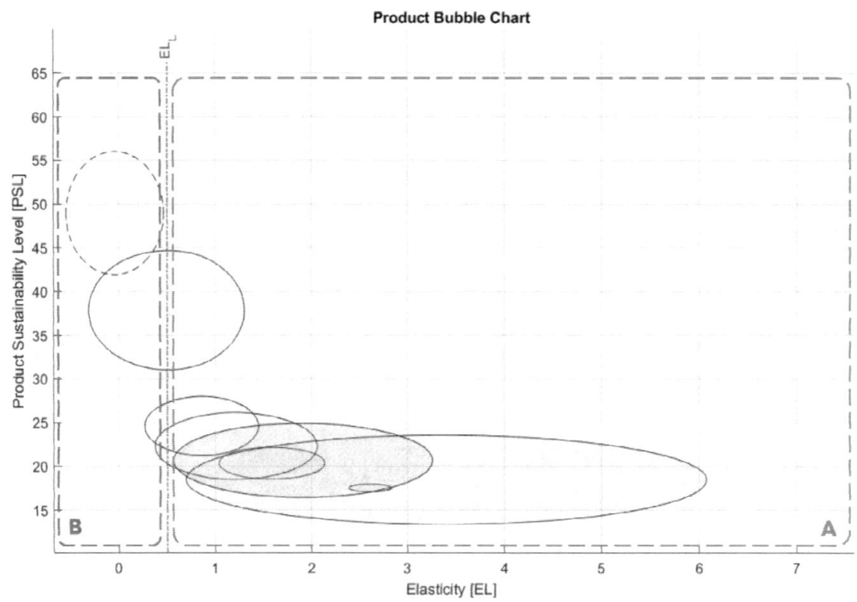

Figure 6. Bubble chart of design solutions with FF=50%.

Final ranking. In this last phase, the final ranking of the acceptable solutions S^{acc}_i is compiled and the PSL values are compared. Since each solution has a PSL^{range}, the final ranking is done through the average value of that range; it is defined as PSL_{ss} as follows (equation (30)):

$$PSL_{ss} = \frac{PSL_{max}^{range} + PSL_{min}^{range}}{2} \tag{30}$$

Where:
PSL_{ss} = single score of Product Sustainability Level.

The acceptable solutions S^{acc}_i are ranked through the value PSL_{ss} calculated for each bubble/solution.

3. Case study

Since the main goal of this paper is the application of DeSA methodology, a case study of an engine bracket in automotive context is chosen. In order to obtain the product sustainability level (PSL), the possibility to freely explore materials and manufacturing processes is given to the designer, while remaining consistent with system functional/performance requirements (i.e., boundary conditions). The main features/inputs of the case study are synthesized in the form below (see Table 6).

Table 6. Case study input data.

Input Design Data		
Design Inputs	**Sustainability Inputs**	**Vehicle Features Inputs**
Primary Shape Design limitations Loadcase Boundary conditions	Materials Data Processes Data	Vehicle Lifetime Vehicle Class Powertrain Characteristics Driving Cycle

3.1. Selection

The first step of the selection phase is the choice of the component primary shape; The Figure 7 presents the initial FEM model of engine bracket for DeSA methodology. As shown in figure, the chosen shape by the designer for the case study analysis is the three-dimensional shape (3D).

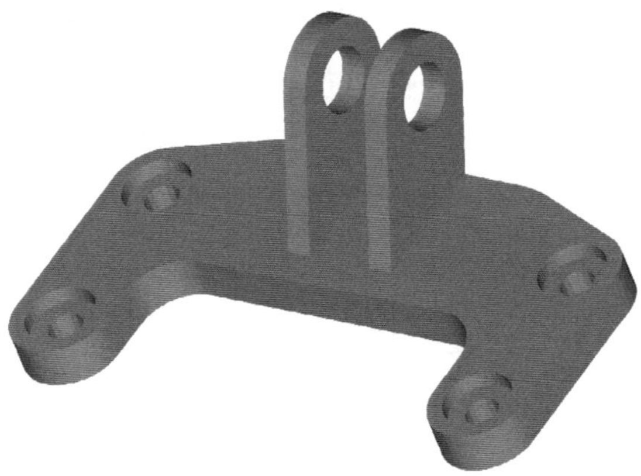

Figure 7. Engine bracket FEM model.

The second step is the definition of production database which consists of the entirety of materials and manufacturing processes available. For this case study, the list of allowable materials is only made of metals and alloys: each material is defined by an ID (i.e., Mxx) and its mechanical (such as density,

Young modulus) and environmental properties (i.e., GWP). Table 7 presents the materials chosen for this case study. Appendix A collects all of the materials data used in the case study analysis.

Table 7. List of allowable materials in production database.

Mat ID	Material Class	Material Subclass	Material Name
M1	Metals and Alloys	Ferrous	High Carbon Steel
M2	Metals and Alloys	Ferrous	Low Alloy Steel
M3	Metals and Alloys	Ferrous	Stainless Steel
M4	Metals and Alloys	Non-Ferrous	Age-Hardening Wrought Al-Alloys
M5	Metals and Alloys	Non-Ferrous	Titanium Alloys

Then, the list of allowable processes is only composed by processes capable of creating three-dimensional (3D) shapes: each process is defined by an ID (i.e., CAxx) and its mechanical and environmental features. Table 8 presents the processes chosen for this case study. Appendix B collects all the processes data used in the case study analysis.

Table 8. List of allowable processes in production database.

Pr ID	Process Class	Process Subclass	Process Name
CA1	Casting	Investment Casting Processes	Investment Casting
DE1	Deformation	Bulk Deformation Processes	Forging
PO1	Powder Methods	Powder Pressing Processes	Pressing and Sintering

The final step is the screening phase, where the objective is determining the list of all design solutions that are feasible from a technical point of view and that satisfy the design requirements at the same time. The application of design constraints and design choices in terms of primary shape, materials and processes will automatically provide several solutions. All the combinations that do not meet the functional requirements are not included in the list.

3.2. Sustainability analysis.
This phase is aimed at evaluating the different design alternatives obtained in the selection step by means of the PSL. In design analysis phase, Finite Element Method (FEM) simulation modelling of all the feasible solutions is provided by the screening step. The loadcase is assessed through the Performance Index (PI), on the basis of the performance level in terms of structural integrity provided by the generic design solution.

The boundary conditions (BC) of engine bracket are presented in Figure 8: the connection with engine and the link between bracket and vehicle chassis are defined by zone A and B, respectively. The connection between engine and bracket is modelled through bolted joints (see Figure 9); the bolts head are modelled with 1D deformable element (RBE3) in order to restrict degree of freedom without adding further stiffness to the FEM model (see the red spider webs in Figure 9). Instead, bolts stem is modelled with 1D solid circular section beam element (see the blue lines in Figure 9). The screws characteristics are shown in

.

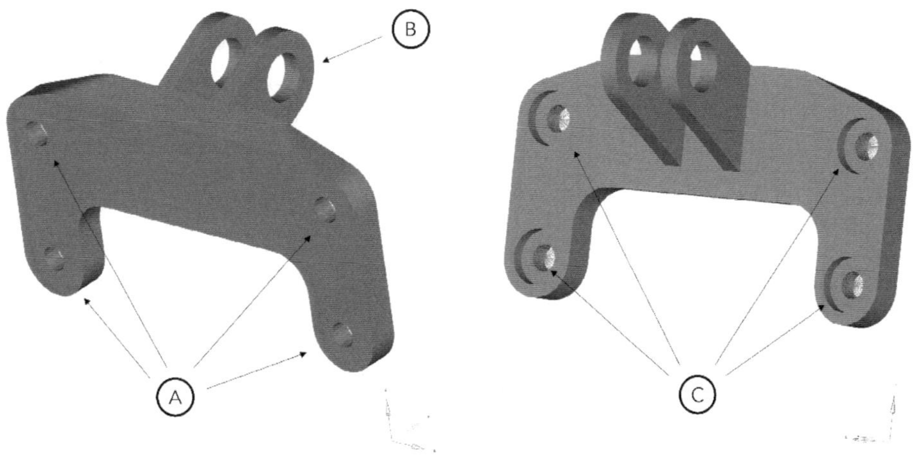

Figure 8. Engine bracket BC zones.

Table 9. Screw's characteristics.

Screw Properties	
Class	8.8
Diameter (D)	10 [mm]
Yield Stress (σ_y)	640 [MPa]

Each screw is pretensioned with a force (i.e., F_{pre}) defined by the following equation (31):

$$F_{pre}=0.6\cdot\sigma_y\cdot A=0.6\cdot\sigma_y\cdot\frac{\pi\cdot D^2}{4}\approx 30160\ N \tag{31}$$

The load is applied to zone C as a pressure vector (according to the global system in Figure 8); one component in x-direction (longitudinal), the second component in y-direction (lateral), and the latter component in z-direction (vertical). The pressure load is calculated considering a static distribution of the engine mass on each bracket used for mounting the engine to the cradle (i.e., F_{eng}) and multiplying this result by load coefficients consistent with the case study. Finally, the resultant forces are divided by the load application surface (i.e., the crowns where the bolts heads are placed, S_{cr}, equal to 265 mm^2) (see equations (32-35)). The engine mass must be consistent with the vehicle used for the case study simulation (in this case, a vehicle with a V4 engine):

$$m_{eng}=100\sim 150\ kg\rightarrow m_{eng}=150\ kg \tag{32}$$

$$F_{eng}=\frac{m_{eng}}{4}\cdot g\approx 375\ N \tag{33}$$

$$\begin{cases} F_x=0.5\cdot F_{eng}\approx 190N \\ F_y=0.5\cdot F_{eng}\approx 190N \\ F_z=1.5\cdot F_{eng}\approx 570N \end{cases} \tag{34}$$

$$\begin{cases} p_x = \dfrac{F_x}{4 \cdot S_{cr}} \approx 0.18\ MPa \\ p_y = \dfrac{F_y}{4 \cdot S_{cr}} \approx 0.18\ MPa \\ p_z = \dfrac{F_z}{4 \cdot S_{cr}} \approx 0.55\ MPa \end{cases} \tag{35}$$

The y-directional motion of zone A (i.e., where the connection between engine bracket and engine is made) is prohibited. Finally, in zone B (i.e., link between cradle and bracket) a cylindrical constraint is applied, locking the displacements respect to the axial and radial axes.

Since each design solution presents data ranges, Latin Hypercube method (LH) is performed to use this information in the design simulations. In this case study, a variable number of simulations (n_{sim}) is defined (variable between 2-5), depending on the solution considered (as shown in Figure 4).

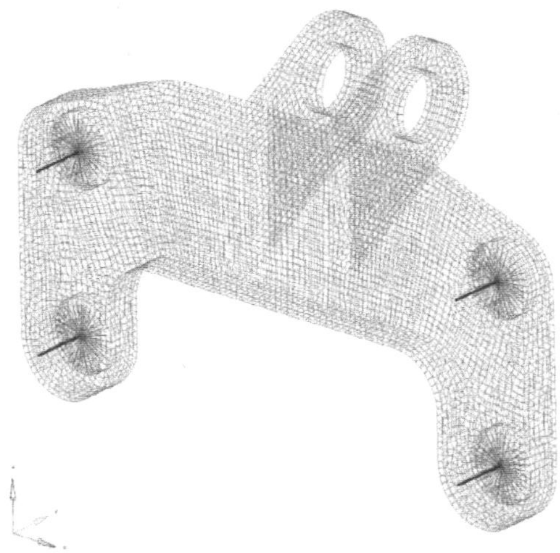

Figure 9. Bolt joints representation.

In the environmental analysis phase, the assessment of the environmental profile of the feasible sub-simulations provided by the design performance analysis section is carried out. The environmental assessment is carried out by means of the LCA impact category GWP according to the ILCD 1.09 impact assessment method. The first step of this phase is the choice of the vehicle where the case study is mounted for hypothesis. The chosen vehicle for the analysis is VOLKSWAGEN Golf 2.0 TSI GTI, whose characteristics are shown in Table 10. Finally, the lifetime of engine bracket is equal to vehicle lifetime (in this case, no substitution is required).

Table 10. List of vehicle features inputs.

Vehicle Features Inputs	
Vehicle Mass	1355 [kg]
Vehicle Lifetime	150000 [km]
Vehicle Class	C
Powertrain	ICEV
Driving Cycle	WLTP
CO_2 Consumption	149 [g/km]

3.3. *Classification.*
In this phase the Product Sustainability Level (PSL) is calculated by means of indexes calculated for each sub-solutions and the final ranking of design alternatives is performed. In the PSL calculation phase, the aggregated score is evaluated using a weighted sum formula to combine the relative importance weights of criteria with performance scores for each design alternative (see equation (23)). The PSL scale factor (K) chosen for the case study is equal to 10; with the presence of the correction coefficient (cor), the value of PSL varies between the values reported in equation (24).

In the Elasticity screening step, the determination of the list of all design solutions that are acceptable from an elasticity point of view is occurred. This screening is assessed through an Elasticity Line (EL_L) and Flexibility Filter (FF); in this case study, EL_L is equal to 1.5 (i.e., safety factor equals to 2.5); FF is equal to 50%, i.e., all solutions that have less than 50% of the elasticity range (EL_{range}) to the right of EL_L will be discarded and not considered in the final ranking phase.

In the Ranking phase, the PSL values of the acceptable solutions will be compared; as mentioned, this ranking is done by mean of PSL_{ss}.

3.4. *Results*
The solutions created in the selection phase is reported in Table 11. Each design solution is identified with an ID defined as follows:

$$\textit{Design Solution ID} \rightarrow n_{ID}\text{-}\textit{Shape}\text{-}MAT_{ID}\text{-}PR_{ID}$$

Where:
n_{ID} = Solution number.
Shape = Primary component shape defined by designer.
MAT_{ID} = Material ID (see Table 7).
PR_{ID} = Process ID (see Table 8).

The application of design constraints and design choices in terms of primary shape, materials and processes created 13 solutions. The LH method created 5 simulations, grouped according to the materials in the list, with a total of 16 sub-simulations. The elasticity screening passed all 13 solutions, divided in two groups:

- Solutions that present an elasticity range always greater than the EL_L value (see green lines in Table 11).
- Solutions that present an elasticity range with values lesser than the EL_L value (see yellow lines in Table 11).

All analyzed solutions in the sustainability analysis phase can be represented through ellipses, in a PSL-EL bubble chart. The engine bracket bubble chart is shown in Figure 10: the full-acceptable solutions S^{acc}_i (in green) are those that present an elasticity value (EL) always greater than the limit imposed by the designer (i.e., EL_L); instead, the FF-acceptable solutions S^{acc}_i (in yellow) are those that can present an elasticity value (EL) lesser than the limit imposed by the designer. The other design alternatives that are not acceptable are discarded; however, in this case study no solution has been discarded by the methodology.

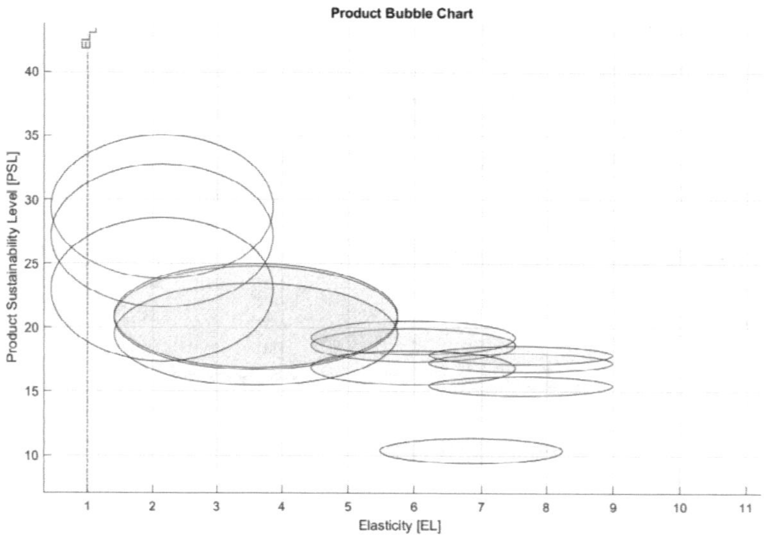

Figure 10. Engine bracket bubble chart, with FF=50%.

Finally, Table 12 comparing the PSL values of the acceptable solutions is compiled. Looking at the first 3 solutions in the ranking (see Table 12), they present the same material type (i.e., Age-Hardening Wrought Al-Alloys, with ID = M4), but the manufacturing process differs from one solution to another. Considering the best solution (with ID = 4), the main life cycle phases (reported in logarithmic values) are presented in Figure 11:

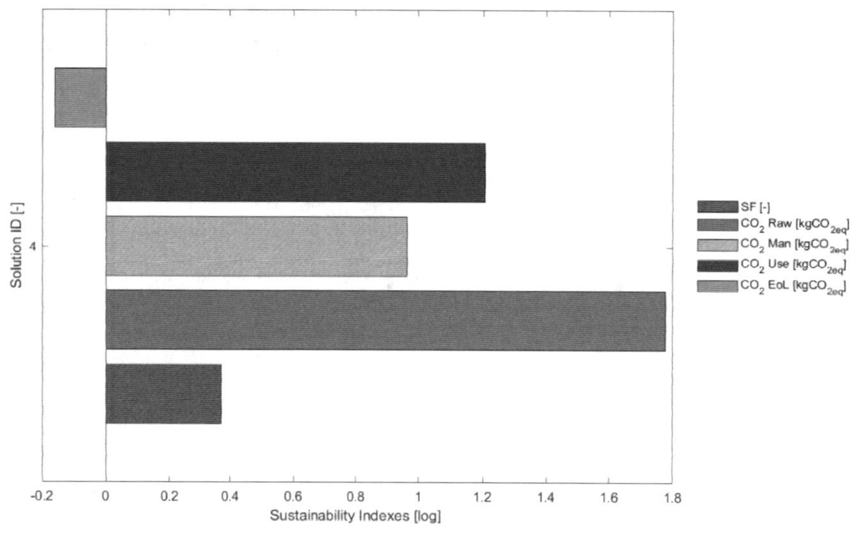

Figure 11. Life cycle phases of best solution (with ID = 4).

The raw material phase, compared to the other phases, has high value, and depends on the component material.

Looking at the generated solutions, they present very large elasticity ranges and away from elasticity line EL_L (with the exception of solutions that touch EL_L, but still show massive ranges). This highlights how the finite element model (with its boundary conditions) used for the case study presents possibilities for improvement.

Table 11. List of solutions created from selection phase.

Design Solution ID	PSL Range	EL Range	Screen
1_Solid_M1_CA1	[17.3701,19.9418]	[4.4379,7.5181]	OK
2_Solid_M2_CA1	[16.559,18.0186]	[6.2095,8.9864]	OK
3_Solid_M3_CA1	[16.7473,24.8011]	[1.4024,5.7481]	OK
4_Solid_M4_CA1	[23.8281,35.0472]	[0.44898,3.8531]	OK, With FF
5_Solid_M1_DE1	[17.9333,20.5535]	[4.4379,7.5181]	OK
6_Solid_M2_DE1	[17.195,18.6049]	[6.2095,8.9864]	OK
7_Solid_M3_DE1	[16.9081,24.9975]	[1.4024,5.7481]	OK
8_Solid_M4_DE1	[21.6142,32.7787]	[0.44898,3.8531]	OK, With FF
9_Solid_M5_DE1	[9.4244,11.3832]	[5.4898,8.2355]	OK
10_Solid_M1_PO1	[15.542,18.2384]	[4.4379,7.5181]	OK
11_Solid_M2_PO1	[14.7071,16.2157]	[6.2095,8.9864]	OK
12_Solid_M3_PO1	[15.4796,23.4742]	[1.4024,5.7481]	OK
13_Solid_M4_PO1	[17.3816,28.6046]	[0.44898,3.8531]	OK, With FF

Table 12. Final ranking table of acceptable solutions.

Design Solution ID	PSLss	PSL_{max}	Rank
1_Solid_M1_CA1	18.6559779	19.9418	8
2_Solid_M2_CA1	17.2888227	18.0186	10
3_Solid_M3_CA1	20.7742002	24.8011	5
4_Solid_M4_CA1	29.437672	35.0472	1
5_Solid_M1_DE1	19.2434075	20.5535	7
6_Solid_M2_DE1	17.8999635	18.6049	9
7_Solid_M3_DE1	20.9527804	24.9975	4
8_Solid_M4_DE1	27.1964416	32.7787	2
9_Solid_M5_DE1	10.4037791	11.3832	13
10_Solid_M1_PO1	16.8902052	18.2384	11
11_Solid_M2_PO1	15.4614015	16.2157	12
12_Solid_M3_PO1	19.4768603	23.4742	6
13_Solid_M4_PO1	22.9930991	28.6046	3

Finally, the best acceptable solution is solution with ID=4, whose characteristics are shown in Table 13.

Table 13. Best solution characteristics.

Best Solution Characteristics	
Solution ID = 4_Solid_M4_CA1	
Shape	Solid (3D)
Material	Age-Hardening Wrought Al-Alloys (ID: M4)
Process	Investment Casting (ID: CA1)

4. Conclusions

In this work, a design and sustainability analysis method is defined and applied to design solutions in the automotive field. The single score indicator Product Sustainability Level (PSL), obtained through the integration of finite element simulations and environmental modeling in the LCA perspective, is based on the multi-objective approach and is evaluated for design solutions generated by DeSA methodology.

An engine bracket is the case study used to explain the method. The combination of shape (chosen by the designer), materials and manufacturing processes (defined through the production database) allowed to observe several design alternatives and compare their level of sustainability, demonstrating the potential of the methodology. From the case study results, the use phase shows an extremely high weight in the PSL calculation. Thus, solutions having the same material, but different processes present almost the same level of sustainability.

The integration of performance and environmental issues into the product's concept design is the methodology contribution: several design solutions (combination of shape, materials, and technologies) are defined, analyzed, and compared using a single score index and engineering charts. However, the current approach presents several limits, summarized into the following key points, but which will be the subject of future developments in methodology:

- the method is limited to the structural analysis of the product to be studied and analysis of the environmental issues.
- only primary processes are considered in the creation of the several design alternatives (secondary and joining processes are not considered).
- a certain amount of data to perform the framework is needed.
- the sustainability index (i.e., PSL) of alternatives is generated, but it does not give any information to improve the same index and to consider the component lightweighting.

Further work development will be the application of the topology optimization to acceptable solutions S^{acc}_i, achieving both product lightweighting and a possible sustainability improvement and completing the overview on performance and sustainability potentialities within the automotive field.

Appendix A

Table A1. Design material data in production database.

ID	Material Class	Material Subclass	Material Name	Density [kg/m^3]	Young Modulus [MPa]	Poisson Ratio [-]	Yield Stress [MPa]
M1	Metals and Alloys	Ferrous	High Carbon Steel	7800-7900	200000-215000	0.285-0.295	400-1160
M2	Metals and Alloys	Ferrous	Low Alloy Steel	7800-7900	205000-217000	0.285-0.295	400-1500
M3	Metals and Alloys	Ferrous	Stainless Steel	7600-8100	189000-210000	0.265-0.275	170-1000
M4	Metals and Alloys	Non-Ferrous	Age-Hardening Wrought Al-Alloys	2500-2900	68000-80000	0.32-0.36	95-610
M5	Metals and Alloys	Non-Ferrous	Titanium Alloys	4400-4800	110000-120000	0.35-0.37	750-1200

Table A2. Environmental material data in production database.

ID	Materia l Class	Material Subclass	Material Name	GWP primary [$kgCO_{2eq}/kg$]	GWP EoL – Energy Recovery [$kgCO_{2eq}/kg$]	Substitution Factor [-]
M1	Metals and Alloys	Ferrous	High Carbon Steel	1.71-1.89	0	0.25
M2	Metals and Alloys	Ferrous	Low Alloy Steel	1.93-2.13	0	0.25
M3	Metals and Alloys	Ferrous	Stainless Steel	4.73-5.23	0	0.25
M4	Metals and Alloys	Non-Ferrous	Age-Hardening Wrought Al-Alloys	12.2-13.4	0	0.15
M5	Metals and Alloys	Non-Ferrous	Titanium Alloys	44.1-48.7	0	0.15

Appendix B

Table B1. Design process data in production database.

ID	Process Class	Process Subclass	Process Name	Mass Range [kg]	Economic Batch Size [-]
CA1	Casting	Investment Casting Processes	Investment Casting	0.001-100	1-10000
DE1	Deformatio n	Bulk Deformation Processes	Forging	0.01-5000	1000-100000
PO1	Powder Methods	Powder Pressing Processes	Pressing and Sintering	0.01-5	5000-5000000

ID	Process Class	Process Subclass	Process Name	Section Thickness [mm]	Tolerance [mm]	Roughness [µm]
CA1	Casting	Investment Casting Processes	Investment Casting	1-75	0.05-0.25	0.5-3.2
DE1	Deformation	Bulk Deformation Processes	Forging	3-250	0.2-1	3.2-12.5
PO1	Powder Methods	Powder Pressing Processes	Pressing and Sintering	1.5-8	0.025-1	1-10

Table B2. Design process/shape compatibility data in production database.

	Primary Shape Type					
Process ID	**1D - Circular**	**1D – Non-Circular**	**2D - Flat**	**2D - Dished**	**3D - Solid**	**3D - Hollow**
CA1	1	1	0	0	1	1
DE1	0	1	0	0	1	0
PO1	1	1	0	0	1	1

Table B3. Design process/material compatibility data in production database.

	Material ID				
Process ID	**M1**	**M2**	**M3**	**M4**	**M5**
CA1	1	1	1	1	0
DE1	1	1	1	1	1
PO1	1	1	1	1	0

Table B4. Environmental process/material compatibility data in production database.

	Material ID - GWP manufacturing [$kgCO_{2eq}/kg$]				
Process ID	**M1**	**M2**	**M3**	**M4**	**M5**
CA1	0.807-0.892	0.819-0.906	0.809-0.894	0.819-0.906	0
DE1	0.247-0.273	0.238-0.263	0.562-0.621	5.24-5.79	1.05-1.16
PO1	2.7-3.27	2.83-3.43	2.89-3.31	20.2-24.4	0

References

[1] Ness B, Urbel-Piirsalu E, Anderberg S and Olsson L 2007 Categorising tools for sustainability assessment. *Ecol. Econ.* **60** 3 pp 498–508
[2] Sala S, Ciuffo B and Nijkamp P 2015 Asystemic framework for sustainability assessment. *Ecol. Econ.* **119** pp 314–32
[3] Del Pero F, Delogu M, Pierini M and Bonaffini D 2015 Life cycle assessment of a heavy metro train. *J. Clean. Prod.* **87** pp 787-799
[4] Mayyas A, Qattawi A, Omar M and Shan D 2012 Design for sustainability in automotive industry: a comprehensive review. *Renew. Sustain. Energy Rev.* **16** pp 1845-1862
[5] Graedel T and Allenby B 1998 *Industrial ecology and the automobile.* (Prentice Hall, New Jersey)
[6] Maclean H L and Lave L B 2003 Life cycle assessment of automobile/fuel options. *Environ. Sci. Technol.* **37** pp 5445-5452
[7] Delogu M, Del Pero F and Pierini M 2016 Lightweight design solutions in the automotive field: environmental modelling based on fuel reduction value applied to diesel turbocharged vehicles. *Sustainability* **8** 11 1167
[8] Kelly J C, Sullivan J L, Burnham A and Elgowainy A 2015 Impacts of vehicle weight reduction via material substitution on life-cycle greenhouse gas emissions. *Environ. Sci. Technol.* **49** pp 12535–12542
[9] Ciacci L, Marselli L, Passarini F, Santini A and Vassura I 2010 A comparison among different automotive shredder residue treatment processes. *Int. J. Life Cycle Assess.* **15** pp 896–906
[10] Delogu M, Zanchi L, Maltese S, Bonoli A and Pierini M 2016 Environmental and economic life cycle assessment of a lightweight solution for an automotive component: a comparison between talc-filled

and hollow glass microspheres-reinforced polymer composites. *J. Clean. Prod.* **139** pp 548–560
[11] Rousseaux P, Gremy-Gros C, Bonnin M, Henriel-Ricordel C, Bernard P, Floury L, Staigre G and Vincent P 2017 "Eco-tool-seeker": a new and unique business guide for choosing ecodesign tools. *J. Clean. Prod.* **151** pp 546–577
[12] Rossi M, Germani M and Zamagni A 2016 Review of ecodesign methods and tools. Barriers and strategies for an effective implementation in industrial companies. *J. Clean. Prod.* **129** pp 361–373
[13] Delogu M, Del Pero F, Romoli F and Pierini M 2015 Life cycle assessment of a plastic air intake manifold. *Int. J. Life Cycle Assess.* **20** pp 1429–1443
[14] Sun X, Liu J, Lu B, Zhang P and Zhao M 2017 Life cycle assessment-based selection of a sustainable lightweight automotive engine hood design. *Int. J. Life Cycle Assess.* **22** pp 1373–1383
[15] Sun X, Meng F, Liu J and McKechnie J 2019 Life cycle energy use and greenhouse gas emission of lightweight vehicle – a body-in-white design. *J. Clean. Prod.* **220** pp 1-8
[16] Cecchel S, Chindamo D, Collotta M, Cornacchia G, Panvini A, Tomasoni G and Gadola M 2018 Lightweighting in light commercial vehicles: cradle-to-grave life cycle assessment of a safety-relevant component. *Int. J. Life Cycle Assess.* **23** pp 2043–2054
[17] Fassi H F, Ourihi R and Elahrach K 2021 An integrated method involving design-manufacturing-environment applied in strutural optimization. *Mat. Proceedings* **38** pp 135-138
[18] Deng Y, Guo Y, Wu P and Ingarao G 2019 Optimal design of flax fiber reinforced polymer composite as a lightweight component for automobiles from a life cycle assessment perspective. *J. Ind. Eco.* **23** pp 986-997
[19] Ghadimi P, Azadni A H, Yusof N M and Mat Saman M Z 2012 A weighted fuzzy approach for product sustainability assessment: a case study in automotive industry. *J. Clean. Prod.* **33** pp 10-21
[20] Delogu M, Maltese S, Del Pero F, Zanchi L, Pierini M and Bonola A 2018 Challenges for modelling and integrating environmental performances in concept design: the case of an automotive component lightweighting. *Int. J. Sustain. Eng.* **11** 2 pp 135-148
[21] Reimer L, Kaluza A, Cerdas F, Meschke J, Vietor T and Herrmann C 2020 Design of Eco-Efficient Body Parts for Electric Vehicles Considering Life Cycle Environmental Information. *Sustainability* **12** 14 5838
[22] Russo D and Matina D 2012 Structural Optimization and LCA for a computer-aided sustainable design. *Proceedings of the ASME 2012 International Design Engineering Technical Conferences and Computers and Information in Engineering Conference. Volume 2: 32nd Computers and Information in Engineering Conference, Parts A and B. Chicago, Illinois, USA. August 12–15* pp. 329-338
[23] Russo D and Rizzi C 2014 An ECO-DESIGN Approach Based on Structural Optimization in a CAD Framework. *Comp. Des. Appl.* **11** 5 pp 579-588
[24] Viana F A 2013 Things you wanted to know about the Latin hypercube design and were afraid to ask. *Proceedings of the 10th World Congress on Structural and Multidisciplinary Optimization*
[25] Helton J C and Davis F J 2003 Latin hypercube sampling and the propagation of uncertainty in analyses of complex systems. *Reliab. Eng. Syst. Saf.* **81** 1, pp 23-69
[26] ILCD handbook, International Reference Life Cycle Data System - General Guide for Life Cycle Assessment
[27] Del Pero F, Delogu M, Berzi L, Dattilo C A, Zonfrillo G and Pierini M 2019 Sustainability Assessment for Different Design Solutions within the Automotive Field. *Procedia Structural Integrity* **24** pp 906-925
[28] Zhou L, Tokos H, Krajnc D and Yang Y 2012 Sustainability performance evaluation in industry by composite sustainability index. *Clean Technol. Environ. Policy* **14** pp 789–803
[29] Vafaei N, Ribeiro R A and Camarinha-Matos L M 2016 Normalization Techniques for Multi-Criteria Decision Making: Analytical Hierarchy Process Case Study. *IFIP Advances in Information and Communication Technology* **470** pp 261-269

IOP Conf. Series: Materials Science and Engineering 1214 (2022) 012046 doi:10.1088/1757-899X/1214/1/012046

Energy harvester duty cycle evaluation for railway vehicle health monitoring

C Russo[a], M Lo Monaco[a] and A Somà[a]

[a] Politecnico di Torino - Department of Mechanical and Aerospace Engineering, C.so Duca degli Abruzzi, Torino,10129, Italy

E-mail: caterina.russo@polito.it

Abstract. The recent growth of the IoT (Internet of Things) technologies makes these connected devices suitable for monitoring and diagnostic in different applications. Through these devices, a wireless sensor network has become a smart solution for monitoring structures, vehicles, and other devices. Each node in the network can be placed in an inaccessible or unsafe location for human intervention and provide a real-time data stream, useful for the diagnostic and maintenance of the structure. In this context, the power node becomes a fundamental problem since the replacement of batteries is a disadvantage both for environmental disposal and for the related costs. Thus, the interest in the so-called AIOT (Autonomous Internet of Things) is growing, and the energy harvester generators can be a possible solution to this problem. In this scenario, an inductive linear generator having a non-symmetrical gravitational suspension is presented. The main characteristics of the generator and the magnetic suspension are introduced with the description of the Matlab/Simulink model that simulates the same behavior. In this work, a first study of the duty cycle of the generator to power a wireless sensor node for industrial application is presented as well. This study is carried out with a particular focus on the acceleration frequency evaluation of railway vehicles to better understand the possible effective power that can be extracted from the harvester. The relevance of this work lies in the fact that the generator sizing cannot be separated from the detailed knowledge of the energy source and of the sensing/monitoring system that must be powered.

1. Introduction

The growing spread of IoT (Internet of Things) and monitoring systems based on micro-electro-mechanical systems (MEMS) also encompass the railway system. Through these connected devices it is possible to have real-time information and data regarding the monitored vehicle or component. These new data offer the opportunity to have a new vision regarding the status of the monitored device providing information now accessible thanks to the rising of these technologies. With a continuous stream of data, it is possible to have knowledge regarding the health of the monitored system that can be crucial for the maintenance, safety, and design field. In fact, these data allow a better knowledge of the monitored system in its working condition, making it possible to the create algorithms of predictive maintenance and to improve the design phase. Moreover, these data can provide fundamental information regarding the working conditions condition and offer an evaluation concerning the safety and the durability of the monitored device. A wireless monitoring network allows a continuous stream of data but also has to comply with some strict characteristics to be a powerful instrument. All these systems should be characterized by a low power demand, a reduced size, a long-life cycle, and wireless connection reliability. The reduced size is required to be easily placed in dangerous or hard-to-reach places and to prevent any interference with the regular functioning the regular functioning of the system monitored. Moreover, a solid wireless connection to the server and through nodes is mandatory to have a reliable network ad an affordable set of data. At last, a long lifetime is required to reduce the cost of maintenance and of battery replacement, which are the main reasons for low power demand. The batteries used in these networks become no more sustainable in terms of cost and for environmental disposal but also in terms of safety. The need of replacing a battery is the main drawback of these systems that have to be designed with the logic of 'place and forget'. Also, for freight vehicles, the design of these monitoring devices remains a challenge due to the lack of power onboard. This is also the reason why these devices still do not achieve a widespread diffusion and so the interest regarding the alternative methods of power supply is growing. In this contest, one of the possibilities is the use of energy harvester devices the energy harvester devices and the process of harvesting the energy losses in the environment.

In literature, there are many research works regarding the design and study of the energy harvester devices in the railway field. In particular, the applications and the characteristics of these devices can be summarized in different ways. For example, in the comprehensive reviews from Bernal et al. [1], Pan et al. [2] and Hosseinkhani et al. [3] the authors focus on the different monitoring systems, dividing them on th bais of how they are powered and what they measure and monitor. In the work of Bosso et al. [4] a comprehensive review of energy harvester used in the railway field is reported, where they classify the energy harvester used energy harvester on the basis of the type. Another classification method is based on the location in which the devices are mounted, as proposed by Chunsheng et al.[5]. The ones mounted on-board are generally composed by an accelerometer, gyroscope, noise sensors, GPS, and temperature sensor and they are used to provide a continuous monitoring system with real-time detection of defects, track irregularities, vehicle dynamic, location, and speed. The other ones are mounted around the track for wheel profile monitoring, wheel impact, and bogie performance evaluation. The main drawback of these systems is the fact that they provide information only in that specific location of the track. From our research group, two possible monitoring systems are proposed, one presented in the work of Aimar et al.[6],[7], which is followed in this work, and followed in this work and the other presented in Bosso et al.[8]. So, in this paper, it will be shown an onboard monitoring system and the prototype realization of an energy harvester for its power supply. In section 2 the description of the numerical model of the energy harvester is proposed with details on the integration in the system of the experimental nonlinear stiffness and the comparison with the experimental measurements. Then in section 3, the monitoring system is described with all its components and the heatmaps and plots of the measured data are presented in section 3.1. Then the evaluation of the load cycle on the energy harvester computed through the experimental data is explained in section 4. The same section shows the of the power the power obtained from the harvester if it is excited with the load curve computed with 5 days of activity. In the last section 5, the duty cycle design for this kind of harvester is presented.

2. Mathematical Model

Roundy and al. [9] propose basic principles of the functionality of vibration energy harvester. These generators consist of a seismic mass m attached with spring and viscous damper to a moving base inside a box (on the left of Figure 1). The analytic model is a 1 d.o.f mass-spring-damper system. The moving magnet has a relative motion z with respect to the base, which receives an imposed displacement y. The vibratory external source makes the seismic mass move out of phase with respect to the box. This relative motion between the mass and the base can drive a suitable transducer to generate electrical energy. Like all inertial devices, the harvester must be set in vibration to generate electrical power.

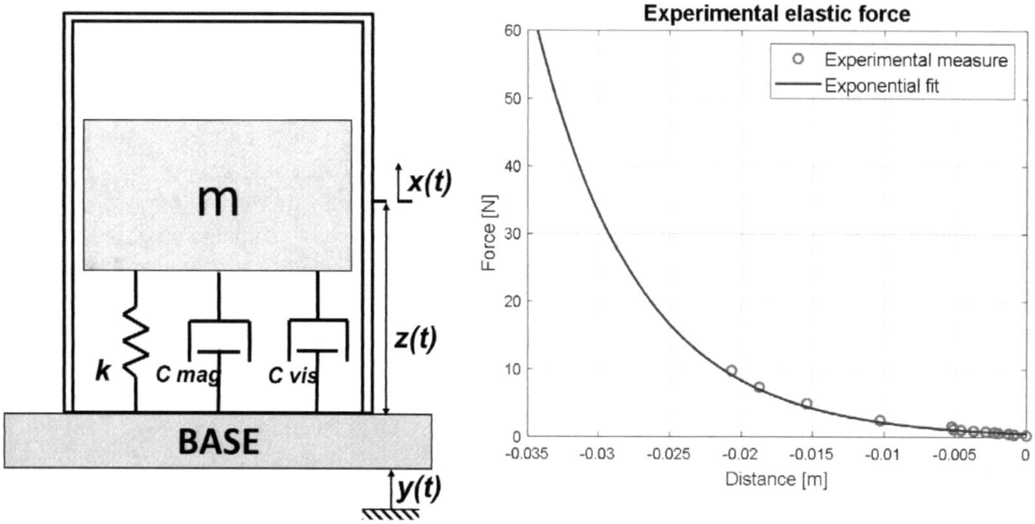

Figure 1: on the left, the model of the harvester with reference system, on the right the electromagnetic force

The magnetic interaction within the two magnets can be studied as a non-linear magnetic spring and the experimental and fitting curve of the elastic force obtained from the suspension is shown on the right of Figure 1. The static measures are carried out by placing different calibrated masses on the moving magnet with a supporting beam having a plate on the top for the application of these masses. By processing these measures, it is possible to find the numerical fit of the elastic force and then compute the stiffness curve to insert it in the numerical model. The equation of motion can be written in terms of the relative motion, considering the starting position as the equilibrium of the magnet,when its weight is equal to the repulsive magnetic force. Thus, the absolute motion coordinate of the moving magnet x is related to the static equilibrium condition of the moving magnet x_0.

$$m\ddot{x} + c_{tot}(\dot{x} - \dot{y}) + k(x - y) = 0 \tag{1}$$

$$z = x - y \tag{2}$$

$$m\ddot{z} + c_{tot}\dot{z} + kz = -m\ddot{y} \tag{3}$$

$$\ddot{y} = -\omega^2 y_0 sin(\omega t) = -Y_0 sin(\omega t) \tag{4}$$

Then the form to introduce this equation in Matlab Simulink is:

$$\ddot{z} = -\frac{c_{tot}}{m}\dot{z} - \frac{k}{m}z + Y_0 sin(\omega t) \tag{5}$$

Where:

- m is the mass of the moving magnet
- c_{tot} is the sum of viscous damping c_{vis} and the magnetic damping c_{mag}
- k is the magnetic non-linear stiffness
- Y_0 is the acceleration amplitude, in this case it is sinusoidal
- ω is the excitation frequency

Figure 2 below shows the complete Simulink model and the blocks that are circled in blue are the most characterizing ones and are explained in Table 1.In detail, about the block used in the numerical simulation, the ones blue circled are reported in Table 1.

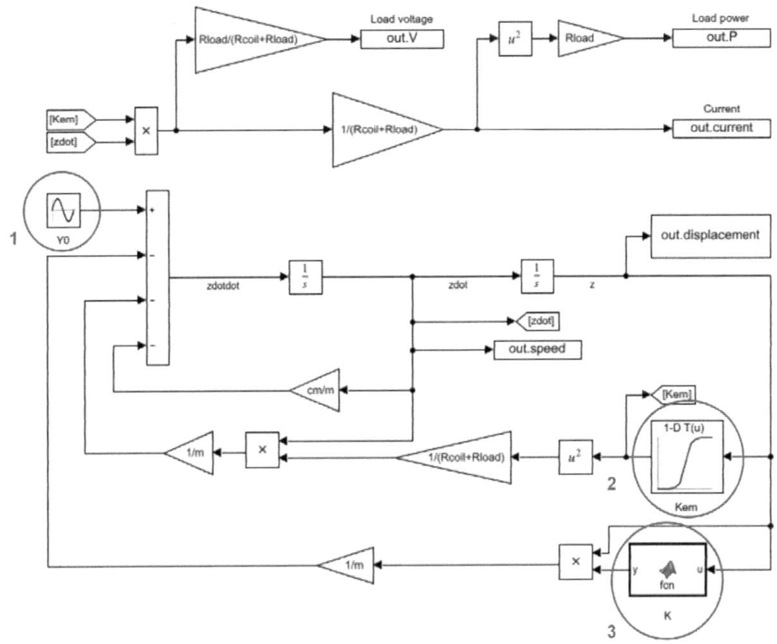

Figure 2: Simulink model of the energy harvester

N	Type	Description
1	Sine Wave	Excitation with Y_0 amplitude and ω frequency
2	1-D Lookup Table	Electromagnetic coefficient-moving magnet displacement relation
3	Matlab Function	Stiffness-moving magnet displacement relation

Table 1: Main Simulink blocks

The results obtained by comparing the numerical simulation results with the experimental ones are promising in terms of Frequency Response Function (FRF). Despite some slight differences in the matching resonant frequency and in the amplitude, a good comparison is obtained, as can be seen in Figure 3. It is important to notice that these differences are very reduced, the maximum difference of the peak resonance is about 0.3 Hz. The height difference in the peak is mainly due to the modelling of the friction damping in the model. Experimental tests show a non linear behavior of this parameter, as the variation of the damping coefficient obtained from the logarithmic decrements is significant. In fact, the variation of the damping coefficient obtained from the logarithmic decrements was huge. This is because of inaccuracies in the realization of the tube causing more friction when the magnet is close to the equilibrium position. Also, there is a dependence of the damping coefficient on the moving magnet speed. When the magnet moves faster and widely during resonance, the damping is at its minimum value vice; versa when the magnet oscillates away from resonance the damping is higher. In this first phase of the study, it is chosen not to implement the non-linearity of the damping friction and it is introduced as a constant value. Then, it is verified that this simplification does not affect heavily the obtained results. A future step of this work is to introduce this nonlinearity to complete the model. More details about these developments are presented in the work of Russo et al. [10].

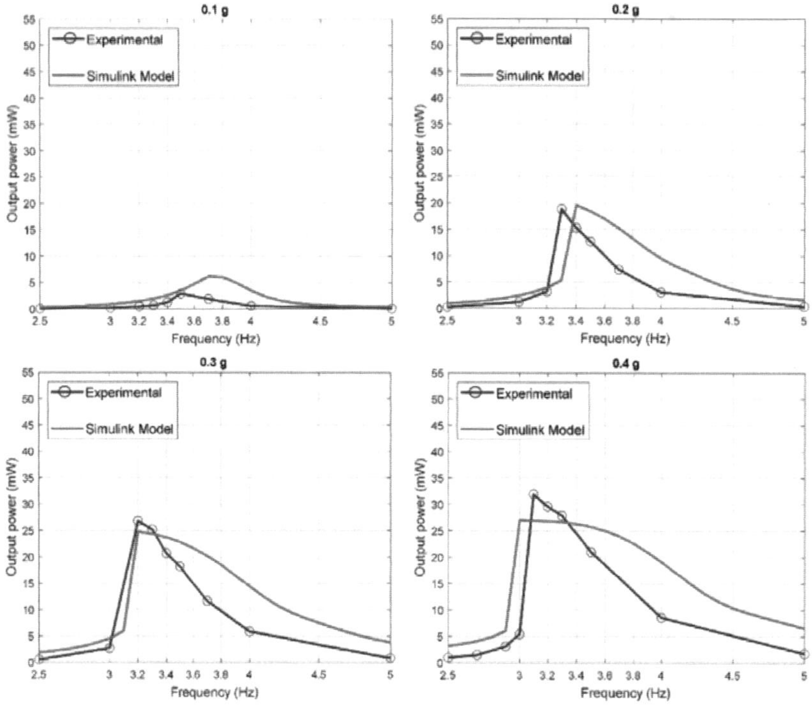

Figure 3: Experimental and numerical FRF's plot

3. The monitoring device

The monitored device developed by our research group [6] is build with these main components that are summarized in the table below:

N	Components	Descripion
1	Microprocessor	ATmega2560
2	Triaxial Accelerometer	16 g
3	Memory	μSD Card
4	GPS module	position of the system
5	Temperature Sensor	
6	Differential pressure sensors	

Table 2: Hardware components description

The data acquisition system is based on a micro-controller ATmega2560, which is an embedded software and firmware developed entirely by our research group. Through the GPS it is possible to have fundamental information regarding the position and the velocity of the train. The GPS signal is sampled with a frequency of 1 Hz. In case of no GPS signal (e.g. inside the railway galleries), the monitoring system keeps sampling the other parameters. In particular, the monitored data are the accelerations in the x-y-z directions, the temperature, and the pressure. All these parameters are sampled at 200 Hz and then the RMS, minimum and maximum value of the acceleration are computed every second and saved on the sd memory. The accelerometer measures allow to have a better measures is possible to have a better understanding of the vehicle dynamics and to detect potential impacts. Going into detail, the z-axis accelerations provide important information on the wheel-rail track interaction;, the x-axis, the longitudinal one, is crucial during the braking operations and the y-axis is the lateral one. Moreover, the vertical acceleration values represent an excellent basis for the correct design of an energy harvester device [11] as presented in this work. The temperature sensor is used to monitor the temperature of the brake block. The sensor is glued into a small groove cast iron brake block and it has a range temperature from -50°C to 500°C, enabling high temperature measures during braking. The pressure sensor, with a range of 0 to 7 bar, is placed on the same attachment points used during the periodical brake test after the laboratory calibration.

3.1. Heatmap of the data monitored
The data obtained from the GPS can be used to plot the real path of the train knowing latitude and longitude. The heatmaps allow to highlight, through a color graduated scale, the trend along the path of a certain data. In particular, the quantities of interest are the altitude, the speed, and the temperature to run a cross analysis for train braking monitoring. Below are reported, as an example, the data obtained during the 9 august 2020 between 9:00 and 14:00.

The altitude values were obtained thanks to an online software called "GPS Visualizer" which interpolates the latitude and longitude values provided by the system.

From the heatmaps, it is possible to see that the speed remains almost constant along all the routes and around 100 km / h. The duration of stops along the way can be obtained by crossing the GPS data with the respective instants of time.

4. Load diagram
The vertical acceleration data allows the definition of the excitation input of the energy harvester. The RMS value of the vertical acceleration, along the y axis, it is to count the number of peaks in a certain range and how often these are repeated in a daily data set.Through this approach it is possible to elaborate the load diagram which shows how many times a certain range of peaks acceleration is repeated in a time period. The process is shown in Figure 5, where at the top the peaks counting for day 09-08-20 is reported as an example. The same process is followed for all the other days that are analyzed. Subsequently, by means of the load diagram it is possible to transform the random acceleration signal in a sum of sinusoidal waves with amplitude values equal to the corresponding peaks range of the diagram and a temporal extension equal to the ratio between the number of cycles in which that amplitude occurs and the total number of cycles. The excitation frequency was considered equal to the resonance frequency of the energy harvester (3.5 Hz), which is the optimum working condition of the device.

Once this sinusoidal input is introduced in the Simulink model of the energy harvester, it is possible to evaluate the electrical quantities of the system and the output power, as it is shown in Figure 7. The obtained RMS value of power is compared with respect to the power generated by the mathematical model of the harvester excited by a sinusoidal with a constant amplitude equal to 0.4 g and a resonance frequency of 3.5 Hz.This point of comparison is the maximum theoretical power that can be generated by the model with the chosen sinusoidal excitation input. The load resistance and the number of turns used as a comparison are the optimal ones of the configuration that produces the maximum theoretical power.

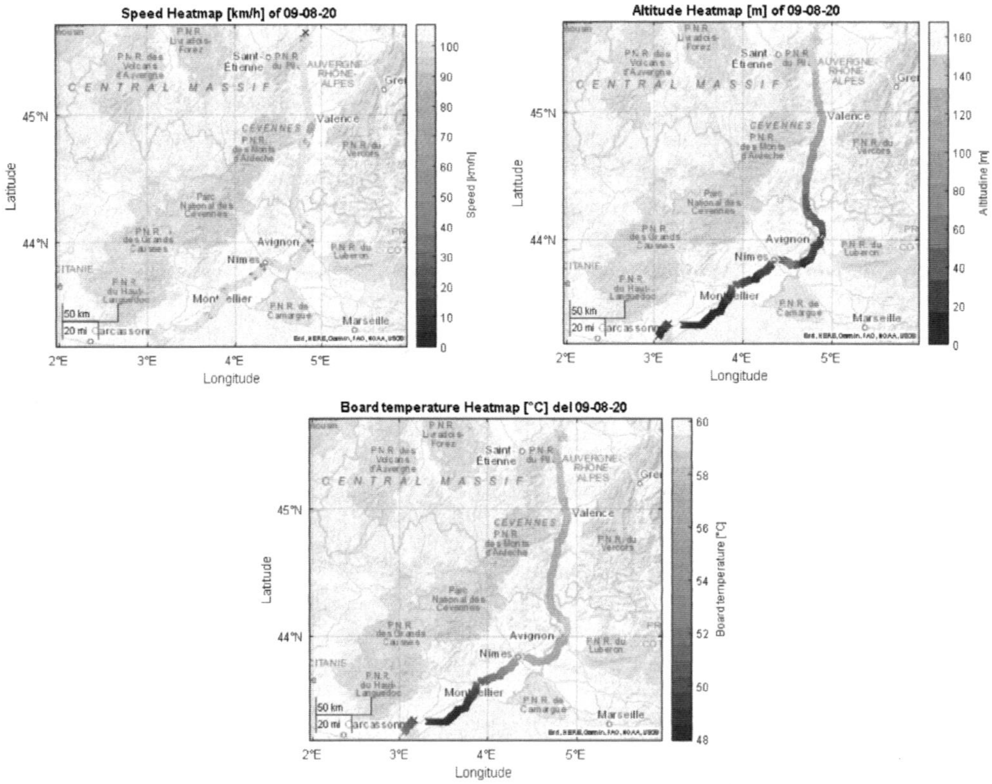

Figure 4: Altitude,speed and velocity heatmaps

The Table 3 below shows the effective RMS power that can be obtained from the 5 days chosen and the theoretical RMS power described before. From these two parameters, it is possible to have a first estimation of the harvester efficiency during a working cycle.

Day	Effective RMS Energy at 3.5 Hz [mWh]	Ideal RMS Energy [mWh]	Efficiency
09-08-20	30.5	50	61%
13-08-20	20	50	40%
04-09-20	20.4	50	41%
06-09-20	17.4	50	35%
16-09-20	10.5	50	21%

Table 3: Energy comparison of five examples days

To have a better knowledge of what happens moving around the resonant frequency,the RMS power at 2.5 Hz and 4.5 Hz is computed as well. As expected, the efficiency of the harvester continues to decrease. The asymmetric decreasing of the efficiency is due to the non-linearity of the device. The stiffness characteristic has a softening behavior, which leads to a sudden drop of the power when the frequency decreases from the resonance. On the contrary, when the excitation frequency has greater values than the resonance, the output power decreases more slowly. Finally, the efficiency is evaluated comparing again the effective output power with the theoretical one in terms of RMS value. The Table 4 below shows the obtained energy values for

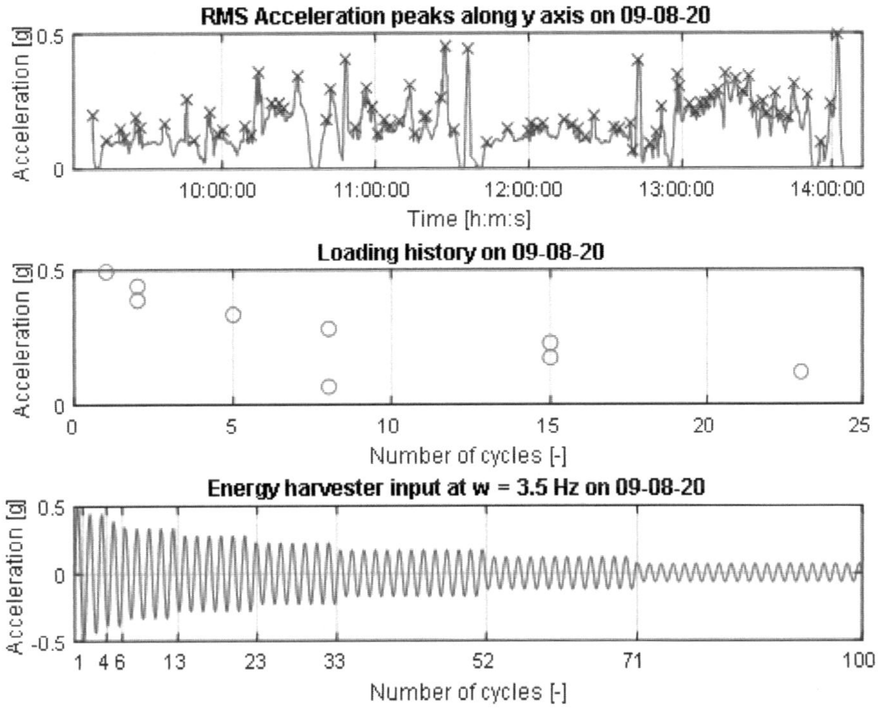

Figure 5: Peaks of vertical RMS acceleration, Load diagram and sum of sinusoidal with proportion obtained from he load diagram

the frequency range during different working days.

Day	RMS energy 2.5 Hz [mWh]	RMS energy 3.5 Hz [mWh]	RMS energy 4.5 Hz [mWh])
09-08-20	3.1	30.5	11.7
13-08-20	4.1	20	10.1
04-09-20	2.7	20.4	9.2
06-09-20	1.7	17.4	7
16-09-20	0.9	10.5	3.9

Table 4: Energy obtained around the resonant frequency

From this evaluation it is possible to point out the importance of a correct design of the harvester. It is necessary to provide a wide range for a wide range of the FRF spectrum so that the efficiency of the harvester remains constant in a wide range of excitation frequencies. Moreover, the duty cycle of the energy harvester should be evaluated. This parameter has become an essential part of the designing phase in order to obtain an efficient energy harvester. Real devices in working conditions always have accelerations or input excitation that are different from the laboratory one and it is mandatory to know how much power the device can effectively generate during its working operation. This concept will be explained in detail in the next

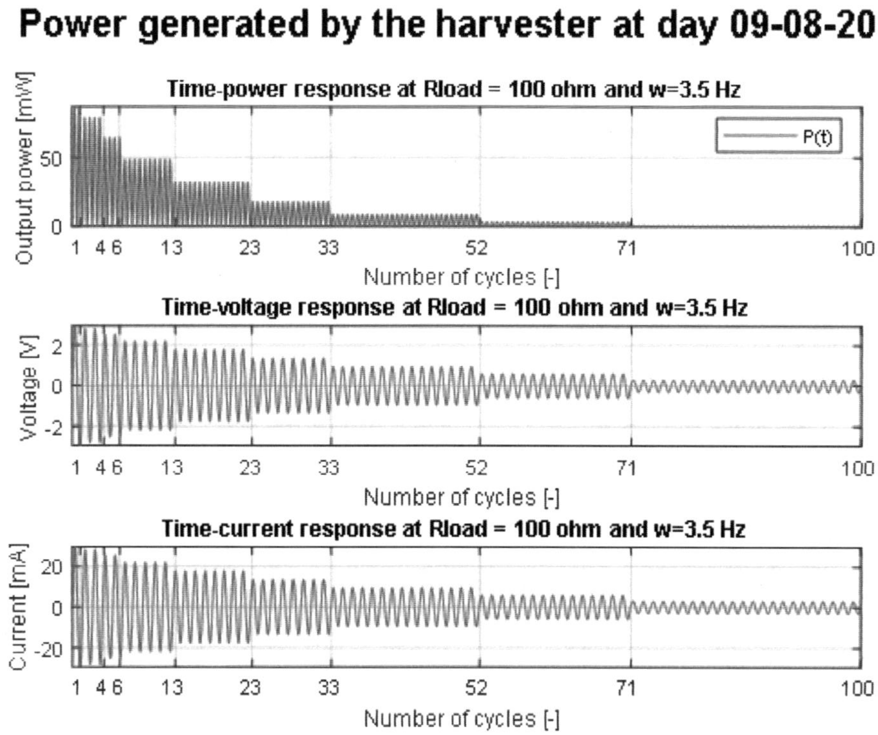

Figure 6: Power, current and voltage obtained during 09-08-20

section.

5. Duty cycle: cycle type and application range

The generator sizing cannot be separated from the detailed knowledge of the sensing/monitoring system that must be powered. The activity of design and dimensioning of the generator must be supported by complete information about the characteristics of the overall autonomous system including its frequency of data transmission, its powering demand during the acquisition data phase and the transmission one, etc. Moreover, it is important to consider the powering request of all the components between the transducer and the sensor. Usually, the following functional blocks can be identified (see Figure 8): current rectifier, charge reservoir (storage battery), one or more sensing devices (sensors),the transceiver device.

In almost every case, the energy generated by the harvester is stored in batteries or supercapacitors before use, requiring a preliminary the preliminary conversion of the alternate current produced by the harvester in continuous current. The charge storing is an expensive activity in terms of energetic efficiency. However, it is often necessary due to some reasons: firstly, to provide continuous supply also in case of irregular power generation; secondly, to reach the prescribed energy threshold needed to supply the utilizer. The energy produced by the harvester and stored in the battery can be used for supplying only when the minimum charge threshold is reached; this lower level of the battery charge is imposed by the energetic demand of the utilizers. Similarly, the time interval when the power flows from the battery to the utilizer must be calculated to know the consumed energy in time unit. The mentioned constraints are considered in defining the duty cycle of the self-powered system and in Figure 9 is represented

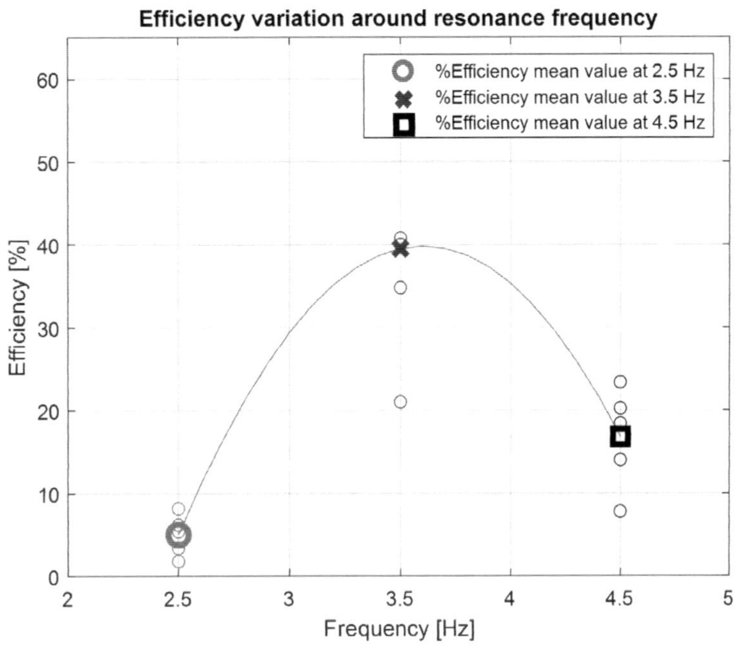

Figure 7: Effciciency plot obtained at 2.5 Hz,3,5 Hz and 4.5 Hz

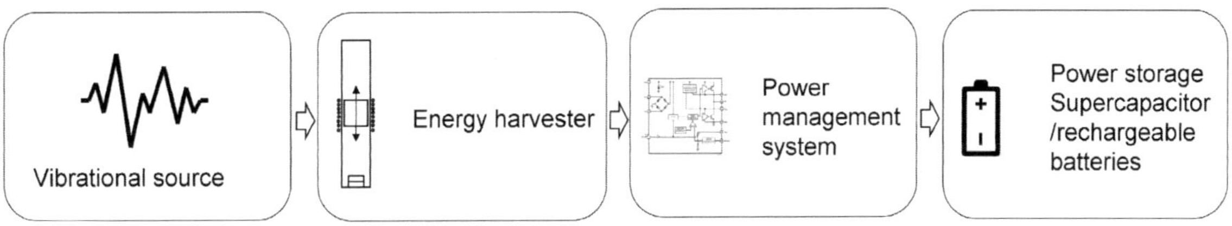

Figure 8: Energy harvester functional blocks

a typical working schematic duty cycle.

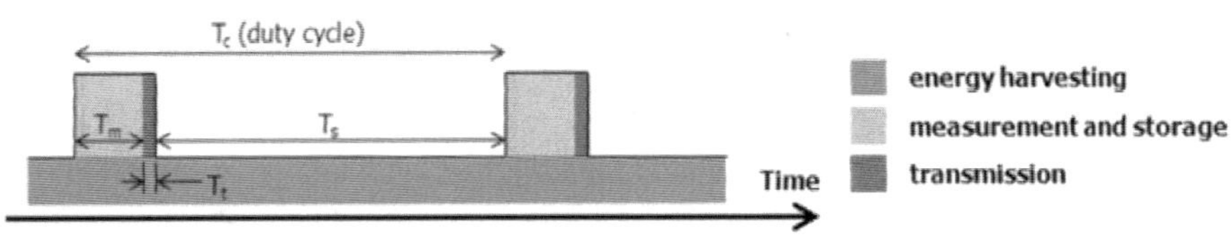

Figure 9: Transmission duty cycle

In general, during the design phase of a self-powered monitoring system, it is mandatory to consider the application in which the systems must work. It is important to define how and what the system must measure and how often it must send the data to the server. This global analysis allow to have a complete understanding of the system that needs to be monitored. The working time (T) of the harvester device and its transmission time period (t) must be evaluated. For the purpose of a long-lasting autonomous device, the ratio between these two quantities should be minor than 1. The time in which the harvester reaches power level has to be greater than the time in which the power is requested. In this case, it is possible to have a harvester

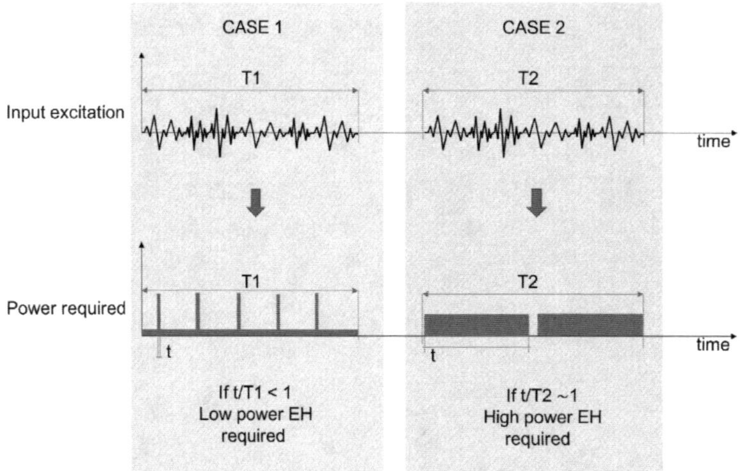

Figure 10: Duty cycle examples for low and high power EH

with a power output magnitude of mW and working for a long period. While, if continuous or more frequent information is required to be sent to the server, the ratio between the two time characteristics tends to 1. In this case, a more powerful energy harvester with output power magnitude of W has to be designed To have a better understanding of this concept see Figure 10). In conclusion, this ratio has to be a fundamental parameter to be consider during the study of an energy harvester. The definition of the duty cycle becomes one of the early phases of the designing process that allows obtain a prototype suitable for the application desired.

6. Discussion

This paper shows the evaluation of an energy harvester power generation in working conditions. The analysis is carried out firstly by the realization thanks to the first realization of a numerical model that simulates the real behavior of a prototype energy harvester. The model is built in Matlab/Simulink environment and leading to a good superimposition of numerical and experimental results. Also, a brief description of the monitoring board design by our research group is described in order to understand the cause of the excitation of the harvester mounted on a rail freight car. Then, the heatmaps of the data obtained from the monitoring device are presented and the vertical acceleration is processed to compute the load diagram. In this way, it was possible to build a sinusoidal wave with power content equivalent to the random excitation accelerations. Five days of train working are analyzed,in which the GPS coordinates are recorded. This process allows to have a raw evaluation of the efficiency of the energy harvester design during a real working operation. This phase is fundamental because it points out the necessity to design a harvester that presents a wider peak in the FRF in order to generate more power also when the resonance is not reached. The non-linearity of the harvester causes the generated power to rapidly drop when the frequency decreases from the resonance (softening stiffness), while it decreases slowly at higher frequencies. Finally, the duty cycle concept is introduced and its importance in the early design phase to understand the necessity of the definition of a duty cycle also in the early design phase to understand how much power has to be generated and if the energy harvester solution is suitable for the application.

References

[1] Bernal E, Maksym S,Colin C 2019 Onboard Condition Monitoring Sensors, Systems and Techniques for Freight Railway Vehicles: A Review *IEEE SENSORS JOURNAL* **19** 4-24

[2] Pan H, Qi L, Zhang Z,Yan J 2021 Kinetic energy harvesting technologies for applications in land transportation: A comprehensive review *Applied Energy* **286** 116518
[3] Hosseinkhani A, Younesian D,Eghbali P, Moayedizadeh A,Fassih A 2021 Sound and vibration energy harvesting for railway applications: A review on linear and nonlinear techniques *Energy Reports* **7** 852-874.
[4] Bosso N, Magelli M, Zampieri N 2020 Application of low-power energy harvesting solutions in the railway field: a review *Vehicle System Dynamics*
[5] Chunsheng L, Shihui L, Colin C, Maksym S 2017 An overview: modern techniques for railway vehicle on-board health monitoring systems *Vehicle System Dynamics* **55** 1045-1070
[6] Aimar M, Somà A 2017 Study and results of an onboard brake monitoring system for freight wagons *Proc. Inst. Mech. Eng. Part F J. Rail Rapid Transit.* **232** 1277–1294
[7] Somà A, Aimar M, Zampieri N 2021 Simulation of the Thermal Behavior of Cast Iron Brake Block during Braking Maneuvers *Applied Science* **11** 5010
[8] Bosso N, Gugliotta A, Magelli M, Zampieri N 2019 Monitoring of railway freight vehicles using onboard systems *Procedia Structural Integrity* **24** 692–705
[9] Roundy S, Wright P.K, Rabaey J 2003 A study of low level vibrations as a power source for wireless sensor nodes *Computer Communications 26*, 1131–1144.
[10] Russo C, Lo Monaco M, Fraccarollo F, Somà A 2021 Experimental and Numerical Characterization of a Gravitational Electromagnetic Energy Harvester *Energies 14*
[11] De Pasquale G, Somà A, Zampieri N 2012 Design, Simulation, and Testing of Energy Harvesters With Magnetic Suspensions for the Generation of Electricity From Freight Train Vibrations *AJournal of Computational and Nonlinear Dynamics* **7**

Analysis of the rider's body movement during the intervention of the Autonomous Emergency Braking system for Motorcycles (MAEB)

Giovanni Cassese[1], Cosimo Lucci[1], Giovanni Savino[1,2] and Niccolò Baldanzini[1]

[1] Dept. of Industrial Engineering, University of Florence, Via di Santa Marta 3, 50139 Florence, Italy

[2] Monash University Accident Research Centre, 21 Alliance Ln, Clayton VIC 3800, Australia

email: giovanni.cassese@stud.unifi.it

Abstract. Among the rider assistance systems for powered-two-wheelers (PTWs) that are currently in the developing stage, autonomous emergency braking (identified by the acronym MAEB - Motorcycle Autonomous Emergency Braking) was shown to be promising to significantly improve the safety of such vehicles. This system, which is already available on passenger cars and trucks (known as AEB), reduces the vehicle speed in the event of a forthcoming collision. The lack of implementation of AEB on standard motorcycles is due to the characteristic capsize instability of PTWs and their complex dynamics, which is, strongly influenced by the motion of the rider. In a recent field-test campaign within the EU funded project "PIONEERS", tests were conducted with common riders as participants to evaluate the intervention of MAEB in urban riding scenarios. A combined analysis of the data recorded from the vehicle, data related to the movement of the rider's body measured through an inertial measurement unit and videos recorded during the test, allowed characterizing the different behaviours of the rider's body in response to the activation of the automatic braking system in straight riding conditions. The results showed that body movement can be used as an indicator of the riders' ability to control the vehicle under automatic braking conditions. In addition, in tests conducted with 0.5 g automatic decelerations, riders showed to be able to recover to natural riding position within the timeframe of the automatic braking activation event. This study defined an innovative method for evaluating the response of motorcyclists to the braking intervention and provides insights into the applicability of MAEB on standard vehicles.

1. Introduction

Powered-two-wheeler (PTW) users are globally increasing in recent years, due to the cost-effectiveness and flexibility in urban environments of this means of transport [1]. However, the increasing number of users is related to the increase of crashes and fatalities due to the lower stability of PTWs and high vulnerability suffered by PTW users [2]. In order to prevent and mitigate the negative outcome of crashes involving PTWs, researchers worked in recent years to introduce on PTWs some of the

Advanced Driver Assistance Systems (ADAS) already available on four-wheeled vehicles. Among the rider assistance systems derived from ADAS, the Motorcycle Autonomous Emergency Braking (MAEB), a technology which autonomously deploys a braking action in pre-crash condition to reduce PTW speed or even prevent crashes, has been indicated as one of the most promising technologies for improving PTW users' safety [3]. In the past, some studies assessed the effectiveness of such a system [4] and its applicability in real-world conditions, mainly with expert riders [5,6] and straight-line riding conditions [7]. A further step in the development of MAEB was recently represented by a new field test campaign performed within the PIONEERS project [8]. In such experiment, extensive field testing involving two PTWs and common riders as participants was conducted to assess the applicability and the acceptability of Automatic Braking (AB) events reproducing unexpected MAEB interventions [9]. The first results of these field tests highlighted that safe MAEB interventions are feasible with decelerations up to 0.5 g [10] and in manoeuvres other than straight-line riding [11]. These important findings suggest that MAEB can be more effective in reducing speed before the crash and reducing injury risks for PTWs users then what was initially shown [12]. Even if the extensive field test campaign of PIONEERS provided evidence of the feasibility of safe MAEB interventions when adopting certain working parameters, concerns may still arise regarding the effect of MAEB intervention on the rider's body, especially when the rider is not prepared for the intervention of the system.

During an unexpected automatic emergency braking event, the rider's body tends to move forward. Since unlike car occupants, the rider is not restrained by seat belts, such forward movement may lead to a loss of control. The understanding of the rider's body response during MAEB intervention is, therefore, a key factor to assess MAEB safe applicability [13]. In addition, the body position and movement of the rider immediately before and during MAEB intervention could be used as indicators of rider capacity to handle MAEB intervention. Such indicators may be used to inform the system intervention regarding rider stability in real-time, in order to maximize the effect of the intervention while ensuring its safety. The objective of the present study is to characterize the body response of the rider during unexpected AB activations in straight-line riding conditions, to further validate the acceptability of MAEB among end-users and to enlarge its field of safe applicability in the future on standard powered-two-wheelers.

2. Methods

In this section, the test protocol employed to field test automatic braking (AB) interventions and the instrumentation used for the tests performed within the PIONEERS project [14] will be briefly described. Then, a summary of the data analysis performed to analyse the body movement in relation to the AB intervention will be presented.

2.1. Field test campaign

The participants for the AB tests were recruited among active riders having a minimum riding experience of two years or 10,000 km ridden and aged between 20 and 65 who volunteered for the tests based on an online survey. The test protocol for this study was approved by the Ethics Committee of the University of Florence (Written opinion N. 46, 20/03/2019).

The test vehicle was a Ducati Multistrada 1260S, a sport-touring motorcycle equipped with Cornering ABS (Anti-lock Braking System), Combined braking and a four-stroke engine with a displacement of 1262 cm^3. The motorcycle was also equipped with outriggers to prevent the motorcycle from lateral fall (Figure 1). The vehicle was provided with a remote-controlled AB system, set to decelerate the motorcycle at a nominal deceleration of 0.5 g for a duration of approximately 1s: this includes a fade-in time to reach the value of deceleration and a constant deceleration time.

During the tests, data from the CAN-bus of the motorcycle, data from external sensors, GPS position, and video from two Go-pro Hero4 cameras installed on the motorcycle were recorded. At the end of the test, subjective data including an adjusted controllability rating scale during the AB intervention was collected using questionnaires.

Figure 1. Test vehicle provided with outriggers.

The test protocol, which was developed based on previous work [15,16], consisted of the following phases: a) a warm-up session to become familiar with the vehicle and the track; b) a familiarization with the AB system, consisting of deploying declared AB interventions in straight-line, c) two test sessions involving pseudo-unexpected AB activations during straight riding and manoeuvring using a nominal AB deceleration of 0.3 g and 0.5 g respectively, d) the completion of a final questionnaire. The AB intervention was deployed with a remote-controlled braking device which was activated by the investigator to produce unexpected automatic decelerations without intervention by the rider as the participants went through different spots of the test track. In the 0.5 g AB deceleration session, each participant tested the pseudo-unexpected AB intervention twice in the straight-line section, with an additional final AB intervention in a different spot of the track. The latter intervention was meant to get the rider less prepared than any other ordinary AB event; this was called "out of the blue" activation. The test protocol of this study was designed to test AB interventions with unprepared participants minimizing any learning effect in the participants. Further description of the test protocol and the field test campaign is available in previous papers [9,16].

3. Data analysis

The methodology used for the data analysis is presented in the scheme shown in Figure 2.

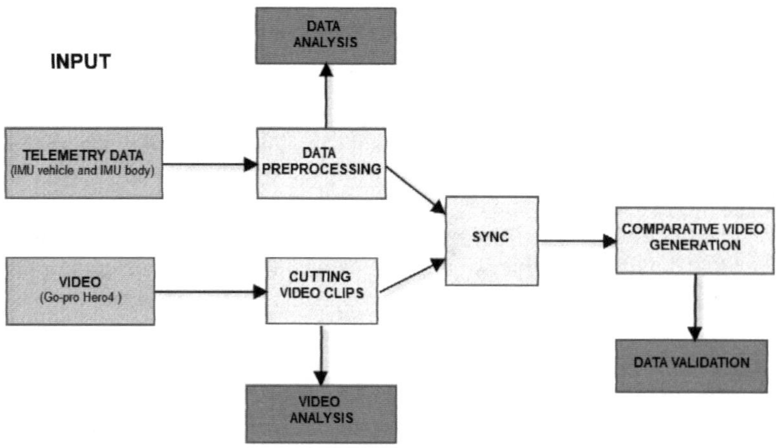

Figure 2. Methodology used for data analysis of the telemetry data and video recording data.

3.1. *Data pre-processing*

Every rider was equipped with an Inertial Measurement Unit (IMU) that measured accelerations and angles of the rider's back on three axes. Data recorded by this sensor were used to describe the dynamics of the rider's body during an AB activation. To give a complete description of the rider's movements during AB events, videos acquired by a Go-Pro Hero4 camera installed on the outrigger at the right-hand side of the vehicle (to provide a side view of the rider's movements) were employed. The combined analysis of data recorded on the vehicle, data related to the rider's body movement and videos recorded during the test, allowed us to describe the different behaviours of the rider in response to the activation of the AB system in straight riding conditions. The analysis focused also on comparing the rider's reactions in pseudo-unexpected AB activations with the "out of the blue" ones, which were assumed to be more difficult to control as completely unexpected [13].

The processing of data collected by the inertial platform unit (IMU) positioned on the back of the rider involved several steps (filtering, vector rotation, synchronisation). Data measured by accelerometer and gyroscopic sensors (accelerations, angular velocities, angles inclination) described the behaviour and the reactions of participants during AB interventions.

3.2. *Video analysis and data validation*

The first phase of the analysis was dedicated to the validation of the data. Using a specifically designed Matlab window displaying the data collected during the AB activation and the corresponding video recorded by the camera, videos and data were compared to have a qualitative validation of data and synchronization. This comparison also allowed us to a simple and intuitive analysis of the rider's behaviour during AB activation from the video together with data in every timestep of the AB intervention. Further validation of the data was made employing the open-source software Kinovea, which was used to analyse the rider's movements from the video. This allowed us to evaluate the accuracy of the data by comparing the participant's movements captured by the camera with the data acquired by the sensors. We chose the variation of the rider's body pitch angle during the AB event as a measure of comparison between the measured data and the videos (see Figure 3). At the time of the AB trigger, the magnitude of the inclination of the rider was positioned to a reference value of 0° and then, for the duration of the automatic braking, the movement of the rider in terms of the angle of inclination of the body (pitch) was compared between videos and data.

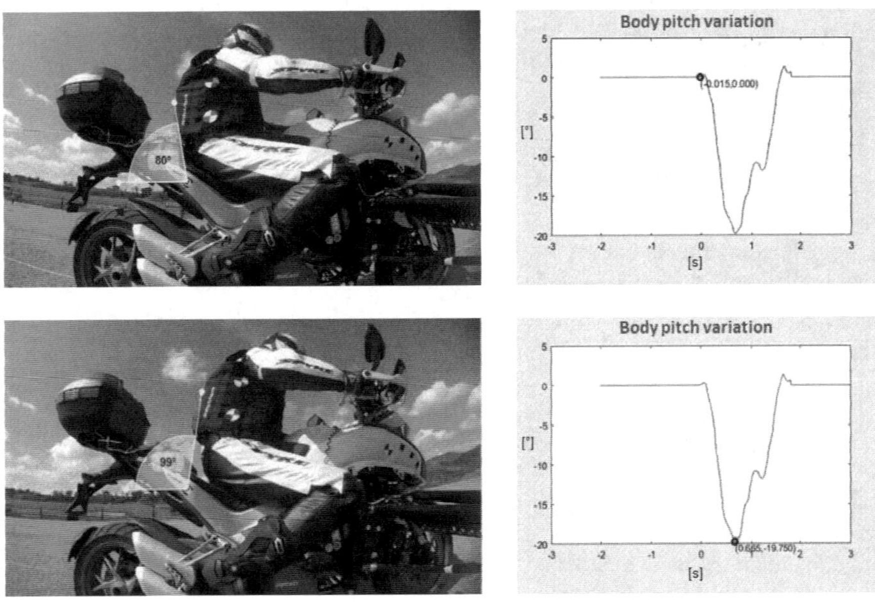

Figure 3. Example of video-validation process.

The analysis of the videos also offered the possibility to make a first analysis of the AB effects on the rider. In this phase, the analysis was focused not only on the movements related to the rider's torso, but also on the dynamics of the entire body of the rider during AB events. This allowed detecting patterns of behaviour during AB that were not detectable by the available sensors. Although there were no sensors to detect the movement of the head, the force exerted on the foot pegs or on the handlebars, this analysis allowed us to identify how riders reacted differently to the deceleration caused by AB. The analysis was carried out on 32 videos, each of which was rated from 1 to 3 by three experts (two of them made a first evaluation, while a third one independently assessed the evaluation agreement) depending on the rider's perceived effort in managing the AB intervention. The videos were also viewed at slower playback speeds to focus more on details. To make these evaluations several elements were considered: the change in torso pitch angle during activation, head movement, stress on legs or arms, shoulder movements and any movement on the saddle. Scores were given related to the increasing difficulty level as reported in Table 1.

Table 1. Levels of disturbance of the rider due to AB determined by the video analysis.

Levels	Description
Level 1	The AB activation does not generate any destabilisation and the rider's reactions to deceleration were insignificant; the rider remains almost unmoved
Level 3	The AB activation is notably destabilising the rider body movements are evident and involve, high body's pitch angles, pelvis rising from the saddle, strain on arms and legs (detectable from video)
Level 2	The AB activation has an influence on the rider that is in between the situations described in Level 1 and Level 3

3.3. *Data analysis*

After the video analysis and data validation, the final phase of the body movement analysis and characterization during AB intervention was performed based on all the data recorded from the vehicle and the IMU positioned on the rider's back. The relevant signals were plotted and compared to perform a quantitative body movements analysis.

4. Results

4.1. *Video analysis*

In Table 2, the results of the qualitative analysis performed by three experts through videos collected during the tests are reported.

Table 2. Results of the video analysis.

	All AB activations		Pseudo-unexpected AB		Out of the blue AB	
	n.	%	n.	%	n.	%
Level 1	16	50.0	11	50.0	5	50.0
Level 2	13	40.6	9	40.9	4	40.0
Level 3	3	9.4	2	9.1	1	10.0

A first important result is that in any video major events of destabilization or imminent danger were found. Even in the three cases classified with the maximum level of disturbance, the rider was always able to control the PTW with ease, but body movements were more relevant compared with what was visible in the rest of the cases examined (see as example Figure 3). In 50% of the 32 AB activations evaluated, the rider did not show any unconventional movement and remained almost unperturbed during the AB intervention. Such events were classified with the lowest score of 1.

Figure 4. Example of level 3 AB reaction.

Separating the "pseudo-unexpected" AB activations from the "Out of the blue" cases (Table 2), the results are still in line with the overall distribution. Among the 22 pseudo-unexpected activations, 50% were classified with the lowest score (level 1), while only two cases were classified with the highest one (level 3). On the other hand, in the cases with the "Out of the blue" AB, the frequency of the level 3 score was slightly higher, even though the low number of activations available (10) has an influence on the variation.

4.2. Data analysis

4.2.1. Test participants. The test campaign involved 31 volunteer participants, non-professionals riders, characterized by different age, body size and riding experience (for further information regarding the sample of participants see [14]). Due to problems in data acquisition, only data from 12 out of 31 participants were considered for the analysis of the body movement in order to guarantee the best reliability. The following analysis is therefore based on data from 36 automatic braking activations in straight line riding conditions (three interventions per participant for a total of 24 pseudo-unexpected AB and 12 "out of the blue" AB).

4.2.2. Automatic Braking (AB) intervention. The Automatic Braking (AB) produced a longitudinal deceleration of the vehicle by acting on the throttle inhibition and the combined braking system. Figure 4 shows a typical deceleration profile (red curve): the trigger signal is followed by an increasing deceleration ramp and then, once a target value (approx. 0.5 g) is reached, there is a constant deceleration phase. The end of the intervention is characterised by an exit ramp, and it returns to the conditions prior to the trigger. The average duration of the event was 1.19s and the AB deceleration produced a mean speed reduction of 15.6 km/h. The longitudinal fade-in jerk of the vehicle during the AB activation phases reached peaks around 20 m/s^3 (see Table 3).

Table 3. AB intervention characterization.

	Mean	SD
AB deceleration [m/s^2]	4.50	0.42
Fade-in jerk [m/s^3]	15.98	2.61
AB duration [s]	1.19	0.02
Initial speed [km/h]	49.1	5.1
AB Speed reduction [km/h]	15.7	2.2

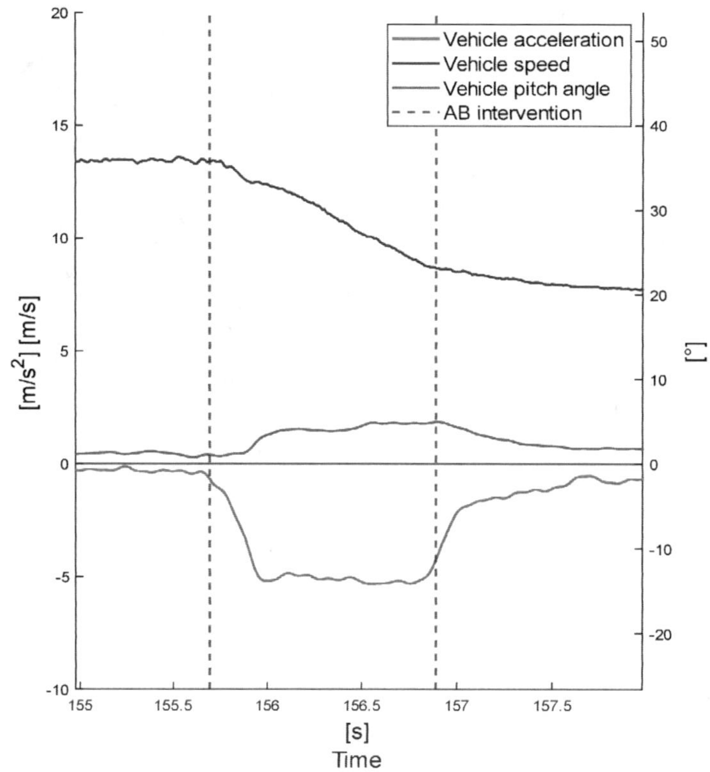

Figure 5. Example of an AB intervention: vehicle deceleration and speed reduction.

4.3. Body movement

The data analysis performed on the 36 AB interventions highlighted two types of typical behaviour of the rider's body during the AB events. Both these types were characterized by a similar initial phase while they differ in the second part of the AB intervention.

4.3.1. Convergent behaviour. As showed in Figure 5, the so-called "convergent behaviour" is characterized by a first phase in which the rider and the PTW decelerate with the same deceleration (red/blue curve). After the automatic braking has been activated and the deceleration of the AB increases with a constant fade-in jerk, there is no relative motion between the body of the rider and the vehicle. This phase starts at the AB trigger and ends when the target AB deceleration is reached. In the second phase, the deceleration of the body and the vehicle diverge: the deceleration of the rider's body continues to increase and registers a maximum peak, almost double the deceleration of the vehicle at the same time. This phenomenon is caused by a relative motion of the rider with respect to the vehicle, observable as an increase in the pitch angle of the torso, as a translation of the rider on the saddle or as a combination of these two movements. The rider at this stage does not assume his/her typical natural riding position and this phase represents the most critical (less stable) time during AB intervention for the vehicle control. In the final phase of the AB intervention, the rider regains his/her natural riding position (while the AB system is still active) and the body deceleration returns to follow the vehicle deceleration. With this type of behaviour, the final part of AB intervention is characterized by a stable condition, with the body of the rider moving with magnitude and trends of deceleration similar to those of the vehicle: this behaviour is therefore called "convergent".

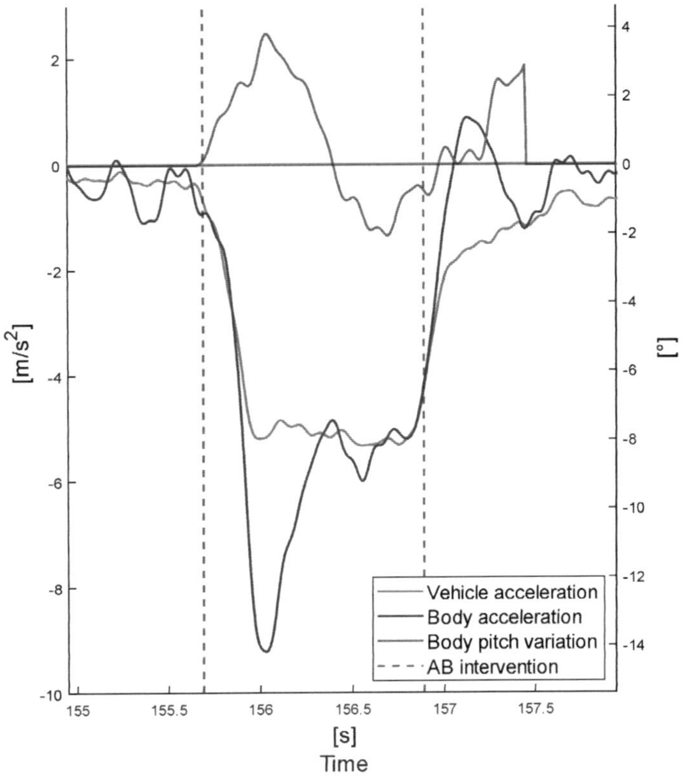

Figure 6. Example of convergent behaviour in response to AB intervention.

4.3.2. Non convergent behaviour. The second type of behaviour detected by the analysis does not present the last phase within the AB event duration (Figure 6): after an initial deceleration peak, other peaks of the same intensity are reached by the body of the rider. The trend of the variation of the pitch angle of the body shows greater peaks than in the "convergent" behaviour and the body remains bent forward for the entire duration of the AB activation. The decelerations of the body and the vehicle reach similar values at the end of the AB activation. The rider never returns to a natural riding position within the duration of the AB intervention, but only when it is concluded. In this case, the two decelerations do not converge within the AB activation time frame, and therefore this type of behaviour was called "non convergent".

4.3.3. Incidence of behaviours. Among the 36 AB activations analysed, 16 (44.4%) recorded a non convergent behaviour: 10 pseudo-unexpected activations (41.7%) and 6 "out of the blue" ones (50%). The other AB interventions were characterized by a "convergent" behaviour of the rider's body.

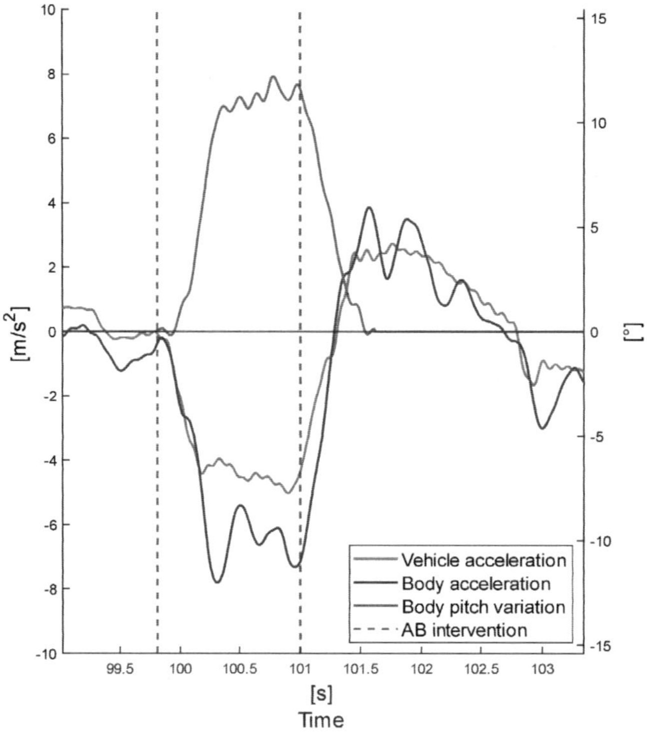

Figure 7. Example of non convergent behaviour in response to AB intervention.

4.3.4. Timing of body movement. The first phase of the response to AB intervention in common to both behaviours, resulted in an average duration of 0.32 seconds: 0.29 seconds in the case of convergent behaviour and 0.36 seconds in the case of non convergent behaviour (Figure 7). If we consider only the activations that have shown a "convergent" behaviour, the least stable phase of the rider (phase 2) had an average value of 0.43 seconds: considering that the entire activation event has an average duration of 1.19 seconds, for about 36% of the activation the rider performed an oscillatory or translational motion with respect to the motorcycle. These times were slightly higher in "out of the blue" braking, with an average duration of 0.46 seconds, compared to the pseudo-unexpected ones with an average of 0.41 seconds.

4.4. Vehicle controllability

Further analysis was carried out by analysing parameters that characterize the lateral dynamics of the vehicle, such as roll angle and lateral acceleration (of both the body and the vehicle). Abnormalities of these parameters in straight riding conditions may have been indicative of instability and incipient loss of control. However, no peculiar deviation in such parameters was detected in any of the 36 activations analysed. This was also confirmed by the results emerging from the subjective questionnaires completed by the participants at the end of the session [10]: 100% of the participants included in the study rated the AB intervention with a controllability rating score lower than three, indicating good vehicle behaviour and easiness of control (for details regarding the controllability rating scale see [16]), with a limited compensation required by the rider to execute the desired trajectory during the automatic braking.

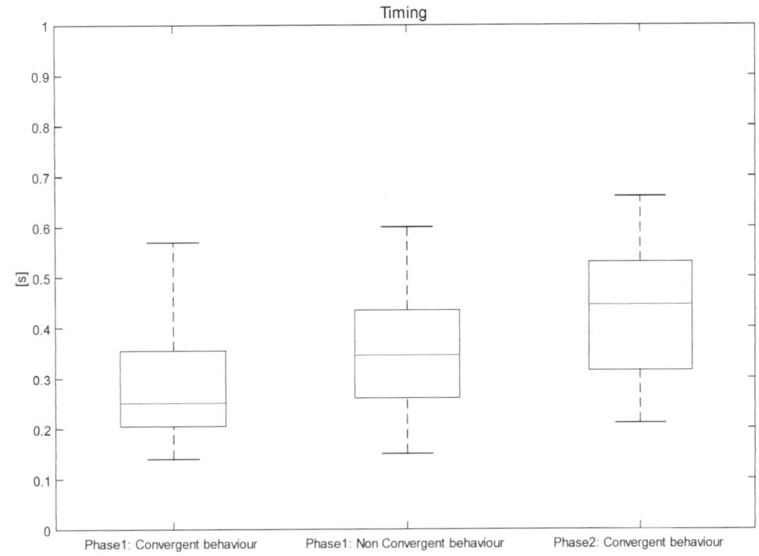

Figure 8. Timing of body movement in response to AB intervention.

5. Discussion

In this paper, further results of a field test campaign carried out within the PIONEERS project [14] were presented. This analysis focused on the rider body movement during the intervention of an Automatic Braking event representing an unexpected intervention of a Motorcycle Autonomous Emergency Braking (MAEB) system in straight-line riding conditions. A new approach was developed to analyse the movement of the body of the rider during the activation of the AB system, based on videos and data recorded during field testing. First, in-depth video analysis of the AB intervention recorded from lateral position made by three experts defined a qualitative classification of the AB intervention based on the effort of riders to cope with the vehicle during the AB intervention. Even if the results coming from this type of analysis are inevitably influenced by the subjective opinion of the experts, they allow an identification of the most relevant cases to be further investigated. At the end of the analysis, no potential loss of control events during the AB intervention were identified. After the video analysis, a quantitative analysis of data collected by vehicle sensors and an IMU positioned on the back of the rider was performed. This analysis allowed characterizing the body movement of the rider during the AB intervention: two types of behaviours defined by different timing required to regain the natural riding position of the body after the AB intervention were identified. These results confirmed the importance of measuring the motion of the rider's torso to define rider state during AB, as highlighted in previous work [17]. Overall, our results were compatible with the limited prior research on this topic [18].

This study has also some limitations. First, due to some recording issues, not all the participants involved in the field test campaign were considered for the analysis. Those included in the study were however a heterogeneous subsample (about 30%) of the whole group, the selection of which was basically random. Also, in previous analyses, such sub-sample did not show any specific bias and analysis performed in other studies did not indicate any different behaviour between our subsample and the overall sample in terms of incipient loss of control [9,10]. A further limitation is represented by the challenge to collect reliable data regarding body movement with simple instrumentation. Several data processing and filtering were required to perform this analysis, for which video recordings collected during the tests were used as a reference in order to double-check the phenomena observed in the data analysis.

Despite these limitations, the analysis highlighted that in most of the cases riders were able to return to a natural riding position during the AB activation event, and also in those cases considered less stable (non convergent behaviour), where participants had more difficulty regaining a natural riding position, any incipient loss of vehicle control was noted as shown in previous studies [9,10]. More critical conditions, such as those occurring in the unexpected activation, did not particularly affect the behaviour of the riders.

6. Conclusions

This paper presented an integrated methodology to characterize the movement of the body of riders in relation to the activation of an active safety system for PTWs such as MAEB, based on videos and quantitative data recorded during field testing. This methodology highlighted a good potential to characterize the response of the rider in case of the intervention of an automatic braking system, identifying different types of behaviours whereby in the specific case the rider reacts to unexpected braking actions. The results of the analysis highlighted two types of behaviour in response to the AB intervention, characterized by different timing required to regain the natural riding position of the body after the intervention. Even if these two behaviours were associated with different levels of disturbance of the rider due to AB intervention, no potential loss of control was detected. The results of this work confirmed, through rider's body movement analysis, that MAEB is likely to be safely applied in real riding conditions, as emerged in previous studies.

Acknowledgement

The authors would like to acknowledge Ducati Motor Holding SPA and Robert Bosch GmbH for their support in this project. The authors would like also to acknowledge Pedro Huertas-Leyva and Mirko Marra for their support during the experimental tests and data preparation.

Funding

This paper is part of a project that has received funding from the European Union's Horizon 2020 research and innovation program under Grant Agreement N° 769054, project PIONEERS (Protective Innovations Of New Equipment for Enhanced Rider Safety).

Disclosure statement: All authors have read and agreed to the submitted version of the manuscript. The authors declare no conflict of interest.

References

[1] World Health Organization 2018 Global status report on road safety 2018.
[2] European Commission 2020 Road safety: Europe's roads are getting safer but progress remains too slow.
[3] Terranova P, Dean M, Gabler H C, Piantini S and Savino G 2020 Active safety systems for motorcycles : where are we ? A novel transnational comparison of applicability in the Australian, American and Italian fleets AAAM Student Symposium (Portland, Oregon Active) .
[4] Savino G, Lot R, Massaro M, Rizzi M, Symeonidis I, Will S and Brown J 2020 Active safety systems for powered two-wheelers: A systematic review Traffic Inj. Prev. 21 78–86.
[5] Merkel N, Pless R, Scheid K and Winner H 2018 Limits of Autonomous Emergency Brake Systems for Powered Two-Wheelers – an Expert Study 12th International Motorcycle Conference (IFZ) (Cologne, Germany) pp 122–44.
[6] Savino G, Pierini M and Baldanzini N 2012 Decision logic of an active braking system for powered two wheelers J. Automob. Eng. 226 1026–36.
[7] Savino G, Pierini M, Thompson J, Fitzharris M and Lenné M G 2016 Exploratory field trial of motorcycle autonomous emergency braking (MAEB): Considerations on the acceptability of unexpected automatic decelerations Traffic Inj. Prev. 17 1–12.
[8] 2020 PIONEERS Project Webpage.
[9] Marra M, Lucci C, Huertas-Leyva P, Baldanzini N, Pierini M and Savino G 2021 The future of the Autonomous Emergency Braking for Powered-Two-Wheelers: field testing end-users' acceptability in realistic riding manoeuvres IOP Conf. Ser. Mater. Sci. Eng. 1038 11.
[10] Lucci C, Baldanzini N and Savino G Does Motorcycle Autonomous Emergency Braking (MAEB) mitigate rider injuries and fatalities? Design of effective working parameters and field test validation of their acceptability, Forthcoming.
[11] Lucci C, Baldanzini N and Savino G 2021 Field testing the applicability of motorcycle autonomous emergency braking (MAEB) during pre-crash avoidance manoeuvre *Traffic Inj. Prev.*
[12] Lucci C, Allen T, Pierini M and Savino G 2021 Motorcycle Autonomous Emergency Braking (MAEB) employed as enhanced braking: estimating the potential for injury reduction using real-world crash modeling *Traffic Inj. Prev.*
[13] Huertas-Leyva P, Savino G and Baldanzini N Upper-body response to automatic braking on motorcycles: assessment of the unexpected nature of repeated exposure to undeclared interventions for an acceptability test, Forthcoming.
[14] Lucci C, Huertas-Leyva P, Marra M, Pierini M, Savino G and Baldanzini N 2020 Autonomous Emergency Braking system for Powered-Two-Wheelers: testing end-user acceptability of unexpected automated braking events deployed in typical pre-crash trajectories *13th International Motorcycle Conference (IFZ)* (Cologne, Germany).
[15] Lucci C, Berzi L, Baldanzini N and Savino G 2019 Remote controlled braking actuation for motorcycle safety system development *2019 IEEE 5th International forum on Research and Technology for Society and Industry (RTSI)* pp 477–82.
[16] Lucci C, Marra M, Huertas-Leyva P, Baldanzini N and Savino G 2021 Investigating the feasibility of Motorcycle Autonomous Emergency Braking (MAEB): design criteria for new experiments to field test automatic braking *MethodsX* **8** 101225.
[17] Merkel N L and Winner H 2019 Measures for the Evaluation of Riders' Adaption to the Changing Vehicle State during Autonomous Emergency Braking Maneuvers on Motorcycles *Proceedings of Dynamics and Control of Single Track Vehicles Conference* (September 2019).
[18] Merkel N L and Winner H 2020 Characteristic Rider Reactions to Autonomous Emergency Braking Maneuvers on Motorcycles *13th International Motorcycle Conference (IFZ)*

(Cologne, Germany).

IOP Conf. Series: Materials Science and Engineering 1214 (2022) 012048 doi:10.1088/1757-899X/1214/1/012048

Development of methods to measure the forces at the rear suspension of a race car during racetrack driving

D Cortivo[1], L Dindo[2], M Vendramin[2] and G Meneghetti[1]

[1] Università degli Studi di Padova – Dipartimento di Ingegneria Industriale, Via Venezia 1, Padova 35131, Italy

[2] Michelotto S.a.s., Via Chiesanuova 27, Padova 35136, Italy

giovanni.meneghetti@unipd.it

Abstract. The forces generated by the tyres of a vehicle are responsible for the maximum achievable performance for a race car. In this work, a geometric matrix method combined with a sensitivity matrix method has been applied to a rear multi-link suspension of a rear-wheel-drive race car to estimate the tyre forces from the measurement of the loads acting on the suspension arms. The geometric matrix method calculates the wheel forces from the equilibrium of the wheel assembly, thus involving all reaction forces exchanged with the suspension arms. The reaction forces have been measured through the application of axial strain gauge bridges on the link arms; however, the lower arm has a complex geometry and exchanges multi-axial forces with the upright, therefore a sensitivity matrix method has been implemented. The strain gauges positions have been identified with FE analyses and after installing the sensors, the calibration of the entire suspension assembly has been performed in a dedicated test rig where known forces at the ideal contact point between the wheel and the ground can be applied. After the calibration and validation in the laboratory, the instrumented suspension has been installed in a race vehicle and multiple racetrack acquisitions have been successfully performed.

1. Introduction

In the race vehicles field, minimizing the lap-time is always a performance target, which means to maximize the lateral and longitudinal accelerations developed in the track. The acceleration limits in cornering and braking are typically set by the performance of the tyres [1]. Measuring the wheel forces during racetrack driving allow to characterize and compare different types of tyres [2], but also to estimate the vertical load transfer and the aerodynamic forces, the friction coefficient of the brakes and the design loads.

In the literature, different methods are proposed to measure the forces acting between the tyre and the ground, which can be distinguished between indirect and direct approaches. Indirect approaches estimate the wheel forces from easily available sensor signals like the vehicle accelerations, the longitudinal velocity and the steering wheel angle. A vehicle dynamics model is required, which is usually implemented in the form of a Kalman Filter [3,4] and the accuracy of force estimation depends on the uncertainties in the parameters of the vehicle model, like inertia properties and aerodynamic coefficients which need to be measured through specific tests. On the other hand, direct approaches, on which this study is focused, estimate the loads by measuring the deformations or the accelerations of the tyre, of the wheel or of the suspension components. Braghin et al. [5] attached three accelerometers in the inner liner of the tyre to estimate the vertical load, while Tuononen [6] estimated the vertical, lateral and longitudinal forces by measuring the tyre carcass displacements with optical sensors. These approaches require to take into account the effects of the tyre speed and internal pressure, increasing the calibration complexity. Measurement systems independent from the tyre operating conditions are the wheel force transducers [7] which are mounted in place of the original wheel, or as an interface with the hub, and then leading to a variation of the upsprung mass that influence the vehicle dynamic response. To maintain unchanged the vehicle static and dynamic characteristics, instrumentation of the vehicle components is required. Cheli et al. [8] measured the wheel forces and moments by applying three strain gauge bridges to the rim, but for motorsport applications where the frequent tyres change is done by replacing the entire wheel, multiple identical measuring systems would be necessary. To overcome these limitations, in a previous work [9], a double-wishbone front suspension of a rear-wheel-drive race vehicle has been instrumented to measure the tyre forces. The axial loads, which equilibrate the tyre forces, have been measured by applying and calibrating a strain gauge bridge on each arm. The tyre forces have been calculated with a geometric matrix method that has been validated with a laboratory test rig before performing racetrack acquisitions. In this paper, the geometric matrix method combined with the sensitivity matrix method is applied to a multi-link suspension. The lower arm of the suspension has a complex geometry and reacts with multi-axial forces to the upright, which have been estimated through a sensitivity matrix method. Then the tyre forces are calculated from the wheel assembly equilibrium through the geometric matrix method.

2. Methods for measuring the suspension forces

The object of the study is a rear left multi-link suspension of a rear-wheel-drive race vehicle. Figure 1 shows the 3D model of the suspension in the static configuration with the acronyms of each arm and the ISO [10] reference system used in this paper. The static configuration is set by the suspension travel of the car at rest with its weight sustained by the wheels.

The arms ARA, CA, SD, and TA (Anti-Rotation Arm, Camber Arm, Spring&Damper, Toe Arm) are tubular components constrained by means of two spherical joints at their ends and will be in the paper referred to link arms. The lower arm (figure 1b) is a tubular arc-welded structure stiffened by folded plates; it is constrained with four spherical joints: two of them are connected to the vehicle frame, one to the ARA arm and one to the upright.

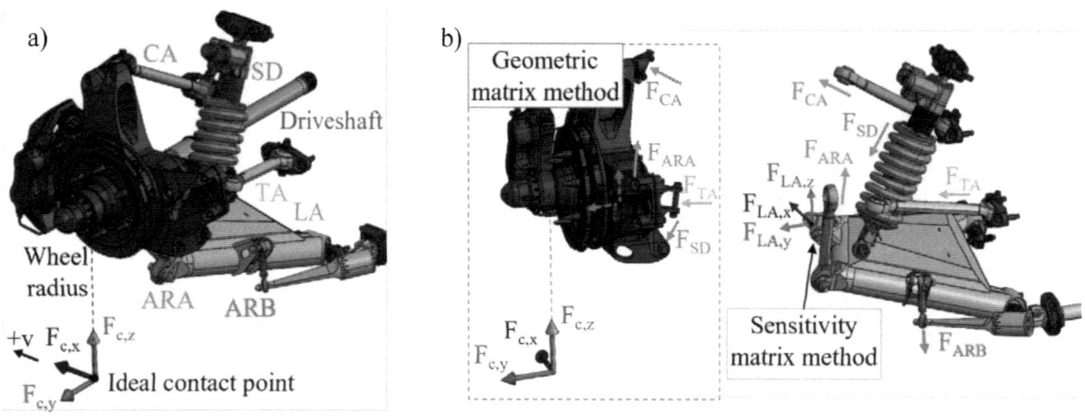

Figure 1. (a) Suspension 3D model in static configuration, (b) exploded view with internal reaction forces and outline of the methods applied; the reference system is the ISO reference system.

The wheel-road forces are assumed to be applied at the ideal contact point, placed in the wheel plane orthogonal to the wheel axle and containing the wheel centre, at a distance equal to the wheel radius in the static configuration. The geometric matrix method described in [9], calculates the tyre forces from the equilibrium of the wheel assembly, which is composed of the following macro-components: wheel, upright and hub. The measurement of all the reactions between the suspension arms and the upright is required to calculate the tyre forces, which become the sole unknown variables in the equilibrium equation. As shown in figure 1b, each link arm force act along the axis defined by the two respective spherical joints and can be measured by applying and calibrating an axial strain gauge bridge. The lower arm and the upright exchange in the common spherical joint a reaction force with variable direction and intensity depending on the applied tyre forces. This reaction force can be expressed as three orthogonal components ($F_{LA,x}$, $F_{LA,y}$, $F_{LA,z}$) that can be measured with the sensitivity matrix method by applying three strain gauge bridges to the component.

2.1. The geometric matrix method

The vector sum of all the reactions developed between the upright and the suspension arms must be in equilibrium with the net force from the three components ($F_{c,x}$, $F_{c,y}$, $F_{c,z}$) that develop between the tyre and the ground.

In equation (1) the method applied to the suspension of figure 1 is shown, where the upright reaction forces are multiplied by the geometric matrices giving the tyre forces. The link arms forces (F_{ARA}, F_{CA}, F_{SD}, F_{TA}) are multiplied by the respective $[G_{LINKS}]$ geometric matrix, which contains the direction cosines of the local axis x_i with respect to the global reference system (X,Y,Z) of the tyre forces. For the sake of clarity, the local axis x_i of each link arm is defined by the centres of the two spherical joints.

For the lower arm a local reference system is defined having axes (x_{LA}, y_{LA}, z_{LA}) coincident with the global reference system when the suspension is in the static configuration. The forces ($F_{LA,x}$, $F_{LA,y}$, $F_{LA,z}$) exchanged between the upright and the lower arm are expressed in its local reference system, the orientation of which vary when the suspension is compressed or extended from the static configuration. The geometric matrix $[G_{LA}]$ contains the director cosines of the local reference system of the lower arm with the global reference system, then it projects the forces ($F_{LA,x}$, $F_{LA,y}$, $F_{LA,z}$) in the global reference system.

$$\begin{Bmatrix} F_{c,x} \\ F_{c,y} \\ F_{c,z} \end{Bmatrix} = [G_{\mathrm{LINKS}}] \cdot \begin{Bmatrix} F_{\mathrm{ARA}} \\ F_{\mathrm{CA}} \\ F_{\mathrm{SD}} \\ F_{\mathrm{TA}} \end{Bmatrix} + [G_{\mathrm{LA}}] \cdot \begin{Bmatrix} F_{\mathrm{LA},x} \\ F_{\mathrm{LA},y} \\ F_{\mathrm{LA},z} \end{Bmatrix} \tag{1}$$

The $[G_{LINKS}]$ and $[G_{LA}]$ are calculated according to equations (2) and (3) and can be numerically computed from the coordinates of the suspension kinematic points. When the suspension is in the static configuration the $[G_{LA}]$ matrix is the identity matrix from the assumptions of the previous paragraph.

$$[G_{\mathrm{LINKS}}] = \begin{bmatrix} \cos(\widehat{x_{\mathrm{ARA}}X}) & \cos(\widehat{x_{\mathrm{CA}}X}) & \cos(\widehat{x_{\mathrm{SD}}X}) & \cos(\widehat{x_{\mathrm{TA}}X}) \\ \cos(\widehat{x_{\mathrm{ARA}}Y}) & \cos(\widehat{x_{\mathrm{CA}}Y}) & \cos(\widehat{x_{\mathrm{SD}}Y}) & \cos(\widehat{x_{\mathrm{TA}}Y}) \\ \cos(\widehat{x_{\mathrm{ARA}}Z}) & \cos(\widehat{x_{\mathrm{CA}}Z}) & \cos(\widehat{x_{\mathrm{SD}}Z}) & \cos(\widehat{x_{\mathrm{TA}}Z}) \end{bmatrix} \tag{2}$$

$$[G_{\mathrm{LA}}] = \begin{bmatrix} \cos(\widehat{x_{\mathrm{LA}}X}) & \cos(\widehat{y_{\mathrm{LA}}X}) & \cos(\widehat{z_{\mathrm{LA}}X}) \\ \cos(\widehat{x_{\mathrm{LA}}Y}) & \cos(\widehat{y_{\mathrm{LA}}Y}) & \cos(\widehat{z_{\mathrm{LA}}Y}) \\ \cos(\widehat{x_{\mathrm{LA}}Z}) & \cos(\widehat{y_{\mathrm{LA}}Z}) & \cos(\widehat{z_{\mathrm{LA}}Z}) \end{bmatrix} \tag{3}$$

2.2. The sensitivity matrix method

The forces on the suspension arms, required for the geometric matrix method, are estimated through the application of strain gauges to each component. The link arms are subjected to pure axial load along their direction, then an axial full-bridge configuration is suitable to measure the exchanged forces at the upright interface. Equation (4) shows the relation, for a generic link arm *i*, between the strain gauge channel CH_i and the respective F_i axial force. The sensitivity s_i is constant and it is calibrated through a uniaxial load testing machine, as described in [9].

$$F_i = \frac{CH_i}{s_i} \tag{4}$$

The lower arm exchanges multi-axial forces: with the vehicle frame, with the upright in the common joint, with the ARA and ARB link arms along their respective directions. To replicate the suspension constraints, as graphically shown in figure 2a, zero displacements and free rotations (UX=UY=UZ=0) are applied at the two centres of the chassis joints, while the displacement along the direction of the ARA arm is set to be null ($Ux_{ARA}=0$). The forces ($F_{LA,x}$, $F_{LA,y}$, $F_{LA,z}$) and the ARB link force are then external forces acting on the lower arm. A sensitivity matrix method (equation (5)) has been applied to the lower arm through the realization of three strain gauge channels ($CH_{LA,x}$, $CH_{LA,y}$, $CH_{LA,z}$), designed to minimize cross-sensitivity. A strain gauge channel has been realized on the ARB link to measure the force F_{ARB}, the sensitivity of the CH_{LA} channels to the force F_{ARB} has been considered by adding a row and a column to the matrix, according to equation (5).

$$\begin{Bmatrix} CH_{LA,x} \\ CH_{LA,y} \\ CH_{LA,z} \\ CH_{ARB} \end{Bmatrix} = [S_{LA}] \begin{Bmatrix} F_{LA,x} \\ F_{LA,y} \\ F_{LA,z} \\ F_{ARB} \end{Bmatrix} = \begin{bmatrix} s_{x,x} & s_{x,y} & s_{x,z} & s_{x,ARB} \\ s_{y,x} & s_{y,y} & s_{y,z} & s_{y,ARB} \\ s_{z,x} & s_{z,y} & s_{z,z} & s_{z,ARB} \\ 0 & 0 & 0 & s_{ARB} \end{bmatrix} \begin{Bmatrix} F_{LA,x} \\ F_{LA,y} \\ F_{LA,z} \\ F_{ARB} \end{Bmatrix} \tag{5}$$

After calibration, the forces ($F_{LA,x}$, $F_{LA,y}$, $F_{LA,z}$, F_{ARB}) can be calculated from the inverse of the sensitivity matrix multiplied by the measured signals ($CH_{LA,x}$, $CH_{LA,y}$, $CH_{LA,z}$, CH_{ARB}).

3. FEA evaluation of the lower arm sensitivity matrix

Finite element simulations have been performed with the software ANSYS® to determine the strain gauges positions. The tubular components and the plates have been discretized with 8-node quadrilateral shell elements, while for the forks and the mountings of the spherical joints 10-node tetrahedral elements have been used; the meshing element size has been set to 2.5mm. The different parts, that in the real component are welded together, have been coupled through beam elements with a circular solid cross-section of 1.25mm and with the same material properties of the lower arm. The arm has been constrained according to paragraph 2.3 and figure 2a, and the forces ($F_{LA,x}$, $F_{LA,y}$, $F_{LA,z}$, F_{ARB}) have been each one individually applied to evaluate the component surfaces subjected to the higher strains.

In figure 2b-d contour plots of maximum and minimum principal elastic strain are shown and the areas subjected to the higher elastic strains for each load case are highlighted with white dashed lines. The strain gauges of each channel have been positioned according to the directions of the principal elastic strains in the respective load case and in order to minimize cross-sensitivity. Figure 3 shows the strain gauges configurations described below:

- $CH_{LA,x}$: two adjacent uniaxial strain gauges aligned with the axis of the tubular component, half-bridge configuration (strain gauges on opposite Wheatstone bridge branches);
- $CH_{LA,y}$: two adjacent XY rosettes with one grid aligned to the axis of the tubular component, full-bridge configuration;
- $CH_{LA,z}$: two adjacent XY rosettes with grids at an angle of 45° to the axis of the tubular component, full-bridge configuration.

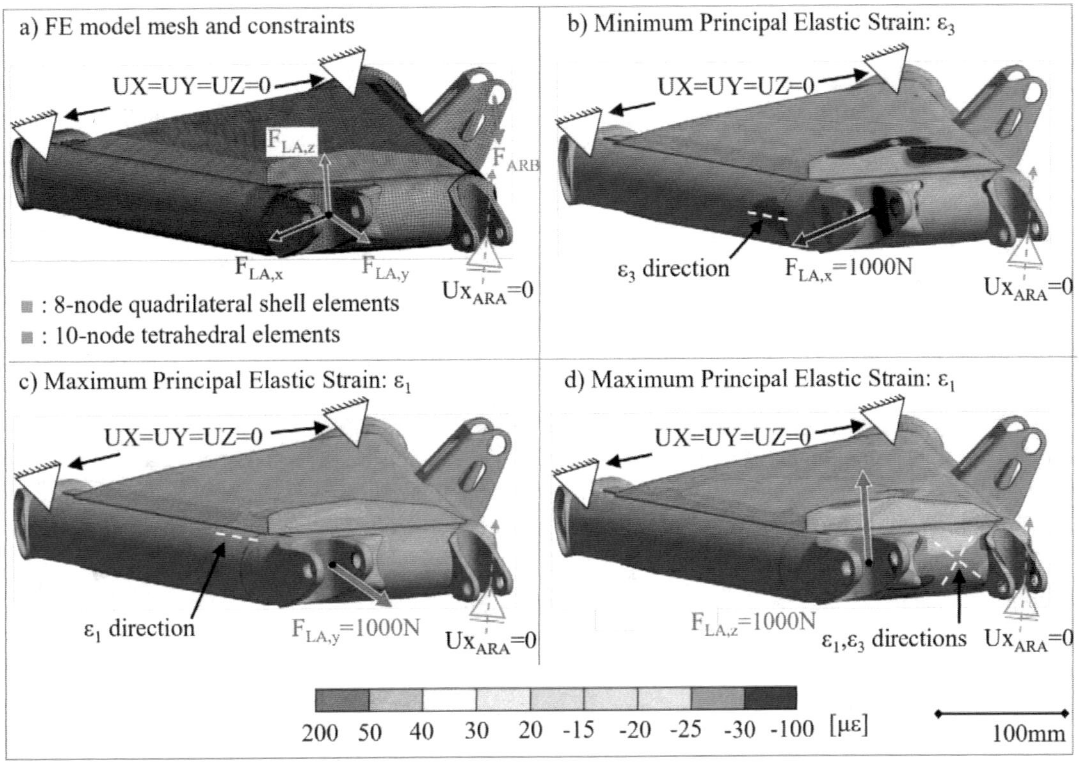

Figure 2. FE model of the lower arm subjected to each force component at the kinematic common point with the upright.

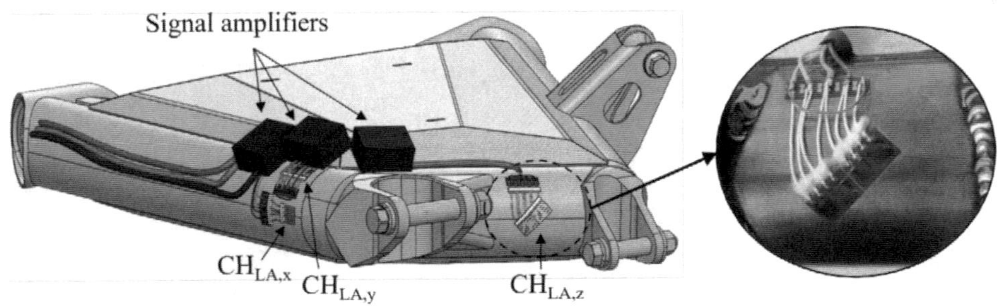

Figure 3. Strain gauges positioning on the lower arm.

The sensitivity matrix has been evaluated from the FE model: each coefficient of equation (6) has been calculated by summing the deformation of each strain gauge grid according to the Wheatstone bridge equation and dividing by the applied force.

$$\begin{Bmatrix} CH_{\mathrm{LA,x}} \\ CH_{\mathrm{LA,y}} \\ CH_{\mathrm{LA,z}} \\ CH_{\mathrm{ARB}} \end{Bmatrix}_{[\mu\varepsilon]} = \begin{bmatrix} -0.056 & +0.036 & +0.010 & -0.017 \\ -0.043 & +0.063 & -0.016 & -0.028 \\ -0.024 & -0.043 & +0.163 & -0.051 \\ 0 & 0 & 0 & s_{ARB} \end{bmatrix} \cdot \begin{Bmatrix} F_{\mathrm{LA},x} \\ F_{\mathrm{LA},y} \\ F_{\mathrm{LA},z} \\ F_{\mathrm{ARB}} \end{Bmatrix}_{[\mathrm{N}]} \tag{6}$$

4. Test rig calibration of the lower arm

The suspension has been instrumented according to the previous paragraphs, 8 strain gauge channels were realized, one for each link arm (ARA, CA, SD, TA, ARB) and three for the lower arm. The link arms have been individually calibrated to measure the axial force with a uniaxial testing machine, while for the lower arm calibration a test rig was re-designed (figure 4) from the one used for the front suspension [9].

4.1. Test rig design

The test rig is shown in figure 4, where the chassis constraints have been replicated on a frame plate constrained to a fixed support, then each suspension arm has been connected at its respective fork and assembled at the upright. The suspension spring/damper has been replaced by a dummy tubular component to set a constant suspension travel and to deal with a load-independent geometric matrix. The length of the rigid dummy can be changed, and the chassis plate is designed with multiple holes to switch between three different suspension travel configurations. The three suspension positions considered are the static, the half-extended and the half-compressed suspension configurations. A plate was mounted in the place of the wheel and it carries the forks that allow the forces ($F_{c,x}$, $F_{c,y}$, $F_{c,z}$) application by four servo-hydraulic actuators. The forks are positioned in order to ensure that the axes of the actuators intersect at the ideal wheel-ground contact point when no load is applied. The two MTS servo-hydraulic actuators in the y direction have a load capacity of ±15kN, while the two ITALSIGMA actuators along the x and z direction have a load capacity of ±10kN; they are controlled through an MTSFlextest FT 60 controller.

To reproduce the forces in a forward acceleration, a positive $F_{c,x}$ force is applied by the respective actuator and the M_y moment around the hub axis is delivered to the driveshaft which is constrained at its ends with two tripods. For the reproduction of the braking events, a pneumatic manual actuator operates on the pump rod through a leverage mechanism (figure 4). The pressurized calliper blocks the disk and a negative $F_{c,x}$ braking force can be applied by the actuator: the M_y moment around the hub axis is delivered mostly to the upright, while a minor component is carried by the driveshaft simulating the engine brake torque.

Lastly, the ARB force can be manually applied and adjusted through a tie rod that connects its link to a fixed support.

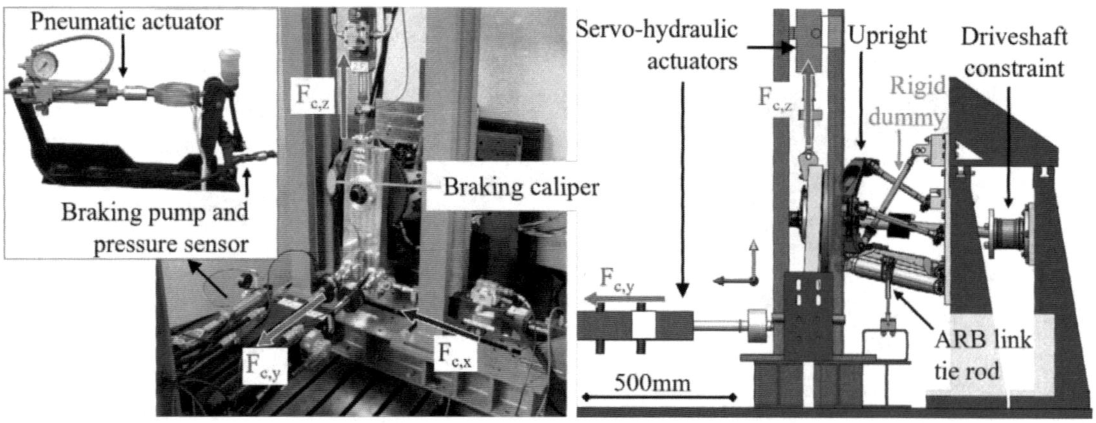

Figure 4: Test rig for the rear instrumented suspension with four servo-hydraulic actuators to apply forces at the ideal wheel-ground contact point.

4.2. Test procedure

A realistic combination of wheel-ground forces has been applied to the suspension, reproducing a racetrack load history in the test rig. The load history, provided by Michelotto S.a.s., is derived from a vehicle simulation on the flipped circuit MotorLand Aragón (Spain). In the flipped circuit the originally left turns become right turns and vice versa, while preserving the length and the radius of curvature, in this way the loads on the left suspension are maximized since it runs more frequently on the outer side. The testing procedure is described below, while, in the following figures 5-7, the forces and signals of the rear suspension are plotted against vertical axes that have been normalized by a constant and common force value.

a) Application of the simulated forces ($F_{c,x}$, $F_{c,y}$, $F_{c,z}$) with the test rig: the loads of the actuators have been acquired by the MTS controller and are shown with continuous lines in figure 5. For the 8 strain gauge channels an IMC CRONOS PL-2 acquisition unit has been used, the three of them applied to the lower arm are plotted in figure 6. The sampling frequency has been set equal to 500Hz.
b) Calculation of the lower arm ($F_{LA,x\text{-}G}$, $F_{LA,y\text{-}G}$, $F_{LA,z\text{-}G}$) applied forces from the equation (1) for each timestep of the acquired signals: the link arms (F_{ARA}, F_{CA}, F_{SD}, F_{TA}) forces are multiplied by the matrix [G_{LINKS}] and subtracted from the ($F_{c,x}$, $F_{c,y}$, $F_{c,z}$) forces, the resultant vector then pre-multiplied by the inverse of the matrix [G_{LA}]. The resultant forces are named ($F_{LA,x\text{-}G}$, $F_{LA,y\text{-}G}$, $F_{LA,z\text{-}G}$), the first of them is plotted in figure 7 with a continuous line.
c) Evaluation of the sensitivity matrix: by multiple linear regression of the ($F_{LA,x\text{-}G}$, $F_{LA,y\text{-}G}$, $F_{LA,z\text{-}G}$, F_{ARB}) forces against the measured lower arm signals ($CH_{LA,x}$, $CH_{LA,y}$, $CH_{LA,z}$, CH_{ARB}) the sensitivity matrix coefficients are evaluated. Naming ($F_{LA,x\text{-}S}$, $F_{LA,y\text{-}S}$, $F_{LA,z\text{-}S}$) the forces calculated from the inverse experimental sensitivity matrix multiplied by the signals ($CH_{LA,x}$, $CH_{LA,y}$, $CH_{LA,z}$, CH_{ARB}), the objective of the multiple linear regression is to minimize the sum of the squared errors between ($F_{LA,x\text{-}S}$, $F_{LA,y\text{-}S}$, $F_{LA,z\text{-}S}$, F_{ARB}) and ($F_{LA,x\text{-}G}$, $F_{LA,y\text{-}G}$, $F_{LA,z\text{-}G}$, F_{ARB}).
d) Calculation of ideal wheel-ground contact point forces ($F_{c,x}$, $F_{c,y}$, $F_{c,z}$) from the instrumented suspension according to equation (1): using as inputs the link arms measured axial forces (F_{ARA}, F_{CA}, F_{SD}, F_{TA}) and the lower arm forces ($F_{LA,x\text{-}S}$, $F_{LA,y\text{-}S}$, $F_{LA,z\text{-}S}$) calculated from the experimental sensitivity matrix. The instrumented suspension forces ($F_{c,x}$, $F_{c,y}$, $F_{c,z}$) are plotted with dashed lines in figure 5 to compare with the forces applied by the actuators, displayed in continuous lines.

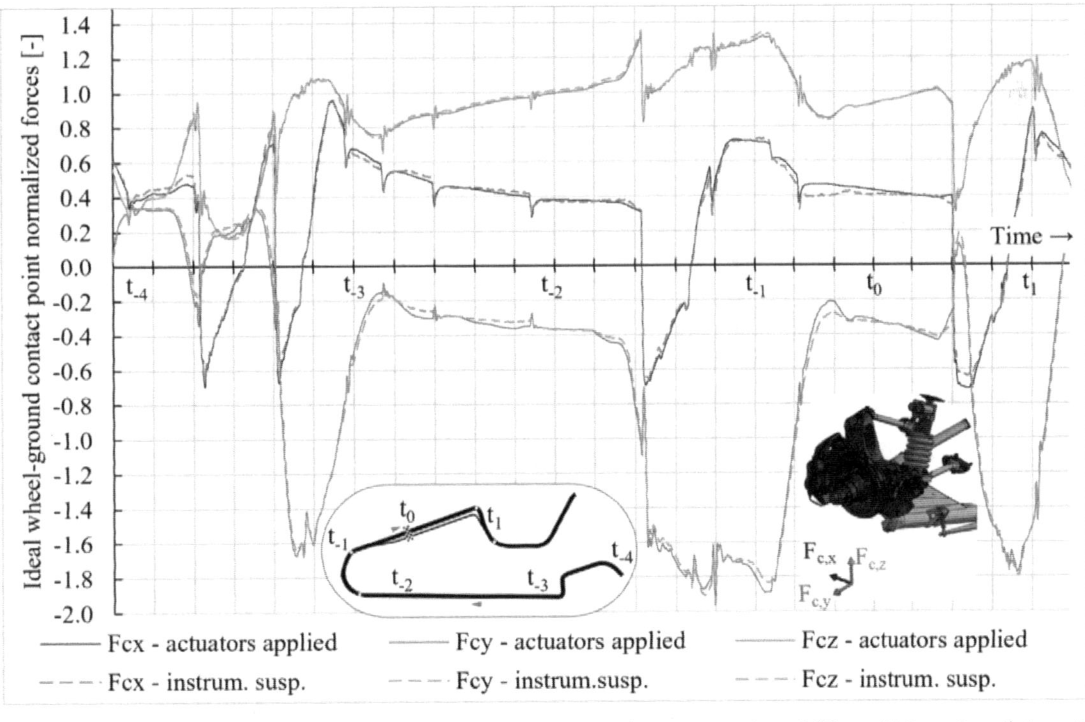

Figure 5: Simulated load history at the ideal wheel-ground contact point of flipped MotorLand Aragón racetrack, comparison between applied forces by the actuators (continuous lines) and measured forces by the instrumented suspension (dashed lines).

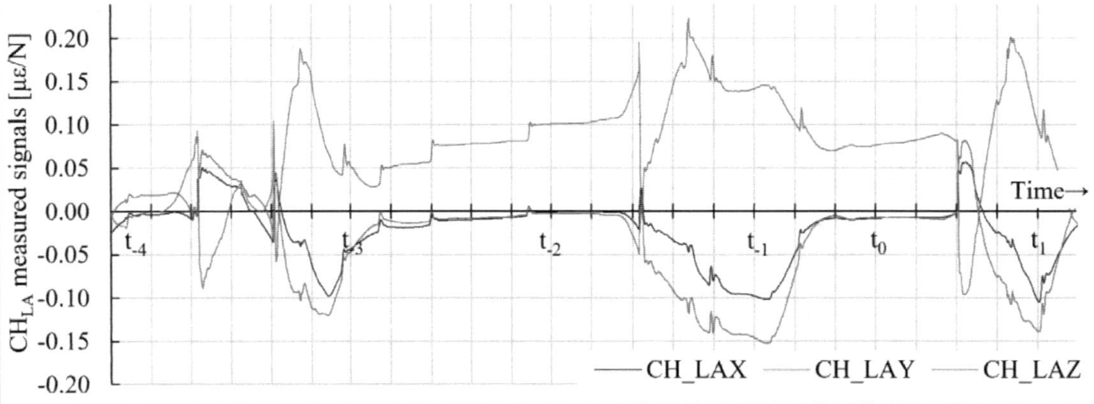

Figure 6: Lower arm CH_{LA} measured signals during application of figure 5 racetrack load history at the ideal wheel-ground contact point. The signals have been normalized by a constant force value, as done in figure 5 and figure 7.

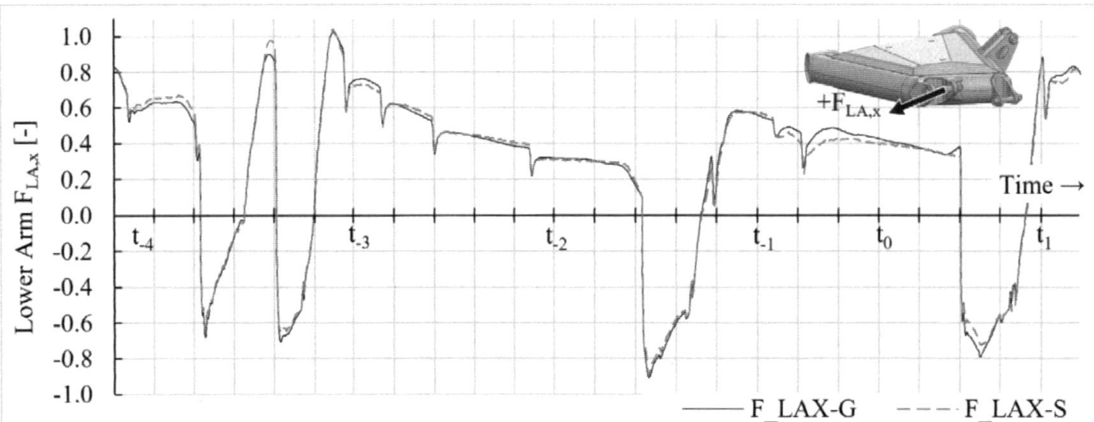

Figure 7: Lower arm $F_{LA,X}$ force during application of figure 5 racetrack load history at the ideal wheel-ground contact point.

4.3. *Test rig results*

The flipped Aragón circuit simulation consists of three right turns where the instrumented suspension is on the outer side, two left turns, four braking events and two straightaways. The load history has been applied with the test rig in the three suspension travel configurations, considering the three different geometric matrices. In order to minimize the errors on the forces estimated by the instrumented suspension, two different sensitivity matrices have been evaluated.

Driving condition: a positive $F_{c,x}$ is applied at the ideal wheel-ground contact point and the M_y moment around the hub axis is carried by the driveshaft; the $F_{LA,x}$ force is mainly positive as can be seen from figure 7. The corresponding Lower arm sensitivity matrix is displayed in equation (7):

$$\begin{Bmatrix} CH_{LA,x} \\ CH_{LA,y} \\ CH_{LA,z} \\ CH_{ARB} \end{Bmatrix}_{[\mu\varepsilon]} = \begin{bmatrix} -0.051 & +0.032 & +0.007 & +0.003 \\ -0.036 & +0.055 & -0.006 & -0.005 \\ -0.032 & -0.039 & +0.129 & -0.033 \\ 0 & 0 & 0 & s_{ARB} \end{bmatrix}_{driving} \cdot \begin{Bmatrix} F_{LA,x} \\ F_{LA,y} \\ F_{LA,z} \\ F_{ARB} \end{Bmatrix}_{[N]} \tag{7}$$

Braking condition: a negative $F_{c,x}$ is applied at the ideal wheel-ground contact point and the M_y moment around the hub axis is delivered mostly to the upright while a minor component is carried by the driveshaft; the $F_{LA,x}$ force is mainly negative as can be seen from figure 7. The corresponding Lower Arm sensitivity matrix is displayed in equation (8):

$$\begin{Bmatrix} CH_{LA,x} \\ CH_{LA,y} \\ CH_{LA,z} \\ CH_{ARB} \end{Bmatrix}_{[\mu\varepsilon]} = \begin{bmatrix} -0.050 & +0.033 & +0.009 & +0.001 \\ -0.044 & +0.057 & -0.008 & -0.005 \\ +0.005 & -0.043 & +0.136 & -0.034 \\ 0 & 0 & 0 & s_{ARB} \end{bmatrix}_{braking} \cdot \begin{Bmatrix} F_{LA,x} \\ F_{LA,y} \\ F_{LA,z} \\ F_{ARB} \end{Bmatrix}_{[N]} \quad (8)$$

The differences between the experimental (equations (7)-(8)) and the numerical (equation (6)) sensitivity matrices, ranging from 10% to 20% for the coefficients in the main diagonal, are believed to be caused by the following factors:

- the strain gauges of the $CH_{LA,y}$ channel are located in the proximity of multiple nearby weld beads that have been FE modelled with beam elements to simplify the analysis, but leading to different strain distribution in the contiguous mesh elements with respect to the real component;
- the lower arm is over-constrained along the direction defined by the two spherical joints at the chassis, leading to reaction forces that depend on the stiffness of the supports, not being modelled in the FE model;
- possible misalignments during the application of the strain gauges due to the complex geometry of the lower arm.

These differences underline the requirement to have performed the experimental calibration of the sensitivity matrix of the lower arm.

The differences between the braking and traction matrices have been found principally in the ($CH_{LA,x}$, $CH_{LA,y}$, $CH_{LA,z}$) channels response to $F_{LA,x}$. They are explained considering the $F_{LA,x}$ force sign: when negative, the load is mainly carried by the side of the upright fork towards the ARA arm since it is stiffened by the gusset plate; when positive, a higher contribution of the load is carried by the front side of the upright fork. This leads to different strain distribution when $F_{LA,x}$ forces with the same intensity but different sign are applied.

The experimental sensitivity matrices have been used for the calculation of the lower arm forces ($F_{LA,x-S}$, $F_{LA,y-S}$, $F_{LA,z-S}$) from the signals ($CH_{LA,x}$, $CH_{LA,y}$, $CH_{LA,z}$, CH_{ARB}), then the forces ($F_{c,x}$, $F_{c,y}$, $F_{c,z}$) have been evaluated according to point (d) of paragraph 4.2. The forces applied by the actuators are plotted in continuous line in figure 5, while the forces estimated from the instrumented suspension are reported in dashed line. The following table 1 summarizes the achieved mean percentage errors on wheel forces estimation.

Table 1: Percent errors for ideal contact point forces estimation in representative racetrack conditions.

Load direction	Mean error
$F_{c,x}$ – traction	5%
$F_{c,x}$ – braking	7%
$F_{c,y}$ – outer side turn	5%
$F_{c,y}$ – inner side turn	10%
$F_{c,z}$	< 2%

5. Racetrack tests

The instrumented rear suspension has been installed in the car to acquire racetrack loads at the MotorLand Aragón circuit. The car is also equipped with five potentiometers which measure the suspension travel and the steering wheel angle, while an accelerometer acquires the longitudinal acceleration and the lateral acceleration of the vehicle in its center of mass. Other acquired signals are the vehicle longitudinal speed and the brakes pressures. All the signals are acquired from the vehicle data logger via the CAN bus, at a frequency of 100Hz; the post-processing procedure has been implemented through the MATLAB® software in an automated routine, which is schematically depicted in figure 8.

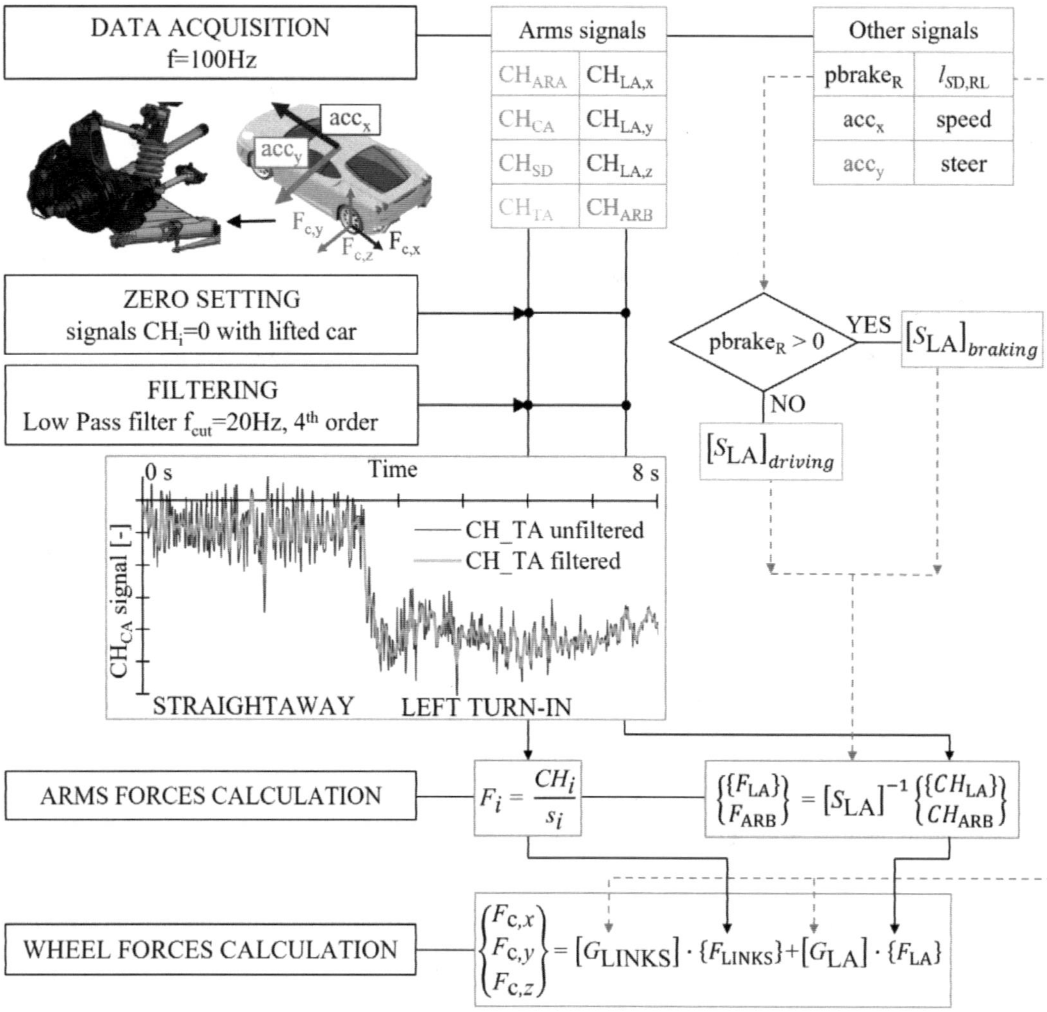

Figure 8: Post-processing procedure for wheel forces calculation from racetrack acquired data. The picture of the car is not representative of the real vehicle.

The strain gauge signals of the suspension are zeroed at the start of every acquisition session with the car lifted from the ground, so the arms are loaded only with the wheel assembly load along the negative $F_{c,z}$ direction: this is set as the zero vertical force value for the instrumented suspension. When the car is lowered to the ground, the instrumented suspension measures a vertical load equal to the sum of the weight of the wheel assembly and the left rear portion of the sprung mass.

The arms signals are then filtered to reduce the signal noise due to the engine vibrations and to the road roughness: a zero-phase digital 4th order Butterworth filter with a cut-off frequency of 20Hz has been applied through the MATLAB® *filtfilt* function. Figure 8 reports the comparison between the unfiltered and the filtered signal for the TA link arm and highlights a halving of the standard deviation.

The calculation of the link arms forces is straightforward with equation (4), however, for the lower arm, the brake pressure signal at the rear callipers ($pbrake_R$) is used to switch between the braking and the driving sensitivity matrices. The wheel forces are then calculated according to equation (1), considering a variable geometric matrix depending on the suspension spring displacement. At each step the geometric matrices are calculated from the signal $l_{SD,RL}$ of the rear left suspension potentiometer and multiplied by the respective arm forces resulting in the wheel forces.

In figure 9 the accelerations and wheel forces acquired during a lap are plotted in the reference system of figure 8. The wheel forces have been normalized by the same force value used in the previous paragraph.

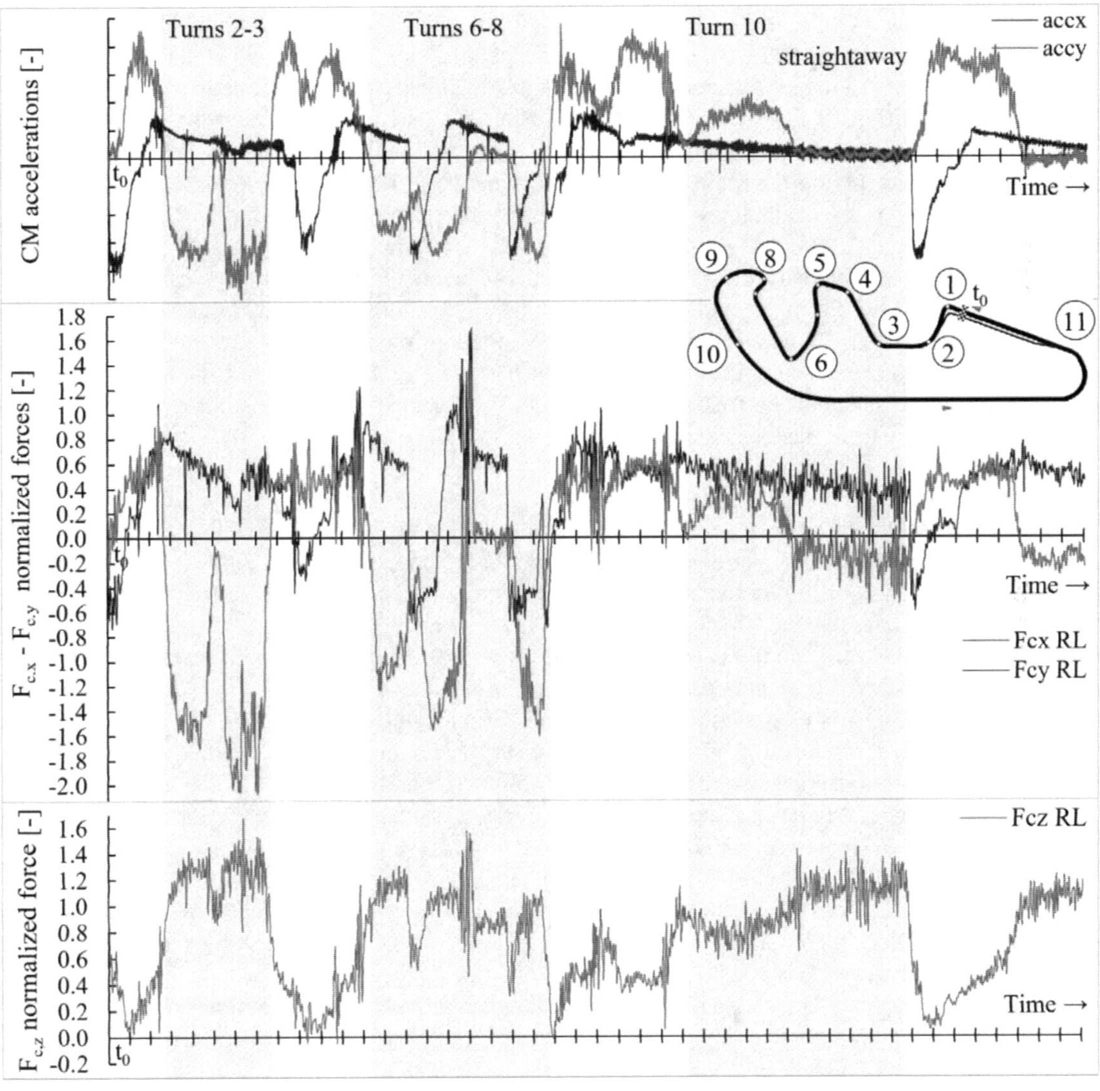

Figure 9: Acquired accelerations and rear left wheel forces from the instrumented suspension for a lap of the MotorLand Aragón circuit. The wheel forces have been normalized by the same force value used in figure 5.

The maximum lateral load $F_{c,y}$ have been measured in the 3rd turn of the circuit where the instrumented suspension is on the outer side, it is associated with the maximum vertical load $F_{c,z}$ of the lap due to the combined lateral and longitudinal load transfer effect. The maximum traction positive F_{cx}

force has been measured at the exit of the 6th turn where the instrumented suspension is on the outer side and thanks to the limited slip differential the stiffer tyre generates the higher longitudinal load. Lastly, the bigger F_{cx} braking force has been measured before the turn 1 of the circuit.

6. Conclusions

In this work, a geometric matrix method combined with a sensitivity matrix method has been developed and experimentally validated to measure the loads acting on a rear multi-link suspension of a rear-wheel-drive race car. The axial loads on the link arms have been measured through the application of axial strain gauges in a full bridge configuration, while two full-bridge and a half-bridge configurations have been applied on the lower arm to measure the three force components exchanged with the upright. The sensitivity matrix has been firstly evaluated through FE analyses and subsequently calibrated in a test rig dedicated to the entire suspension assembly. The calibration has been performed by applying a simulated load history of the MotorLand Aragón circuit at the ideal contact point between the wheel and the ground. Two different sensitivity matrices have been evaluated depending on the sign of the $F_{c,x}$ force applied at the ideal contact point, namely for driving and braking conditions, respectively. The experimental coefficients on the main diagonal differs between 10% and 20% from the numerically evaluated ones. After the calibration of the sensitivity matrix of the lower arm, the wheel forces estimated from the instrumented suspension have been compared with the actual applied ones to calculate the errors achieved in a racetrack acquisition. The errors on the estimated tyre forces are lower than 5% for mid-corner, turn-exit and in the straightaways, while they are lower than 7% for braking and turn-in events. After the laboratory tests for the validation, the instrumented suspension has been installed in the vehicle and multiple racetrack acquisitions have been successfully performed. The instrumented suspension will be used in future works for the characterization of the tyre and brake performance starting from racetrack acquisitions.

References

[1] Guiggiani M 2018 *The Science of Vehicle Dynamics* (Cham: Springer International Publishing)
[2] Farroni F 2016 T.R.I.C.K.-Tire/Road Interaction Characterization & Knowledge - A tool for the evaluation of tire and vehicle performances in outdoor test sessions *Mechanical Systems and Signal Processing* **72–73** 808–31
[3] Kim J 2009 Identification of lateral tyre force dynamics using an extended Kalman filter from experimental road test data *Control Engineering Practice* **17** 357–67
[4] Doumiati M, Charara A, Victorino A, Lechner D and Dubuisson B 2012 *Vehicle Dynamics Estimation using Kalman Filtering: Experimental Validation* (John Wiley and Sons)
[5] Braghin F, Brusarosco M, Cheli F, Cigada A, Manzoni S and Mancosu F 2006 Measurement of contact forces and patch features by means of accelerometers fixed inside the tire to improve future car active control *Vehicle System Dynamics* **44** 3–13
[6] Tuononen A 2009 On-board estimation of dynamic tyre forces from optically measured tyre carcass deflections *International Journal of Heavy Vehicle Systems* **16** 362
[7] Weiblen W, Kockelmann H and Burkard H 1999 Evaluation of Different Designs of Wheel Force Transducers (Part II) *SAE Technical Papers*
[8] Cheli F, Braghin F, Brusarosco M, Mancosu F and Sabbioni E 2011 Design and testing of an innovative measurement device for tyre–road contact forces *Mechanical Systems and Signal Processing* **25** 1956–72
[9] Meneghetti G, Cortivo D, Dindo L and Vendramin M 2020 Sviluppo di metodi per la misura dei carichi al gruppo sospensione anteriore di una vettura sportiva durante la guida in pista *Atti del 49° Convegno Nazionale AIAS - Associazione Italiana per l'Analisi delle Sollecitazioni*
[10] Pacejka H 2012 *Tire and Vehicle Dynamics* (Elsevier)

Development of a hybrid racing car model and optimization of the torque delivery strategy by means of genetic algorithms

J Bertucco[1], L Dindo[1], M Vendramin[1] and G Meneghetti[2]

[1] Michelotto S.a.s., Via Chiesanuova 27, 35136, Padova, IT
[2] University of Padova - Department of Industrial Engineering, Via Venezia 1, 35131, Padova, IT

josef.bertucco@michelotto.eu

Abstract. This paper proposes an optimization of torque strategy of All Wheel Drive hybrid vehicle taking part to Le Mans Hypercar championship. Using the software VI-Grade CarRealTime an existing reference model has been considered, and new co-simulated model with CarRealTime and Simulink with custom torque distribution has been defined. Subsequently, a simpler vehicle model has been developed only in Simulink to speed up the optimization problem, while conserving the same torque strategy of the co-simulated model. Regarding the optimization problem, two analysis are proposed. The first one is based on single-objective optimization that reduces the fuel consumption, by optimizing the use of the electric powertrain along the track; the second one is a multi-objective optimization that minimizes the fuel consumption and variation of State of Charge (SoC) using the same decision variable of single-objective optimization. The same penalty function is adopted in both cases to have a final SoC equal to the initial one. This analysis allow to design the size of battery and fuel tank as low as possible.

1. Introduction

Since 2020/21 season a new category in the World Endurance Championship (WEC) has been introduced, named Le Mans Hypercar (LMH), as replacement of LMP1-H category. For this group two powertrain solutions are available: one Rear Wheel Drive (RWD) with an Internal Combustion Engine (ICE), and the other with a hybrid powertrain, with ICE RWD plus an Electric powertrain (E) at the Front Wheel Drive (FWD) [1].

In both layouts the same power curve must be respected, as shown in Figure 1. In standard condition (i.e. atmospheric pressure 1010 mbar, ambient temperature 20°C, relative humidity 0% and no Balance of Performance applied) the maximum power achievable is 500 kW.

Moreover for the hybrid solution, the maximum power deliverable by the electric motor is 200 kW. Even though the ICE solution with a single engine it is easier from a design perspective, the hybrid solution gives huge advantages in the vehicle dynamics, thus justifying the adoption of an additional powertrain. In facts, this solutions aims to reduce the lap-time, increasing the overall powertrain efficiency thanks to the high efficiency of the electric powertrain. For hybrid powertrain, the FIA rules imposes that no electric power can be used when the vehicle speed is lower than 120 km/h with dry tack condition and 150 km/h with wet condition [1].

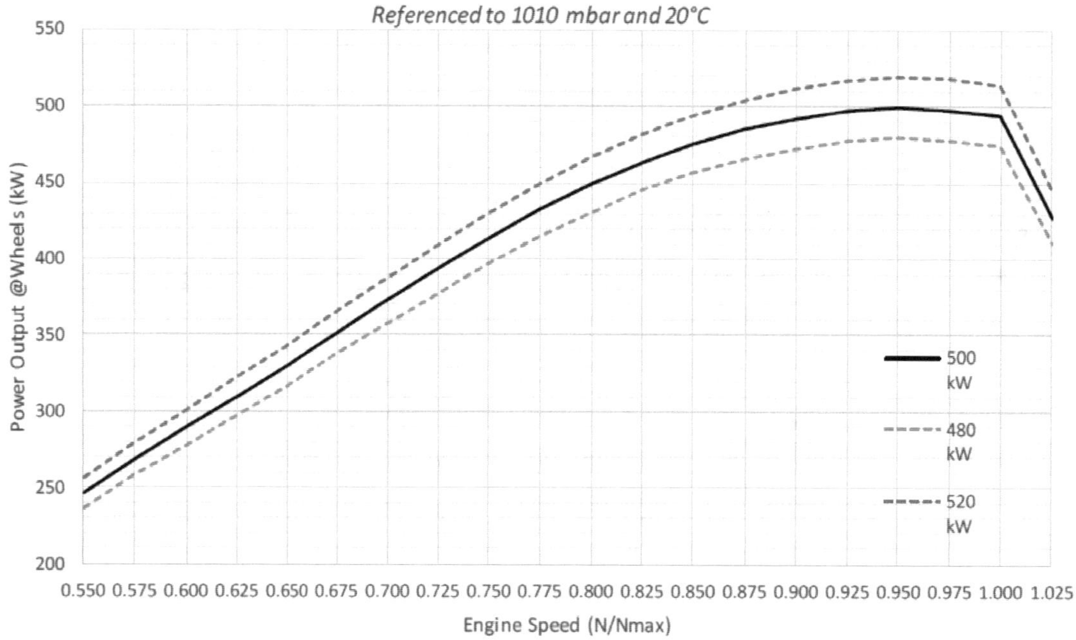

Figure 1. Powertrain power from FIA regulation (Appendix 4.b)[1]. Black line standard condition power curve, red dot curve Max adjustment by BOP (Balance of Performance), blue dot curve min adjustment by BOP.

The hybrid powertrain has been preferred, because an All Wheel Drive (AWD) vehicle gives more control in the corners and allows an improved vehicle strategy during an endurance race, allowing to demand the required power to a powertrain rather than the other (for example if a powertrain reaches overheating conditions, it is possible moves the request to the other). A more complex powertrain layouts as which adopted, require a more complex traction strategy as well. Nevertheless, a hybrid solution gives the possibility to apply different energy strategies, which has beneficial effects during an endurance race, for example in terms of lap time. Many solutions can be discovered with a hybrid powertrain, but in this study only a few are taken into account, in particular those useful in an early stage of the car development.

Given the powertrain architecture, FIA regulations gives the freedom to choose the capacity of the Electric Energy Storage (ES) and fuel tank, thus the energy available for each powertrain and thus different distributions of the torque between the two powertrain systems can be used. Therefore an optimization of the ES and fuel tank capacity is needed, from which different torque strategies are derived.

The paper is structured as follows: in first section all vehicle models used are explained, then the optimization problem is presented and eventually results are discussed.

2. Vehicle Modelling

2.1. CarRealTime Model

As mentioned in the Introduction, the first model was defined in the software VI-Grade CarRealTime (CRT). In fact this software offers an easy implementation of all physical characteristics of the vehicle, translating into a fast and robust lap time simulation, which is used as reference lap time for all next steps. This software simulates a 14 DOF (Degree of Freedom) vehicle model: 6 degrees for rigid free body, 1 for each wheel rotation and 1 DOF between each wheel and the body. The kinematic suspension is defined with a table that define a function between the wheel travel against suspension travel.

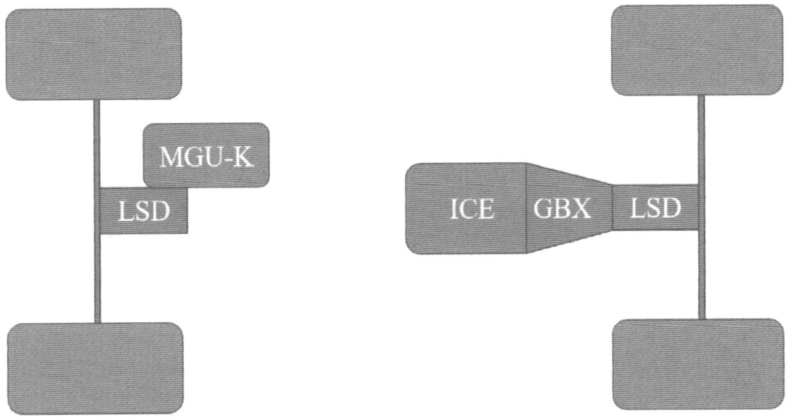

Figure 2. LSD= Limited Slip Differential, MGU-K= Motor Generator Unit - Kinematic (i.e. electric motor), ICE= Internal Combustion Engine, GBX= Gearbox.

For this study, this part of vehicle modelling is assumed as a given reference model. In fact, CRT is able to simulate a hybrid powertrain, as shown in Figure 2, according to the hybrid configuration indicated by FIA (See article 6.3 [1]). However, CRT does not allow a custom torque distribution between powertrains, thus for example if the driver requires 50% of the available torque, then 50% of torque of all powertrains will be provided. This limitation of CRT requires a custom torque distribution development, which has been obtained in this paper combining CRT with Simulink.

2.2. Co-Simulated Model

For the purpose of this study it is necessary to define a custom torque distribution of the two powertrains. The intention is to maintain the vehicle model in CRT (including the powertrain) while in Simulink model only determines how to split the target torque between the two powertrains.

To do that, CRT offers a Simulink library that easily permits to transfer the variables from CRT to Simulink and vice versa. The following variables are transferred from CRT to Simulink:

- Target Torque: torque necessary to follow the vehicle speed profile;
- Max ICE Torque: maximum torque available from engine torque curve at rated engine speed;
- Min ICE Torque: minimum torque available from engine torque curve at rated engine speed;
- Engine ratio: the total ratio from engine to wheel (primary engine ratio · gearbox actual ratio · rear differential ratio);
- Max E-Torque: maximum torque available from e-motor torque curve at rated motor speed;
- Min E-Torque: minimum torque available from e-motor torque curve at rated motor speed;
- Engine speed, measured in rpm;
- E-motor speed, measured in rpm;
- Vehicle path: the position along the track, measured along the curvilinear coordinate, measured in m;
- Vehicle speed, measured in km/h;
- Front longitudinal slips to implement a basic slip ratio control;

2.2.1. Torque distribution. Figure 3 shows the Simulink subsystem that, given the target torque from CRT as input, calculates the torque to the electric powertrain and to the ICE. To understand the torque strategy, it is important highlight that in the LMH regulations three important limitations are given for hybrid powertrains:

- Electric powertrain can be used only when vehicle speed is greater than 120 km/h, or 150 km/h when there are wet conditions (see article 5.3.2 [1]);
- The total amount of power delivered by the whole powertrain must be lower than a power curve defined by FIA, shown in Figure 1;
- The maximum battery power must be lower than 200 kW (see article 5.3.2 [1]);

Starting from these boundary conditions given by the rules, an algorithm to split the torque between the two powertrains has been proposed. Before exposing it, it is preferable clarify that hereafter when performance of the powertrain is limited by the physics, it means that the maximum power available is given by the specifications of the motor for the E-powertrain or from the ICE for the internal combustion powertrain, that could be different from what indicated by the FIA rules, depending by the choice of the powertrains. The strategy workflow is the following:

- The total target torque is initially assigned to the electric torque. If the target torque is greater than the maximum deliverable E-torque, then the E-torque will be equal to the maximum value. Note that the limit of the electric powertrain is in turn given by two limits:
  - The limit of the motor itself, given by the physics of the motor, that must be lower than the torque curve given by the producer;
  - The power limitations imposed by rules, that if the power is positive (traction phase) must be equal to the maximum battery power - 200 kW - minus the efficiency of the motor and inverter. On the other hand if the power is negative (brake phase) the maximum power at the motor will be the limit of the battery power plus the efficiency of the motor and inverter;
- The E-torque can be controlled with a switch that reset the torque amount if:
  - The SoC is out of the bounds;
  - When user decided to de-activate it via a control variable. This control variable will be the control variable that the optimization will use to decide when activate the e-powertrain;
- The difference between the target torque and the E-torque (zero if the target torque is lower than maximum E-torque) is given to the ICE torque;
- A block check the vehicle speed: if it is lower than 120 km/h, the ICE torque will be equal to the target torque. If vehicle speed is greater than 120 km/h, no changes are made.
- A slip ratio control is made on the front wheel. If the maximum slip ratio between the two front wheels is greater than a constant, then the E-torque will be reduced by a factor given by the ratio between the constant and the actual maximum slip ratio. If the slip ratio is lower no action are done;
- *FIA Power limit*: in this block is computed the maximum power by FIA rules. To do that it is necessary to use the ICE speed that gives the power limit at rated speed by means of a look-up table. If the total power of the powertrain is lower than the FIA limit, no actions are taken. Conversely, if the total power of the powertrain is bigger than FIA limit, the ICE power will be reduced in order that the sum of E power and ICE power comply the FIA power limit. No modification are applied to the E-power;
- The torque assigned to the ICE is limited to the maximum torque achievable by the physics of the engine;

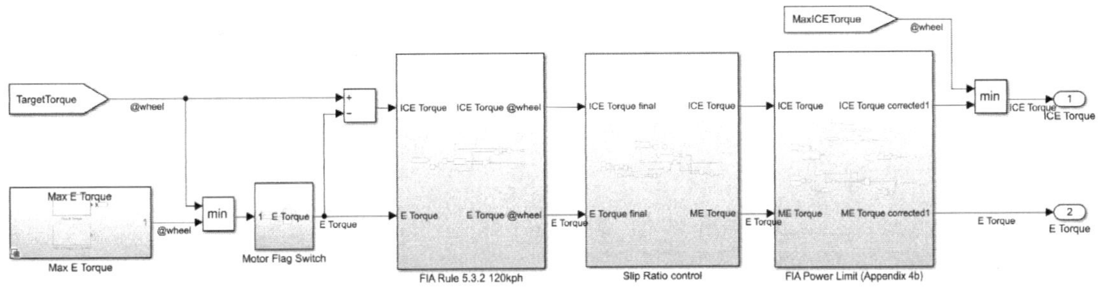

Figure 3. Torque strategy definition in Simulink.

2.3. Simulink model

The models presented in the previous sections presents intrinsic limitations for the scope of this paper, and in particular:

- The first model (*CRT model*) does not offer the possibility to the user to define a custom torque split between the powertrains;
- The second model (*Co-simulated model*) cannot be used for optimizations because the simulation time is not compatible for an optimization problem, where thousands of simulation are needed and thus the total amount of time would not be acceptable;

For this reason another model was defined to run an optimization in a reasonable time, having the same torque strategy explained in the co-simulated model.

To this aim all vehicle dynamics previously treated in CRT was transferred into Simulink. At this point is important to remember that this study is simplified and focused only on the longitudinal dynamics. Furthermore, since the optimization problem does not change the power profile but only how it is split between the front and rear powertrain, therefore it is possible use the CRT variables in Simulink, where lateral dynamics are not modelled. Since this study it is not focused in the reduction of lap-time, the comparison with previous models can be done overlapping the speed profiles. This assumption is acceptable because if the vehicle characteristics and power profile of the whole powertrain does not change among the models, then the vehicle speeds as first approximation are not be affected. Since the vehicle speed is assumed an invariant between the models, it will be used as reference variable for the driver modelling of the Simulink model.

2.3.1. Longitudinal vehicle model. The vehicle model chosen for the Simulink implementation is a model commonly known as "single-track model", used in many books of vehicle dynamics [2,3,4]. The equation that describes the longitudinal dynamics of the vehicle is expressed in equation (1).

$$F_x = m \cdot a_x = F_{x,\mathrm{front}} + F_{x,\mathrm{rear}} + F_{\mathrm{aero}} \tag{1}$$

Furthermore, to take into account some lateral dynamics effects, in particular the effect of the front steering wheels, the equation of the longitudinal vehicle has been changed to:

$$F_x = m \cdot a_x = F_{x,\mathrm{front}} \cdot \cos\delta - F_{y,\mathrm{front}} \cdot \sin\delta + F_{x,\mathrm{rear}} + F_{\mathrm{aero}} \tag{2}$$

where δ is the steer angle at wheel, express as mean of toe angle of front wheels, see equation (3)

$$\delta = \frac{|toe\ angle_{\text{front left}}| + |toe\ angle_{\text{front right}}|}{2} \quad (3)$$

These lateral data are taken from the results of the co-simulated model. As afore mentioned, because of the power profile along the track is not affected - in first approximation - by the power split, it is acceptable use steer and lateral force data from the same simulation made with the co-simulated model.

2.3.2. Driver modelling. A simple driver is modelled in Simulink, to replace the more complex CRT driver. The control is performed via a PID controller, where the input error variable is the difference between the actual speed and the reference speed, given by a target velocity profile made with the CRT model with a powertrain of 720 kW. The choice of using a more powerful engine is to ensure that the Simulink model can only follow the CRT model in order to remove any upper bound to the performance of the optimized solution. The PID is tuned with the dedicated Simulink Toolbox *Simulink Design Optimization*. In Figure 4 is shown the vehicle speed profile obtained with the Simulink model compared to the co-simulated model.

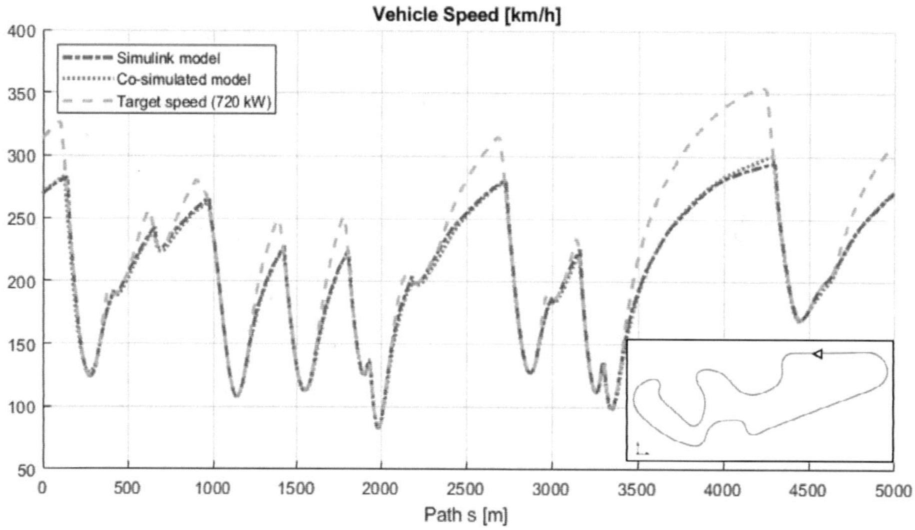

Figure 4. Vehicle speed comparison at Aragon Motorland Circuit. Yellow dashed line represents the vehicle speed target, obtained with a 720 kW powertrain. Orange dot line is the vehicle speed profile of the co-simulated model. Blue dash-dot line is the vehicle speed of the Simulink model. This comparison showed that the difference between Simulink simple model and the Co-simulated model are acceptable, below the 2% of maximum difference.

3. Optimization problem

The focus of this study moves now to the optimization problem. It is worth recalling that the aim of this optimization is to reduce the fuel consumption, by defining where the electric powertrain is to be used, taking into account that the energy available from the battery is equal to the net recover energy net of the e-powertrain after losses. A second optimization in addition to fuel consumption reduction is the battery size minimization, by reducing the difference between the maximum and minimum value of SoC in a lap. The adopted approach is as follows: initially perform a single-objective optimization to define the fuel per lap consumption and have an indication of the maximum battery capacity necessary to cover the optimized lap; secondly a multi-objective optimization is performed to identify the relation between the fuel-consumption and the battery capacity, the battery capacity being set from the single-objective optimization. The optimization has been made in MATLAB using the toolbox *Optimization Toolbox*, coupled with *Parallel Computing Toolbox* and *Global Optimization Toolbox*, has been used.

3.1. Single-objective problem

The first optimization aims to minimize the fuel consumption. The decision variable used for this study is the activation/de-activation of the electric powertrain, transferring the torque request from the electric motor to the ICE (see Section 2.2.1). The variable domain is limited to the straights of the track, that in this study are intended as part of the track where the performance is power-limited and not tire-limited. With this definition are defined in each track a number of straights where it is up to the optimizer define where to use the e-powertrain. In fact, the algorithm consider that in each straights, if the conditions permits it, e-powertrain must be used at the beginning of the straights, and the optimizer decide where to stop the use of the e-powertrain.

3.2. Multi-objective problem

In this optimization the goal is to minimize two variables during a lap: the fuel consumption and the difference between the maximum SoC and the minimum SoC. This optimization gives a correlation between the fuel consumption and the battery size, and thus it is possible define the best e-powertrain activation in different race strategy condition, i.e. the e-power profile that minimize the fuel consumption, or the SoC use, depending on which is fixed.

3.3. Genetic algorithm

The genetic algorithm has been used because the decision variable used for this study is not a continuous variable, and this make impossible to use gradient-optimization based algorithms such as "fmincon" present in MATLAB. To better illustrate the necessity for a genetic algorithm, Figure 5 shows a typical profile activation of the e-powertrain.

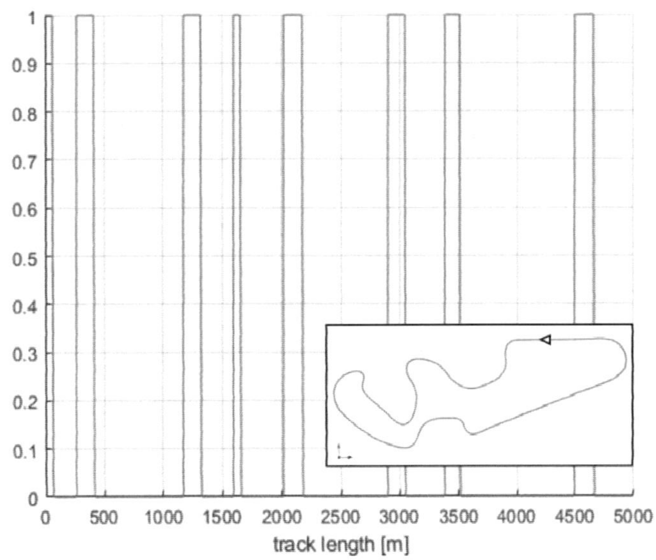

Figure 5. Example of e-powertrain activation, Aragon Motorland circuit. If the variable is zero the e-motor must be deactivated. If the variable is one, the e-power can be used, respecting the conditions given by the torque distribution controller, see Section 2.2.1.

4. Results

4.1. Single-objective optimization

The results of the first optimization problem are reported in Figure 6.

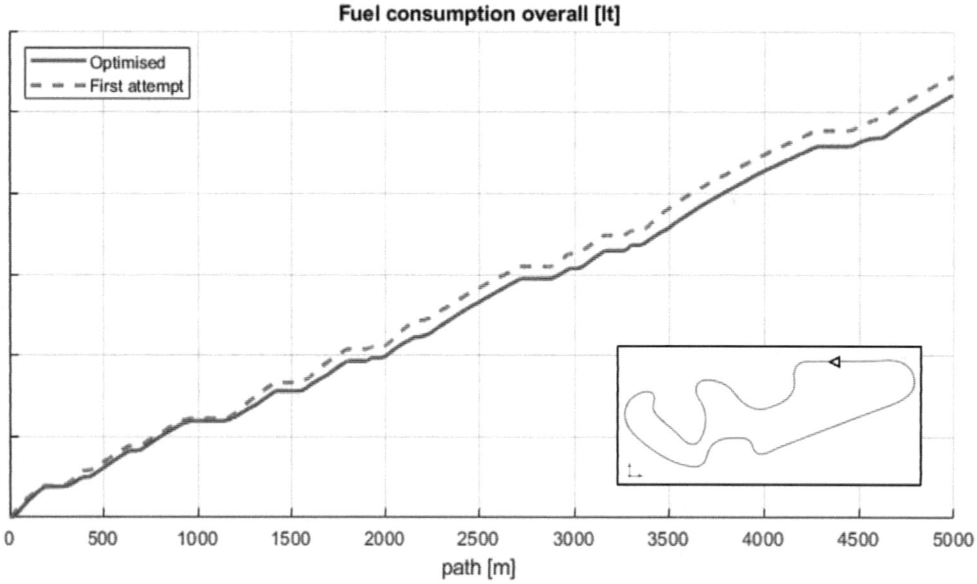

Figure 6. Fuel consumption comparison. Orange dashed line is a first attempt where e-powertrain activation is defined by user. Blue line represent the fuel consumption after optimization. Simulation performed on Aragon Motorland circuit.

The fuel reduction obtained after optimization (set to 20 generations with a population size of 200) is about 4.2%. This reduction was obtained without modifying the total power profile along the track and maintaining the final SoC equal to the initial.

4.2. Multi-objective optimization

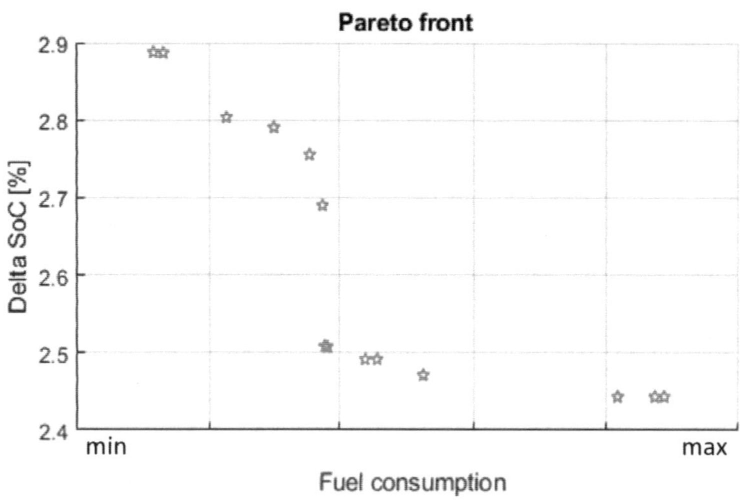

Figure 7. Pareto front of the multi-objective optimization. In x-axis the fuel consumption of a lap, increasing from left to right. In y-axis the difference between maximum and minimum SoC value in a lap, expressed as percentage of total capacity of the battery. Results obtained at Aragon Motorland circuit.

As expected, Figure 7 shows that the correlation between the two objective variables to minimize are inverse dependent. The figure shows the result of an optimization performed by means of 20 generations and 200 individuals per generation. This combination, like in the single-objective function, is the result of a trade-off between accuracy and simulation time for the whole optimization. Another test was run with a bigger population and more generations (250 individuals and 896 generations); as a result it converged before max generations limit was reached (reached at 117 iterations) thus demonstrating that results are acceptable with the first choice of generations and individuals, as suggested in [5].

5. Conclusion

In this work, a method for torque strategy definition has been proposed to minimize the fuel consumption and battery usage. VI-Grade CarRealTime has been used to define an initial velocity profile target and vehicle dynamics behaviour. A second model in co-simulation between CRT and Simulink has been developed to define a custom torque strategy in the hybrid powertrain. Finally a Simulink vehicle model has been defined to reduce the computational effort in the optimization process. More precisely the optimization problem has been split in two optimization: the former, consisting in a single-objective optimization, defined the minimum fuel consumption and a target battery size; the latter, consisting in a multi-objective optimization, defined the correlation between the minimum battery size and the minimum fuel consumption. Thanks to these two processes it was possible to define a target for battery and fuel tank sizing.

References

[1] FIA 2021 *2021 Le Mans Hypercar Technical Regulations* (fia.com)
[2] Limebeer D J N and Massaro M 2018 *Dynamics and Optimal Control of Road Vehicles* (Oxford University Press)
[3] Genta G 1997 *Motor vehicle dynamics: modeling and simulationvol 43* (World Scientific)
[4] Jazar R N 2017 *Vehicle dynamics: theory and application* (Springer)
[5] Alajmi A and Wright J 2014 *International Journal of Sustainable Built Environment* 3 18–26 ISSN 2212-6090

IOP Conf. Series: Materials Science and Engineering 1214 (2022) 012050 doi:10.1088/1757-899X/1214/1/012050

Thermo magnetic FEM simulation of a PM synchronous motor with input data from telemetry driving cycles

A Somà, F Mocera and S Venuti

Politecnico di Torino, Department of Mechanical and Aereospace Engineering, Corso Duca degli Abruzzi 24, 10129 Torino, Italy

E-mail: aurelio.soma@polito.it

Abstract. Nowadays, the requirements to reduce greenhouse gas emissions and to provide a healthy and more habitable environment, has led to the development of several sustainable alternative for eco-mobility. Since the improvement of internal combustion engine has reached a steady state point in terms of overall efficiency, the increasingly stringent requirements imposed by international normative standards are leading automotive companies to find other alternatives to reduce pollution. Thanks to the exponentially growth of power electronics, the huge interest on research of high energy and power density batteries and the more integration of the embedded systems, the central role of the electric drive has taken over on most of vehicles applications. The wide use of permanent magnets synchronous motors for electric vehicles application has rapidly spread out, thanks to their capability to provide high torque and efficiency with low weight and size. Since the natural behaviour of permanent magnets to demagnetize under severe conditions, both coupled thermal and magnetics, the necessity to understand and predict the phenomena is mandatory. This paper carries out a performance analysis in duty cycle, given by real CAN and GPS readings in a studied pathway, with discussion of numerical and graphic technical evaluations. Finite element software has also been used for coupled electromagnetic and thermal calculation to set the magnets working point and establish the temperature distribution within the motor itself during the whole thermal transient. A particular overview is done on the dependence of material used and different cooling solutions adopted.

1. Introduction

Earth's changing climate has several impacts on individual mobility[1], logistics and agriculture[2]. In the last years many companies have been announcing their conversion to produce only zero-emission vehicles. This strong shift has been made possible thanks to the consumer sentiment, buffered also by the extended financial incentives, the growth in charging infrastructures and the expansion on the affordable models available. With the recent progress in the power electronics control[3], AC motors are suppressing almost all the applications in which electric motors are involved. Research in magnetic materials has led to the use of rare-earth magnets that enhances the motor power, but at the same time they are more prone to demagnetize under thermal stress[4]. To develop more power, a strong attention needs to be paid on how batteries[5] fed the motor and on their energy capacity and current discharge rate which enhance the peak torque achievable, taking into account also the inherent stress[6] along the charge and discharge cycle. In the context of the Industry 4.0, which has been representing the fourth industrial revolution, refers to the continuous communication between machines and network to prevents faults, programming maintenance and allowing real time remote control. To integrate all components constituting this chain, it is essential the development of a platform in which all the information are collected, so the role of big data management and the IoT become crucial. Furthermore, once these data are stored, simulation covers a central role in solving problems or accurately predict the maintenance needed. This paper carries out a performance analysis of a full electric vehicle test bench using an Arduino based electronic platform as acquisition system[7]. To accomplish this goal, starting from the mechanical characteristics declared by the manufacturer, a permanent magnet synchronous motor (shortened PMSM) was designed following a reverse engineering procedure. Furthermore, data acquired by the electronic platform are collected and filtered, separating the ones referring to the vehicle from the motor ones. So, resorting to coupled GPS and data visualization the correct functioning of the motor is examined. If particular functioning conditions are observed, these data become the input for the duty cycle FEM simulation in Ansys MotorCad with the previously designed motor. The investigation aims to control thermal and magnetic behavior avoiding demagnetization fault, which determines costs related to magnets substitution and to machine down-time. Finally, the calculation of a simple performance index is proposed with the aim of describing how far magnets are from their critical condition and if heat is properly managed.

2. Coupled electromagnetic model

The electric motor investigated in this work is a PMSM which characteristics to be replicated are:

Table 1. Manufacturer's technical specifications.

	Value	Unit
N. Pole	6	-
n_{nom}	9000	rpm
P_{nom}	43	kW
C_{nom}	45	Nm
BEMF	180	V_{RMS}
I_{nom}	150	A_{RMS}

Dealing with electric motors the design consists of a fixed stator and a rotating rotor, held in place by the use of bearings and separated by a thin layer of air called air-gap.

First the stator design is carried out by considering an integral slot motor, in a way in which the harmonics produced by each winding are in phase with each other, so the total number of slots is found considering one coil per pole per phase. Choosing a number of magnet poles N_m=6 and by considering a three phase alternating currents, the number of slots constituting the stator lamination N_s=18 was calculated. Therefore, the value of air gap length comes out from a tradeoff between the cogging torque and torque produced. A good compromise is obtained with[8] :

$$g=\frac{N_s}{2\,N_m} \tag{1}$$

Where g is the amount of stator yoke required for each magnet pole.

Considering a commercially available magnet length of 40 mm, the magnetic flux density in the air gap is determined by the magnets thickness, an increase in this dimension reduces the motor inductance. A desired condition is a higher inductance coupled to a high flux density. From the simulations, to match motor specifications, magnets need to be 4 mm thick with an arc length of 150 °Electrical. The overall dimensions of the motor designed are:

Table 2. Geometry properties of the designed motor.

Data	Value (mm)
Φ_{ext}	220
Φ_{bore}	126
Airgap	1.5
Φ_{shaft}	40

Where drawing dimensions are shown in figure 1 and figure 2.

To complete the stator design a supposed lamination material has been chosen by considering its magnetics properties. Thus, was calculated dimensions of stator teeth and radial back iron length. In addition, since the space are constrained by the destination place, knowing relation of the torque:

$$T=kD^2L \tag{2}$$

length is consequently adjusted to fit the requirement, obtaining a motor with 3 axial slice of magnets covering a length of 120 mm while the overall motor length reach 280 mm considering the space in which are placed windings overhang. Since all these aforementioned data are hypothesis based on simulations output, it is necessary to consider an uncertainty on the data so as the material achieved. To conclude this treatment about the electromagnetic design, reducing the torque ripple[9] involves a shifting of a row from each other by an angle determined from the following relations:

$$\alpha_{skew}=\frac{N_m}{N_s}\cdot 180=60°E \tag{3}$$

Since the correspondence between electrical degree and mechanical degree is:

$$\theta_M=\theta_E\cdot\frac{2}{N_m} \tag{4}$$

Where N_m is the number of magnetic poles, the skewing angle in mechanical degree is found considering also the number of magnetic rows:

$$\alpha_{skew_M}=\alpha_{skew_E}\cdot\frac{2}{N_m}\cdot\frac{1}{N_{rows}}=6.67°M \tag{5}$$

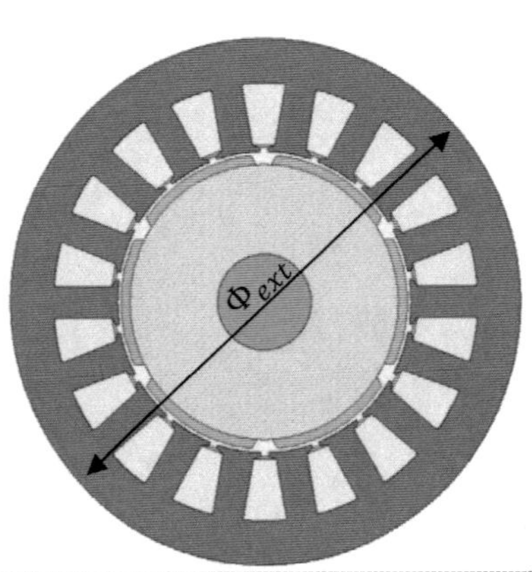

Figure 1. Radial section of PMSM designed

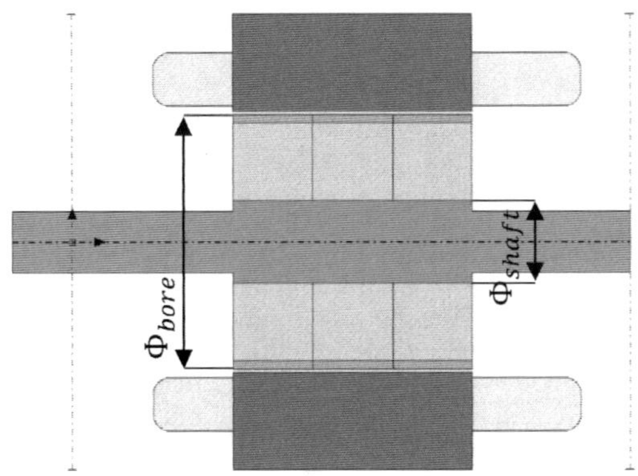

Figure 2. Axial section of the PMSM designed.

2.1. Electromagnetic analysis of motor

Before reaching the final configuration, many attempts were done maintaining as constants internal geometric data, because other parameters could determine the performance of the machine in terms of power and voltage, such lamination material, thickness, winding configuration, ferromagnetic material and temperature conditions. Temperature takes effect on the flux density remanence B_r. In fact, when increasing T, B_r decreases becoming zero at Curie's temperature. The following simulations are conducted with an internal temperature of 80°C at which the motor withstand after a starting transient.

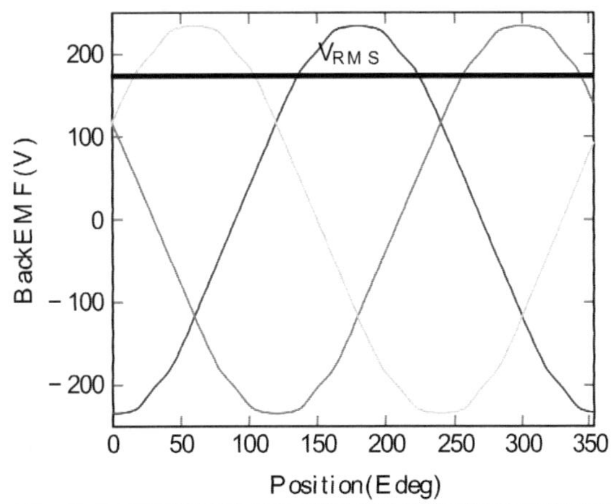

Figure 3. Back electromotive force produced by the motor

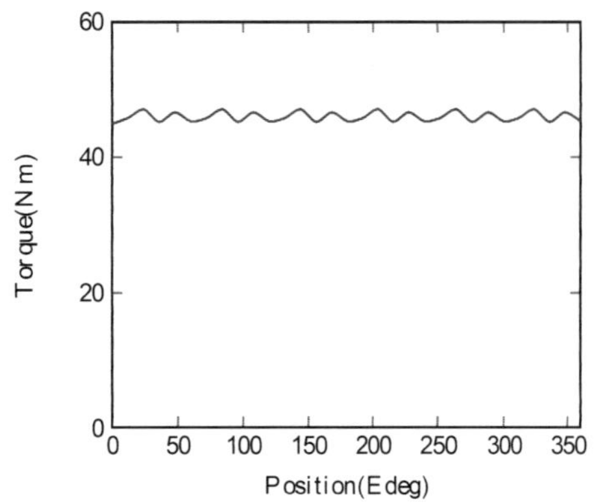

Figure 4. Total torque developed from each slice contribution

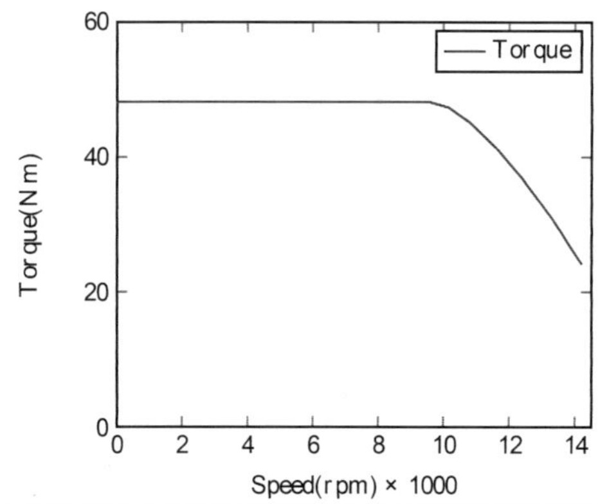

Figure 5. Torque envelope as function of the speed

All the relevant parameters are in line with the specifications and the cogging torque is arranged on 4% thanks to the magnets skewing. The efficiency obtained at nominal point working condition is η=96% using the following configuration of winding:

Table 3. Winding definition

	Value	Unit
N. Turns	2	-
Layers	2	-
N. Strands	70	-
Cond/slot	280	-
Φ_{wire}	0.885-0.8	mm

3. Electronic acquisition system to telemetry monitoring

With the spread of Industry 4.0, necessity in systems remote monitoring through telemetry platforms becomes important, in order to reduce machine downtime and to predict failure, in the view of a continuous improvement in the design of projects. In this paper the electronic system used is based on Arduino components and consists of:

- Arduino Mega
- Can Bus shield
- IMU 9-axis motion sensor
- GPS shield

The boards are used to read vehicle information available on the CAN-Bus communication line and to save hexadecimal data on a µSD card for offline post processing. The inertial measurement unit, with a data fusion algorithm, couples accelerometer, gyroscope and magnetometer readings, making available linear and gravity acceleration on each axis composing the 3-axis frame. Moreover, the gyroscope measures the angular velocity about the Euler's angle. Finally, to make significant the data acquired, is useful to link them with GPS coordinates, in a way to justify or make further investigations on particular vehicle behaviours.

IOP Conf. Series: Materials Science and Engineering 1214 (2022) 012050 doi:10.1088/1757-899X/1214/1/012050

This box is mounted on board of a "Fiat Panda" electric prototype which is propelled by the motor above described. This is a test bench allowing the study and the optimization of design control strategies. Concerning the vehicle, its overall specifications are reported in the following table:

Table 4. Vehicle specifications

	Value	Unit
Power	43	kW
Battery capacity	16	kWh
Nominal Voltage	640	V
Max speed	120	km/h
Average range	150	km

The acquisition platform is mounted on the vehicle, by connecting it to the CAN-Bus line and 12V power system. The cycle covered is stored and data are post processed with Matlab, where data are converted from hexadecimal to decimal values, a proper scaling factor and offsets are applied to make them meaningful. Finally, data are plotted to visualize information and catching functional anomalies or relevant working conditions.

3.1. Duty cycle definition

The working cycle[10] studied consists in a mixed route path, to test the vehicle in different conditions: urban, highway and uphill. To better understand the covered path a map showing the geographic data like altitude, latitude and longitude is shown in Figure 6:

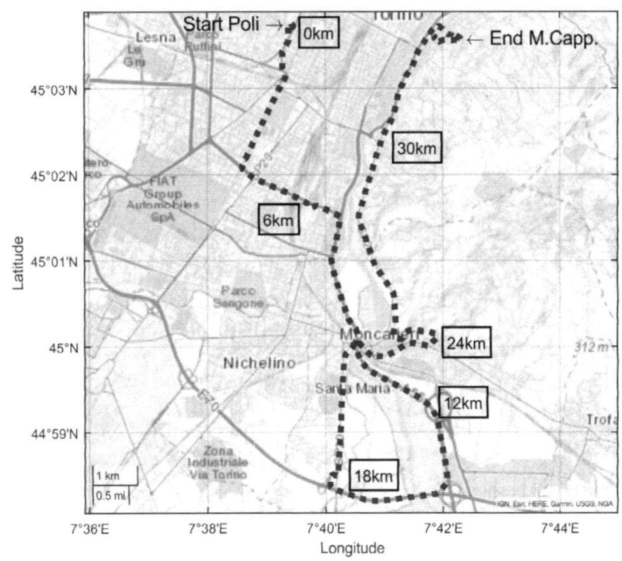

Figure 6. Driving cycle GPS path.

GPS readings report the following altitude path:

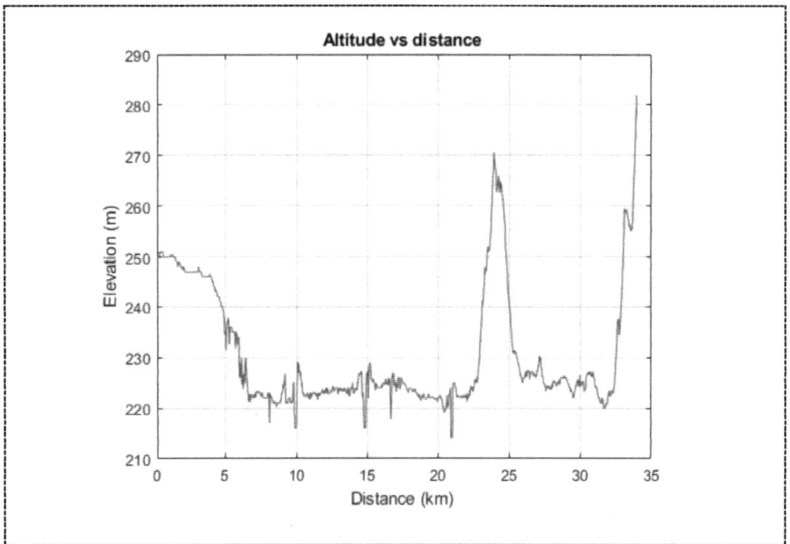

Figure 7. Altitude vs distance recorded path profile

A first skim may be sufficient to produce thoughts on power consumption and recovered energy: in particular are expected two regeneration periods coupled with two peak-power demands. The main characteristics of the covered cycle are shown below:

Table 5. Duty cycle characteristics

	Value	Unit
Space covered	34	km
Altitude difference	60	m
Average speed	36	km/h
Cycle time	60	min

By analyzing the data acquired a practical visualization on the above predicted data has been considered interesting. The following picture shows the motor power with color shades on which is superimposed the road slope represented by the markers area, in a way to analyze how the energy is used and evaluating also if the control strategy (speed or torque control based) is optimal for the prototype application.

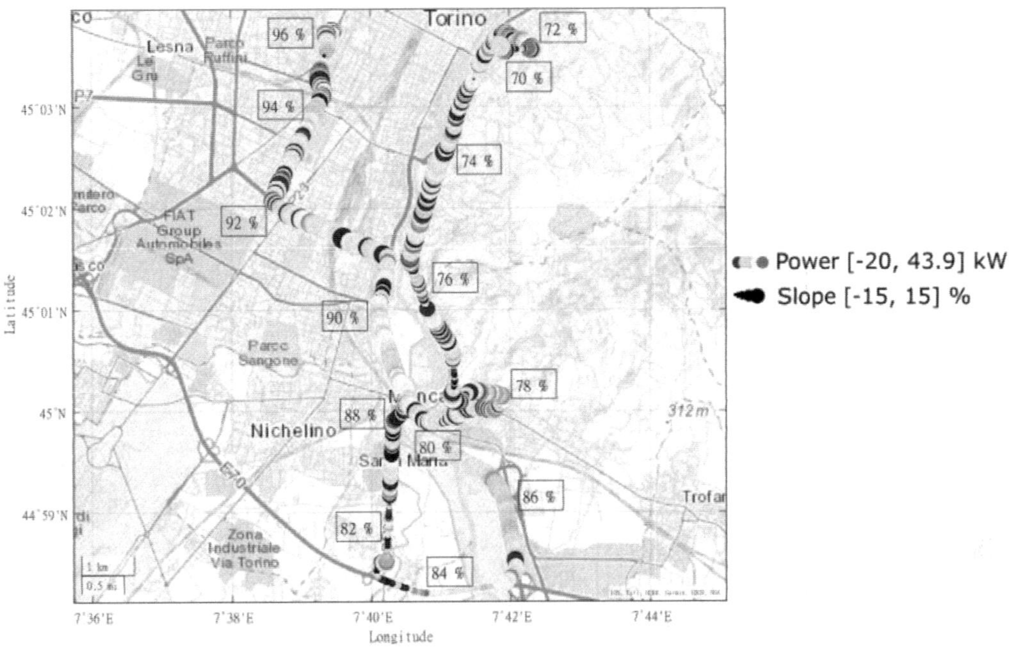

Figure 8. Power and road slope superimposed path with reference to battery SOC (%)

The above figure shows the power delivered to the motor and recovered[11], linked with altitude information and the battery-pack state of charge. To obtain a visualizable graphic GPS-altitude data are not used. Instead, a better information that describes the road slope can be given by the magnetometer readings. Acquiring the two components of the gravity acceleration we can trace the slope, furthermore since these data are so noisy, it was necessary to apply filters making smoother the data. This plot may confirm the above stated hypothesis, if the energy recovered is in correspondence of a downhill either near continuous starts and stops due to a traffic lights. In particular, knowing the covered space from the previous picture (figure 6) we can calculate the energy consumption as:

$$\text{Energy}=\Delta \text{SOC}\cdot \text{BatteryCapacity}=26\%\cdot 16\ \text{kWh}=4.16\ \text{kWh} \tag{6}$$

Allowing the calculation of the battery life in terms of average space coverable, which result is reported in table 4. To understand the correct functioning of the motor, the focus is now moved to data acquired and post-processed in Matlab. First is useful to have a look to the motor characteristics data like currents, temperature and speed:

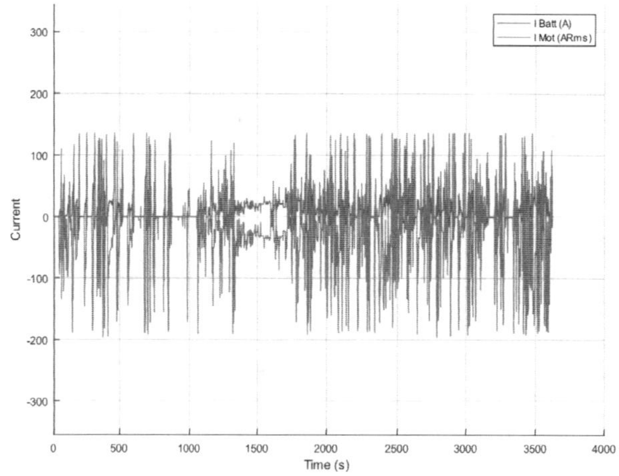

Figure 9. Motor and battery currents plot.

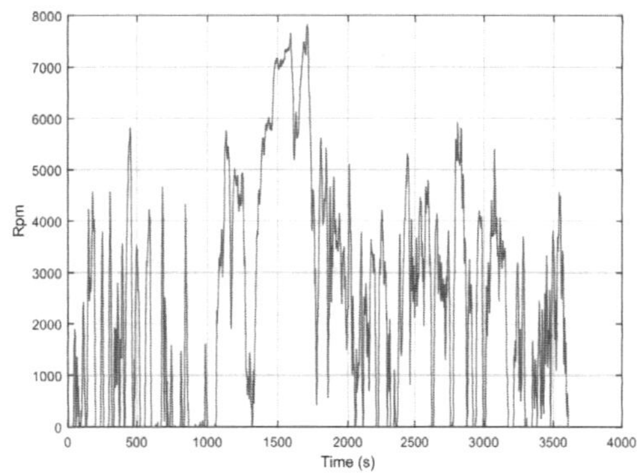

Figure 10. Motor shaft speed.

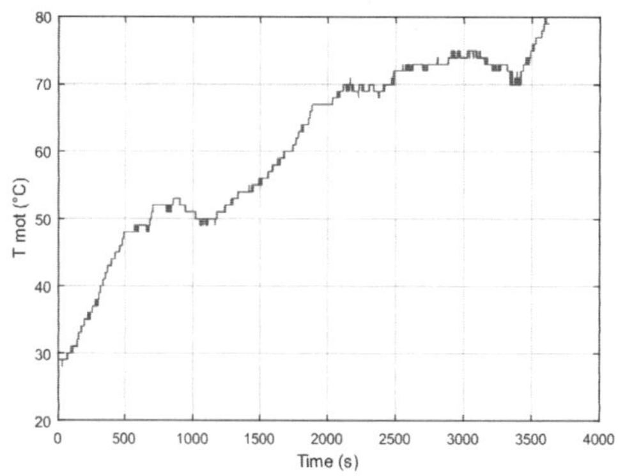

Figure 11. Temperature profile with time.

Looking at the above graphs, reported in order of visualization: in the first are plotted motor and battery currents, with opposite signs by convention: assuming positive value when battery current flows to the motor, negative value for the motor current when it receive the battery one. When torque is requested by the driver, the power electronics loops a certain amount of current to the motor, which is function of the speed. Since motor produces back electromotive force (BEMF) working against the applied voltage, the current flowing through the winding is:

$$I_{mot}=\frac{V_{drive}\text{-BEMF}}{R_{winding}} \tag{7}$$

This means that a rise in motor speed makes I_{motor} closer to $I_{battery}$, as it can be seen in the middle period (highway path) where figure 10 reports high value of motor speed. In addition, a further evaluation regards the temperature, the figure 11 shows a continuously rising suggests that no thermal equilibrium occurs within the motor and that thermal management deserves a review. Back to the figure 9, it can show motor health in the sense that if demagnetization takes place, due to thermal either uncontrolled magnetic field produced by extremely high currents, motor loose its capacity in BEMF production, so contrary to as we used to see before, at high speed motor current doesn't get closer to battery current.

3.2. MotorCad FEM simulation

To understand possible faults or to simulate effects of a thermal design review Ansys MotorCad is used, a FEM-based software which allows to evaluate design improvements effects on the reviewed motor project. From the design discussed in Section 2, acquired data have been used to visualize motor performance. Data input of the model are the motor currents and speed as a function of time and the outputs are all motor data at each step with corresponding losses, used then as thermal simulation input.

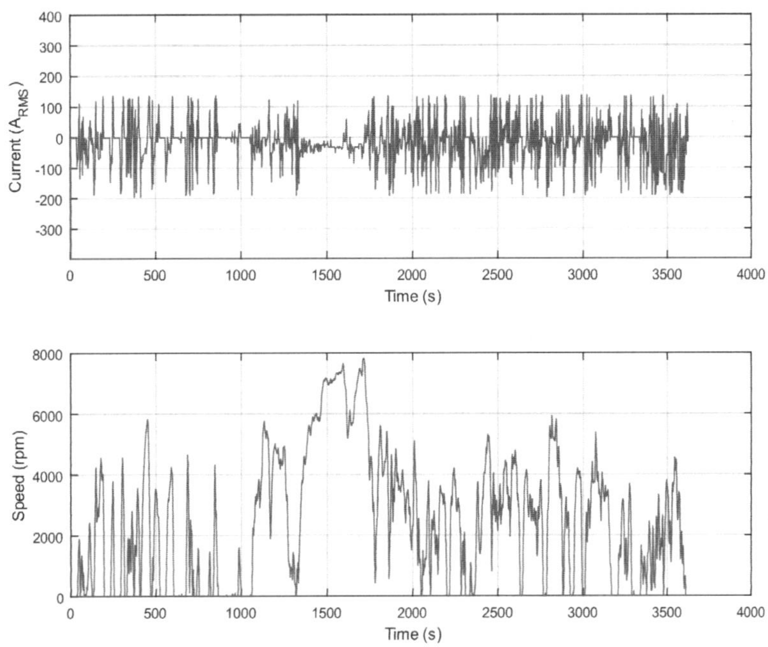

Figure 12. Motor current and motor shaft speed for MotorCad duty cycle simulations.

These data allow for calculation of point to point performance. From the calculated shaft power and knowing battery output power it is possible to estimate the overall efficiency:

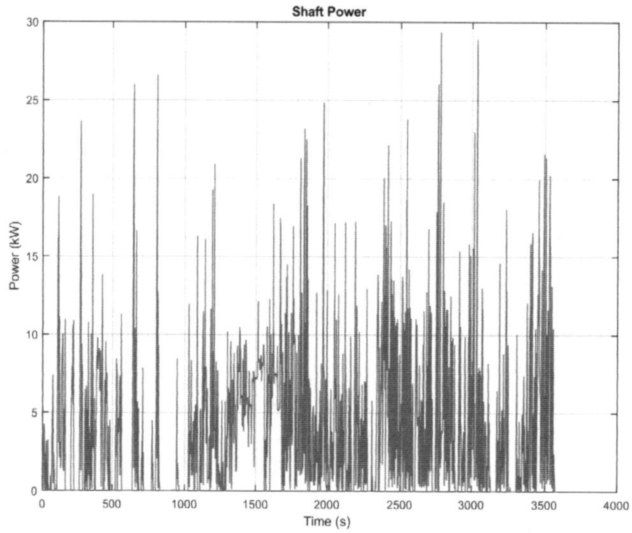

Figure 13. Shaft output motor power calculated with MotorCad simulation.

The overall efficiency calculated as product of $\eta_g=\eta_{electronics}\cdot\eta_{motor}=70\%$ where the electronics efficiency includes all the power thermal losses (wires, EPM, inverter). A similar result can be found using manufacturer's K_t to find torque, so using speed to calculate shaft power. In the following table are reported data that describes the complete cycle of the motor:

Table 6. Output duty cycle simulation

	Value	Unit
Average efficiency	88.6	%
Electrical input energy	3250	Wh
Shaft motoring energy	3100	Wh
Recovered energy	695	Wh
Shaft-generated energy	730	Wh
Total loss	190	Wh
% Motor	75.7	%
% Generator	25.3	%

Visualizing all the working points on the efficiency map of the designed motor, considerations about the motor size and the proper design for the application can be made.

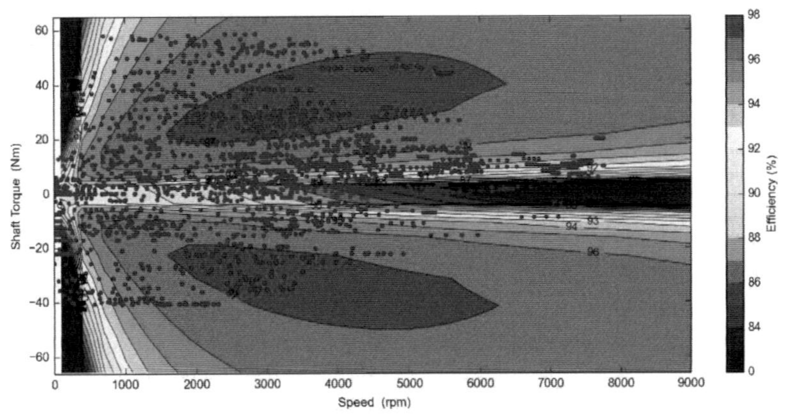

Figure 14. Motor efficiency map with duty cycle working points.

As can be seen just a little amount of the total points are in the high efficiency region as the combination of torque and speed. The other part are in a lower efficiency area, this is the reason of the average efficiency reported in table 6. Most of the working points are settled in the low efficiency region of the map, with a combination of low torque and high speed, this suggests that a further review should move points through high efficiency improving losses and energy consumption, this could be made by use of a different control strategy like a variable phase angle instead a constant one fixed to zero.

3.3. MotorCad thermal review simulation

The advantage of this software is the easy way with which multiphysics simulations are made, by passing from mechanical to thermal one and evaluating the effect of their mutual interaction. Losses calculated in the duty cycle simulation are used as inputs for the thermal model.

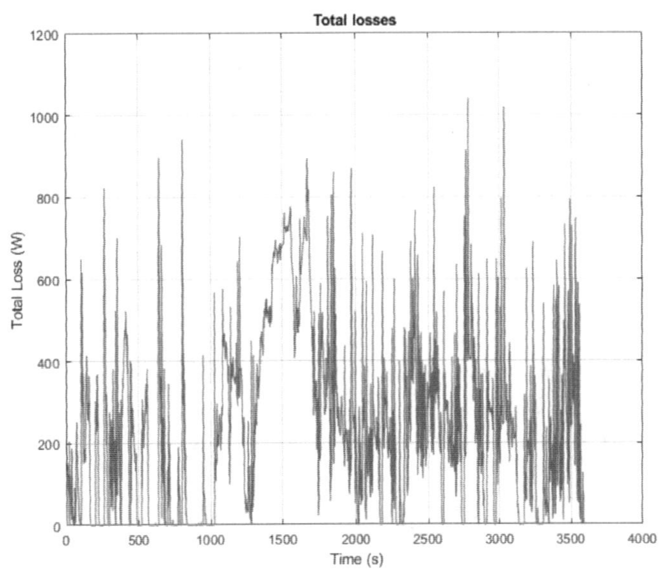

Figure 15. Motor losses calculated.

As stated from the Steinmetz's equation losses depend on the current frequency:

$$P_w = k \cdot f^a B^b \tag{8}$$

Where k,a,b are the Steinmetz's coefficients depending on interpolation from material B-H curve and the calculated P_w is expressed as power per unit of mass $\frac{W}{kg}$. Concerning material[12], the requests are high magnetic flux permeability and low losses. Since the power losses are proportional to $P_w=RI^2$ the lower is the current, the lower will be losses. So, to reduce eddy currents in the laminations electrical steel is widely used, in which the presence of silicon increases resistivity reducing the current produced, following the Ohm's law $I=\frac{V}{R}$. In this work, M350-50A electrical steel is used for stator and rotor laminations. The cooling system is assigned to housing water jacket with spiral ducts in which passes a cooling fluid consists in a mixture of water and ethylene glycol 50/50, chosen thanks to their good capability in heat absorption, furthermore presence of glycol rises the boiling temperature and decreases the freezing point. A first analysis consists in the evaluation of a general scenario at the end of the driving cycle, by knowing the point in which is installed the single NTC thermistor critical issues are investigated.

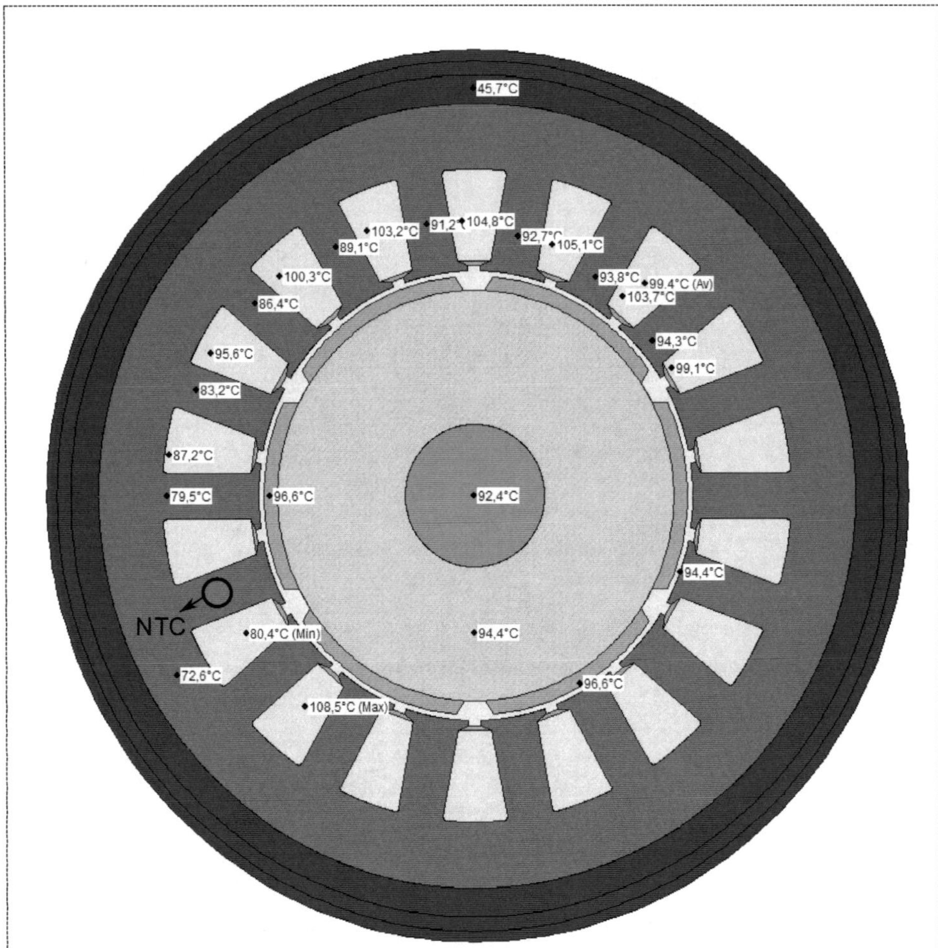

Figure 16. Radial section temperature analysis.

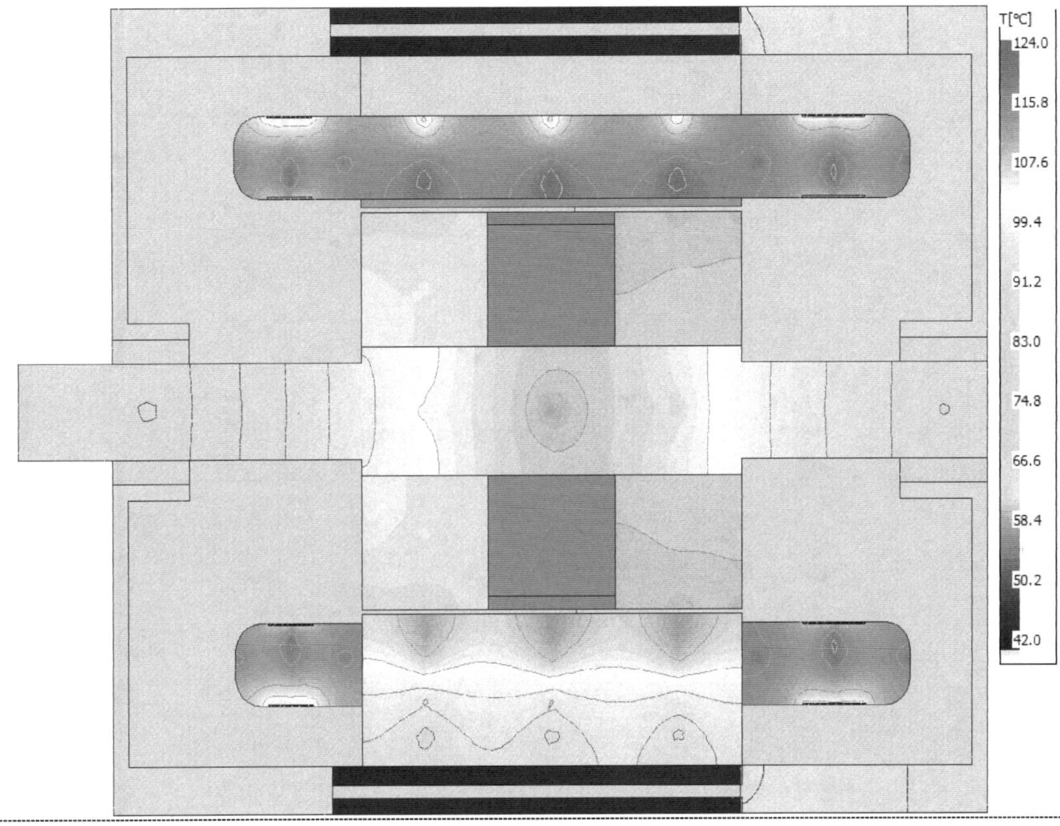

Figure 17. Radial section temperature analysis.

As it can be observed, even if we register temperature of 80°C on the external ring of the stator, in contact with cool water, internal temperature can be higher also of 20°C as the figure 14 shows. Another proof that the sensor is not well positioned is in the right figure, in which the central slice of magnets is hotter than the sides, due to the fact that exchange continuously heat with air in the endcaps region. There is an appreciable temperature difference from a side slice to a central one, of about 10°C. Generally speaking, with the used magnets N30UH 180°C class, these temperatures are not a problem, but if the using time is increased, either the power demand changes, the situation could be different. So, a general trend is to reach a steady state condition, where all the heat generated can be totally transferred to the fluid. This desired condition is achieved with a re-designed[13] cooling solution, looking for a fluid speed allows a turbulent flow and ensure high values of convective heat transfer coefficient h in the cooled region. First aluminium alloy for the inner chamber of the ducts is used, then reducing height of the channel ensures an increase in flow speed and increases structural resistance. Also the space between ducts is slightly increased, in this way endcap length can be reduced so as distance from the channel to the external surface of the stator lamination avoiding structural damage. The final geometrical configuration is:

Table 7. Radial and axial water jacket dimensions

	Value	Unit
WJ channel-lam	5	mm
WJ channel height	4	mm
WJ channel width	33	mm
WJ channel spacing	25	mm

The result achieved is an increased heat exchange surface, with enhanced heat exchange which, after a first transient, allows thermal equilibrium. As the following transient simulation[14] shows.

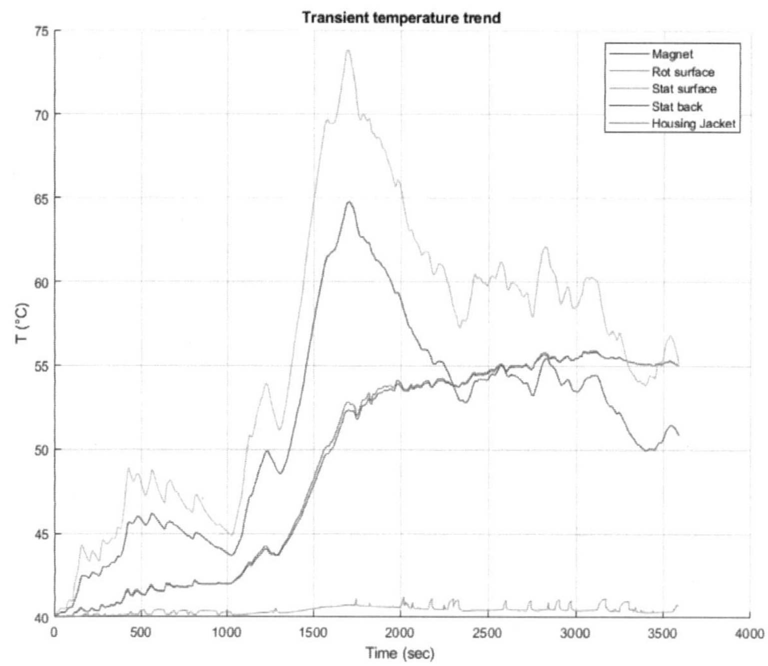

Figure 18. Duty cycle thermal analysis with reviewed cooling design

4. Demagnetization analysis[15]

Sintered NdFeB magnets are subjected to irreversible demagnetization starting from temperature of 120°C, many causes can lead to this fault: i.e. short circuits or wrong thermal management. In this section the worse condition verified in the previous duty cycle is analysed finding a safety coefficient to evaluate how far the magnets are from fault condition. When a magnet is subjected to an external magnetic field, its behaviour is dictated by the hysteresis cycle and its working point moves along a curve called recoil line[16] parametrized with the temperature. In correspondence of the maximum temperature (for N30UH 180°C), the lines show a knee at which irreversible demagnetization occurs. Instead for temperatures lower than maximum the knee is located to negative values both of B and H. The phenomenon takes place in a sense that the remaining value of magnetic field density B_r is reduced to a value in which the magnet is not able anymore to produce BEMF. Demagnetization[17] may occurs with a coupled effect of impulsive currents and high temperature, considering that heat is generated in part in the coils and another part in laminations due to eddy current losses, cooling these hot spots represents the highest priority in thermal management. In this work is discussed a fault analysis starting from a critical point identified in the cycle. In this case, a critical point is found with the following data:

Table 8. Characteristics of critical working point

	Value	Unit
n	2090	rpm
Torque	58.7	Nm
I_{peak}	274.3	A
Power	25	kW
Temp	70	°C
Loss	635	W

In order to find a simple and representative value that defines how far is the worst condition at which magnets need to withstand. The operating point is simulated coupling electromagnetic and thermal FEM, taking into account the working temperature. With the resulting B and H it is calculated the permeance coefficient P_c defined as:

$$P_c = -\frac{B_m}{\mu_0 H_m} \tag{9}$$

Table 9. Magnetic results at critical working point

	Value	Unit
B	0.498	T
H	441.7	kA/m
P_c	0.9	-

Demagnetization curve, at working temperature of this condition shows a $H_{knee}=1430\frac{kA}{m}$, defining the starting point of irreversible demagnetization.

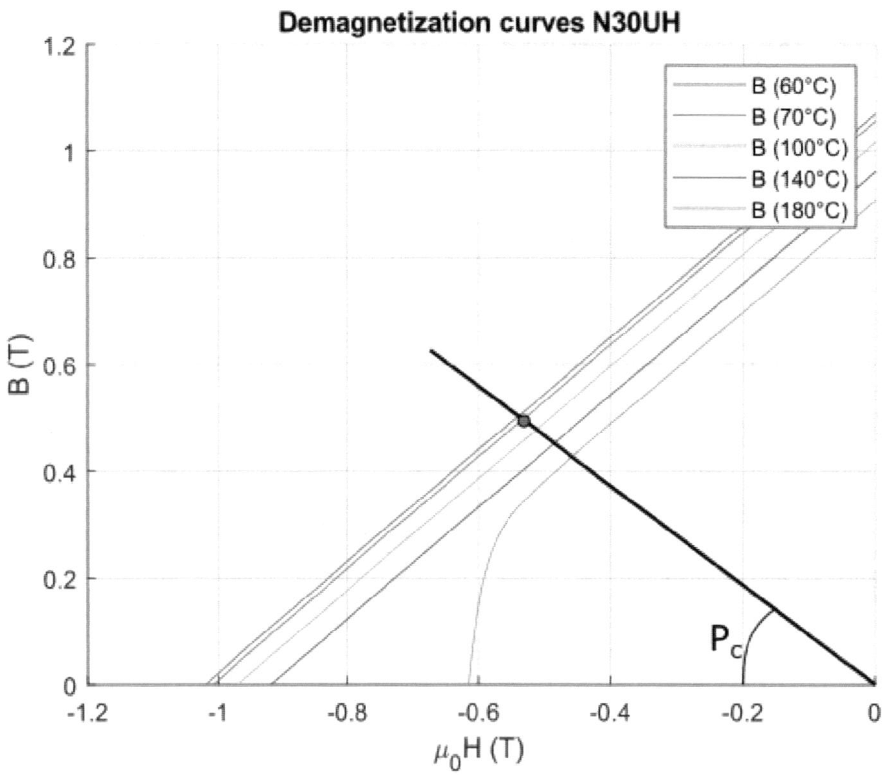

Figure 19. Demagnetization curves of magnet N30UH.

As it can been appreciated this kind of load in terms of current and heat is not critical. Instead moving to the 180°C characteristic, considering these electromagnetic conditions, since the permeance line crosses the knee point, the magnet would results more prone to demagnetize. However, the safety coefficient which represents qualitatively the distance from fault is:

$$Sf = \frac{H}{H_{knee}} = 3.2 \tag{10}$$

This safety factor suggests electric motor works in a safe condition in the most loaded point of the cycle.

5. Conclusion

This work points out a method for continuous monitoring of a full electric vehicle equipped with a permanent magnets synchronous motor, by acquiring data with an electronic Arduino based system connected to the vehicle CAN-bus and by using them as input data for a reverse engineer motor from manufacturer's overall characteristics. Data have shown an unbalanced thermal management and the paper propose a re-designed cooling jacket which improve heat exchange. Furthermore, a method to analyse potentially critical working point and evaluating possible faults has been developed. Thermo-magnetic coupled FEM simulations have made it possible to detect eventual failures and calculate a value which states distance from fault. In this work the existence of a rotor retaining sleeve structure, useful to withstand the high centrifugal forces generated when the motor turns at high speed, was neglected. The high number of degrees of freedom available at the design stage of electric motor has made difficult the preliminary design, for this reason it has been necessary the use of mathematical

proportionality equations[8]. This work has pointed out the importance on the right choose of materials, in fact, many attempts to reach an acceptable efficiency have been made. The final configuration reports a higher efficiency than manufacturers specifications, considering however that the purpose of this study was matching performance neglecting costs. The analysis have shown a motor particularly responsive to heat with sudden rises in temperature recorded with the electronic tool, which behaviour was considered dependent on a underestimation of heat exchange characteristic parameters, so it was considered useful the proposal of improvements to be taken into account in the subsequent models. Finally, the last evaluation was done on demagnetization, without observing any critical scenario, that could have been occurred if the critical point would have happened at higher temperatures.
The future research will focus on more innovative cooling solutions regarding also the rotor and find materials with minor losses in a way to expand the high efficiency working range.
Further improvement may regard the implementation of a simile technique as a standard analysis for IoT remote control, allowing fault prediction or to help the service in detection of electric vehicle overall failures.

References

[1] Mocera F and Somà A 2021 A Review of Hybrid Electric Architectures in Construction, Handling and Agriculture Machines *New Perspectives on Electric Vehicles* ed Prof Marian Găiceanu (Intechopen)
[2] Mocera F and Somà A 2020 Analysis of a Parallel Hybrid Electric Tractor for Agricultural Applications *Energies* **13**(12) 3055
[3] B K Bose 2009 Power Electronics and Motor Drives Recent Progress and Perspective *IEEE Transactions on Industrial Electronics* **56**(2) pp 581-588
[4] Tae-Kyoung B, Kyung-Hun S, Jeong-In L, Jong-Hyeon W, Han-Wook C and Jang-Young C 2021 Design of high-speed permanent magnet synchronous machines considering thermal demagnetization and mechanical characteristic of permanent magnet *AIP Advances* **11** 025129
[5] Mocera F, Vergori E and Somà 2020 A Battery Performance Analysis for Working Vehicle Applications in *IEEE Transactions on Industry Applications* **56**(1) pp 644-653
[6] Mocera F, Somà A and Clerici D 2020 Study of aging mechanisms in lithium-ion batteries for working vehicle applications *Fifteenth international conference on ecological vehicles and renewable energies (EVER)* (Monte-Carlo, Monaco)
[7] Vergori E, Mocera F and Somà A 2018 Battery Modelling and Simulation Using Programmable Testing Equipment *Computers* **7**(2) p 20
[8] Duane C Hanselman 2006 *Brushless Permanent Magnet Motor Design* Magna Physics Publishing
[9] Urresty J, Riba J, Romeral L and Garcia A 2010 A simple 2-D finite-element geometry for analyzing surface-mounted synchronous machines with skewed rotor magnets *IEEE Transactions on Magnetics* **46** pp 3948-54
[10] Mocera F and Somà A 2018 Working Cycle requirements for an electrified architecture of a vertical feed mixer vehicle *Procedia Structural Integrity*, **12** pp 213-223
[11] Lu D, Ouyang M, Gu J and Li J 2014 Instantaneous optimal regenerative braking control for a permanent-magnet synchronous motor in a four-wheel-drive electric vehicle *Proceedings of the Institution of Mechanical Engineers, Part D: Journal of Automobile Engineering* **228**(8) pp 894-908
[12] Berardi G and Bianchi N 2017 Rotor losses reduction in high speed PM generators for organic rankine cycle systems *IEEE Energy Conversion Congress and Exposition (ECCE)* **55**(6) pp 5800-08
[13] Huang J, Shervin S , Miller R, Rizzo D, Sebeck K, Shurin S, Wagner J et Al. 2019 A Hybrid Electric Vehicle Motor Cooling System- Design, Model, and Control *IEEE Transactions on Vehicular Technology* **68**(5) pp 4467-78
[14] Fan J, Zhang C, Wang Z, Dong Y, Nino C E, Rehman A and Strangas E 2010 Thermal Analysis

of Permanent Magnet Motor for the Electric Vehicle Application Considering Driving Duty Cycle. Magnetics, *IEEE Transactions* **46** pp 2493 – 96
[15] Sjökvist S 2014 Demagnetization Studies on Permanent Magnets *Uppsala Universitet*
[16] Egorov D, Petrov I, Pyrhönen J, Link J and Stern R 2017 Linear recoil curve demagnetization models for ferrite magnets in rotating machinery 43rd Annual Conference of the IEEE Industrial Electronics Society pp 2046-2052
[17] Zhilichev Y 2008 Analysis of Permanent Magnet Demagnetization Accounting for Minor B-H Curves *IEEE Transactions on Magnetics* **44**(11) pp 4285-88

AUTHOR INDEX